CARBANIONS

LIVING POLYMERS

and

ELECTRON TRANSFER PROCESSES

CARBANIONS
LIVING POLYMERS
AND
ELECTRON TRANSFER
PROCESSES

MICHAEL SZWARC, F.R.S.

State University of New York Polymer Research Center
at the College of Forestry
Syracuse, New York

WITH A FOREWORD BY
DONALD J. CRAM
University of California at Los Angeles

INTERSCIENCE PUBLISHERS

A DIVISION OF JOHN WILEY & SONS
NEW YORK · LONDON · SYDNEY · TORONTO

CHEMISTRY

LIBRARY OF CONGRESS CATALOG CARD NUMBER 67–13964
SBN 470 84305 5
PRINTED IN THE UNITED STATES OF AMERICA

To the memory of my parents

Foreword

As in all endeavors, chemistry has its scholars, investigators, craftsmen, prospectors, speculators, adventurers, entrepreneurs, statesmen, and chroniclers whose combined functions give movement and substance to the science. In the early stages of development of a branch of chemistry occasionally one person integrates some of these activities. But as the field expands, the hard economics of time drive investigators into specialization, both of subject matter and function. Polymer chemistry, in general, and carbanion chemistry, in particular, are now vast fields with hundreds of practitioners, objectives, techniques, and approaches that range from almost pure physics to synthetic organic chemistry. Ideally, a monograph possesses enough breadth to encompass a field, enough depth to critically evaluate its current state, and enough insight to anticipate and define the goals of the future. Ambitious, indeed, is the man who, understanding these things, writes a monograph embracing two vital fields, one of practical and the other of theoretical significance. Professor Szwarc is such a man, and *Carbanions, Living Polymers, and Electron Transfer Processes* is such a monograph. The functions of scholar, investigator, and chronicler are nicely blended in this book, as are the arts and crafts and theory of physical, organic, and applied polymer chemistry.

Carbon with a partial or full negative charge is the unifying theme, with polymers and their models, the vehicle. The first chapter provides orientation in the general field of anionic addition polymerization and provides a context in which living polymers are placed. The structural, stereochemical, and kinetic problems are stated. The second chapter deals with the environmental factors that control molecular weight and its distribution, and the means by which block and other "tailored" structural features of polymers are generated. Chapter III is concerned with the thermodynamics of carbanion addition and of polymerization. The interplay of techniques and fundamentals is stressed. The next chapter details techniques applicable in kinetic studies of carbanion reactions and most useful from the point of view of extraction of mechanistic information. These four chapters provide a background for discussions of the next several chapters that deal with reaction mechanism. Chapter V is a wide-ranging discussion of ions, ion-pairs, rates of dissociation and association of ions, their structure, their solvation, their reactivities, and their thermodynamic properties. Radical-ions, particularly radical-anions and solvated electrons, are similarly treated in Chapter VI, along with electron transfer phenomena. These two chapters build a foundation for the remaining six in which physical-chemical concepts are applied to the anionic polymerization process. Initiation, propagation, copolymerization, and termination are given explicit treatment, as are the special problems encountered with organolithium compounds, both simple and polymeric.

The character of this book springs more from what is found at the paragraph level than in the formal organization. Professor Szwarc is equally conversant with quantum and molecular models. The concepts and tools of the physical chemist merge with those of the structural organic chemist in many of the chapters. His quest for knowledge of the structure, properties, and reactivities of short-lived ionic species represents a confluence of streams of ideas and techniques to which all the chemical disciplines contribute. Stereochemical and kinetic probes of mechanism stand side by side with spectroscopic and conductivity results obtained on static systems. Equally impressive is the evolutionary flavor that emerges enough times from the print to remind the reader of the historical dimension of chemistry. The enthusiasm of the author for his subject is always in evidence, and his use of words and imagery are pleasing and natural. For example, in discussing changes with temperature of equilibrium constants between contact and solvent-separated ion pairs, the author states: "The same solvent at two different temperatures represents, after all, two different media. For example, tetrahydrofuran at $+25$ and $-70°C$ has different densities, viscosities, dielectric constants, etc. One may say, with some justification, that it is only the label on the bottle that remains the same, but the content is different at different temperatures."

A monograph oriented toward research that draws from so many disciplines must be written by a practitioner of those disciplines, and here the author shines. He knows what he is writing about because he led the way in many of the subjects treated. This is particularly true in his discussions of living polymers and the treatment of ion-pair phenomena. This book should be of value to those investigators who wish to look behind the curtain of empiricism and find the beginnings of a science of anionic polymerization. For creative readers, the book points to many unsolved and intriguing problems. For the scholarly, the book reflects the critical judgment of a talented and seasoned investigator.

July 1968   DONALD J. CRAM
                                         University of California at Los Angeles

Preface

Developments in the chemistry of carbanions and ion-radicals have been astonishingly rapid during the last ten or fifteen years. A comprehensive survey of the whole subject in a single volume is virtually impossible and, therefore, attention in this book is focused on the mechanistic aspects of this fascinating field.

In his recent book *Fundamentals of Carbanion Chemistry*, Donald Cram extensively reviewed molecular rearrangements and proton-transfer reactions involving carbanions. Consequently, these topics are treated lightly here, and the emphasis is shifted to electron transfer reactions, and to the addition of carbanions to olefinic and unsaturated substrates. Anionic polymerization provides an excellent example of such a process and the application of living polymers facilitates the relevant research and greatly extends its scope. Indeed, the new vista in synthetic chemistry opened through the advent of living polymers is impressive and the pertinent achievements are reviewed in the second chapter of this book. Living polymers provide, also, new approaches to thermodynamic problems, and this subject is treated in Chapter III.

Kinetic studies of carbanions and ion-radicals require special techniques and call for many elaborate precautions in experimentation. It is desirable to discuss all of these problems together, and this is done in Chapter IV.

Studies of organic processes involving ionic species provide ample demonstration of the importance of ionic aggregation. It is imperative to know the fundamentals of the structure of free ions, ion-pairs, and their agglomerates in order to understand the patterns of reactions in which these species participate. Therefore, a general discussion of the behavior of ions, ion-pairs, and their agglomerates in non-aqueous solutions is presented in Chapter V.

The subject of ion-radicals and related entities is too large to be covered fully in one chapter. Therefore, Chapter VI focuses on problems of electron affinities, electron transfers, disproportionation and dimerization of ion-radicals, and others.

Chapters VII–IX are devoted to a detailed discussion of the interaction of carbanions with vinyl monomers and other unsaturated compounds. The effects arising from such phenomena as ion-pairing, solvation, and complexation are emphasized. It is hoped that the reader will be able to generalize the outlined principles and apply them to other organic reactions of ionic and ionizable species.

Chapters X and XI digress from the main theme. Nevertheless, the reactions discussed there are interesting to the student of ionic processes, and this justifies their inclusion in this volume. The ideas outlined in Chapter XI are highly speculative, but they may be important for some systems perhaps not yet known.

Some processes leading to the destruction of carbanions are discussed in the last chapter. However, this subject is by no means exhausted and provides only a few glimpses into this vast field.

In conclusion, I wish to express my thanks to my students and co-workers whose work and enthusiasm greatly contributed to this volume. I am especially grateful to Dr. M. Levy who pioneered most of the studies carried out in this laboratory, to Dr. J. Smid who contributed so much to the understanding of the role of ions and ion-pairs in these reactions, and to Dr. J. Jagur-Grodzinski who elaborated the details of many electron transfer reactions. My thanks go to Professors F. S. Dainton, K. J. Ivin, R. M. Fuoss, and T. G Fox, and to Drs. H. A. Skinner, R. Waack, H. W. McCormick, and J. M. Pearson who helped by critically reading the individual chapters. I acknowledge also the help of Dr. J. F. Henderson in writing Chapter II.

Finally, I wish to express my gratitude to my colleague and friend, E. C. Jahn, Dean of the State University College of Forestry, for his help and encouragement throughout the years. Also, I gratefully acknowledge the financial support of the National Science Foundation which made our studies possible.

July 1968

M. SZWARC
Syracuse

Contents

Chapter I

Introduction

1. POLYMERIZATION PROCESSES AND THEIR CLASSIFICATION

The process by which small molecules are converted into large ones, attaining eventually macromolecular dimensions, is known as polymerization. It is customary to differentiate between two types, viz., condensation and addition polymerization. This classification was founded on criteria that evolved through various stages, concurrently with the development of polymer chemistry. The mechanistic distinction, based on the diversity of mechanisms governing the progress of these processes, seems to be the most appropriate criterion for the classification because it stresses the essential differences between these two reactions. On the other hand, the distinction between addition and condensation *polymers* is questionable and perhaps even undesirable. Although products of addition polymerization often differ from those which result from condensation polymerization, it is possible to visualize systems which form the same macromolecules by both routes. For example, polycondensation of β-hydroxypropionic acid yields a polyester which could also be synthesized by the addition polymerization of β-propiolactone initiated by a strong base. The ambiguity, illustrated by this example, proves the point. It is advantageous, however, to retain the distinction between condensation and addition polymerizations.

Condensation polymerization proceeds only with bifunctional or polyfunctional monomers. It is a stepwise process consisting of a series of chemically identical reactions. In each of them a covalent bond is formed in the course of interaction involving two functional groups associated with two different molecules. Thus, monomers are combined into dimers, the latter form trimers or tetramers, and, as the reaction progresses, still larger units are eventually produced. Such a progress decreases continually the number of molecules present in the system and simultaneously increases their average molecular weight. It is described by the equation:

$$P_i + P_j \rightarrow P_{i+j}$$

where P_i, P_j, and P_{i+j} denote an i-mer, j-mer and $(i+j)$-mer, i and j being any

1

integers, not excluding unity. This equation is the most characteristic feature of polycondensation.

The salient feature of addition polymerization is the participation of reactive centers in the reaction leading to evolution of polymeric molecules. The polymerization proceeds by a consecutive accretion of monomeric units to reactive centers, and on completion of each addition cycle a new reactive center is regenerated at the end of the enlarged macromolecule. This process yields usually linear polymers, each chain continuing its growth until the reactivity of the growth-propagating end group is lost or the supply of monomer is exhausted. The growth-initiating centers may be produced gradually in the course of polymerization, or all of them may be formed at the onset of the reaction. Their activity may be retained until the conversion is completed, or their decay may continually take place in the course of polymerization. These are the factors that determine the concentration of growing centers at any stage of the process, and therefore they influence profoundly the kinetics of conversion of a monomer into a polymer.

In many condensation polymerizations, small molecules are eliminated whenever a new linkage is formed. For example, formation of each ester bond in the polycondensation of a diol with a dicarboxylic acid takes place with the simultaneous elimination of one water molecule. It was thought once that such an elimination reaction is the most characteristic feature of condensation polymerizations, differentiating it from addition polymerization. This view cannot be maintained. For example, polymerization of 1-bromo-2-dimethyl-amino ethane undoubtedly represents a polycondensation process,

$$n \, Br \cdot CH_2CH_2 \cdot N(CH_3)_2 \rightarrow Br{-}[CH_2 \cdot CH_2 \cdot \overset{+}{N}(CH_3)_2]_{n-1}{-}CH_2 \cdot CH_2 \cdot N(CH_3)_2$$
$$Br^-$$

although no small molecules are eliminated in the course of this reaction. Similarly, polymerization of urethanes, a process of great industrial importance, exemplifies a polycondensation yielding what is commonly known as a typical condensation polymer, viz.,

$$n\{HO \cdot R \cdot OH + OC \cdot NR'N \cdot CO\} \rightarrow$$
$$HOR \cdot [O \cdot CO \cdot NHR'NH \cdot CO \cdot OR]_{n-1}O \cdot CO \cdot NH \cdot R' \cdot NCO.$$

Again, no small molecules are produced in this polymerization. On the other hand, polymerization of Leuch's anhydride (NCA) into a polypeptide is a chain-propagated polyaddition,

$$\sim\!\!NH_2 + \underset{\underset{\displaystyle CH(R)}{OC\diagdown{}_{NH}}}{\overset{O{-}CO}{|\quad\ |}} \longrightarrow \ \sim NH \cdot CO \cdot CH(R) \cdot NH_2 + CO_2.$$

Nonetheless, a molecule of CO_2 is eliminated in each consecutive propagation step.

Addition polymerizations are classified according to the nature of the growth-propagating center. Thus, we classify a process as a radical polymerization when the growth-propagating end group possesses one unpaired electron. The presence of a reactive, positively or negatively charged group on the end of a polymeric molecule imparts an ionic character to the respective polyaddition process. Coordination polymerization ensues when a reactive coordination complex, attached to the end of a developing polymer, is responsible for its growth.

Distinction between radical and ionic polymerization is unequivocal, although in some systems initiation may simultaneously induce ionic and radical growth. On the other hand, the demarcation line between ionic and coordination polymerization is ill-defined. Cations and anions must be involved in any ionic system and, depending on the conditions of the reaction, the ions may be free or associated into neutral ion-pairs. A strongly associated ion-pair may be considered as a complex and, vice versa, a highly polar complex may be treated as an associated ion-pair. Clearly, these are the border cases which may be arbitrarily classified as ionic or coordination types of polymerization. Even a more confusing situation may be created when growth-propagating complexes exist in equilibrium with ion-pairs or dissociated free ions. All these species may contribute to polymerization, albeit each to a different degree. Such systems behave in a complex manner, particularly if the character of the growth-propagating ends changes during the lifetime of an individual polymeric molecule. The growth of such a macromolecule is governed then by one type of mechanism in the earlier stages of its formation, while another mode controls it in later periods. Each of these processes may impart a different characteristic property to the resulting segment of a polymeric molecule. For example, one mode of addition may result in an isotactic arrangement while another gives a syndiotactic linkage; or in diene polymerization a *cis*-isomer may be produced by the activity of one center while a *trans* could result from the growth of another. The sequence of segments in such a polymer cannot be described by a Markoff chain, i.e., the probability of finding a particular type of segment at position n is not solely determined by the sequence of all the preceding $n-1$ units, but it is governed by some outside events associated with the momentary character of the growing center which is no longer perceptible in the final product.

In coordination polymerization, the reacting monomer molecule is probably first associated with the complex—the coordination catalyst—and thereafter is inserted into the growing chain, thus forming a covalent bond with a preceding unit. This behavior is often considered as the shibboleth of coordination polymerization. In radical polymerization, addition of a monomer to a growing chain occurs directly in one step, and it was believed that ionic polymerization proceeds in a similar way. However, a two-step addition may not be uncommon in many typical ionic polymerizations. A molecule of monomer may first become loosely associated with the counterion of a growing polymer, and be bounded at this stage by, say, charge-transfer forces. Its insertion into a polymeric molecule, and formation of a covalent bond, takes place then in another kinetically distinguished reaction. Evidence favoring the two-step propagation in ionic polymerization will be presented later; however, it should be stressed at this juncture that a two-stage propagation need not necessarily be a unique feature encountered in coordination polymerization only.

In ionic and in coordination polymerizations, the growth-propagating end group is associated with another species: with a counterion in ionic propagation, and with a coordination complex in coordination polymerization. The bond linking the last unit of a growing macromolecule with its escort always possesses some ionic character. It is customary to classify the process as ionic if this bond is mainly electrovalent, whereas, if its character is essentially co-valent, the polymerization is included in the coordination class. This is not a rigorous division, although it may be useful.

Ionic polyadditions are naturally subdivided into polymerizations involving positive or negative centers. In this classification, one is concerned with the polarity of the last atom, or group of atoms, of the terminal recurring unit in the growing polymer. The best known examples of these two types of reactions are the carbonium ion and the carbanion polymerizations of vinyl monomers.

The distinction between cationic and anionic polymerization seems to be unambiguous. The cationic process ensues when the last atom of the growing chain carries a positive charge or forms the positive end of a dipole, whereas, if the charge is negative, or the atom in question forms the negative end of a dipole, the resulting polymerization is classified as anionic. However, it is possible that the rate-determining step of a cationic polymerization may involve the respective negative counterion, and an analogous situation may be encountered in some anionic polyadditions. For example, a monomer approaching an anionically growing polymer may initially interact with its positive counterion instead of its negatively charged end. Such an interaction results in the formation of a labile intermediate; and, then, either its stationary concentration or its rate of formation may determine the rate of propagation. The classic anionic propagation is a nucleophilic reaction, but, if pre-association with the positive counterion is rate-determining, the reaction acquires an electrophilic character. Reverse cases may be observed in some cationic polymerizations. Consequently, it has been suggested by Danusso (1) that ionic polymerizations should be subdivided into positive and negative cationic polyadditions and negative and positive anionic polymerizations. Of course, such a classification does not apply to reactions proceeding through a concerted push–pull mechanism, when the approaching monomer is polarized simultaneously by both poles of the growing center, each acting on the respective end of the added molecule.

2. HOMOGENEOUS AND HETEROGENEOUS POLYMERIZATIONS. STEREOSPECIFICITY AND ORGANIZATION

Like other chemical reactions, polymerization may proceed in a homogeneous or heterogeneous system. Heterogeneity may arise from various causes. The polymer may be insoluble in the reaction medium and then it precipitates while still growing. Precipitation does not lead to a new type of polymerization, neither does it change the fundamental character of the investigated process, although some new peculiar features may appear. For example, in radical polymerization, precipitation may trap the free radicals and this interferes with their bimolecular termination. Nevertheless, the reaction retains all the characteristics of a radical polymerization.

The insolubility of the initiator is another cause of heterogeneity. For example, a complex of boron trifluoride with water ($BF_3 \cdot H_2O$) is insoluble in hexane. Polymerization of isobutene induced by this initiator and carried out in hexane is, therefore, heterogeneous. This state of the system may affect its kinetics; nevertheless, the carbonium ion character of the reaction remains its most distinct feature.

Many coordination polymerizations involve insoluble initiators of a rather complex nature. Although this heterogeneity influences the behavior of the system, basically the reaction should not substantially differ from a coordination polymerization involving a similar complexing agent which is soluble in the reaction medium. It had been claimed (2) that the insolubility of a coordination complex is a necessary condition needed for the formation of a stereospecific polymer. This hypothesis can no longer be upheld because examples of homogeneous stereospecific polyadditions are now well known. It happens that most of the available highly stereospecific initiators form insoluble and infusible solids. This seems to be, however, an incidental and insignificant feature of these systems and not the rather fundamental reason for their stereospecific controlling power (3). The group that ascertains a particular stereo-placement of the added monomeric molecule, or which determines the stereoconfiguration of the preceding monomeric segment if its configuration becomes fixed only *after* the addition of the next monomer, cannot exert its influence beyond, say, a 20-Å radius. Hence, the species responsible for stereo-specificity need not be larger than, at most, 40 Å in diameter and, therefore macroscopic surface, or an insoluble catalyst, is not needed. A similar argument, denying the necessity of heterogeneity in stereospecific polymerization, was forwarded by Patat and Sinn (4).

Macroscopic surface may play an essential role in inducing stereospecificity of the resulting polymer, if many monomeric molecules have to be arrayed into a well-defined pattern *prior* to their polymerization. For example, there were suggestions that monomers are first adsorbed on Ziegler–Natta catalysts and then the crystal lattice organizes them, i.e., its order is reflected in the orientation of the adsorbed layer. Subsequent polymerization perpetuates the pattern and produces, therefore, a stereospecific polymer. Although this idea is most interesting, and systems operating in that way may be found in future, there is no doubt that such a mechanism does not apply to the ordinary Ziegler–Natta type of process.

There are techniques which allow one to array the monomeric molecules *prior* to their polymerization. For example, polymerization of a crystalline monomer in the solid state may yield a polymer having a structure which reflects the order imposed upon the monomer by the crystal lattice. The most striking example of such a phenomenon was described by Okamura (5) who produced natively oriented polymeric fibers by γ-ray irradiation of single crystals of trimethylene oxide.

If a polymer precipitates while it is formed, the orientation of the incoming monomer may be controlled by the precipitated macromolecular chain. Examples of such phenomena are encountered in the spontaneous polymerization of gaseous *p*-xylylene on a solid poly-*p*-xylylene film (6) and in the Ziegler–Natta type of polymerization of ethylene on a catalyst deposited on natural fibers (7).

Urea and other compounds of similar nature crystallize in the presence of hydrocarbons, forming canal complexes. These are crystalline structures built up, say, from urea molecules arranged in a tubular fashion with the hydrocarbon guest-molecules trapped inside the tubes. The guest molecules form, therefore, a linear pattern, and, e.g., in the case of butadiene, the individual molecules acquire a *trans*-conformation which maintains contact between the 4-carbon atom of one molecule and the 1-carbon of the following one. Consequently, the polymerization of butadiene entrapped in such a complex yields all *trans*-1,4-polybutadiene (8). The regularity of this polymer reflects again the order initially imposed upon the monomer, because in the absence of urea a random polymer is formed under otherwise identical conditions.

The effect of the initially ordered orientation of monomer on its polymerization was discussed by Kabanov et al. (61). For example, polymerization of liquid crystals of methacrylyl-hydroxy-benzoic acid yields stereo-regular polymer, although atactic polymer is formed when the polymerization proceeds in solution. Some unusual phenomena observed in the course of polymerization of salts of vinyl pyridinium ions (62) also suggest that their initial orientation, and the resulting agglomeration, play an important role in the process.

A stereospecific polymer may organize adjacent molecules of monomer and thus induce a pattern in the resulting polymer. Such replica polymerization has been discussed (9). Similar phenomena take place in many biological systems where templates induce order in the macro-molecules produced. Interesting examples of synthetic templates, which operate in vinyl-type addition polymerization, have been recently described by Kämmer and Ozaki (10). Although not much is known about stereospecificity of the resulting product the template ensures a unique molecular weight for the formed polymer.

Finally, some most interesting findings of Liquori and his associates (53) should be mentioned. It is known that polymers may form complexes if their structures complement each other. The double helix of DNA, proposed by Watson and Crick (54), is the classic example of such a complex. Here the two polymeric molecules associate with each other in a parallel fashion and, therefore, this structure does not extend through three dimensions. Recently, an interesting and novel structure was proposed for nucleohistones by Zubay (55). In his model DNA molecules form an array across which the α-helical proteins are arranged; thus, a crisscrossed network is formed.

It is known that methyl methacrylate may be polymerized into an isotactic polymer under one set of conditions, or it may form, under different conditions, a syndiotactic polymethyl methacrylate (56). It is also possible to produce a third form, which was assumed to be a stereo-block. Studies of solution behaviour of the stereospecific polymethyl methacrylates revealed a peculiar behaviour of mixtures of their isotactic and syndiotactic polymers (57,58). Liquori (59) was impressed by the similarity of organizations revealed by such a mixture and by the "stereo-block." He proposed, and essentially proved, that a 1:2 stereo-complex is formed between an isotactic and syndiotactic polymethyl methacrylate. The chains of syndiotactic polymers, being placed across the arrays of isotactic polymethyl methacrylates, fill up the grooves, and form a crisscrossed network similar to that proposed by Zubay for nucleohistones.

This observation suggests that polymerization of methyl methacrylate carried out in the presence of a stereospecific polymer may indeed affect and regulate the process. The intriguing observations of Melville and Watson (60) should, therefore, be reexamined in the light of these findings.

3. BASIC STEPS OF ADDITION POLYMERIZATION

Three basic steps participate in most addition polymerizations: (*1*) An initiation, i.e., a process forming the reactive centers from which macromolecules evolve. This stage of the reaction may be referred to as the birth of a polymeric molecule. (*2*) The initiation is followed by a propagation, or growth, step during which monomer molecules are consecutively added to the active centers, yielding eventually a linear array of interconnected monomer segments of a polymer. (*3*) Finally, the activity of the growing end group of a macromolecule may be lost. This process is known as the termination of polymerization and may be described as the death of a macromolecule. The resulting product, incapable of continuing its spontaneous growth, is known, therefore, as a dead polymeric molecule. A macromolecule may irrevocably lose its ability to grow. Such a process represents a genuine termination. Alternatively, the capacity to grow may be transferred from the growing polymer molecule to another species, and then a new macromolecule eventually evolves from the latter. Such a reaction is referred to as a chain transfer—it terminates the growth of one macromolecule by converting it into a dead polymer, but it does not interrupt the kinetic chain of the reaction. Chain transfer may involve dead polymer molecules, and then new growing chains appear on their side. This process yields, therefore, branched chain macromolecules.

Most of the conventional addition polymerizations involve some termination and, therefore yield dead polymer molecules. Termination is inevitable in radical polymerization, at least in systems which do not become too viscous. It arises from a rapid and unavoidable interaction between two radicals in which both species become annihilated through their combination or disproportionation. Termination is also inevitable for those growing macromolecules which are essentially labile and undergo a spontaneous transformation into stable and unreactive species. This case is illustrated by some carbonium ion polymerization. For example, in the polymerization of isobutene initiated by the $BF_3 \cdot H_2O$ complex termination results from intramolecular proton transfer,

$$\sim CH_2C(Me)_2^+, \ (BF_3OH)^- \rightarrow \sim CH{=}C(Me)_2 + BF_3 \cdot H_2O,$$

or

$$\sim CH_2 \cdot C^+(CH_3){-}CH_3 \rightarrow \sim CH_2 \cdot C(CH_3){=}CH_2 + BF_3 \cdot H_2O.$$

In the polymerization of styrene initiated by HCl in nitrobenzene, the collapse of the ion-pair represents another type of unavoidable termination,

$$\sim CH_2CH(Ph)^+, Cl^- \rightarrow \sim CH_2CHCl(Ph).$$

A chain-transfer agent or a terminator may lead to still another type of termination. For example, in the presence of low-molecular weight alcohol, the anionic polymerization of ethylene oxide is terminated by a reaction capable of initiating the growth of another macromolecule.

$$\sim\!CH_2CH_2O^-,Na^+ + ROH \rightarrow \sim\!CH_2CH_2OH + RO^-,Na^+$$

$$RO^-,Na^+ + \text{ethylene oxide} \rightarrow \text{new polymer.}$$

Polymers possessing reactive but stable end groups may remain active, at least for the time needed to complete the polymerization, if the experimental conditions are judiciously chosen. This requires a rigorous exclusion of impurities and other substances which may act as terminators or chain-transfer agents. Under these conditions the macromolecules finally produced retain their ability of spontaneous growth and, therefore, they are referred to as living polymers (11) in contradistinction to dead polymer molecules which have lost this ability. In a system forming living polymers, polymerization comes to an end when the monomer, initially present in the reacting mixture, becomes exhausted. The characteristic feature of such a system is revealed on addition of further monomer. Not only does spontaneous polymerization reappear, but the reaction increases the size of the existing macromolecules, i.e., their number average molecular weight becomes higher.

Living polymers are often encountered in anionic polymerizations and their characteristic features will be discussed in a later section of this chapter.

4. EARLY STUDIES OF ANIONIC POLYMERIZATION

The first reports of the phenomenon known today as anionic polymerization appeared at the end of the last century when several authors reported the formation of gums and resins under the influence of alkali metals. Conscientious activity in this field started at the beginning of this century and in 1910 a patent was issued in England to Matthews and Strange (12) describing the polymerization of dienes induced by alkali metals. In the following year Harries (13) published his pioneering work on isoprene polymerization and three years later Schlenk (14) described the formation of high-molecular weight polymers prepared in ether solution by reacting styrene or 1-phenylbutadiene with sodium dust. However, the modern concepts of polymerization were not appreciated in those days, and it was not until 1920 that Staudinger (15) recognized the chain character of addition polymerization. He was probably also the first to understand the anionic character of formaldehyde polymerization initiated by bases such as sodium methoxide. Strangely enough, eight years later it was argued (16) that polymerization initiated by alkali metals was propagated by radicals! This most interesting mistake is instructive and it is beneficial, therefore, to examine the origin of this error.

In the middle 1920's Ziegler in Germany and Lebedev in Russia began systematic studies of polymerization processes initiated by alkali metals. Both investigators recognized that the process is a chain reaction, described in those days as an attack of a C—Na bond on C=C double bond (17). This description is essentially equivalent to the modern representation of anionic propagation as shown, e.g., by the following equation:

$$\sim\!\!\overset{\diagdown}{\underset{\diagup}{C}}^-,Na^+ + CH_2:CH(Ph) \rightarrow \sim\!\!\overset{\diagdown}{\underset{\diagup}{C}}\cdot CH_2\cdot CH(Ph)^-,Na^+.$$

According to Ziegler (18) the polymerization was initiated by disodium compounds, e.g.

$$2Na + CH_2:CH\cdot CH:CH_2 \rightarrow Na\!-\!CH_2\cdot CH:CH\cdot CH_2\!-\!Na$$

or

$$2Na + CH_2:CH(Ph) \rightarrow Na\!-\!CH_2\cdot CH(Ph)\!-\!Na,$$

and in his subsequent papers (19,20) he furnished further evidence to support his claim. However, many observations reported in those days could be accounted for by mechanisms involving radical-ions, and, in fact, this approach was proposed in 1930 by Schlenk and Bergmann (21). The concept of radical-ions was yet unknown, and therefore Schlenk and Bergmann represented the initiating species as radicals, e.g., $Na\!-\!CH_2\cdot CH:CH\cdot CH_2^{\cdot}$. This representation led them to believe that the resulting polymerization is propagated through a radical mechanism (21), while it is recognized nowadays that radical-ions rapidly dimerize (11,22) and consequently the resulting polymerization is propagated by dianions.

In the period 1935–45 the concepts of addition polymerization were firmly established as a result of intensive studies of radical processes. Ionic polymerizations attracted the attention of a few investigators only, and consequently development in this field languished. The success of radical theories caused even some regression, e.g., Schulz (23) in 1938 and Bolland (24) as late as 1941 were still upholding Schlenk and Bergmann's ideas attributing a radical mechanism to the polymerization of butadiene initiated by metallic sodium.

Nevertheless, slow progress was made. Ziegler correctly explained some peculiarities of the sodium-initiated polymerization, and the heterogeneous nature of this reaction was established by the elegant experiments of Abkin and Medvedev (25). These workers carried out polymerization of butadiene in a closed vessel shown in Figure 1. The reaction was initiated by bringing the monomer into contact with metallic sodium placed in arm *A*, and the progress of the reaction was then followed manometrically. After a while, the monomer was transferred by distillation to arm B, and the connected stopcock was closed. The constancy of the pressure demonstrated the absence of any reaction, indicating that the reactive species were destroyed or left behind in arm A. When the monomer was retransferred into A the reaction resumed with its previous rate. Abkin and Medvedev concluded, therefore,

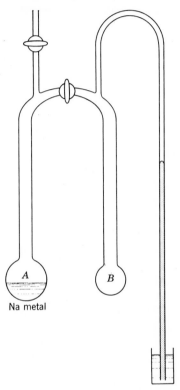

Fig. 1. The apparatus used by Abkin and Medvedev to demonstrate the long-lived nature of the species responsible for anionic polymerization of butadiene.

that the reacting species are long-lived. They were not destroyed by the distillation of the monomer from arm A into B and, therefore they could not be radicals. The S-shaped conversion curves showed that a slow initiation was followed by a relatively rapid propagation. This observation verifies the earlier conclusion of Ziegler.

In the following years, several processes were recognized as anionic polymerizations. Blomquist (26) described the polymerization of nitroolefins and implied that this is an anionic process initiated by hydroxyl ions. Beaman (27) recognized the anionic character of methacrylonitrile polymerization initiated in ether solution by Grignard reagents or by triphenylmethylsodium. This reaction was shown to be extremely rapid, and produced a high-molecular weight polymer when initiated by metallic sodium in liquid ammonia at $-75°C$. The lack of inhibition by oxygen or by the sodium salt of quinone was correctly brought forward as an argument against the radical hypothesis. The extremely high molecular weight (about 100,000) of the product obtained in liquid ammonia at $-75°C$ shows that under those conditions propagation is much faster than termination.

Sanderson and Hauser (28) postulated an anionic mechanism for the polymerization of styrene initiated by sodamide in liquid ammonia. The addition of an NH_2^- ion to the C=C double bond was assumed to initiate the chain process and a proton transfer from NH_3 to the growing carbanion was postulated as the termination step which regenerates the initiating species. The molecular weight of the resulting polymer (about 3000) was independent of the sodamide concentration and of the time of polymerization, confirming, therefore, the proposed mechanism. Sanderson and Hauser reported also that sodamide does not initiate the polymerization of styrene in ether solution, and that addition of α-methylstyrene or butadiene to the liquid ammonia solution of the polymerizing styrene lowers the molecular weight of the product. It seems that the degree of dissociation of sodamide in ether solution is too low to produce a sufficiently high concentration of NH_2^- ions which, as was later shown by Higginson and Wooding (29), are responsible for the initiation step. Lowering of the molecular weight of the product on addition of α-methylstyrene or butadiene is puzzling. It may be that a relatively rapid termination of the $\sim\sim C(CH_3)(Ph)^-$ ions, which apparently are more basic than polystyryl$^-$ ions, and therefore more readily protonated, accounts for the observed retardation.

In 1949 Landler (30) reported on his studies of polymerization initiated by Grignard reagents and interpreted the results in terms of anionic polymerization. In the following year he published (31,32) the first studies of anionic copolymerization, calculating the reactivity ratios by the conventional method, i.e., by comparing the composition of the copolymer with the proportion of the respective monomers in the feed. For the system styrene–methyl methacrylate polymerized by metallic sodium in liquid ammonia, the respective reactivity ratios were found to be 0.123 and 6.4. The unreliability of the classical method of determining reactivity ratios in anionic copolymerization is now well appreciated and therefore it is not surprising that Landler's results were disputed later by Tobolsky and O'Driscoll (33).

Finally, in 1950, Walling, Briggs, Cummings, and Mayo (34) published a most significant paper describing the copolymerization of styrene and methyl methacrylate induced by a variety of initiators. Some of their results are shown in Table I, and the profound differences in copolymer compositions

TABLE I

The Composition of Styrene–Methyl Methacrylate Copolymer Obtained
with Different Initiators

Initiator	T, °C	% Styrene in the copolymer for a 1:1 styrene: methyl methacrylate feed
Benzoyl peroxide	60	51
$SnCl_4$	30	99
Na	30	1
K	30	1

led them to conclude that the polymerization initiated by sodium or potassium must be carbanionic in nature. In fact, their observation has been one of the strongest evidence for the truly anionic nature of sodium initiated polymerization. Consequently, it was suggested (34) that the composition of a styrene–methyl methacrylate copolymer provides a test which should reveal the character of the propagation step induced by an unknown initiator.

It is interesting to read in the same paper (34) the following statement: "Carbanionic polymerization seems to be the most limited in the scope of the three types..." (radical, cationic, and anionic). This sentence reflected the sentiments of those days. Obviously, the truly explosive developments which took place in the field of anionic polymerization only a few years later were not anticipated!

5. EARLY KINETIC STUDIES OF ANIONIC POLYMERIZATION

The first thorough kinetic study of anionic polymerization was reported in 1952 by Higginson and Wooding (29). Pursuing the ideas developed in their preliminary publication (35) they postulated the following mechanism for the styrene polymerization initiated by potassium amide in liquid ammonia:

$$K^+, NH_2^- \rightleftarrows K^+ + NH_2^-$$

$$NH_2^- + CH_2:CH(Ph) \rightarrow NH_2 \cdot CH_2 \cdot CH(Ph)^- \text{ initiation,}$$

$$\sim\!\!CH_2 \cdot CH(Ph)^- + CH_2:CH(Ph) \rightarrow \sim\!\!CH_2 \cdot CH(Ph) \cdot CH_2 \cdot CH(Ph)^- \text{ propagation.}$$

$$\sim\!\!CH_2 \cdot CH(Ph)^- + NH_3 \rightarrow \sim\!\!CH_2 \cdot CH_2(Ph) + NH_2^- \text{ termination.}$$

The presence of one terminal NH_2 group in each polymeric molecule was a strong argument for the proposed initiation. This fact, and the absence of unsaturation in the polymer, proved that no chain transfer to monomer occurs in the process. A possible, although highly improbable, termination involving a K^+ ion, i.e.,

$$\sim\!\!CH_2\cdot CH(Ph)^- + K^+ \rightarrow \sim\!\!CH_2\cdot CH(Ph)^-, K^+$$

was discarded, since potassium amide is not consumed in the polymerization. Moreover, such a termination should lead to a decrease in the molecular weight of the product on the addition of K^+ ions, contrary to their observations. However, the addition of K^+ ions substantially decreases the rate of polymerization, indicating that either of the following equilibria:

$$K^+, NH_2^- \rightleftarrows K^+ + NH_2^-$$

or

$$K^+, NH_2^- + CH_2:CH(Ph) \rightleftarrows K^+ + NH_2\cdot CH_2\cdot CH(Ph)^-$$

is involved in the initiation step. The participation of the latter equilibrium is ruled out, since such a process should introduce an additional $[CH_2:CH(Ph)]^{1/2}$ factor into the rate expression contrary to experimental evidence. In fact, the rate of the polymerization was found to be proportional to the square of the monomer concentration and to the square root of the amide concentration, in full agreement with the proposed mechanism. Further quantitative evidence supporting the mechanism was obtained from studies of the number average degree of polymerization, DP_n, of the produced polystyrene. This was shown to be proportional to the monomer concentration and independent of that of the initiator (K^+, NH_2^-).

At a constant monomer concentration, the temperature dependence of DP_n permits calculation of the difference between the activation energies of the propagation (E_p) and the termination (E_t) steps. Thus, $E_p - E_t$, was found to be -4 ± 1 kcal/mole, showing that termination requires a higher activation energy than propagation. This relationship, $E_t > E_p$, is often found in carbonium ion polymerization.

The temperature dependence of the rate of the overall reaction leads to an overall activation energy $E = E_i + E_p - E_t + \frac{1}{2}\Delta H_{KNH_2 diss.} = 9 \pm 2$ kcal/mole. The heat of dissociation of potassium amide ($\Delta H_{KNH_2 diss.}$) was determined independently from the temperature coefficient of the equilibrium constant of $K^+, NH_2^- \rightleftarrows K^+ + NH_2^-$, and found to be 0 ± 2 kcal/mole. Hence, the activation energy for initiation (E_i) has the value of 13 ± 4 kcal/mole.

The high activation energy of this initiation is significant. It indicates that at least some of the ammonia molecules, solvating the NH_2^- ions, must be desolvated in the transition state. This conclusion is plausible in view of the low heat of the dissociation of K^+,NH_2^-, and demonstrates that the solvation energy of the NH_2^- ion by ammonia is indeed considerable.

The studies of Higginson and Wooding are important for, at least, two reasons. (*1*) They established conclusively the participation of the three conventional steps, initiation, propagation, and termination, in anionic polymerization. (*2*) They drew attention to the problem of free ions and ion-pairs as possible distinct intermediates in such reactions. The dielectric constant of liquid ammonia is high and solvation power rather large. Hence, in this medium Higginson and Wooding found that free ions are intrinsically more reactive initiators of styrene polymerization than ion-pairs. In fact, their data not only prove that NH_2^- ions, and not K^+,NH_2^- ion-pairs, initiate the polymerization, but that the propagation involves the free $\sim\!CH_2\cdot CH(Ph)^-$ ions and not the corresponding ion-pairs. In drawing this conclusion, it is tentatively assumed that the ratios of the rate constants of the propagation and termination steps are unlikely to be the same for ion-pairs and for free ions.

The importance of free ions and ion-pairs in ionic polymerization will be discussed comprehensively in later chapters. In solvents of lower dielectric constant than ammonia, e.g., in tetrahydrofuran, the free ions may form only a minute fraction of the growing polymers. Nevertheless, their reactivity might be so high that most of the polymerized monomer adds to these species, the remaining polymers contributing only a little to the overall growth. As the dielectric constant of the reaction medium decreases, the fraction of free ions diminishes, and eventually their significance in the polymerization may become negligible. In such systems, growth involves ion-pairs.

An early example of propagation involving ion-pairs was provided by the studies of Gee, Higginson, and Merrall (36), who investigated the polymerization of ethylene oxide initiated by sodium methoxide in dioxane. The extremely low solubility of sodium methoxide in this solvent made it necessary to add some methanol to the reaction mixture, and the latter acted, therefore, as a chain-transfer agent. The propagation is described by the equation:

$$\sim\!CH_2\cdot CH_2\cdot O^-,Na^+ + \overset{\displaystyle O}{\overset{\displaystyle /\backslash}{CH_2\!\!-\!\!CH_2}} \rightarrow \sim\!CH_2\cdot CH_2\cdot O\cdot CH_2\cdot CH_2\cdot O^-,Na^+$$

and this process is coupled with chain-transfer,

$$P_n\cdot CH_2\cdot CH_2\cdot O^-,Na^+ + P_m\cdot CH_2\cdot CH_2\cdot OH \rightleftarrows$$
$$P_n\cdot CH_2\cdot CH_2\cdot OH + P_m\cdot CH_2\cdot CH_2\cdot O^-,Na^+.$$

The system is composed, therefore, of two types of chains. The growing ones having the $-CH_2\cdot O^-,Na^+$ end group and the "dormant" possessing the $-CH_2\cdot OH$ ending. The reversible chain-transfer equilibrium converts the growing polymers into dormant ones and vice versa.

The activation energy of propagation was determined as 17.8 ± 1 kcal/mole, the temperature independent factor being 2×10^8 liter mole^{-1} sec^{-1} which indicates that the degree of solvation of the reagents is not significantly changed as the system passes through its transition state.

The data of Gee, Higginson, and Merrall led to the following conclusions:

1. The rate constant for propagation, k_p, is independent of the molecular weight of the polymer if the degree of polymerization exceeds 2.

2. The magnitude of k_p depends on the counterion; e.g., the growing ends associated with K$^+$ were found to propagate the polymerization faster than those associated with Na$^+$.

3. The solvation of the ion-pairs by hydroxylic groups profoundly affects the magnitude of k_p; e.g., increasing the concentration of OH groups from 0.6 to 3.15 M raises the respective rate constant of propagation from 2.4 to 4.7×10^{-5} liter mole^{-1} sec^{-1}. These data refer to experiments performed at 30°C and for a concentration of growing ends of about 3.5×10^{-2} M. It seems that solvation favors formation of the solvent-separated ion-pairs, and that the reactivity of the latter exceeds that of contact ion-pairs. This conclusion is supported by the observation that a free alkoxide anion adds more readily to an epoxide than its ion-pair (37).

6. LIVING POLYMERS

Extensive studies of radical polymerization, and particularly the investigations carried out in the period between 1935–1950, established the basic mechanism of polyaddition (38). As in many other chain reactions, a termination is involved in this process and its existence was unquestionably demonstrated in all the investigated systems (39). A termination step has to be included in a radical mechanism, because radicals interact with each other and become annihilated through bimolecular collisions. However, it is not imperative to postulate termination for other polyadditions, although the great success of the conventional polymerization scheme created the impression that a terminationless polymerization is highly improbable, or perhaps even impossible, especially if the process involves vinyl or vinylidene monomers.

Polymerization schemes free of termination had been considered in earlier days. For example, the first kinetic treatment of polymerization, proposed by Dostal and Mark (40) in 1935, did not involve termination. Similarly, termination was not visualized in Ziegler's mechanism of the sodium-initiated polymerization of butadiene. In fact, the need for a termination step was not fully appreciated in those days. Later, several examples of terminationless polymerization were considered by Flory (41), who also discussed ramifications of such schemes.

However, it was not until 1956 that Szwarc and his associates (11,22) re-
discovered the terminationless polymerization of vinyl monomers and
clearly demonstrated all the potentialities of such systems. Their original
experiments were performed in an all-glass apparatus depicted in Figure 2.
The device was thoroughly evacuated and then sealed off under vacuum. The
initiator, a green solution of sodium naphthalene dissolved in 50 cc of tetrahydrofuran, was admitted into reactor (A) by crushing a breakseal of the
storage ampoule (B). Thereafter, 10 cc of rigorously purified styrene, stored
in ampoule (C), was slowly added into the reactor. As soon as the first drops

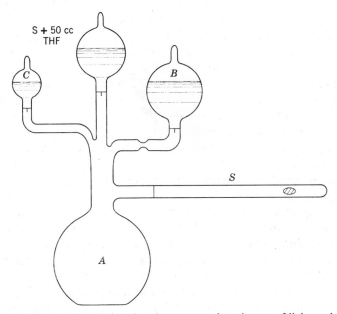

Fig. 2. The apparatus designed to demonstrate the existence of living polymers.

of monomer came into contact with the initiator solution its color changed
abruptly from green to cherry-red, and then, within a second or two, the
quantitative polymerization of styrene was completed. Nevertheless, although
the reaction was over, the cherry-red color persisted, suggesting that, perhaps,
the reactive species had remained in solution.

By turning the reactor through 90°, the long side arm (S) was placed verti-
cally and was filled with the polymerized solution. An iron weight, enclosed
in a glass envelope, was raised with the help of a magnet to a marked line and
then allowed to fall freely. The time of fall, about 5 sec, provided an estimate
of the viscosity of the investigated solution. Thereafter, the reactor was
restored to its original position, and a fresh batch of 10 cc of styrene dissolved

in 50 cc of tetrahydrofuran was added to the polymerized mixture. Again, within seconds, the added monomer polymerized but the color of the solution was not appreciably affected, although its viscosity increased considerably. As shown by the falling weight method, the time of falling increased to about 50 sec as compared to 5 sec found previously. The higher viscosity of the solution could not be attributed to an increase in the concentration of the polymer; in fact, the latter remained unaltered, viz., 0.2 g/cc. This proved that the molecular weight of the polymer formed in the first stage of the experiment increased upon addition of fresh styrene. Hence, the original polymer, produced during the first polymerization, retained its ability to grow and became longer when a second batch of styrene was added. In view of this, the polymer formed in such a system has been described as a living polymer to differentiate it from a conventional dead polymer which cannot spontaneously resume its growth upon addition of fresh monomer.

To prove that *all* of the polystyrene formed in the first stage of the process was living, a modified experiment was performed. Instead of styrene, isoprene was added to the reactor in the second stage of the experiment. The total mass of the resulting polymer, determined after its precipitation, proved that the added isoprene had polymerized quantitatively. However, the isolated material, after being dissolved in toluene, could not be precipitated from such a solution by isooctane, whereas a 50:50 mixture of homopolystyrene and homopolyisoprene *could* be separated by this procedure, polystyrene being precipitated quantitatively. This result proved that no polystyrene was left after addition of isoprene, i.e., *all* of the original polymers were living and eventually produced block polymers of polystyrene–polyisoprene.

The characteristic features of living polymers are clearly revealed by these experiments. Living polymers may resume their growth whenever monomer is added to the system. They do not die but remain active and wait for the next prey. If the monomer added is different from the one previously used, a block polymer results. This, indeed, is the most versatile technique for synthesizing block polymers and its details and applications will be discussed in the next chapter.

Living polymers do not become infinitely long. Any system producing living polymers contains a finite amount of monomer. The system also contains some specified concentration of growing centers, or living polymers, and consequently all of the available monomer becomes partitioned among them. Hence, the number average degree of polymerization, $\overline{DP}_n$, is simply given by the ratio, (total no. of moles of added monomer)/(total no. of moles of living polymers).

The lack of natural death, i.e., of spontaneous termination, does not imply immortality either. The reactive end groups of living polymers may be annihilated by suitable reagents, a process known as "killing" of living polymers. It is desirable, indeed, to distinguish between a "killing" reaction and a spontaneous termination. The latter is governed by the law of probabilities and its course is set by the conditions existing *during* polymerization. On the other hand, the moment of the "killing" reaction is determined by the free choice of the experimenter, usually after completion of the polymerization. Moreover, he is free also to choose at will the reagent which converts active end groups into dead ones. For example, in the anionic polymerization of styrene propagated by carbanions, addition of water, or of any other proton-donating substance, converts the active $\sim\!\!CH(Ph)^-$ end groups into

inactive $\overset{\diagdown}{\underset{\diagup}{C}}\!\!-\!\!H$, whereas addition of carbon dioxide converts $\overset{\diagdown}{\underset{\diagup}{C}}{}^-$ into

$\overset{\diagdown}{\underset{\diagup}{C}}\!\cdot\!COO^-$ and eventually into $\overset{\diagdown}{\underset{\diagup}{C}}\!\cdot\!COOH$. Similarly, a terminal hydroxyl group may be formed upon addition of ethylene oxide. It is important to realize that all these end groups cannot propagate styrene polymerization, and in this sense a dead polymer is produced in each case. We should stress that this technique allows us to introduce valuable functional end groups into a macromolecule, giving, therefore, a novel and interesting product. The scope of this method will be also considered in the following chapter.

Ideally, living polymers should retain their activities forever, if not subjected to a "killing" reaction. However, in any real system this is not the case. Some slow side reactions which annihilate the growing ends are always encountered, and these set an upper limit to the durability of living polymers. Nevertheless, if the rates of these reactions are sufficiently slow to permit completion of our task, the system may be correctly classified as composed of truly living polymers.

7. EXAMPLES OF LIVING POLYMERS

Perhaps, the earliest example of living polymer was provided by Ziegler. In the course of his studies of sodium initiated polymerization of butadiene, it was noted (17) that the molecular weight of polybutadiene left in contact with metallic sodium increased upon addition of fresh monomer. This result suggests that at least some living polybutadiene was present in the investigated mixture, and it continued its growth as the monomer was supplied. However, the heterogeneous nature of the system, the possibility of a renewed initiation (an excess of metallic sodium was still available), and the feasibility of cross-linking or branching made these experiments not as conclusive as could be desired. Consequently, they did not make the impact which otherwise could be expected, and did not lead to further development of this field.

The results of Abkin and Medvedev (25) may be also interpreted as a manifestation of living polymers. Polymerization of butadiene *was* resumed when monomer, previously distilled out of the system, was reintroduced. Alternatively, it is possible that a termination reformed an efficient initiator which started new chains when fresh monomer was supplied. Such a situation has been observed, e.g., in the studies of Evans and Meadows (42) who polymerized isobutene with $BF_3.H_2O$ complex. Twelve successive additions of isobutene were performed in a row and on each successive addition, the reaction was reinitiated, its kinetic course exactly duplicating the first run. The system did not contain, however, living polymers because tetramers and pentamers only were produced, as a result of efficient chain transfer to monomer.

Anionic polymerizations often provide convenient systems for the synthesis of living polymers. Polymerization of ethylene oxide is one of the earliest examples (41). Unfortunately, to overcome the insolubility of the initiators it was necessary to add some alcohol to the reacting mixture which acts as a chain transfer agent. This interrupts the growth of living polymers and converts them, at least temporarily, into dead or dormant species.

Polymerization of ε-caprolactam, pyrrolidone, and of other similar monomers can be also performed under conditions leading to the formation of living polymers growth of which is propagated through an anionic mechanism. This subject will be dealt with in Chapter XI.

Finally, one of the best examples of anionic living polymers is provided by the polymerization of the N-carboxyanhydride of amino acids (NCA). It was shown that the growing groups of such polymers $\sim\sim CO\cdot CH(R)NH_2$, may propagate the polyaddition without being destroyed, that the molecular weight of the polymer increases as more monomer is supplied, and that block polymers are formed when the newly added monomer differs from the previous one. A comprehensive discussion of this subject is given in Chapter X.

Anionic polymerization is not unique in generating living polymers. Examples of living polymers formed in cationic systems are now available. The earliest and the best documented case is provided by the cationic polymerization of tetrahydrofuran (43). Bawn and his co-workers (43) showed that $Ph_3C^+, SbCl_6^-$ initiates polymerization of this cyclic ether. They proved that the process involves a proton transfer.

$$Ph_3C^+, SbCl_6^- + \begin{matrix} H_2C-CH_2 \\ O \\ H_2C-CH_2 \end{matrix} \longrightarrow Ph_3CH + \begin{matrix} H_2C-CH_2 \\ \overset{+}{O} \\ HC-CH_2 \end{matrix}$$
$$SbCl_6^-$$

followed by the reaction,

$$
\begin{array}{ccc}
\underset{\substack{H_2C-CH_2 \\ | \quad\;\; O \\ H_2C-CH_2}}{} + \underset{SbCl_6{}^-}{\substack{H_2C-CH_2 \\ HC{}^+\!\!\diagdown_{O}\diagup CH_2}} & \longrightarrow & \underset{SbCl_6{}^-}{\substack{H_2C-CH_2 \;\; H_2C-CH_2 \\ | \quad\; O-HC\diagdown_{O}\diagup CH_2 \\ H_2C-CH_2}}
\end{array}
$$

forming an oxonium ion. The latter reacts rather slowly with another molecule of tetrahydrofuran, producing a new oxonium ion,

$$
\overset{+}{O}\!-CH_2\!\cdot\!CH_2\!\cdot\!CH_2\!\cdot\!CH_2O\!-
$$

$$SbCl_6{}^-$$

which perpetuates a relatively rapid propagation by continuous repetition of the above step. The lack of termination was demonstrated by thermodynamic studies, by increase in the molecular weight of the product upon monomer addition, and through formation of block polymers (44).

Most carbonium ion polymerizations do not yield living polymers because the propagating carbonium ions are too reactive and either rapidly decompose or isomerize into stable, nonpropagating ions. However, in some systems these undesirable reactions are inhibited at extremely low temperatures, while the propagation may remain still relatively rapid. Hence, living polymers may be formed (45,46) under these conditions.

In a typical coordination polymerization of the Ziegler–Natta type, the most common termination involves formation of the hydride with simultaneous ejection of a dead polymer containing terminal $C=C$ bond. In some catalyst systems, this reaction may be avoided, at least for a reasonable time, and living polymers are then formed. This is manifested by an increase in molecular weight of the product as the conversion proceeds, and by the formation of block polymers when another monomer is supplied (47).

Finally, two more examples of living polymers will be given. Polymerization of diazomethane initiated by BF_3 leads to $\sim\!\!\sim CH_2CH_2CH_2BF_2$—a perfectly stable species (48). Diazomethane coordinates with the $—BF_2$ moiety giving a complex $\sim\!\!\sim CH_2BF_2 \rightarrow CH_2N_2$, which decomposes spontaneously with the evolution of nitrogen,

$$
\sim\!\!\sim\{CH_2BF_2 \rightarrow CH_2N_2\} \rightarrow \sim\!\!\sim CH_2CH_2BF_2 + N_2.
$$

Hence, the product of the reaction is a living polymer, it resumes its growth whenever diazomethane is added to the system.

Polymerization of chloral may be initiated by organotin compounds (49). The propagation is due to an insertion process, viz.,

$$\text{\textasciitilde CH—O—Sn(CH}_3)_3 + \text{CH}{=}\text{O} \rightarrow \text{\textasciitilde CH·O·CH—O—Sn(CH}_3)_3,$$
$$\qquad\;|\qquad\qquad\qquad\quad\; |\qquad\qquad\;\;\; |\quad\; |$$
$$\;\;\text{CCl}_3\qquad\qquad\qquad\text{CCl}_3\qquad\text{CCl}_3\;\;\text{CCl}_3$$

and it resembles the initiation which involves a reaction of $(CH_3)_3Sn$—OR with the C$=$O double bond of chloral. The resulting organometallic compound is stable if impurities, such as CO_2 or moisture, are excluded and it continues polymerization whenever fresh monomer is supplied.

8. STABILITY OF LIVING POLYMERS

Living polymers are endowed with reactive, but also intrinsically stable, end groups. They must not interact mutually like free radicals, should not isomerize or decompose easily, and the system must not involve any agents which may convert living polymers into inert species. These conditions impose several restrictions upon systems in which living polymers operate:

1. High temperatures must be avoided and, in fact, some living polymers must be kept at extremely low temperatures to avoid their decomposition. For example, a living polymethyl methacrylate is only stable below $-60°C$.

2. Solvents have to be judiciously chosen. For example, living polymers endowed with carbanionic groups cannot be produced in proton-donating solvents, such as alcohols, liquid ammonia, etc.

3. Many living polymers react readily with oxygen, carbon dioxide, or moisture. Consequently, their studies have to be performed in equipment isolated from the surrounding atmosphere. The work has to be performed either in carefully evacuated apparatus or in equipment filled with rigorously purified inert gases.

4. Damaging impurities have to be meticulously excluded from all reagents and solvents, and the walls of the equipment which come in contact with living polymers have to be rigorously purged. The degree of purity needed in such a work is really astonishing. Two examples are given to emphasize this point.

Consider preparation of a 10% solution of living polystyrene having molecular weight of about 100,000. Moisture is one of the most damaging impurities and one may easily verify that 2 ppm of water in the system are sufficient to kill 10% of all the living polymers. Another example. Say that we wish to introduce a $10^{-5}M$ solution of living polystyrene into a cubic cell 1 cm high. Such a solution contains 6×10^{15} carbanions. Let us assume that one molecule of water adsorbed on the wall occupies an area of 10 Å². If the surface of the cube is covered by a unimolecular layer of moisture, then the amount of adsorbed water is sufficient to kill *all* of the living polymers!

9. LIVING AND DORMANT POLYMERS

In many systems involving living polymers, two or more species may co-exist, one of which does not participate, or participates only to a small extent, in the propagation. A few examples serve to illustrate this situation:

1. In ionic polymerization, ion-pairs and free ions may coexist and, in principle, both contribute to the reaction; however, free ions are enormously more reactive than ion-pairs. Thus, the ions may be considered as the living and the ion-pairs as the dormant polymers.

2. The anionic polymerization of ethylene oxide carried out in the presence of small amounts of alcohols has been discussed previously. In view of the equilibrium,

$$P_nO^-,Na^+ + P_mOH \rightleftarrows P_mO^-,Na^+ + P_nOH$$

the ion-pairs form the living polymers, whereas the hydroxyl compounds are the dormant ones.

3. Living polystyrene, $\sim\!\!S^-$, forms reversibly a complex with anthracene

$$\sim\!\!S^- + A \rightleftarrows \sim\!\!S^-,A$$

and, because the latter does not grow, it forms a dormant polymer which is reversibly converted into a living one (50).

4. It is probable (51) that a lithium salt of living polystyrene in benzene solution forms a dormant dimer $(\sim\!\!S^-,Li^+)_2$ which is in equilibrium with the living monomeric species,

$$(\sim\!\!S^-,Li^+)_2 \rightleftarrows 2\sim\!\!S^-,Li^+.$$

Apparently, only the latter is responsible for the growth step. An even more complex situation is encountered in the presence of lithium butoxide. According to Roovers and Bywater (52), the following equilibria are involved:

$$(\sim\!\!S^-,Li^+)_2 \xrightarrow{\text{LiOBu}} (\sim\!\!S^-,Li^+;BuOLi)_2$$
$$\updownarrow \qquad\qquad\qquad \updownarrow$$
$$2\sim\!\!S^-,Li^+ \qquad\quad 2(\sim\!\!S^-,Li^+;BuOLi)$$

and only non-associated $\sim\!\!S^-,Li^+$ and $\sim\!\!S^-,Li^+;BuOLi$ contribute to the growth, although the latter is less reactive than the former.

5. Anionic polymerization of methyl methacrylate is assumed to involve two types of chains (see Chapter XII). If the last three monomeric segments are linked through isotactic linkages, the chain is open and grows fast; however, if this is not the case, a ring is formed and the polymer grows slowly.

These examples suffice to illustrate the character of dormant polymers. Either they do not propagate polymerization, or only propagate it very slowly. However, they are not dead. Their conversion into living polymers is governed by some reaction which is usually independent of propagation, e.g., a proton transfer, dissociation of ion-pairs into free ions or, more generally, conversion of an associated form into a dissociated one, etc. The last example, viz., the anionic polymerization of methyl methacrylate, illustrates an interesting case when conversion is associated with propagation. This implies that at least two distinct types of propagation operate in such a system. An isotactic placement of an open chain retains its character of living polymer, while a syndiotactic placement converts it into a dormant chain, which is in turn revived when three monomer molecules add consecutively in an isotactic fashion to this slowly growing species.

References

(1) F. Danusso, *Makromol. Chem.*, **35A**, 116 (1960).
(2) N. G. Gaylord and H. Mark, *Makromol. Chem.*, **44**, 448 (1961).
(3) M. Szwarc, *Advan. Chem. Phys.*, **2**, 147 (1959).
(4) F. Patat and H. Sinn, *Angew. Chem.*, **70**, 496 (1958).
(5) S. Okamura, K. Hayashi and M. Nishi, *J. Polymer Sci.*, **58**, 125 (1962); **60**, S26 (1962).
(6) See for a review L. A. Errede and M. Szwarc, *Quart. Rev. (London)*, **12**, 301 (1958).
(7a) National Leed Co., Belg. Pat. 575,599 (1959).
(7b) H. D. Crangy, Ph.D. Thesis, New York State College of Forestry, Syracuse, New York (1966).
(8a) J. F. Brown and D. M. White, *J. Am. Chem. Soc.*, **82**, 5671 (1960).
(8b) D. M. White, *J. Am. Chem. Soc.*, **82**, 5678 (1960).
(9) M. Szwarc, *J. Polymer Sci.*, **13** 317 (1954).
(10) H. Kämmerer and Sh. Ozaki, *Makromol. Chem.*, **91**, 1 (1966).
(11) M. Szwarc, *Nature*, **178**, 1168 (1956).
(12) F. E. Matthews and E. H. Strange, Brit. Pat. 24790 (1910)
(13) C. Harries, *Ann. Chem.*, **383**, 213 (1911).
(14) W. Schlenk, J. Appenrodt, A. Michael, and A. Thal, *Chem. Ber.*, **47**, 473 (1914).
(15) H. Staudinger, *Chem. Ber.*, **53**, 1073 (1920).
(16) W. Schlenk and E. Bergmann, *Ann. Chem.*, **463**, 1 (1928).
(17) K. Ziegler and K. Bähr, *Chem. Ber.*, **61**, 253 (1928).
(18) K. Ziegler, H. Colonius, and O. Schäfer, *Ann. Chem.*, **473**, 36 (1929); K. Ziegler and O. Schäfer, *Ann. Chem.*, **479**, 150 (1930).
(19) K. Ziegler, F. Dersch, and H. Woltham, *Ann Chem.*, **511**, 13 (1934).
(20) K. Ziegler, L. Jakob, H. Wolltham, and A. Wenz, *Ann. Chem.*, **511**, 64 (1934).

(21) W. Schlenk and E. Bergmann, *Ann Chem.*, **479**, 42, 58, 78 (1930).

(22) M. Szwarc, M. Levy, and R. Milkovich, *J. Am. Chem. Soc.*, **78**, 2656 (1956).

(23) G. V. Schulz, *Ergeb. Exakt. Naturw.*, **17**, 405 (1938).

(24) J. L. Bolland, *Proc. Roy. Soc. (London), Ser. A*, **178**, 24 (1941).

(25) A. Abkin and S. Medvedev, *Trans. Faraday Soc.*, **32**, 286 (1936).

(26) A. T. Blomquist, W. J. Tapp, and J. R. Johnson, *J. Am. Chem. Soc.*, **67**, 1519 (1945).

(27) R. G. Beaman, *J. Am. Chem. Soc.*, **70**, 3115 (1948).

(28) J. J. Sanderson and C. R. Hauser, *J. Am. Chem. Soc.*, **71**, 1595 (1949).

(29) W. C. E. Higginson and N. S. Wooding, *J. Chem. Soc.*, **1952**, 760.

(30) Y. Landler, *Rec. Trav. Chim.*, **68**, 992 (1949).

(31) Y. Landler, *Compt. Rend.*, **230**, 539 (1950).

(32) Y. Landler, *J. Polymer Sci.*, **8**, 63 (1952).

(33) K. F. O'Driscoll and A. V. Tobolsky, *J. Polymer Sci.*, **37**, 363, (1959).

(34) C. Walling, E. R. Briggs, W. Cummings, and F. R. Mayo, *J. Am. Chem. Soc.*, **72**, 48 (1950).

(35) M. G. Evans, W. C. E. Higginson, and N. S. Wooding, *Rec. Trav. Chim.*, **68**, 1069 (1949).

(36) G. Gee, W. C. E. Higginson, and G. T. Merrall, *J. Chem. Soc.*, **1959**, 1345.

(37) G. Gee, W. C. E. Higginson, P. Levsley, and K. J. Taylor, *J. Chem. Soc.*, **1959**, 1338.

(38) P. J. Flory, *Principles of Polymer Chemistry*, Cornell Univ. Press, Ithaca, New York, 1953.

(39) C. H. Bamford, W. G. Berb, A. D. Jenkins, and P. F. Onyon, *The Kinetics of Vinyl Polymerization by Radical Mechanism*, Butterworths, London, 1958.

(40) H. Dostal and H. Mark, *Z. Physik. Chem.*, **B29**, 299 (1935).

(41) P. J. Flory, *J. Am. Chem. Soc.*, **65**, 372 (1943).

(42) A. G. Evans and G. W. Meadows, *Trans. Faraday Soc.*, **46**, 327 (1950).

(43) C. E. H. Bawn, R. M. Bell, and A. Ledwith, Communicated at the Anniversary Meeting of the Chemical Society, Cardiff, 1963; *Polymer (London)*, **6**, 95 (1965).

(44) M. P. Dreyfuss and P. Dreyfuss, *Polymer (London)*, **6**, 93 (1965); *J. Polymer Sci. A*, **4**, 2179 (1966).

(45) J. P. Kennedy and R. M. Thomas, *Makromol. Chem.*, **64**, 1 (1963).

(46) J. P. Kennedy, L. S. Minckler, and R. M. Thomas, *J. Polymer Sci.*, **2**, 367 (1964).

(47a) G. Bier, *Angew. Chem.*, **73**, 186 (1961).

(47b) E. G. Kontos, E. K. Easterbrook, and R. D. Gilbert, *J. Polymer Sci.*, **61**, 69 (1962).

(48) C. E. H. Bawn, A. Ledwith, and P. Matthies, *J. Polymer Sci.*, **34**, 93 (1959).

(49) A. J. Bloodworth and A. G. Davies, *Proc. Chem. Soc. (London), Ser A*, **1963** 315.

(50) S. N. Khanna, M. Levy, and M. Szwarc, *Trans. Faraday Soc.*, **58**, 747 (1962).

(51) D. J. Worsfold and S. Bywater, *Can. J. Chem.*, **38**, 1891 (1960).

(52) J. E. L. Roovers and S. Bywater, *Trans. Faraday Soc.*, **62**, 1876 (1966).

(53) A. M. Liquori, G. Anzuino, V. M. Coiro, M. D'Alagni, P. de Santis, and M. Savino, *Nature*, **206**, 358 (1965).

(54) J. D. Watson and F. H. C. Crick, *Nature*, **171**, 964 (1953).

(55) G. Zubay, in *The Nucleohistones*, Holden-Day, San Francisco, 1964, p. 95.

(56) T. G. Fox. B. S. Garrett, W. E. Goode, S. Gratch, J. F. Kincaid, A. Spell, and J. D. Stroupe, *J. Am. Chem. Soc.*, **80**, 1768 (1958).

(57) W. H. Watanabe, C. F. Ryan, P. C. Fleisher, and B. S. Garrett, *J. Phys. Chem.*, **65**, 896 (1961).

(58) N. Beredjick, R. A. Ahlbeck, T. K. Kwai, and H. E. Ries, *J. Polymer Sci.*, **46**, 268 (1960).

(59) V. Crescenzi, M. D'Alagni, A. M. Liquori, L. Picozzi, and M. Savino, *Ric. Sci.*, **33**, 123 (1963).
(60) H. W. Melville and W. F. Watson, *J. Polymer Sci.*, **11**, 299 (1953).
(61) V. A. Kabanov, K. V. Aliev, O. K. Kargina, T. I. Petrikeeva, and V. A. Kargin, *Intern. J. Pure A. Chem.*, **16**, 1079 (1967).
(62) V. A. Kabanov, T. I. Petrikeeva, and V. A. Kargin, *Proc. Acad. Sci. USSR*, **168**, 1350 (1966).

Chapter II

Application of Living Polymers to Synthetic Polymer Chemistry

1. INTRODUCTION

The discovery of living polymers had a great impact upon the synthetic techniques of polymer chemistry. The lack of termination opened a number of interesting and novel avenues for organic work. It permits the synthesis of polymers having nearly a Poisson molecular weight distribution, i.e., of materials which may be treated for certain purposes as being virtually mono-disperse. It leads to preparation of unusual and unique block polymers, star- and comb-shaped polymers, etc. It allows us, also, to introduce a variety of desired functional groups at one or both ends of polymeric chains, both in homopolymers and block polymers. Most of all, the use of living polymers provides an enormous degree of control of the architecture of a polymer and permits the synthesis of complicated macromolecules according to rigid specifications imposed by a scientific or a technological demand. The principles governing the behavior of living polymers were outlined by this writer in several publications (1) and elaborated in more recent review articles (2). Their application to practical problems appears to be successful, and a number of achievements may now be claimed.

In this chapter, the various opportunities for polymer synthesis will be critically examined. Emphasis will be laid upon present gains, technical difficulties, and theoretical limitations. Clear understanding of these problems should pave the way for further improvements of the available methods and should lead to the extension of these techniques to new fields of organic polymer chemistry.

2. POLYMERS OF UNIFORM MOLECULAR WEIGHT

Most polymerization processes do not give rise to macromolecules of uniform size. The heterogeneity of the product results from the statistical

27

character of chemical reactions. For example, consider a polymerization in which the growing chain has only two alternatives: either it may add another monomer molecule and thus increase its length, or it may be terminated, i.e., converted into a dead polymer. If the probabilities of these two events remain constant during the course of reaction, say p for the addition step and $q = 1 - p$ for the termination, then the probability P_j of j-mer formation is given by

$$P_j = p^{j-1}(1 - p).$$

The mole fraction of j-mers in the polymerized material is equal, therefore, to P_j, whereas the weight fraction is given by W_j:

$$W_j = jp^{j-1}(1 - p)/\sum_{i=1}^{\infty} \{ip^{i-1}(1 - p)\} = jp^{j-1}(1 - p)^2.$$

Hence, the number average degree of polymerization is (see ref. 3),

$$\langle j \rangle_n = 1/(1 - p)$$

and the weight average becomes

$$\langle j \rangle_w = (1 + p)/(1 - p).$$

The ratio $\langle j \rangle_w/\langle j \rangle_n$ is therefore $1 + p$. For a high molecular weight polymer $p \approx 1$, hence $\langle j \rangle_w/\langle j \rangle_n$ is approximately 2. Of course, if all the macromolecules in a polymer sample are of identical size, then $\langle j \rangle_w = \langle j \rangle_n$ and $\langle j \rangle_w/\langle j \rangle_n = 1$. Thus, it became customary to measure the uniformity of a polymer sample by the ratio $\langle j \rangle_w/\langle j \rangle_n$. The closer its value to unity, the more homogeneous with respect to their size are the macromolecules composing the investigated polymer.*

Two points should be stressed, however. The ratio $\langle j \rangle_w/\langle j \rangle_n$ does not determine the molecular weight distribution of a sample in a unique way. A precise description of polydispersity demands a complete specification of some distribution function, such as P_j or W_j. For example, consider a mixture of 37.2 mole % of 100-mers and 62.8 mole % of 10-mers. Its number average degree of polymerization is $\langle j \rangle_n = 43.5$ and the ratio $\langle j \rangle_w/\langle j \rangle_n = 2$. Obviously, such a material differs from a sample also having $\langle j \rangle_n = 43.5$, but conforming to the most probable molecular weight distribution, a term used in the current literature to denote the type of distribution discussed in

---

* Other functions were proposed as measures of the uniformity of a polymer sample. Schulz (67) recommends the expression $\langle j \rangle_w/\langle j \rangle_n - 1$, to which he refers as the dishomogeneity factor, while Zimm (68) prefers the ratio $[(\langle j \rangle_z - \langle j \rangle_w)/\langle j \rangle_w]^{1/2}$, known as the relative dispersion factor. $\langle j \rangle_z$ is the so-called z-average, defined through the third moment. The square root of Schulz's dishomogeneity factor was utilized by Lowry (136) as an index of polymer uniformity.

the preceding paragraphs. Nevertheless, both polymers have the same weight average degree of polymerization and correspond to the same ratio $\langle j \rangle_w / \langle j \rangle_n = 2$.

There is an alternative method which permits us to completely describe the molecular weight distribution, namely, by specifying all the moments. A moment of order n is defined as $Q_n = \sum_0^\infty j^n P_j$, where P_j is the mole fraction of j-mers. In these terms, $\langle j \rangle_n = Q_1/Q_0$, $\langle j \rangle_w = Q_2/Q_1$, $\langle j \rangle_z = Q_3/Q_2$, etc. For many polymerization systems it is relatively easy to find the analytical expressions for all the moments, whereas it could be difficult to derive the distribution function directly. Therefore, it is important to show that the distribution function, P_j, may be calculated in a unique way from the known moments Q_n. The proof was provided by Bamford and Tompa (69), and the method is based on the expansion of the distribution function in terms of a series of Laguerre polynomials. In addition to the exponential distribution, which was discussed previously, its more general form is of considerable interest. The relation between moments and P_j for such a distribution was discussed by Billmeyer and Stockmayer (70).

A ratio $\langle j \rangle_w / \langle j \rangle_n$ close to unity does not imply a really high degree of uniformity of the respective polymer. For example, a mixture of 100-mers and 50-mers corresponds to a $\langle j \rangle_w / \langle j \rangle_n$ ratio of 1.05, although as much as 18 mole % of the total material has one-half the molecular weight of the remaining macromolecules. When the mole % of the low molecular weight component of the mixture described above is reduced to 5%, the ratio $\langle j \rangle_w / \langle j \rangle_n$ becomes as low as 1.01.

The usefulness of the $\langle j \rangle_w / \langle j \rangle_n$ ratio as a reliable measure of polydispersity is further hampered by technical limitations associated with the determination of the number and weight average molecular weights of an investigated polymer. The number average DP (degree of polymerization) of a high molecular weight polymer may hardly be determined with an accuracy better than 3%, and the accuracy of the weight average determination probably does not exceed 2%. Therefore, the ratio $\langle j \rangle_w / \langle j \rangle_n$ is usually uncertain to within 5% and, hence, a result giving $\langle j \rangle_w / \langle j \rangle_n = 1.05$ is experimentally indistinguishable from the ideal case of $\langle j \rangle_w / \langle j \rangle_n = 1.000$. To determine the homogeneity of a really uniform polymer, one needs to develop some other techniques. For example, the temperature interval within which a polymer sample precipitates on cooling its solution in a poor solvent is apparently a very sensitive measure of its uniformity. The onset of precipitation is preceded by a large increase in the apparent $\overline{M}_w$ of the sample as measured by light scattering, a phenomenon known as the critical opalescence (137). Suppose that this substantial increase of the apparent $\overline{M}_w$ is observed at temperature $T_p + \Delta T$, while the precipitation is complete at T_p. As the sample becomes more uniform, ΔT decreases, being of the order of 2 to 3°C for a conventionally fractionated sample. An extremely high degree of uniformity is needed to decrease ΔT to 0.1°C.* McIntyre et al. (138) made some attempt to utilize this technique for characterization of the mono-dispersity.

It is hoped that gel permeation chromatography may provide the necessary tool, but more work is necessary to obtain the required calibration data. A very extensive study of this problem was recently reported by Tung and co-workers (35). The limited resolution of the chromatographic column distorts the results, and consequently special mathematical techniques had to be developed to account for this disturbance. Tung described three approaches which permit one to solve the respective integral equations. It was stressed, however, (35c) that the narrower the distribution, the more stringent are the experimental requirements for ascertaining the reliability of the calculations.

The molecular weight distribution function may be also determined from ultracentrifugation data, especially if the sedimentation technique is applied. However, here again studies of highly uniform samples are scanty, although this technique was thoroughly investigated by McCormick (15).

More refined methods to handle the sedimentation-diffusion equilibria of polymer solution have been developed recently by Scholte (186), who applied Fujita's method of concentration gradient centrifugation. This approach permits one to calculate four molecular weight averages, viz., M_n, M_w, M_z, and M_{z_1}, and it appears that molecular weight distribution can be determined with a reasonable degree of resolution.

The most reliable data about molecular weight distribution are probably obtained through careful fractionation of the investigated sample, followed by determination of molecular weights of the collected fractions. The best examples of such studies are provided by the work of Schulz et al. (187) and of Wyman (188).

Let us consider now the molecular weight distribution of a polymer pro-duced in a system in which all termination processes are rigorously excluded. The simplest situation exists in a reactor containing a monomer which is instantly mixed with growing species, having negligible size compared with the polymer ultimately produced. In such a system all the growing chains (living polymers) have equal chance to add the polymerized monomer. The reaction proceeds until the conversion is complete (assuming an irre-versible propagation) and then, as was pointed out by Flory (4), the resulting polymer has a Poisson molecular weight distribution, i.e.,

$$P_j = e^{-\gamma}\gamma^{j-1}/(j-1)!$$

* This method was mentioned to the present writer by Professor Debye. Its reliability is, however, questioned by some workers. For example, doubt has been expressed by Drs. H. W. McCormick and L. H. Tung of Dow Chemicals (private communication from Dr. McCormick).

and

$$W_j = [\gamma/(\gamma + 1)]je^{-\gamma}\gamma^{j-2}/(j - 1)!$$

where γ is the ratio $[M]_0/$[living polymers], i.e., the average number of monomer units added to each living polymer. If the degree of polymerization of the initial living polymer is assumed to be 1 (a permissible simplification), then

$$\langle j \rangle_n = \gamma + 1$$

and

$$\langle j \rangle_w = \gamma + 1 + \{\gamma/(\gamma + 1)\}.$$

Hence, the ratio $\langle j \rangle_w/\langle j \rangle_n$, which is equal to $1 + [\gamma/(\gamma + 1)^2]$, is only slightly greater than unity, e.g., for $\gamma = 100$ its value is 1.01 only. The resulting distribution is illustrated in Figure 1, where the weight fraction, W_j, is plotted as a function of j for $\gamma = 50, 100,$ and 500.

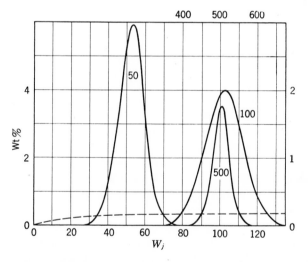

Fig. 1.   Poisson distribution for $\gamma = 50, 100,$ and 500. W_j plotted versus j. The W_j scale for the curve $\gamma = 500$ is given on the right. The dashed curve gives the "most probable" distribution for $\langle j \rangle_n = 101$ (after ref. 4).

The degree of uniformity attained in polymers having a Poisson molecular weight distribution increases greatly with their size. For example, for $\langle j \rangle_n = 100$, about 70% by weight of the polymer is composed of chains having their molecular weights within 10% of the average value. The proportion of such chains increases to 96% for a degree of polymerization of 500. The ratios $\langle j \rangle_w/\langle j \rangle_n$ for these materials are 1.01 and 1.002, respectively. In fact, it would be impossible to detect the heterogeneity of the latter polymer with the presently available techniques and, therefore, we may consider such a material as being virtually monodispersed.

In the absence of templates, the Poisson molecular weight distribution represents the highest possible limit of a polymer's homogeneity that may be achieved in any polymerization process involving linear macromolecules growing in one direction only.* Replication involving templates produces, of course, polymers of unique size and structure. Such materials are formed in biological systems, e.g., in the cellular synthesis of enzymes. Examples of nonbiological polymerization involving templates have been reported (5). For a recent review of this subject the reader may consult Schulz (139).

Let us now summarize the conditions necessary for attaining the Poisson molecular weight distribution in a prepared polymer:

(*1*) Termination, or chain-transfer, must be rigorously prevented during the course of preparation, i.e., the system should contain living polymers only.

(*2*) All the initiating species should be present and *active* at the onset of the reaction.

(*3*) A perfectly uniform spatial monomer concentration and temperature must be maintained throughout the whole reaction vessel. These need not be constant during the time of polymerization, but if they vary the change must occur simultaneously, and to the same degree, in each volume element of the polymerizing material.

(*4*) The rate constants of each consecutive addition of a monomer molecule to a growing chain should be independent of its size. In particular, the rate constant of the addition of the first molecule, which is often referred to as the initiation rate constant, k_i, should not be smaller than the rate constant of the subsequent additions, k_p.

(*5*) Condition *4* implies that all the growing chains behave identically with respect to their growth. This is not always the case. For example, in some systems free ions and ion-pairs may act as the growth propagating end groups, each growing at a different rate. Such a situation leads to a broadening of the molecular weight distribution even if a rapid, but not infinitely rapid, exchange converts one propagation center into another (see Section 7).

(*6*) Propagation must be irreversible, i.e., the rate of depropagation should be vanishingly small.

Szwarc and his co-workers (6–8) have pointed out that anionic polymerization of some vinyl monomers, which leads to the formation of living

---

* Under otherwise identical conditions, linear polymers growing simultaneously from both ends always give a product of more uniform size than those possessing one growing end only (see Section 4). In principle, it should also be possible to improve the homogeneity of a polymer if the propagation rate constant (or more generally the rate of propagation) decreases with increasing size of an individual macromolecule.

polymers, is particularly adaptable for the synthesis of polymers having a narrow molecular weight distribution. This goal was eventually achieved by several groups of workers (9–11), and polymer samples showing a dispersity ratio, $\langle j \rangle_w / \langle j \rangle_n$, as low as 1.012 have been produced. For the sake of illustration, a molecular weight distribution of a polystyrene prepared by Wenger and Yen (10) is shown in Figure 2.

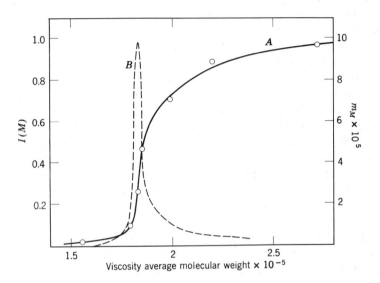

Fig. 2. Integral (A) and differential (B) weight distribution for polystyrene initiated with sodium biphenyl (after ref. 10). $I(M)$ is the total weight fraction of polymers up to molecular weight M.

The spectacular progress made in the preparation of uniform polymers should not distract our attention from the fact that even the best samples produced by the anionic technique have fallen short of expectation, the distribution being broader than predicted by the Poisson law. Obviously, some of the conditions listed above are not fulfilled by the practical systems. We shall therefore investigate the way in which the molecular weight distribution of the resulting products is distorted by failure to maintain one or the other of the required conditions.

3. FACTORS BROADENING THE MOLECULAR WEIGHT DISTRIBUTION. THE EFFECT OF IMPURITIES

A successful preparation of nearly monodispersed polymers requires meticulous elimination of all damaging impurities from the reacting system.

The reagents, solvents, and walls of the reactor have to be rigorously purged, and it is imperative to remove any trace of moisture, oxygen, and of other hydroxylic compounds, particularly when the system involves growing carbanions. Detailed description of some techniques used in practical applications were reviewed by Fetters (182) and are outlined in Chapter IV. The low molecular weight living oligomers are perhaps the best purging agents if the resulting dead material can be removed from the ultimate product. Alternatively, one may use other organometallic compounds, especially those which are reactive but not capable of initiating polymerization. For example, fluorenylsodium or lithium is a most convenient purging agent for the final purification of styrene or α-methylstyrene, because this reactive salt does not initiate polymerization of either monomer.

It has been claimed (10) that anionically polymerized systems contain some relatively inert impurities which are not destroyed during the purging operation, and which eventually cause a slow killing of living polymers during actual preparation. This writer feels that such a killing most probably arises from a slow reaction of living polymers with solvent. The purging operation cannot eliminate this difficulty; improvement of the procedure then requires a change of solvent or some judicial adjustments of the conditions maintained in the process, e.g., lowering the temperature of polymerization or increasing the monomer concentration.

Studies of electronic spectra of some anionic polymerization systems revealed that some growing end groups may change with time (134). For example, polyisoprenyllithium isomerizes rapidly in THF. Although the polybutadienyl carbanion cannot isomerize, the spectrum of polybutadienyl-potassium in THF changes significantly after prolonged storage. With lithium as the counterion, the isomerization appears to be pseudo-reversible. The original carbanion reforms when more isoprene is added to the system, indicating that the isomerized end group is capable of adding monomer and, thus, the original carbanion is again produced. Such a change does not occur when sodium or potassium is the counterion. Spontaneous isomerization of carbanion end groups of living polymers to some unreactive forms is kinetically similar to a pseudo-unimolecular termination of living polymers by solvent.

It is obvious that impurities which terminate the growth of living polymers broaden the molecular weight distribution of the final product. Polymers which were terminated before they had attained their expected size contribute to the low molecular weight fraction, whereas those which retained their activity until completion of the conversion became longer than expected because the monomer not utilized by the others contributed to their own length. The form of the resulting distribution function depends on the conditions maintained in the process, and we shall consider now some special cases often encountered in practice.

Polymerization may be carried out in a reactor containing a solvent and an initiator (e.g., a low molecular weight living polymer) to which a monomer is added slowly with efficient stirring. Such a system is referred to as a continuous feed polymerization. Impurities present in the reactor and the solvent are destroyed by the initiator *before* the onset of polymerization. However, the loss of initiator does not affect the molecular weight distribution of the final high molecular weight product, although it increases the degree of polymerization.* On the other hand, the distribution is affected by any damaging impurities present in the slowly added monomer because the living polymers are terminated continually *in the course* of their growth. The resulting distribution was derived by Szwarc and Litt (12), who assumed that killing takes place simultaneously with the addition of monomer which is rapidly consumed. The amount of initiator present at the onset of the reaction is denoted by I_0 and the total amount of monomer added at time t by M. Let f denote the mole fraction of impurities in the monomer, it follows that after the addition of M moles of monomer the amount of living polymers retained in the reactor is reduced to $I = I_0 - fM$. Subsequent addition of dM moles of monomer kills fdM moles of the growing macromolecules, having DP equal to j, and simultaneously increases the degree of polymerization of the survivors by $dj = dM/(I_0 - fM)$. Hence, the degree of polymerization of living polymers present in the reactor at that stage of the process is

$$j = \int_0^M (I_0 - fM)^{-1}dM = -f^{-1} \ln (1 - fM/I_0)$$

and

$$dx_j/dj = f \exp (-fj),$$

where $dx_j = fdM/I_0$ is the mole fraction of dead polymers having a degree of polymerization confined to the interval j up to $j + dj$. Note that dx_j/dj, i.e., the molecular weight distribution, depends on f only, while the ratio M/I_0 determines its highest degree of polymerization, j_M:

$$j_M = -f^{-1} \ln (1 - fM/I_0).$$

Thus, for $M < M_{max}$, where $M_{max} = I_0/f$,

$$dx_j/dj = f \exp (-fj) \qquad \text{for } 0 < j < j_M,$$
$$x_j = (1 - fM/I_0) \qquad \text{for } j = j_M,$$
$$x_j = 0 \qquad \text{for } j > j_M.$$

The resulting molecular weight distribution is shown in Figure 3. The peak for $j = j_M$ represents the fraction of living polymers that survived monomer

* This is the reason why the so-called "reliability coefficient" stressed in Wenger's papers (e.g., 10) does not apply to the present problem.

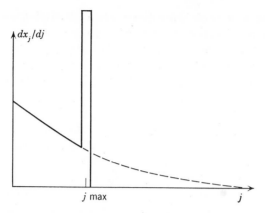

Fig. 3.   Distribution observed in a polymerization in which monomer contaminated with terminating agents is added to a living polymer solution. The peak represents the still living polymers. The distribution shown by the dashed line is obtained when all the polymers are "killed" during the monomer addition.

addition. It is obvious that all the polymers are killed when $M = M_{max}$, and, thereafter, further addition of monomer no longer contributes to the polymerization.

TABLE I

Degree of Dispersion $\langle j \rangle_w / \langle j \rangle_n$ of Polymers Obtained in a Process in which Monomer Containing Mole Fraction f of Impurities is Added Continually to the Initiator

fM/I_0 [a]	$\langle j \rangle_w / \langle j \rangle_n$	fM/I_0 [a]	$\langle j \rangle_w / \langle j \rangle_n$
0.0	1.000	0.6	1.299
0.1	1.028	0.7	1.382
0.2	1.075	0.8	1.492
0.3	1.122	0.9	1.647
0.4	1.168	1.0	2.000
0.5	1.228		

[a] Fraction of killed living polymers.

Table I shows how the dispersity coefficient, $\langle j \rangle_w / \langle j \rangle_n$, of polymer thus produced changes with $x = fM/I_0$, i.e., with the fraction of killed polymers. It suffices to kill 10% of the living polymers to increase this coefficient to 1.028. For $M = M_{max}$, this ratio becomes 2, and the resulting polymer then conforms to the "most probable" molecular weight distribution, i.e., the distribution characterizing products of a conventional radical polymerization.

In the derivation of the above equations it was assumed that, at any stage of the process, polymers still living at that time are perfectly monodispersed, all having a degree of polymerization given by j_M. The validity of this approximation may be tested. Denote by P_j^* the concentration of living j-mers and by M the constant, stationary concentration of the monomer (the condition $M = $ constant does not affect the generality of the argument). It follows that

$$dP_j^*/dt = kMP_{j-1}^* - (1 + f)kMP_j^*.$$

Substituting dZ for $kMdt$, and remembering that P_1^* at time 0 is I_0, while $P_j^* = 0$ for $j > 1$, we find

$$P_j^*/I_0 = Z^{j-1} \exp\{-(1 + f)Z\}/(j - 1)!.$$

Therefore, in spite of continuous termination due to impurities, the living polymers that survived the monomer addition have a Poisson molecular weight distribution. Because $dP_j/I_0 = f(\ln A)^{f-1}A^f dA/A(j-1)!$, where $\ln A = -Z$ and P_j is the concentration of dead j-mers, one finds

$$P_j/I_0 = \{f/(1 + f)^j\}\{1 - A^{1+f} \sum_{s=0}^{j-1} \{[-(1 + f)\ln A]^s/s!\}.$$

When $M = M_{\max}$, and all the polymers are killed, the dispersity coefficient becomes virtually equal to 2, i.e.,

$$\langle j \rangle_w/\langle j \rangle_n = 2 - f(1 + f) \approx 2 \qquad \text{if } f \ll 1.$$

Let us consider in turn a system in which all the monomer, the instantly initiating species, and the terminating impurities are rapidly mixed together and allowed to react in a well-stirred reactor. We shall refer to such a setup as a batch process. The following factors affect the distribution in this system:

(1) The kinetics of the termination processes, viz., whether it is a spontaneous, unimolecular, or pseudo-unimolecular reaction of living polymers, or a bimolecular interaction of impurities with the growing polymers.

(2) The ratio of k_p/k_t, i.e., of the propagation constant, to that of the killing process. Obviously, it is desirable that $k_p M$ be much greater than k_t (or k_t [impurity]), even when 90% of the monomer has been polymerized.

The shape of the distribution curve also depends on

(3) the ratio M_0/I_0, viz., of the initial concentrations of monomer and initiator, and, if termination involves impurities, on the ratio $T_0/I_0 = f$. The latter gives the amount of impurities present in the system at the onset of the process. If high molecular weight polymer is formed, M_0/I_0 must be large.

For spontaneous termination, the molecular weight distribution can be calculated from the following equation:

$$dx_j/dj = (k_t/k_p)M^{-1} \exp(-k_t t)$$

where M is the concentration of the monomer at time t, and the degree of polymerization of living polymers then is j. Because

$$M = M_0 \exp \{-(k_p I_0/k_t)[1 - \exp(-k_t t)]\}$$

we have

$$dx_j/dj = (k_t/k_p M_0) \exp \{-k_t t + (k_p I_0/k_t)[1 - \exp(-k_t t)]\}.$$

In such a process, conversion of monomer into polymer is never quantitative. The fraction of unreacted monomer remaining in the system at the completion of the reaction is given by

$$M_\infty/M_0 = \exp(-k_p I_0/k_t).$$

For practical reasons, we are interested in conditions which make $k_p I_0/k_t = \alpha$ large, e.g., $\alpha \geqslant 3$, because only then does 95% or more of the monomer polymerize. The distribution is characterized by the parameter $\beta = k_t/k_p M_0$. Note that for the lowest values of j $dx_j/dj = \beta$. As j increases dx_j/dj increases too, reaching its maximum value of $\beta \exp(\alpha - 1 - \ln \alpha)$ for $j \approx (1/2\beta)[1 - \exp(1 - \alpha)] \ln \alpha$, and then steeply decreases to zero. The peak of the distribution curve is high for a large α, its height being given approximately by $\beta \exp(\alpha - 1)$. For example, for $\alpha = 4$, it exceeds by a factor of 20 the value of dx_j/dj for $j = 1$. The degree of polymerization of the most abundant fraction is about $1/\beta$, e.g., it attains a value of $j = 100$ for $M_0 = 1$ mole/liter when the termination constant k_t is 100 times smaller than the propagation constant. Examples of distribution functions for living polymers which are spontaneously terminated are shown in Figure 4.

For a bimolecular termination, the fraction of killed polymers having the degree of polymerization j (i.e., terminated at time t) is given by the equation

$$dx_j/dt = k_t I_0(1 - x_j)(f - x_j),$$

where $f = T_0/I_0$ is the mole ratio of impurities to the initiator at the onset of the reaction. Obviously $dj = k_p M dt$, and therefore,

$$dx_j/dj = (k_t I_0/k_p)M^{-1}(1 - x_j)(f - x_j).$$

The concentration of the monomer, M, is

$$M = M_0 \exp \left\{ k_p I_0 \int_0^t y \, dt \right\}$$

and y, the fraction of polymers still living at time t, is

$$y = (1 - f)\{1 - f \exp[-k_t I_0(1 - f)t]\}^{-1}.$$

For $t = 0$, $dx_j/dj = k_t f I_0/k_p M_0$ and $y = 1$. The fraction y decreases to its ultimate value of $1 - f$ for $f < 1$ and to 0 for $f > 1$. For f substantially larger than 1, the problem is reduced to that of a spontaneous, unimolecular termination with a pseudo-unimolecular constant given by $k_t T_0$. However, if $f < 1$, a fraction $(1 - f)$ of living polymers survives at the end of the process; this represents a cutoff in the respective molecular weight distribution curve, which rises steeply and eventually becomes vertical at the j cutoff value.

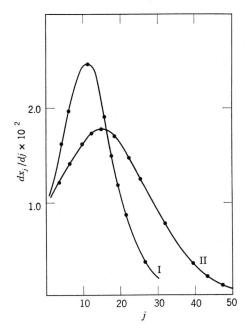

Fig. 4. Batch polymerization of living polymers in which a spontaneous first-order termination takes place: (I), $\alpha = 3.0$, $\beta = 10^{-2}$; (II) $\alpha = 2.5$, $\beta = 10^{-2}$.

The derivations given in the preceding paragraphs involved the implicit, simplifying assumption that at any stage of the reaction all the living polymers are of equal size. A more general and very elegant treatment of such problems was outlined by Coleman, Gornick, and Weiss (13). They assume that a homogeneous addition polymerization has the following properties:

(1) Initiator is monofunctional and instantly initiates polymerization, i.e., there is no further initiation during the polymerization.
(2) There is no chain-transfer or depropagation.
(3) Termination is irreversible.
(4) The spatial concentration of all the reagents is uniform throughout the whole polymerizing mass.

Two reactions only are considered:

$$L_j + M \xrightarrow{k} L_{j+1},$$

$$L_j \xrightarrow{\gamma} D_j,$$

where L_j denotes a living j-mer, D_j a dead j-mer, and M is the monomer. The rate coefficients k and γ need not be constant and may be functions of time. For example, if termination is due to impurity X, then $\gamma = k_X[X]$ and it varies with concentration of X. However, neither k nor γ depends on j, i.e., all the polymeric molecules react identically whatever their size.

Let N_t^* and N_t denote the concentrations of living and dead polymers, respectively, at time t. Of course, $N_t^* + N_t = N_0^*$. Furthermore, denote by $P_j^*(t), P_j(t)$, and $Q_j(t)$ the mole fractions at time t of living j-mers, dead j-mers, and all the j-mers, respectively. Obviously,

$$Q_j(t) = P_j^*(t) + P_j(t),$$

$$\sum_1^\infty P_j^*(t) = N_t^*/N_0^* \quad \text{and} \quad \sum_1^\infty P_j(t) = N_t/N_0^*.$$

The proposed mechanism involving propagation and termination only leads to the differential equations

$$dP_1^*(t)/dt = -\{k[M]_t + \gamma\}P_1^*(t),$$

$$dP_j^*(t)/dt = -\{k[M]_t + \gamma\}P_j^*(t) + k[M]_t P_{j-1}^*(t)$$

and

$$dP_j(t)/dt = \gamma P_j^*(t),$$

with the initial conditions $P_1^*(0) = 1$, $P_j^*(0) = 0$ for $j > 1$ and $P_j(0) = 0$ for all j. As expected, $dN_t^*/dt = -\gamma N_t^*$, and therefore,

$$N_t^* = N_0^* \exp(-\psi_t),$$

where $\psi_t = \int_0^t \gamma d\tau$ and $d\psi_t/dt = \gamma$.

Let us introduce an auxiliary variable θ_t defined as

$$\theta_t = \int_0^t k[M]_\tau d\tau,$$

i.e., $d\theta_t/dt = k[M]_t$. It can be shown that the above differential equations lead to

$$P_j^*(t) = \theta_t^{j-1} \exp[-(\psi_t + \theta_t)]/(j-1)!$$

and

$$P_j(t) = \int_0^t \{\gamma \theta_\tau^{j-1} \exp[-(\psi_\tau + \theta_\tau)]/(j-1)!\} d\tau.$$

Hence, the problem of finding the molecular weight distribution is solved if k and γ are given as functions of time, because $Q_j(t) = P_j^*(t) + P_j(t)$.

An interesting parameter of this treatment is the ratio

$$R_t = \gamma/k[\text{M}]_t = d\psi_t/d\theta_t.$$

If R_t is a constant independent of time, we deal with a special case, basically the one discussed by Szwarc and Litt (12), when the ratio of probabilities of growth and termination is time independent. Such a case will be referred to as the *stationary-ratio* polymerization. Note, however, that the above derived expressions for $P_j^*(t)$ and $P_j(t)$ are perfectly general and valid whether the stationary-ratio assumption holds or not. For $\gamma = 0$ there is no termination and the result is reduced to a Poisson distribution for $P_j^*(t)$ with parameter θ. Of course, $P_j(t)$ is then 0 for all j.

Because θ_t and ψ_t are one-to-one functions of time, θ may be regarded as a function of ψ, namely, θ_ψ. This permits us to rewrite the equation for $P_j(t)$:

$$P_j(t) = \int_0^{\psi_t} \{\theta_\psi^{j-1} \exp\left[-(\psi + \theta_\psi)\right]/(j-1)!\}d\psi.$$

The case of continuous-feed polymerization, with termination caused by impurities present in the monomer, corresponds to the condition

$$\lim_{t \to \infty} \psi_t = \infty.$$

At infinite time all the polymers are dead and the limiting distribution is given by

$$Q_j(\infty) = \int_0^\infty \{\theta_\psi^{j-1} \exp\left[-(\psi + \theta_\psi)\right]/(j-1)!\}d\psi.$$

For

$$\lim_{t \to \infty} \psi_t = \psi_\infty < \infty \qquad \text{and} \qquad \lim_{t \to \infty} \theta_t = \theta_\infty < \infty$$

we deal with a batch polymerization in which the terminating impurity is at a lower concentration than the initiator. In such a case

$$Q_j(\infty) = \theta_\infty^{j-1} \exp\left[-(\psi_\infty + \theta_\infty)\right]/(j-1)!$$

$$+ \int_0^\infty \{\theta_\psi^{j-1} \exp\left(\psi + \theta_\psi\right)/(j-1)!\}d\psi.$$

Having these results, Coleman et al. (13) derived the equation for the respective moments, i.e.,

$$\langle j^m \rangle_t = \sum_1^\infty j^m Q_j(t).$$

Thus, the number average degree of polymerization was found to be

$$\langle j^1 \rangle_t = 1 + \int_0^{\theta_t} \exp\left(-\psi_\theta\right) d\theta$$

and

$$\langle j^2 \rangle_t = 1 + 3 \int_0^{\theta_t} \exp\left(-\psi_\theta\right) d\theta + 2 \int_0^{\theta_t} \theta \exp\left(-\psi_\theta\right) d\theta.$$

In the special case of the stationary-ratio polymerization,

$$d\psi_t / d\theta_t = R = \text{constant},$$

and since $\psi_0 = \theta_0 = 0$, one finds $\psi_t / \theta_t = R = \text{constant}$. This gives

$$\langle j^1 \rangle_t = 1 + R^{-1}\{1 - \exp\left(-R\theta_t\right)\}$$

and

$$\langle j^2 \rangle_t = 1 + 3R^{-1}\{1 - \exp\left(-R\theta_t\right)\} + 2R^{-2}\{1 - \exp\left(-R\theta_t\right)\}$$
$$- 2\theta_t R^{-1} \exp\left(-R\theta_t\right).$$

For

$$\lim_{t \to \infty} \theta_t = \infty,$$

which applies to batch polymerization in which the concentration of impurities exceeds that of initiator, the limiting values for $t = \infty$ are

$$\lim_{t \to \infty} \langle j^1 \rangle_t = 1 + R^{-1},$$

$$\lim_{t \to \infty} \langle j^2 \rangle_t = (1 + R^{-1})^2 + R^{-1}(1 + R^{-1}),$$

and then $\langle j \rangle_w / \langle j \rangle_n = 1 + 1/(1 + R)$, i.e., for small R the dispersity ratio approaches 2. However, if the concentration of initiator exceeds that of impurities, then $\lim_{t \to \infty} \theta_t$ cannot be ∞ if we demand $R = \text{constant}$, and the above equations are then not valid. These limiting results, derived for $R = \text{constant}$, were previously obtained by Orofino and Wenger (135). Let us stress again that they apply to a system in which all the living polymers are terminated.

The assumption of stationary ratio is valid only if the ratio of concentrations of monomer to impurity remains constant in time. This is the case when both bimolecular rate constants, k_X and k, are equal, i.e., their ratio $\rho = k_X / k$ is unity. For $\rho > 1$, the impurity (X) disappears more rapidly than the monomer, whereas the impurity accumulates when $\rho < 1$.

The formalism developed by Coleman et al. leads to elegant expressions for the kinetics of batch polymerization following the mechanism,[*]

$$d[M]_t / dt = -k[M]_t N_t^*$$

and

$$d[X]_t / dt = -k_X [X]_t N_t^*.$$

[*] In the following treatment k and k_X are constant independent of time.

Let us put $\alpha_t = [M]_t/[M]_0$, $\beta_t = [X]_t/[X]_0$, $\sigma = [X]_0/N_0^*$, and $\rho = k_X/k$. Hence,

$$d \ln \alpha_t/dt = -kN_t^* \quad \text{and} \quad d \ln \beta_t/dt = -k_X N_t^*,$$

with the initial conditions $\alpha_0 = \beta_0 = 0$. This gives

$$\beta_t = \alpha_t^\rho,$$

and, therefore, the parameter $R_t = R_0 \alpha^{\rho-1}$; the polymerization is then classified as the stationary-ratio type only if $\rho = 1$, i.e., $k = k_X$. In view of the stoichiometric requirement,

$$N_t = N_0^* - N_t^* = [X]_0 - [X]_t,$$

we find for a general case (any ρ)

$$\beta_t = (1 - \sigma)/\{\exp [(1 - \sigma)k_X N_0^* t] - \sigma\}$$

and

$$\alpha_t = (1 - \sigma)^{1/\rho}\{\exp [(1 - \sigma)k_X N_0^* t] - \sigma\}^{-1/\rho}.$$

These equations describe the overall kinetics of such a polymerization and give the concentrations of the monomer and of the impurities as explicit functions of time. With the help of previously derived equations they lead to the moments and, therefore, to the molecular weight distribution. Our main interest is to find the expression for the moments when $\sigma \ll 1$, i.e., when impurities are highly dilute. Thus,

$$\langle j^1 \rangle_t = 1 + ([M]_0/N_0^*)(1 - \alpha_t)$$

$$\langle j^2 \rangle_t = \langle j^1 \rangle_t^2 + \langle j^1 \rangle_t - 1 + 2\{[M]_0/N_0^*\}^2 \sum_{i=1}^\infty \sigma^i \int_{a_t}^1 da' \int_{a'}^1 (1 - \alpha^\rho)^i d\alpha.$$

The ratio $\langle j \rangle_w/\langle j \rangle_n = r_t$ depends of course on σ—the proportion of impurities in the monomer. We find that

$$r_t(\sigma) - r_t(0) = 1 + \langle j^1 \rangle_t^{-1} - \langle j^1 \rangle_t^{-2},$$

where $r_t(0)$ is the ratio of the weight to number average degree of polymerization in the absence of impurity ($\sigma = 0$, i.e., Poisson distribution). For very small σ,

$$r_t(\sigma) = r_t(0) + (1 - \langle j^1 \rangle_t^{-1})^2 C(\alpha_{t,\rho})\sigma + 0(\sigma^2),$$

where $C = 1 - 2(1 - \alpha_t)^{-2}\{(1 - \alpha_t^{\rho+2})/(\rho + 2) - \alpha_t(1 - \alpha_t^{\rho+1})/(\rho + 1)\}$.

To convey the meaning of these results a plot of

$$\{r_\infty(\sigma) - r_\infty(0)\}/(1 - \langle j \rangle_{n,\infty}^{-1})^2$$

as a function of σ is given in Figure 5. Two cases are illustrated: $\rho = 1/2$ (the impurities disappear faster than the monomer), and $\rho = 2$ (the impurities accumulate in the process).

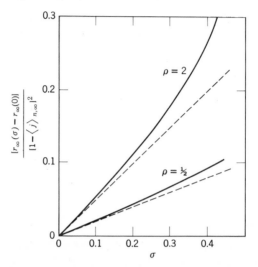

Fig. 5. Plot of $\{r_\infty(\sigma) - r_\infty(0)\}/(1 - \langle j \rangle_{n,\infty}^{-1})^2$ versus σ. The upper curve corresponds to $\rho = 2$, i.e., when the impurities accumulate in the process, and the lower line corresponds to $\rho = \frac{1}{2}$, i.e., when the impurities disappear faster than monomer. Broken lines indicate the linear approximation (after ref. 13).

4. MOLECULAR WEIGHT DISTRIBUTION OF LIVING POLYMERS PRODUCED BY BIFUNCTIONAL INITIATORS

In the preceding section, we were concerned with processes leading to living polymers endowed with one growing end per chain, i.e., derived from monofunctional initiators. It is possible to produce living polymers with both ends capable of growing, and in principle such macromolecules should have a narrower distribution of chain sizes than that formed from living polymers with one growing end only. The resulting distribution may be derived by modifying the mathematical treatment outlined by Flory (3) who investigated the molecular weight distribution of polymers formed by termination involving the coupling of two growing chains. The basic assumption states that the fate of each growing end is independent of that of the other. Hence, if P_i denotes the probability of attaining a length i through growth of one end only, then the probability of producing a macromolecule of length n through a simultaneous growth of both ends is

$$Q_n = \sum P_i \cdot P_k \quad \text{for all } i + k = n.$$

For example, if P_i and P_k are given by the most probable distribution, then

$$Q_n = \sum_{i+k=n} \{p^i(1-p)p^k(1-p)\} = (n+1)p^n(1-p)^2.$$

This leads to the dispersity ratio, $\langle j \rangle_w / \langle j \rangle_n$ of 1.5, i.e., smaller than the ratio 2 that characterizes the ordinary, most probable distribution. Similarly, if P_i and P_k are given by a Poisson distribution, Q_n corresponds to an even narrower distribution.

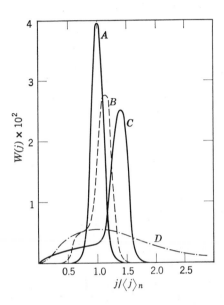

Fig. 6. Differential distribution function for bifunctional systems $\langle j \rangle_n = 101$ (after ref. 135). $W(j)$ is the weight fraction of polymers having the degree of polymerization j.

Curve	Mole fraction of added impurities	$\langle j \rangle_w / \langle j \rangle_n$
A	0.000	1.01
B	0.005 (two growing ends)	1.05
C	0.005 (one growing end)	1.23
D	0.020 (two growing ends)	1.44

The molecular weight distribution of living polymers growing simultaneously from both ends in a medium containing terminating impurities was discussed by Figini (14) and independently by Orofino and Wenger (135). The results depend on the relative rates of growth and termination, as illustrated in Figure 6. Note the development of a peak in the distribution curve at a j value of about one-half that corresponding to the average degree of polymerization.

It is assumed that the terminating agent attacks the growing ends indiscriminately and, hence, the reaction produces three types of polymers: dead polymers, polymers having only one growing end, and polymers possessing two growing ends. If killing is completed during the early stages of the reaction, polymerization results from the growth of two types of living polymers—the mono- and the bifunctional ones. The product then forms a mixture of two polymers, both virtually uniform in their molecular weight distribution, but one having twice the average molecular weight of the other.

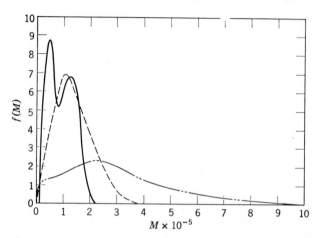

Fig. 7.   Molecular weight distributions of polystyrenes. Sample S 51 was produced by a difunctional initiator. Notice the two peaks in the distribution function of this sample (after ref. 11).

		M_w	M_n	M_w/M_n
————	S 51	84,000	64,000	1.31
– – – –	S 46	158,000	137,000	1.15
— · · —	S 43	295,000	140,000	1.99

$f(M)$ is the weight fraction of polymers having molecular weight M.

In fact, such a situation arises when a partial killing of a bifunctional initiator takes place during its preparation. A striking example of such a case was reported by Brower and McCormick (11) (see Figure 7), who initiated the styrene polymerization in dioxane by adding an electron-transfer initiator (sodium naphthalene) to the monomer. The growth is slow in dioxane, and therefore termination of the bifunctional living polymers, which were formed by electron-transfer initiation (1), was completed at a very early stage of polymerization.

In conclusion, the use of a monofunctional, in preference to a bifunctional initiator is beneficial when the impurity level is relatively high or when the spontaneous termination is not negligible. If a preparation is carried out in a meticulously purged system, a narrower distribution could be expected with a bifunctional initiator, especially if the sample finally prepared could be fractionated, and hence freed from the fraction derived from the partially killed initiator which has half the molecular weight of the bulk of the polymer.

5. EFFECT OF INEFFICIENT MIXING

We may now turn to the next essential requirement for a successful preparation of polymers of uniform size—the condition of constant spatial monomer concentration throughout the whole volume of the reactor. For a relatively slow propagation it should be possible to mix all the reagents at the onset of the reaction and eventually to maintain identical conditions in each volume element of the reactor during the course of polymerization. In practice, the experimenter still faces some problems. The temperature of the reacting mixture must be uniform; if not, local depletion of monomer, arising from the uneven rate of reaction, would lead to broadening of the distribution. Hence, stirring is necessary even if the reagents have been initially well mixed. Unfortunately, stirring and heat transfer become more difficult to cope with as the solution becomes more viscous, which makes the preparation of high molecular weight polymers even more difficult, especially if the rate of propagation is high.

Let us consider a specific example of a system in which the monomer and living oligomer may be mixed before propagation takes place. Polymerization of a relatively dilute solution of living α-methylstyrene is prevented at 25°C by its high equilibrium concentration (see Chapter III). Thus, the monomer and its low molecular weight living oligomer may be mixed together without causing further polymerization, and, thereafter, the well-homogenized mixture could be chilled to the required low temperature. Under such conditions, equilibrium is shifted in favor of polymerization which should then yield a perfectly uniformly sized product because all the damaging impurities were destroyed *in situ* during the preceding period. This method was successfully utilized by Sirianni, Worsfold, and Bywater (9), by McCormick and Friedrich (36), and subsequently by Wenger (16). It is obvious that an efficient heat transfer is then required to ensure the formation of a superior product.

When polymerization is rapid, it is necessary to supply the monomer gradually, i.e., its solution must be added continually or intermittently to the well-stirred reaction mixture. This again introduces some difficulties. Living polymers dissolved in the layers close to the points at which monomer is supplied are preferentially "fed," and hence grow longer than those located farther away. The mathematical consequences of poor mixing have been developed by Figini and Schulz (17) and by Litt (18).

Figini and Schulz visualized mixing as an exponential growth of a droplet of monomer introduced into a solution of living polymers. On its expansion, the droplet engulfs the growing polymers, which react with monomer. The molecular weight distribution of the resulting product depends, therefore, on the rate of propagation relative to that of expansion. If the former is much greater than the latter, a small fraction of living polymers consumes all the monomer; they become enormously longer than other chains and the ratio $\langle j \rangle_{w,1}/\langle j \rangle_{n,1}$ of the initial product (arising from an individual droplet) becomes very large. In the reverse case, nearly all the polymers acquire an approximately equal share of monomer.

Under typical polymerization conditions, $\langle j \rangle_{w,1}/\langle j \rangle_{n,1}$ was estimated to be between 30 and 500. As more monomer is added, the living polymers react over and over again, and the dispersity of the distribution gradually decreases. The final result, arising from the addition of z droplets, was computed by a summation method. Polymerization of the monomer contained in each individual droplet is considered as an independent event. Assume now that the living polymers which reacted with only one droplet of monomer are characterized by the averages $\langle j \rangle_{n,1}$ and $\langle j \rangle_{w,1}$. After reacting with two droplets the averages are then

$$\langle j \rangle_{n,2} = 2\langle j \rangle_{n,1},$$

and

$$\langle j \rangle_{w,2} = \langle j \rangle_{w,1} + \langle j \rangle_{n,1}.$$

In general, the population which reacted with k droplets has the average degrees of polymerization given by

$$\langle j \rangle_{n,k} = k\langle j \rangle_{n,1}$$

and

$$\langle j \rangle_{w,k} = \langle j \rangle_{w,1} + (k - 1)\langle j \rangle_{n,1},$$

leading to

$$\langle j \rangle_{w,k}/\langle j \rangle_{n,k} = 1 + U_1/k,$$

where $U_1 = \langle j \rangle_{w,1}/\langle j \rangle_{n,1} - 1$. The number of molecules of each size was computed on the assumption that the chance of reaction of a living polymer with monomer contained in a droplet is independent of its previous history. Thus, the final summation gives

$$\langle j \rangle_n = \langle j \rangle_{n,1}(zN/N_0)/[1 - (1 - N/N_0)^z]$$

and

$$\langle j \rangle_w = \langle j \rangle_{n,1}[1 + U_1 + (z - 1)N/N_0].$$

Here, N_0 is the total number of living polymers, N gives the number of living polymers that react with each droplet, and z is the number of droplets introduced. Denoting zN/N_0 by γ, one may approximate the above equations for a large z, i.e.,

$$\langle j \rangle_n = \langle j \rangle_{n,1} \gamma / [1 - \exp(-\gamma)]$$

and

$$\langle j \rangle_w = \langle j \rangle_{n,1} (1 + U_1 + \gamma - N/N_0).$$

The distribution eventually approaches the Poisson type as z increases, and the deviation arises from the fact that monomer is added in droplets rather than in the form of individual molecules.

The expanding droplet model was modified by Litt (18), who pointed out that, as the drop hits the surface of a stirred liquid, it quickly becomes spread by the velocity gradient into a flat ribbon. As the ribbon spreads thinner and thinner, its surface area increases, and thus more and more living polymers come into contact with the monomer. The mathematical treatment hinges on the assumption that the concentration of monomer at the interface depends on its concentration in the bulk of the layer. Litt considers two cases: (a) the interface concentration is proportional to that in the bulk; (b) the interface concentration remains constant.

On the basis of the first assumption, the degree of polymerization of a living polymer that arrived at the interface at time t is

$$j_t = \int_t^\infty (k_p M_0/v) \exp\{-k_p KGd^2 N_0 t^2/2v\} dt.$$

Here G is the rate of shear, d the diameter of the droplet, v its volume, M_0 initial concentration of monomer, and N_0 the concentration of living polymers. K represents a geometrical factor which depends on surface concentration of the reagents and k_p is the rate constant of propagation. The final results are published in the original paper in the form of graphs giving $\langle j \rangle_{n,1}$ and $\langle j \rangle_{w,1}$ as functions of those parameters that determine the rate of spreading of the ribbon.

The second assumption gives

$$j_t = k_p M_s (t_f - t),$$

where $t_f = \{2v/k_p K' Gd^2 N_0\}^{1/2}$ and eventually the ratio $\langle j \rangle_{w,1}/\langle j \rangle_{n,1}$ is calculated to be 1.33.

According to Litt, the first assumption seems to be more realistic than the second and therefore the former was used in the final computations, which were carried out by applying the summation method of Figini and Schulz. Thus,

$$\langle j \rangle_w / \langle j \rangle_n = (1 + U_1 + \gamma - N/N_0)[1 - \exp(-\gamma)]/\gamma,$$

where the symbols have the previously defined meaning, and U_1 is obtained from the graph given by Litt (18).

An attempt was made to test the various models by comparing experimental data with the results of calculations. Lack of the necessary parameters and the uncertainties of the experimental conditions, which prevailed during the polymerizations singled for the comparison, make such tests of doubtful value.

What, therefore, are the best conditions for monomer addition? Dilution of the added monomer may appear advantageous, but this method magnifies the dangers arising from the presence of impurities. It is attractive to introduce the monomer in the form of its vapor kept at the lowest practical partial

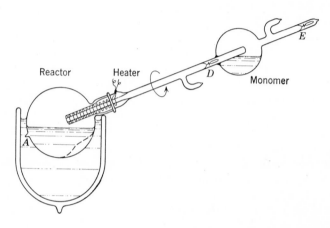

Fig. 8. Apparatus by means of which the monomer may be introduced in the form of a vapor into a stirred reactor. D and E, breakseals (after ref. 19).

pressure. Such a method was described, e.g., by Cowie, Worsfold, and By-water (19), and the diagram of an ingenious apparatus described in their paper is shown in Figure 8. However, regardless of the method of addition, efficient stirring is imperative.

6. EFFECT OF SLOW INITIATION ON MOLECULAR WEIGHT DISTRIBUTION

We may now turn to the next problem—the effect of slow initiation on the molecular weight distribution of the resulting product. The simplest kinetic scheme involves two rate constants:

$$P_0^* + M \rightarrow P_1^*, \qquad\qquad k_i$$

and

$$P_j^* + M \rightarrow P_{j+1}^*, \qquad\qquad k_p$$

for all $j \geq 1$. In such a system, the initiating species is never quantitatively used up. In fact, if the ratio $[P_0^*$ at $t = \infty]/[P_0^*$ at $t = 0]$ is denoted by $1 - f$, then f may be calculated (2a) from the equation

$$M_0/P_{0,0}^* = -(k_p/k_i)[f + \ln (1 - f)] + f,$$

where M_0 and $P_{0,0}^*$ are the initial concentrations of monomer and of initiating species P_0^*. Table II gives the numerical values of f for various ratios

TABLE II

k_p/k_i	$M_0/P_{0,0}^*$	f
0.1	5	> 0.99
0.5	5	> 0.99
1.0	5	0.99
2.0	5	0.97
2.0	8	0.99
5.0	5	0.81
5.0	7	0.90
5.0	20	0.99
10.0	5	0.67
10.0	10	0.82
10.0	20	0.95
100.0	10	0.38
100.0	50	0.69
100.0	100	0.84

$M_0/P_{0,0}^*$ and k_p/k_i. The resulting average degrees of polymerization obtained at $t = \infty$ were calculated by Litt (18):

$$\langle j \rangle_n = -(k_p/k_i)\{1 + [\ln (1 - f)]/f\}$$

and

$$\langle j \rangle_w = -(k_p/k_i)\{[\ln (1 - f)]^2/[f + \ln (1 - f)] + 2\}.$$

The molecular weight distribution obtained in polymerization that is slowly initiated but not terminated was derived by Gold (20). His treatment starts with the following set of differential equations:

$$dP_0^*/dt = -k_i M P_0^*$$

$$dP_1^*/dt = (k_i P_0^* - k_p P_1^*)M$$

. . . .

$$dP_j^*/dt = k_p(P_{j-1}^* - P_j^*)M \qquad \text{for all } j > 1.$$

M denotes the concentration of the monomer, P_0^* that of the initiating species, and P_j^* that of a living j-mer. The rate constant of propagation is assumed to be constant and independent of j for all living j-mers including P_1^*. In addition,

$$-dM/dt = \{k_i P_0^* + k_p \sum_{1}^{\infty} P_j^*\}M.$$

The initial conditions for $t = 0$ are: $P_0^*(0) = I_0$ and $P_j^*(0) = 0$ for all $j \geq 1$. Solution of these equations leads to the values of $P_j^*(\alpha_t)$ which are implicit functions of time, because $\alpha_t = P_0^*(t)/P_0^*(0)$ denotes the fraction of the initiator present at time t and therefore is a unique function of time.

The values of P_j^* were obtained by summation of a derived infinite power series involving a *positive* parameter u defined by the equation

$$u/(1 - r) = \ln [P_0^*(0)/P_0^*(t)].$$

This *limits* the treatment to systems for which k_p/k_i, denoted by r, is greater than 1. This is not a serious limitation because only these systems are of practical interest. It should be stressed, nevertheless, that the series used in Gold's treatment diverge for $q = r/(1 - r) \geq 1$. Fortunately, his results remain correct even for those values of q, despite the nonrigorous procedure used in their derivation. Having P_j^*, Gold proceeded to compute the moments. Unfortunately, the final results were not obtained in a closed form and, therefore, extensive tables and graphs had to be given in his paper from which $\langle j \rangle_n$, $\langle j \rangle_w$, and $\langle j \rangle_w/\langle j \rangle_n$ could be calculated as functions of the auxiliary variable u. Inspection of Figure 9, reproduced from Gold's paper, shows that the ratio $\langle j \rangle_w/\langle j \rangle_n$ initially increases with u, reaches its maximum (always less than 1.4), and then asymptotically tends to unity.

The shortcoming of Gold's treatment has been overcome by an elegant approach reported by Nanda and Jain (140). The system of differential equations given by Gold was presented in a matrix notation, thus greatly simplifying the algebra. The mole fraction P_j^* of a j-mer is given as a function of a modified time, τ, defined by the equation

$$t = \int_0^\tau d\tau/k_p M_\tau.$$

Thus,

$$P_j^* = P_0^*(0)\alpha \exp(-\tau)(1 - \alpha)^{-j}\left\{\sum_{k=j}^{\infty} (w^k/k_i)\right\},$$

where $\alpha = k_i/k_p$ and $w = \tau(1 - \alpha)$. This expression, in contradistinction to that given by Gold, is valid for *all* values of α, both greater and smaller than unity. Nanda and Jain also succeeded in providing a general expression for any moments, and in particular they obtained $\langle j \rangle_n$ and $\langle j \rangle_w$ in closed forms:

$$\langle j \rangle_n = [\alpha\tau - 1 + \alpha + (1 - \alpha) \exp(-\alpha\tau)]/[\alpha - \alpha \exp(-\alpha\tau)]$$

and

$$\langle j \rangle_w = \frac{\alpha^2\tau^2 - 2\alpha\tau + 2 + 3\alpha^2\tau - 3\alpha + \alpha^2 - (2 - 3\alpha + \alpha^2) \exp(-\alpha\tau)}{\alpha[\alpha\tau - 1 + \alpha + (1 - \alpha) \exp(-\alpha\tau)]}.$$

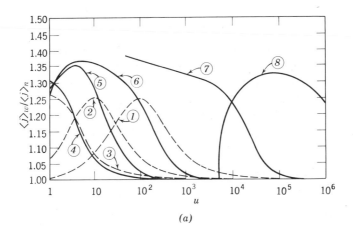

(a)

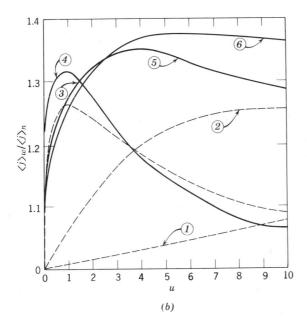

(b)

Fig. 9.   (a) The behavior of the dispersion index $\langle j \rangle_w / \langle j \rangle_n$ as polymer yields increase. An increase in u is equivalent to increasing conversion of monomer into polymer (after ref. 20). $u = -R \ln I/I_0$ where I denotes the concentration of initiator at time t, and I_0 is its initial concentration. $R = 1 - k_p/k_i$. The numbers on the curves refer to the different values of k_p/k_i, viz., 10^{-2}, 10^{-1}, 0.5, 2, 10, 10^2, 10^4, and 10^6. (b) Enlargement of section of graph a.

For $\alpha\tau \ll 1$, a condition encountered in all cases of practical interest, the averages may be given by the power series of $\alpha\tau$:

$$\langle j \rangle_n = (\tau/2) \frac{(1 + 2/\tau) - \alpha\tau/3 + \cdots}{1 - \alpha\tau/2 + \cdots}$$

and

$$\langle j \rangle_w = (2\tau/3) \frac{(1 + 9/2\tau + 3/\tau^2) - \alpha\tau/4 + \cdots}{(1 + 2/\tau) - \alpha\tau/3 + \cdots}.$$

In such a case $\langle j \rangle_w / \langle j \rangle_n \rightarrow 4/3$ when $\tau \gg 1$. This is the situation encountered in a polymerization leading to a high molecular weight polymer when only a minute amount of the initiator, P_0^*, is consumed.

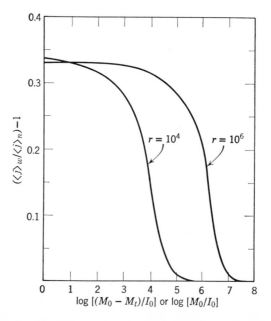

Fig. 10.   The variation of $(\langle j \rangle_w / \langle j \rangle_n) - 1$ with conversion $(M_0 - M_t)$. The variation of the ultimate value of $(\langle j \rangle_w / \langle j \rangle_n) - 1$ with the starting monomer concentration M_0 is represented by the same set of curves. The curves correspond to different values $r = k_p / k_t$. Note that the dishomogeneity coefficient remains constant for a wide range of conversion, i.e., $\langle j \rangle_w / \langle j \rangle_n \approx 1.33$ (after ref. 140).

For $\alpha\tau \gg 1$ and $\tau \gg 1$, the general expressions for $\langle j \rangle_n$ and $\langle j \rangle_w$ show that both averages tend to τ and their ratio $\langle j \rangle_w / \langle j \rangle_n$ approaches unity. This is essentially the case when initiation is virtually completed in the early stage of polymerization that eventually yields a high molecular weight polymer.

For a given τ the ratio $\langle j \rangle_w / \langle j \rangle_n$ is the highest when $\alpha \rightarrow 0$. For example, for $\tau = 6$, the ratio is 1.375 while $\langle j \rangle_n = 4$. One should not conclude, however, that the high values of $\langle j \rangle_w / \langle j \rangle_n$ are limited to the early stages of the reaction, and that the ratio approaches unity at its completion. In fact, under some conditions, the ratio $\langle j \rangle_w / \langle j \rangle_n$ may remain nearly constant for a wide range of conditions (see Figure 10). Thus, its value is ~ 1.33 when $(2\alpha M_0 / I_0)^{1/2} \ll 1$ and approaches unity when $\alpha M_0 / I_0 \gg 1$. An interesting and instructive diagram is shown in Figure 11. For various values of

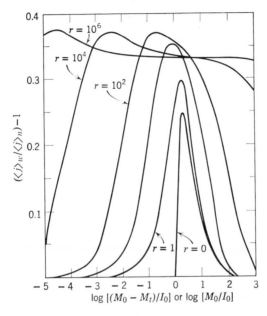

Fig. 11. Variation of $\{(\langle j \rangle_w / \langle j \rangle_n) - 1\}$ with conversion $(M_0 - M_t)$. The variation of the ultimate value of $\{\langle j \rangle_w / \langle j \rangle_n - 1\}$ with the starting monomer concentration M_0 is represented by the same set of curves, and shown for different values of r (after ref. 140).

$r = k_p / k_i$, the dishomogeneity coefficient, $\langle j \rangle_w / \langle j \rangle_n - 1$, is given as a function of relative conversion, $(M_0 - M_t)/I_0$. This relation is valid if the amount of the polymerized monomer is smaller than the amount of the initially available initiator. Its inspection clearly illustrates the conditions necessary for producing narrow molecular weight polymers.

Finally, it should be remarked that the mathematical approach of Nanda and Jain could prove to be useful in solving more complicated problems.

The treatments outlined above assume a constant initial concentration of the monomer. The generalization, applicable to the systems in which monomer is added in the course of reaction, was developed by Figini (32).

Other initiation schemes are also possible, but it is improbable that they are encountered in systems leading to the uniform size polymers. One, which deserves a more detailed discussion may apply to the polymerization of Leuch's anhydrides (NCA) to polypeptides. In such a scheme, initiation is bimolecular,

$$M_1 + M_1 \rightarrow M_2 \qquad k_i$$

followed by propagation,

$$M_n + M_1 \rightarrow M_{n+1} \qquad k_p \qquad \text{for } n > 1.$$

For a special case, $k_i = k_p = k$, and for the initial conditions $M_1 = C_0$ and $M_n = 0$ for all $n > 1$ at time zero, one finds

$$\langle j \rangle_n = \exp(p) \qquad \text{and} \qquad \langle j \rangle_w = 1 + 2p,$$

giving

$$\langle j \rangle_w / \langle j \rangle_n = (1 + 2p)/\exp(p).$$

Here $p = \{1/[M_1]_0\} \int_0^t k M_1 dt$. As $t \rightarrow \infty$, the distribution becomes independent of $[M_1]_0$:

$$\lim_{t \rightarrow \infty} \langle j \rangle_n = e, \qquad \lim_{t \rightarrow \infty} \langle j \rangle_w = 3, \qquad \text{and} \qquad \lim_{t \rightarrow \infty} \{\langle j \rangle_w / \langle j \rangle_n\} = 3/e.$$

For a more general case, $k_i < k_p$, one finds

$$\langle j \rangle_n = \exp(p)/\cos[p(k_p/k_i - 1)^{1/2}],$$

and, if $k_p/k_i > 10^4$, the approximate forms are valid:

$$\langle j \rangle_n = 1/\cos \alpha, \qquad \text{where } \alpha = p(k_p/k_i)^{1/2}.$$

At infinite time, $\langle j \rangle_n$ and $\langle j \rangle_w$ become independent of $[M_1]_0$:

$$\lim_{t \rightarrow \infty} \langle j \rangle_n = 1/\beta \qquad \text{and} \qquad \lim_{t \rightarrow \infty} \langle j \rangle_w = (\pi - 2)/\beta,$$

where $\beta = (k_i/k_p)^{1/2}$.

The problem of slow initiation becomes critical when lithium alkyls are used to initiate anionic polymerization in hydrocarbon solvents. In this connection, it is interesting to recall the observation of Holden and Milkovich (47). They found that polymerization initiated by a secondary alkyllithium compound gives a more uniform polymer than that obtained on initiation with n-butyllithium. This work implies that tertiary or secondary derivatives are faster initiators than primary ones (141). To avoid the difficulties arising from a slow initiation, it was proposed to add the monomer in two stages (see, e.g., ref. 21). The first stage is designed to complete the initiation step while the molecular weight of the polymer is still low and, thus, to restrict variation in molecular sizes to an initially formed portion of the macromolecules. It should be stressed, however, that this method is useful only for the systems in which the depropagation of living polymers is sufficiently rapid, because then the polymerized monomer again becomes available for further initiation:

living polymers ⇌ monomer,

initiator + monomer → living polymers.

In the absence of depropagation, no improvement in distribution is possible by restricting the reaction first to a fraction of the monomer and then adding the remaining portion later. This point is not appreciated by some investigators although it is obvious if one examines the kinetic scheme discussed by Dostal and Mark (22). Depropagation has a high activation energy; hence it is advisable to maintain a higher temperature at this stage of the process (21). Alternatively, one may first initiate polymerization of a small amount of monomer which has a low ceiling temperature, e.g., of α-methylstyrene, if its living oligomer acts as a rapid initiator. Thereafter, the bulk of the desired monomer could be added (19,23).

So far we have discussed kinetic schemes for which the initiation is first order in respect to initiator and the propagation is first order in respect to living polymers. There are, however, systems which are more complex. For example, polymerizations proceeding in benzene, or in other hydrocarbons, involve agglomerated species. Butyllithium is known to be hexameric in nonpolar solvents, and the lithium salt of living polystyrene is dimeric in benzene (see Chapter VIII). As shown by Worsfold and Bywater (31), the initiation is then 1/6 order in butyllithium and propagation is 1/2 order in living polystyrene. Under such conditions the initiator is converted quantitatively into living polymers only when the monomer concentration exceeds a critical value of $0.84 \, [P_0^*(0)]^{4/3} \cdot k_p/k_i$. If both reactions are 1/2 order, the critical concentration of monomer is $1.60 P_0^*(0) \cdot k_p/k_i$, and, when both steps are of zero order, M must be greater than $(1 - k_p/k_i)P_0^*(0)$.

The initiation by alkyllithium is slow because of its agglomeration. Polar additives dissociate the agglomerates and therefore speed up the initiation more than propagation. Such an effect is observed when a small amount of tetrahydrofuran is added to styrene polymerized by butyllithium in benzene (184).

7. SIMULTANEOUS GROWTH OF TWO OR MORE SPECIES. LIVING AND DORMANT POLYMERS

Throughout this treatment, it was assumed that all the living polymers grow with the same rate constant. This is often not the case. For example, anionic polymerization of styrene in tetrahydrofuran involves ion-pairs and free ions, and the latter are about 1000 times as reactive as the former

(23–25). Hence, even if the free ions form only 1% of all the living polymers, they polymerize 90% of the monomer, whereas the remaining 99% of living polymers, present in the form of ion-pairs, is responsible for only 10% of the polymerization. A similar situation exists in other ethereal solvents, e.g., in tetrahydropyran (26–28) or methyltetrahydrofuran (28), but not in dioxane. An analogous problem is encountered in other systems. For example, living anionic polymers having a lithium counterion are associated in hydrocarbon solvents. The associated species are inert and do not propagate polymerization. Hence, only a minute fraction of the nonassociated polymers contributes to the growth (29–31), and therefore, at any stage of the reaction, living polymers are divided into two classes—those which grow and those which are dormant.

Participation of dormant and living polymers (45) in the propagation leads to broadening of the molecular weight distribution, the degree of which depends on the rate of exchange. A mathematical treatment of this problem was given by Coleman and Fox (64c), by Figini (33) and independently by Szwarc and Hermans (34). Consider the system

$$P_n^* + M \rightarrow P_{n+1}^* \qquad k_p,$$

and

$$P_n^* \underset{k_{-1}}{\overset{k_1}{\rightleftarrows}} P_n.$$

The asterisk denotes a living polymer and the nonasterisked symbol represents a dormant polymer which does not contribute to growth. The kinetics of propagation is then given by the following set of differential equations:

$$-dM/dt = k_p M \sum P_n^*,$$

$$dP_n^*/dt = k_p M (P_{n-1}^* - P_n^*) + k_{-1} P_n - k_1 P_n^*,$$

$$dP_n/dt = k_1 P_n^* - k_{-1} P_n,$$

coupled with the initial conditions maintained at zero time:

$$P_1^* = a, \qquad P_1 = b, \qquad k_1 a = k_{-1} b,$$

$$P_n^* = P_n = 0 \qquad \text{for all } n > 1,$$

and

$$M = M_0.$$

To evaluate the average degrees of polymerization, $\langle j \rangle_n$ and $\langle j \rangle_w$, the following sums have to be determined:

$$\Lambda_0 = \sum_1^\infty (P_n^* + P_n) = \text{total number of polymer molecules,}$$

$$\Lambda_1 = \sum_1^\infty n(P_n^* + P_n) = \text{total weight of polymer molecules,}$$

$$\Lambda_2 = \sum_1^\infty n^2(P_n^* + P_n) = \text{second moment.}$$

Then, $\langle j \rangle_n = \Lambda_1/\Lambda_0$ and $\langle j \rangle_w = \Lambda_2/\Lambda_1$.

To present results in a concise form, denote $k_{-1}/(k_1 + k_{-1})$ by m and $k_pM/(k_1 + k_{-1})$ by g, and use du for k_pMdt. Two cases will now be considered:

(1) A continuous supply of monomer maintains constant its stationary concentration. Then

$$\Lambda_0 = a + b = a/m,$$

$$\Lambda_1 = a + b + kat = (a/m)(1 + u),$$

$$\Lambda_2 = (a/m)\{1 + [3 + 2g(1 - m)]u + u^2 - 2g^2m(1 - m)[1 - \exp(-u/gm)]\}.$$

The systems of practical interest are those giving a product of high molecular weight, i.e., when $u \gg 1$ and therefore $u^2 \gg 3u$. In such a case,

$$\langle j \rangle_w/\langle j \rangle_n = 1 + 2(1 - m)g/u - 2m(1 - m)(g/u)^2[1 - \exp(-u/gm)].$$

This shows that the dispersity ratio may differ significantly from unity, even when u is large, provided g is also large.

(2) All the monomer is added at the onset of the reaction. Then the effective rate constant of propagation, k_pM, becomes a function of time, since

$$M = M_0 \exp(-\lambda t) \qquad \text{when } \lambda = k_pa,$$

and $\Lambda_0/a = 1/m$, $\Lambda_1/a = 1/m + (M_0/a)[1 - \exp(-\lambda t)]$, and

$$\Lambda_2/a = 1/m + M_0/a(3 + 2mM_0/a)[1 - \exp(-\gamma t)]$$

$$+ (M_0/a)^2[(\lambda - \beta)/(\gamma - \lambda)][1 - \exp(-2\lambda t)]$$

$$- (M_0/a)^2 2\lambda^2\alpha[1 - \exp -(\lambda + \gamma)t]/\gamma(\gamma^2 - \lambda^2)$$

where $\gamma = \alpha + \beta$, $\alpha = k_1$, and $\beta = k_{-1}$.

The treatments of Coleman and Fox (64) and of Figini (33) are more general. Both species are assumed to contribute to growth but each at a different rate. The resulting equations resemble the one presented above, but they contain two rate constants.

The effect of exchange between the rapidly growing free ions and slowly propagating ion-pairs was investigated by Löhr, and Schulz et al. (41) and then again by Figini (42). The polymerization of styrene in tetrahydrofuran was performed in the absence and in the presence of sodium tetraphenylboride. The large difference between the reactivity of the free polystyryl ions, $\sim\!\!\sim\!S^-$, and of the $\sim\!\!\sim\!S^-,Na^+$ ion-pairs leads to a substantial deviation of the

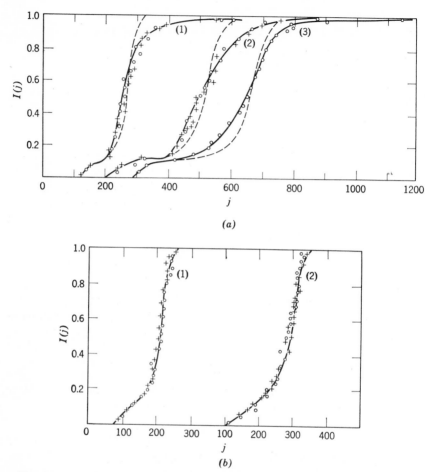

Fig. 12. Experimental molecular weight distribution of anionically polymerized polystyrene in a flow system at high Reynolds numbers (adequate rate of mixing of living polymers with the monomer). (a) Polymerization in the absence of sodium tetraphenylboride, i.e., in a system in which two types of growing species (free ions and ion pairs) participate in the reaction. The broadening arises from their unequal rate of growth. [M] = 0.57 M, [living polymers] = 1.1×10^{-3} M.

Curve	Reynolds number	Conversion, %	$(\langle j \rangle_w/\langle j \rangle_n) - 1$
1	14,000	21	0.08
2	14,000	39	0.075
3	12,000	49	0.06

The dashed line represents theoretical molecular weight distribution for the infinite rate of exchange. (b) Polymerization in the presence of sodium tetraphenylboride—the concentration of free ions is depressed and only ion pairs propagate.

Curve	Conversion, %	[Living polymers]
1	17.4	0.9×10^{-3} M
2	45	1.65×10^{-3} M

The experimental curve coincides with the calculated Poisson distribution.

molecular weight distribution from the expected one. This is seen in Figure 12a. On addition of sodium tetraphenylboride, the dissociation of ion-pairs is repressed and, for a sufficiently high concentration of the salt, the concentration of the free $\sim$S$^-$ ions becomes negligible. Under such conditions, only one species, the ion-pair, propagates the polymerization and, consequently, a polymer having a sharp distribution of molecular weights is obtained. This dramatic change of molecular weight distribution is seen by comparing parts (a) and (b) of Figure 12.

Some puzzling observations may be explained by the participation of ions and ion-pairs in the propagation of sodium salts of living polystyrene in tetrahydrofuran. The viscosity average molecular weight of living polystyrene was higher when prepared at $-76°C$ than at room temperature, all other conditions being constant (43). Most probably, the broadening resulted from a substantial decrease in the rate of exchange between ions and ion-pairs at low temperature.

The observed deviations of the molecular weight distribution from the Poisson type were utilized to calculate the rate of exchange between ions and ion-pairs (41,42). Although the published calculations led to reasonable results, there is some doubt as to whether the accuracy of this technique is sufficient to allow a reliable determination of rate constants. Let us remark, also, that two mechanisms may contribute to the exchange: (1) a direct dissociation of ion-pairs, coupled with the association of the free ions, e.g.,

$$\sim S^-, Na^+ \rightleftarrows \sim S^- + Na^+,$$

and (2) a bimolecular reaction of ions with ion-pairs, e.g.,

$$\sim S^-, Na^+ + \sim S^- \rightleftarrows \sim S^- + \sim S^-, Na^+.$$

Apparently, no attention was paid in the past to the fact that active and nonactive living polymers coexist in hydrocarbon solvents when Li$^+$ is the counterion (see, however, ref. 142). Therefore, the resulting product should, most probably, have a broader molecular weight distribution than otherwise expected. On the other hand, it has been shown that polymers of a relatively uniform size are obtained under these conditions. This indicates a relatively fast rate of exchange as compared with the rate of growth, which is slow under these conditions. More extensive studies of this problem would be advisable.

Another example of an interplay between living and dormant polymers is found in the studies of Gee et al. (44), who polymerized ethylene oxide by initiating the reaction with sodium alkoxide in the presence of small amounts of alcohol. The alcohol was required to solubilize the catalyst. The system is described by

$$\sim CH_2CH_2O^-, Na^+ + CH_2\overset{O}{\overset{\diagup\diagdown}{-}}CH_2 \rightarrow \sim CH_2 \cdot CH_2 \cdot O \cdot CH_2 \cdot CH_2 \cdot O^-, Na^+,$$

$$\sim O^-, Na^+ + \sim OH \rightleftarrows \sim OH + \sim O^-, Na^+.$$

A controlled adjustment of the concentration of dormant polymers may serve to regulate at will the molecular weight distribution of the resulting product (142).

The problem of "tailoring" the molecular weight distribution of polymers formed in anionic terminationless processes was discussed by Eisenberg and McQuarrie (46). They proposed to adjust the $\langle j \rangle_w / \langle j \rangle_n$ ratio by programing the addition of an initiator to a slowly polymerized mixture of living polymers and unpolymerized monomer. Various hypothetical cases were treated and the relevant mathematical relations were derived. The treatment refers to an ideal system, and all the complicated factors discussed in the preceding sections were ignored. A simpler mathematical treatment of a similar problem was reported by Litt and Richards (143) who proposed to calculate the absolute rate constant of an ideal polymerization of living polymers from the value $\langle j \rangle_w / \langle j \rangle_n$.

A peculiar situation is created if a living polymer possesses two growing ends which may reversibly associate and form a dormant polymer. In such a case the concentration of open, living polymers depends on their degree of polymerization—the longer the polymer the higher is the probability of finding it in an open state. Thus the apparent rate constant of propagation increases with increasing degree of polymerization, i.e., the longer the polymer the faster its growth. This is a unique case when the degree of polymerization must affect the rate of propagation.

8. EFFECT OF DEPROPAGATION ON MOLECULAR WEIGHT DISTRIBUTION

We may pass now to the last cause of the broadening of the molecular weight distribution of living polymers, namely, the reversibility of the propagation step. This problem was initially treated by Brown and Szwarc (48) and later elegantly and thoroughly elaborated by Miyake and Stockmayer (49). A living polymer, which was initially composed of uniform size macromolecules, spontaneously broadens its molecular weight distribution because of its ability to depropagate. This problem is treated in Chapter III, where we shall discuss the equilibrium established between living polymers and their monomer and show that the ultimate equilibrium in such a system demands a most probable molecular weight distribution of the living polymers, i.e., the ratio $\langle j \rangle_w / \langle j \rangle_n$ must ultimately approach its limiting value of 2.

At present we are interested in knowing how rapidly such a change takes place when the initially formed macromolecules were all of identical size. This question was considered by Brown and Szwarc (48) who pointed out that $\langle j \rangle_n$ remains constant in such systems. Therefore, the rate of the process is reflected in the rate of change of the weight average molecular weight or of some higher moment. The following result was obtained by assuming a constant, although minute, concentration of the residual monomer,

$$d\langle j\rangle_w/dt = \alpha k_d/\langle j\rangle_n(1 - x_1\langle j\rangle_n),$$

where α has a value between 1 and 2, k_d is the rate constant of depropagation, and x_1 is the mole fraction of the lowest living polymer. This relation may be expected intuitively. The initial molecular weight distribution is preserved if $k_d = 0$ and its rate of broadening is accelerated as k_d increases. Moreover, the longer the polymers, the slower the change of their distribution since each step of the process involves dissociation of a single monomeric unit at the end of one polymeric molecule, coupled with addition of the dissociated monomer to another growing end.

A profound and more deeply penetrating study of this problem was presented by Miyake and Stockmayer (49). How does the molecular distribution of the product change with conversion and time if depropagation takes place simultaneously with propagation while termination and chain transfer are absent? The following kinetic scheme was proposed:

$$I + M \underset{k_{-i}}{\overset{k_i}{\rightleftarrows}} P_1^*,$$

$$P_j^* + M \underset{k_d}{\overset{k_p}{\rightleftarrows}} P_{j+1}^*,$$

and it leads to the relevant set of differential equations. It is interesting to note that the problem is identical mathematically with that dealing with the kinetics of formation of multimolecular adsorption layers on a surface represented by a Brunauer–Emmett–Teller model. If the instantaneous concentration of the monomer $M(t) = M_0$ is constant all the time (this may be achieved by continuous addition of the monomer into a stirred reactor at an appropriate rate), then an exact solution may be obtained by means of Laplace transforms. The transforms are defined by the equations:

$$i(s) = \int_0^\infty I(t) \exp(-st)dt,$$

$$j(s) = \int_0^\infty J(t) \exp(-st)dt,$$

and
$$m_j(s) = \int_0^\infty P_j^*(t) \exp(-st)dt,$$

where $I(t)$ is the concentration of the initiator at time t, $j(t)$ that of all the growing centers, and $P_j^*(t)$ the concentration of j-mers. Obviously, $J(t) = I(t) + \sum_1^\infty P_j^*(t)$ and its value varies with time if the initiator is supplied during the reaction. The solutions allow one to calculate the Laplace transforms of the various momenta, $Q_P(t)$, defined as

$$Q_P(t) = \sum_1^\infty j^p P_j^*(t),$$

and therefore, the required distribution function may be eventually obtained. For the simplest case, when

$$k_i = k_p \quad \text{and} \quad k_{-i} = k_d$$

one finds

$$P_j^*(t) = \int_0^t \varphi_j(t - t')J(t')dt' \quad \text{for } j > 0.$$

The auxiliary function $\varphi_j(t)$ is given by:

$$\varphi_j(t) = t^{-1} \exp\{-(k_d + k_p M_0)t\}(k_p M_0/k_d)^{j/2}$$
$$\times \{jB_j(u) - (k_p M_0/k_d)^{1/2}(j + 1)B_{j+1}(u)\}$$

where $B_j(u)$ is the modified Bessel function of the order j, and u abbreviates $2t(k_d k_p M_0)^{1/2}$. The moments Q_j are then given by:

$$Q_j(t) = \int_0^t \phi_j(t - t')J(t')dt',$$

where

$$\phi_0(t) = (k_p M_0/k_d)^{1/2}t^{-1} \exp\{-(k_d + k_p M_0)t\}B_1(u),$$

$$\phi_1(t) = k_p M_0 - k_d \int_0^t \phi_0(t')dt',$$

$$\phi_2(t) = k_p M_0 + k_d \int_0^t \phi_0(t')dt' - 2(k_d - k_p M_0) \int_0^t \phi_1(t')dt', \quad \text{etc.}$$

When *all* the monomer is supplied at the onset of the reaction,

$$J(t) = I_0 = \text{constant}$$

and then

$$P_j^*(t) = I_0 \int_0^t \varphi_j(t')dt',$$

$$Q_p = I_0 \int_0^t \phi_p(t')dt',$$

and

$$I_t = I_0\left\{1 - \int_0^t \phi_0(t')dt'\right\}.$$

Now when the monomer is added to the system continually to maintain a constant concentration (M_0) during the process and if $k_d > k_p M_0$, the polymers will grow to a finite average length and will attain at $t = \infty$ a state of equilibrium with their monomer. Their distribution then approaches the most probable molecular weight distribution. This case is of little practical significance because, for synthetic purposes, one always chooses the conditions $k_d \ll k_p M_0$. In the latter instance, final equilibrium cannot be attained; the polymers will continue to grow on and on as long as the constant concentration of monomer, $M_0 > k_d/k_p$, is maintained. The respective moments are then given by the equations:

$$Q_1(t) = I_0(k_p M_0 - k_d)t + 0(1),$$

$$Q_2(t) = I_0(k_p M_0 - k_d)^2 t^2 + 0(t), \qquad 0(t) \to 0 \text{ as } t \to \infty,$$

$$Q_3(t) = I_0(k_p M_0 - k_d)^3 t^3 + 0(t^2), \qquad \text{etc.}$$

This case is of practical interest. The plot of $\langle j \rangle_w / \langle j \rangle_n$ versus $k_p M_0 t$, shown in Figure 13, is most instructive. For the ratio $A = k_d/k_p M_0 < 1$,

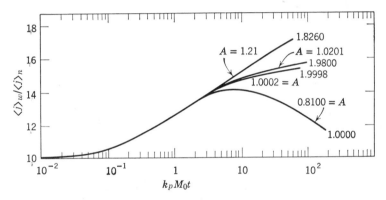

Fig. 13.   The ratio of weight average to number average chain length, $\langle j \rangle_w / \langle j \rangle_n$, in the reversible living batch polymerization as a function of $k_p M_0 t$. The curves correspond to different values of $A = k_d/k_p[M]_0$, namely, $A = 1.21$, 1.021, 1.0002, and 0.81. The numbers at the end of each curve give the limiting values (for $t = \infty$) of the $\langle j \rangle_w / \langle j \rangle_n$ ratios.

the ratio $\langle j \rangle_w / \langle j \rangle_n$ first reaches a maximum; but ultimately its value decreases asymptotically to 1. This means that eventually all the initiator is converted into living polymers and, as they become sufficiently long, the differences in their molecular weights become insignificant. For $A > 1$, the ratio $\langle j \rangle_w / \langle j \rangle_n$ approaches asymptotically a ratio close to, but always slightly smaller than, 2.

When $k_p \neq k_i$, the solutions are more complex but still presentable in an explicit form. It is interesting that for a range of k_d,

$$k_{-i} + (k_i^{1/2} - k_p^{1/2})^2 M_0 < k_d < k_{-i} + (k_i^{1/2} + k_p^{1/2})^2 M_0,$$

the functions $\varphi_j(t)$ and $\phi_j(t)$ become oscillatory in nature. This means the growth process speeds up and slows down in a back and forth manner.

The other extreme case considered by Miyake and Stockmayer is a batch polymerization in which all the monomer and initiator were supplied at the onset of the process. The simplified conditions, $k_i = k_p$ and $k_{-i} = k_d$, were assumed and the general equations lead, of course, to the Poisson distribution for $k_d = 0$. The final distribution attained at $t = \infty$ was derived, but no explicit solutions could be given for the distribution attained at time t.

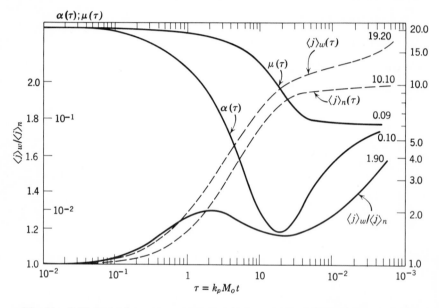

Fig. 14. Initiator and monomer concentrations, number average and weight average chain lengths, and the weight average to number average length ratio as functions of time for reversible living batch polymerization. $A = k_d/k_pM_0 = 0.1$, $B = I_0/M_0 = 0.1$, $\tau = kM_0t$. $\alpha(\tau) = [\text{initiator}]_t/[\text{initiator}]_0$, $\mu(\tau) = [\text{monomer}]_t/[\text{monomer}]_0$.

Therefore, numerical solutions were constructed for guidance. The symbols used in these calculations are as follows:

$$k_d/k_pM_0 = A \qquad\qquad \alpha(\tau) = I(\tau)/I_0$$
$$I_0/M_0 = B \qquad\qquad \mu_j(\tau) = P_j^*(\tau)/M_0$$
$$\tau = kM_0t \qquad\qquad \mu(\tau) = M(\tau)/M_0.$$

The first example, calculated for $A = k_d/k_pM_0 = 0.1$ and $B = 0.1$, is shown in Figure 14. The concentration of the monomer decreases monotonically and rapidly approaches its equilibrium value, while the concentration of the initiator falls *below* its equilibrium value and then increases, reaching the equilibrium concentration asymptotically. The ratio $\langle j \rangle_w / \langle j \rangle_n$ goes through a maximum, then a minimum, and eventually reaches its final value. This strange behavior arises from a more rapid approach of $\langle j \rangle_n$ than of $\langle j \rangle_w$ to their ultimate values. In fact, the polymerization proceeds through three stages:

(1) Proper polymerization when the concentration of the monomer decreases rapidly and the initiator is consumed. At this stage, the reaction resembles an irreversible polymerization leading to the Poisson distribution.

(2) A transient state when the concentration of the monomer remains

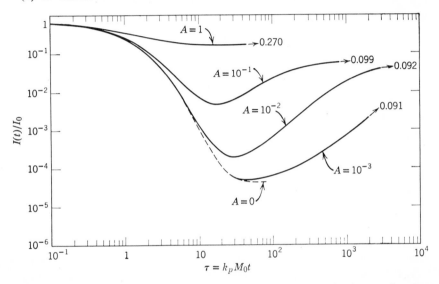

Fig. 15. Initiator concentration in reversible living batch polymerization as a function of time, for $A = 1$, 0.1, 0.01, and 0.001, and $B = 0.1$; broken curve is for $A = 0$ and $B = 0.1$. $A = k_d / k_p M_0$ and $B = [\text{initiator}]_0 / [\text{M}]_0$. The numbers following the arrows give the fraction of initiator left at time ∞.

virtually constant at approximately its equilibrium value, while the initiator is regenerated by the depropagation of the lowest molecular weight living polymers. This is clearly seen in Figure 15.

(3) Final redistribution of the molecular weight, arising from depropagation, causing the final increase of $\langle j \rangle_w / \langle j \rangle_n$.

Taking data found for styrene polymerization, i.e., $k_p = 10^3 M^{-1} \sec^{-1}$, $k_d = 2 \times 10^{-4} \sec^{-1}$, $I_0 = 2 \times 10^{-3} M$, and $M_0 = 2M$, the characteristic times of each of these three steps were calculated. Thus, $t_1 \sim 4.6$ sec, $t_2 \sim 8.1$ sec, whereas $t_3 \sim 80$ years! Hence, the polymerization giving the narrow molecular weight polymer could be completed in seconds; the broadening of the molecular weight distribution would take more than a century.

Depropagation is not the only mechanism by which the originally narrow molecular weight distribution may be broadened. An attack of a growing end on an internal segment of another chain may lead to a similar result. The kinetics of this type of broadening molecular weight distribution was investigated by Hermans (50).

In conclusion, living polymers provide an excellent system for producing, even on a large scale, polymers of uniform size. In principle, a Poisson molecular weight distribution may be obtained, although to reach this goal is not easy. Spectacular progress was made in this field during the last decade, but improvements are possible and more work is still needed. In addition, the problem stimulated much mathematical work, by posing challenging puzzles, as has been amply demonstrated in the preceding sections.

All the difficulties and limitations discussed in this section explain why the early literature of this subject was most confused and sometimes contradictory. Many of the problems were not fully comprehended by early investigators, so that failures were frequently attributed to the wrong causes. It is hoped that this discussion will help to dispel much of the past misunderstandings.

9. REPORTED POLYMERS OF NARROW MOLECULAR WEIGHT DISTRIBUTION AND THEIR CHARACTERIZATION

A few types of polymers of narrow molecular weight distribution were reported. Most of the work was performed with styrene, and samples having molecular weights ranging from less than 10,000 up to 500,000 are now available. Low molecular weight polystyrenes, $M \leqslant 10,000$, having $\langle j \rangle_w / \langle j \rangle_n \leqslant 1.1$ were described by Altares, Wyman, and Allen (144). Anionic polymerization initiated by butyllithium and performed in benzene was employed in their preparation. The viscosity-molecular weight relation in benzene at 25°C. gives $[\eta] = 1.0 \times 10^{-3} M_n^{-0.5}$ for $500 < \overline{M}_n \leqslant 10,000$ and $8.5 \times 10^{-5} \overline{M}^{0.75}$ for higher molecular weights. A single expression, covering the whole range of molecular weights (500 up to 1.5×10^6) was obtained for theta conditions, viz., in cyclohexane at 34.5°C., namely $[\eta] = 8.4 \times 10^{-4} M^{0.5}$. Samples prepared by McCormick et al. (11), by Wenger and Yen (10), and by Cowie, Worsfold, and Bywater (19) were discussed previously. Commercially prepared samples are also available now.

Narrow molecular weight distribution has been claimed for polybutadiene (145) and polyisoprene (146) prepared from the respective living polymers. However, these materials were not sufficiently well characterized to substantiate the claims. Better results were reported recently (192).

Living polymers of propylene sulfide were prepared (106b) but their molecular weight distribution was rather broad. Similar results were obtained with living polyacrylonitriles. Much better results were obtained with living polytetrahydrofuran prepared by cationic polymerization (147), although the resulting molecular weight distribution is still far from the Poisson type. A relatively narrow sample of polyisoprene was claimed by Medvedev's group (148) which used a Ziegler catalyst for the preparation.

Very successful preparations were reported for poly-α-methylstyrene (9,149) and perhaps for polymethyl methacrylate (16,150,151,193). Anionic methods were applied for both syntheses.

It was shown by Cori that the glucose-phosphate (Cori ester) may be polymerized to amylose by phosphorylase. Studies of Husemann and her colleagues (152) showed that this synthesis may be performed under conditions where all the polymers are initiated at the same time and grow simultaneously without being terminated. Consequently, the resulting amylose has a Poisson molecular weight distribution—a fact documented by thorough characterization of the product. It is probable that the same conditions exist in living plants when cellulose is formed. This is shown by the recent studies of Marx-Figini and Schulz (153) which demonstrate that the native cellulose has a unique molecular weight.

In addition to the methods already discussed (see p. 29), the narrowness of molecular weight distribution may also be assessed by some indirect methods, e.g.,

(*1*) Anderson (154) and Pennings (155,156) observed that linear polyethylene of broad molecular weight distribution becomes separated on crystallization. The low molecular weight fraction forms simple crystals containing extended chains, whereas the high molecular weight fraction yields crystals built up by folded chains.

(*2*) Evaporation of highly dilute solutions of polystyrene of molecular weight about one million left nearly spherical particles which could be photographed in an electron microscope (157). The distribution of their diameters could be correlated with the molecular weight distribution of the original polymer.

10. UTILIZATION OF UNIFORM SIZE POLYMERS

Molecular weight distribution affects many physical and mechanical properties of polymers. Systematic study of this field, which is important in the development of polymer engineering, was prevented by the technical difficulties encountered in acquiring sufficiently large samples of uniform size macromolecules. Conventional fractionation techniques were obviously too laborious, and therefore impractical. The advent of anionic polymerization, with the practical application of living polymers, removed this objectionable feature of the study by making it possible to produce large quantities of reasonably homogeneous polymers. The first investigation of this type was reported by McCormick, Brower, and Kin (37).

It was known that many physical properties of polymers follow the relation proposed by Flory (38):

$$\text{property} = a + b/\langle j \rangle_n.$$

In fact, the studies of Merz, Nielsen, and Buchdahl (39) confirmed such a relation for polystyrene films made from fractionated and unfractionated

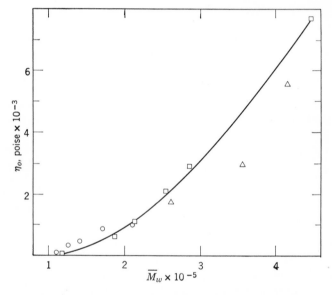

Fig. 16.   A plot of zero shear melt viscosity against weight average molecular weight for (○) anionic, (□) isothermal, and (△) thermal polystyrene.

polymers. On the other hand, there are properties which are functions of the weight average molecular weight—melt viscosity, for example, appears to be given by the relation

$$\eta(\text{melt}) \sim \{\langle j \rangle_w\}^q,$$

where $q = 3$–4 (see ref. 40). It could be expected, therefore, that the uniform size polymer would provide optimal properties at lowest melt viscosity, an obvious advantage from the engineering point of view.

The studies of McCormick and colleagues (37) dispelled this hope. They have shown that at zero rate of shear the melt viscosity is indeed determined by $\langle j \rangle_w$, whatever the distribution (Figure 16). However, at high rate of shear, the uniform size polymer has a higher melt viscosity than the conventional polymer of the same weight average degree of polymerization (Figure 17). Apparently the low molecular weight fraction acts as a kind of lubricant at a high rate of shear. One may note that at high rate of shear the melt flow of polystyrene closely approaches Newtonian behavior as the molecular weight distribution narrows. This observation was confirmed by Rudd (160). We may recall, also, the studies of Stratton (158) and of Porter, Cantow, and Johnson (159), who found that for a narrow molecular weight distribution polyisobutylene, the Newtonian flow region extends to much higher shear rates and shear stresses than for the ordinary material. In fact,

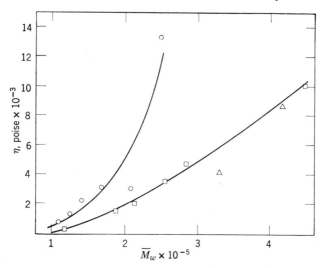

Fig. 17. A plot of high shear melt viscosity against weight average molecular weight for (○) anionic, (□) isothermal, and (△) thermal polystyrene.

this region was extended by a factor of ten when the carefully fractionated polyisobutylene samples were compared with the unfractionated polymer.

Studies of tensile strength demonstrated again the importance of molecular weight distribution (Figure 18). It appears that the narrower the distribution, the poorer the tensile strength at a given $\langle j \rangle_n$, but the higher this property at a given $\langle j \rangle_w$. The modulus associated with the maximum relaxation time is inversely proportional to the molecular weight for ordinary polymers. However, extensive studies of Tobolsky (161,162) demonstrated that for narrow molecular weight distribution polystyrenes this property becomes independent of the molecular weight. The stress relaxation and creep of such polymers were investigated (119), and the master curves for three samples of different molecular weight were reported. The steady-state shear compliance is almost independent of the molecular weight for the narrow distribution

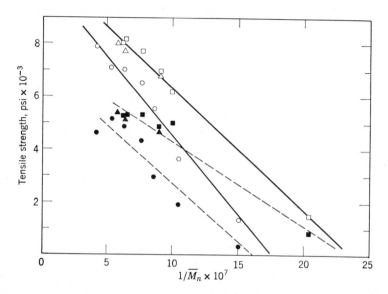

Fig. 18.   A plot of tensile strength versus the reciprocal of number average molecular weight for ($\bigcirc$, $\bullet$) anionic, ($\square$, $\blacksquare$) isothermal, and ($\triangle$, $\blacktriangle$) thermal polystyrene. Open symbols for injection-molded specimens and solid symbols for compression-molded specimens.

polystyrenes, although it increases with the molecular weight for the broad distribution material. Again, interesting differences are observed when the birefringence/stress ratio is determined for the conventional and for the narrow-distribution polystyrene (163). This ratio is one hundred times as great for the former material than for the latter.

The effect of molecular weight distribution on the rate of creep was investigated also by Plazek (189). Creep compliance of narrow molecular weight polystyrene was measured from 10^{-10} to 10^{-4} cm^2/dyne.

Availability of uniform size polymers also permitted establishing in a more satisfactory way various relations between, e.g., viscosity, sedimentation constant, etc., and $\langle j \rangle_n$. Examples of such studies are given in references 15–19 and 164 and in the most recent papers of Berry (190,191). A review of this subject was published recently by Fox and Berry (202). Many more investigations of this type are to be expected in the near future.

All these observations are most interesting, but they are restricted for the time being to polystyrenes. It is desirable to extend these studies to polymers of other types in order to establish the scope of generalizations which have been noted. The availability of living polymers of uniform size also makes it possible to prepare more complex macromolecules having unique dimensions and structure. This subject will be discussed in the following sections.

11. CLASSIFICATION OF BLOCK POLYMERS

Linear macromolecules are classified as homopolymers if all their segments are identical or virtually identical; otherwise such species are described as copolymers. In a random copolymer, segments of different types are distributed along the chain in a haphazard way without forming a regular pattern. In an alternating copolymer, they follow consecutively, giving rise to a regular structure, $\ldots$A.B.A.B.A.B.$\ldots$. Still, in other polymers, units of one type form a long sequence followed by a long sequence of units of the other type; such macromolecules are referred to as block polymers or block copolymers.

To facilitate discussion of block polymers, it is advisable to adopt a notation which describes in a concise way the various possible structures of such molecules. Thus, a symbol {A} will be used to denote a sequence (block) of units of type A, and a subscript may be given to specify the degree, or the average degree, of polymerization of such a block. Thus, $\{A_n\}$ represents a sequence of n, or on the average n, segments of type A, forming a single block of a block polymer.

A block polymer may be composed of two or more blocks. Thus,

$$\{A\}\{B\} \quad \text{or} \quad \{A_n\}\{B_m\}$$

denotes a two-block polymer, whereas symbols such as

$$\{A\}\{B\}\{A\} \quad \text{or} \quad \{A\}\{B\}\{C\}$$

refer to ter-block polymers, etc. A block polymer may involve two, or more, types of units, e.g., the first of the ter-polymers mentioned above is built up from two types of units, whereas the other involves three different types of segments.

The blocks described in this survey form homopolymers. One may visualize, however, a copolymeric block, e.g., $\{A\}\{A,B\}$, which represents a two-block polymer composed of a block of A linked to a block of a random copolymer of A and B. A complete specification of such a macromolecule should include information about the length of each block and the composition of the copolymeric block.

It should be stressed that the overall composition and the total length of a macromolecule are definitely not sufficient to describe a block polymer. For example, the two ter-block polymers

$$\{A_n\}\{B_{2m}\}\{A_n\} \quad \text{and} \quad \{B_m\}\{A_{2n}\}\{B_m\}$$

have identical length and composition, i.e., they are "isomeric." However, their properties may be enormously different. A specific example of such "isomeric" block polymers, namely A = styrene and B = butadiene, will be discussed later; while the former is technologically most valuable, the latter seems to be useless.

Finally, tapered structures deserve discussion. In some preparations, a polymer is formed which possesses a block composed of pure A at one end, and a pure block of B on the other. These are linked through a copolymeric block, the composition of which gradually changes from that of pure A to that of pure B. A symbol $\{A \rightarrow B\}$ is proposed for such a unit, and hence the ter-block polymer described here is denoted by $\{A\}\{A \rightarrow B\}\{B\}$.

The advent of stereoregular polymers led to the concept of stereoblocks, e.g., a sequence of isotactic units may be followed by a sequence of segments arranged in a syndiotactic manner, or a sequence of all d-oriented units may be followed by a sequence of l units. Alternatively, an all cis-polybutadiene block may be followed by a block of 1,2-linked dienes. Many stereoblock polymers have been described in the recent literature and some have been thoroughly characterized. Short stereoblocks are most common in ordinary "atactic" polymers. In fact, a perfectly atactic, i.e., statistically random, polymer is unlikely to be formed in most processes.

12. SYNTHESIS OF BLOCK POLYMERS

The preparation of block polymers may be afforded through many techniques. A review of this subject was published by Smets and Hart (51), who were interested mainly in a radical type of polymerization. Ionic methods have been discussed, however, in a later paper from the Smets laboratory (78).

Basically, two methods lead to the formation of block polymers. If an end of a homopolymer A remains, or becomes active, and if it is capable of initiating the polymerization of a monomer B, the resulting product is a block polymer $\{A\}\{B\}$. Alternatively, two independently prepared homopolymers of A and of B may be linked, and thus an $\{A\}\{B\}$ block polymer is formed again.

Living polymers are the most suitable materials for the synthesis of block polymers by either method. Their ends are active and capable of initiating polymerization of properly chosen monomers. Therefore, the simplest and the most straightforward method leading to the preparation of block polymers is based on the following procedure (1,2): A living polymer, prepared from monomer A, is placed in a reactor to which monomer B is added. If the active end of the living poly-A initiates polymerization of B, a block of B may grow from the end of a block of A. This technique is superior to many other methods used in the synthesis of block polymers. Whenever termination and chain-transfer are rigorously excluded, a product free of homopolymers of either A or B is obtained. In fact, contamination of block polymers by homopolymers is a serious drawback of nearly all the other synthetic methods that yield block polymers. The presence of homopolymers greatly impedes the characterization of the prepared block polymer, and it is often extremely difficult to resolve such a mixture into pure components, particularly when the product is composed of blocks of nonuniform size.

The living polymer technique offers other important advantages to a synthetic polymer chemist. In the preceding sections, we discussed the conditions under which a uniform size living polymer may be formed. If such conditions are maintained during the preparation of block polymers, it becomes possible to produce macromolecules composed of blocks having a predetermined size. In fact, the length of each block may be easily adjusted by a proper feeding of the desired amounts of A, and then of B, into the reactor. Thus, an experimenter, or a manufacturer, may rapidly prepare a whole spectrum of block polymers the sizes of which vary according to a predetermined scheme. Thus, systematic academic or technological studies of their properties as functions of the lengths of individual blocks become feasible and reliable.

It is possible to prepare living polymers with both ends active (1,2). These materials are most valuable for the preparation of ter-block polymers of the type {A}{B}{A} if a living poly-B may initiate polymerization of A. Again, having all the necessary conditions fulfilled, the sizes of {A} and {B} may be varied at will and uniform polymers $\{A_n\}\{B_m\}\{A_n\}$ could be produced.

A living poly-A does not initiate polymerization of every monomer B. For example, living, anionically growing polystyrene may initiate polymerization of methyl methacrylate, but living polymethyl methacrylate— again of the anionic type—cannot initiate polymerization of styrene (52,165). Similarly, polymerization of ethylene oxide may be initiated by living polystyrene, but obviously an alcoholate—a terminal active group of a living polyglycol—cannot initiate polymerization of a vinyl hydrocarbon (53). It is possible, however, to choose a pair of two monomers such that a living poly-A initiates polymerization of B and, vice versa, a living poly-B initiates polymerization of A. In such a system, an interesting spectrum of "isomeric" block polymers may be produced. The following example outlines the procedure and its potentialities.

Let two reservoirs containing, say, a mole of monomer A and of monomer B be attached to a reactor in which a living polymer may be formed. The polymerization may be performed by adding slowly all the A and thereafter all the B. This yields a two-block polymer of 50:50 composition, viz.,

A....................A.B....................B

and of a particular degree of polymerization which is determined by the amount of initiator present at the onset of the reaction.

Now, the procedure may be repeated by changing only the modes of additions, keeping, however, all the other variables unaltered. Let us add one-half a mole of A to begin with, and then one-half a mole of B. The remaining one-half mole A would be added next, followed by one-half mole B. As a result of these additions a different block polymer is expected, namely,

A.........A.B.......B.A.........A.B........B

The new macromolecule should have the same composition and molecular weight as the previous one, but a different distribution of monomers along the chain. The generalization of this procedure is obvious, and thus one may vary in a great number of ways the distribution of A and B in the polymer, nevertheless keeping constant its overall composition and its total molecular weight.

Studies of the effect of distribution upon the properties of block polymers have therefore become feasible. Such studies were carried out by Schlick and Levy (54). They prepared block polymers of styrene and isoprene, maintaining the 1:1 mole ratio of monomers in the polymers but varying their distribution along the chain. Two series of block polymers were investigated, one having $DP \sim 100$, the other of about 1000. In each series, the monomers were distributed among 3, 5, 7, and 9 blocks, and in addition "copolymers" were prepared by a slow, dropwise addition of the monomer mixture to the reactor. Furthermore, they investigated polyblocks terminated with polystyrene and others, possessing terminal blocks of polyisoprene. It was shown that the solubility and viscosity of these polymers depends on their structure, i.e., at constant composition and molecular weight the distribution affects the properties of macromolecules. Similar studies were reported by Korotkov et al. (55), who block polymerized isoprene and butadiene. Thorough studies of ter-block polymers have been described by Baer (56).

It is frequently observed that ionic polymerization performed in a mixture of two monomers greatly favors one component over the other (57). The preferred monomer is polymerized first, almost exclusively, and the other reacts only after the supply of the preferred monomer becomes virtually exhausted. Such a situation is encountered in a mixture of butadiene and styrene polymerized with the aid of alkyllithium initiators in hydrocarbons (58–60). The monomers were analyzed as the reaction proceeded. The results showed that, as long as butadiene was present in the feed, only a minute amount of styrene polymerized, although this monomer formed 55% of the initial mixture and its proportion in the feed increased during the first stages of the reaction. The final product, obtained at the completion of polymerization, was composed of macromolecules, uniform in their composition, a fact which in conjunction with the other finding clearly proves its block structure (59). Additional evidence was reported by Zelinski (60), who fractionated the ultimate polymer and demonstrated in this way the absence of homopolymers. He also degraded the product with osmium tetroxide, a reagent that attacks polybutadiene but not polystyrene, and isolated homopolystyrene from the degraded material. The isolated polystyrene did account for 80% of all the initial styrene present in the feed. Hence, the presence of polystyrene blocks in the original material is unequivocally demonstrated.

Further evidence confirming the formation of block polymers in this system comes from the spectrophotometric studies of Johnson and Worsfold (61). Polybutadienyl anions, but not polystyryl ions, were observed in the polymerizing solution as long as butadiene was still present in the feed, although the concentration of styrene exceeded that of butadiene.

All these facts indicate that polymerization initiated by alkyllithium in a styrene–butadiene mixture yields a ter-block polymer,

$$\{butadiene\}\{butadiene \rightarrow styrene\}\{styrene\}.$$

The existence of the midde tapered block is proved by the shape of the conversion curve (59), although its length is not yet known. Of course, the terminal blocks are not free of contamination. A trace of styrene has to be present in the polybutadiene block, and a trace of butadiene should be incorporated into the polystyrene block.

An interesting situation may be created in ionic polymerization if a suitable complexing agent is present in the system. Two types of growing ends then participate in such a propagation—the complexed and the uncomplexed. It is often observed that each type of growing end may have its own characteristic mode of monomer addition, e.g., one may favor isotactic placement, whereas the other leads to syndiotactic addition. Due to the dynamic nature of this system, the complexing agent is exchanged continually between the different units and, if the exchange is not too rapid, the resulting macromolecule forms a stereoblock polymer. Such a phenomenon was observed in the polymerization of methyl methacrylate initiated by fluorenyllithium in a mixture of toluene and dioxane (63). The average length of the stereoblocks is determined by the rates of each mode of polymerization and by the rate of exchange; the latter depends on the concentration of the complexing agent. A mathematical analysis of such polymerization was presented by Coleman and Fox et al. (64), and the molecular weight distribution of the resulting stereoblocks was calculated by Figini (65). Complexing agent may be added at a later stage of the process, e.g., a diene polymerized by a lithium alkyl in a hydrocarbon solvent forms a 1,4 polydiene. Thereafter a small amount of ether is added, the polymerization continued, and 1,2-linked units are then produced (166).

Living polymers are also valuable reagents for synthetic work involving the "linking up" technique. Several variants of this method may lead to block polymers. For example, an anionically growing polystyrene may be converted to an α,ω-dicarboxylic acid (53), and its condensation with molecules of nylon terminated by amino groups should then yield a block polymer of styrene and nylon. A polyblock may be obtained if the nylon molecules possess amine groups on both of their ends. Hayashi and Marvel (85) explored this method. They prepared block copolymers by refluxing a toluene solution of a mixture of polystyrenedicarboxylic acid and either polybutadiene glycols or hydrogenated polybutadiene glycols. Direct melt condensation was unsuccessful due to the incompatibility of the polymers. However, block polymers were obtained by using urethan as the linking agent. Uranech, Short, and Zelinski (112) have described many methods for preparing polymers with functional end groups which may then be condensed to yield block copolymers.

Alternatively, the application of bifunctional linking agents may convert living polymers into block polymers. For example, {styrene}{CH$_2$CH$_2$O}$^-$,Na$^+$ block polymer may be converted by adipyl chloride into a

$$\{styrene\}\{CH_2CH_2O\}CO(CH_2)_4CO\{OCH_2CH_2\}\{styrene\}$$

ter-block polymer. Again, a polyblock is formed if a bifunctional block polymer, such as Na$^+$,$^-$\{OCH$_2$CH$_2$\}\{styrene\}\{CH$_2$CH$_2$O\}$^-$,Na$^+$ is used in the synthesis (53). In a similar way, a styrene–butadiene block polymer was linked up by p-CH$_2$Cl·C$_6$H$_4$·CH$_2$Cl (110).

An interesting and unusual technique of linking living polymers has been described by Berger, Levy, and Vofsi (111). Polystyryl dianions were prepared in tetrahydrofuran from α-methylstyrene tetramer, after which most of the solvent was evaporated. Polytetrahydrofuran cations were prepared at 0°C using (C$_2$H$_5$)$_3$O$^+$,BF$_4$-complex as the initiator. After two hours, the living polystyrene anions were added to the living polytetrahydrofuran cations. The product was poured into methanol; a semisolid precipitate thus formed was washed with methanol and dried *in vacuo*. The infrared spectrum of the product showed that it consisted of both polystyrene and polytetrahydrofuran. The product could not have been a mixture of homopolymers because then the precipitate formed on pouring it into methanol would have consisted of polystyrene only. Three fractions of the block copolymer prepared by precipitation with methanol or water contained both polystyrene and polytetrahydrofuran, although the first fraction was richer in styrene than the other two. The same treatment of a mixture of homopolymers yielded two fractions, one consisting of polystyrene only and the other of polytetrahydrofuran. Attempts to prepare multiblocks

$$\{THF_m\}\{S_n\}\{THF_m\}\{S_n\}\{THF_m\}\ldots$$

by using difunctional living polytetrahydrofuran yielded insoluble gels which were difficult to characterize. It would be interesting to apply this method to other pairs of living cationic and anionic monomers.

The principles outlined here may be applied to the preparation of new types of polymers having unusual structures. Two examples suffice to illustrate the method:

(1) Interaction of α-methylstyrene with metallic potassium leads eventually to a tail-to-tail linked dimer, $K^+, C^-(CH_3)(Ph) \cdot CH_2 \cdot CH_2 \cdot C^-(CH_3)(Ph), K^+$. Its reaction with I_2 was shown (62) to produce a polymer,

$$-C(CH_3)(Ph) \cdot CH_2 \cdot CH_2 \cdot C(CH_3)(Ph) \cdot C(CH_3)(Ph) \cdot CH_2 \cdot CH_2 \cdot C(CH_3)(Ph)-,$$

having head-to-head tail-to-tail linked α-methylstyrene units. This polymer proved to be distinct from the conventional head-to-tail poly-α-methylstyrene.

(2) The reaction of the above dimer with adipyl chloride gives an alternating copolymer (109),

$$-C(CH_3)(Ph) \cdot CH_2 \cdot CH_2 \cdot C(CH_3)(Ph) \cdot CO \cdot (CH_2)_4 CO \cdot C(CH_3)(Ph) \cdot CH_2 \cdot CH_2 \ldots$$

In principle, these reactions provide examples of polycondensations which are applied to living polymers. The high reactivity of the latter makes them most useful and valuable for such procedures.

13. REMARKS ON THE PREPARATION OF BLOCK POLYMERS

In spite of the apparent simplicity of the living polymer technique, the preparation of block polymers is extremely difficult in practice. Studies of block polymers of styrene and methyl methacrylate illustrate the various pitfalls facing the experimenter. The first preparation of such a block polymer led to killing of the living polymethyl methacrylate (92), probably caused by some impurities still present in the monomeric methyl methacrylate and by a relatively high temperature of reaction (20 to 30°C). More careful purification of the reagents and lowering of the temperature to $-60°C$ led eventually to stable living polymers of this monomer (113,114). Nevertheless, earlier investigators were still unable to produce truly linear block polymers of styrene and methyl methacrylate (66). In fact, light-scattering studies carried out in solvents with the same refractive index as the polymethyl methacrylate blocks showed that the molecular weight of the polystyrene fragments attached to the polymethyl methacrylate was considerably higher than that of the living polystyrene from which the block polymer was derived (165). Presumably, some of the polystyryl anions reacted with the ester function of the growing polymethyl methacrylate and this led to branched structures. Subsequently, Freyss, Rempp, and Benoit (98) reported that this side reaction could be avoided by adding a small amount of 1,1-diphenylethylene to the living polystyrene. This monomer does not homopolymerize but the anion derived from it rapidly initiates methyl methacrylate polymerization without attacking the ester functions. Light scattering and fractional precipitation experiments proved that the resulting block copolymers were highly uniform with respect to both composition and molecular weight. This therefore illustrates the useful role of the detailed characterization of a polymer in developing useful synthetic procedures. Indeed, it is impossible to overestimate the value of this approach. Too many times polymers have been prepared and their structure deduced merely from synthetic considerations, with no detailed physical characterization ever attempted. This point should be kept in mind when we discuss some of the polymers described below.

To illustrate the scope of these methods (without claiming, however, that such a review is thorough or complete), let us survey briefly the literature describing the formation of some block polymers by anionic polymerization or by application of some other living polymers. The preparation of block polymers of ethylene oxide and acrylonitrile was described by Furukawa et al. (79); their procedure utilized anionic polymerization. This was accomplished by treating polyethyleneglycol in toluene with sodium metal to produce alkoxide anions. Acrylonitrile was then added slowly to the system. Similar work was described by Novitskaya and Konkin (80). Morphological studies of block polymers composed of polyacrylonitriles and polymethacrylonitriles and synthesized by anionic technique were reported by Rempp et al. (81). Anionic block polymerization of styrene or isoprene with cyclic octamethyl tetrasiloxanes was described by Morton and Rembaum (84). Living polystyrene or polyisoprene was the initiator for the cyclic compound, the technique being analogous to that used in styrene–ethylene oxide block polymerization (53).

Interesting block polymers, synthesized through the same approach and involving episulfides, have been reported by Sigwalt and his colleagues (104), who showed that propylene episulfide polymerizes easily and rapidly under the influence of an ionic catalyst such as butyllithium, sodium amide, potassium hydroxide, or metallic sodium. Their work confirms the earlier observations of Ohta et al. (105). Polymers of the highest molecular weight were obtained with films of metallic sodium ($\eta = 2.65$ [100 cm^3/g] determined in benzene at 25°C). Polymerization may also be initiated by sodium naphthalene in tetrahydrofuran (106), although it is not clear whether or not electron transfer initiation takes place. This suggested, therefore, that block polymers may be prepared and, indeed, such materials were reported by Boileau and Sigwalt (107) and by Baker (108). They produced blocks of polypropylene episulfide and polyethylene episulfide, as well as blocks of polystyrene and polypropylene episulfides. Some attempts were made to prepare block polymers with blocks of elemental sulfur. Apparently, sulfur may be polymerized on living polystyrene if the reaction is performed at $-80°C$ (108). The product was shown to contain 23% sulfur but the length of the sulfur blocks was not estimated. In view of the high floor temperature of sulfur, this result is intriguing and indicates that sulfur blocks must have a complex structure and do not form simply linear polymeric chains.

Both perfluoroethylene oxide and perfluoroethylene episulfide add to living polymers and form the respective block polymers (115). This observation is supported by the fact that infrared spectra of these products are not affected by extraction. Similar reactions were observed when hexafluoroacetone was used as a monomer (115).

Block polymers of styrene and 2-vinylpyridine were studied by Sigwalt and his colleagues (116). A block polymer is undoubtedly formed when vinylpyridine is added to living polystyrene; however, it is doubtful whether a linear block polymer is produced when styrene is added to living 2-vinyl-pyridine (117). These studies were extended by Sigwalt's group to monomers other than styrene, e.g., dihydronaphthalene (88).

Block polymerization involving formaldehyde was described by Smith et al. (83), by Baker (108) and by Bastien (183). The success depends on the purity of the monomer, and a delicate "balance" of conditions is necessary to obtain formaldehyde suitable for the preparation of the block polymers. The formaldehyde blocks were grown on anionic living polystyrene, and both

$$\{styrene\}\{CH_2O\} \qquad and \qquad \{CH_2O\}\{styrene\}\{CH_2O\}$$

polymers were obtained. These results confirm the previous claim of Noro et al. (120) and contradict the findings of Carter and Michelotti (121), who reported that formaldehyde gives only terminal formate groups; probably impurities present in the aldehyde vitiated their results. In addition, living polyisoprene, polymethyl methacrylate, and polyacrylonitrile were used as initiators of formaldehyde polymerization, and the respective block polymers were isolated. To prevent unzipping of the polyformaldehyde blocks, the respective polymers were "capped" by acetylation of the original hydroxyl end groups. A block polymer $\{A\}\{B,C\}$ was also described, where A is styrene and B and C are formaldehyde and phenyl isocyanate, respectively. The materials produced in this study were characterized by determining their tensile strength, elongation, stiffness, etc.

For the reader's convenience, a list of references concerned with some typical block polymerization processes has been compiled in Table III.

Finally, let us consider some block polymerization processes which utilize a coordination catalyst. It has been pointed out by Natta (93) that the lifetime of olefin chains growing on Ziegler–Natta catalysts may be sufficiently long to justify their treatment as living polymers and, hence, it should be possible to produce block polymers by consecutive feeding of

TABLE III[a]

Preparation of Block Polymers {A}{B} or {B}{A}{B} Using
Living Polymer Technique

Living polymer A	Monomer B	References
Styrene	α-Methylstyrene	56,86
	1-Vinylnaphthalene	87
	1,2-Dihydronaphthalene	88
	p-Bromostyrene	89
	2-Vinylpyridine	88,89
	4-Vinylpyridine	89
	Butadiene	76
	Isoprene	1,47,54
	Methyl methacrylate	56,89,90,92,97
	Hexyl methacrylate	91
	Isopropyl acrylate	91
	Acrylonitrile	78,90,91
	Methacrylonitrile	89,91
	Ethylene oxide	53,56,74
Isoprene	Styrene	1,47,54,55
	Butadiene	55
	Ethylene oxide	89
α-Methylstyrene	Styrene	56,86
	Methyl methacrylate	51
Methyl methacrylate	Butyl methacrylate	91
	Hexyl methacrylate	91
	Isopropyl acrylate	91
	Acrylonitrile	90,91

[a] A more extensive list has been compiled recently by Morton and Fetters (194).

monomers. For example, a block polymer of propylene–ethylene was produced when a catalyst prepared from titanium trichloride and triethylaluminum in xylene was employed (94). Propylene was introduced first, then after five minutes it was flushed out with nitrogen, and the latter replaced in turn by ethylene. Bier and his coworkers (95) described similar systems producing polymers which retained their activity for at least fifteen hours at room temperature; and even more stable catalyst–living polymer complexes, lasting for several days, were reported by Kontos et al. (96). According to Bier, the product is free of homopolymers. This conclusion was drawn after examination of the molded articles made from block polymers. The articles were more transparent and stretchable than those made from a blend of the respective homopolymers. However, such an observation may simply indicate that the block polymer, present together with a mixture of polyethylene and polypropylene, acts as an emulsifying agent and facilitates the formation of a highly dispersed blend of the homopolymers.

14. SPECIFIC BLOCK POLYMERS AND THEIR PROPERTIES

Let us compare a copolymer with the respective parent homopolymers. The new features of a copolymer arise from the fact that the *local* environment of nearly every segment of such a macromolecule (i.e., the conditions existing within a radius of 5–10 Å from its core) differs from the environment experienced by the segments of the corresponding homopolymers. Therefore, any property of such a chain which is determined by molecular interactions of segments 10–20 Å long, is profoundly affected by the composition of a copolymer. Moreover, since the interactions between two adjacent, but not identical, monomeric units are important, the properties of copolymers are not given by any simple additive rules.

Block polymers, and particularly those built up from long sequences of identical monomeric units, retain, on the whole, the segmental environments of the parent homopolymers. In this respect, a block polymer resembles a blend of two homopolymers; thus many characteristic features of an *isolated* molecule of a block polymer are additive and can be calculated by averaging the properties of the component blocks (199). The heat of solution, or of dilution, of a block polymer is perhaps the best example of such a property. To a lesser degree, the same applies to the average end-to-end distance and to the radius of gyration, although a closer examination of this problem suggests the possibility of some deviations from simple additivity. For example, studies of Benoit and his associates (167,200) indicate a degree of incompatibility of the blocks, even in a highly dilute solution. Apparently, segments of type A tend to avoid those of type B, and vice versa, and this leads to the expansion of the macromolecule and to an increase in its end-to-end distance.

The effects arising from polymer-polymer interactions become perceptible and important in concentrated solutions, and even more so in the solid state. These interactions create some new properties by virtue of which a block polymer becomes basically different from the corresponding polymer blend.

Homopolymers are, on the whole, incompatible. Their concentrated solutions separate into two phases, and the same tendency leads to a heterogeneous structure of a solid polyblend. This tendency is responsible also for the microseparation of block polymers, but the bonds linking the two blocks do not allow them to separate on a macroscopic scale. The two, or more, incompatible chains are forced, therefore, to coexist in colloidal domains, and thus some specific colloidal relations have to be expected. This leads to a specific topology of a swollen gel or a solid polymer composed of block polymers.

The first extensive studies of the morphology of such systems were reported by Skoulios and his colleagues (71,72). They investigated {A}{B} block polymers built up from polystyrene and polyethylene oxide, of total molecular weight of about 14,000. These were dissolved in nitromethane—a good solvent for polyethylene oxide but poor for polystyrene, or, alternatively, in butyl phthalate, which dissolves the latter polymer but precipitates the former. The gels produced were examined using the low angle x-ray scattering technique. The results proved that in nitromethane cylindrical micelles were formed, whereas lamellae were observed in butyl phthalate. These structures are presented schematically in Figure 19, the domains of mesomeric phases ranging from 100–300 Å, depending on the amount of added solvent. In concentrated solutions, cylindrical micelles showed hexagonal packing, whereas the lamellae were equidistantly spaced. The polyethylene oxide

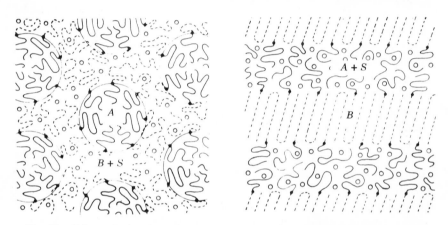

Fig. 19.   Schematic drawing of structures formed by block polymers of styrene and ethylene oxide. Right side: cylindrical micelles packed hexagonally formed in nitromethane; left side: lamellae formed in butyl phthalate. (A) blocks of polystyrene (drawn in solid line), (B) blocks of polyethylene oxide (drawn in dashed lines), (S) solvent molecules denoted by small circles.

layers of these lamellae were crystalline, and those of polystyrene were amorphous and swollen by the solvent. A comprehensive review of this subject has been published by Sadron (73). Recent studies of this type of solid block polymer have been reported by Kovacs et al. (82) and by Skoulios et al. (99).

The block polymers of styrene and ethylene oxide were prepared by adding ethylene oxide to living polystyrene (53) or by condensing polyethylene oxide, terminated by hydroxyl groups, with polystyrene having —COCl end groups (74). The latter polymer was prepared by reacting living polystyrene with $COCl_2$ (phosgene).

Some unusual properties of the $\{OCH_2 \cdot CH_2\}\{styrene\}\{CH_2CH_2O\}$ block polymers were described by Richards and Szwarc (53). Their solutions in methyl ethyl ketone or in tetrahydrofuran turned only slightly cloudy when large amounts of water were added, but in no case was a precipitate formed. On the other hand, a gelatinous colloidal precipitate appeared when a drop of water was added to a clear solution of the block polymer in benzene. Apparently, a nonhydrated polymer is soluble in benzene, while the hydrated chains of ethylene oxide coagulate or form micelles.

A closer examination of this phenomenon showed that it is more complex than at first appears. Addition of a drop of water to a benzene solution of polyethylene oxide does not precipitate the polymer, nor is a precipitation or emulsion observed when water is added to a solution of a mixture of the homopolymers of styrene and ethylene oxide in benzene. This observation again illustrates the difference in behavior of a block polymer and of the blend of the respective homopolymers. It indicates also the value of block polymers as efficient emulsifying agents. In the example cited above, the micelles of polyethylene oxide, swollen by water, disperse this liquid throughout the hydrocarbon phase. The blocks of polystyrene prevent coagulation because the solubility of this polymer in benzene is high. However, addition of salt (e.g., sodium iodide) leads to salting out of water and the benzene solution becomes clear again. Block polymers are expected to be efficient homogenizing agents useful in blending incompatible polymer solutions or solid polymers (75,201); this should lead to the formation of new and perhaps useful materials. An example of homogenizing some immiscible solutions of polymers in oil with block or graft polymers has been discussed by Molau (118).

A spectacular phenomenon arising from the ordering of incompatible blocks of a block polymer was described by Vanzo (76). By initiating anionic polymerization of a styrene–butadiene mixture with butyllithium, he produced block polymers of a very high molecular weight. Their concentrated solutions in ethylbenzene exhibited irridescent colors which depended on the concentration of the polymer and changed reversibly when stress was applied. As in the case of the polystyrene–polyethylene oxide blocks, the incompatible polybutadiene and polystyrene blocks separate on a microscale and form equidistant layers which, being sufficiently thick, cause the interference of visible light. Addition of solvent changes the spectrum of the reflected light. For example, Figure 20 gives the observed spectra for solutions containing 10, 20, and 30% of ethylbenzene. Carefully dried films retain the layer structure, and their electron micrographs are reproduced in Figure 21. The patterns vary according to the degree of monodispersity of the blocks, and the spacing should vary with their molecular weight. Unfortunately, technical difficulties prevented systematic studies of this subject.

Highly significant, and technologically most valuable, are the ter-block polymers of the type

{styrene}{isoprene}{styrene}   or   {styrene}{butadiene}{styrene}

These were prepared by a number of workers, but their industrial value has been investigated only recently by Milkovich and his colleagues (77). The above ter-block polymers combine the elastomeric properties of rubber with the thermoplastic properties of polystyrene and yield a new class of materials —*the thermoplastic rubbers.*

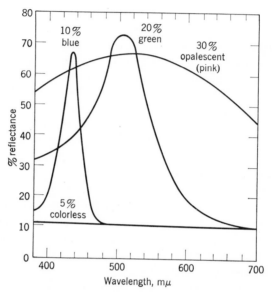

Fig. 20.   Visible spectrum of reflected light from layers of styrene–butadiene block copolymer at various concentrations in ethylbenzene.

The preparation of thermoplastic rubbers follows the usual procedures of anionic block polymerization. Living polystyrene is prepared in hydrocarbon solvents by initiating the polymerization with alkyllithium. The use of secondary alkyllithiums offered some unexpected advantages (47) and yielded products of narrow molecular weight distribution. The living polymers are thereafter fed with isoprene and thus living two-block polymers,

$$\{styrene\}\{isoprene\}^-, Li^+$$

result. The two-block polymers are eventually combined into ter-block polymers either by using a linking agent (e.g., CH_2Br_2) or by feeding styrene again. Thus, the ter-block polymer

$$\{styrene\}\{isoprene\}\{styrene\}$$

is formed.

At temperatures above the glass transition point of polystyrene, these materials have the flow characteristics of conventional nonvulcanized rubber.

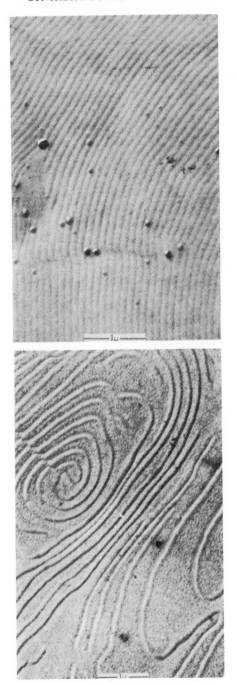

Fig. 21. Surface structure of styrene–butadiene block polymer film.

The latter, as is well known, are tacky materials of low strength and are of little utility until subjected to a vulcanization treatment. This improves their strength but renders them insoluble in hydrocarbons. In this respect, the new thermoplastic rubbers are basically different. Under service conditions (i.e., below the glass transition temperature of polystyrene) the latter chains act both as physical cross-links and as reinforcing fillers so that the valuable properties of vulcanized rubber are obtained directly without rendering the material insoluble.

The technological and commercial advantages are obvious. Such rubbers may form dispersions in oil or greases which improve the stringiness or tack and, being soluble, improve the adhesion of such compositions to metal surfaces. They are useful, therefore, in paints, etc. Moreover, the used material, i.e., the scrap, may be reutilized like any other thermoplastic, in contrast to the scrap of vulcanized rubber, which is useless.

In order to obtain the optimum "self-vulcanizing" properties, as well as satisfactory processing characteristics, the molecular weights of the blocks must be carefully controlled. Here, the advantage of the living polymer technique is obvious. Moreover, since the optimal results are obtained when the blocks have a narrow molecular weight distribution (47), again the living polymer technique is uniquely suited to yield such products.

The high degree of elasticity of these materials is believed to be due to two causes: (a) the degree of contact between the rubbery and plastic phases is high, and (b) the plastic phase is highly dispersed with most of the plastic being in isolated islands rather than in the matrix. The stress–strain curves of these elastomeric thermoplastics show that the initial moduli are high, although a pronounced yield point eventually occurs (168). It appears, therefore, that some of the polystyrene is present in the matrix rather than in islands, perhaps as a kind of network. Films cast from heptane, a nonsolvent for polystyrene, ought to increase the fraction of polystyrene present in the islands. This conclusion is supported by the observed stress–strain relations, viz., the initial modulus becomes lower and the material exhibits a less pronounced yielding. Birefringence measurements indicate that the segmental mobility of these materials is high. Their incomplete recovery appears to be associated with the process of yielding and is not due to permanent flow since retraction was complete on heating. In conclusion, it appears that the highest degree of elasticity is achieved when the fraction of the plastic phase in the islands is as high as possible.

The difference in the properties of two-block and ter-block polymers is remarkable, as may be seen from Table IV. The ter-block polymers show strength, whereas the two-block polymers have low cohesion. Furthermore, while

{styrene}{isoprene or butadiene}{styrene}

TABLE IV

Physical Properties of Uncoupled and Coupled Styrene–Isoprene Block Polymers

	S-I block[a]			S-I-R-I-S block		
Sample	Tensile strength, psi	Elongation, %	Set, %	Tensile strength, psi	Elongation, %	Set, %
A	50	630	50	510	1220	40
B	38	200	0	2225	1260	40
C	115	1290	25	2750	1180	20
D	95	1360	40	2100	1310	20
E	30	300	40	1050	1460	15
F	32	580	50	1490	1350	30

[a] The styrene–isoprene block polymer (S-I) samples were prepared as shown below. Ter-blocks were formed by the coupling procedure.

	S-I				Intrinsic viscosity of coupled block polymer
Sample	Polystyrene, $\overline{M}_v \times 10^{-3}$	Polyisoprene, $\overline{M}_v \times 10^{-3}$	Intrinsic viscosity	Coupling agent	
A	30	107	0.85	1,10-Dibromodecane	0.97
B	25	80	0.70	1,4-Dibromobutane	0.98
C	24	88	0.74	1,2-Dibromoethane	1.03
D	26	96	0.80	1,3-Dibromopropane	1.06
E	30	77	0.71	1,3-Dibromopentane	0.93
F	25	70	0.69	α,α'-Dichloro-p-xylene	0.86

Sheets were pressed at 120°C for 5 min (conventional rubber press) microdumbbell specimens were tested on Instron—crosshead speed 1.2 in./min.

ter-block polymers are industrially valuable, the isomeric

{isoprene}{styrene}{isoprene}

polymers are of little technological interest.

These results emphasize once again the importance of texture, which may be controlled by composition, length, and distribution of the blocks. Apparently many mechanical properties of polymers are determined by the dimensions of the colloidal domains, and not only by their molecular structure. There are evidently some critical sizes for these domains and these may be associated with sharp maxima or minima in mechanical properties. In addition, their topological interrelations are extremely important in modifying such properties, and the distribution of blocks permits us, at least in some measure, to control this feature of the solid polymer. Extensive and systematic work in this field may uncover a wealth of new phenomena which could be important academically and industrially.

It is advisable to pursue this discussion further, even if presently it is highly speculative. The importance of colloidal domains in determining the mechanical properties of a polymer becomes more apparent when other phenomena of somewhat similar nature are considered. A comparison of laminar and turbulent flow may serve this purpose. In laminar flow, the motions of the individual molecules of flowing liquid are not intercorrelated; each moves freely from one layer to another independently of the movements of the other molecules. On the other hand, a large number of molecules travel together in turbulent flow, forming a coherent cluster, their movements being intercorrelated in each cluster. Hence, the onset of turbulent flow is associated with a *decrease* in the entropy of the system. It is obvious, therefore, that the conditions governing the relations between clusters, and the determination of their sizes, are extremely important for the hydrodynamics of turbulent flow.

Some mechanical properties may be associated with phenomena which, in some way, resemble those occurring in turbulent flow. This is particularly pertinent for the dynamic properties that govern the changes taking place in the investigated material in response to rapidly varying external forces, impact strength being perhaps the most outstanding example. In view of this, it is not surprising that the sizes of colloidal domains, their distribution, and the conditions existing on their boundaries are significant in shaping the properties of solids composed of block polymers. This problem undoubtedly will require much detailed and thorough investigation before it can be fully understood.

Some more examples are available to show the effects of colloidal structure and texture on the properties of materials. For instance, the same block polymer of butadiene and styrene may form a rubbery film or a rigid film, depending on the nature of the solvent from which the film is deposited. The different behavior of block polymers and polyblends with respect to the torsional deformation of polymeric rods was utilized by Tobolsky et al. (122) as an analytical tool which permits one to distinguish between the two materials.

15. STAR-SHAPED, GRAFT, AND COMB-SHAPED POLYMERS

Syntheses of star-shaped and comb-shaped polymers (see Figure 22) provide further examples of the versatility of the living polymer technique in preparative polymer chemistry.

A molecule of a *star-shaped polymer* possesses a number of arms radiating

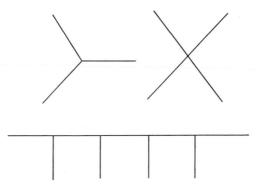

Fig. 22. Schematic representation of star- and comb-shaped polymers.

from a common center. Two distinct routes lead to their formation, and the application of living polymers provides a great degree of control upon their architecture (2a):

(1) One approach involves application of multifunctional initiators. For example,

$$C\left(CH_2 \cdot C_6H_4 \cdot \underset{\underset{CH_3}{|}}{\overset{\overset{CH_3}{|}}{C}}^-, K^+\right)_4$$

may initiate the polymerization of styrene, and four polystyryl chains are expected to grow from such a center. If the proper conditions are maintained, all the arms should be of the same length and all the star-shaped molecules would be identical. A similar product is expected to be formed in the polymerization of ethylene oxide initiated by the trisodium salt of triethanolamine, $N(CH_2CH_2O^-, Na^+)_3$. The reaction should yield a star-shaped polymer endowed with three arms, while the previous initiator had been expected to produce a molecule with four arms.

(2) Alternatively, it is possible to begin the synthesis with living polymers endowed with one growing end each, and link them through a polyfunctional terminating agent. For example, a monodispersed living polystyrene possessing Li^+ counterions reacts with

$$ClCH_2 \overset{}{\underset{CH_2Cl}{\bigcirc}} CH_2Cl$$

and yields a three-armed, star-shaped polymer (100). Analogous reactions involving $SiCl_4$ (101), $(PNCl_2)_3$, or Si_2Cl_6 (123) should produce star-shaped polymers with four or six arms, respectively. However, closer examination proved that the products consisted of mixtures of star-shaped polymers with different numbers of arms (195). It is possible that the difficulties arise, at least in part, from the relatively low reactivity of the lithium derivatives or from metal–halogen interchange reaction (169). They could be eliminated if potassium compounds were used in the synthesis.

This contention was confirmed by Yen (102). The polymerization of styrene was initiated by cumylpotassium, and the living polymers were terminated by $1,2,3,4-C_6H_2(CH_2Cl)_4$. The results were most encouraging. Fractionation and ultracentrifugation demonstrated a high degree of uniformity of the resulting product and all of the molecules had four arms of nearly identical length. The ease of coupling of potassium derivatives, when compared with the reactions of lithium compounds, suggests that the use of potassium derivatives as initiators may lead to other highly uniform polymers of the desired structure. The reliability of Yen's synthesis was confirmed by studies of the physical properties of the resulting products. Extensive study of the relations between the properties of star-shaped polymers and their degree of dispersity were reported by Decker and Rempp (103, see also 196).

The technique described here may be extended, and thus even more complex macromolecules may be synthesized according to predetermined patterns. The following example is given to illustrate the possible approaches. Under suitable conditions, monodispersed living polymers may be terminated by phosgene to yield a polymer possessing two equally long sequences of monomeric units linked to a carbonyl group, i.e., $CO(A\ldots\ldots A)_2$. This macromolecule may be reacted with another uniform size living polymer, either of the same type as the previous one but having a different degree of polymerization, or built up from other monomeric units. This may lead to a three-armed polymer:

$$\underbrace{(A\ldots\ldots A)_2\overset{\displaystyle \overset{OH}{|}}{C}(A\ldots\ldots A)}_{n\qquad\qquad m} \quad \text{or} \quad \underbrace{(A\ldots\ldots A)_2\overset{\displaystyle \overset{OH}{|}}{C}(B\ldots\ldots B)}_{n\qquad\qquad m}$$

Conversion of the hydroxyl group into a bromide, followed by coupling with still another living polymer, should produce such star-shaped polymers as

$$\begin{matrix} (B\ldots\ldots B) & & (A\ldots\ldots A) \\ & \diagdown\ \diagup & \\ & C & \\ & \diagup\ \diagdown & \\ (C\ldots\ldots C) & & (A\ldots\ldots A) \end{matrix}$$

Other variants of this method are easily conceived.

If the reactive groups are part of a polymeric molecule, then the coupling processes described above lead to *grafting* of living polymer on another polymer. Many examples of such reactions are reported in the recent literature; the grafting of living polystyrene on polymethyl methacrylate attracted much attention. Pioneering work in this field was done by Rempp and his colleagues (124,125). Other examples are provided by the work of Waack (126), Leavitt (127), and many others (see also 197).

Grafting of living polystyrene on methylated xylans was reported by O'Malley and Marchessault (185). The resulting materials were thoroughly characterized. Each potential grafting site reacted at least once. The expected decrease in viscosity of branched polymer, as compared with linear, was noted.

The use of monodispersed living polymers permits placing branches of identical size along the chain. Such graft polymers are referred to as *comb-shaped polymers*. Note, however, that the conventional grafting of living polymers leading to comb-shaped polymers provides a method for introducing a number of *identical* branches, but *randomly* distributed along the main chain. It is possible, at least in principle, to produce comb-shaped polymers with identical branches attached to the main chain at *regular* intervals (2). In the previous section, it was mentioned that one may link up living polymers endowed with two growing ends with bifunctional linking agents. Let us combine monodispersed polymers of this type with, e.g., phosgene. This should yield the following linear polymer:

$$\ldots\ldots |\!\!-\!\!-\!\!-\!\!|.CO.|\!\!-\!\!-\!\!-\!\!|.CO.|\!\!-\!\!-\!\!-\!\!|.CO.\ldots\ldots$$
$$\quad\quad\quad n \quad\quad\quad n \quad\quad\quad n$$

i.e., a macromolecule possessing reactive carbonyl groups placed at equal intervals along the chain. These groups may react quantitatively with a monodispersed living polymer B, and then one obtains a molecule such as

Again, it is easy to conceive a great number of variants of this method.

Alternatively, one could produce reactive spots on a main polymer chain from which branches may grow. This is a typical procedure leading to radical-type grafting. However, if the active spots lead to the formation of living polymers, one may reap all the benefits associated with their unusual properties. Three examples of such a procedure may be quoted. Leavitt (128) copolymerized styrene and *p*-bromostyrene, lithiated the bromine-substituted benzene rings, and used the resulting macromolecule as an initiator for anionic branching. Dondos and Rempp (129,170) prepared poly-1,1-diphenylpropylene, converted with sodium naphthalene the —CH(Ph$_2$) into —C(Ph)$_2$$^-$,Na$^+$, and utilized the latter macromolecule as a poly-initiator. Cuddihy et al. (130) reacted polyvinylnaphthalene with sodium and then treated the resulting polysodium compound with ethylene oxide. Again, a branch polymer formed. The latter workers also described some unusual rheological properties of their graft polymers which differ from those of the respective polyblends. This may be seen from an inspection of Figure 23.

Chemical reactions involving polymers as substrates often lead to some peculiar difficulties which arise from the degree of accessibility of the functional groups. These groups may be buried in the polymer and therefore become inaccessible to reagents. The problem becomes even more serious

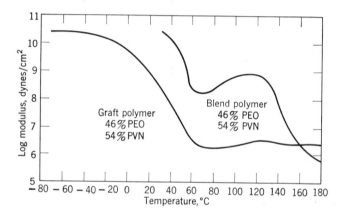

Fig. 23.   Plots of modulus vs. temperature for a blend and graft polymer of polyvinyl-naphthalene (PVN) and polyethylene oxide (PED).

when the chosen reagent is also polymeric in nature, because polymers, on the whole, are not compatible. A good illustration of such difficulties are provided by the work of Rempp's group (131). Grafting of living polystyrene on polymethyl methacrylate leads to a "saturation" phenomenon. Only a small fraction of the ester groups may be reacted even if an excess of living polystyrene is present. It has been suggested that the partially grafted polymer collapses in the reaction medium, forming an inner shell of polymethyl methacrylate and an outer shell of polystyrene. Only the ester groups which are on the "surface" of the collapsed structure are available for further grafting.

Star-shaped and comb-shaped branched polymers of narrow molecular weight distribution prepared by the living polymer technique have been of great value as model compounds. For example, much progress has been realized by means of these polymers in relating viscoelastic properties to molecular parameters. This is one of the major aims of polymer physics. Fox and his colleagues have been especially interested in the use of monodisperse branched polystyrenes for studying the relation between zero-shear viscosity and the intermolecular friction constant on one hand, and the molecular parameters such as the dimensions of the polymer coil on the other. The latter is affected by both molecular weight and branching. Some of this work was summarized recently by Fox who concluded that the zero-shear viscosity of a polymer depends directly on its average chain dimensions and that this dependence is not affected by branching if the degree of branching and molecular weight are not too high (171). Using monodisperse, star-shaped polybutadienes, Kraus and Gruver observed that the relation between viscosity and chain dimensions does depend upon the degree of branching when the molecular weight is greater than about 60,000 (172). Without these model branched polymers, these advances in our knowledge of viscoelasticity would not yet have been made and the living polymer technique represents the only practical method for the synthesis of such materials.

Little information about the physical properties of graft polymers is yet available. However, the dilute solution behavior of graft polymers indicates that interactions between the chemically different sequences are pronounced (173). These interactions occur because the chemically different chains tend to be incompatible. In this respect, graft polymers resemble block polymers. It is interesting to note that the intensity of these interactions in dilute solution and, hence, the conformations and dimensions of the polymer molecules depend upon the nature of the solvent. Selective solvation of one of the chains by solvent promotes a kind of segregation of graft polymer because of repulsions between unlike chains. For example, when unsolvated backbone is kept in solution by solvated grafts, formation of "monomolecular micelles" has been observed (173). Such structures exist when polydiphenylpropene containing polystyrene grafts is dissolved in mixtures of dioxane and cyclohexane. At high cyclohexane content, each molecule resembles a small particle of gel surrounded by solvated grafts which prevent it from precipitating. The formation of such structures is accompanied by tremendous shrinkage of the polymer molecule.

16. POLYMERS WITH FUNCTIONAL END GROUPS

Living polymers possess reactive groups on one or both of their ends. In anionic polymerization of vinyl monomers these are carbanions, which as is well known, may easily be converted into other functional groups. For example, their reaction with carbon dioxide leads to terminal carboxyl groups and ethylene oxide yields terminal hydroxyl groups. Rempp and Loucheux

(132) showed how terminal amino groups may be introduced, etc. Many more examples of such reactions which permit us to introduce still other end groups, e.g., halogens, nitriles, aldehydes, mercapto, etc., are reported by Uraneck et al. (112). Application of 4-dimethylaminobenzaldehyde yields polymers with end groups containing both hydroxyl and tertiary amines (176). These procedures are particularly valuable when α,ω-bifunctional polymers are formed because these may undergo further reversible or irreversible polymerization linking polymeric molecules into larger structures.

While straightforward in principle, the synthesis of polymer molecules with one or two reactive functional end groups per molecule is difficult in practice. It is essential that none of the living polymer terminates by isomerization, by reaction with solvent, or by reaction with impurities. Moreover, the reaction between the carbanions and the other reagent must be free of side reactions and must go to completion. This last condition is sometimes difficult to realize, especially when either ethylene oxide or carbon dioxide is the reagent (174). For example, reactions between carbanions and these reagents yield oxide and carboxyl anions, respectively. The carboxyl anions are much more prone to association than are carbanions with the result that the polymer solutions become very viscous as the substitution proceeds. In fact, gels are formed when living polymers with reactive groups at both ends are used. Because of this increase in viscosity, the mixing of ethylene oxide or carbon dioxide with living polymer must be as complete as possible before the reaction progresses too far. Otherwise, the hydroxy and carboxy contents will be lower than theoretical, unless the reaction can be allowed to take place over a long period of time. One approach to instantaneous mixing described by Short is to let the living polymer solution and the reagent flow into a T-shaped reaction tube under conditions which lead to turbulent mixing at the junction point (175). The degree of carboxylation was higher when this technique was used than when the living polymer was poured onto Dry Ice or when gaseous carbon dioxide was injected into the living polymer solution. If the reaction is slow, the product may react with the unreacted living polymer. For example, in carboxylation the reaction of $-COO^-$ with living polymer yields ketones and even t-alcohols (174).

Polymer molecules having reactive functional groups at both ends are much preferred to those which are monofunctional. The former polymers are of great interest in condensation reactions provided the functionality is exactly 2. If the functionality is even slightly less than 2, the degree to which the condensation can be carried will be drastically reduced. It was reported earlier that polymers with reactive end groups can be used to prepare block polymers by means of condensation, e.g., polystyrenedicarboxylic acid with polymeric glycols (85). Of even greater interest is the possibility of preparing low molecular weight liquid polymers with reactive end groups, which after casting can be chain extended by condensation and vulcanized to yield solid rubbers. If the functional end groups are also involved in the vulcanization, free polymer chains will be absent in the network structure and the network chains will be uniform in length. Hence, such networks will be more ordered than those which can be obtained by sulfur vulcanization. There are several references to the preparation of such materials, including those of low molecular weight polybutadienes containing terminal hydroxyl groups which have been chain extended and then crosslinked by diisocyanates (176–178). Low molecular weight polybutadiene glycols have also been extended and cross-linked by reaction with stoichiometric amounts of the oxides or sulfides of triaziridinyl, e.g., 1 mole of hexa-2-methylaziridinyl triphosphotriazine per mole of hydroxyl group (179). Chain extension occurs during the first stage of the vulcanization which is followed by cross-linking. To yield rubbers with high resilience down to low temperatures, the low molecular weight polymers of diolefins should be high in 1,4 content, i.e., free of side-chain unsaturation. Consequently, the polymerizations must be carried out in hydrocarbon solvents using dilithium compounds such as 1,2-dilithiumdiphenylethane as initiators (180,181,198).

On the whole, it is not easy to ascertain quantitatively the effect of terminal groups on the rheological and mechanical properties of polymers. Usually different samples have to be prepared in order to introduce the different investigated end groups, and even then it is not certain whether the end group, or some other change, e.g., molecular weight distribution, is responsible for the observed effect. The use of living polymers removes this difficulty. A batch of living polymers, say, living polystyrene, may be prepared and the product divided into several aliquots. Each aliquot is then reacted with another reagent, e.g., one carboxylated, the other hydroxylated, still another converted into an ordinary hydrocarbon by proton transfer, etc. We know, therefore, that the products are identical in every respect, having the same composition, molecular weight, molecular weight distribution, etc., and differ *only* in the nature of their end groups. Comparison of their properties then gives an unequivocal indication of end-group effects. Such a study was reported by Richards and Szwarc (53), who prepared α,ω-disubstituted polystyrenes and showed that the viscosity-concentration dependence of their solutions, but not the intrinsic viscosity, was affected by the end groups. With increasing concentration, an increasing number of polymer molecules becomes intermolecularly hydrogen bonded. This leads, therefore, to a higher viscosity of concentrated solutions of the bifunctional polymers than that observed for identical polymers which do not possess functional groups. The effect of terminal ion-pairs on the viscosity of polymer solution is even more dramatic, as was shown by studies of Brody et al. (133). The bifunctional polymers may undergo *intra-* and *inter*-molecular associations, and a change from one type of association to the other may profoundly affect the flow properties of the respective material. Thus, reversible, temperature-dependent associations which convert intramolecular bonds into intermolecular ones may lead to interesting viscosity index improvers.

In conclusion, it should be stressed again that the living polymer technique offers enormous control of the synthesis of polymers. It permits many modifications in the architecture of macromolecules and it yields, if properly applied, products composed of virtually identical species.

References

(1) (a) M. Szwarc, M. Levy, and R. Milkovich, *J. Am. Chem. Soc.*, **78**, 2656 (1956).
(b) M. Szwarc, *Nature*, **178**, 1168 (1956).

(2) (a) M. Szwarc, *Advan. Chem. Phys.*, **2**, 147 (1959). (b) M. Szwarc, *Makromol. Chem.*, **35**, 132 (1960). (c) M. Szwarc, *Proc. Roy. Soc. (London)*, *Ser. A*, **279**, 260 (1964).

(3) P. J. Flory, "Principles of Polymer Chemistry," Cornell University Press, Ithaca, N.Y., 1953, p. 334.

(4) P. J. Flory, *J. Am. Chem. Soc.*, **62**, 1561 (1940).

(5) H. Kämmerer and S. Ozaki, *Makromol. Chem.*, **91**, 1 (1966).

(6) R. Waack, A. Rembaum, J. D. Coombes, and M. Szwarc, *J. Am. Chem. Soc.*, **79**, 2026 (1957).

(7) H. Brody, M. Ladacki, R. Milkovich, and M. Szwarc, *J. Polymer Sci.*, **25**, 221 (1957).

(8) M. Szwarc, *Advan. Chem. Phys.*, **2**, 147 (1959).

(9) A. F. Sirianni, D. J. Worsfold, and S. Bywater, *Trans. Faraday Soc.*, **55**, 2124 (1959).

(10) F. Wenger and S. P. S. Yen, *Makromol. Chem.*, **43**, 1 (1961).

(11) F. M. Brower and H. W. McCormick, *J. Polymer Sci. A*, **1**, 1749 (1963).

(12) M. Szwarc and M. Litt, *J. Phys. Chem.*, **62**, 568 (1958).

(13) B. D. Coleman, F. Gornick, and G. Weiss, *J. Chem. Phys.*, **39**, 3233 (1963).

(14) R. V. Figini, *Makromol. Chem.*, **44–46**, 497 (1961).

(15) H. W. McCormick, *J. Polymer Sci.*, **36**, 341 (1959).

(16) F. Wenger, *J. Am. Chem. Soc.*, **82**, 4281 (1960).

(17) (a) R. V. Figini, *Z. Physik. Chem. (Frankfurt)*, **23**, 224 (1960). (b) R. V. Figini and G. V. Schulz, *ibid.*, 233. (c) R. V. Figini and G. V. Schulz, *Makromol. Chem.*, **41**, 1 (1960).

(18) M. Litt, *J. Polymer Sci.*, **58**, 429 (1962).

(19) J. M. G. Cowie, D. J. Worsfold, and S. Bywater, *Trans. Faraday Soc.*, **57**, 705 (1961).

(20) L. Gold, *J. Chem. Phys.*, **28**, 91 (1958).

(21) F. Wenger, *Makromol. Chem.*, **36**, 200 (1960).

(22) H. Dostal and H. Mark, *Z. Physik. Chem.*, **B29**, 299 (1935).

(23) T. Shimomura, K. J. Tölle, J. Smid, and M. Szwarc, *J. Am. Chem. Soc.*, **89**, 796 (1967).

(24) D. N. Bhattacharyya, C. L. Lee, J. Smid, and M. Szwarc, *Polymer*, **5**, 54 (1964); *J. Phys. Chem.*, **69**, 612 (1965).

(25) (a) H. Hostalka, R. V. Figini, and G. V. Schulz, *Makromol. Chem.*, **71**, 198 (1964). (b) H. Hostalka and G. V. Schulz, *Z. Physik. Chem. (Frankfurt)*, **45**, 286 (1965).

(26) W. K. R. Barnikol and G. V. Schulz, *Makromol. Chem.*, **68**, 211 (1963); **86**, 298 (1965).

(27) F. S. Dainton, G. C. East, G. A. Harpell, N. R. Hurworth, K. J. Ivin, R. T. LaFlair, R. H. Pallen, and K. M. Hui, *Makromol. Chem.*, **89**, 257 (1965).

(28) M. Fisher, M. van Beylen, J. Smid, and M. Szwarc, to be published.

(29) F. J. Welch, *J. Am. Chem. Soc.*, **81**, 1345 (1959).

(30) K. F. O'Driscoll and A. V. Tobolsky, *J. Polymer Sci.*, **35**, 259 (1959).

(31) D. J. Worsfold and S. Bywater, *Can. J. Chem.*, **38**, 1891 (1960).

(32) R. V. Figini, *Z. Physik. Chem. (Frankfurt)*, **38**, 341 (1963).

(33) R. V. Figini, *Makromol. Chem.*, **71**, 193 (1964).

(34) M. Szwarc and J. J. Hermans, *J. Polymer Sci. B*, **2**, 815 (1964).

(35) (a) L. H. Tung, *J. Appl. Polymer Sci.*, **10**, 375 (1966). (b) L. H. Tung, J. C. Moore, and G. W. Knight, *ibid.*, 1261. (c) L. H. Tung, *ibid.*, 1271.

(36) H. W. McCormick and R. E. Friedrich, U.S. Pat. 3,069,405 (1962).

(37) H. W. McCormick, F. M. Brower, and L. Kin, *J. Polymer Sci.*, **39**, 87 (1959).

(38) P. J. Flory, *J. Am. Chem. Soc.*, **67**, 2048 (1945).

(39) E. H. Merz, L. E. Nielsen, and R. Buchdahl, *Ind. Eng. Chem.*, **43**, 1396 (1951).

(40) (a) R. S. Spencer and R. E. Dillon, *J. Colloid Sci.*, **4**, 241 (1949). (b) R. S. Spencer, *J. Polymer Sci.*, **5**, 591 (1950).

(41) (a) G. Löhr and G. V. Schulz, *Makromol. Chem.*, **77**, 240 (1964). (b) R. V. Figini, G. Löhr and G. V. Schulz, *Polymer Letters*, **3**, 985 (1965). (c) R. V. Figini, H. Hostalka, K. Hurm, G. Löhr, and G. V. Schulz, *Z. Physik. Chem. (Frankfurt)*, **45**, 269 (1965).

(42) R. V. Figini, paper presented at the IUPAC Meeting, Prague, Sept. 1965.

(43) R. Waack, Ph.D. Thesis, Syracuse University, Syracuse, N.Y., 1959.

(44) G. Gee, W. C. E. Higginson, and G. T. Merrall, *J. Chem. Soc. (London)*, **1959**, 1345.

(45) S. N. Khanna, M. Levy, and M. Szwarc, *Trans. Faraday Soc.*, **58**, 747 (1962).

(46) A. Eisenberg and D. A. McQuarrie, *J. Polymer Sci. A-1*, **4**, 737 (1966).

(47) G. Holden and R. Milkovich, U.S. Pat. 3,231,635 (1966).

(48) W. B. Brown and M. Szwarc, *Trans. Faraday Soc.*, **54**, 416 (1958).

(49) A. Miyake and W. H. Stockmayer, *Makromol. Chem.*, **88**, 90 (1965).

(50) J. J. Hermans, *J. Polymer Sci. C*, **4**, 345 (1966).

(51) G. Smets and R. Hart, *Fortschr. Hochpolymer.-Forsch.*, **2**, 173 (1960). See also A. E. Woodward and G. Smets, *J. Polymer Sci.*, **17**, 51 (1955), and G. Smets, Bunsen-Discussionstagung, Ludwigshafen (1965).

(52) R. K. Graham, D. L. Dunkelberger, and W. E. Goode, *J. Am. Chem. Soc.*, **82**, 400 (1960).

(53) D. H. Richards and M. Szwarc, *Trans. Faraday Soc.*, **55**, 1644 (1959).

(54) S. Schlick and M. Levy, *J. Phys. Chem.*, **64**, 883 (1960).

(55) A. A. Korotkov, L. A. Shivaev, L. M. Pyrkov, V. G. Aldoshin, and S. Ya. Frenkel, *Vysokomolekul. Soedin.*, **1**, 157 (1960).

(56) M. Baer, *J. Polymer Sci.*, *A*, **2**, 417 (1964).

(57) S. Bywater, *Fortschr. Hochpolymer.-Forsch.*, **4**, 66 (1965).

(58) (a) G. V. Rakova and A. A. Korotkov, *Dokl. Akad. Nauk SSSR*, **119**, 982 (1958). (b) A. A. Korotkov and N. N. Chesnokova, *Vysokomolekul. Soedin.*, **2**, 365 (1960).

(59) I. Kuntz, *J. Polymer Sci.*, **54**, 569 (1961).

(60) R. P. Zelinski, U.S. Pat. 2,975,160 (1961). See also R. N. Cooper, U.S. Pat. 3,030,346 (1962).

(61) A. F. Johnson and D. J. Worsfold, *Makromol. Chem.*, **85**, 273 (1965).

(62) D. H. Richards, D. A. Salter, and R. L. Williams, *Chem. Commun.*, **1966**, No. 2, 38.

(63) (a) T. G Fox, B. S. Garrett, W. E. Goode, S. Gratch, J. F. Kincaid, A. Spell, and J. D. Stroupe, *J. Am. Chem. Soc.*, **80**, 1768 (1958). (b) T. G Fox, W. E. Goode, S. Gratch, C. M. Huggett, J. F. Kincaid, A. Spell, and J. D. Stroupe, *J. Polymer Sci.*, **31**, 173 (1958).

(64) (a) B. D. Coleman and T. G Fox, *J. Chem. Phys.*, **38**, 1065 (1963). (b) B. D. Coleman, T. G Fox, and M. Reinmöller, *J. Polymer Sci. B*, **4**, 1029 (1966); see also (c) B. D. Coleman and T. G Fox, *J. Am. Chem. Soc.*, **85**, 1241 (1963); *J. Polymer Sci., C*, **4**, 345 (1963).

(65) R. V. Figini, *Makromol. Chem.*, **88**, 272 (1965).

(66) W. Bushuk and H. Benoit, *Can. J. Chem.*, **36**, 1616 (1958).

(67) G. V. Schulz, *Z. Physik. Chem.*, **B47**, 155 (1940).

(68) B. H. Zimm, *J. Chem. Phys.*, **16**, 1099 (1948).

(69) C. H. Bamford and H. Tompa, *J. Polymer Sci.*, **10**, 345 (1953); *Trans. Faraday Soc.*, **50**, 1097 (1954).

(70) F. W. Billmeyer and W. H. Stockmayer, *J. Polymer Sci.*, **5**, 121 (1950).

(71) A. Skoulios, G. Finaz, and J. Parrod, *Compt. Rend.*, **251**, 739 (1960).

(72) A. Skoulios and G. Finaz, *ibid.*, **252**, 3467 (1961).

(73) C. Sadron, *Angew. Chem.*, **75**, 472 (1963).

(74) G. Finaz, P. Rempp, and J. Parrod, *Bull. Soc. Chim. France*, p. 262 (1962).

(75) M. Szwarc, Brit. Pat. 852,823 (1960).

(76) E. Vanzo, *J. Polymer Sci. A-1*, **4**, 1727 (1966).

(77) (a) R. Milkovich (Shell); S. African Pat. Application 642,271 (1964); Brit. Pat. 1,000,090 (1965). (b) R. Milkovich, G. Holden, E. T. Bishop, and W. R. Hendricks Brit. Pat. 1,035,873 (1966).

(78) P. Claes and G. Smets, *Makromol. Chem.*, **44–46**, 212 (1961).

(79) J. Furukawa, T. Saegusa, and N. Mise, *ibid.*, **38**, 244 (1960).

(80) M. A. Novitskaya and A. A. Konkin, *Vysokomolekul. Soedin.*, **7**, 1719 (1965).

(81) J. C. Galin, J. Herz, P. Rempp, and J. Parrod, *Bull. Soc. Chim. France*, **1966**, 1120.

(82) (a) A. J. Kovacs and B. Lotz, *Kolloid-Z. Z. Polymere*, **209**, 97 (1966). (b) B. Lotz, A. J. Kovacs, G. A. Bassett, and A. Keller, *ibid.*, 115 (1966).

(83) W. E. Smith, F. R. Galiano, D. Rankin, and G. J. Mantell, *J. Appl. Polymer Sci.*, **10**, 1659 (1966).

(84) (a) M. Morton, A. Rembaum, and E. E. Bostick, *J. Polymer Sci.*, **32**, 530 (1958). (b) M. Morton and A. Rembaum, U.S. Pat. 3,051,684 (1962).

(85) K. Hayashi and C. S. Marvel, *J. Polymer Sci., A*, **2**, 2571 (1964).

(86) G. Meshitsuka, M. Kamachi, and K. Hirota, *Kobunshi Kagaku*, **17**, 641 (1960).

(87) E. Franta and P. Rempp, *Compt. Rend.*, **254**, 674 (1962).

(88) G. Champetier, M. Fontanille, A. C. Korn, and P. Sigwalt, *J. Polymer Sci.*, **58**, 911 (1962).

(89) E. Franta and P. Rempp, *Compt. Rend.*, **254**, 674 (1962).

(90) G. Champetier, M. Fontanille, and P. Sigwalt, *ibid.*, **250**, 3653 (1960).

(91) R. K. Graham, J. R. Panchak, and M. J. Kampf, *J. Polymer Sci.*, **44**, 411 (1960).

(92) M. Szwarc and A. Rembaum, *J. Polymer Sci.*, **22**, 189 (1956).

(93) G. Natta, *ibid.*, **34**, 531 (1959).

(94) G. Natta, E. Giachetti, and I. Pasquon, Fr. Pat. 1,220,573 (1959).
(95) (a) G. Bier, G. Lehmann, and H. J. Leugering, *Makromol. Chem.*, **44–46**, 347 (1961). (b) G. Bier, *Angew. Chem.*, **73**, 186 (1961).
(96) E. G. Kontos, E. K. Easterbrook, and R. D. Gilbert, *J. Polymer Sci.*, **61**, 69 (1962).
(97) (a) J. R. Urwin and J. M. Stearne, *Makromol. Chem.*, **78**, 194 (1964). (b) *Ibid.*, **78**, 204 (1964). (c) *European Polymer J.*, **1**, 227 (1965).
(98) D. Freyss, P. Rempp, and H. Benoit, *J. Polymer Sci. B*, **2**, 217 (1964).
(99) E. Franta, A. Skoulios, P. Rempp, and H. Benoit, *Makromol. Chem.*, **87**, 271 (1965).
(100) F. Wenger and S. P. S. Yen, 141st ACS Meeting, Washington, D.C., *Abstracts*, **3**, 163 (1962); T. A. Orofino and F. Wenger, *J. Chem. Phys.*, **67**, 566 (1963).
(101) M. Morton, T. E. Halminiak, S. D. Gadkery, and F. Bueche, *J. Polymer Sci.*, **57**, 471 (1962).
(102) S. P. S. Yen, *Makromol. Chem.*, **81**, 152 (1965).
(103) D. Decker and P. Rempp, *Compt. Rend.*, **261**, 1977 (1965); **262**, 726 (1966).
(104) S. Boileau and P. Sigwalt, *ibid.*, **252**, 882 (1961).
(105) M. Ohta, A. Kaudo, and R. Ohi, *Nippon Kagaku Zasshi*, **75**, 985 (1954).
(106) (a) S. Boileau, J. Coste, J. M. Raynal, and P. Sigwalt, *Compt. Rend.*, **254**, 2774 (1962). (b) S. Boileau, G. Champetier, and P. Sigwalt, *Makromol. Chem.*, **69**, 180 (1963).
(107) S. Boileau and P. Sigwalt, *Compt. Rend.*, **261**, 132 (1965).
(108) W. P. Baker, U.S. Pat. 3,225,120 (1965).
(109) R. Rempp, private communication.
(110) G. Finaz, Y. Gallot, J. Parrod, and P. Rempp, *J. Polymer Sci.*, **58**, 1363 (1962).
(111) G. Berger, M. Levy, and D. Vofsi, *J. Polymer Sci. B*, **4**, 183 (1966).
(112) C. A. Uraneck, J. N. Short, and R. P. Zelinski, U.S. Pat. 3,135,716 (1964).
(113) F. Wenger, *Chem. Ind.*, **1959**, 1094.
(114) M. Baer, *J. Polymer Sci. A*, **1**, 2171 (1963).
(115) Du Pont, Brit. Pat. 981,361 (1965).
(116) M. Fontanille and P. Sigwalt, *Compt. Rend.*, **251**, 2947 (1960).
(117) C. L. Lee, J. Smid, and M. Szwarc, *Trans. Faraday Soc.*, **59**, 1192 (1963).
(118) G. E. Molau, *J. Polymer Sci. A*, **3**, 4131 (1965).
(119) J. J. Aklonis and A. V. Tobolsky, *J. Appl. Phys.*, **36**, 3483 (1965).
(120) K. Noro, H. Kawazura, T. Moriyama, and S. Yoshioka, *Makromol. Chem.*, **83**, 35 (1965).
(121) J. H. Carter and F. W. Michelotti, ACS Sept. Meeting, Chicago, *Preprints Polymer Div.*, 614 (1964).
(122) A. Rembaum, F. R. Ells, R. C. Morrow, and A. V. Tobolsky, *J. Polymer Sci.*, **61**, 155 (1962).
(123) (a) J. A. Gervasi and A. B. Grosnell, Camille Dreyfus Laboratory, Preprint No. 88 (1965). (b) A. B. Grosnell, J. A. Gervasi, and A. Schindler, *ibid.*, No. 102 (1965). (c) J. A. Gervasi and A. B. Grosnell, *J. Polymer Sci. A-1*, **4**, 1391 (1966).
(124) P. Rempp, V. I. Volkov, J. Parrod, and Ch. Sadron, *Bull. Soc. Chim. France*, **1960**, 919.
(125) Y. Gallot, P. Rempp, and H. Benoit, *Compt. Rend.*, **253**, 989 (1961).
(126) R. Waack, U.S. Pat. 3,235,626 (1966).
(127) F. C. Leavitt, U.S. Pat. 3,234,196 (1966).
(128) F. C. Leavitt, U.S. Pat. 3,234,193 (1966).
(129) A. Dondos and P. Rempp, *Compt. Rend.*, **254**, 1426 (1962).
(130) E. Cuddihy, J. Moacanin, and A. Rembaum, *J. Appl. Polymer Sci.*, **9**, 1385 (1965).

(131) Y. Gallot, P. Rempp, and J. Parrod, *J. Polymer Sci. B.*, **1**, 329 (1963).
(132) P. Rempp and M. H. Loucheux, *Bull. Soc. Chim. France*, (**1958**), 1497.
(133) H. Brody, D. H. Richards, and M. Szwarc, *Chem. Ind.* (*London*), **45**, 1473 1958.
(134) S. Bywater, A. F. Johnson, and D. J. Worsfold, *Can. J. Chem.*, **42**, 1255 (1964).
(135) T. A. Orofino and F. Wenger, *J. Chem. Phys.*, **35**, 532 (1961).
(136) G. G. Lowry, *J. Polymer Sci. B*, **1**, 489 (1963).
(137) (a) P. Debye, H. Coll, and D. Woermann, *J. Chem. Phys.*, **32**, 939; *ibid.*, **33**, 1746 (1960). (b) P. Debye, D. Woermann, and B. Chu, *ibid.*, **36**, 851 (1962).
(138) (a) D. McIntyre, A. Wims, and M. S. Green, *J. Chem. Phys.*, **37**, 3019 (1962). (b) D. McIntyre, A. Wims, and J. K. O'Mara, *Polymer Preprints*, **6**, No. 2, 1037 (1965).
(139) G. V. Schulz, *Pure Appl. Chem.*, **12**, 85 (1966).
(140) V. S. Nanda and R. K. Jain, *J. Polymer Sci. A*, **2**, 4583 (1964).
(141) H. L. Hsieh, *Rubber Plastic Age*, **46**, 394 (1965).
(142) M. Szwarc, *Polymer Preprints*, **8**, No. 1, 22 (1967).
(143) M. Litt and H. D. Richards, *J. Polymer Sci. B*, **5**, 867 (1967).
(144) T. Altares, D. P. Wyman, and V. R. Allen, *J. Polymer Sci. A*, **2**, 4533 (1964).
(145) J. T. Gruver and G. Kraus, *J. Polymer Sci. A*, **2**, 797 (1964).
(146) A. A. Korotkov, N. N. Chesnokova, and L. B. Trukhamnova, *Vysokomolekul. Soedin*, **1**, 46 (1959).
(147) M. P. Dreyfuss and P. Dreyfuss, *Polymer*, **6**, 93 (1965).
(148) E. V. Zabolotskaya, V. A. Khodzhemirov, A. R. Gentmakher, and S. S. Medvedev, *Vysokomolekul. Soedin.*, **6**, 76 (1964).
(149) D. E. Burge and D. B. Bruss, *J. Polymer Sci. A*, **1**, 1927 (1963).
(150) A. Roig, J. E. Figueruelo and E. Llano, *J. Polymer Sci. B*, **3**, 171 (1965).
(151) A. Roig, J. E. Figueruelo and E. Llano; Preprint IUPAC Symposium on Macromolecular Chemistry, Prague, 1965, p. 548.
(152) (a) E. Husemann, B. Fritz, R. Lippert, and B. Pfannemüller, *Makromol. Chem.*, **26**, 199 (1958). (b) E. Husemann, *Makromol. Chem.*, **35**, 239 (1959).
(153) G. V. Schulz, *Pure Appl. Chem.*, **12**, 85 (1966).
(154) F. R. Anderson, *J. Polymer Sci. C*, **8**, 275 (1965).
(155) R. Koningsveld and A. J. Pennings, *Rec. Trav. Chim.*, **83**, 552 (1964).
(156) A. J. Pennings and A. M. Kiel, *Kolloid Z. Z. Polymere*, **205**, 160 (1965).
(157) M. J. Richardson, *Proc. Roy. Soc.* (*London*), *A*, **279**, 50 (1961).
(158) R. A. Stratton, *J. Colloid Interface Sci.*, in press.
(159) R. S. Porter, M. J. R. Cantow, and J. F. Johnson, *Proc. Intl. Congr. Rheol. 4th Providence*, **2**, 479 (1963).
(160) J. F. Rudd, *J. Polymer Sci.* **60**, 8 (1962).
(161) A. V. Tobolsky, J. J. Aklonis, and G. Akovali, *J. Chem. Phys.*, **42**, 723 (1965).
(162) A. V. Tobolsky, R. Schaffhauser and R. Böhme, *J. Polymer Sci. B*, **2**, 103 (1964).
(163) J. F. Rudd and E. F. Gurnee, *J. Polymer Sci. A*, **1**, 2857 (1963).
(164) V. R. Allen and T. G Fox, *J. Chem. Phys.*, **41**, 337 (1964).
(165) P. Rempp, *Polymer Preprints*, **7**, 141 (1966).
(166) I. Kuntz, U.S. Pat. 3,140,278 (1964).
(167) H. Benoit, Robert A. Welch Symposium on Polymers, 1966.
(168) E. Fischer, K. H. Grundy and J. F. Henderson, *J. Polymer Sci.*, in press.
(169) J. A. Gervasi and A. B. Grosnell, Camille Dreyfuss Laboratory, Preprints No. 88 and 102 (1965).
(170) A. Dondos, *Bull. Soc. Chim. France*, **1963**, 2762.
(171) T. G Fox in *Structure and Properties of Polymers*, A. V. Tobolsky, Ed. (*J. Polymer Sci. C*, **9**), Interscience, New York, 1966, p. 35.

(172) G. Kraus and J. T. Gruver, *J. Polymer Sci. A*, **3**, 105 (1965).
(173) A. Dondos, P. Rempp, and H. Benoit, *J. Polymer Sci. B*, **4**, 293 (1966).
(174) D. P. Wyman, V. R. Allen, and T. Altares, *J. Polymer Sci. A*, **2**, 4545 (1964).
(175) J. N. Short, U.S. Pat. 3,225,089 (1965).
(176) R. P. Zelinski, H. L. Hsieh, and C. W. Strobel, U.S. Pat. 3,109,871 (1963).
(177) Polymer Corporation Ltd., Brit. Pat. 946,300 (1964).
(178) E. J. Goldberg, U.S. Pat. 3,055,952 (1962).
(179) C. A. Uraneck, H. L. Hsieh, and O. G. Buck, *J. Polymer Sci.*, **46**, 535 (1960).
(180) Phillips Petroleum, Brit. Pat. 895,980 (1962).
(181) Phillips Petroleum, Brit. Pat. 964,478 (1964).
(182) L. J. Fetters, *J. Res. Natl. Bur. Std.*, **70A**, 421 (1966).
(183) B. N. Bastien, Canadian Pat. 746,555 (1966).
(184) S. Bywater and D. J. Worsfold, *Can. J. Chem.*, **40**, 1564 (1962).
(185) J. J. O'Malley and R. H. Marchessault, *J. Phys. Chem.*, **70**, 3235 (1966).
(186) Th. G. Scholte, *J. Polymer Sci. A*, **7**, 461 (1967).
(187) G. V. Schulz, K. C. Berger, and A. G. R. Scholz, *Ber. Bunsengesel. Phys. Chem.*, **69**, 856 (1965).
(188) D. P. Wyman, Technical Report ASD-TDR-62110 from Melon Institute (1963).
(189) D. J. Plazek, *J. Phys. Chem.*, **69**, 3480 (1965).
(190) G. C. Berry, *J. Chem. Phys.*, **44**, 4550 (1966).
(191) G. C. Berry, *J. Chem. Phys.*, **46**, 1338 (1967).
(192) N. Calderon and K. W. Scott, *J. Polymer Sci. A*, **3**, 551 (1965).
(193) D. L. Glusker, I. Lysoff, and E. Stiller, *J. Polymer Sci.*, **49**, 315 (1961).
(194) M. Morton and L. Fetters, in *Macromolecular Reviews*, vol. 2, A. Peterlin, M. Goodman, S. Okamura, B. H. Zimm, and H. F. Mark, Eds., Interscience, New York, p. 71.
(195) T. Altares, D. P. Wyman, V. R. Allen, and K. Meyersen, *J. Polymer Sci. A*, **3**, 4131 (1965).
(196) D. Decker-Freyss and P. Rempp, *Compt. Rend.*, **261**, 1977 (1965).
(197) P. Rempp, *Polymer Preprints*, **7**, 141 (1966).
(198) W. B. Reynolds, U.S. Pat. 3,074,917 (1963).
(199) A. Dondos, D. Froelich, P. Rempp, and H. Benoit, *J. Chim. Phys.*, in press.
(200) H. Benoit and D. Froelich, *Makromol. Chem.*, **92**, 224 (1966).
(201) G. Riess, J. Kohler, C. Tournut, and A. Banderet, *Rev. Gén. du Caoutchouc*, **3**, 361 (1966).
(202) T. G Fox and G. C. Berry, *Advan. Polymer Sci.*, **5**, 261 (1967).

Chapter III

Thermodynamics of Propagation

1. GENERAL PRINCIPLES

The first basic approach to the thermodynamics of addition polymerization was presented in 1948 by Dainton and Ivin (1) and developed in their review paper (2) published ten years later. In their exposition, they stressed the significance of the propagation step in polymerization, emphasizing its critical role in the whole process. This is the step whereby the macromolecule is gradually formed by the sequence of reactions

$$\text{\small$\sim\sim$}M_n\text{---}X + M \rightleftarrows \text{\small$\sim\sim$}M_{n+1}\text{---}X$$

converting the monomer M into a polymer molecule $\sim\sim M_n \cdot X$ possessing the active end group X. For large values of n, i.e., for a high molecular weight polymer, the reactivity of X is independent of n and then the above equation may be symbolically represented by

$$M_f \rightleftarrows M_s,$$

where M_f and M_s denote, respectively, a free monomer molecule and a monomer segment located somewhere in the midst of a long chain of a polymer molecule $\sim\sim M_n \cdot X$. The free energy change, ΔF_p, accompanying the addition of one mole of monomer to a high molecular weight polymer is given, therefore, by the difference $\Delta F_p = \Delta F(M_s) - \Delta F(M_f)$ where $\Delta F(M_f)$ and $\Delta F(M_s)$ are, respectively, the molar free energies of the monomer and of a monomer segment of a high molecular weight polymer (both referring to the conditions prevailing in the system). ΔF_p is thus independent of both the nature of X and the magnitude of n, although its value depends on the nature of the solvent and, to some extent, on the concentrations of the monomer and polymer. Their presence affects $\Delta F(M_f)$ and $\Delta F(M_s)$ by modifying the environment in which the process takes place. Finally, it should be noted that the free energy of propagation for any monomer-polymer system is constant if the thermodynamic conditions are defined, e.g., pure liquid monomer $\rightarrow$ amorphous solid polymer. Its value is unaffected by the mechanism of polymerization, provided that the state and conformation of the polymer do not change.

The last corollary deserves some comment. The nature of the growing end and the mechanism of polymerization affect the rates of propagation and depropagation. However, thermodynamic functions are always independent of the reaction path and are determined uniquely by the initial and final states of the system. Hence, any conceivable process, even if impractical, may serve to calculate ΔF_p. Let us consider, e.g., a high molecular weight polymer with the terminal X and Y groups linked by a long chain of n monomer segments

$$X \cdot M \cdot M \underbrace{\cdots \cdots \cdots \cdots \cdots}_{\text{long chain}} M \cdot M \underbrace{\cdots \cdots \cdots \cdots \cdots}_{\text{long chain}} M \cdot M \cdot Y \cdot$$

The conversion of an n-mer into an $(n + 1)$-mer may be imagined to occur in two steps: (1) fission of an M—M bond located in the middle of the chain; (2) insertion of a monomer molecule into the broken linkage coupled with the formation of two new M—M bonds. Because the end groups are far away from the reaction center, their nature cannot influence the free energy of the process which thus is independent of the magnitude of n and of the nature of X and Y. This is true even if X and Y are some unreactive end groups.

The conventional thermodynamic equation

$$\Delta F_p = \Delta H_p - T \Delta S_p$$

relates the free energy of propagation to the heat and entropy change of this process. Depending on the signs of ΔH_p and ΔS_p, four distinct situations may occur:

(1) For negative ΔH_p and ΔS_p (most commonly the case encountered in addition polymerization), ΔF_p becomes positive above a certain critical temperature, $T_c = \Delta H_p / \Delta S_p$, known as the "ceiling temperature" of the system. The value of T_c depends on the concentrations of the monomer and of the polymer as well as on the nature of the solvent, if the latter is present in the system. Of course, the high molecular weight polymer cannot be formed above the ceiling temperature. For any monomer–polymer system of this class, the maximum ceiling temperature corresponds to the process which converts pure liquid monomer into a crystalline polymer.

(2) For positive ΔH_p and ΔS_p, the system can polymerize only above the critical floor temperature, $T_f = \Delta H_p / \Delta S_p$. The phenomenon of floor temperature is exhibited by some cyclic monomers. On polymerization these lose part of the binding energy of the ring associated with the closeness of some atoms or groups which become separated upon forming the linear structure. Polymerization of such monomers may increase the entropy of the system, if the loss of monomer translational and rotational entropy is more than compensated by the entropy of internal rotation around those bonds which were rigid in the original cyclic molecule but allow free rotation in the newly

formed chain. The polymerization of octameric sulfur molecules, S_8, into polymer chains of plastic sulfur exemplifies such a process.

It should be remarked that dilution of the monomer decreases the entropy of polymerization and, hence, even if its value were positive for pure monomers it eventually becomes negative at a sufficiently high dilution. Therefore, for any system showing the phenomenon of floor temperature, polymerization becomes impossible at any temperature below a certain critical monomer concentration.

(3) When ΔH_p is positive and ΔS_p negative, no polymerization is possible under any conditions.

(4) When ΔH_p is negative and ΔS_p positive, polymerization may occur at all temperatures.

In some systems, the same polymer may be produced from different monomers, e.g., plastic sulfur can be formed by polymerizing S_6 as well as S_8 molecules. In such systems, a polymer may be stable with respect to one monomer but labile with respect to another one. For example, linear siloxane $(—Me_2SiO—)_n$, which could be formed from the cyclic trimer $(—Me_2SiO—)_3$, decomposes into the cyclic tetramer $(—Me_2SiO—)_4$ and the reaction $4(—Me_2SiO—)_3 \rightarrow 3(—Me_2SiO—)_4$ proceeds, therefore, through a polymer intermediate. Oligomers having larger rings are also formed (63).

The equilibria between linear polymers and a variety of cyclic oligomers are encountered in the polymerization of formaldehyde. A recent comprehensive review of this subject was published by Kern et al. (59). Formation of cyclic oligomers was discussed by Jaacks (60), and the miscibility of the resulting solutions by Iwabuchi, Kern, and Jaacks (61). The pertinent thermodynamic data were published by Bevington (62).

Similar phenomena may be observed in some copolymerizations. For example, tetrafluoroethylene, C_2F_4, and trifluoronitrosomethane, CF_3NO, copolymerize spontaneously into a $1:1$ copolymer,

$$\sim[—CF_2 \cdot CF_2 \cdot O \cdot N(CF_3) \cdot —]_n \sim.$$

Its degradation produces, however, a pair of inert monomers, i.e.

$$\sim[—O \cdot CF_2 \cdot CF_2 \cdot N(CF_3)—]_n \sim \rightarrow n(CF_2O) + n(CF_2:NCF_3),$$

which are thermodynamically more stable than the copolymer.

2. EXPERIMENTAL DETERMINATION OF CEILING TEMPERATURE

The ceiling temperature of a polymerizable system may be obtained from studies of its polymerization kinetics at a series of rising temperatures. For many polymerizations, the sum of the activation energies for the initiation and propagation greatly exceeds the activation energy for the termination step and, therefore, the initial rate of the overall reaction increases with temperature. However, as the ceiling temperature is approached, the depropagation step, which is too slow at the lower temperatures to contribute significantly to the rate, becomes important and the rate then decreases steeply as shown in Figure 1. Extrapolation to zero rate gives, therefore, the required

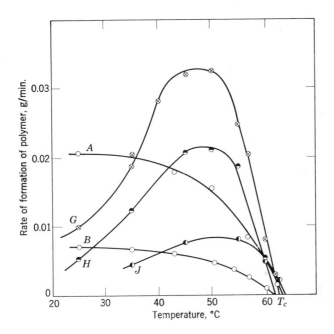

Fig. 1. Rate of formation of polybutene sulfone from monomer mixtures containing 9.1 mole-% 1-butene. (*A*) and (*B*) photochemical initiation at two different intensities; (*G*) and (*H*) initiation by silver nitrate at two different concentrations; (*J*) initiation by benzoyl peroxide.

ceiling temperature. The accuracy of the method depends on the steepness of the descending line and usually it is about 2–3°C.

The kinetic technique was extensively used by Dainton and his associates (3,4) particularly in their studies of the copolymerization of olefins with sulfur dioxide to 1:1 polysulfones. To check their results, they compared the heats of polymerization calculated from the ceiling temperatures determined at various partial pressures of olefin and sulfur dioxide with those directly obtained by calorimetry. The agreement between these two sets of data is shown in Table I. Further verification (5) was provided by determining

TABLE I
Heats of 1:1 Copolymerization of Olefins and SO_2 to Polysulfones

Olefin	$-\Delta H$ (from T_c), kcal/mole	$-\Delta H$ (from calorimetry), kcal/mole
1-Butene	20.7 ± 1.4	21.2 ± 0.1
cis-2-Butene	20.8 ± 0.7	20.1 ± 0.1
trans-2-Butene	19.3 ± 0.5	18.7 ± 0.1
1-Hexadecene	19.2 ± 1.2	19.9 ± 0.1

calorimetrically the respective ΔS values and comparing them with $\Delta S = \Delta H/T_c$.

Alternatively, one might investigate the reaction between polymer free radicals and monomer vapor. Chains of polymer molecules, kept in a monomer atmosphere, are ruptured by a suitable photochemical technique, thus producing the respective polymer radicals. The polymeric radicals grow and consume monomer, if its vapor pressure exceeds the respective equilibrium value, p_c, whereas in the reverse case depolymerization takes place and free monomer is liberated. Consequently, the pressure in the system

TABLE II
Heats of Polymerization Determined from Ceiling Temperatures and from Calorimetry

Monomer	$-\Delta H$ (from T_c), kcal/mole		$-\Delta H$ (from calorimetry), kcal/mole	
Methyl methacrylate	13.4 ± 0.5	(a, b)	13.9 ± 0.3	(c)
Ethyl methacrylate	14.4 ± 0.6	(d)	14.1	(e)
Methacrylonitrile	15.3 ± 1.0	(f)	—	
α-Methylstyrene	8.15	(g)	8.42	(h)
ϵ-Caprolactam	3.6	(i)	3.25	(j)

(a) S. Bywater, Trans. Faraday Soc., **51**, 1267 (1955).

(b) K. J. Ivin, ibid., p. 1273.

(c) S. Ekegren, O. Ohrn, K. Granath, and P. O. Kinell, Acta Chem. Scand., **4**, 126 (1950).

(d) R. E. Cook and K. J. Ivin, Trans. Faraday Soc., **53**, 1132 (1957).

(e) K. Iwai, J. Chem. Soc. Japan, Ind. Chem. Sect., **49**, 185 (1946).

(f) S. Bywater, Can. J. Chem., **35**, 552 (1957).

(g) S. Bywater and D. J. Worsfold, J. Polymer Sci., **26**, 299 (1957).

(h) D. E. Roberts and R. S. Jessup, J. Res. Natl. Bur. Std., **46**, 11 (1951).

(i) A. B. Meggy, J. Chem. Soc., **1953**, 796.

(j) S. M. Skuratov, A. A. Strepikheev, and E. N. Kanarskaya, Kolloidn. Zh., **14**, 185 (1952).

decreases when $p > p_e$ but increases when $p < p_e$ and, thus, at each temperature the respective equilibrium pressure may be found by interpolation. This method was used by Ivin (6) in his studies of the polymethyl methacrylate–gaseous methyl methacrylate system and it was subsequently extended to other systems listed in Table II. A similar approach, adapted to equilibria established in solutions, was developed by Bywater (7).

3. THERMODYNAMIC STUDIES OF POLYMERIZATIONS INVOLVING LIVING POLYMERS

The equilibrium techniques of Ivin and of Bywater, discussed above, can be applied elegantly to systems involving living polymers. The ability of these species to grow implies, in accordance with the principle of microscopic reversibility, their ability to degrade. Hence, in a living polymer–monomer system, the following equilibria must eventually be established:

$$P_{n_0}^* + M \rightleftarrows P_{n_0+1}^*, \qquad \dots K_{n_0}$$

$$P_{n_0+1}^* + M \rightleftarrows P_{n_0+2}^*, \qquad \dots K_{n_0+1}$$

$$\dots$$

$$P_{n_0+j-1}^* + M \rightleftarrows P_{n_0+j}^*, \qquad \dots K_{n_0+j-1}$$

$$\dots$$

In these equations, $P_{n_0+j}^*$ denotes a living $n_0 + j$-mer, M the monomer, and $P_{n_0}^*$ the lowest living polymer which may grow but not degrade. For example, in the cumylpotassium–α-methylstyrene system, $n_0 = 1$ and $P_{n_0}^*$ is cumylpotassium. It is possible to convert α-methylstyrene into a tail-to-tail dimer

$$K^+, {}^-C(Me)(Ph) \cdot CH_2 \cdot CH_2 \cdot C^-(Me)(Ph), K^+ = ({}^-\alpha\alpha^-)$$

and therefore in the system ${}^-\alpha\alpha^- + \alpha$ (α denoting α-methylstyrene) the dimer acts as $P_{n_0}^*$.

Whenever $K_{n_0} = K_{n_0+1} = \cdots = K_{n_0+j} = \cdots = K_\infty$, the equilibrium concentration of the living $n_0 + j$-mer, $P_{n_0+j}^*$, may be expressed in terms of equilibrium concentrations of the monomer, M_e, and of the living n_0-mer, $P_{n_0}^*$, namely,

$$P_{n_0+j}^* = P_{n_0}^*(K_\infty M_e)^j.$$

Tobolsky (8) has shown that in such a case M_0, P_{total}^*, and M_e are related by the following equation:

$$\{M_0 - M_e\}/P_{\text{total}}^* = K_\infty M_e/\{1 - K_\infty M_e\}$$

where M_0 is the initial concentration of the monomer which was added to a solution of the living n_0-mers present initially at concentration P_{total}^*.

Evidently, $P_{total}^* = \sum_0^\infty P_{n_0+j}$ and $(M_0 - M_e)$ is the amount of monomer polymerized per unit volume of reacting solution. The total volume of the solution was assumed implicitly to be constant and not affected by the polymerization—a reasonable assumption for dilute solutions. The problem of volume contraction may, however, be circumvented (9) by expressing all the concentrations in moles per unit mass of solution.

Tobolsky's approach requires some modification if K_j varies with j. In a more general treatment, described recently by Szwarc (10), the equalities $K_j = K_\infty$ are assumed to be valid only for j's exceeding some value $s + 1(s > 0)$, whereas for $j \leqslant s$ the respective K_j's may differ from K_∞ and from each other. In such a system, K_∞ may be found from the equation

$$(M_0 - R_s - M_e)/(P_{total}^* - Q_s^*) = K_\infty M_e/(1 - K_\infty M_e)$$

where the symbols R_s and Q_s^* are defined by the ratios Q_s^*/P_{total}^* and R_s/M_0. The former represents the mole fraction of all the living polymers having degrees of polymerization less than $n_0 + s$, whereas the latter denotes the fraction of the initially introduced monomer which becomes incorporated into the above polymers of $DP \geqslant n_0 + s$ (DP = degree of polymerization). All the remaining symbols retain their previous meaning.

For a high average degree of polymerization, the ratios Q_s^*/P_{total}^* and $(P_{total}^* - Q_s^*)/(M_0 - R_s - M_e)$ tend towards zero and, thus, $K_\infty = \lim (M_e^{-1})$. The determination of K_∞ is reduced, therefore, to an analytical problem of finding the ultimate concentration of the monomer which coexists in equilibrium with its high molecular weight living polymer. In some systems, notably α-methylstyrene, the analysis presents no difficulties because the equilibrium concentration of monomer is relatively high, e.g., in a tetrahydrofuran solution of living poly-α-methylstyrene, $M_e = 0.8$ mole/liter at 0°C. This system was thoroughly investigated by McCormick (11), by Worsfold and Bywater (12), and by Vrancken, Smid, and Szwarc (13). Their combined results are presented in a plot of log K_∞ against $1/T$ as shown in Figure 2. The agreement between all three sets of data is excellent and the best line through all the experimental points gives $\Delta H = -7.5$ kcal/mole and $\Delta S = -26.5$ eu.

McCormick (11) and Worsfold and Bywater (12) initiated the polymerization of α-methylstyrene by using sodium naphthalene, brought the system to equilibrium at the desired temperature, and then "killed" the living polymers. They showed that at 0°C, the concentration of the monomer attains its equilibrium value in less than 16 hr, and they observed no further polymerization over periods lasting for as long as 124 hr. Furthermore, it was shown that the equilibrium concentration of monomer was independent of the concentration of living ends, i.e., of the amount of catalyst applied.

Various problems may be encountered in thermodynamic investigations of living polymers. An instructive example is provided by the studies of the styrene–living polystyrene system as reported recently by Bywater and Worsfold (14). At 0°C, the equilibrium concentration of styrene was expected to be about 10^{-7} mole/liter which is too low to be determined by conventional analytical techniques. The system was, therefore, investigated in the temperature range of 100–150°C, where the equilibrium concentrations were expected

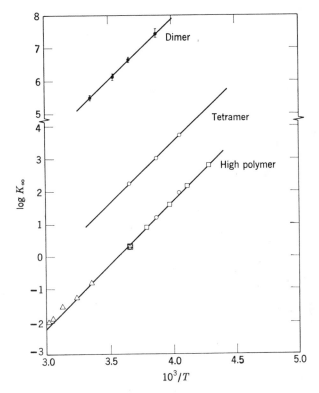

Fig. 2. Equilibrium constant for polymerization of α-methylstyrene to living oligomers and to a high molecular weight polymer. (○) Vrancken, Smid, and Szwarc; (□) Worsfold and Bywater; (△) McCormick.

to rise to 10^{-4} to 10^{-3} mole/liter. At these concentrations ultraviolet spectrophotometric techniques are applicable. The temperatures of 100–150°C are well above those normally considered suitable for anionic polymerizations. In fact, in tetrahydrofuran under these conditions one observes a rapid "isomerization" of living polystyrene as manifested by the appearance of a new absorption band at $\lambda_{max} = 560$ mμ and by a decay of the optical density at $\lambda_{max} = 340$ mμ, the absorption band characteristic of living polystyrene. To

avoid these difficulties, the equilibrium was studied in hydrocarbon solvents
i.e., in benzene and in cyclohexane, using butyllithium to initiate the polymerization. At these elevated temperatures, the half-life of the reaction was
found to be of the order of a few seconds in cyclohexane, and even less in benzene and, hence, in less than a minute the system approached, within 1%, its
equilibrium state. This was important because some slow side reactions still
take place even in these solvents, e.g., the characteristic 334 mμ absorption
peak of lithium-containing living polystyrene, $\sim\!\!\sim$CH$_2$.CH(Ph)$^-$, Li$^+$, decreased during the experiments and a small absorption band appeared near

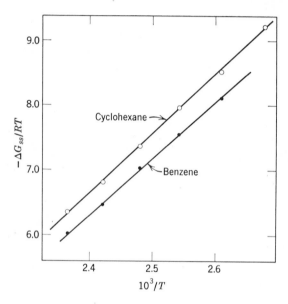

Fig. 3.   Equilibrium constant of living polystyrene propagation in cyclohexane and
benzene; $\ln K = \Delta G_{ss}/RT$ plotted versus $1/T$.

450 mμ. Fortunately, the half-life of this side reaction was nearly a hundred
times longer than that of the polymerization and, thus, this disturbance was of
no significance unless the solution was heated for an excessively long time.
The equilibrium was quenched by adding a small amount of water at the
temperature of the experiment. Thereafter, the mixture was cooled rapidly
to minimize any thermal polymerization of the residual styrene.

   The results, collected in Table III, are shown graphically in Figure 3 as a
plot of log K_∞ against $1/T$. Benzene and cyclohexane are good and poor
solvents, respectively, for polystyrene. Their effects upon the equilibrium
polymerization constants are apparent from the displacement of the lines as
shown in Figure 3. The difference in the respective K_∞'s arises mainly from

TABLE III
Equilibrium Concentration of Styrene in Contact with Living Polystyrene[a]

T,°C	M_e in benzene mole/liter $\times 10^5$	M_e in hexane mole/liter $\times 10^5$
100	—	3.97
110	12.0	7.79
120	20.7	13.6
130	34.2	23.8
140	59.0	41.6
150	90.8	65.0

[a] From S. Bywater and D. J. Worsfold, *J. Polymer Sci.*, **58**, 571 (1962).

the polymer–solvent interactions which affect the free energy of the polymer segments. A more extensive discussion of this subject will be found in Section 11 of this chapter.

The living polymer technique has been applied recently to studies of the equilibrium tetrahydrofuran $\rightleftarrows$ polytetrahydrofuran. This polymerization proceeds by a cationic mechanism (15), the propagation step being described by the equation:

$$\text{~~~—O·(CH}_2)_4\text{—}\overset{+}{O}\underset{X^-}{\overset{\displaystyle \text{CH}_2\text{—CH}_2}{\diagup \diagdown}}\underset{\text{CH}_2\text{—CH}_2}{\Big|} + O\overset{\displaystyle \text{CH}_2\text{—CH}_2}{\underset{\text{CH}_2\text{—CH}_2}{\diagup \diagdown \Big|}} \rightleftarrows$$

$$\text{~~~—O·(CH}_2)_4\text{—O—(CH}_2)_4\text{—}\overset{+}{O}\underset{X^-}{\overset{\displaystyle \text{CH}_2\text{—CH}_2}{\diagup \diagdown}}\underset{\text{CH}_2\text{—CH}_2}{\Big|}.$$

Various agents may initiate such a polymerization, e.g., Ph_3C^+, $SbCl_6^-$, and their action was investigated by Bawn and co-workers (16) who were the first to recognize the formation of living polymers in the tetrahydrofuran system. The kinetics of tetrahydrofuran polymerization initiated by Et_3O^+,BF_4^- was investigated by Vofsi and Tobolsky (17) who determined the monomer equilibrium concentration to be $2.6M$ at 0°C. They showed that the initial rate of polymerization decreases linearly with the initial monomer concentration (see Figure 4) and hence the equilibrium concentration of the monomer was obtained by extrapolating the initial rate to zero. Conceptually this approach is similar to that developed by Ivin (6).

A most thorough investigation of the tetrahydrofuran system was reported recently by Dreyfuss and Dreyfuss (18). They initiated the polymerization by the decomposition of benzenediazonium hexafluorophosphate (PhN_2^+, PF_6^-), which provides a Ph^+ ion and an extremely stable PF_6^- counterion. It seems that the stability of the counterion accounts for the simplicity of this system which is admirably suited for kinetic and thermodynamic studies. The existence of living polymers was demonstrated in a variety of ways, e.g., by increase in the molecular weight of the polymer upon further addition of

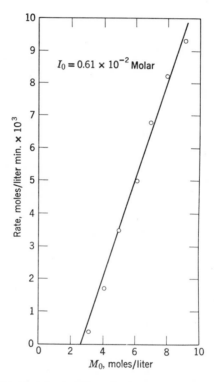

Fig. 4. Rate of cationic polymerization of tetrahydrofuran to polyether at 0°C as a function of monomer concentration.

monomer, by formation of block polymers, etc. The equilibrium concentration of monomer was determined at various temperatures and the plot of $\log K_e$ against $1/T$ is shown in Figure 5. These authors presented recently a comprehensive review of tetrahydrofuran polymerization (56).

Dioxalone may be polymerized in methylene dichloride solution by catalytic action of perchloric acid, and the resulting living polymer, or macroring, remains in equilibrium with its monomer. This system was studied by Plesch and Westermann (57) who extrapolated the kinetic results, obtained at constant temperature but different monomer concentrations, to zero rate.

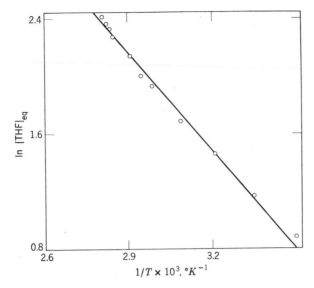

Fig. 5. Equilibrium constant for tetrahydrofuran polymerization to polyether as function of temperature (medium pure tetrahydrofuran).

They confirmed their findings by investigating the dependence of molecular weight of the polymer on the monomer concentration. Thus, they derived the values $\Delta H_{ss} = -5.1$ kcal/mole and $\Delta S_{ss} = -18.6$ eu, leading to a ceiling temperature of $1.5°C$ for a $1M$ monomer solution. Direct calorimetric study was used to check the derived value of ΔH_{ss}.

Extension of this work to 1.3-dioxepane (58) gave the following values: $\Delta H_{ss} = -3.5$ kcal/mole, $\Delta S_{ss} = -11.7$ eu, and T_c (for $1M$ solution) $= +26°C$.

4. USE OF RADIOACTIVE TRACERS IN THE DETERMINATION OF THE EQUILIBRIUM CONCENTRATIONS OF MONOMERS

For many systems the equilibrium concentration of monomer is expected to be extremely low and one may attempt to use the radioactive tracer technique to determine its value. The method requires some refinements because even a small isotope effect might affect considerably the final result. The following calculation illustrates this point.

Let x_0 and x_e denote the respective initial and final (equilibrium) concentrations of the monomer and a_0 and a_f their initial and final specific activities expressed as mole fraction of the radioactive tracer. Consider an isotope effect which makes the rate of addition of the tracer molecule f-times greater than that of the non-radioactive monomer. Hence, the increase, da, in the specific activity of the residual monomer, arising from the polymerization of $-dx$ moles of monomer ($dx < 0$), is given by

$$a + da = a(x + f\,dx)/(x + dx)$$

where x and a respectively denote the concentration and specific activity of the monomer at some stage of polymerization. Rearrangement of this equation leads to

$$da/a = (f - 1)\,dx/x$$

which on integration gives

$$\log (a_f/a_0) = (f - 1) \log (x_e/x_0).$$

Denoting the fraction of the total activity recovered in the residual monomer by γ, so that

$$\gamma = x_e a_f/x_0 a_0,$$

one finds,

$$\log \gamma = f \cdot \log x_e - f \cdot \log x_0.$$

Hence, a plot of $\log \gamma$ against $\log x_0$ should result in a straight line having a slope, $-f$, and an intercept, $f \cdot \log x_e$.

In any real system, f is expected to deviate from unity by not more than 10%. Two numerical examples will show how much the results are affected by such a deviation. Consider, e.g., a polymerization which occurs upon mixing a $1M$ solution of monomer with living polymers so that the resulting reaction reduces eventually the monomer concentration to its equilibrium value of 10^{-7} mole/liter. Using the derived formulae, we find

$$a_f/a_0 = 5.0 \quad \text{for} \quad f = 0.90$$

$$a_f/a_0 = 1.2 \quad \text{for} \quad f = 0.99.$$

Hence, even for an isotope effect as low as 1%, the specific activity of the residual monomer would increase by about 20%. The enrichment of monomer with radioactive tracer affects the apparent x_e values. For example, had the correction not been introduced in the calculation, erroneous values of $x_e = 5.0 \times 10^{-7}$ mole/liter or 1.2×10^{-7} mole/liter (for $f = 0.90$ and 0.99, respectively) would be obtained instead of the correct value of $x_e = 1.0 \times 10^{-7}$ mole/liter.

In deriving these equations, exchange between monomer and living polymer was ignored. This is permissible when the rate of depropagation is much slower than the rate of propagation. However, as the system approaches equilibrium, the depropagation step becomes comparable to the propagation step and the validity of the approximation may then be questioned. Fortunately, the final results are only slightly distorted by this factor, because the specific activity of the monomer varies slowly with its concentration. In the example given, for $f = 0.90$, the specific activity of the monomer increased fourfold as its concentration decreased from 1 to 10^{-6} mole/liter. At this stage, propagation is still ten times faster than depropagation. Further decrease in the monomer concentration to the limiting value of 10^{-7} mole/liter increases its specific activity only by an additional 25%. Hence, at this last reaction stage, the specific activity of the growing ends (and not of the polymer

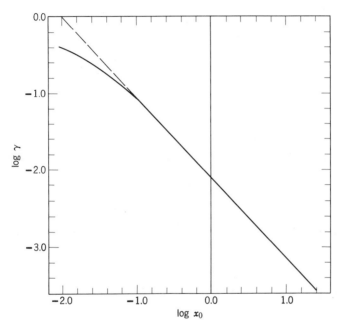

Fig. 6.   Dependence of the recovered activity fraction, γ, of a radioactive monomer, remaining in equilibrium with its living polymer, on its initial concentration, x_0. $K_e = 100$ moles/liter, $P^* = 0.01$ mole/liter, $f = 1.05$.

as a whole) is similar to the expected final specific activity of the monomer so that the exchange cannot be of any great importance, if the mixture is not left for too long before "killing" the polymers.

To complete this discussion, let us consider the plot of $\log \gamma$ against $\log x_0$ as shown in Figure 6. Evidently, our treatment applies only when x_0 is much greater than x_e. However, when x_0 becomes comparable to x_e, the theoretical straight line becomes curved as illustrated in Figure 6. For ratios x_0/x_e of 1000 or more, the deviations from the theoretical line are expected to be small and the tracer method may be used whenever $x_0/[\text{growing ends}]$ is sufficiently large, e.g., greater than 100.

It should be stressed that this technique may be utilized for the determination of small isotope effects in polymerization. In fact, it is more reliable for this purpose than for determining the equilibrium concentration of monomer.

5. EQUILIBRIA BETWEEN MONOMER AND LIVING OLIGOMERS

Although for high molecular weight polymers, the equilibrium propagation constant is independent of the degree of polymerization, this need not be the case for low molecular weight oligomers, i.e., dimers, trimers or tetramers, or for macromolecules which involve head-to-head or tail-to-tail linkages. In fact, recent studies of oligomers of α-methylstyrene (13) revealed great deviations of such equilibrium constants from the K_∞ values.

Prolonged contact of α-methylstyrene with potassium or sodium-potassium alloy in tetrahydrofuran yields a living, tail-to-tail dimer, i.e.,

$$K^+, {}^-C(CH_3)(Ph) \cdot CH_2 \cdot CH_2 \cdot C(CH_3)(Ph)^-, K^+ = K^+, {}^-\alpha\alpha^-, K^+.$$

His structure was conclusively established by carboxylation of the dimer which produced a mixture of stereoisomers of 2-5-dimethyl-2,5-diphenyladipic acids (19).

It will be shown later (see p. 123) that the addition of α-methylstyrene to both ends of the dimer proceeds much faster than the further growth of the resulting tetramer. Consequently, a living tetramer, denoted by T_2, may be prepared by the addition of the required amount of α-methylstyrene to the living α-methylstyrene dimer ($K^+, {}^-\alpha\alpha^-, K^+$). The structure of T_2 follows from the method of its preparation, namely,

$T_2 =$
$$K^+, {}^-C(CH_3)(Ph) \cdot CH_2 \cdot C(CH_3)(Ph) \cdot CH_2 \cdot CH_2 \cdot C(CH_3)(Ph) \cdot CH_2 \cdot C(CH_3)(Ph)^-, K^+$$

or symbolically, $T_2 = {}^-|{-}{\scriptstyle\text{I}}{-}|{-}{\scriptstyle\text{I}}{-}{\scriptstyle\text{I}}{-}|{-}{\scriptstyle\text{I}}{-}|{}^-$.

On the other hand, relatively brief contact of a dilute tetrahydrofuran solution of α-methylstyrene with metallic sodium gives an isomeric living tetramer (13) denoted by T_1. Studies of its formation (20) indicated for this species, a tail-to-tail, head-to-head, tail-to-tail structure, i.e.,

$T_1 =$
$$Na^+, {}^-C(CH_3)(Ph) \cdot CH_2 \cdot CH_2 \cdot C(CH_3)(Ph) \cdot C(CH_3)(Ph) \cdot CH_2 \cdot CH_2 C(CH_3)(Ph)^-, Na^+$$

or symbolically, $T_1 = {}^-|{-}{\scriptstyle\text{I}}{-}{\scriptstyle\text{I}}{-}|{-}|{-}{\scriptstyle\text{I}}{-}{\scriptstyle\text{I}}{-}|{}^-$. All these oligomers show the characteristic absorption spectrum of benzyl carbanions, i.e., $\lambda_{max} = 340$ mμ and $\epsilon = 1 \times 10^4$ (per end), leaving no doubt that these species contain terminal —$CH_2 \cdot C(CH_3)(Ph)^-$ units.

Formation of tetramers by the action of metallic sodium on ether solutions of α-methylstyrene had been described previously by Bergmann et al. (21) who proposed the following structure for this compound

$$Na^+, {}^-C(CH_3)(Ph) \cdot CH \cdot C(CH_3)(Ph) - CH_3$$
$$\overset{|}{Na^+}, {}^-C(CH_3)(Ph) \cdot CH \cdot C(CH_3)(Ph) - CH_3.$$

This work needs further verification. Bergmann's conclusion is based on the assumption that through some rearrangement of unknown nature, a dimer

$$C(CH_3)(Ph) : CH \cdot C(CH_3)(Ph) \cdot CH_3$$

is initially formed, and that it in turn reacts further with sodium giving the proposed product.

The equilibrium constants for the addition of monomer to low-molecular weight living oligomers have been determined by a method developed by Vrancken, Smid, and Szwarc (13). In a solution containing a constant, although very low, concentration of living oligomers which may grow but not decompose, the equilibrium concentration, M_e, of an added monomer may be determined as a function of its initial concentration, M_0. A typical result is shown in Figure 7 where M_e is plotted as a function of M_0 for a constant concentration of living tetramer, T_1. In fact, such a plot proves the stability of the tetramer, T_1, because the resulting curve passes through the origin indicating that T_1 does not decompose into monomer and a trimer. The decomposition of a tetramer would be expected, if one of the terminal monomeric units were linked in a head-to-tail fashion. Since dissociation is not observed, this provides an additional argument favoring the proposed tail-to-tail, head-to-head, tail-to-tail structure of T_1. The stability of T_1 may be contrasted with the behavior of tetramer T_2 which is stable only in the presence of an equilibrium concentration of monomer amounting, e.g., at 25°C, to 0.02 mole/liter of α-methylstyrene for the concentration of 0.04 mole/liter of living ends.

Having established the equilibrium concentration of the monomer, M_e, as a function of its initial concentration, M_0, one may proceed to calculate the respective equilibrium constants. Vrancken et al. (13) accomplished this as follows: they assumed that equilibrium had been established between monomer and *all* of the living j-mers, $j \geqslant n_0$, that each end of the two-living ended polymer reacts independently of the fate of the other end, and that *all* of the equilibrium constants have identical values, i.e., $K_1 = K_2 = \cdots = K_\infty$. The apparent constant, K_1', was calculated from Tobolsky's relation

$$(M_0 - M_e)/P_{total}^* = K_1' M_e (1 - K_1' M_e)^{-1}.$$

In this equation, P_{total}^* denotes the total concentration of living ends which is

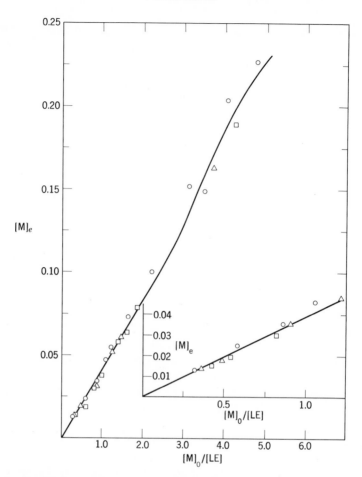

Fig. 7. Equilibrium concentration M_e of α-methylstyrene in contact with its living tetramer as a function of its initial concentration M_0. Solvent, THF; temperature, 0°C; concentration units, moles/liter. Results of three series of experiments, each corresponding to about $0.05M$ concentration of living tetramer.

equal to the initial concentration of the growing ends of the living n_0-mer. Keeping $P^*_{total} = (P^*_{n_0})_0$ constant in a series of experiments, they gradually increased M_0. Every pair of values, M_0 and its respective M_e, enabled them to calculate the apparent constant K'_1, and hence K'_1 may be determined as a function of M_0. Had the assumption, $K_1 = K_2 = \cdots = K_\infty$ been valid, then each pair of values, M_0 and M_e, would have given the same $K'_1 = K_1$, i.e., a plot of K'_1 against M_0 should then produce a straight line parallel to the M_0 axis. The authors (13) found, however, that such a plot was curved, as shown in Figure 8, invalidating the assumption requiring *all* K_j to be identical.

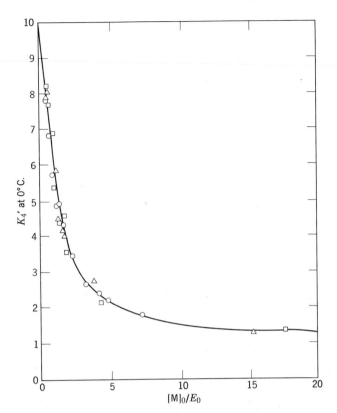

Fig. 8. Apparent equilibrium constant, K_4' for the system α-methylstyrene–living α-methylstyrene tetramer in THF at 0°C. (○) Experiments series 4; (□) experiments series 5; (△) experiments series 6. E_0 denotes P^*_{total}.

Nevertheless, the first equilibrium constant, K_1,

$$\text{P}^*_{n_0} + \text{M} \rightleftarrows \text{P}^*_{n_0+1}, \qquad \ldots K_1$$

may be determined from such a plot by extrapolating the experimental curve to $M_0 = 0$. In this way, Vrancken et al. determined the equilibrium constant for the processes

$$^-\alpha\alpha^- + \alpha \leftrightarrows \text{Trimer}, \qquad \ldots K_D$$

and

$$^-\text{T}_1^- + \alpha \rightleftarrows \text{Pentamer}, \qquad \ldots K_T$$

the pertinent data being collected in Table IV.

It should be noted that the function $K_1' = f(M_0)$ asymptotically approaches K_∞ for a high ratio M_0/P^*_{total} and, hence, the latter constant may also be determined from the plots shown in Figure 8. The same value of K_∞ should

TABLE IV
The Equilibrium Constant of the Growth Process for
Living α-Methylstyrene Oligomers[a]

T,°C	K_D, liter/mole	K_{T1}, liter/mole	K_∞(extrap. K_D'), liter/mole	K_∞(extrap. K_{T1}'), liter/mole	K_∞(lit.), liter/mole
25.4	240 ± 30	—	0.8 (?)	—	0.41
10.0	450 ± 50	—	0.85	—	0.9
0	750 ± 100	9.4	1.24	1.33	1.35
−15.0	1750 ± 300	20	3.3	3.3	3.3
−26.5	—	40	—	6.8	6.8

[a] K_D refers to $^-\alpha\alpha^- + \alpha \rightleftarrows {}^-\alpha\alpha\alpha^-$; K_{T1} refers to $^-T_1^- + \alpha \rightleftarrows {}^-T_1\alpha^-$; K_∞ refers to high-molecular weight poly-α-methylstyrene; K_∞(lit.) data from combined results of H. W. McCormick (11) and D. J. Worsfold and S. Bywater (12).

be obtained from the data derived from studies either of the dimer or of the tetramer. The results listed in the fourth and fifth columns of Table IV, in fact, confirm this conclusion. Moreover, as shown in the last column of this table, the K_∞ values thus derived agree well with those reported by McCormick (11) and by Worsfold and Bywater (12).

The procedure applied to determine K_1 may be utilized to calculate K_2

$$P_{n_0+1}^* + M \rightleftarrows P_{n_0+2}^*, \qquad \dots K_2.$$

By assuming that $K_2 = K_3 = \cdots = K_\infty$ and accepting the K_1 value previously derived, K_2 may be calculated (22). The apparent K_2' is

$$K_2' = M_e^{-1} - (1/2)K_1\{\sqrt{1 + 4P_{total}^*/K_1 M_e(M_0 - M_e)} - 1\}.$$

Extrapolation of K_2' to $M_0 = 0$ should then give the correct value for K_2, while K_∞ should again be given by the asymptotic value of K_2' for a large M_0/P_{total}^*. An extension of such calculations might give values for K_3, K_4, etc. However, the experimental errors accumulate rapidly in such calculations, rendering the method useless for the determination of K_3 and unreliable even for K_2, if the individual errors of experimental M_e values exceed 2%.

The preceding discussion brings us to the general problem of the reliability of such determinations of equilibrium constants. The reproducibility of the data may be seen by examining plots of M_e as a function of M_0 for a constant P_{total}^* as shown, e.g., in Figure 7. By this test, the results seem to be satisfactory. However, much more severe demands on the reproducibility and the reliability of the data are imposed by plotting K_1' as a function of M_0 and Figure 8 illustrates this point. While K_1' is a slowly varying function of M_0 for large values of M_0/P_{total}^*, the curve rises steeply as M_0/P_{total}^* approaches zero. The steepness of such curves increases as the ratio K_1/K_∞ becomes larger and, hence, the reliability of the data must be decidedly greater in studies of the living dimer ($K_D/K_\infty \approx 500$) than in comparable studies of the living tetramer ($K_{T1}/K_\infty \approx 6$). Moreover, the relative accuracy of M_e determinations diminishes as M_0 decreases, whilst studies in the low range of M_0 are crucial for the determination of K_1. Exceptional care is therefore essential in these experiments, details of which are given in reference 13.

For $K_1 > K_\infty$, the equation of Tobolsky gives the apparent constant K_1' smaller than K_1 and, hence, extrapolation proceeds through a set of points lying *below* the value of K_1. In testing the reliability of extrapolation, it is desirable to use an alternative method of calculation giving the apparent K_1'' greater than K_1 and thus, one may approach K_1 from above. This was done by Vrancken et al. (13) for the region $M_0/P_{\text{total}}^* < 1$ by assuming $K_2 = K_3 = \cdots = K_\infty = 0$. This leads to the equation

$$K_1'' = (M_0/M_e - 1)/\{P_{\text{total}}^* - (M_0 - M_e)\} \text{ leading to } K_1'' > K_1 \text{ for any } M_0.$$

The limiting value K_1 is given again by $\lim K_1''$ for $M_0/P_{\text{total}}^* \to 0$ but extrapolation involves now a set of points greater than K_1. Using this approach it was possible to obtain a reliable estimate of the accuracy for each determination of K_1, see, e.g., reference 13. This discussion shows again the futility of calculating K_3 and exposes the great uncertainties involved in the calculation of K_2. Although Vrancken et al. attempted in their preliminary publication (22) to determine K_2 for the T_1 tetramer system, later work indicated that the claimed values are doubtful. It seems, nevertheless, that for that system K_2 may be slightly greater than K_∞, while K_1 is about six times as large. Kinetic work, described later, indicates that the increase in the K_1 value, when compared with K_∞, arises from a smaller depropagation rate constant, whereas the propagation rate constant, $T_1\alpha + \alpha \to T_1\alpha\alpha$ differs insignificantly from that observed for a high molecular weight polymer.

Finally, it should be stressed that the nomenclature used in this section refers to consecutive additions of monomers to the same end. It is assumed implicitly that the equilibrium constants for the reactions $\alpha T_1 + \alpha \rightleftarrows \alpha T_1\alpha$ and $T_1 + \alpha \rightleftarrows T_1\alpha$ are virtually identical, and by defining P_{total}^* as the total concentration of *living* ends, one incorporates this assumption into the calculations.

6. KINETIC TECHNIQUES FOR THE DETERMINATION OF THE EQUILIBRIUM CONSTANT OF PROPAGATION

The equilibrium constant of any process is related to the rate constants of the forward and backward reactions, namely,

$$K = k_{\text{forward}}/k_{\text{backward}}.$$

Hence, the equilibrium propagation constant may be calculated if the absolute propagation and depropagation rate constants are determined. Various techniques used in the determination of the absolute rate constants of anionic propagation and depropagation steps will be discussed in Chapter VII. At this stage, only one kinetic method will be described; this gives simultaneously the propagation and depropagation rate constants as well as the equilibrium constant. The method is illustrated by the reaction, whereby upon addition of α-methylstyrene to the living α-methylstyrene dimer, $^-\alpha\alpha^-$, the latter yields reversibly the respective living α-methylstyrene trimer, i.e.,

$$^-\alpha\alpha^- + \alpha \underset{k_{-1}}{\overset{k_1}{\rightleftarrows}} {}^-\alpha\alpha\alpha^-, \qquad \ldots K_D.$$

The system, $^-\alpha\alpha^- + \alpha \rightleftarrows {}^-\alpha\alpha\alpha^-$, was investigated by Lee, Smid, and Szwarc (23) who studied its kinetics by using a stirred-flow reactor. The great simplicity of this device arises from the fact that the concentrations of all the reagents, products, and intermediates remain constant in the reactor (24). The reagents are fed into the reactor at a constant rate, i.e., v cc/sec. The incoming solution contains reagent A at concentration $C_{A,0}$, B at $C_{B,0}$, etc. Simultaneously v cc/sec of a solution containing the products flow out. Efficient stirring in the reactor maintains the composition of the reacting solution homogeneous throughout its entire volume, V, and, hence, the composition of the outflowing liquid is identical with that maintained in the reactor. In the living dimer → living trimer system, the reaction is quenched instantly in the outgoing liquid by leading it into wet tetrahydrofuran.

The simplicity of such a system is evident from the fact that one needs only to determine, for different rates of flow, the concentration of the residual monomer in the quenched reaction fluid in order to get all the required constants. The initial concentration of the monomer, x_0, and that of the living dimer, y_0, are chosen at will and kept constant in each series of experiments. The reactor is operated for a few minutes to allow the system to reach its stationary state and then a sample of the outgoing liquid is withdrawn and analyzed for α-methylstyrene. The result gives, therefore, the stationary concentration of the monomer, x_t, in the reactor. The stationary concentrations of the dimer, y_t, and of the trimer, z_t, are then determined by the stoichiometry of the process, i.e.,

$$y_t = y_0 - (x_0 - x_t) \quad \text{and} \quad z_t = (x_0 - x_t).$$

The amount of a reagent consumed in the reactor is given by the mass balance, i.e., by the difference between the amounts flowing in and out, and the balance equation permits one to calculate the required rate constants. For monomer, the mass balance becomes:

$$v(x_0 - x_t) = V\{2k_1 x_t y_t - k_{-1} z_t\}$$

or

$$v(x_0 - x_t) = V\{2k_1 x_t [y_0 - (x_0 - x_t)] - k_{-1}(x_0 - x_t)\}.$$

The factor 2 arises from the presence of two living ends in each dimer, both of them being capable of reacting with monomer. If t, the residence time of the liquid in the reactor, is defined as the ratio V/v, one arrives at the following expression

$$1/t = 2k_1 x_t\{y_0/(x_0 - x_t) - 1\} - k_{-1}.$$

Hence, a plot of $1/t$ against $x_t\{y_0/(x_0 - x_t) - 1\}$ should give a straight line as shown in Figure 9, with a slope equal to $2k_1$. The intercept on the $1/t$ axis

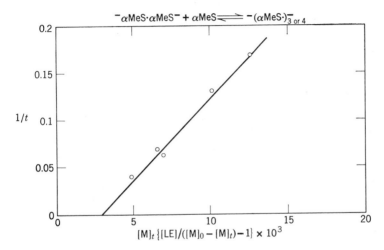

Fig. 9. Graphic solution of the balance equation for the reaction $^-\alpha\alpha^- + \alpha \rightleftarrows {}^-\alpha\alpha\alpha^-$. α denotes α-methylstyrene; $^-\alpha\alpha^-$ its living dimer; and $^-\alpha\alpha\alpha^-$ its living trimer. $1/t$ is the reciprocal of the resident time in the stirred-flow reactor.

is equal to $-k_{-1}$ while that on the other axis is related to the equilibrium constant, i.e., $K_D = (1/2) \, (\text{intercept})^{-1}$. These constants, obtained for different initial conditions of the reaction, are given in Table V, and their self-consistency demonstrates the reliability of the method.

Addition of two or more monomer units to a particular end of the dimer was ignored in deriving the above equation. This is justified because the rate of addition to the right end of the trimer,

$$^-C(CH_3)(Ph) \cdot CH_2 \cdot CH_2 \cdot C(CH_3)(Ph) \cdot CH_2 \cdot C(CH_3)(Ph)^-,$$

was found to be much slower than that to the left end. The latter reaction is accounted for in the equation by introducing the factor $2k_1$ instead of k_1. In this procedure it is implied that the rate of the reaction at one end of the dimer is not affected by the fate of the other end, an assumption which seems to be justified by the data.

TABLE V

Constants for the Reversible Reaction $^-\alpha\alpha^- + \alpha \underset{k_{-1}}{\overset{k_l}{\rightleftarrows}} {}^-\alpha\alpha\alpha^- \cdots K^a$

in Tetrahydrofuran at 25°C.   Counterion: K^{+a}

P^*_{total}, $M \times 10^3$	M_0, $M \times 10^3$	k_1, liter/mole, sec	k_{-1}, sec^{-1}	K_D, liter/mole
3.0	3.0	17.2	0.050	346
4.4	2.9	17.3	0.052	333
4.5	1.1	16.7	0.049	339
6.5	2.9	18.9	0.062	304
14.0	3.0	15.4	0.046	333
	Average:	17.1	0.052	331

[a] Data of Lee, Smid, and Szwarc (23).

The kinetic method produced results concordant with those derived from the direct static approach (13), and leads to a value of 331 liter/mole for K_D as compared with $K_D = 240 \pm 30$ liter/mole obtained by extrapolation of K'_D to $M_0 = 0$. In view of the steepness of the extrapolation, the kinetic value seems to be more reliable.

7. FACTORS AFFECTING THE EQUILIBRIUM CONSTANT OF PROPAGATION

Steric strain in the polymer and the rigidity of its chain are the two most important factors affecting the equilibrium propagation constant. Bulky substituents on the C=C group of a vinyl monomer are the main cause of steric strain in its polymer and the hindrance becomes considerable when two substituents are attached to the same carbon atom. The strain is manifested by a relatively low heat of polymerization, e.g., $-\Delta H$ for ethylene polymerization is ~24 kcal/mole, for styrene it amounts to 17–18 kcal/mole, while polymerization of α-methylstyrene evolves only 7.5–8 kcal/mole. Inspection of models shows that substituents on alternate carbon atoms of a —C—C—C— chain interfere with each other to a greater extent than those located on adjacent atoms. For example, a direct linkage of two *tert*-butyl groups gives a relatively strainless hydrocarbon, namely, 2,2,3,3-tetramethylbutane

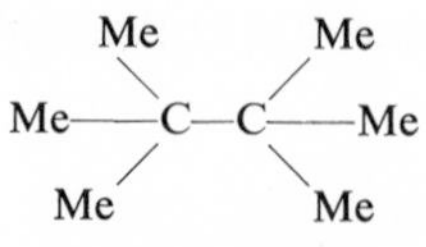

$$
\begin{array}{ccc}
\text{Me} & & \text{Me} \\
\diagdown & & \diagup \\
\text{Me}\!-\!\!-\!\text{C}\!-\!\text{C}\!-\!\!-\!\text{Me} \\
\diagup & & \diagdown \\
\text{Me} & & \text{Me}
\end{array}
$$

whereas their linkage through a CH_2 group leads to severe strain

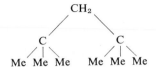

These considerations explain why a head-to-head and tail-to-tail polyiso-butene $\sim\!\!CMe_2\cdot CH_2\cdot CH_2\cdot CMe_2\cdot CMe_2\cdot CH_2\cdot CH_2\cdot CMe_2\!\!\sim$ is expected to be much more stable than the conventional head-to-tail polyisobutene, $\sim\!\!CH_2\cdot CMe_2\cdot CH_2\cdot CMe_2\!\!\sim$. The same reasons partially explain the stability of the living α-methylstyrene tetramer, T_1, $^-C(Me)(Ph)\cdot CH_2\cdot CH_2\cdot$ $C(Me)(Ph)\cdot C(Me)(Ph)\cdot CH_2\cdot CH_2\cdot C(Me)(Ph)^-$ which, as remarked previously, does not decompose into lower oligomers, whereas the living tetramer, T_2, $^-C(Me)(Ph)\cdot CH_2\cdot C(Me)(Ph)\cdot CH_2\cdot CH_2\cdot C(Me)(Ph)\cdot CH_2\cdot C(Me)(Ph)^-$ exists only in equilibrium with the monomer and the respective trimer and dimer.* The same factors account for the lower heat of polymerization of α,α'-substituted vinylidene monomers as compared with that of analogous $\alpha\text{-}\beta$-substituted olefins.

Steric strain is maximal when the addition of monomer causes the substituents of the preceding unit to become crowded by bulky groups on *both* sides. For example, in the conversion of living α-methylstyrene dimer into trimer

$$^-C(Me)(Ph)\cdot CH_2\cdot CH_2\cdot C^-(Me)(Ph) + CH_2:C(Me)(Ph) \rightleftarrows$$
$$\underset{1}{^-C}(Me)(Ph)\cdot\underset{2}{CH_2}\cdot\underset{3}{CH_2}\cdot\underset{4}{C}(Me)(Ph)\cdot\underset{5}{CH_2}\cdot\underset{6}{C}(Me)(Ph)^- \qquad \cdots K_{D,1}$$

the strain between the substituents on carbon atoms 4 and 6 may be reduced by a conformational adjustment of the substituents on carbon 4. Subsequent addition of another monomer unit, viz.,

$$^-C(Me)(Ph)\cdot CH_2\cdot CH_2\cdot C(Me)(Ph)\cdot CH_2\cdot C(Me)(Ph)^- + CH_2:C(Me)(Ph) \rightleftarrows$$
$$\underset{1}{^-C}(Me)(Ph)\cdot\underset{2}{CH_2}\cdot\underset{3}{CH_2}\cdot\underset{4}{C}(Me)(Ph)\cdot\underset{5}{CH_2}\cdot\underset{6}{C}(Me)(Ph)\cdot\underset{7}{CH_2}\cdot\underset{8}{C}(Me)(Ph)^- \qquad \cdots K_{D,2}$$

introduces strain between substituents on carbon atoms 6 and 8 which cannot be released, because the groups on carbon 6 are now squeezed between those located on carbons 4 and 8. The same type of strain results upon each subsequent monomer addition. One expects, therefore, the equilibrium constant $K_{D,1}$ to exceed $K_{D,2}$ while the latter should be similar in magnitude to K_∞. Studies by Vrancken et al. (13) confirmed this supposition. However, the expected decrease in the heat of polymerization was not detected. On the other hand, a direct calorimetric study indicated a substantial increase in the heat of the reaction $^-\alpha\alpha^- + \alpha \rightleftarrows {}^-\alpha\alpha\alpha^-$ as compared with that evolved in the polymerization (53).

* The structure of the tetramer was investigated recently by NMR technique. The results obtained by two groups were contradictory (64,65).

Electrostatic repulsion between the two charged ends (or the respective C^-, K^+ dipoles) of the living α-methylstyrene dimer may also contribute to its heat of polymerization, because the electrostatic energy of the system decreases as $^-\alpha\alpha^-$ grows to $^-\alpha\alpha\alpha^-$. On these grounds, one expects the equilibrium constant $K_{D,1}$ to be larger than the analogous $K_{T1,1}$ constant,

$^-C(Me)(Ph)\cdot CH_2\cdot CH_2\cdot C(Me)(Ph)\cdot C(Me)(Ph)\cdot CH_2\cdot CH_2\cdot C(Me)(Ph)^-$
$$+ \ CH_2:C(Me)(Ph) \ \rightleftarrows$$
$^-C(Me)(Ph)\cdot CH_2\cdot CH_2\cdot C(Me)(Ph)\cdot C(Me)(Ph)\cdot CH_2\cdot CH_2\cdot C(Me)(Ph)\cdot$
$$CH_2\cdot C(Me)(Ph)^- \qquad \cdots K_{T1,1}.$$

This indeed was found by Vrancken et al. (13), although here again the expected decrease in the heat of polymerization was not observed. The importance of the electrostatic effect could be tested by comparing the equilibrium constant $K_{D,1}$ ($^-\alpha\alpha^- + \alpha = \ ^-\alpha\alpha\alpha^-$) with that of the cumyl-potassium-α-methylstyrene process which gives the relevant dimer containing one living end only. Since no electrostatic contribution is associated with the free energy of the latter reaction, its heat might be expected to be lower than that of the former. Preliminary studies of the cumylpotassium system have been carried out and, surprisingly, the equilibrium constant of this reaction was found to be even higher than $K_{D,1}$. The calorimetric data indicate also that the addition of α-methylstyrene to cumene potassium is more exothermic than the addition to the α-methylstyrene dimer (53). One must conclude, therefore, that electrostatic repulsion, if any, results from the interaction of dipoles—not of free charges—and apparently its contribution to the free energy of the addition reaction seems relatively small. Further studies of this problem would be desirable.

Steric strain may also decrease the entropy of the polymer chain by increasing its rigidity and, hence, it could increase the value of $-\Delta S_p$. However, the observed changes in the entropy of polymerization are relatively insignificant, e.g., $-\Delta S_p$ for styrene polymerization to a solid polymer amounts to ~ 25 eu, whereas the corresponding value for α-methylstyrene was found to be ~ 26 eu. This is indeed a small change when compared with a 10-kcal/mole decrease in the heat of polymerization. The growth of oligomers should probably lead to a smaller change in the entropy of the system than the growth of analogous high polymers, and this factor may partially account for their higher equilibrium constants.

Finally, it should be stressed that, in contradistinction to high molecular weight polymers, the equilibrium growth constants of low molecular weight oligomers depend on the mechanism of the reaction and on the nature of the reactive end groups. This point was recognized and stressed by Dainton and Ivin (2).

8. EFFECT OF STEREOSPECIFICITY ON THE MONOMER–LIVING POLYMER EQUILIBRIA

In a dilute solution of living polymers the equilibrium concentration of the monomer is determined solely by the temperature and nature of the solvent if the structure of the resulting polymer is uniquely established. However, the structure of the polymer depends on the nature of the growing ends and on the conditions prevailing during the process. For example, isotactic, syndiotactic, or a variety of atactic polymers may be formed from the same vinyl monomer if suitable catalysts are employed. Similarly, polymerization of a diene may lead to 1,2- or *cis*- or *trans*-1,4 polymers, depending on the conditions of polymerization. Let us consider, therefore, how the equilibrium established between living polymers and their monomer is affected by such phenomena.

Consider addition of a vinyl monomer to a growing chain having a random structure. Each individual placement may occur in an isotactic or syndiotactic fashion; let $-\Delta F_i$ and $-\Delta F_s$ denote the free energies associated with the formation of the respective linkages. Let us assume that the probability of finding an isotactic linkage in a long polymeric chain is constant, say p, i.e., we deal with a Bernoulie chain. Hence, the free energy of such a polymer (per mole of segment) is given by

$$p\Delta F_i + (1 - p)\Delta F_s - RT\{p \ln p + (1 - p) \ln (1 - p)\}$$

and for a system at equilibrium p has to fulfill the condition

$$p/(1 - p) = \exp (\Delta F_i - \Delta F_s)/RT.$$

If the reaction yields only isotactic polymer due, e.g., to some kinetic restrictions, then the equilibrium constant K_i

$$\text{monomer} \rightleftarrows \text{isotactic polymer}$$

is given by the usual thermodynamic equation $RT \ln K_i = \Delta F_i$. Similarly, for a system in which only the syndiotactic placement is possible the equilibrium constant K_s

$$\text{monomer} \rightleftarrows \text{syndiotactic polymer}$$

is given by $RT \ln K_s = \Delta F_s$. Hence, at true thermodymamic equilibrium

$$p/(1 - p) = K_i/K_s.$$

This treatment implies that the free energy of a linkage is not affected by the nature of the neighboring linkages. A more general treatment was discussed by Coleman and Fox.

For $\Delta F_i \gg \Delta F_s$, p is virtually unity and the resulting polymer is entirely isotactic due to the thermodynamic conditions. Similarly, a thermodynamically stable syndiotactic polymer is formed when $\Delta F_s \gg \Delta F_i$. For $\Delta F_i = \Delta F_s$ an atactic polymer is formed for which $p = \frac{1}{2}$.

The above thermodynamic conditions may be represented by a kinetic scheme (Scheme I) in which the rate constants of the monomer addition are determined entirely by the type of placement being formed, and not by the nature of the last linkage, whereas the rate constants of depropagation depend on the nature of the last linkage and are not affected by the type of penultimate linkage.

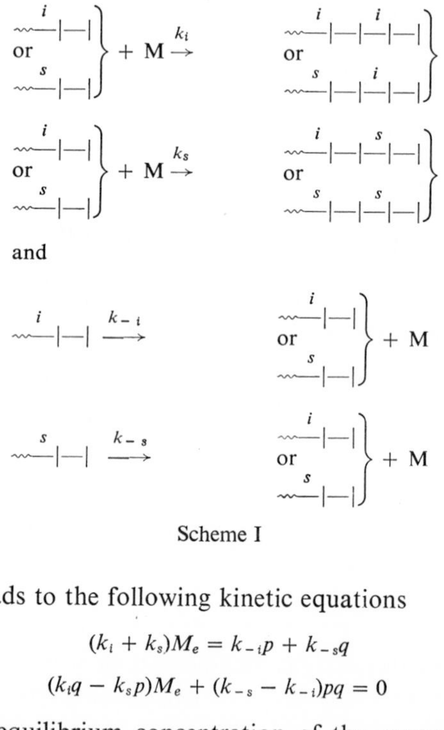

$$
\left.\begin{array}{c} i \\ \sim\!\!-\!|\!-\!| \\ \text{or} \\ s \\ \sim\!\!-\!|\!-\!| \end{array}\right\} + \mathrm{M} \xrightarrow{k_i} \left\{\begin{array}{c} i \qquad i \\ \sim\!\!-\!|\!-\!|\!-\!|\!-\!| \\ \text{or} \\ s \qquad i \\ \sim\!\!-\!|\!-\!|\!-\!|\!-\!| \end{array}\right.
$$

$$
\left.\begin{array}{c} i \\ \sim\!\!-\!|\!-\!| \\ \text{or} \\ s \\ \sim\!\!-\!|\!-\!| \end{array}\right\} + \mathrm{M} \xrightarrow{k_s} \left\{\begin{array}{c} i \qquad s \\ \sim\!\!-\!|\!-\!|\!-\!|\!-\!| \\ \text{or} \\ s \qquad s \\ \sim\!\!-\!|\!-\!|\!-\!|\!-\!| \end{array}\right.
$$

and

$$
\begin{array}{c} i \\ \sim\!\!-\!|\!-\!| \end{array} \xrightarrow{k_{-i}} \left.\begin{array}{c} i \\ \sim\!\!-\!|\!-\!| \\ \text{or} \\ s \\ \sim\!\!-\!|\!-\!| \end{array}\right\} + \mathrm{M}
$$

$$
\begin{array}{c} s \\ \sim\!\!-\!|\!-\!| \end{array} \xrightarrow{k_{-s}} \left.\begin{array}{c} i \\ \sim\!\!-\!|\!-\!| \\ \text{or} \\ s \\ \sim\!\!-\!|\!-\!| \end{array}\right\} + \mathrm{M}
$$

Scheme I

Such a scheme leads to the following kinetic equations

$$(k_i + k_s)M_e = k_{-i}p + k_{-s}q$$

$$(k_i q - k_s p)M_e + (k_{-s} - k_{-i})pq = 0$$

where M_e is the equilibrium concentration of the monomer and p and q denote the probabilities of finding an isotactic or syndiotactic linkage, respectively, in the polymeric chain. Naturally, $p + q = 1$.

The first equation results from the necessity of balancing the propagation and depropagation, the second from the condition requiring the rate of rupture of isotactic (or syndiotactic) linkages to be the same as their rate of formation. The condition of random structure implies that a fraction p of all the polymers has an isotactic terminal linkage and a fraction q has a terminal syndiotactic linkage. Such a line of reasoning leads to the conclusion that on rupture of either linkage the probability of producing a new terminal isotactic linkage is p, and for a terminal syndiotactic linkage the probability is q.

Solution of the above kinetic equations leads to the equilibrium concentration of the monomer, M_e,

$$M_e = (K_i + K_s)^{-1},$$

and again to the previously derived result,

$$p/q = p/(1 - p) = K_i/K_s,$$

where $K_i = k_i/k_{-i}$ and $K_s = k_s/k_{-s}$. Whenever one of the modes of propagation is inoperative (due, e.g., to the catalyst's preference for one type of placement only), the respective pseudo-equilibrium concentration of the monomer is given by $1/K_i$ or $1/K_s$. In the absence of such a restriction the equilibrium concentration of monomer decreases.

9. MOLECULAR WEIGHT DISTRIBUTION IN MONOMER–LIVING POLYMER EQUILIBRIA

In discussing the equilibria between a monomer and its living polymer, two aspects of the problem should be considered: (1) establishment of an equilibrium, or more correctly of a stationary state, between monomer and growing ends; (2) establishment of an equilibrium between all the growing polymers. The first process approaches its equilibrium state fairly rapidly, whereas the other is usually very slow and requires a long time to produce the ultimate equilibrium distribution.

It was shown (25) that for a solution of living polymers, the ultimate equilibrium molecular weight distribution is of the Flory type (26a), i.e., the mole fraction of the $(n_0 + j)$-mer is given by $(K \cdot M_e)^j \cdot (1 - KM_e)$, if all the equilibrium constants are identical. In accordance with our symbolism, $P_{n_0}^*$ defines the lowest living n_0-mer which may grow but not degrade, and its mole fraction in the equilibrated mixture should be $(1 - KM_e)$. For a high average degree of polymerization $1 - KM_e \ll 1$ and, therefore, only a minute fraction of polymers exists as n_0-mers.

Whenever a high molecular weight polymer is produced, the propagation rate constant, k_p, must be much greater than k_d, the depropagation constant. Furthermore, the initial concentration of the monomer, M_0, must be substantially larger than its equilibrium concentration M_e, and $M_0/(P_{n_0}^*)_0 \gg 1$. Polymerization in such a system usually leads to a Poisson molecular weight distribution (26b) when nearly all the polymeric molecules have a degree of polymerization close to $M_0/(P_{n_0}^*)_0$. Although such a system is not yet in its true equilibrium state, the polymerization is essentially completed, and the concentration of the monomer differs only insignificantly from that attained at the ultimate equilibrium. In fact, its value, M'_e, is determined approximately by the condition of the stationary state, i.e.,

$$k_p M'_e \left\{ \sum_0^\infty P^*_{n_0+j} \right\} = k_d \left\{ \sum_1^\infty P^*_{n_0+j} \right\}$$

leading to $M'_e = (1/K)(1 - P^*_{n_0,t}/P^*_{total})$ which is only slightly greater than M_e, because $P^*_{n_0,t}$ is expected to be only a small fraction of P^*_{total}. A numerical example shows how insignificant the difference is between M'_e and M_e. Let us consider a system where initial monomer concentration is 1 mole/liter, total concentration of living polymers equals 0.01 mole/liter, and the propagation equilibrium constant $K = 99$ liters/mole. The calculated ultimate $M_e = 0.01 M$, corresponding to $P^*_{n_0,eq} = 10^{-4} M$. For the Poisson distribution $P^*_{n_0,t} = 0$ and $M'_e = 0.0101$ mole/liter, i.e., only 1% higher than the ultimate M_e.

The redistribution of molecular weight in the living polymer system represents an interesting kinetic problem. In this reaction, neither the concentration of the polymerized material nor its number average molecular weight changes. Hence, had the process been studied in a dilatometer or in an osmotic cell, no change would be observed. However, the weight average molecular weight, and all the higher average molecular weights, increase as the distribution changes from the Poisson to the Flory type. The differential equation giving $d(\overline{DP}_w)/dt$ was discussed by Brown and Szwarc (25).* Assuming the steady state approximation, they found

$$d(\overline{DP}_w)/dt = 2k_d/(\overline{DP}_n)[1 - x_0(\overline{DP}_n)]$$

where (DP_w) and (DP_n) denote, respectively, weight average and number average degree of polymerization, k_d is the depropagation rate constant, and x_0 the mole fraction of the lowest living n_0-mer. A more rigorous treatment leads to

$$d(\overline{DP}_w)/dt = k_d\{[(\overline{DP}_w) - 1]/(\overline{DP}_n)\} \cdot (x_{0,eq} - x_0)$$

or

$$d(\overline{DP}_w)/dt \simeq k_d\{\overline{DP}_w)/(\overline{DP}_n)\} \cdot (x_{0,eq} - x_0)$$

where $x_{0,eq}$ is the mole fraction of the living n_0-mer at the ultimate equilibrium state. Since $x_{0,eq} = 1/(\overline{DP}_n)$, the latter equation becomes identical with the former if $(\overline{DP}_w)/(\overline{DP}_n) = 2$, i.e., for a system being near to its ultimate equilibrium state. These equations make it obvious that the redistribution of polymer chains is an extremely slow process at a high degree of polymerization, because the rate is inversely proportional to $\overline{DP}_n$.

* See p. 63 and reference 66 for a more general treatment.

The redistribution of molecular weights, discussed above, results from reactions taking place at the polymeric *ends* only, viz.,

$$P_n^* + M \rightleftarrows P_{n+1}^*.$$

However, in some systems the redistribution of monomer segments may result from other reactions, such as trans-esterification, trans-amidation or trans-etherification. Consider, e.g., a solution of living polyglycols

$$\sim CH_2CH_2O \cdot CH_2 \cdot CH_2O^-, Na^+$$

As the result of a reaction,

$$\overbrace{\sim CH_2CH_2O}^{l\ units} \cdot \overbrace{CH_2CH_2O\sim}^{k\ units} + \overbrace{^-OCH_2 \cdot CH_2\sim}^{j\ units} \rightarrow$$

$$\overbrace{\sim CH_2CH_2O^-}^{l\ units} + \overbrace{\sim CH_2CH_2O \cdot CH_2 \cdot CH_2O\sim}^{j+k\ units}$$

an $(l + k)$-mer and a j-mer are converted into an l-mer and a $(j + k)$-mer. The ultimate equilibrium distribution resulting from this reaction is again of a Flory type; however, the kinetics of this process are different from those previously considered. Its mathematical treatment has been reported recently by Hermans (27).

10. SHARPNESS OF TRANSITION PHENOMENA IN EQUILIBRIUM POLYMERIZATION

In his approach to the thermodynamics of propagation, Dainton was concerned with a process transforming free monomer into monomer segments of a *high* molecular weight polymer. The free energy change of this reaction is given by the equation

$$\Delta F_p = \Delta F(M_s) - \Delta F(M_f),$$

where the molar free energy of the monomer, $\Delta F(M_f)$, involves the concentration term $RT \ln (M_f)$, whereas no equivalent term appears in the free energy, $\Delta F(M_s)$, of the monomer segments, because the segments are part of already existing macromolecules. This gives to the process the characteristic feature of a physical aggregation such as, e.g., crystallization. The concentration of polymer affects ΔF_p, indirectly only, inasmuch as it determines the deviation of the monomer's activity from its ideal value by influencing the monomer–solvent–segment and segment–segment interactions. In this terminology, the ceiling temperature has a uniquely defined value determined by the equation $\Delta F_p = 0$ for constant thermodynamic conditions of the system, i.e., for specified concentrations of monomer and polymer and for the particular nature of the solvent.

Alternatively, we may look at the state of ultimate equilibrium of a system containing a monomer the polymerization of which involves no termination and yields living polymers. In such a discussion, the modes of initiation of the polymerization and the concentration of the initiator or of the lowest living oligomer must be specified.

This approach has been developed by Tobolsky (9,28), who considered three distinct systems differing in their modes of initiation:

$$
\begin{aligned}
\text{I.} \quad & I + M \rightleftarrows IM^* && \cdots K_1 \\
& \left.\begin{aligned}
IM^* + M &\rightleftarrows IM_2^* \\
&\vdots \\
IM_{n-1}^* + M &\rightleftarrows IM_n^*
\end{aligned}\right\} && \cdots K
\end{aligned}
$$

$$
\begin{aligned}
\text{II.} \quad & M \rightleftarrows M^* && \cdots K_1 \\
& \left.\begin{aligned}
M^* + M &\rightleftarrows M_2^* \\
&\vdots \\
M_{n-1}^* + M &\rightleftarrows M_n^*
\end{aligned}\right\} && \cdots K
\end{aligned}
$$

$$
\begin{aligned}
\text{III.} \quad & M + M \rightleftarrows M^* && \cdots K_1 \\
& \left.\begin{aligned}
M_2^* + M &\rightleftarrows M_3^* \\
&\vdots \\
M_{n-1}^* + M &\rightleftarrows M_n^*
\end{aligned}\right\} && \cdots K
\end{aligned}
$$

A system containing a living n_0-mer which may grow but not degrade is exemplified by Case I. Polymerization of cyclic monomers initiated by spontaneous ring opening provides an example of Case II. Finally, a hypothetical thermal polymerization of a vinyl monomer arising from bimolecular initiation and proceeding without termination illustrates Case III.

We may limit the discussion to Case I without losing sight of the generality of behavior which characterizes all three processes. The system is determined by the initial concentrations of monomer, M_0, initiator, P_0^*, and the equilibrium constants K_1 and K. Its qualitative behavior is not affected by the simplifying assumption demanding K_1 to be equal to K and therefore, for the sake of clarity, this is accepted in the present discussion.

The state of equilibrium is characterized by the equilibrium concentration of the monomer, M_e, and the number average degree of polymerization, $\bar{\jmath}$, the latter being defined by the equation $(\bar{\jmath}) = \sum_1^\infty j \cdot P_j^* / \sum_1^\infty P_j^*$. M_0, P_0, M_e, and $\bar{\jmath}$ are related by the following equations derived from first principles, viz.,

$$
(\bar{\jmath}) = (M_0 - M_e)/(P_0^* - P_{0,eq}^*) = 1/(1 - KM_e)
$$

$$
P_0^* = (\bar{\jmath})P_{0,eq}^*
$$

and

$$
(\bar{\jmath})^2 KP_0^* + (\bar{\jmath})[K(M_0 + P_0^*) - 1] - 1 = 0.
$$

The equilibrium constant K is a function of temperature, T; hence, the equilibrium concentration of monomer, M_e, the degree of polymerization, $\bar{\jmath}$, and the fraction of polymerized monomer, $(M_0 - M_e)/M_0$, are all unique functions of T for constant M_0 and P_0^*. For the α-methylstyrene–living α-methylstyrene system, these functions are shown in Figures 10 and 11. The steep decrease in $\bar{j}$ or in $(M_0 - M_e)/M_0$ within a narrow temperature range is the striking feature of these curves. This behavior characterizes all equilibrium

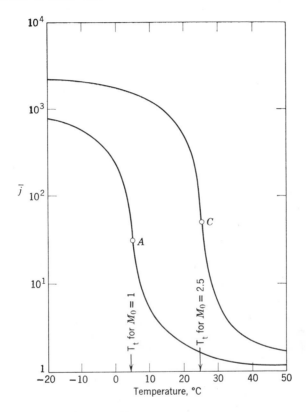

Fig. 10. Degree of polymerization, $\bar{j}$, versus temperature in °C for equilibrium polymerization of α-methylstyrene: (A) initial concentration of monomer, 1 mole/kg.; (C) initial concentration of monomer, 2.5 moles/kg; concentration of living ends, 0.001 mole/kg.

polymerizations and the sudden increase in $\bar{j}$ on lowering the temperature shows the similarity between these phenomena and other transitions based on physical aggregation.

For equilibrium polymerizations, the sharpness of transition has a well-defined physical meaning and may be defined mathematically by, e.g., $d\bar{j}/dT$ in the transition region. This derivative may be evaluated by differentiating the quadratic equation for $\bar{j}$ with respect to T and substituting

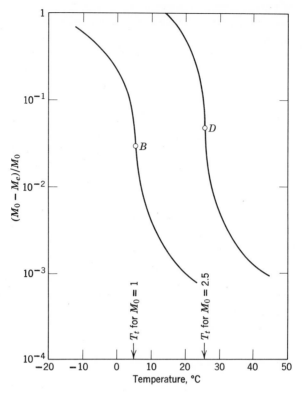

Fig. 11. Amount of polymerized monomer, $(M_0 M_e)/M_0$ versus temperature for equilibrium polymerization of α-methylstyrene. (B) $M_0 = 1$ mole/kg or (D) $M_0 = 2.5$ moles/kg; concentration of living ends = 0.001 mole/kg.

$(d\bar{j}/dK)_T \cdot (dK/dT)$ for $d\bar{j}/dT$. At the transition temperature, the result may be set in the approximate form

$$d\bar{j}/dT = (\Delta H_p/RT^2)/KP_0^*[2 - (KP_0^*)^{1/2}]$$

by using the simplifying relations

$$\bar{j}_{\text{transition}} \approx (KP_0^*)^{-1/2}$$

and

$$(M_0 - M_e)_{\text{transition}} \approx (P_0^*/K)^{1/2}.$$

The latter are plausible approximations because at the transition temperature, and above it, the initial monomer concentration, M_0, is nearly equal to its equilibrium concentration, i.e., $M_0 \approx M_e \approx 1/K$. In view of the last relation, the ceiling temperature of the system is defined as the temperature at which $K(T) = 1/M_0$.

It is interesting to compare Dainton's definition of ceiling temperature with that of Tobolsky. Both have the same mathematical form. Dainton's definition refers to the single chemical change: free monomer $\rightarrow$ monomer segment of a *high* polymer, and takes no account of the building process leading to a macromolecule. On the other hand, Tobolsky refers to the whole stepwise building reaction and gives full consideration to the initiation processes. Therefore, the sharpness of the transition which is irrelevant in Dainton's definition is pertinent in Tobolsky's treatment. However, Dainton's definition is applicable to systems in which equilibrium between various living polymers has not yet been established, e.g., the photodegraded systems studied by Ivin(6), or the pseudo-equilibrium situation when a Poisson distribution is attained as a consequence of rapid polymerization of living polymers. On the other hand, Tobolsky's definition applies only when complete equilibrium is established among *all* the polymeric molecules as well as between polymer and monomer.

Finally, the factors affecting the sharpness of the transition should be examined. The expression giving $[dj/dT]_{P_0^* M_0}$ shows clearly that the transition is sharper, the greater ΔH_p (the heat of polymerization) and the smaller P_0^* (the number of initiating species). The effect of ΔH_p is obvious: increase in its absolute value narrows the temperature range in which K varies from values much larger than $1/M_0$ to those much smaller. The effect of P_0^* is equally easy to comprehend. The amount of polymerized monomer increases on lowering the temperature of the system. For lower P_0^*, the latter is distributed among fewer chains, giving a greater increase in j.

In a similar way, one may calculate the degree of polymerization, $\bar{j}$, as a function of M_0 for constant temperature and constant P_0^*. The results are presented in Figure 12 and show again a sharp transition occurring within a critical region of M_0. The sharpness of the transition is defined by $(dj/dM_0)_T$ in the vicinity of the critical concentration M_0 and is given approximately by the equation $[dj/dM_0]_T = 1/2P_0^*$.

The greater sharpness of the ceiling temperature arising from an increase in the size of the polymer has its analogy in the behavior of the melting point of a crystal (2). The melting temperature is sharp for large crystals but it becomes a broad range for small crystallites because the contributions of the surface, edges, and apexes to the free energy of the system become significant when compared with the bulk free energy. The similarity to the effect due to the presence of polymer chain ends is obvious.

These effects were observed in experiments reported by Dainton's group. For example, they were clearly demonstrated in the study of the cationic polymerization of α-methylstyrene (29). The formation of low molecular weight polymers above the ceiling temperature was attributed to a higher heat of polymerization when oligomers are formed. All of the kinetic studies by

Dainton's group showed that the rate of polymerization does not drop abruptly to zero at the ceiling temperature but that in its vicinity the curve, giving rate as a function of temperature, becomes asymptotic to the T axis. This again points to a broadening of the transition arising from a decrease in the molecular weight of the resulting polymer.

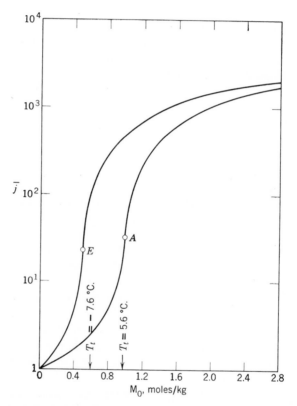

Fig. 12. Degree of polymerization, $\bar{J}$, versus initial monomer concentration of α-methylstyrene. Equilibrium polymerization at $+5.6$ or $-7.6°C$; concentration of living ends $= 0.001$ mole/kg.

Finally, the data published by Gee (30) permit one to evaluate the sharpness of a transition involving floor temperature. Gee studied the temperature dependence of the viscosity of liquid sulfur and observed its sudden, steep increase at a critical temperature followed by its decrease at still higher temperatures. He developed the first, relatively complete theory of equilibrium polymerization of liquid sulfur (30) from which he estimated the chain length of the polymeric sulfur at various temperatures. His results have been confirmed recently by experimental measurements of the magnetic susceptibility of liquid sulfur (50) and of its electron spin resonance (51).

Gee's theory was unified by Tobolsky and Eisenberg (31) and further improved by Tobolsky et al. (52) who removed the original restriction demanding a multiple number of 8 sulfur atoms in each chain. This theory demands the existence of a transition (floor) temperature and excellently accounts for its sharpness as determined from Gee's results. A similar situation is observed with liquid selenium (32).

11. DEVIATIONS FROM THE LAWS OF IDEAL SOLUTIONS

It has been remarked in the preceding sections that the equilibrium concentration of monomer in solution of its living polymer is affected by the nature of the solvent and by the polymer concentration, because these factors influence the activities of the components. A quantitative treatment of these effects, based on Scott's modification of the standard lattice theory of polymer solutions (33), has been outlined recently by Bywater (34).

Polymerization of one mole of a liquid monomer to form a solid polymer is associated with a decrease in the free energy of the system given by the equation

$$-\Delta F_{\text{liquid, solid}} = RT\{\ln a_p - \ln a_m\}$$

where a_m and a_p denote, respectively, the activities of the free monomer and of the monomer segments in the polymer chain when they are at equilibrium in the investigated solution. The activities refer to liquid monomer and solid polymer as their respective standard states. The lattice theory of polymer solutions gives these activities in terms of the volume fractions of the components, i.e., ϕ_s, ϕ_m, and ϕ_p, and of the three interaction parameters, $\mu_{s,m}$, $\mu_{s,p}$, and $\mu_{m,p}$ of the respective pairs: solvent–monomer, solvent–polymer, and monomer–polymer. Thus,

$$-\Delta F/RT = (\mu_{s,p} - \mu_{s,m})\phi_s - 1 - \ln \phi_m + \mu_{m,p}(\phi_m - \phi_p)$$

which may be converted into ΔF_p (the free energy of polymerization of a $1M$ monomer solution to a $1M$ solution of polymeric segments) by adding to the equation the respective free energy of a $1M$ solution of the monomer and of the polymeric segments, i.e.,

$$\Delta F_m^*/RT = \mu_{s,m}\phi_s^* + \phi_s^*(1 - \phi_s^*)^{-1}\ln \phi_s^* + \ln \phi_m^*$$

and

$$\Delta F_p^*/RT = \mu_{s,p}\phi_s^* - 1 + \phi_s^*(1 - \phi_s^*)^{-1}\ln \phi_s^*,$$

ϕ^* denoting the respective volume fractions of the monomer, solvent, and polymer in the relevant $1M$ solutions. This leads to the equation

$$\ln K = -\Delta F_p/RT = \ln (\phi_m^*/\phi_m) + (\mu_{s,p} - \mu_{s,m})(\phi_s - \phi_s^*) + \mu_{m,p}(\phi_m - \phi_p)$$

giving the equilibrium propagation constant in terms of the mole fractions of the components and the relevant interaction parameters.

When $\mu_{s,p} = \mu_{s,m}$, and if either $\mu_{m,p} = 0$ or ϕ_m and ϕ_p tend to zero, the equation for ln K reduces to ln $K = -\ln(\phi_m/\phi_m^*)$. Moreover, if $\phi_m/\phi_m^* = [M]_e$, as would be expected for such a system, the laws of ideal solution apply, i.e., ln $K = -\ln M_e$ or $KM_e = 1$.

For experiments carried out in a highly diluted solution of living polymer, ln K is given by the simplified equation

$$\ln K = \ln(\phi_m^*/\phi_m) + (\mu_{s,p} - \mu_{s,m})(\phi_s - \phi_s^*) + \mu_{m,p}\phi_m.$$

Deviations from ideal behavior arise from three causes:

(1) ϕ_m is not proportional to the monomer concentration in which event $\phi_m/\phi_m^* \neq M_e$.

(2) The interaction of solvent with monomer is different from that with polymer, i.e., $\mu_{s,m} \neq \mu_{s,p}$.

(3) The monomer–polymer interaction is not negligible, i.e., $\mu_{m,p} \neq 0$.

The deviations due to variation of the partial volume of monomer with dilution are expected to be negligible in not too concentrated solutions. For example, if the partial volume of the monomer remains constant up to $1M$ solution and $M_e < 1$ mole/liter, then $\phi_m/\phi_m^* = M_e$.

The inequality $\mu_{s,m} \neq \mu_{s,p}$ accounts for the dependence of M_e on the nature of the solvent. Its magnitude is shown, e.g., by the results of Bywater and Worsfold (14) who found the equilibrium concentration of styrene to be different in hexane and in benzene solutions (see p. 112 and Figure 3). The effect of solvent is not eliminated even at an extremely low equilibrium concentration of monomer; in fact, in such a case, the deviation from ideal behavior is given by $(\mu_{s,m} - \mu_{s,p})(1 - \phi_s^*)$.

The effect of solvent may be separated into heat and entropy terms. The former given by the difference in the molar heats of dilution of the monomer and polymer solutions from the standard solvent volume fractions ϕ_s^* to their equilibrium fractions ϕ_s. The entropy term is constant for many solvents. However, it becomes of great importance when the random configuration of the polymer chain, assumed to prevail in "normal" solutions, is drastically changed by the solvent, e.g., when a polymer acquires a helical structure in the investigated solution.

Finally, the last term, $\mu_{m,p}(\phi_m - \phi_p)$ is unimportant if ϕ_m and ϕ_p are small. However, it becomes significant in concentrated polymer solution as shown, e.g., by the studies of Vrancken et al. (13). Their investigation of the α-methyl-styrene–living poly-α-methylstyrene system in tetrahydrofuran solution demonstrated that for a *high* concentration of living ends the equilibrium concentration of the monomer increases initially and thereafter *decreases* as more and more monomer is added to the system. This strange behavior, illustrated by Figure 13, contrasts with that shown in Figure 7, where, for low concentrations of living ends, M_e increases monotonically with M_0 and eventually reaches a limiting value characteristic of an equilibrium involving a high molecular weight living polymer. The decrease in M_e shown in Figure 13 results from an increase in the *total* concentration of polymer. For example, upon the addition of dead poly-α-methylstyrene to a solution of living poly-α-methylstyrene which was in equilibrium with its monomer, the

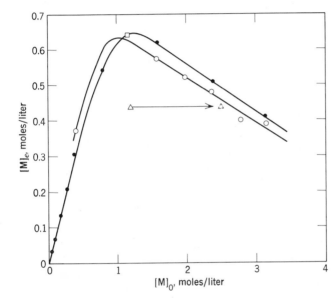

Fig. 13. Equilibrium concentration M_e of α-methylstyrene, in contact with its living polymer in THF at 0°C, plotted as a function of its initial concentration M_0 (●) series 1, [LE] = 0.068; (○) series 2, [LE] = 0.014; (□) series 2, cumene added; (△△) series 2, dead polymer added.

equilibrium concentration of the latter decreased from 0.63 mole/liter (the square point in Figure 13) to about 0.42 mole/liter (the triangle point in the same figure). By adding the mass of the dead polymer to M_0, the effective M_0 is calculated and this, as shown by the dotted triangle, corresponds to the anticipated value of M_e. In terms of the derived equation, the effect arises from an increase in the term $\mu_{m,p} \cdot \phi_p$ caused by the addition of polymer or of monomer which eventually polymerized. The addition of cumene, on the other hand, has a negligible influence on the M_e value (13).

The decrease in M_e upon addition of monomer superficially seems to contradict the law of mass action. Addition of monomer increases its *activity* in equilibrated solution if the concentration of living polymer (in g/liter) is low. The increase, however, becomes negligible as the activity approaches its limiting value. On the other hand, if the amount of added monomer is sufficiently large to increase substantially the concentration of the resulting polymer, then this raises the activity coefficient of the monomer and causes the observed decrease in the equilibrium monomer concentration.

The effect of polymer concentration on M_e has also been shown in the study of Tobolsky et al. (35). Unfortunately, their experimental technique was not sufficiently refined, and the wide scatter of experimental points prevented quantitative deductions from their data. Comment is necessary with respect to one statement made in their paper. The authors assumed that the "true" value of M_e could be derived by linear extrapolation from the experimental M_e's, obtained at high polymer concentrations, up to the point of intersection with the line $M_e = M_0$. However, such an extrapolation is not valid because, as has been shown by Vrancken et al. (13), the results depend on the concentration of living ends. The proper determination of the "true" M_e requires studies at low concentrations of living ends, as was done by Worsfold and Bywater (12) and by McCormick (11).

In a recent paper, Ivin and Leonard (36) considered the effect of a soluble polymer in a liquid monomer on the free energy of polymerization. The process may be performed in three steps: (*1*) remove a mole of monomer from the solution ($-\Delta \bar{G}_M$), (*2*) convert the monomer into polymer ($\Delta G_{l,s}$), and (*3*) dissolve the polymer in the solution ($\Delta \bar{G}_p$). In view of the equilibrium established between the monomer and polymer

$$-\Delta \bar{G}_M + \Delta G_{l,s} + \Delta \bar{G}_P = 0.$$

Accepting the conventional Flory–Huggins treatment, one deduces then the free energy of polymerization, $\Delta G_{l,s}$, of a liquid monomer into solid monomer to be

$$\Delta G_{l,s} = RT\{\ln \phi_1 - (\ln \phi_2)/n + 1 - 1/n + \chi(\phi_2 - \phi_1)\}.$$

For a high molecular weight polymer, i.e., for a large n, this equation is reduced to

$$\Delta G_{l,s} = RT\{\ln \phi_1 + 1 + \chi(\phi_2 - \phi_1)\}.$$

This equation is applicable to a system such as liquid tetrahydrofuran-dissolved polytetrahydrofuran which was discussed in an earlier part of this section.

The thermodynamic relations are particularly simple for a polymer which is insoluble in a solution of the monomer. For such a system

$$\Delta G_{l,m} + RT \ln a_m = \Delta G_{s,p}$$

and this equation applies to a living polymer which is in equilibrium with its monomer. The subscripts l,m and s,p refer to liquid monomer and solid polymer, respectively.

An interesting example of such a system is a mixture of isobutene–sulfur dioxide in contact with the respective polysulfone (the alternating copolymer of these two monomers) which is insoluble in the liquid monomers. For each composition of such a mixture there is a ceiling temperature, T_c, at which the

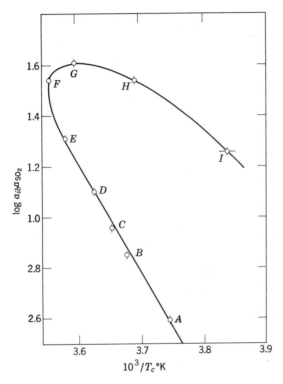

Fig. 14.   Temperature dependence of the product of the activities of isobutene, a_{ib}, and SO$_2$, a_{SO_2}, in equilibrium with their respective copolymers.

system is at equilibrium. One expects a monotonic relation between $RT \ln \{a_{ib} \cdot a_{SO_2}\}$ and the respective T_c, because different values of the activities product should correspond to different ceiling temperatures. However, the studies of Cook, Ivin, and O'Donnell (37a) led to a most surprising result. Using the activity coefficients determined in a previous work (37b), they constructed a curve, shown in Figure 14, giving $\log \{a_{ib} \cdot a_{SO_2}\}$ as a function of $1/T_c$. This plot reveals that the respective function is not singled value and this means that different polymers are formed at different compositions of the mixture.

The copolymers may differ in their composition. However, elementary analysis showed that this is not the case; all the polymers had a composition corresponding to 1:1 ratio of the monomers. Alternatively, they may differ in their structure or conformation. Both changes could be reflected in the crystal structure if the polymer is crystalline.

The authors preferred the latter explanation. They argue that the difference in the dielectric constant of the medium, when a hydrocarbon-rich mixture is compared with one rich in sulfur dioxide, affects the ratio of *gauche* to *trans* conformations of the polymer chain. Such effects were observed in other systems, e.g., in 1,2-dichloroethane (38). Hence, the minute amount of living polymer, which is still in solution, should be intrinsically different in both media. Nevertheless, the precipitated polymer might be ultimately the same, regardless of the mixture from which it was deposited, if the relaxation time is not too long. Since the glass transition of the polysulfone is rather high, it is possible that the ultimate equilibrium structure of the solid has not been attained, and therefore samples precipitated from solutions having different compositions would exhibit a different x-ray pattern. This indeed has been observed (39).

One more possibility may be considered. The equilibrium is attained with the polymer in the solid phase. However, this phase may be swollen by the liquid mixture and the degree and type of swelling would be different when the solid is in contact with a liquid rich in hydrocarbon or rich in SO_2. The degree of swelling affects the chemical potential of the swollen polymer and, hence, the respective ceiling temperature. According to this explanation, the solid polymer should be in equilibrium with its solution, and again its structure (conformation) has to be different, depending upon whether it is in the hydrocarbon-rich or SO_2-rich liquid.

In conclusion, studies of the equilibria between living polymers and their monomers provide means of determining monomer–polymer interaction parameters, and the interaction parameters of monomer and polymer with a solvent.

12. EFFECT OF PRESSURE ON THE EQUILIBRIUM CONSTANT OF PROPAGATION

Polymerization leads to a contraction in the volume of the system so that the equilibrium of a monomer–polymer system shifts in the direction of the reaction as the hydrostatic pressure increases. This was demonstrated by Weale (40) in his studies of α-methylstyrene polymerization under high pressure. The ceiling temperature increased from 61°C at 1 atm to 170°C at 6480 atm.

A polymerization which cannot be observed under normal pressure might become practical at elevated pressures. Consider, e.g., the case of a polymerization prohibitively slow at low temperatures and thermodynamically impossible at higher temperatures. Because the ceiling temperature should rise at higher pressure, polymerization could become feasible under these conditions.

Consider now a polymerization which is thermodynamically impossible at normal pressure because the reaction proceeds with a positive ΔH and a negative ΔS. At higher pressures, the term $P\Delta V$ may balance the positive ΔH and the process would then be feasible. The problems arising from pressure effects have been reviewed in a recent paper by Ivin (41) (see also reference 42).

Equilibrium between living polymer and its monomer offers excellent opportunities for studies of systems which cannot be polymerized under normal pressure. For example, steric strain prevents the polymerization of 1,1-diphenylethylene at atmospheric pressure, but the reaction might take place at a sufficiently high pressure. It should be possible to study such an equilibrium in the following device. Two optical cells, introduced into a chamber maintained at high pressure, are inserted into the optical paths of a two-beam spectrophotometer. One contains a solution of 1,1-diphenylethylene, the other has an identical solution to which some living dimer of 1,1-diphenyl-ethylene has been added. The change in the optical density of the monomer could then be examined at different temperatures and pressures and this provides information about the thermodynamics of the polymerization reaction. This description is given to focus the reader's attention on the potentialities of such studies.

13. ISOTOPE EFFECT IN EQUILIBRIUM POLYMERIZATION

The large decrease in the heat of polymerization of α-methylstyrene when compared to that of styrene is attributed to the steric repulsion between the bulky groups which become crowded in the polymer. The importance of repulsion between the nonbonded atoms was stressed by many workers, e.g., in a recent paper by Bartell (55). In overcrowded molecules even a small additional compression or a small decrease in the degree of compression may lead to appreciable effects.

In hydrocarbons steric repulsion is caused by the interaction between the nonbonded hydrogen atoms. Substitution of deuterium for hydrogen should decrease the repulsion energy because deuterium atoms take less space than hydrogen atoms as a result of their lower amplitude of vibration. This idea has been recently tested by Fetters, Pummer, and Wall (54), who investigated the effects of deuteration on the equilibrium established between living high molecular weight ($DP \sim 10^3$) α-methylstyrene and its monomer. For their studies they chose α-trideuteromethyl-β,β-dideutero-styrene, α-methyl-2,3,4,5,6-pentadeutero-styrene, and perdeutero-styrene. The equilibrium studies were performed in tetrahydrofuran at temperatures ranging from -20 to $+35°C$. The equilibrium constants were determined in the usual way, viz., $K = 1/M_e$.

The results are presented graphically in Figure 15 as plots of $-\Delta F_{ss}/RT$ versus $1/T$. The lower line refers to the ordinary nondeuterated α-methylstyrene, and the results obtained with α-trideuteromethyl-β,β-dideutero-

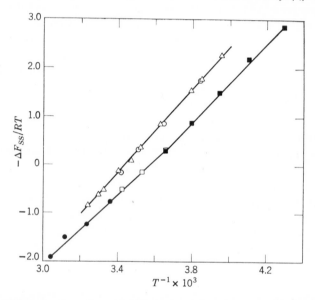

Fig. 15.   Polymerization of α-methylstyrenes. Free energy function versus reciprocal temperature. (△) α-Trideuteromethyl-β,β-dideuterostyrene, (○) perdeutero-α-methyl styrene, (□) 2,3,4,5,6-pentadeutero-α-methylstyrene, (●) α-methylstyrene (3), (■) α-methylstyrene (4).

styrene and with perdeutero-α-methylstyrene lie on the upper line. α-Methylstyrene deuterated in the ring behaved like nondeuterated styrene (see Figure 15).

Inspection of Figure 15 demonstrates that deuteration of the methyl group, but not of the ring, increases the equilibrium constant. Moreover, the exothermicity of polymerization seems to be higher for the monomer deuterated in the methyl group than for the nondeuterated α-methylstyrene. The exothermicity of α-trideuteromethyl-β,β-dideuterostyrene was determined to be -8.6 ± 0.2 kcal/mole as compared with 7.0 kcal/mole found for the ordinary α-methylstyrene. However, the increase in the exothermicity of polymerization of the deuterated monomer is compensated by a decrease in the entropy of polymerization, $\Delta S = -29.6$ eu. For the nondeuterated monomer the entropy of polymerization is only -24.8 eu.

14. EQUILIBRIUM COPOLYMERIZATION

Copolymerization of monomers A and B involves steps

$$\sim\text{A} + \text{B} \underset{k_{-12}}{\overset{k_{12}}{\rightleftarrows}} \sim\text{AB}$$

and

$$\sim\text{B} + \text{A} \underset{k_{-21}}{\overset{k_{21}}{\rightleftarrows}} \sim\text{BA}$$

in addition to the homopropagation of A and B, respectively. For some systems it has been possible to determine directly the absolute rate constants of copolymerization (43–45) and in two cases even the rate constant of depropagation (46,47). The addition of α-methylstyrene to living poly-1-vinylnaphthalene was shown to proceed reversibly (46) and under those experimental conditions the polymerization of further units of α-methylstyrene was virtually prevented. The reaction was studied in a stirred-flow reactor using radioactive monomer to simplify the analysis of the residual monomer. The method, developed in studies of the reaction α-methylstyrene dimer + monomer $\rightleftarrows$ trimer (23) (see p. 123), led to the evaluation of k_{12}, k_{-12}, and K_{12}.

The additon of styrene (S) to a polymer possessing the $\text{CH}_2 \cdot \text{C(Ph)}_2^-,\text{Na}^+$ end group was also investigated in a stirred-flow reactor (47). The process involves two steps

$$\sim\text{C(Ph)}_2^-,\text{Na}^+ + S \underset{k_{-21}}{\overset{k_{21}}{\rightleftarrows}} \sim\text{C(Ph)}_2 \cdot S^-,\text{Na}^+, \qquad K_{21}$$

and

$$\sim\text{C(Ph)}_2 \cdot S^-,\text{Na}^+ + S \overset{k_{11}}{\longrightarrow} \sim\text{C(Ph)}_2 \cdot S \cdot S^-,\text{Na}^+, \text{etc.}$$

The balance equations led to the relation

$$\{(S_0 - S_t)/S_t^3 \cdot D_t^-\}\{1 + (k_{11}/k_{-21})S_t\} = K_{21}k_{11}^2 t^2,$$

where t denotes the residence time, S_0 the initial concentration of styrene, and S_t and D_t^- the concentration of styrene and of $\sim\text{C(Ph)}_2^-,\text{Na}^+$, respectively, in the stirred reactor. From this relation the value of K_{21} was found to be about 5×10^{-2} liter/mole.

In a system containing living polymers, the propagation and depropagation steps continue until an equilibrium is established with respect to *all* components. Therefore, for a living copolymer of A and B, the ultimate distribution of the comonomers in the chain is given by the equilibrium determined by the partition functions of the AA, BB, and AB linkages. Calculation of the ultimate equilibrium distribution has been recently reported by Alfrey and Tobolsky (48) who pointed out that for an infinitely long polymer, the mathematical formalism of the problem is identical with that developed by Ising (49) in his treatment of ferromagnetism.

Denoting by N_A and N_B the number of A and B units and by M_{AB} the number of AB linkages, one finds that

$$(N_A - M_{AB})(N_B - M_{AB})/(M_{AB})^2 = (f_{AA} \cdot f_{BB}/f_{AB}) \exp(\Delta E_{AB}/RT)$$

f_{AA}, f_{BB}, and f_{AB} denote the partition functions of the respective linkages and $\Delta E_{AB} = 2E_{AB} - E_{AA} - E_{BB}$ where E's are the energies of the respective bonds.

The distribution of sequence lengths is given by the equations

$$n_x^A = N_A p_{AA}^{x-1}(1 - p_{AA})^2$$
$$n_x^B = N_B p_{BB}^{x-1}(1 - p_{BB})^2$$

where n_x^A and n_x^B are the numbers of A or B segments having length x, and p_{AA} and p_{BB} are defined by

$$p_{AA} = (N_A - M_{AB})/N_A \quad \text{and} \quad p_{BB} = (N_B - M_{AB})/N_B.$$

It should be stressed that the initial copolymer formed by the addition of two monomers to an ionic initiator is expected to be a block polymer because in ionic polymerization it is common to find k_{AA} and k_{BA}, i.e., the rate constants of A addition to units terminated by A or B respectively, to be much greater than either k_{BB} or k_{AB}. This rule is not perfectly general and exceptions are known. The redistribution of monomers to form the ultimate equilibrium copolymer would be usually a slow process, see, e.g., Section 8, p. 132, and probably it will not be achieved in any practical experiment.

References

(1) F. S. Dainton and K. J. Ivin, *Nature*, **162**, 705 (1948).
(2) F. S. Dainton and K. J. Ivin, *Quart. Rev.* (*London*), **12**, 61 (1958).
(3) F. S. Dainton and K. J. Ivin, *Discussions Faraday Soc.*, **14**, 199 (1953).
(4) R. E. Cook, F. S. Dainton, and K. J. Ivin, *J. Polymer Sci.*, **26**, 351 (1957).
(5) F. S. Dainton, J. Diper, K. J. Ivin, and D. R. Sheard, *Trans. Faraday Soc.*, **53**, 1269 (1957).
(6) K. J. Ivin, *ibid.*, **51**, 1273 (1955).
(7) S. Bywater, *ibid.*, p. 1267.
(8) A. V. Tobolsky, *J. Polymer Sci.*, **25**, 220 (1957); **31**, 126 (1958).
(9) A. V. Tobolsky and A. Eisenberg, *J. Am. Chem. Soc.*, **82**, 289 (1960).
(10) M. Szwarc, *Proc. Roy. Soc.* (*London*), *A*, **279**, 260 (1964).
(11) H. W. McCormick, *J. Polymer Sci.*, **25**, 488 (1957).

(12) D. J. Worsfold and S. Bywater, *ibid.*, **26**, 299 (1957).

(13) A. Vrancken, J. Smid, and M. Szwarc, *Trans. Faraday Soc.*, **58**, 2036 (1962).

(14) S. Bywater and D. J. Worsfold, *J. Polymer Sci.*, **58**, 571 (1962).

(15) H. Meerwein, D. Delfs, and H. Morschel, *Angew. Chem.*, **72**, 927 (1960).

(16) C. E. H. Bawn, R. M. Bell, and A. Ledwith, *Polymer (London)*, **6**, 95 (1965). Communicated at the Anniversary Meeting of the Chemical Society, Cardiff, 1963.

(17) D. Vofsi and A. V. Tobolsky, *J. Polymer Sci.*, **3A**, 3261 (1965).

(18a) M. P. Dreyfuss and P. Dreyfuss, *Polymer (London)*, **6**, 93 (1965). (b) M. P. Dreyfuss and P. Dreyfuss, *J. Polymer Sci.*, **4**, 2179 (1966). (c) C. E. H. Bawn, C. Fitzsimmons, and A. Ledwith, *Proc. Chem. Soc.*, **1964**, 391. (d) D. Sims, *J. Chem. Soc. (London)*, **1964**, 864. (e) C. E. H. Bawn, R. M. Bell, C. Fitzsimmons, and A. Ledwith, *Polymer*, **6**, 661 (1965).

(19) C. E. Frank et al., *J. Org. Chem.*, **26**, 307 (1961).

(20) C. L. Lee, J. Smid, and M. Szwarc, *J. Phys. Chem.*, **66**, 904 (1962).

(21) E. Bergmann, H. Taubadel, and H. Weiss, *Chem. Ber.*, **64**, 1493 (1931).

(22) A. Vrancken, J. Smid, and M. Szwarc, *J. Am. Chem. Soc.*, **83**, 2772 (1961).

(23) C. L. Lee, J. Smid, and M. Szwarc, *ibid.*, **85**, 912 (1963).

(24a) K. G. Denbigh, *Trans. Faraday Soc.*, **40**, 352 (1944). (b) B. Stead, F. M. Page, and K. G. Denbigh, *Discussions Faraday Soc.*, **2**, 263 (1947).

(25) W. B. Brown and M. Szwarc, *Trans. Faraday Soc.*, **54**, 416 (1958).

(26) P. J. Flory, "Principles of Polymer Chemistry," Cornell University Press, Ithaca, N.Y., 1953; (a) p. 318; (b) p. 336.

(27) J. J. Hermans, *Makromol. Chem.*, **87**, 21 (1965).

(28) A. V. Tobolsky and A. Eisenberg, *J. Colloid. Sci.*, **17**, 49 (1962).

(29) F. S. Dainton and R. H. Tomlinson, *J. Chem. Soc.*, **1953**, 151.

(30) G. Gee, *Trans. Faraday Soc.*, **48**, 515 (1952).

(31) A. V. Tobolsky and A. Eisenberg, *J. Am. Chem. Soc.*, **81**, 780 (1959).

(32) A. Eisenberg, *J. Polymer Sci. B.*, **1**, 33 (1963).

(33) R. L. Scott, *J. Chem. Phys.*, **17**, 268 (1949).

(34) S. Bywater, *Makromol. Chem.*, **52**, 120 (1962).

(35) A. V. Tobolsky, A. Rembaum, and A. Eisenberg, *J. Polymer Sci.*, **45**, 347 (1960).

(36) K. J. Ivin and J. Leonard, *Polymer (London)*, **6**, 621 (1965).

(37a) R. E. Cook, K. J. Ivin and J. H. O'Donnell, *Trans. Faraday Soc.*, **61**, 1887 (1965). (b) P. B. Ayscough, K. J. Ivin, and J. H. O'Donnell, *ibid.*, **61**, 1601 (1965).

(38) A. Wada, *J. Chem. Phys.*, **22**, 198 (1954).

(39) K. J. Ivin and J. H. O'Donnell, *Trans. Faraday Soc.*, in press.

(40) J. G. Kilroe and K. E. Weale, *J. Chem. Soc.*, **1960**, 3849.

(41) K. J. Ivin, *Pure Appl. Chem.*, **4**, 271 (1962).

(42) K. E. Weale, *Quart. Rev. (London)*, **16**, 267 (1962).

(43) D. N. Bhattacharyya, C. L. Lee, J. Smid, and M. Szwarc, *J. Am. Chem. Soc.*, **85**, 533 (1963).

(44) M. Shima, D. N. Bhattacharyya, J. Smid, and M. Szwarc, *ibid.*, **85**, 1306 (1963).

(45) M. Shima, J. Smid, and M. Szwarc, *J. Polymer Sci. B.*, **2**, 735 (1964).

(46) J. Stearne, J. Smid, and M. Szwarc, *Trans. Faraday Soc.*, **60**, 2054 (1964).

(47) E. Ureta, J. Smid, and M. Szwarc, *J. Polymer Sci. A.*, **4**, 2219 (1966).

(48) T. Alfrey and A. V. Tobolsky, *ibid.*, **38**, 269 (1959).

(49) E. Ising, *Z. Physik.*, **31**, 253 (1925).

(50) J. A. Poulis, C. H. Massen, and P. van der Leeden, *Trans. Faraday Soc.*, **58**, 474 (1962).

(51) J. A. Poulis and W. Derbyshire, *ibid.*, **59**, 559 (1963).

(52) J. A. Poulis, C. H. Massen, A. Eisenberg, and A. V. Tobolsky, *J. Am. Chem. Soc.*, **87**, 413 (1965).

(53) H. A. Skinner and M. Szwarc, to be published.

(54) L. J. Fetters, W. J. Pummer, and L. A. Wall, *J. Polymer Sci. A*, **4**, 3003 (1966).

(55) L. S. Bartell, *Tetrahedron*, **17**, 177 (1962); *Iowa State J. Sci.*, **36**, 137 (1961).

(56) P. Dreyfuss and M. P. Dreyfuss, *Advan. Polymer Sci.*, **4**, 528 (1967).

(57) P. H. Plesch and P. H. Westermann.

(58) P. H. Plesch.

(59) W. Kern, H. Deibig, A. Grefer, and V. Jaacks, *Pure Appl. Chem.*, **12**, 371 (1966).

(60) V. Jaacks, *Makromol. Chem.*, **99**, 300 (1966).

(61) S. Iwabuchi, V. Jaacks, and W. Kern, *Makromol Chem.*, **100**, 276 (1967).

(62) J. C. Bevington, *Quart. Rev.*, **6**, 141 (1952).

(63) C. L. Lee and O. K. Johannson, *J. Polymer Sci. A-1*, **4**, 3013 (1966).

(64) M. N. Berger, J. J. K. Boulton, B. W. Brooks, and M. J. Evans, *Chem. Commun.*, **1967**, 8.

(65) R. L. Williams and D. H. Richards, *Chem. Commun.*, **1967**, 414.

(66) A. Miyake and W. H. Stockmayer, *Makromol.Chem.*, **88**, 90 (1965).

Chapter IV

Experimental Techniques in Kinetic Studies of Anionic Polymerization

1. REACTION VESSELS

Any experimental work involving living polymers, and particularly kinetic studies of their formation and growth, requires meticulous cleaning of the apparatus used in such investigations. It is desirable to carry out the experiments on a high-vacuum line, to eliminate, as far as possible, greased stopcocks and replace them by breakseals and constrictions, and to destroy by some suitable means all damaging contaminants adsorbed on the walls of the apparatus. The growing ends and the initiators of anionic polymerization are extremely reactive and are readily deactivated by moisture, oxygen, carbon dioxide, proton-donor impurities, etc. In many studies, it is necessary to work with highly diluted reagents; in fact, in some investigations one deals with 10^{-6} or even $10^{-7}M$ solutions of organometallic compounds. In such a case, a trace of impurities which is insignificant in other studies may destroy all the reagent available in the system.

Consequently, the conventional "flaming in vacuum" technique is not sufficient for precise experiments, and it is necessary to wash the whole setup with a concentrated solution of the investigated organometallic compound (1–3). Such a "cleaning" solution is kept in an evacuated ampoule equipped with a breakseal and fused to the apparatus used. The latter is evacuated, thoroughly flamed, if possible, and then sealed off the vacuum line. Thereafter, the breakseal is crushed and the whole unit is rinsed with the "cleaning" solution which is allowed to stand for a while in contact with the walls. Eventually, by proper tilting of the apparatus, the solution is returned to the original ampoule and the walls of the unit are then washed with pure solvent. This is achieved by chilling the walls from the outside, thus causing condensation of the solvent at the desired place. The latter being evaporated from the "cleaning" solution is free of any damaging impurities. Finally, the contents of the ampoule are frozen with liquid nitrogen and the ampoule is sealed off from the unit. The apparatus thus cleaned may be used for studies of the most diluted solutions.

One may avoid the above procedure by introducing an excess of the organometallic compound, thus hoping that it will destroy the damaging impurities present in the system. Such a method, albeit permissible in some studies, cannot be recommended. Destruction of the organometallic compound produces other impurities which may affect profoundly the experimental results. For example, the resulting salts or oxides may complex with the living polymers and thus affect their reactivities and even their state of aggregation. There is ample evidence that such interactions have vitiated many studies and invalidated quite a few data reported in the literature. The sensitivity of some systems to impurities is illustrated by the discussion given in the Appendix of Slates' thesis (34) and in some other recent publications (35,36). It is obvious that decomposition products may contribute to the conductivity of solutions of organometallic compounds and, therefore, their presence is highly undesirable in conductance studies.

Some investigators recommend coating of the inner glass walls of the reactor by hydrophobic —OSiMe$_3$ groups which are formed by exposing the vessel to the vapor of Me$_3$SiCl. This treatment has to be followed by prolonged evacuation in a high vacuum to secure the removal of the excess Me$_3$SiCl and the HCl formed. Although such a procedure is occasionally useful, it cannot be recommended for studies of anionic polymerization because alkalis and organometallic compounds may react with the silanes.

Greased stopcocks must not be used in contact with the reagents or solvents, although they may be tolerated in lines connecting the apparatus with a high-vacuum pump or in those permitting the admission of gases needed for pressurizing the reagents. But even then, it is advisable to use Teflon stopcocks, because truly vacuum-tight types are now available. Teflon stopcocks may also be used for introducing the reagents; they are slowly attacked by most organometallic compounds and then become coated with a protective, black layer which prevents further reaction. All-metal stopcocks are less useful—the oxide layer covering their surface may destroy some organometallic compounds.

The use of hypodermic syringes and self-sealed rubber gaskets is widely accepted in industrial kinetic research. This technique offers flexibility and is expedient. It may be recommended for removal of samples needed for analysis, but, in this writer's opinion, it should not be applied in precise kinetic studies, especially if the investigation is performed with $10^{-4}M$, or even more dilute, solutions.

2. PURIFICATION OF SOLVENTS AND MONOMERS AND PREPARATION OF INITIATORS

Various purification methods for solvents and monomers can be found in the original papers [see, e.g., refs. 1–3].

It is recommended to store the purified and fractionated solvent in an evacuated flask attached to a high-vacuum line. The solvent should be constantly in contact with a sodium-potassium alloy, and it is desirable to add to it a small amount of benzophenone or anthracene, because the resulting organometallic compound reacts more efficiently with impurities than the alkali metal. Moreover, the appearance of the characteristic color serves as a sort of indicator demonstrating the absence of damaging impurities. The solvent is distilled directly, whenever needed, into the required apparatus, the distillation being carried out on a high-vacuum line.

Final purification of monomers involves contact with reagents which destroy moisture and other undesirable impurities. Calcium hydride is frequently used. A liquid monomer forms with the hydride a slurry which is stirred magnetically for 24 hr in an evacuated container. The hydrogen evolved during the reaction peels off the resulting crust of calcium hydroxide and thus a fresh hydride surface is available for the drying process. The gas evolution also maintains the porosity of the hydride. Eventually the monomer is distilled under high vacuum into the appropriate ampoules which are stored in a deep-freezer. If necessary, it is diluted with a suitable solvent to prevent its spontaneous polymerization.

Some monomers may be dried with alkali metals, because most of them react only slowly with the metal in the absence of polar solvents. It should be remembered, however, that the surface of the alkali metal may be quickly coated with polymer, or some other product of reaction, and this prevents its further action. Therefore, the metal has to be freshly distilled, and whenever possible liquid sodium-potassium alloy should be employed instead of a solid metallic mirror. The purified monomer is distilled, under vacuum, into ampoules; any polymer or oligomer formed in the reactor remains there. A loss of even 30% of monomer is of no importance in academic studies.

Occasionally, one may use for the final purification an organometallic compound which does not initiate the polymerization of the investigated monomer. For example, in the purification of styrene or its derivatives sodium fluorenyl may be used to destroy the damaging impurities.

Recently special drying agents which also remove traces of oxygen have become commercially available; for example, the BTS catalyst produced by Badische Anilin und Soda Fabrik. These are particularly useful for drying gaseous monomers. For purification of gaseous or volatile monomers, molecular sieves may also be useful, providing they do not induce their polymerization (4).

Solid monomers, if stable at their melting temperature, may be purified by zone refining. This is a most promising technique and it may be desirable to develop it even for monomers which are liquid at room temperature.

For most purposes high-vacuum sublimation of recrystallized material seems to be adequate.

The purity of solvents and monomers should be checked by a sensitive gas chromatograph, e.g., using a hydrogen flame detector. This permits detection of impurities down to 0.01%.*

Absolutely pure materials do not exist. A material is considered to be pure when the level of impurities, which may affect the experimental results, is too low to produce a detectable effect. The ultimate test for purity is therefore functional; if different methods of preparation and purification lead to identical results then these may be considered genuine and characteristic for the "pure" material.

To illustrate the difficulties encountered in some purifications, let us consider the problem of the purification of 1,1-diphenylethylene (5). This hydrocarbon reacts with alkali metals giving the dianion, $^-C(Ph)_2 \cdot CH_2 \cdot CH_2 \cdot C(Ph)_2^-$, which is in equilibrium with minute amounts of its parent $CH_2 : C(Ph)_2^{\cdot -}$ radical-ions. ESR studies of the dianion solution showed signals which eventually were identified as those of ketyl, $Ph_2C{=}O^{\cdot -}$, and of the radical-anion of biphenyl, $Ph \cdot Ph^{\cdot -}$. They arose from the presence of traces of biphenyl and of benzophenone both often contaminating even the highly purified samples. The benzophenone is formed as a result of the spontaneous oxidation of 1,1-diphenylethylene; even short exposure of the hydrocarbon to air is sufficient to produce 1 part per million of the ketone— enough to give a discernible ESR signal in the presence of alkali metal. To remove the ketone, one reacts the purified hydrocarbon with its dimeric dianion which converts the ketone into ketyl. Thereafter, the hydrocarbon is distilled off under high vacuum, at room temperature, into a fresh solution of the dianion. This procedure, repeated four times, gives 1,1-diphenylethylene free of benzophenone, i.e., a $0.1M$ solution of its dimeric dianion shows no ESR signal in the most sensitive ESR spectrometer. Of course, the purified samples should be kept under vacuum and never be exposed to the atmosphere.

No method was found to remove the traces of biphenyl. This impurity is introduced in the course of preparation of the commercial 1,1-diphenylethylene. The conventional synthesis involves an intermediate alcohol which is formed by reaction of PhMgBr with ethyl benzoate, and the respective Grignard reagent is always contaminated with biphenyl. Therefore, the hydrocarbon had to be prepared by reaction of benzophenone with CH_3MgI, and then freed of benzophenone by the above described procedure. Such 1,1-diphenylethylene was found to be truly "pure."

---

* This method fails if the impurity has a retention time close to that of the substance analyzed. Fortunately, the undesirable impurities often are more volatile than the monomer or solvent.

The trace of benzophenone present in the supposedly pure 1,1-diphenyl-ethylene led to an unusual observation which caused much speculation in the literature. Leftin and Hall (6) described spectacular color changes when 1,1-diphenylethylene reacted with sodium or potassium in tetrahydrofuran. With excess hydrocarbon the solution is blue, adding an excess of alkali metal changes the color to red; blue appears again when an excess of hydrocarbon is added to the red solution, etc. However, it was admitted that some samples, presumably more carefully purified, did not display these intriguing color changes. In spite of this remark the authors implied that the blue species is the radical-anion of 1,1-diphenylethylene. These phenomena were observed earlier in our laboratory and proved to be caused by traces of benzophenone. The blue ketyl is present with an excess of hydrocarbon; excess alkali converts all of the hydrocarbon into the red dimeric-dianion which masks the blue. A large excess of the hydrocarbon, i.e., the addition of more ketone, intensifies the blue color, etc.

Another example of an unfortunate mistake, arising from the presence of biphenyl impurity in 1,1-diphenylethylene, is found in a recent paper by Evans (37) who reported kinetic studies of the dimerization of radical-anions of diphenyl ethylene. The radical-ion was claimed to be formed from its parent hydrocarbon by reduction with potassium. Its spectrum was given in the paper and on closer examination one finds it to be identical with that of the radical-anion of biphenyl. Hence, the authors followed destruction of biphenyl $^-$ and not dimerization of radical-ions of 1,1-diphenylethylene.

Alkali metals used in the preparation of initiators should be vacuum distilled or even redistilled. Commercial sodium or potassium metals are acceptable starting materials. Unfortunately, metallic lithium cannot be heated in a Pyrex vessel, because it reacts with glass causing its fracture. Special steel vessels have to be used in such a distillation. Commercial lithium metal is sufficiently pure for most purposes. It contains, however, ca. 1% of sodium and unpublished reports claimed that some kinetic results are profoundly affected by the sodium contamination. Careful checking of this point would be desirable. Lithium metal may be, perhaps, prepared by vacuum decomposition of lithium azide or by evaporation of ammonia from purified solutions of metallic lithium in liquid ammonia. The former approach is most advantageous if one wishes to introduce into a reactor an *exact* amount of pure alkali metal. For example, the desired amount of sodium, or cesium, azide may be weighed, introduced into a reactor, and then quantitatively decomposed in high vacuum by gentle heating. The latter technique, deposition of metal from its liquid ammonia solution, may again be used in quantitative work, however, one may always question whether the deposit contains some amide and the completeness of the ammonia removal. Metallic rubidium and cesium are prepared most conveniently *in situ* by reacting the respective chlorides with metallic calcium or alternatively by pyrolyzing in high vacuum the respective azides.

Potassium may be advantageously used in the form of a liquid sodium-potassium alloy. The prepared organometallic compound contains less than 1% of sodium with respect to potassium, even if the proportion of the former in the alloy exceeds 50%.

Colloidal dispersions of lithium or sodium, although most useful for some synthetic work, are not recommended for kinetic studies; many of the additives present in these materials affect the reaction.

Organolithium initiators such as butyllithium and ethyllithium are available commercially. It seems that commercial butyllithium, supplied by the Foot Co., may be used without further purification, if the original container was not open to the air. Purification of butyllithium is difficult because of its low volatility and its easy decomposition at elevated temperatures hampers distillation. Sinn (7) recommends a high-vacuum distillation in which the distilled material flows through a heated trough. In this way one avoids accumulation in the heated liquid of decomposition products which seem to be responsible for its further auto-catalytic decomposition.

Ethyllithium is a solid compound, and the commercial material may be easily recrystallized from benzene. A useful crystallization procedure is described by Morton et al. (8). The properties of the resulting living polymer are apparently not affected by the nature of the alkyl group present in the initiator, provided that the initiation step used up all of the initiator or that its excess has been destroyed. Therefore, in precise kinetic studies recrystallized ethyllithium may be preferred to the butyllithium.

Two additional organolithium compounds merit mentioning. Fluroenyl-lithium has been used in kinetic studies of methyl methacrylate polymerization (9). The presence of the fluorenyl moiety in the eventually formed polymeric chain is easily demonstrated by its ultraviolet absorption. Salts of the fluorenyl carbanion do not initiate the polymerization of styrene (10). This compound is most conveniently prepared from fluorene and ethyl- or butyllithium (9c). Reaction of the hydrocarbon with the metal may lead to complications arising from the formation of a radical-anion.

1,1-Diphenylhexyllithium is prepared by adding butyllithium to 1,1-diphenylethylene (11). It initiates polymerization slowly, but its use frequently precludes side reactions encountered with other initiators.

Preparation of living polymers possessing the sodium counterion and only one growing end per chain is most conveniently achieved by initiating the reaction with benzylsodium (12). Commercially available benzylsodium appears to be unsatisfactory and the compound has to be prepared from dibenzylmercury and metallic sodium. The reaction proceeds readily in tetrahydrofuran solution at $-20°C$ and within a few hours about 70%

conversion may be attained. At room temperature, some unknown reaction converts benzylsodium into another organosodium compound (13) which shows an absorption maximum at 485–490 mμ, whereas pure benzylsodium absorbs at $\lambda_{max} = 350$ mμ. Upon standing, some mercury deposits on the walls of ampoules containing the filtered solution. The sodium compounds probably react with the unconverted dibenzylmercury, i.e.,

$$Ph \cdot CH_2^-, Na^+ + Hg(CH_2 \cdot Ph)_2 \rightarrow Hg + Ph \cdot CH_2 \cdot CH_2 \cdot Ph + Ph \cdot CH_2^-, Na^+.$$

Benzylsodium is a slow initiator of styrene polymerization, while the α-methylstyrene analog initiates rapidly. Therefore, it is imperative to convert $Ph \cdot CH_2^-, Na^+$ into $Ph \cdot CH_2 \cdot CH_2 \cdot C(Me)Ph^-, Na^+$ by adding to the cold solution a small excess of α-methylstyrene. It may be remarked that the reaction of dibenzylmercury with metallic lithium proceeds smoothly in tetrahydrofuran giving benzyllithium, a concentrated solution of which is stable at room temperature.

Preparation of diphenylmethylsodium and of triphenylmethylsodium may be accomplished by reacting the respective chlorides in tetrahydrofuran with metallic sodium. The reaction proceeds smoothly and the resulting products show no peculiarities; however, these are slow initiators.

Cumylpotassium, prepared by Ziegler's method (14) from methyl cumyl ether and potassium metal, is an excellent initiator giving living polymers, having the potassium counterion and possessing only one growing end per chain. This initiator is stable and shows no changes on storage at room temperature. After preparation its solution should be chilled to about $-70°$C and then filtered or decanted from precipitated potassium methoxide.

Cumylcesium is a suitable initiator to produce living polymers having one growing end per chain associated with the Cs$^+$ counterion. Its preparation is similar to that of cumylpotassium; however, care must be taken not to expose the reagents for more than 6–8 hr to the action of metallic cesium, because side reactions then take place, which are revealed by the appearance of new absorption peaks. The mixture should be cooled to $-80°$C for 2 hr before decanting the clear solution through a glass-sintered plate. The cooling enhances the coagulation of the precipitated cesium methoxide. The conversion is about 90%.

It should be stressed that methyl cumyl ether must be fractionated just before being reacted with the metal. The ether is intrinsically unstable, and on standing, even in a freezer, it spontaneously decomposes into α-methylstyrene and methanol. In the course of preparation of cumyl potassium or cesium, α-methylstyrene dimerizes and forms the dianions which in turn yield living polymers endowed with two growing end groups.

Frequently, it is advantageous to prepare a living polystyrene possessing the desirable counterion and then to add the monomer being investigated, thus producing the required living polymer in the form of a block polymer. For slow initiators, it is advisable to use α-methylstyrene instead of styrene because the low ceiling temperature of its propagation step permits a complete conversion of the initiator into $\sim\!CH_2 \cdot C(Me)(Ph)^-, M^+$ even with a relatively small excess of monomer. Before adding the monomer being investigated, the solution is chilled to -40 or $-50°C$ for a few hours to complete the polymerization of residual α-methylstyrene.

Living polymer possessing counterions of one kind may be converted into a polymer possessing another counterion by applying an exchange technique. For example, by adding sodium tetraphenylboron to living polystyrene having K^+ counterion, insoluble K^+, BPh_4^- is precipitated, and $\sim\!S^-, K^+$ converted into $\sim\!S^-, Na^+$. It is difficult, however, to carry out the process stoichiometrically and therefore the resulting solution usually contains some excess $Na^+BPh_4^-$.

Polymers possessing two growing ends are prepared by electron-transfer initiation. This reaction may involve as the electron-donor an alkali metal or a suitable organometallic compound such as sodium naphthalene or biphenyllithium (see Chap. VI). It is often essential to prepare a living polymer of low molecular weight and then the setup described in Figure 1 may be useful. Tetrahydrofuran (THF) is placed in reservoir B and the monomer in bulb C. After chilling B and C with solid carbon dioxide, the apparatus is evacuated and the alkali metal deposited on the inner surface of the cold finger A. Thereafter, the unit is sealed off at the constriction. The cold finger A is filled with a solid carbon dioxide–methanol mixture and the content of B is gently heated to about $25°C$. Tetrahydrofuran then begins to circulate through the system, being evaporated in B, condensed on A, and returned to B through the narrow tube linking A with B. If potassium or cesium is deposited on A the blue color of the alkali metal solution appears soon in B. The bulb C is then heated to the temperature required to maintain the desired partial pressure of the monomer in the stream of THF vapor.

This device maintains a very low concentration of monomer in the condensed liquid which is in contact with the metal. However, the final concentration of the living polymer retained in B may be quite high. Using this procedure, a living trimer of styrene or a living dimer of α-methylstyrene can be easily prepared.

The device described above is also useful for preparing organometallic compounds when prolonged contact of metal with the substrate is undesirable.

Most of the organometallic compounds are unstable. Therefore, it is advisable to prepare them just when needed and to perform the experiment not later than 24 hr after preparation. If necessary, they may be stored in a deep-freezer, or preferentially in a solid carbon dioxide bath, for a few days.

The formation of hydrides is a quite common side reaction which leads to an undesirable product. For example, stored sodium naphthalene is converted slowly into sodium hydride and naphthylsodium. The presence of a hydride may be easily detected by the evolution of hydrogen gas formed upon the addition of water to the solution of organometallic compound.

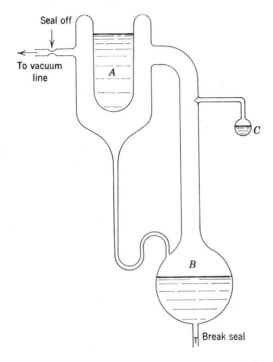

Fig. 1.   Apparatus for preparation of extremely low molecular weight living polymers (dimers or trimers).

Finally, it should be pointed out that living polymers may be more readily prepared in one solvent than in another, e.g., tetrahydrofuran is often the most convenient solvent for such preparations. However, if the study requires use of another solvent, the following procedure may be adopted. The living polymer is prepared in tetrahydrofuran which is then evaporated under a high vacuum and the desirable solvent is distilled into the dry mass. The product is dissolved and then the latter solvent is also evaporated and a fresh amount of it is then distilled in. This operation removes traces of the first solvent, if any were left, in the prepared sample (4). Sometimes solvation by THF is so strong that one or two molecules remain permanently bound to the organometallic compound. To check whether this is the case one removes an aliquot of the solution, destroys the organometallic compound (e.g., by adding a few drops of water), and then analyzes the resulting liquid by VPC for THF.   A negative result proves that *all* THF has been removed.

By using a solvent of high melting point, e.g., dioxane or benzene, one may freeze out the solution and then freeze-dry the living polymer. This method provides a powdery, dry material which often is more stable than its solution and therefore useful for storage. The powdery material may be dissolved in the desired solvent whenever needed.

3. PROPERTIES OF SOME SOLVENTS USED IN KINETIC STUDIES OF ANIONIC POLYMERIZATION

In many mechanistic studies it is necessary to have a detailed knowledge of the density, viscosity, and dielectric constant of the solvent over a wide range of temperatures. The available data are collected and presented in Tables I–X. Comments pertaining to specific solvents are given at the bottom of each table.

TABLE I

Physical Properties of Dimethoxyethane (DME) (b.p. $= 87$–$88°C$)

$T, °C$	Density, g/cc	$\eta \times 10^3$, poise	Dielectric constant, D
25	0.859	4.55	7.20
10	0.874	5.30	7.60
5	0.879	5.55	7.75
0	0.883	6.10	8.00
-10	0.893	6.70	8.45
-20	0.903	7.80	8.85
-25	0.908	8.60	9.05
-30	0.913	9.30	9.30
-40	0.923	11.30	9.85
-50	0.932	13.80	10.45
-60	0.942	16.9	11.05
-70	0.952	21.0	11.75
-75	0.957	—	12.15

$\partial \ln D / \partial \ln T = -1.28$; $D = -2.83 + 2950/T$; $\log \eta = -3.773 + 425/T$; $d \ln v/dT = 0.00114$; $v =$ molar volume of the liquid.
From C. Carvajal, K. J. Tölle, J. Smid, and M. Szwarc, *J. Am. Chem. Soc.*, **87**, 5548 (1965).

TABLE II

Physical Properties of Tetrahydrofuran (THF) (b.p. $= 65.66°C$)

$T,°C$	Density, g/cc	$\eta \times 10^3$, poise	Dielectric constant, D
25	0.880	4.61	7.39
10	0.894	5.42	7.88
0	0.904	6.08	8.23
−10	0.914	6.90	8.60
−20	0.924	7.91	9.00
−30	0.934	9.16	9.43
−40	0.945	10.75	9.91
−50	0.955	12.8	10.43
−60	0.966	15.5	10.98
−70	0.978	19.1	11.58

$\partial \ln D/\partial \ln T = -1.16$; $D = -1.495 + 2660/T$; $\log \eta = -3.655 + 393/T$; $d \ln v/dT = 0.001085$; $v =$ molar volume of the liquid.
From C. Carvajal, K. J. Tölle, J. Smid, and M. Szwarc, *J. Am. Chem. Soc.*, **87**, 5548 (1965); confirmed by D. J. Metz and A. Glines, *J. Phys. Chem.*, **71**, 1158 (1967).

TABLE III

Physical Properties of 2-Methyltetrahydrofuran (MeTHF) (b.p. $= 78°C$)

$T,°C$	Density, g/cc	$\eta \times 10^3$, poise	Dielectric constant, D
25	0.848	4.57	6.24
10	0.862	5.36	6.63
0	0.871	6.01	6.92
−10	0.880	6.80	7.22
−20	0.889	7.77	7.55
−30	0.898	8.98	7.91
−40	0.908	10.51	8.30
−50	0.917	12.47	8.72
−60	0.926	15.04	9.19
−70	0.935	18.47	9.70
−75	0.940	20.63	9.97

$\partial \ln D/\partial \ln T = -1.125$; $D = -1.14 + 2200/T$; $\log \eta = -3.635 + 386/T$; $d \ln v/dT = \alpha = 0.00109$.

TABLE IV

Physical Properties of Tetrahydropyran (THP) (b.p. $= 86°C$)

$T,°C$	Density, g/cc	$\eta \times 10^3$, poise	Dielectric constant, D
25	0.878	7.64	5.61
20	0.883	8.26	5.71
10	0.894	9.73	5.90
0	0.904	11.61	6.12
-10	0.910	14.03	6.35
-20	0.919	17.2	6.59
-30	0.928	21.4	6.84
-40	0.938	27.3	7.12
-45	0.942	31.0	~ 7.26

A plot of $\ln D$ versus $\ln T$ is slightly curved. $(d \ln D/d \ln T) = -0.97$.

$D = 0.11 + 1640/T$; $\log \eta = -4.05 + 580/T$; $d \ln v/dT = 0.00108$; $v =$ molar volume of the liquid.

Unpublished data of C. Sutphen and M. Szwarc; confirmed by P. Siegwalt.

TABLE V

Physical Properties of Diethylether (b.p. $= 34.6°C$)

$T,°C$	Density, g/cc	$\eta \times 10^3$, poise	Dielectric constant, D
25	0.707	2.33	4.34
20	0.711	2.43	4.43
10	0.721	2.66	4.62
0	0.731	2.88	4.84
-10	0.741	3.21	5.09
-20	0.751	3.62	5.36
-30	0.761	4.10	5.68
-40	0.771	4.63	5.99
-50	0.781	5.08	6.39
-60	0.791	6.14	6.87
-70	0.801	7.41	7.42

$\partial \ln D/\partial \ln T = -1.33$. D is not linear with $1/T$, the curve is concave. $\log \eta = -3.74 + 330/T$; $d \ln V/dT = 0.00141$.

Unpublished data of C. Sutphen and M. Szwarc.

TABLE VI
Physical Properties of Tetrahydrofuran–Dioxane Mixtures at 25°C

Dioxane, vol. %	Density, g/cc	Dielectric constant, D	$\eta \times 10^3$, poise
0	0.880	7.29	4.60
10	0.895	6.76	4.86
23	0.910	6.10	5.24
33	0.923	5.62	5.60
47	0.947	4.81	6.41
68	0.975	3.82	7.78
100	1.028	2.21	11.96

From M. van Beylen, D. N. Bhattacharyya, J. Smid, and M. Szwarc, *J. Phys. Chem.*, **70**, 157 (1966).

TABLE VII
Physical Properties of Benzene–THF Mixtures at 20°C

THF, mole fraction	Dielectric constant, D	η, cp
0.0	2.283	0.649
0.0545	2.51	0.638
0.1371	2.83	0.624
0.1994	3.10	0.613
0.3060	3.57	0.596
0.4309	4.17	0.576
0.6033	—	0.549
0.6231	5.09	—
0.7386	5.84	0.530
0.8704	6.65	0.509
1.0000	7.587	0.491

From S. Bywater and D. J. Worsfold, *J. Phys. Chem.*, **70**, 162 (1966).

TABLE VIII
Dielectric Constant of Dioxane–Benzene Mixtures at 25°C

Mole % dioxane	Dielectric constant, D
100	2.21
75.3	2.23
61.1	2.245
59.6	2.246
53.0	2.250
45.3	2.255
41.2	2.260
25.5	2.270
0.0	2.274

Determined by W. P. Purcell and J. A. Singer, *J. Phys. Chem.*, **69**, 4097 (1965).

TABLE IX
Physical Properties of Pyridine (b.p. = 115°C)

T, °C	Density, g/cc	$\eta \times 10^3$, poise	Dielectric constant, D
−40	1.047	31.3	17.0
−30	1.032	24.5	16.2
−20	1.0225	19.6	15.5
−10	1.013	15.9	14.8
0	1.003	13.15	14.2
10	0.994	11.0	13.6
20	0.984	9.31	13.1
25	0.979	8.59	12.80
30	0.975	7.96	12.55

$D = 4530/T - 2.40$; $\log \eta = -4.079 + 600/T$; $d \ln v/dT = \alpha = 0.000978$.

Unpublished data of C. Sutphen and M. Szwarc.

TABLE X
Physical Properties of Methylene Chloride

T, °C	Density, g/cc	$\eta \times 10^3$, poise	Dielectric constant, D
25	1.317	4.58	9.01
15	1.335	4.98	9.39
5	1.352	5.44	9.81
−5	1.369	6.05	10.26
−15	1.378	6.71	10.74
−25	1.397	7.59	11.26
−35	1.403	8.61	11.82
−45	1.409	9.86	12.43
−55	1.452	11.5[a]	13.10
−65	1.474	13.6[a]	13.83
−75	1.491	16.3[a]	14.64
−85	1.509	19.9[a]	15.53

$\partial \ln D/\partial \ln T = -1.21$; $D = 3320/T - 2.13$.

$\log \eta = -3.430 + 325/T$.

$d \ln V/dT = 0.001178$.

S. O. Morgan and H. H. Lowry, *J. Phys. Chem.*, **34**, 2385 (1930). C. Sutphen and M. Szwarc, unpublished data.

4. DETERMINATION OF CONCENTRATIONS OF GROWING (LIVING) ENDS

Many quantitative studies of the kinetics of anionic polymerization have involved species possessing carbanions, e.g., $\sim CH(Ph)^-,Na^+$, as their growing ends. Therefore, special attention will be paid to the methods used for determination of the concentrations of these groups. Determination of the concentrations of other growing ends, such as $\sim O^-$, $\sim S^-$, $\sim NH^-$, or $\sim COO^-$, is relatively easy because those groups are reasonably stable being much less reactive than the carbanions. Ordinary analytical techniques are, therefore, applicable in their analysis.

Carbanions are very strong bases and, if, conjugated with aromatic or olefinic moieties, they absorb light in the near ultraviolet or in the visible portion of the spectrum. For example, $\sim CH(Ph)^-,Na^+$, the growing end of living polystyrene, has an intense absorption band at about 340 mμ with a tail extending into the visible region; light absorption due to this tail makes living polystyrene solution cherry-red.

Many techniques can be used to determine the concentration of living polystyrene and a detailed discussion of the various methods will serve to illustrate the numerous problems encountered in the quantitative analysis of this and of other similar living end groups.

The basic nature of carbanions suggests that titrimetric acid-base techniques may be useful in the determination of their concentration. For example, an aliquot of living polystyrene solution may be added to water, or aqueous tetrahydrofuran, and the liberated base, e.g., NaOH, titrated with a suitable acid. In practice, such an aliquot is sealed in a glass ampoule which, thereafter, is crushed under water. However, this method does not discriminate between carbanions (living ends) and other bases which may be present in the investigated solution. For example, moisture or traces of alcohol introduced with the reagents in the course of the preparation of living polymer, or water adsorbed on the walls of the reaction vessel, destroy a fraction of the living ends and produce an equivalent amount of base which then contributes to the titration. This erroneously increases the apparent concentration of living ends.

To determine the concentration of living polymer in a stock solution it is often necessary to withdraw aliquots (in suitable ampoules) each having a known volume and the *same* concentration as the bulk of the solution. This operation may be performed by using the device depicted in Figure 2. Ampoule A containing the prepared solution is sealed to a dividing vessel B having a breakseal S at the bottom and several side tubes (up to 8) on its periphery. Calibrated special ampoules C are connected to these side-arms, and if necessary two ampoules may be linked to each arm. The connecting tubes are also calibrated having a cross-section corresponding to about 0.1 cm². The dividing vessel, B, is then evacuated, flamed in high vacuum and thereafter sealed off from the line. The breakseal on ampoule A is crushed and its contents transferred quantitatively to B by washing ampoule A with

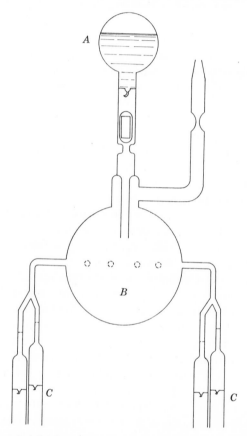

Fig. 2. Apparatus for dividing under vacuum a solution of a sensitive organometallic compound or living polymer into a series of aliquots, each having a known volume and all of the same concentration.

condensed solvent. This is done by cooling the washed part of the apparatus with a cloth soaked in a methanol-solid carbon dioxide mixture. The solvent then condenses on the chilled surface and dissolves the residue left on it. After freezing the solution, ampoule A is sealed off. By tilting bulb B, each ampoule is filled approximately to the mark scratched on the calibrated tube. The apparatus is then left for a while until, as a result of some condensation of the solvent on the walls, the narrow tubes are clean. Thereafter, each ampoule is quickly sealed off and detached from the main bulb. The ampoules are kept for about a halfhour in a thermostat at 25°C and the volume of the liquid is determined by observing the position of the meniscus with respect to the mark. It is important to note that in sealing off the ampoules, the contents cannot be cooled because this would lead to distillation of the solvent from one ampoule to another and the resulting changes of concentration would introduce rather large errors. However, the small amount of water and/or CO_2 formed by pyrolysis of the solvent vapor on the hot glass also introduces an error in analysis. The latter errors, due to "killing," were shown to be less than 1% when 10 ml or more of a 0.05M solution were withdrawn (15).

The difficulties of sealing off may be eliminated by using Teflon or all-metal stopcocks, and the former are preferred if they are truly vacuum-tight. Such arrangements are particularly advantageous in a large-scale operation involving larger amounts of liquid or many standard samples. A detailed and reliable description of these techniques is given in a paper by Brower and McCormick (16).

In spite of its shortcomings, the acid titration of all the bases is useful, because if the results of such a titration coincide with those obtained by a more discriminating method one may conclude that all the living polymers remained intact, i.e., within the experimental uncertainty of 1–2%, no accidental killing had occurred in the course of preparation.

A convenient titration method, applied by Brower and McCormick (16), uses the color of the organometallic compound as its own indicator. The solution to be analyzed of living polystyrene or poly-α-methylstyrene is introduced into a buret leading to a closed Erlenmeyer flask containing a standard solution of propanol or butanol. Prior to analysis the flask has to be evacuated and flamed, then specially purified nitrogen or argon is admitted, and finally the standard butanol solution is introduced. The presence of traces of oxygen causes discoloration of the "killed" solution, which interferes with the observation of the end point. The method is useful when the living polymer solution is intensively colored, e.g., it is advantageous for the titration of living polyvinylnaphthalene, dimeric dianions of 1,1-diphenylethylene or for relatively concentrated solutions of living polystyrene, poly-α-methylstyrene, polyvinylpyridine, etc. It is rather unreliable for the titration of living polydienes, e.g., living polybutadiene or polyisoprene, and of course useless for those living polymers which do not absorb in the visible portion of the spectrum.

The above method discriminates between living ends and other bases formed by hydrolysis or solvolysis. However, sodium hydride, or other hydrides, may react with alcohols and their presence introduces an error.

Hydrides, as has been mentioned previously, are frequently formed in stored solutions of living polymers (17) and of other organometallic compounds. Because they evolve hydrogen on addition of water, the formation of this gas may be used for their quantitative analysis.

In the writer's laboratory, the titration of growing ends is performed by reacting the living polymers with methyl iodide. The aliquot to be analyzed is kept in an evacuated flask, the solution is chilled and stirred magnetically. The flask is connected through a stopcock to an evacuated bulb containing methyl iodide. By opening the stopcock, methyl iodide vapor is slowly introduced and the stopcock is closed as soon as the color disappears. Thus, excess methyl iodide is avoided, this being important because it reacts, although much more slowly, with other bases which may be present in the solution. All the volatile ingredients (solvent, excess CH_3I, etc.) are then evaporated. The flask with the dry or resinous residue is detached from the line and, after adding water and benzene (to dissolve the polymer which is insoluble in water), the inorganic iodide is titrated in the conventional way. Fluorescent light, commonly used in many laboratories, decomposes silver iodide thus leading to too high results. The titration must be performed, therefore, in dimmed light. The analysis described above is rapid and accurate for solutions more concentrated than about $0.01M$. Variations of this procedure were described to the writer by other workers in the field, e.g., butyl bromide may be used instead of methyl iodide, etc.

In the titration of living polymers with methyl iodide, or other halides, one must ascertain the absence of any halogen salts in the analyzed solution. For example, some commercially prepared organometallic compounds, such as lithium compounds or cumylpotassium, contain a substantial amount of halide impurities. Obviously, the analysis gives then too high results for the concentration of living ends.

The course of the reaction, $\sim\sim S^- + CH_3I \rightarrow I^- +$ other products, is not well established. It appears that the process may involve electron transfer giving $\sim\sim S\cdot + CH_3\cdot + I^-$. Hence, not all of the methyl groups formed by the reaction are attached to the titrated polymer. Consequently, the technique based on counting the polymer terminated by radioactive $C^{14}H_3I$ gives results which are too low.

An accurate, albeit rather time-consuming, method for determining the concentration of living polymers was described by Trotman and Szwarc (18). The reaction of living ends with Michler's ketone leads to a leuco base which eventually is oxidized to the dye, the concentration of the latter is then determined spectrophotometrically. Strict adherence to the details described in the original paper is essential.

Application of radioactive tracers for the determination of the concentration of living ends offers many interesting possibilities. For example, an aliquot of a living polymer may be rapidly sucked, by crushing a break-seal, into an evacuated flask containing an excess of radioactive carbon-dioxide frozen at its bottom. This arrangement provides very rapid mixing of the solution with the gas—an important feature of this method—because, otherwise, the primary carboxylated product may react with non-neutralized living polymers. For example,

$$\sim CH(Ph)\cdot COO^- + \sim CH(Ph)^- \rightarrow \sim CH(Ph)\cdot \overset{\overset{\displaystyle \bar{O}}{|}}{\underset{\underset{\displaystyle \bar{O}}{|}}{C}}\cdot CH(Ph)\sim$$

and eventually, on hydrolysis of the primary product, a ketone is formed. This complication may become serious when one deals with low-molecular weight two-ended living polymers, because the intramolecular reaction involving both ends may be quite rapid.

The carboxylated polymer may be converted into acid and then freed of solvent. The residue is then weighed and thereafter assayed, e.g., in a liquid scintillation counter. However, this procedure is rather lengthy and apparently no more accurate than titration with methyl iodide.

In principle, other labeled compounds may be used instead of carbon dioxide. However, carbon dioxide (being a gas) may be easily removed from the reaction mixture, whereas, this may become a problem when one deals with non-volatile terminating agents. Purification of labeled polymer by precipitation is not recommended, because the low molecular weight fraction may be lost in this operation, and, therefore, one arrives at a too low value for the concentration of living ends.

Spectrophotometric techniques provide the most satisfactory means of determining the concentration of living ends as well as of many other organo-metallic compounds. They are most valuable, in fact unique, in analyses of diluted solutions of living polymers, viz., 10^{-7} to $10^{-4}M$, or in experiments in which a series of solutions having variable concentrations of living ends have to be compared.

Most of these materials have strong absorption peaks in the visible or the near ultraviolet region of the spectrum, the linear decimal extinction coefficients being of the order of 10^4. The respective spectra and the tables of extinction coefficients are given in the following section where we discuss also the effect of the solvent and of the counterion upon λ_{max} and ϵ.

Recording spectrophotometers are most useful in determining optical density. These measurements require a frequent check of the base line which sometimes may change during an experiment. The difficulty may be avoided if the investigated spectrum shows an absorbance maximum and minimum which are not too far apart, or if there are two relatively close maxima, of greatly different intensities. In such a case, the difference in the respective optical densities is proportional to the concentration of the analyzed species and is independent of the position of the base line.

Finally, the concentration of living polymers may be calculated by determining its molecular weight and comparing it with the theoretical value calculated from the ratio of monomer to initiator. A very accurate procedure based on this principle has recently been described by Barnikol and Schulz (19) who checked their results by titrating the solution with ethyl bromide. Their remark, that below $10^{-3}M$ concentration, the titration with ethyl bromide tends to give too low values, is not surprising. Any "killing," which becomes more probable as the polymerization proceeds, has a great effect on the titration result, but it affects only slightly the molecular weight of the product. It is, therefore, more correct to consider the results of Barnikol and Schulz as evidence of a careful preparation of a living polymer solution, in which "killing" was essentially avoided, and not as an analytical method for determining the concentration of living ends.

5. DETERMINATION OF EXTINCTION COEFFICIENTS AND CHECK ON THE VALIDITY OF BEER'S LAW

Spectrophotometric techniques require knowledge of the relevant extinction coefficients and the following device is useful in such a study. Ampoule A containing an aliquot of concentrated living polymer solution* is sealed to the apparatus shown in Figure 3. It consists of a 2-mm thick, quartz optical cell B equipped with a 1.9-mm spacer C, a flask D with a calibrated narrow neck which serves to determine accurately the total volume of the solution, a round-bottomed flask E with a magnetic stirrer in which the titration of living polymers is carried out, and a sealed ampoule F containing pure deaerated methyl iodide. The apparatus is evacuated through the connecting tube, flamed, and eventually sealed off at constriction P_1. The breakseal on ampoule A is crushed and the solution of living polymers introduced into E. Ampoule A is then sealed off at constriction P_2 after being washed with condensed solvent. By rinsing all of the apparatus with the living polymer solution any moisture adsorbed on the walls is removed. All the contents are then collected in E and the remaining parts of the equipment are washed by cooling the walls and condensing the solvent. From E, the liquid is quantitatively transferred to D, brought to the desired temperature, and its volume is measured by determining the position of the meniscus in the narrow neck.

---

* The titration of diluted solutions is less reliable and therefore the extinction coefficient is determined for concentrated (about $0.1M$) solutions.

The solution is then introduced into the optical cell B, mixed well by pouring it in and out a few times, and eventually its spectrum is recorded. Thence, it is again quantitatively transferred to E, cooled, and stirred magnetically. The breakseal on the methyl iodide ampoule F is crushed and, as soon as the solution is bleached, the distillation of methyl iodide is interrupted by cooling F. Flask E is then cut off, solvent and excess methyl iodide are evaporated, and its contents are titrated for inorganic iodide, following the procedure described in the preceding section.

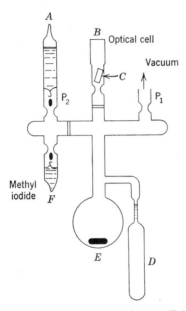

Fig. 3. Apparatus for determining the extinction coefficients of living polymers and other sensitive organometallic compounds; applicable to a relatively concentrated solution.

The procedure may be improved by eliminating the analytical difficulties arising from the uncertainty of the end point determination. Instead of ampoule F containing an excess of methyl iodide, a small ampoule containing an exactly weighed amount of deaerated water, butanol, stearic acid, or other suitable proton-donor, is attached to flask E. The amount of proton-donor must be insufficient to destroy all the living ends, and thus, after its content is introduced into flask E, the optical density of the remaining polymer may be redetermined. The difference in optical densities, in conjunction with the calculated moles of the "killing" agent, gives the value of the extinction coefficient.

The decimal extinction coefficients of living polymers, radical-anions and other organo-alkali compounds of interest are collected in Tables XI. To assess the effect of counterion and solvent upon the λ_{max} and ε, the pertinent data for some of the more thoroughly investigated systems are collected in Table XIB.

Two methods may be used to check the validity of Beer's law. A known volume of concentrated solution of living polymers, or of another organo-metallic compound being investigated, is kept in an ampoule P equipped with a suitable optical cell and a breakseal. The optical density of that solution is determined and then the ampoule is fused to a double vessel shown in Figure 4. A known, large volume of solution of the same compound is sealed in compartment B. The whole unit is well rinsed with the latter solution which is eventually quantitatively returned to bulb B. Thereafter, the remaining parts of the setup are washed with solvent, using the condensation technique described on p. 151. The solvent is then distilled into bulb A, frozen, and bulb

TABLE XIA

Spectra and Extinction Coefficients of Living Polymers
(If not marked, the data refer to sodium counterion
and tetrahydrofuran solution at 25°C)

Living polymer derived from	λ_{max}, μ	$\varepsilon \times 10^{-4}$	Ref.
Styrene	342	1.20 (1.4)	a
α-Methylstyrene	352	1.2	b
Vinyl mesitylene	360	—	b
2-Vinyl pyridine	315 (red)	1.04	c
4-Vinyl pyridine	315 (yellow)	1.04	c
Vinyl biphenyl	405	—	i
1-Vinyl naphthalene	558	0.65	d
2-Vinyl naphthalene	410	0.91	d
9-Vinyl anthracene	392 and 372	—	j
1,1-Diphenyl ethylene	470	2.6	h
Butadiene (Li$^+$, cyclohexane)	275	0.83	e
Isoprene (Li$^+$, cyclohexane)	270	0.69	f
Methyl methacrylate (Li$^+$, THF)	335	0.24	g

a. D. N. Bhattacharyya, C. L. Lee, J. Smid, and M. Szwarc, *J. Phys. Chem.*, **69**, 612 (1965).

b. D. N. Bhattacharyya, J. Smid, and M. Szwarc, *J. Polymer Sci. A*, **3**, 3099 (1965).

c. C. L. Lee, J. Smid, and M. Szwarc, *Trans. Faraday Soc.*, **59**, 1192 (1963).

d. F. Bahsteter, J. Smid, and M. Szwarc, *J. Am. Chem. Soc.*, **85**, 3909 (1963).

e. A. F. Johnson and D. J. Worsfold, *J. Polymer Sci. A*, **3**, 449 (1965).

f. D. J. Worsfold and S. Bywater, *Can. J. Chem.*, **42**, 2884 (1964).

g. D. M. Wiles and S. Bywater, *J. Polymer Sci. B*, **2**, 1175 (1964).

h. E. Ureta, J. Smid, and M. Szwarc, *J. Polymer Sci. A1*, **4**, 2219 (1966).

i. A. Rembaum, J. Moacanin, and E. Cuddihy, private communication.

j. A. Rembaum and A. Eisenberg, in "Macromolecular Reviews," Vol. 1, A. Peterlin, M. Goodman, S. Okamura, B. H. Zimm, and H. F. Mark, Eds., Interscience, New York, 1967.

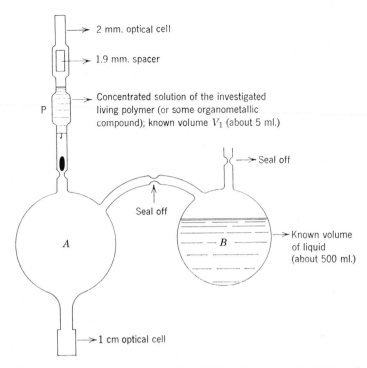

Fig. 4. Apparatus for checking the validity of Beer's law by Method I.

TABLE XIB

Effect of Counterion and Solvent on the Spectrum of Living Polystyrene

Solvent	Counterion	λ_{max}, μ	$\varepsilon \times 10^{-4}$	Ref.
Tetrahydrofuran	Li$^+$	337 (338)	1.00	a (b)
Dioxane	Li$^+$	336	1.02	a
Benzene	Li$^+$	335	1.3	c
Cyclohexane	Li$^+$	328	1.35	d
Tetrahydrofuran	Na$^+$	342 (343)	1.20 (1.18) (1.4)	a (b)
Dioxane	Na$^+$	339	1.21	a
Benzene	Na$^+$	332	—	e
Tetrahydrofuran	K$^+$	343 (346)	1.20	a (b)
Dioxane	K$^+$	340	1.21	a
Cyclohexane	K$^+$	330	1.32	—
Tetrahydrofuran	Rb$^+$	340	1.2	a
Dioxane	Rb$^+$	341	1.23	a
Tetrahydrofuran	Cs$^+$	345	1.3	a
Dioxane	Cs$^+$	342	1.24	a
Cyclohexane	Cs$^+$	333	1.25	—

a. D. N. Bhattacharyya, C. L. Lee, J. Smid, and M. Szwarc, *J. Phys. Chem.*, **69**, 612 (1965).

b. S. Bywater, A. F. Johnson, and D. J. Worsfold, *Can. J. Chem.*, **42**, 1255 (1964).

c. D. J. Worsfold and S. Bywater, *Can. J. Chem.*, **38**, 1891 (1960).

d. A. F. Johnson and D. J. Worsfold, *J. Polymer Sci. A*, **3**, 449 (1965).

e. J. E. L. Roovers and S. Bywater, *Trans. Faraday Soc.*, **62**, 701 (1966).

TABLE XIC

Spectra and Extinction Coefficients of Some Radical-Ions
(If not stated to the contrary, counterion: sodium; solvent: tetrahydrofuran)

Radical-anion derived from	λ_{max}, μ	$\varepsilon \times 10^{-4}$	Refs.
Benzene[1]	286	—	a
	435	—	
Toluene[1]	294	—	a
	435	—	
Pyrazine	249	0.47	b
	364	0.30	
Biphenyl	400 (407, Li$^+$)[2]	4.0 (3.6)[3]	c,d,e
	630 (650, Li$^+$)[2]	1.25 (1.20)[3]	
Bipyridyl (4.4)	383	—	b
	580	—	
Naphthalene	291	1.71	c,d,f
	325	1.54	
	775	0.24	
	820 (815)	0.25 (0.32)	
Triphenylene	406	1.32 (1.43)	d,g
	525	0.46	
	667	0.57	
	723	0.55	
Phenanthrene	415	0.87	h
	445	0.91	
	654	0.11	
	945	0.25	
Anthracene	262	7.8	c,d,f
	325	3.5 (3.75)	
	366	2.4 (2.45)	
	550	0.32	
	595	0.44	
	658	0.70	
	720 (715)	1.00 (0.96)	
9,10-Dimethyl anthracene	330	2.13	i
	376	2.50	
	725	0.93	
Pyrene	320	1.14	f,i
	384 (383)	1.35 (1.48)	
	490	4.95 (4.1)	
	730	0.40	
	1010	0.31	
Perylene	318	2.27 (1.18)	d
	577	5.90	
Tetracene	286	6.38	d,f
	401	5.72 (2.09)	
	711	1.45 (0.7)	
	800	1.25 (1.43)	

(*continued*)

TABLE XIC (*continued*)

Radical-anion derived from	λ_{max}, μ	$\varepsilon \times 10^{-4}$	Refs.
Diphenyl acetylene[1]	417	—	c,j
	447 (445)	2.7	
	580	0.78	
	860	1.17	
Tetraphenylethylene	370	~2.0	k
	660	1.13	
Acrylamide[4]	275	1.05	l
Methacrylamide[4]	275	0.415	l
Acrylic acid[4]	265	⩾1.6	l
Acrylonitrile[4]	255	0.68	l

[1] Recorded at $-80°C$ in dimethoxyethane; potassium as counterion.

[2] Free ion in hexamethyl phosphoramide.

[3] Spectrum recorded at $-196°C$ in 2-methyl tetrahydrofuran; lithium as counterion.

[4] Aqueous solution, pulse radiolysis.

References

a. C. L. Gardner, *J. Chem. Phys.*, **45**, 572 (1966).

b. J. W. Dodd, F. J. Hopton, and N. S. Hush, *Proc. Chem. Soc.*, **1962**, 61.

c. M. Szwarc and J. Jagur-Grodzinski, private communication.

d. J. Jagur-Grodzinski, M. Feld, S. L. Yang, and M. Szwarc, *J. Phys. Chem.*, **69**, 628 (1965).

e. K. H. J. Buschow, J. Dieleman, and G. J. Hoijtink, *Mol. Phys.*, **7**, 1 (1963).

f. G. J. Hoijtink, N. H. Velthorst, and P. J. Zandstra, *ibid.*, **3**, 533 (1960).

g. G. J. Hoijtink, *ibid.*, **2**, 85 (1959).

h. P. Balk, S. De Bruijn, and G. J. Hoijtink, *Rec. Trav. Chim.*, **76**, 907 (1957).

i. D. Gill, J. Jagur-Grodzinski, and M. Szwarc, *Trans. Faraday Soc.*, **60**, 1424 (1964).

j. D. Dadley and A. G. Evans, *J. Chem. Soc.*, B **1967**, 418.

k. R. C. Roberts and M. Szwarc, *J. Am. Chem. Soc.*, **87**, 5542 (1965).

l. K. Chambers, E. Collinson, F. S. Dainton, and W. Seddon, *Chem. Commun.*, **1966**, 498.

TABLE XID

Spectra and Extinction Coefficients of Dinegative Anions of Aromatic Hydrocarbons

Dianion derived from	λ_{max}, μ	$\varepsilon \times 10^{-4}$	Refs.
Anthracene	335	5.1 (3.8)	c,f
	605	1.85 (1.3)	
9,10-Dimethyl anthracene	600	2.1	c
Perylene	260	3.1	c,f
	321	1.8	
	552 (548)	3.2 (3.2)	
	701	1.4 (1.9)	
Tetracene	357	4.4	f
	619	1.4	
Tetraphenylethylene	485	3.7	k
Diphenylethylene (dimer)	470	5.4	c

For references, see those of Table XIC.

B is sealed off. As a result of this operation, the walls of ampoule A and the solvent distilled into it are free of any impurities which could destroy the living polymers. The breakseal on ampoule P is crushed and its contents are transferred quantitatively to A. Thus, the original solution may be diluted by a known factor, as large as 100, without inducing any "killing." The optical density is then redetermined, if necessary, in a longer optical cell.

Alternatively, a larger volume of dilute solution of the compound being

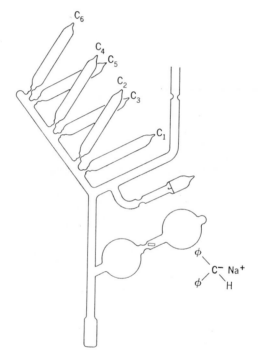

Fig. 5.   Apparatus for checking the validity of Beer's law by Method II.

investigated is introduced into the apparatus shown in Figure 5. After determining the optical density, one distills about half of the solvent into the empty ampoule C_1, which is sealed off when the apparatus is cooled with liquid nitrogen. The optical density is redetermined and the solution is concentrated again by distilling about half of the remaining solvent into the empty ampoule C_2. This procedure is continued until the solution is concentrated to the minimum feasible volume (usually not less than 1/20 of its original volume), the optical density is determined, and the concentrated solution is then collected in the last ampoule. The volumes of the liquid collected in each ampoule are accurately measured, e.g., by weighing, and thus the dilution factors are calculated. For the sake of illustration, the results obtained by this technique for diphenylmethylsodium are shown in Figure 6.

A useful method which permits one to check the validity of Beer's law over a dilution range as large as a factor of 1000 has been described recently by Slates and Szwarc (32,33). It is based on the assumption that the extinction coefficients of hydrocarbons, and of other stable and inert compounds, are constant and independent of concentration, whereas those of

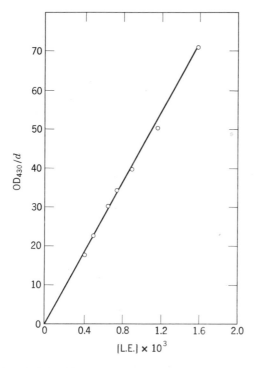

Fig. 6. Plot of optical densities per unit length of the cell versus concentration as determined by Method II: The experiment performed for a solution of sodium salt of diphenylmethyl carbanions in THF.

organometallic compounds, ion pairs, radical ions, etc., may vary because dilution may affect their degree of association or their ionic dissociation. The apparatus used is shown in Figure 7 and is composed of two flasks, A and B, the latter being linked to three optical cells having optical paths of 0.2 cm, 1 cm, and 10 cm. A 0.19 cm spacer permits one to reduce the optical path of the narrowest cell to 0.01 cm. Thus, a thousandfold variation in optical density may be easily determined. The apparatus is washed and dried in the conventional way (see pp. 151 and 172) and then filled with a relatively concentrated solution of the investigated organometallic compound and, say, of some nonvolatile aromatic hydrocarbon which does not react with the former. It is essential that the chosen compounds have distinct not overlapping absorption peaks, e.g., tetracene$^-$,Na$^+$ radical-ions mixed with tetracene form a convenient pair for such studies. Now, the optical spectrum is determined in the narrowest cell (optical path 0.01 cm).

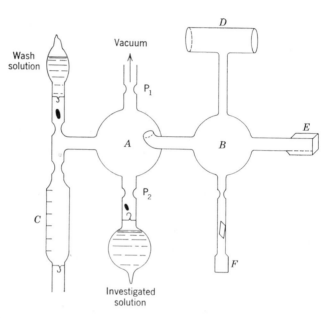

Fig. 7.   Apparatus for the determination of relative changes in extinction coefficients by means of a standard compound.

Thereafter most of the solution is transferred from flask B to A leaving in flask B only one-third or one-fourth of the original liquid. By cooling B, we transfer the volatile solvent from A to B and thus dilute by a factor of 3 or 4 the content of the latter. The spectrum is redetermined, and thereafter the dilution procedure is repeated. In this way, a series of spectra are obtained for decreasing concentrations (a final dilution may be as high as a factor of about 1000), and the constancy of the ratio (OD of the investigated compound)/(OD of the chosen standard) demonstrates the validity of Beer's law for the former.

A very ingenious method for checking Beer's law was described recently by Garst and his co-workers (20). Their apparatus, depicted in Figure 8, consists of two matching optical cells which contain the same volumes of organometallic solution, if the respective menisci coincide with the marks on the connecting tubes. The concentrations of the investigated compound may be varied by pouring the liquid from one cell to the other and then redistilling the solvent to the half-empty cell until the meniscus reaches the mark. After thorough mixing of the contents, the optical densities are redetermined, and if Beer's law is obeyed, their sum should remain constant. Unfortunately, the method is not very sensitive to small changes of ϵ. For example, if ϵ increases by 10% on tenfold dilution, then $D_1 + D_2$ increases by only

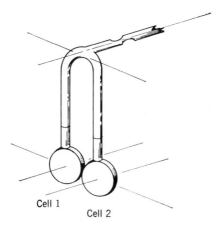

Cell 1

Cell 2

Fig. 8.   Apparatus for the determination of relative changes in extinction coefficients upon dilution.

0.5% when the originally equimolar solutions are diluted tenfold in one cell and concentrated by a factor of 1.9 in the other.

The device described by Garst is most useful for studies of systems such as

$$nA \rightleftarrows B, \ldots, K$$

where A, but not B, absorbs light. It may be shown then that the sum of optical densities, D_1 and D_2, is a linear function of the sum $D_1^n + D_2^n$, namely,

$$D_1 + D_2 = (-nK/\epsilon^{n-1} \, l^{n-1})(D_1^n + D_2^n) + T\epsilon \, l,$$

where l is the optical path (identical for both cells), ϵ the extinction coefficient of A, and T the total amount of A in either form (monomeric and n-meric). To illustrate the usefulness of the method the results obtained by Garst et al. for the system,

$$2Ph_2CO^-,Na^+ \rightleftarrows Na^+,{}^-OC(Ph)_2 \cdot C(Ph)_2O^-,Na^+,$$

are presented graphically in Figure 9. The ketyl absorbs at 630 mμ, while the dimeric pinacol does not. The plot $D_1 + D_2$ vs. $D_1^2 + D_2^2$ gives a straight line from the slope of which the dimerization equilibrium constant may be calculated after an independent determination of the extinction coefficient of the

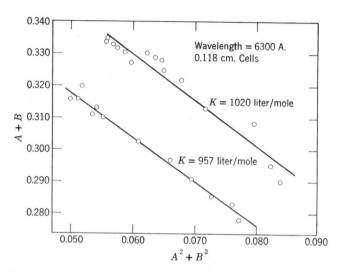

Fig. 9. Spectrophotometric study of association (or dissociation) equilibria by Garst method. Deviations of sodium benzophenone ketyl, from Beer's law at 6300 Å (room temperature). In the calculation of dimerization constants (K) the extinction coefficient was taken to be 12,000. A denotes optical density through cell 1; B denotes optical density through cell 2.

ketyl. In fact, the required extinction coefficient may be determined from the intercept of the line, but due to the long extrapolation this procedure is unsatisfactory.

6. DETERMINATION OF MONOMER CONCENTRATION

A variety of techniques may be used in determining the concentration of monomer in quenched solutions, two of which are especially recommended: (*1*) gas phase chromatography, and (*2*) spectrophotometry.

The first technique is the more reliable and the analysis may be performed in an elegant way by using an internal standard added to the solution being analyzed. For monomers composed of only C, H, and O, it is convenient to choose a standard of the same number of carbon atoms as the monomer, e.g., ethylbenzene for styrene, cumene for α-methylstyrene or stilbene for 1,1-diphenylethylene. If a hydrogen flame detector is used, the peak areas on the chromatogram are then identical for equal amounts of standard and monomer. Hence, the molar ratio (monomer)/(standard), is given by the ratio of areas under the respective peaks.

The spectrophotometric technique is, on the whole, more expedient than the chromatographic method, particularly if many samples containing the same polymer have to be analyzed. However, care must be taken in choosing a suitable absorption peak in the monomer spectrum and a reference cell, containing polymer solution of the same concentration as the sample, has to be inserted in the second beam of the spectrophotometer. The validity of Beer's law must be established for each monomer, and it is desirable to check once or twice the reliability of the spectrophotometric method by comparing the results with those obtained by VPC. It should also be remembered that oxidation of monomer or of polymer, or some other side reactions, may produce substances with strong light absorption and the presence of these substances vitiates the results of a spectrophotometric analysis. This often occurs when alkaline solutions are left overnight and are analyzed thereafter. Hence, it is desirable to acidify each solution before it is exposed to oxygen and to analyze it as soon as possible.

The spectrophotometric technique permits continuous monitoring of monomer concentration during polymerization. For details, see p. 183.

7. EXPERIMENTAL TECHNIQUES USED IN KINETIC STUDIES OF BATCH POLYMERIZATION

No technical difficulties are encountered when the progress of a slow anionic polymerization is followed, even when the reaction leads to the average addition of only one or less monomeric molecule to each growing chain. In the latter case, it is most convenient to determine the concentration of free monomer in solution at various times, and this may be accomplished either by withdrawing aliquots from the reactor and analyzing them for monomer or by applying spectrophotometric techniques.

In a slow reaction, viz. half-lifetime 15 min or more, aliquots may be most satisfactorily withdrawn from the reactor by syringes with long (about 6-in.) needles. A typical reactor used for such studies is shown in Figure 10. The main reaction flask, kept in a thermostat, is equipped with a magnetic stirrer and is connected to two Teflon stopcocks. The reagents are introduced into the evacuated reactor by crushing the breakseals on the respective ampoules, and the reactor is then filled with purified nitrogen or argon maintained at a pressure slightly higher than atmospheric. The Teflon stopcocks are outside the thermostated bath, because they may leak if heated, or stick if chilled. At the desired time intervals, aliquots are withdrawn by inserting the syringe needle through one of the stopcocks. Alternatively, self-sealing rubber stoppers may be used which are punctured by the needle. The reaction in the withdrawn sample is quenched immediately, e.g., by introducing the aliquot into an aqueous or alcoholic THF solution, and then the liquid is analyzed for monomer.

In a spectrophotometric study, the reagents are mixed in a reactor to which an optical cell is attached. Again, no major problems are encountered when the reaction is slow. However, for a fast reaction, special arrangements are required to assure rapid mixing of the reagents at the onset of the process. Two setups used in such studies will be described.

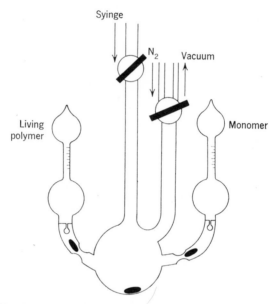

Fig. 10.   Batch reactor used in studies of relatively slow reactions.

For a reaction completed in about 3–15 min the apparatus shown in Figure 11 may be useful. The optical cell A is attached to a reactor having the shape of the inverted letter V. Ampoule B containing a living polymer solution is sealed to one arm of the reactor while one or more ampoules C, each containing monomer solution are linked to the other arm. The whole unit is evacuated, flamed, purged of impurities if necessary (see p. 151), and eventually sealed off the line. By crushing the appropriate breakseals, the solutions are let into the respective arms of the reactor and the empty ampoules may then be sealed off the reactor. At the desired moment, the contents are shaken vigorously, the reactor turned upside down thus filling the optical cell, and the latter is inserted into the cell compartment of a spectrophotometer equipped with a recorder and set at the desired wavelength. With some practice, these operations may be completed in 15 sec and the progress of the reaction may therefore be followed from that moment. After completion of the reaction, the mixture may be collected in one arm of the reactor, fresh monomer solution introduced into the other, and the whole procedure repeated. The reactor is less clumsy if only one ampoule with monomer is attached each time; other ampoules are attached, when needed, to the sidearms equipped with breakseals. The space between the attached ampoule and the breakseal is evacuated and then the unit is sealed off the vacuum line. The concentration of growing ends may be determined at the end of each

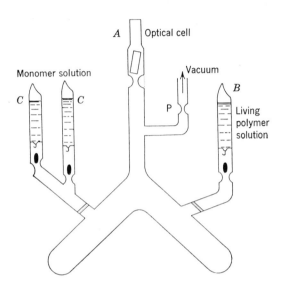

Fig. 11. Reactor used to follow spectrophotometrically a relatively slow reaction (60 sec or more).

experiment or, if desirable, even in the course of the reaction, by shifting the monochromator of the spectrophotometer to the maximum absorption wavelength of the growing ends. Figure 12 shows a typical recorder tracing obtained in an experiment performed by this technique. The peaks giving the concentration of living ends are clearly seen. The latter are measured twice with the monochromator turning towards longer wavelength and then run in the reverse direction. This facilitates the determination of optical density at λ_{max}.

In performing a series of experiments with decreasing concentrations of living ends it is convenient to attach empty ampoules to the arm containing living polymers. After completion of each experiment, the bulk of the solution is transferred to the empty ampoule leaving only a small portion in the reactor. Then, the solvent is distilled from the ampoule into the reactor and, eventually, the ampoule containing the solid residue is sealed off. In this way the solution may be diluted each time by a factor of 10 or more, without increasing the volume of the liquid in the reactor. Removal of the spacer in the optical cell or attachment of two cells having different path lengths allows one to determine accurately the concentration of living ends in the concentrated and in the diluted solutions.

In some studies, infrared absorption may be used. For example, such a technique was applied by Idelson and Blout (21) in their investigation of the kinetics of NCA polymerization to polyamino acids (see Chapter X). The infrared absorption spectra of NCA show two characteristic bands at 1860 and

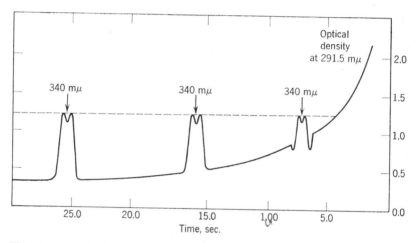

Fig. 12.   Recorder tracing of typical propagation of living polymer. Note: peak of living polymers is taken in the forward and backward direction, viz, it appears twice. This facilitates the determination of its optical density at λ_{max}.

1790 cm^{-1} which are attributed to the C=O stretching vibrations. On conversion of the monomer to polymer they disappear and, therefore, the progress of the reaction may be followed by a decrease in the optical density at those wavelengths. An additional band at 1735 cm^{-1}, present in the infrared spectrum of γ-benzyl-L-glutamate NCA, arises from the stretching vibration of the carbonyl group of the benzyl ester moiety. Its intensity is not affected by polymerization, and therefore it conveniently serves as an "internal standard." Idelson and Blout compared the results of their kinetic studies of γ-benzyl-L-glutamate NCA polymerization followed by infrared spectrophotometry with those obtained by the ordinary gasometric technique. The agreement was most satisfactory.

The evolution of carbon dioxide in NCA polymerization,

$$\sim CO-CH(R)-NH_2 + CO-O \rightarrow \sim CO-CH(R)-NH-CO-CH(R)-NH_2 + CO_2,$$
$$\underset{\underset{NH-CO}{|}}{\overset{|}{R \cdot CH}} \underset{}{\overset{|}{}}$$

allows one to monitor the progress of polymerization (22) and various techniques used for this purpose will be described later (see Chapter X).

The kinetics of propagation of living polymers carried out at low temperatures (12) may be investigated in the apparatus shown in Figure 13. This is a modification of the setup described earlier. The living polymers and the wash solution are kept in bulbs A and B. The unit is evacuated, flamed, and then sealed off at constriction H. After purging its walls with the wash solution, following the procedure described earlier, bulb B with the used purging

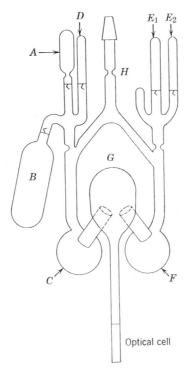

Fig. 13. Reactor used in studies of polymerization kinetics at very low temperature.

solution is sealed off. Living polymer solution is then admitted to C by crushing the breakseal on ampoule A, and the monomer transferred from E to F. The latter is diluted by distilling some solvent from C to F. The whole unit is immersed in a large Dewar flask containing the cooling medium. Another Dewar flask equipped with quartz windows and filled with a cooling liquid maintained at the same temperature as that of the large Dewar flask is placed in the cell compartment of a spectrophotometer. When the reagents reach the desired temperature, the apparatus is quickly withdrawn from the large Dewar flask, turned upside down, thus mixing the reagents in bulb *G*. Upon its inversion, the optical cell is filled up with the reacting solution, and then it is placed in the small Dewar flask and the progress of the reaction is monitored by the recorder.

Note that the homopropagation step in a living polymer system is a first order reaction and, therefore, its progress may be followed from any desirable moment.

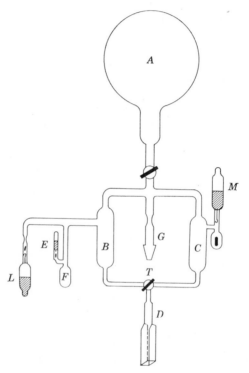

Fig. 14.   Apparatus for the investigation of propagation kinetics of living polymers at extremely low concentrations of living ends.

A more sophisticated system (3a) may be designed for faster reactions. The apparatus shown in Figure 14 consists of a 2-liter flask *A* containing specially purified nitrogen at approximately atmospheric pressure, two containers *B* and *C* which are connected through a three-way stopcock *T* to the optical cell *D*, and ampoules *E*, *M*, and *L* containing the solution of living polymers, monomer, and a concentrated solution of living polymer, respectively. The latter is used to destroy moisture and other impurities adsorbed on the walls of the evacuated apparatus, excluding of course flask *A*. After this purging operation, all of the solution is returned to ampoule *L*, using the previously described washing technique, the solvent is distilled into the lower part of ampoule F (to dilute the living polymer) and then the ampoule L with its contents is sealed off. Thereafter the living polymer solution is introduced into B and the monomer solution into C, the latter being diluted by distilling some solvent from B to C. Nitrogen is then introduced into both containers by opening the stopcock on flask A. The optical cell is now inserted into the

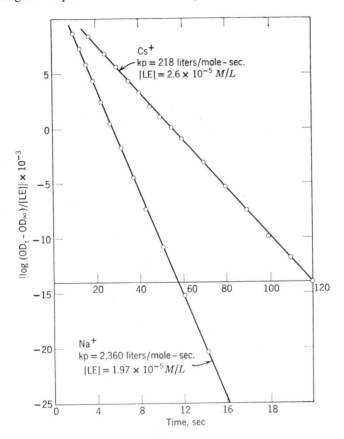

Fig. 15.   Unimolecular plot of $\log \{(OD_t - OD_\infty)\}/[\text{living ends}]$ versus time of reaction. Fast reaction: propagation of sodium salt of living polystyrene in THF; slow reaction: propagation of cesium salt of living polystyrene in THF.

spectrophotometer which is set for the required wavelength and the recorder is switched on. Swift opening of stopcock T allows both solutions to flow rapidly into the evacuated optical cell and the resulting turbulence mixes efficiently the reagents. In this device, the concentration of the monomer is monitored one second after the onset of polymerization and the reactions having a half-life of about 3–4 sec may be easily investigated. The concentration of living ends is determined spectrophotometrically at the end of the experiment, usually not later than 10–20 sec after the onset of the reaction. For the sake of illustration a plot of $\{\log(OD_t - OD_\infty)\}/[\text{living ends}]$ vs. time is shown in Figure 15. The slope of the resulting straight line gives the absolute value of the propagation rate constant.

8. FLOW TECHNIQUES: THE CAPILLARY FLOW METHOD

Studies of still faster reactions require adaptation of flow techniques; the capillary flow method will be considered first and the stop-flow technique will be reviewed in a later section.

Although the capillary flow technique is well known, having been described for the first time by Hartridge and Roughton (23,24) in 1923, its application to the investigation of anionic polymerization raises a few problems. The flow apparatus developed in the writer's laboratory (25) is shown in Figure 16. Two calibrated, cylindrical reservoirs A and B contain the solutions of monomer and of living polymers, respectively. The reservoirs are surrounded by wider tubes through which circulates water maintained at a constant temperature by a thermostat. Narrow tubes link the bottom of each reservoir with 1-mm wide capillaries forming the arms of a T-shaped, three-way Teflon stopcock T. The outlet of the vertical third arm of the stopcock is immersed in a beaker containing wet tetrahydrofuran or methyl iodide dissolved in dry tetrahydrofuran. The Teflon barrel of the stopcock has specially drilled 1-mm holes which form a mixing chamber. The liquid from each reservoir is pressed into the stopcock by dry and rigorously purified nitrogen which is introduced through tubes linked to the bottom of each vessel. This device assures a constant rate of flow for each of the liquids despite their varying head-levels, i.e., changes of hydrostatic pressure are balanced by increase in the pressure of the gas present above the liquid. The pressure of the incoming nitrogen may be regulated at will, but it is kept constant during each experiment. To minimize any possible fluctuations in the gas pressure a 3-liter flask, which acts as a barostat, is inserted in the nitrogen line.

In the course of an experiment, the reagents flow through the stopcock (where their mixing takes place) into the vertical capillary where the polymerization proceeds. The reaction is quenched at the outlet of the capillary when the mixture contacts the wet solvent in the beaker. The reaction time is determined by the dimensions of the vertical capillary and by the adjustable pressure. In each series of experiments the residence time may be varied within a factor of 5 or 6. It seems to be impractical to reduce the residence time below 0.03 sec, and it is not advisable to use this technique when the time is longer than 2 sec.

The mixing of the reagents in the stopcock appears to be satisfactory, if the rate of flow is not too low. Any faulty mixing could be recognized visually by the appearance of streaks in the capillary. These are readily perceptible, because one of the solutions is usually colored, whereas the other is colorless. In some runs, "living" polystyrene solution has been mixed with a solution

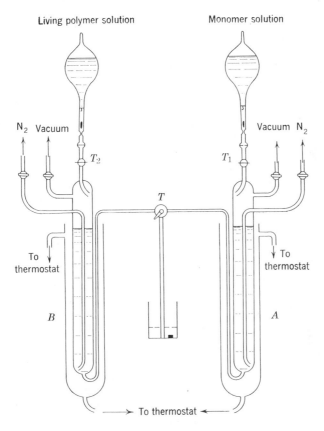

Fig. 16. Capillary flow apparatus used in kinetic studies of anionic polymerization.

of THF containing an equivalent amount of water. "Killing" of "living" ends had been completed in the barrel of the stopcock because the emerging solution was colorless. This proves that mixing takes place in less than 0.003 sec. The ultimate "killing" of the polymer solution extruded into the beaker is also extremely fast; the time interval required to complete this reaction depends upon the rate of dispersion of the reagents in the wet solvent. The reaction may be followed visually by observing the length of the colored "tail" of the liquid extruded from the capillary, because the solution becomes colorless when virtually all of the living ends are destroyed. Thus, a shorter "tail" indicates faster "killing," and the length of the "tail" seems to shorten as the rate of flow increases. Apparently, the negative pressure resulting from the suction action of the extruded liquid speeds up the mixing of the reagents with the surrounding solution. Further reduction in "killing" time is achieved by placing the bottom of the beaker against the outlet of the capillary. It has been estimated that in such a setup the living polymers are destroyed in less than 0.001 sec.

Solutions of living polymers and of monomer are freshly prepared for each series of experiments and sealed in glass ampoules equipped with break-seals. These are fused to the respective reservoirs and the whole apparatus is then pumped for a while to remove moisture adsorbed on the walls. Thereafter, the reagents are admitted to the appropriate reservoirs by crushing the breakseals and manipulating Teflon stopcocks T_1 and T_2. Nitrogen is then allowed to fill up the remaining space above the liquids and the system is left for a while in order to bring the reagents to the desired temperature.

To determine the concentration of living ends, an aliquot of living polymer solution is withdrawn through the vertical capillary into a closed flask having no contact with the surrounding atmosphere and containing methyl iodide in dry solvent. The resulting sodium iodide is then titrated by following the procedure described earlier (page 168). Thereafter, a second aliquot of a *mixture* of living polymers and of monomer is withdrawn and titrated in the same way. In fact, similar aliquots are withdrawn and titrated at the beginning and at the end of each series of experiments, the average value of the respective titrations being used in the final calculation of the rate constant. In a reliable series of experiments, the results of all three titrations should not differ by more than 5%.

Four or five runs usually are performed with each batch of reagents. In such a series of experiments, it is recommended to vary in a non-systematic way the residence time of each run. The concentration of monomer in the killed solution is determined by the usual techniques (VPC or spectrophotometry; see p. 180), and from the rates of flow of each liquid the composition of the mixture and the residence time are calculated.

The stationary flow of liquid is not established immediately after opening the three-way stopcock, and this may introduce some error in the calculation of the residence time. However, the stationary flow seems to be established in less than a second and, because each experiment lasts for 2 or 3 min, this disturbance is of no importance. It certainly is not profitable to eliminate this problem by using a more sophisticated procedure.

The polymerization in a flow system proceeds virtually under adiabatic conditions. The temperature of the reagents rises, therefore, because polymerization is exothermic. To minimize this effect, the heat capacity of the solution should be chosen in such a way that the resulting increase in temperature does not exceed 0.5°C.

Finally, it should be stressed that the polymer may gradually deposit on the walls of the vertical capillary and thus reduce its volume. This effect is particularly disturbing when the concentration of monomer is high, and it is necessary, therefore, to clean the capillary thoroughly after each run, e.g., by using a pipe cleaner for this purpose.

A modified flow technique used in polymerization studies in Mainz has been described recently by Schulz et al. (26). Since then their device has been substantially improved and it is hoped that its detailed description will be published in the near future.

In a flow system, the progress of polymerization may be conveniently followed by the rise in temperature. This method has been developed by Löhr and Schulz but it is not yet described in the literature. Its successful operation requires careful thermostating of the reagents. Calibrations needed in this method have to be performed for each system because the heat of polymerization depends on the solvent used and the heat capacity of the reacting liquid depends on its composition.

9. CALCULATION OF RESULTS OBTAINED BY THE CAPILLARY FLOW METHOD

Calculation of the results requires knowledge of the type of flow established in the vertical capillary. Two extreme cases will be considered: (*1*) plug flow and (*2*) laminar flow.

In plug flow, each volume element of the flowing liquid is assumed to move with the same, uniform vertical velocity independent of its position in the capillary. Moreover, it is assumed that diffusion or convection of reagents or products from one volume element to another is negligible and, therefore, its effects may be disregarded in calculations.

The kinetics of a reaction taking place in plug flow are described by the conventional differential equations in which the residence time t gives the duration of the process, t being the ratio (reactor's volume)/(rate of flow). Such expressions apply approximately to kinetics of processes taking place in turbulent flow, because the turbulence mixes efficiently the reagents in a radial direction, homogenizing the concentrations within each cross-section, whereas mixing in a vertical direction is much less pronounced. Note, however, that there is a gradient of velocities even in a turbulent flow, giving the ratio $u_{max}/\bar{u} = 1.25$. This affects the apparent rate constant and decreases it by about 5%.

In laminar flow, each concentric ring layer flows with a different velocity,

u_r, arising from a radial velocity gradient. Let us denote by r the distance of a ring layer of thickness dr from the center of the capillary. The layer moves with vertical velocity u_r. At distance r, the velocity gradient of the liquid is given by

$$-du_r/dr = pr/2\eta$$

where p is the pressure gradient and η the viscosity of the fluid.

Hence, $u_r = (p/4\eta)(R^2 - r^2)$, if the liquid adjacent to the walls of the capillary is stationary, and the capillary has radius R. The volume of the liquid flowing through the capillary in 1 sec. is,

$$q = \int_0^R 2\pi r \cdot u_r \, dr = \pi R^4 p/8\eta.$$

The "average" residence time, t, retains its previous meaning, viz.

$$t = (\text{volume of capillary})/(\text{rate of flow}) = 8V\eta/\pi R^4 \cdot p = 8L\eta/R^2 \cdot p$$

V denoting the volume of the capillary and L its length. Similarly, the average linear velocity $\bar{u}$, is defined as

$$\bar{u} = (\text{rate of flow})/\pi R^2 = pR^2/8\eta,$$

and hence $u_r = 2\bar{u}(1 - r^2/R^2)$. Therefore, the maximum velocity, in the center of the tube, is twice the average velocity.

The average concentration, $\bar{C}_t$, of a product in the solution emerging from the capillary outlet is given, by the integral,

$$\bar{C}_t = \int_0^R 2\pi r u_r C_r \, dr/\pi R^2 \bar{u},$$

where C_r is the concentration of the product at the end of the capillary in a circular ring located at a distance r from the axis of the tube. This concentration is calculated from the conventional kinetic equation keeping in mind that the reaction in the lamina proceeds for a time, $t_r = L/u_r$. The latter statement implies that diffusion of the reagents, whether radial or axial, may be ignored.

For first order kinetics, e.g., for the concentration change of monomer in an ordinary homopolymerization of living polymers,

$$C_r = C_0 \exp(-k_1 C_{LE} L/u_r) = C_0 \exp(-\alpha/u_r),$$

where $\alpha = k_1 C_{LE} L$; k_1 is the rate constant of polymerization and C_{LE} the constant concentration of living ends. Similarly, for a second order reaction, e.g., for a copolymerization followed by an extremely slow homopolymerization of the second monomer,

$$C_r = \{k_2 L/u_r + 1/C_0\}^{-1},$$

if the initial concentration of both reagents $\sim\!\!A^-$ and B is identical. For different concentrations of the reagents, viz. $[\sim\!\!A^-] = C_0'$ and that of the monomer being C_0, C_r is given by

$$C_r = C_0(C_0' - C_0)/\{C_0' \exp [k_3(C_0' - C_0)L/u_r] - C_0\}.$$

For the first order reaction, integration gives

$$\bar{C}_t = C_0 \int_0^R 2\pi r \, dr \cdot u_r \exp (-\alpha u_r)/\pi R^2 \bar{u}$$

or

$$\bar{C}_t/C_0 = e^{-m}[1 - m + G(m)],$$

where $m = 4\alpha\eta/pR^2 = 1/2k_1 C_{LE}t$ and $G(m) = m^2 e^{-m} \int_m^\infty e^{-x} \, dx/x$.

Numerical values of $\bar{C}_t/C_0$ as a function of m are given in Table XII and presented graphically in Figure 17. From the curve one reads m for the

TABLE XII

Values of the Auxiliary Variable m Needed for the Calculation of the Rate Constant in the Flow Technique Method for a First Order Reaction

m	0.000	0.005	0.010	0.020	0.030	0.040	0.050	0.075	0.100	0.150
C/C_0	1.000	0.990	0.981	0.961	0.943	0.925	0.908	0.868	0.832	0.765
m	0.200	0.250	0.300	0.350	0.400	0.450	0.500	0.550	0.650	0.750
C/C_0	0.704	0.652	0.600	0.555	0.515	0.477	0.443	0.412	0.357	0.310
m	0.850	1.000	1.150	1.350	1.50	1.75	2.00	2.25	2.50	3.00
C/C_0	0.274	0.219	0.179	0.138	0.113	0.082	0.060	0.044	0.028	0.018

experimentally given value of $\bar{C}_t/C_0$. A plot of $2m/$[living ends] vs. the conventional residence time t gives, therefore, a straight line of slope k_1 going through the origin.

For the second order reaction with $[LE] = [M]_0 = C_0$, integration gives

$$\bar{C}_t/C_0 = 1 - a + 2a^2 \ln (1 + a^{-1}),$$

where $a = (1/2)k_2 C_{LE}t$. Numerical values of $\bar{C}_t/C_0$ as a function of a are given in Table XIII and presented graphically in Figure 18. Again, having experimentally determined $\bar{C}_t/C_0$ one reads from the graph (Figure 18) the respective values of a. A plot of $2a/$[LE] vs. t then gives a straight line with a slope k_2.

For a second order reaction in which the initial concentration of $\sim\!\!A^-$ is C_0' and that of monomer C_0, one derives the expression

$$\bar{C}_t/C_0 = 2(1 - C_0/C_0') \int_0^1 x \, dx\{\exp (g/x) - C_0/C_0'\},$$

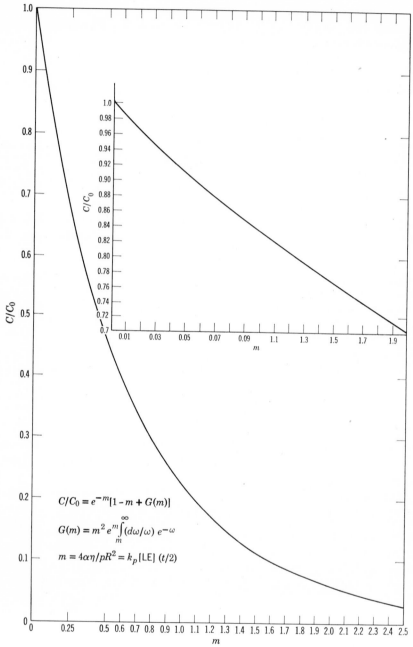

$$C/C_0 = e^{-m}[1 - m + G(m)]$$

$$G(m) = m^2 e^m \int_m^\infty (d\omega/\omega)\, e^{-\omega}$$

$$m = 4\alpha\eta/pR^2 = k_p[LE]\,(t/2)$$

Fig. 17.   Curve giving C/C_0 as a function of m for the calculation of kinetic results
obtained in laminar flow for first-order monomer conversion.

TABLE XIII

Values of the Auxiliary Variable α Needed for the Calculation of the Bimolecular Rate Constant ($A_0 = B_0$) in the Flow Technique Method (Laminar Flow)

α	$\bar{C}_t/C_0$	α	$\bar{C}_t/C_0$	α	$\bar{C}_t/C_0$
0.000	1.0000	0.70	0.4664	2.60	0.1997
0.025	0.9546	0.80	0.4380	2.80	0.1877
0.050	0.9152	0.90	0.4105	3.00	0.1782
0.075	0.8800	1.00	0.3863	3.50	0.1571
0.10	0.8480	1.20	0.3457	4.00	0.1405
0.15	0.7917	1.40	0.3129	4.50	0.1272
0.20	0.7433	1.60	0.2858	5.00	0.1160
0.30	0.6639	1.80	0.2630	6.00	0.1010
0.40	0.6009	2.00	0.2438	7.00	0.0810
0.50	0.5493	2.20	0.2269	8.00	0.0758
0.60	0.5062	2.40	0.2124	9.00	0.0651
				10.00	0.0620

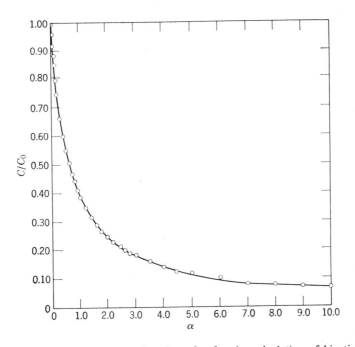

Fig. 18. Curve giving C/C_0 as a function of α for the calculation of kinetic results obtained in laminar flow for second-order monomer conversion $C_0 = C_0'$. $C_0/C_0' = 0.1$.

TABLE XIVA. Values of the Auxilary Variable g for Different Values of C_0/C_0' Needed for the Calculation of a Bimolecular Rate Constant in Laminar Flow.

g	C_0/C_0'				
	0.1	0.2	0.3	0.4	0.5
0.05	0.9014	0.8911	0.8783	0.8620	0.8404
0.10	0.8186	0.8019	0.7815	0.7563	0.7239
0.20	0.6835	0.6599	0.6320	0.5986	0.5576
0.30	0.5770	0.5507	0.5203	0.4850	0.4431
0.40	0.4907	0.4639	0.4337	0.3992	0.3594
0.50	0.4197	0.3937	0.3648	0.3324	0.2958
0.60	0.3606	0.3361	0.3091	0.2793	0.2462
0.70	0.3110	0.2882	0.2635	0.2364	0.2068
0.80	0.2691	0.2482	0.2256	0.2013	0.1750
0.90	0.2335	0.2144	0.1940	0.1732	0.1489
1.00	0.2030	0.1858	0.1675	0.1481	0.1274
1.20	0.1545	0.1405	0.1258	0.1105	0.0944
1.40	0.1183	0.1071	0.0955	0.0834	0.0709
1.60	0.0911	0.0822	0.0730	0.0635	0.0537
1.80	0.0705	0.0634	0.0561	0.0487	0.0411
2.00	0.0547	0.0491	0.0434	0.0375	0.0316
2.50	0.0295	0.0264	0.0232	0.0200	0.0167
3.00	0.0161	0.0144	0.0126	0.0109	0.0091
3.50	0.0089	0.0079	0.0070	0.0060	0.0050
4.00	0.0050	0.0044	0.0039	0.0033	0.0028
4.50	0.0028	0.0025	0.0022	0.0019	0.0016
5.00	0.0016	0.0014	0.0012	0.0011	0.0009

where $g = (1/2)k_3(C_0' - C_0)t$. Numerical values of $\bar{C}_t/C_0$ as functions of g for various values of C_0/C_0' are given in Tables XIVA and B and presented graphically for the case $C_0/C_0' = \frac{1}{2}$ in Figure 19. Again, having $\bar{C}_t/C_0$ and knowing C_0/C_0' one reads from the respective curve the appropriate value of g, and then a plot of $2g/(C_0' - C_0)$ vs. t gives a straight line having a slope of k_3.

Comparison of the kinetic treatments of plug flow and laminar flow shows that for $\bar{C}_t/C_0$ approaching unity both methods give identical results. This is seen in Figure 20 where the dotted line gives the values of $\bar{C}_t/C_0$ which should be observed in plug flow for the same average residence time t. Hence, for very low per cent conversion, i.e., for a short time of polymerization and/or for a low concentration of living ends, the calculated rate constants are reliable whatever method is used for their derivation.

In conclusion, it should be stressed that for the same residence time t the fraction of conversion $1 - \bar{C}_t/C_0$ is always larger for plug flow than for laminar flow, all the other parameters remaining the same. Hence, a set of

TABLE XIVB. Auxiliary Variable g for a Bimolecular Reaction $C_0 \neq C_0'$ in Laminar Flow

g	C_0/C_0'			
	0.6	0.7	0.8	0.9
0.05	0.8104	0.7656	0.6910	0.5385
0.10	0.6807	0.6201	0.5277	0.3672
0.20	0.5063	0.4395	0.3486	0.2160
0.30	0.3927	0.3304	0.2513	0.1467
0.40	0.3128	0.2576	0.1905	0.1072
0.50	0.2540	0.2057	0.1492	0.0819
0.60	0.2092	0.1673	0.1196	0.0645
0.70	0.1742	0.1380	0.0975	0.0519
0.80	0.1463	0.1150	0.0805	0.0424
0.90	0.1238	0.0966	0.0672	0.0351
1.00	0.1054	0.0818	0.0566	0.0294
1.20	0.0775	0.0597	0.0409	0.0210
1.40	0.0578	0.0443	0.0301	0.0154
1.60	0.0437	0.0333	0.0225	0.0115
1.80	0.0333	0.0253	0.0171	0.0086
2.00	0.0255	0.0193	0.0130	0.0066
2.50	0.0135	0.0102	0.0068	0.0034
3.00	0.0073	0.0055	0.0037	0.0018
3.50	0.0040	0.0030	0.0020	0.0010
4.00	0.0022	0.0017	0.0011	0.0006
4.50	0.0012	0.0009	0.0006	0.0003
5.00	0.0007	0.0005	0.0004	0.0002

experimental values of $\overline{C}_t/C_0$ obtained for respective times t leads to a higher value for the rate constant if the flow is assumed to be laminar, than it would have been if the calculations were performed on the assumption of plug flow.

10. STOP-FLOW TECHNIQUE

The stop-flow technique was developed by Chance (27,28) who used it in a most successful way in his studies of enzymatic reactions. The reagents are introduced into two reservoirs, or syringes, and in the course of an experiment they are pressed through a suitable mixing chamber into a short capillary located in a spectrophotometer. As the liquid flows through the observation post, the spectrum of the reacting mixture may be recorded. The recorder tracing refers, therefore, to the stage attained by the process in a time required for the fluid to travel from the mixing chamber to the optical cell. This interval may be as short as 0.002 sec. The monochromator is then set at the desired wavelength, the recorder switched on, and the flow stopped

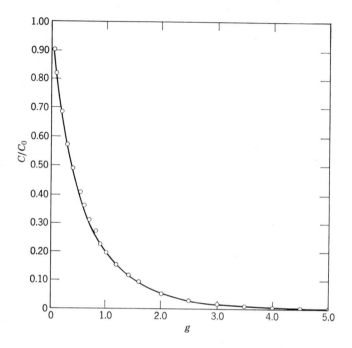

Fig. 19.   Curve giving C/C_0 as a function of g for the calculation of kinetic results obtained in laminar flow for second-order monomer conversion $C_0/C_0' = \frac{1}{2}$.

suddenly, e.g., by turning a stopcock or another suitable device. It is imperative to stop the flow suddenly, because mixing becomes inefficient if the flow is stopped slowly. The progress of the reaction is followed from that moment by the change in the optical density, as illustrated in Figure 21.

The whole procedure may be repeated with the monochromator set at another wavelength, and thus the rates of change of optical density at various wavelengths may be investigated. At the end of the reaction, the whole spectrum may be scanned again, thus providing the absorption spectrum of the final products.

For very fast reactions, the liquids are pressed from the syringes by a piston operated by a synchronic electric motor. A special obstacle stops dead the piston and the impact switches on an oscilloscope which monitors the variation in optical density. Technical details of this method may be found in reference 29, and many further improvements have been introduced recently by Chance and Gibson (38).

Adaptation of this technique to studies of anionic polymerization requires several precautions because contact of the reagents with moisture, air, etc., has to be avoided. The devices used in a capillary flow technique may be

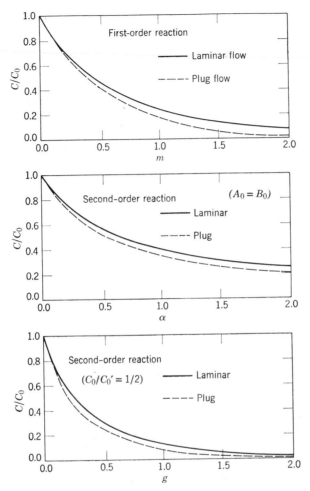

Fig. 20. Comparison of monomer conversions obtained in laminar and plug flows for the same residence time.

beneficially applied to the stop-flow technique. Use of glass syringes leads sometimes to difficulties arising from the etching action of alkali solutions which causes the sticking of pistons, while metal syringes may lead to an undesirable killing of living polymers and to the contamination of the solution by salts.

To calculate the rate constant from the results obtained by the stop-flow technique, one needs to calibrate the optical cell and to determine the required extinction coefficients. This is relatively easy when one observes the decay of the reagent or the formation of a stable final product. However, some

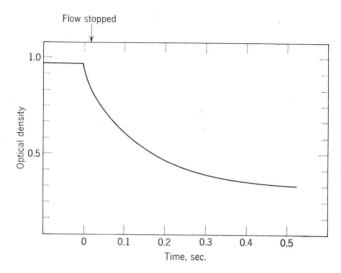

Fig. 21.   Recorder tracing of change in optical density with time in stop-flow technique.

difficulties arise when the absorption of a transient intermediate is followed. To obtain the desired data, one needs to follow the reaction at different wavelengths and frequently it is advantageous to choose some extreme conditions which simplify the kinetics of the process. Moreover, it is imperative in such studies to establish carefully the stoichiometry of the investigated reaction.

11. ADVANTAGES AND SHORTCOMINGS OF FLOW TECHNIQUES

Flow techniques provide one with the opportunity of following fast reactions and in this respect the stop-flow technique is superior to the capillary flow technique because it allows the study of faster reactions. The advantages of the former technique are particularly obvious when the initial rate of the process is investigated, because the progress of the reaction may be continually followed starting from 0.01 sec after its onset. The continuity of the tracing is of the greatest benefit when the resulting curve is used for integration, this being of importance in processes having complex kinetics, e.g., when a reaction involves intermediates which are in equilibrium with the reagents and products. On the other hand, the disadvantage of the stop-flow technique arises from its spectrophotometric character. The reagents and/or products must have a convenient absorption spectra, if possible, high linear extinction coefficients, and most of all, the respective spectra must not overlap too closely. The last condition is usually the most difficult to fulfill; it sometimes forces the investigator to choose a system which otherwise might not be the best for his purpose, and frequently contemplated research cannot be carried out by this method because of too great an overlap of the spectra. Minor changes in the structure of the reagents might lead to spectral differences which make the research feasible or vice versa. For example, on addition of styrene to living poly-1-vinylnaphthalene a complex is formed which shows an absorption maximum at 440 mμ. Since the maximum absorption of the living poly-1-vinylnaphthalene is at 558 mμ, the spectrophotometric study of the kinetics of complex formation is feasible. On the other hand, the absorption maximum of living poly-2-vinylnaphthalene corresponds to 409 mμ, and consequently studies of complex formation are impractical for the latter system, at least by this method.

Impurities and some side reactions might cause great confusion in any spectrophotometric study, and the stop-flow technique is most susceptible to this. It is not infrequent that the impurity initially present in the system, or formed by some side reactions, shows a strong absorption in the investigated region of the spectrum. The utmost care has to be taken to assure the reliability of the observed spectral changes, and it is desirable to investigate the process with reagent prepared in various ways, or at least purified by different techniques, to ascertain the validity of the results.

From a hydrodynamic point of view, a cylindrical optical cell is probably the most desirable because a most uniform flow may be achieved in such a device and the danger of "dead pockets," viz. places in which the flow is stagnant, is excluded. On the other hand, a cylindrical capillary cell has rather awkward optics. The optical path length is not constant and there are undesirable reflections and refractions. It is often necessary to insert slits in front and behind the capillary, but these devices greatly reduce the amount of light falling on the detector and diminish, therefore, the sensitivity of the spectrophotometer. Moreover, to avoid the undesirable large curvature of the capillary, one has to increase its diameter because it seems to be impractical to work with a capillary narrower than 2 mm i.d. In studies of anionic polymerization and of related processes, one deals with solutions having linear extinction coefficients of about 10^4 and, consequently, extremely dilute solutions (less than $10^{-3}M$) have to be used, if the optical path is 1 mm. long. This not only limits the scope of the investigation but introduces many technical difficulties caused by the enhanced importance of impurities. Hence, it was found in the writer's laboratory that a flat cell sealed to a capillary is, after all, advantageous in such studies. By proper choice of the inlet, one may introduce enough turbulence to secure a homogeneous composition of the liquid in the vicinity of the beam of searching light. Any diffusion may distort the results at later stages of the reaction; however, the results obtained during the first 5 or 10 sec are quite reliable, and if a slower reaction is to be investigated, one may always design a suitable static system (see p. 186).

Some of these difficulties may be overcome by using a nontransparent capillary, e.g., a capillary drilled in a stainless steel block. The searching light passes through the holes drilled perpendicularly to the capillary which are closed by suitable quartz windows or short quartz rods (Figure 22). The

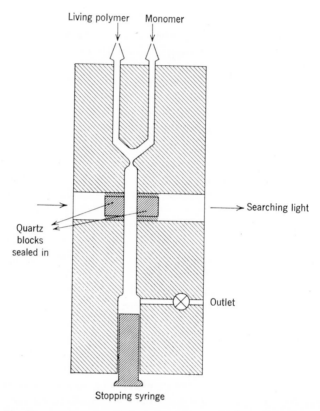

Fig. 22. Design of optical windows for a stop-flow apparatus.

problem of gaskets and sleeves, which should be flexible, may be troublesome when one intends to vary widely the temperature of the cell; however, it was demonstrated that practical solutions exist even for temperatures as low as −180°C. An alternative solution was developed by Gibson (38). His cell, also drilled in a metal block, is shown in Figure 23, which is self-explanatory.

Finally, if the early stages of the reaction lead to rapid changes in optical density, a proper recorder with a fast response should be chosen. The oscilloscope is obviously the most desirable device, particularly, if the reaction is completed within a few seconds. It is often advantageous, by inserting the proper circuit, to convert the signal displayed on the oscilloscope into its log. Thus, optical density (and not transmission) is recorded.

The capillary flow technique has the advantage that the concentration of the reagents or products may be directly determined in an unambiguous way, e.g., by gas chromatography. Moreover, this method allows one, at least in principle, to determine the concentrations of all the stable components of the reacting mixture at the chosen time, t. However, each experiment gives only one point on the time-conversion curve, whereas the whole conversion curve

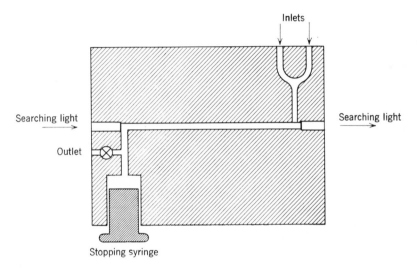

Fig. 23.   Alternative design of optical windows for a stop-flow apparatus.

is obtained by a stop-flow technique. Therefore, when the reaction is studied by the capillary flow technique, a set of experiments—at least 4 or more—has to be performed to obtain enough information to draw such a curve.

The initial concentrations of the reagents are calculated from the known concentrations of the feed solutions and the rates of the respective flows. Denoting the latter by f_1 and f_2, we find the dilution factors to be $f_1/(f_1 + f_2)$ and $f_2/(f_1 + f_2)$. In the system utilizing gas pressure to drive the liquids the rates of flow, f_1 and f_2, are determined by operating the system for 2–3 min, before stopping the flow, and then reading the volumes of the liquids removed from each reservoir. This approach assumes implicitly that the rates of flow at the moment preceding stoppage are the same as the average rates of flow, or at least that their ratio has not changed. It seems that in some runs a speck of some solid or semi-solid material may partially block one of the jets of the mixing chamber and cause a substantial distortion of the antici-pated initial concentrations.

This difficulty is removed entirely in systems in which the liquids are delivered from syringes having pistons driven by a motor or any other suitable device, e.g., hydraulic press. Hence, the flow ratio is determined by the ratio of cross-sections of the respective syringes and therefore it remains constant, independent of the rate of flow. Moreover, the dilution factor in such a system is independent of the viscosities of the solutions fed into the mixing chamber, whereas in the gas driven system the result is affected by their ratio. This effect interferes with the planning of the experiment—the initial concentrations may be very different from those which the investi-gator wished to have.

The use of syringe permits one to perform stop-flow experiments with minute amounts of reagents, if the initial time (time needed for the reagents to reach the observation post) need not be accurately determined. This follows from the fact that one does not need to measure the rate of flow of the liquid; it is sufficient to know that it is faster than some minimum value. The waste of reagents arises from the necessity to attain a steady flow and to determine accurately its rate. This, of course, is required in the capillary flow technique.

The largest drawback of the capillary flow technique arises from some uncertainty about the character of the flow, because, as has been previously pointed out, the calculation of rate constants depends on this factor. Fortu-nately, as has been mentioned in Section 9, the rate constants calculated on the basis of different flow types converge when the conversion is low, and this provides a method of testing the character of the flow and the reliability of the calculated results.

The lowest conversion is achieved for the shortest residence time and, hence, for the fastest flow. Judging from the Reynolds numbers such flows are usually turbulent, and hence for these experiments it is plausible to perform the calculation by assuming a plug-flow (mass-flow). For slower flows, the calcu-lations performed in the same way are justified, if they lead to the same rate constants as those obtained with the fastest flow, and the self-consistency of such a set of data assures their reliability. However, this approach presupposes the knowledge of the kinetic behavior of the system, but, fortunately, for some systems the kinetics of the process is indisputable.

All these uncertainties do not appreciably affect the values of the calculated rate constants, at least for most cases encountered in practice. Geacintov et al. (25) carried out such calculations for several experiments in which the anionic polymerization of styrene was investigated. The spread of rate con-stants, reported in their paper, is not greater than about 30%.

Some artifacts of the capillary flow technique might be mentioned to warn the investigator. For example, an attempt was made to investigate the pseudo-first order rate constant in the anionic homopropagation of styrene at a constant concentration of living ends but for a widely different initial concentration of the monomer. The results, shown in Table XV, indicated a small decrease, by about a factor of 2, in the apparent k_p when the initial monomer concentration increased by a factor of 160. Apparently, the increase in the viscosity of the solution, resulting from a considerable degree

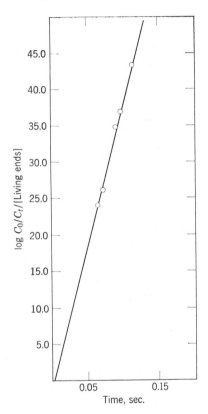

Fig. 24.   Example of result obtained by capillary-flow technique for a first-order reaction, homopolymerization of 1-vinyl-naphthalene. Temperature, 25°C. [LE] = 2.17×10^{-3} mole/liter; [M] = 2.60×10^{-3} mole/liter.

of polymerization attained at higher monomer concentration, changed the character of the flow and reduced the extent of the reaction, thus giving the false impression of a decrease in k_p. It is also possible that the increase in the molecular weight of the polymer could lead to its enhanced adsorption on the wall of the capillary thus reducing the true residence time of the reagents.

To illustrate the reproducibility and reliability of the capillary flow technique, two examples of kinetic studies—one referring to a first order reaction, the other to a second order process—are shown graphically in Figures 24 and 25.

TABLE XV[a]

Apparent Change in k_p for the Anionic Homopropagation Step of Styrene in Tetra-hydrofuran Arising from Increase in Initial Concentration of Monomer[b]

Concentration living ends, mole/liter × 10^3	Initial concentration styrene, mole/liter × 10^3	k_p, (liter/mole) sec^{-1}
11.0	1.0	790
11.0	10.0	630
11.0	160.	400
3.0	3.4	955
3.0	53.0	610

[a] From the Ph.D. Thesis of D. N. Bhattacharyya, Syracuse University, Syracuse, N.Y., 1964.

[b] The observed trend seems to be an artifact of the capillary flow technique.

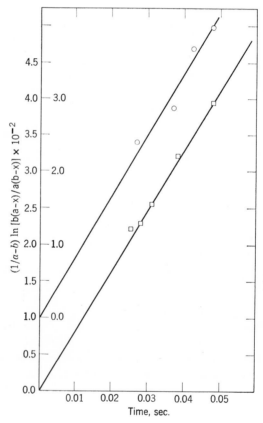

Fig. 25.   Examples of results obtained by capillary flow technique for a second-order reaction, copolymerization of living polystyrene with (□) (outer ordinate) 1-vinyl- and (○) (inner ordinate) 2-vinyl-naphthalene.

12. STIRRED-FLOW REACTOR

The stirred-flow reactor is a device well-known to any chemical engineer and its adaptation to kinetic studies was advocated by Denbigh and his associates (29–31). The principle is exceedingly simple. The reagents are fed continually into the reactor at a constant rate, i.e., v ml/sec, and the reacting liquid is simultaneously withdrawn at the same rate. Thus, the volume, V, of the reacting fluid remains constant. Efficient stirring in the reactor assures a uniform and homogeneous composition of the reacting mass throughout its whole volume, and therefore the outflowing liquid has the same composition as that established in the reactor. The reaction is quenched instantly in the liquid withdrawn, and then a sample of the quenched solution is analyzed. Hence, the results give the composition of the fluid in the reactor.

The reactor is not in a stationary state at the beginning of an experiment. In fact, the initial composition of the liquid is given by the composition of the feed. After a time, the duration of which is determined by the residence time, viz. $t_r = V/v$, and by the rate of the reaction, the system approaches its stationary state sufficiently closely to permit the concentration of all the reagents and products to be considered constant within the whole volume of the reactor and invariant in time.

The industrial stirred-flow reactors are large and reach their stationary state within hours or even days. This is of little importance when the production is continued for weeks or months. The majority of the experimental kinetic reactors described in the literature are still too large for a convenient operation, e.g., those used by Denbigh had a volume of about 0.5 liter. Such a size leads to a slow operation; it takes a long time to reach the stationary state and consequently large amounts of reagents are wasted before the crucial analysis can be performed. This difficulty discourages many investigators from adopting the stirred-flow technique for their kinetic studies.

All these difficulties may be avoided if the reaction is fast. The reactor may be small, and indeed the types developed in the writer's laboratory have volumes of 2–7 ml. This reduces the residence time to a few seconds (about 1–30 sec) and shortens the approach to the stationary state to ca. a few minutes. Consequently, only 10–100 ml of the reacting solution have to be used for each experiment, and this reduces the need for reagents to an acceptable level.

The reactors used in the writer's laboratory are all-glass vessels with two inlet capillaries and an outlet capillary sealed directly to the main chamber, as shown in Figure 26. The reactor is made from a piece of glass tubing sealed on both ends in such a way as to form two conical bearings to support an inner rotor. The rotor is similarly made from glass tubing of a slightly smaller diameter with an iron bar sealed in it. The ends of the rotor are also conically shaped and fit into the conical cavities of the reactor's ends. Its cylindrical surface is grooved, forming channels sloped in opposite directions which push the liquid hither and thither. To insert the rotor, the reactor is cut in two and, after placing the rotor, both parts are fused together, neither too tightly nor too loosely in order to secure a proper motion of the rotor. The volume of the reactor is given by the difference in the volumes of the outer vessel and of the

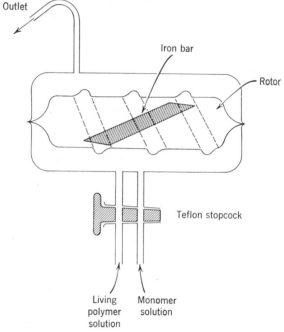

Fig. 26.   Stirred-flow reactor used in kinetic studies of anionic polymerization.

inner rotor, and with proper care it may be reduced to 2 ml. The inlet capillaries are connected to a double Teflon stopcock, and through it to the reservoirs which contain the reagents. The reservoirs and the supplying system used in the capillary flow or the stop-flow techniques may be utilized for the stirred-flow reactor. The outlet capillary is as short as possible and it delivers the reacting liquid to the vessel containing the quenching solution.

The operation is easy and in performing a set of runs, each with a different rate of flow and corresponding, therefore to a different residence time, it is not necessary to empty the reactor. The flow may be stopped whenever desired, then adjusted to a new rate, and a fresh sample may be collected after a while when the system reaches its new stationary state. To check whether the stationary state has been attained, two samples may be collected at different times without interrupting the flow. The identity of their composition proves that the desirable conditions have been attained.

The rotor is driven magnetically and the number of revolutions per unit time is determined stroboscopically. It was found that 20–30 revolutions within residence time seem to be sufficient to assure the proper mixing of the reacting fluid. The reagents are introduced into the lower part of the reactor at one of its ends and the outlet is located in its upper part at the other end. This makes it easy to drive air out of the reactor and to avoid the presence of any air bubbles.

13. MOLECULAR WEIGHT DISTRIBUTION IN STIRRED-FLOW REACTORS

Distribution of molecular weights of a product arising from a polymerization of low molecular weight living oligomers to a high molecular weight polymer is of the "most probable" type, if performed in a stirred-flow reactor providing any termination is prevented. The system is described by the following set of balanced equations, in which P_0 denotes the lowest living oligomer which, it is assumed, is the only species introduced in the feed.

$$v([P_0]_0 - [P_0]_t) = Vk_p[P_0]_t[M]$$

$$-v[P_1]_t = Vk_p[M]\{[P_1]_t - [P_0]_t\}$$

$$\vdots$$

$$-v[P_j]_t = Vk_p[M]\{[P_j]_t - [P_{j-1}]_t\}$$

$$\ldots, \text{etc.}$$

The symbols v, V, and $[M]$ denote the rate of flow, the volume of the reactor, and the stationary concentration of the monomer, respectively. P_j denotes a living polymer possessing j monomeric units more than P_0. The solution of this set of equations gives the following results:

$$[P_0]_t = [P_0]_0 \cdot (k_p[M]t + 1)^{-1}$$

$$[P_1]_t = [P_0]_0(k_p[M]t + 1)^{-1}\{k_p[M]t/(k_p[M]t + 1)\}$$

$$\vdots$$

$$[P_j]_t = [P_0]_0(k_p[M]t + 1)^{-1}\{k_p[M]t/(k_p[M]t + 1)\}^j$$

$$\ldots, \text{etc.}$$

Hence the mole fraction of a j-mer is given by the geometrical series $(k_p[M]t + 1)^{-1} \cdot \{k_p[M]t/(k_p[M]t + 1)\}^j$ and, therefore, the ratio of weight average molecular weight to number average molecular weight is 2.

References

(1) D. J. Worsfold and S. Bywater, *Can. J. Chem.*, **38**, 1891 (1960).

(2) F. Wenger, *Makromol. Chem.*, **37**, 143 (1960).

(3a) D. N. Bhattacharyya, C. L. Lee, J. Smid, and M. Szwarc, *J. Phys. Chem.*, **69**, 612 (1965).

(3b) M. Matsuda, J. Jagur-Grodzinski, and M. Szwarc, *Proc. Roy. Soc. (London)*, *Ser. A*, **288**, 212 (1965).

(4) T. H. Bates, J. V. F. Best, and T. F. Williams, *Nature*, **188**, 469 (1960).

(5a) J. Jagur, M. Levy, M. Feld, and M. Szwarc, *Trans. Faraday Soc.*, **58**, 2168 (1962).

(5b) J. Jagur-Grodzinski and M. Szwarc, *Proc. Roy. Soc. (London)*, *Ser. A*, **288**, 224 (1965).

(5c) G. Spach, H. Monteiro, M. Levy, and M. Szwarc, *Trans. Faraday Soc.*, **58**, 1809 (1962).

(6) H. P. Leftin and W. K. Hall, *J. Phys. Chem.*, **66**, 1457 (1962).

(7) H. Sinn, C. Lundborg, and O. T. Onsager, *Makromol. Chem.*, **70**, 222 (1964).

(8) M. Morton, A. Rembaum, and J. L. Hall, *J. Polymer Sci. A*, **1**, 461 (1963).

(9a) D. L. Glusker, I. Lysloff, and E. Stiles, *ibid.*, **49**, 315 (1961).

(9b) T. E. Hogen-Esch and J. Smid, *J. Am. Chem. Soc.*, **88**, 307 (1966).

(9c) D. L. Glusker, E. Stiles, and B. Yoncoskie, *J. Polymer Sci.* **49**, 297 (1961)

(10) N. S. Wooding and W. C. E. Higginson, *J. Chem. Soc.*, **1952**, 774.

(11) R. Waack and M. A. Doran, *J. Phys. Chem.*, **67**, 148 (1963).

(12) T. Shimomura, K. J. Tölle, J. Smid, and M. Szwarc, *J. Am. Chem. Soc.*, **89**, 796, (1967).

(13) R. Asami, M. Levy, and M. Szwarc, *J. Chem. Soc.*, **1962**, 361.

(14) K. Ziegler and H. Dislich, *Ber.*, **90**, 1107 (1957).

(15) A. Vrancken, J. Smid, and M. Szwarc, *Trans. Faraday Soc.*, **58**, 2036 (1962).

(16) F. M. Brower and H. W. McCormick, *J. Polymer Sci. A*, **1**, 1749 (1963).

(17) G. Spach, M. Levy, and M. Szwarc, *J. Chem. Soc.*, **1962**, 355.

(18) J. Trotman and M. Szwarc, *Makromol. Chem.*, **37**, 39 (1960).

(19) W. K. R. Barnikol and G. V. Schulz, *ibid.*, **68**, 211 (1963).

(20) J. F. Garst, D. Walmsley, C. Hewitt, W. R. Richards, and E. R. Zabolotny, *J. Am. Chem. Soc.*, **86**, 412 (1964).

(21) M. Idelson and E. R. Blout, *ibid.*, **79**, 3948 (1957).

(22) A. Patchornik and Y. Shalitin, *Anal. Chem.*, **33**, 1887 (1961).

(23) H. Hartridge and F. J. W. Roughton, *Proc. Roy. Soc. (London)*, *Ser. A*, **104**, 376 (1923).

(24) F. J. W. Roughton and B. Chance, in "Investigation of Rates and Mechanisms of Reactions," Part II, Chap. 14, Interscience, New York, 1963.

(25) C. Geacintov, J. Smid, and M. Szwarc, *J. Am. Chem. Soc.*, **84**, 2508 (1962).

(26) R. V. Figini, H. Hostalka, K. Hurm, G. Löhr, and G. V. Schulz, *Z. Physik. Chem. (Frankfurt)*, **45**, 269 (1965).

(27) B. Chance, *J. Franklin Inst.*, **229**, 455, 613, 737 (1940).

(28a) Q. H. Gibson and F. J. W. Roughton, *Proc. Roy. Soc. (London)*, *Ser. B*, **143**, 310 (1955).

(28b) Q. H. Gibson and C. Greenwood, *Biochem. J.*, **86**, 541 (1963).

(29) E. F. Caldin, "Fast Reactions in Solution," Blackwell Sci. Publ., Oxford, England, 1964.

(30) K. G. Denbigh and F. M. Page, *Discussions Faraday Soc.*, **17**, 145 (1954).

(31) B. Stead, F. M. Page, and K. G. Denbigh, *ibid.*, **2**, 263 (1947).

(32) R. V. Slates and M. Szwarc, *J. Phys. Chem.*, **69**, 4124 (1965).

(33) P. Chang, R. V. Slates, and M. Szwarc, *ibid.*, **70**, 3180 (1966).

(34) R. V. Slates, M.Sc. Thesis, Syracuse, N.Y. (1967).

(35) J. E. L. Roovers and S. Bywater, *Trans. Faraday Soc.*, **62**, 1876 (1966).

(36) M. Fisher, M. van Beylen, J. Smid, and M. Szwarc, to be published.

(37) A. G. Evans and J. C. Evans, *Trans. Faraday Soc.*, **61**, 1202 (1965).

(38) B. Chance, Q. H. Gibson, R. H. Eisenhardt, and K. K. Lonberg-Holm, *Rapid Mixing and Sampling Techniques in Biochemistry*, Academic Press, New York, 1964.

Chapter V

Ions, Ion-Pairs, and Their Agglomerates

1. INTRODUCTION

Ionic polymerization, as indicated by its name, is propagated by macro-molecules endowed with reactive ionic end groups. It is, therefore, essential to know the detailed structure of these groups if one desires to understand the intricacies of these processes. The problem is far from simple. Ionic species may exist in a great variety of forms—as free ions, ion-pairs, triple ions, quadrupoles, and as still higher aggregates. Because their reactions are studied in the liquid phase, most of these entities may be coordinated with solvent molecules, and in some media the same ionic species may exist in two or more distinct states of coordination. The nature of the solvent determines the type of solvation, and hence a change of medium may profoundly affect the course of the reaction.

As a rule, the different ionic forms of a growing polymer are in dynamic equilibrium with each other. Because all the components of such a system may contribute to the propagation step, albeit each to a different degree, the kinetics of ionic polymerization may be quite complex and could be affected in a most unexpected way by changes of temperature or concentrations of the reagents. Furthermore, the ionic end groups of growing polymers may strongly intereact with other ionic, polar or polarizable substances present in the investigated system. This makes ionic polymerization extremely susceptible to minute amounts of some impurities—a most annoying feature of the reaction. The investigator may often be unaware of the presence of an adventitious substance which strongly affects the course of the studied process and, therefore, artifacts and erroneous observations are frequently encountered in the literature.

The laws governing the behavior of electrolyte solutions apply equally to those systems in which ionic polymerization occurs. Therefore, it is advisable to review the theories of electrolyte solutions and to examine critically the meaning and justification of the proposed concepts. Such an approach may be helpful to those who are interested in the details of ionic polymerization, as well as to other investigators who are concerned in general with ionic reactions.

2. CONDUCTIVITY OF IONOPHORES AND IONOGENS

The pronounced differences in the behavior of "weak" and "strong" electrolytes led Bjerrum (1) to reject the Arrhenius idea of an equilibrium established in a salt solution between the neutral, undissociated molecules and their respective ions. In fact, a typical salt is not a covalently bonded species; even in the solid phase its crystal is composed not of molecules, but of ions which form the lattice. Bjerrum assumed, therefore, that ionophores—substances virtually built up of ions—are completely dissociated in solutions. In contradistinction, only partial dissociation takes place in solutions of ionogens, i.e., compounds made up of neutral molecules which are capable of producing ions by reacting with suitable solvents, e.g., $CH_3COOH + H_2O \rightleftarrows CH_3 \cdot COO^- + H_3O^+$. In the latter system, the parent molecules and the resulting ions coexist in equilibrium with each other, as visualized by Arrhenius.

The classification of electrolytes into ionophores and ionogens, introduced by Fuoss (2) in 1955, is satisfactory for most cases, although there are substances which do not belong to either class. For example, crystals of nitrogen pentoxide are composed of NO_2^+ and NO_3^- ions (3); however, these ions associate into neutral, covalently bonded molecules of N_2O_5 when the solid is dissolved in low polarity solvents. The neutral N_2O_5 molecules exist also in the gas phase and on heating they decompose into *neutral* $NO_2 + NO_3$ radicals (4). A somewhat similar situation is observed for ammonium chloride. On sublimation, ionic crystals of NH_4^+,Cl^- dissociate into a mixture of neutral, gaseous molecules, viz., $HCl + NH_3$; and the same phenomenon is observed when the crystals are dissolved in some nonpolar solvents. In both systems the gain in lattice energy is responsible for the formation of ions in the solid state.

There is an important class of compounds which exists simultaneously in two forms: as covalently and as electrovalently bonded molecules. For example, in polar solvents some organic halides, which are basically covalent, coexist in equilibrium with their electrovalent form, i.e., $R \cdot X \rightleftarrows R^+,X^-$. The latter, R^+,X^- ion-pair, may dissociate into the respective free R^+ and X^- ions. Such a situation was first visualized by Ziegler and Wollschitt (5) who studied the equilibria between the ionic and the covalently bonded forms of $Ph_3C—Cl$ in liquid sulfur dioxide.

Equilibria involving covalent and ionic structures are probably important for some organometallic compounds. Intermediate situations may also be encountered, and then the binding is best described in quantum mechanical language as having some ionic and some covalent character. This representation was originally advocated by Pauling (6). A review of the behavior of such compounds, with special reference to their electrolytic dissociation, has been published recently by Monk (7).

Equilibrium between the two forms of a molecule, the covalent and the ionic, deserves further elaboration. The underlying ideas may be clarified by the energy–distance diagrams shown in Figure 1. Figure 1a represents the most common case of a gaseous molecule XY. Its ground state dissociates into X + Y radicals, while the dissociation of the excited state, X^+,Y^-, gives a pair of ions, $X^+ + Y^-$, the energy of which is much above that of the radicals X + Y because the ionization potential of X is usually larger than the electron affinity of Y. The energy level of X^+,Y^- is often sufficiently high with respect to that of XY to make the equilibrium concentration of the former species negligible. Any interaction between these two states makes the energy gap even greater.

In a suitable solvent, which interacts strongly with free ions and ion-pairs, the energy relation may be reversed, as shown in Figure 1b. In such a case, the solvated ion-pair forms the stable state and dissociates into solvated ions, whereas the covalent form is less stable; its concentration depends on the energy gap $X^+,Y^- \rightleftharpoons XY$, as well as on the difference in the respective entropies of formation.

An interesting situation arises under those conditions which stabilize strongly the free ions but not the ion-pairs. This is shown in Figures 1c and 1d. In the former, resonance splitting leads to an essentially covalent ground state which dissociates into solvated ions, whereas the excited state, described essentially as X^+,Y^- dissociates into free radicals. In the latter, the strong interaction may create two minima in the lower curve giving rise to two forms XY and (X^+,Y^-) as well as to the excited form X^+,Y^-.

It is also possible to visualize a case in which the two curves cross each other twice, because the ionic forces decrease slowly as $1/r^2$, whereas those responsible for covalent bonding fall off rapidly as $1/r^6$. Such a situation was considered by Weiss (151).

Depending on the conditions prevailing in the investigated system, the concentrations of the various species may, or may not, be comparable. In the former case, one deals with observable equilibria of the type $XY \rightleftharpoons X^+,Y^-$ or with even more complex situations involving three species, as depicted in Figure 1d. Alternatively, when the concentration of the excited species is very low, its presence may not be directly demonstrated, although it could influence the course of some reactions by being a fleeting intermediate involved in the process.

Bjerrum's hypothesis of complete dissociation of ionophores led to the modern theory of electrolytes developed by Debye and Hückel (8). To account for the behavior of free ions in solution, they substituted a continuous "ionic atmosphere" for the potential field generated by all the individual ions

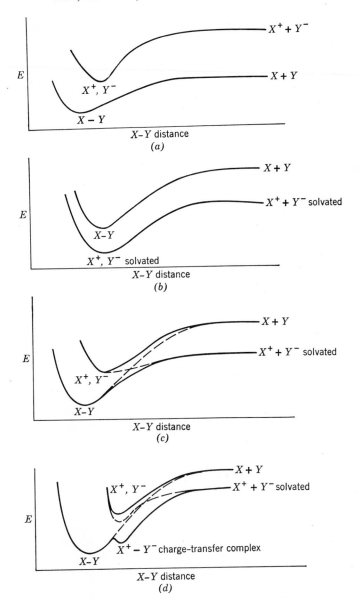

Fig. 1.   Energy diagrams illustrating the dissociation of a covalent molecule XY and an ion-pair X^+, Y^- in the gas phase (a) and in solution (b, c, and d). The formation of a charge-transfer complex (case d) was suggested by E. Grunwald.

located in the neighborhood of a reference ion. This allowed them to calculate the activity coefficients of the ions and the influence of the "ionic atmosphere" on their mobility in an external electric field.

It should be realized, however, that the Debye–Hückel theory is an approximation which cannot be extended to concentrated solutions, particularly in low dielectric constant solvents. In fact, Guggenheim has shown (9) that the Debye–Hückel approach is acceptable only if $z_1 z_2 \varepsilon^2 / aDkT < 2.8$. The reason for the limitation of the theory to dilute solutions is inherent in the model and in its mathematical treatment. The potential of an ionic atmosphere is determined by the Poisson equation and its solution gives an exponential relation between the charge of a reference ion and the potential generated around it. Such a relation is incompatible with the important principle of electrostatics, viz., the linear superposition of electrostatic fields. To retain the principle, it is necessary to approximate the exponential by the first two terms of the respective Taylor series, and thus the Debye–Hückel relations are derived. Hence, in this approach the approximation is a *necessity* and not merely a mathematical convenience, the "exact" solution being in principle not acceptable. Interestingly, for 1:1 electrolytes (more generally for $n:n$ electrolytes), the third term of Taylor's expansion vanishes, making, thus, the Debye–Hückel approximation particularly useful for treatment of such systems.

The Debye–Hückel theory is extremely successful in dealing with dilute, aqueous solutions of strong electrolytes. It leads, in perfect agreement with experiments, to a linear dependence of the activity coefficient of ions on the square root of ionic strength, and it requires a linear relation between the equivalent conductance, Λ, of an extremely dilute solution of an ionophore and the square root of its concentration. The last result vindicates the classic findings of Kohlrausch (10). His precise experiments, concerned with aqueous solutions of strong electrolytes, demonstrated the square root dependence of the equivalent conductance on electrolyte concentration and contradicted the Arrhenius theory, accepted in those days. According to the latter, the conductance of dilute electrolyte solutions should vary linearly with concentration of the investigated salt.

Critical examination of the relaxation phenomena occurring during the movement of an ion in an electric field eventually led Onsager (11) to further improvements of the theory of conductance. For dilute electrolytes his results are embodied in the equation

$$\Lambda = \Lambda_0 - (\alpha \Lambda_0 + \beta)C^{1/2}$$

where Λ_0 denotes the limiting conductance and α and β are constants determined by temperature, by the dielectric constant and viscosity of the solvent, and by the valence type of the salt. The Onsager equation rigorously determines the tangent of the conductance curve at zero concentration. His treatment, which resembles that of an ideal gas, reduces the ions to charged, dimensionless particles endowed with mass, and these are assumed to interact with each other only by virtue of their charges. Such a law must fail at finite concentrations when the dimensions of ions cannot be neglected, although, by analogy with an ideal gas law, the theory represents correctly the situation for *any* electrolyte at limiting conditions of high dilution.

There are circumstances when the ideal gas law, or laws of an ideal electrolyte solution, must fail even at high dilution. The ideal gas law is strictly valid for any gas at a sufficiently low pressure, if its molecules do not *dissociate*, but it does not apply to a dissociating gas, e.g., $I_2 \rightleftarrows 2I$. A similar restriction arises in dilute electrolyte solution when the dielectric constant of the solvent is relatively low. In such systems, as Bjerrum (12) has pointed out, the *association* phenomena cannot be neglected, and then, in addition to free ions, the solution must contain ion-pairs formed by coupling of two oppositely charged ions.

Ion pairs represent thermodynamically distinct species coexisting in equilibrium with free ions. Their neutrality makes them nonconducting and their formation decreases, therefore, the electrolytic conductance of an investigated solution. Under the influence of an electric field their dissociation is enhanced and, hence, the concentration of ion-pairs should decrease with the strength of the applied field. This abnormal increase in conductivity becomes perceptible at higher potentials and is known as the Wien effect. The phenomenon provides, therefore, an indication of the existence of ion-pairs, although, the distortion of the ionic "atmosphere," as discussed by Debye and Hückel, contributes also to an increase in conductance under the influence of a strong electric field. However, while the latter effect is proportional to the square of the field intensity (152), the classic Wien effect varies with its first power, hence, the two effects may be distinguished.

The presence of ion-pairs may be revealed by other, more direct methods. For example, the optical spectrum, and particularly the Raman spectrum, of an ion-pair might differ from that of a free ion (for further details, see Section 13). Solutions of organic radical-ions may reveal their association with the counterion by the additional splitting of the respective ESR spectra (13). This phenomenon is treated in Section 21. Finally, the IR band due to vibration of the cation in respect to the anion has been observed recently (165).

The concept of ion-pairs needs further elaboration. Its utility is limited in the same way as the concept of a molecule. The molecules "exist" in that temperature region in which their binding energy is greater than kT. At sufficiently high temperature, the concept becomes meaningless, because then only fleeting collisions between the free atoms are observed. The electrostatic binding energy of two oppositely charged ions is given by the Coulombic term $z_1 z_2 e^2/aD$, where $z_1 e$ and $z_2 e$ are their charges, a their separation, and D the dielectric constant of the medium (we shall discuss later the question of what is the appropriate value for D). Hence, the concept of ion-pairs is perfectly justified, if $z_1 z_2 e^2/aD$ is much greater than kT. However, in solvents of high dielectric constants, and for sufficiently large ions, such a concept may become ambiguous, if not superfluous. Indeed, a recent treatment of relaxation by Fuoss and Onsager (14) revealed the existence of an additional term in the conductance equation which was overlooked in previous treatments because of the too drastic approximation of the Boltzmann distribution. This term is proportional to $Cf_{\pm}^2$ ($f_{\pm}$—the activity coefficient of the ions) and, hence, its introduction removes the necessity of explicitly postulating an equilibrium between free ions and their associated pairs, if the coupling is caused entirely by electrostatic forces. This provides an alternative way of describing such an ionic agglomeration. Nevertheless, because the concept of ion-pairs is useful from a chemical point of view, particularly for those systems which show extensive association in dilute solutions, and because other forces may also contribute to ionic agglomeration, further discussion of this subject is advisable. Association due to forces other than electrostatic cannot, of course, be accounted for by the above-mentioned treatment of Fuoss and Onsager (14).

3. THE CONCEPT OF ION-PAIRS

The concept of ion-pairs was introduced by Bjerrum (12) on the basis of the following reasoning. Let us consider spherical ions of radius $\frac{1}{2}a$ immersed in a continuous medium of dielectric constant D. The probability, $W(r)$, of finding two oppositely charged ions at a distance r from each other may be calculated from the number of cations present in a spherical shell of thickness dr and radius r surrounding a reference anion. Hence,

$$W(r)dr = (4\pi r^2 dr/v) \exp(-u/kT),$$

where the first factor gives the volume of a spherical shell of radius r and thickness dr as a fraction of the total average volume, v, available for each cation (i.e., $v = 1/n$ where n is the number of cations in 1 cc of solution). The exponential represents the Boltzmann factor arising from the electrostatic interaction energy $u = -e^2/rD$. The resulting "distribution" function has a minimum for $r = r_c = e^2/2DkT$ and, therefore, it appears that a cation located at a distance $r < r_c$ tends to approach the reference anion and form a pair, while the one being outside the sphere $r = r_c$ escapes farther and farther from its partner. This representation, shown in Figure 2, led Bjerrum to propose that all pairs separated by a distance smaller than r_c should be counted as bounded ion-pairs, the remaining ones being considered as free ions. The fraction, α, of all the ions forming ion-pairs is given, therefore, by the integral*

$$\alpha = \int_a^{r_c} (4\pi r^2 dr/v) \exp(e^2/rDkT),$$

* Note the lack of normalization factor. Such a factor could not be introduced because the integrated function tends to infinity with $r \to \infty$.

its lower limit a giving the distance of the closest approach of two ions (the separation distance at contact). The final result of computation is given in terms of a function $Q(b)$, namely,

$$\alpha = (4\pi\beta^3/v)Q(b),$$

where $\beta = e^2/DkT$, $b = e^2/aDkT$, and $Q(b) = \int_2^b Y^{-4} \exp Y dY$. The numerical values of $Q(b)$ were tabulated by Fuoss and Kraus (15).

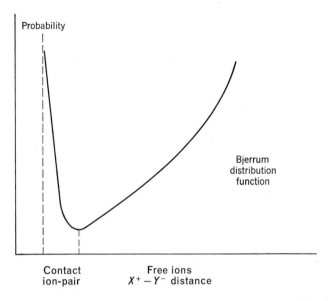

Fig. 2.   Bjerrum distribution function for the X^+,Y^- ion-pair. Note the lack of convergency for large X^+,Y^- distances.

The equilibrium constant of ion-pair association, K_A, is now defined in the conventional way, $K_A = \alpha/c(1 - \alpha)^2 f_\pm^2$, where $f_\pm$ is the activity coefficient of the ions as derived by Debye and Hückel. For a high dilution and in units of liter/mole,

$$K_A \approx (4\pi\beta^3 N/1000)Q(b).$$

The above expression was tested experimentally by Fuoss and Kraus (15) who determined K_A for tetraisoamyl ammonium nitrate in dioxane–water mixtures. As shown in Table I, for all of the investigated mixtures, the equation gives a constant distance a of about 6.4 Å in spite of a 10^{15}-fold variation in the K_A! This remarkable result was hailed as a proof of the reliability of the theory.

A closer inspection of Bjerrum's treatment reveals several unsatisfactory

TABLE I

Association Constants for Tetraisoamyl Ammonium Nitrate in
Dioxane–Water Mixtures

H_2O, %	D	$K_A{}^{-1}$	a, Å
0.60	2.38	2.0×10^{-16}	6.01
1.24	2.56	1.0×10^{-14}	6.23
2.35	2.90	1.0×10^{-12}	6.36
4.01	3.48	2.5×10^{-10}	6.57
6.37	4.42	3.0×10^{-8}	6.65
9.50	5.84	1.65×10^{-6}	6.45
14.95	8.5	1.00×10^{-4}	6.50
20.2	11.9	9.0×10^{-4}	6.70
53.0	38.0	0.25	6.15

a calculated from the Bjerrum–Fuoss equation,
$$K_A = (4\pi\beta^3 N/1000)Q(b), \qquad b = e^2/aDkT, \qquad \beta = e^2/DkT.$$

features. For large r's his "distribution" function diverges—an obvious indication of some oversimplification of the problem. This weakness of Bjerrum's approach was remedied by Fuoss (16). For a sufficiently large value of r one finds cations which, according to Bjerrum's treatment, are formally associated with the reference anion, although another cation, located closer to the reference ion, had been previously counted as its partner. To remove this redundance, Fuoss proposed to count as partner of a reference anion only that cation which is its closest neighbor, provided the latter is not even closer to another unassociated anion. In this way all the ions become paired, each anion having one, and only one, cation associated with it and vice versa. The pairing technique was ingeniously described by Fuoss through the following mental experiment.

Let us freeze out any momentary configuration of ions in a solution. Each anion is then surrounded by a concentric sphere of radius $a/2$. Imagine now all the spheres simultaneously expanding with a constant rate. At some stage of this expansion, a sphere surrounding the i-th-anion reaches the first neighboring cation which has not yet been "grabbed" by the expanding sphere of another anion. These two are then paired and removed from further counting. Continuation of the process eventually provides one and only one partner for each anion and vice versa.

The distribution function $P(r)$ derived from this model, gives the probability of finding a pair thus defined having its ions separated by a distance r. It is obvious that $P(r)dr$ must be proportional to N—the total number of ions present in volume V occupied by the investigated solution. It also has to be proportional to $4\pi r^2 dr/V$, to the Boltzmann factor exp $(e^2/rDkT)$, and to the probability $f(r)$ of having no other unpaired ion within the sphere of radius r. The probability of having an unpaired cation in a spherical shell of radius x

and thickness dx surrounding an anion, while another cation is placed at a distance r, is given by $\{(N - 1)/N\}P(x)$. The factor $(N - 1)/N$ arises from the fact that the function P refers to a system containing N randomly distributed ions, whereas now we deal with $N - 1$ ions only, one of them being known to be located at a distance r. Hence,

$$f(r) = 1 - [(N - 1)/N] \int_a^r P(x)dx.$$

Therefore,

$$P(r)dr = 4\pi r^2 dr(N/V) \exp(e^2/rDkT)\left\{1 - [(N - 1)/N] \int_a^r P(x)dx\right\}$$

and the solution of this integral equation gives

$$P(r) = 4\pi r^2(N/V) \exp\left\{\beta/r - (4\pi N/V) \int_a^{R_\infty} x^2 \exp(\beta/x)dx\right\}$$

where $\beta = e^2/DkT$ and the order of magnitude of R_∞ is determined by the condition $4\pi R_\infty^3/3 \approx V$. In actual calculation, we may assume $R_\infty = \infty$ because R_∞ is enormously greater than a. Of course, $P(r)$ is a proper probability function fulfilling the necessary condition

$$\int_a^\infty P(r)dr = 1;$$

its shape being shown in Figure 3.

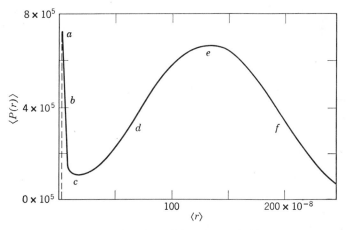

Fig. 3. Fuoss distribution function giving the probability of finding an ion-pair X^+,Y^- separated by a distance r (16). The area under a,b,c gives the fraction of ion-pairs, that under c,d,e,f represents the fraction of free ions. Point c corresponds to the critical distance a.

The problem of determining the equilibrium constant of associated ions is now reduced to the question of what is the appropriate value of the critical distance r_c which permits us to divide the pairs into two classes. The pairs which are separated by a distance smaller than r_c are then referred to as ion-pairs, whereas others, being farther away, are treated as free ions. Following Bjerrum's reasoning, Fuoss defined r_c as that value of r for which $P(r)$ is at a minimum. Thus, the value $r_c = e^2/2DkT$ is again arrived at.

Further improvement of the above treatment is due to Poirier and DeLap (17). In deriving the probability function, one should require that the anion located at a distance r from the fixed cation, which is counted as its partner, should not be paired with another cation. The respective factor, which was omitted by Fuoss, was then introduced, and this slightly modified the distribution curve. Moreover, Poirier and DeLap derived a more general expression for the equilibrium constant of the association by permitting the parameter r_c to have any arbitrary value and not restricting it to the particular condition $r_c = e^2/2DkT$.

All these improvements of the theory involve the unsatisfactory and artificial device imbedded in the concept of the critical distance, r_c. In this type of approach the pairs which are not in physical contact with each other are still counted as bound ion-pairs. Fortunately, in water and in other solvents of a relatively high dielectric constant, viz., greater than 30, the critical distance defined as $r_c = e^2/2DkT$ has a magnitude of molecular dimensions, i.e., at ambient temperature $r_c = 3$–10 Å. Hence, for such systems the proposed classification may be at least plausible on physical grounds. Any two ions separated by less than 10 Å could perhaps be considered as a unit, whereas all the remaining ions should be treated as free, their interactions being determined by the activity coefficients derived from the Debye–Hückel theory. However, the critical distance becomes unreasonably large in solvents of low dielectric constant, e.g., in benzene at 25°C, $r_c \approx 120$ Å, and a pair separated by 100 Å obviously cannot be treated as one unit. Furthermore, strict adherence to the Bjerrum–Fuoss formalism leads to an unreasonable peculiarity of the system—in a solvent of sufficiently high dielectric constant $r_c = e^2/2DkT$ may become equal to, or smaller than, a, and then no ion-pairs could be defined in such a medium.

The difficulty described herein is of general interest and it is desirable to investigate its origin. Whenever a molecular assembly is described by a parameter, or parameters, which allow the classification of its component systems into two thermodynamically distinguishable species, the distribution function given in terms of this parameter (or parameters) must be represented by two, essentially separated lobes. This is shown in Figure 4a, the distance r separating the investigated species serving as the relevant parameter. In such a case, one may neglect all the species having intermediate configurations because they are so few. On the other hand, the distinction between the two classes becomes impractical, and perhaps even undesirable, when the domains overlap as shown in Figure 4b. In such a case, it is still useful to find the distribution function giving the *density*, say, of cations around a reference anion, because such a function permits us to account for the properties of the studied systems. But, it is useless to pair the ions and to treat them as if they were divided into two or more classes, each representing a distinct component of the system in the proper thermodynamic sense. The logical difficulties

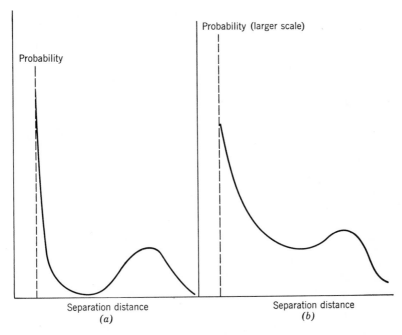

Fig. 4. Thermodynamic justification for the classification of dissociating molecules into two distinct species: (a) the distinction is justified; (b) the distinction is not justified.

inherent in the concept of ion-pairs as defined by Bjerrum and by Fuoss may perhaps be avoided. For example, a treatment suggested by Reiss (18) does not introduce at all the equilibrium notation nor does it involve the mass law. It accounts directly for the concentration dependence of conductance and, when applied to situations which justify the concept of ion-pairs, it leads to results virtually identical with those derived from the previously discussed models. This is most gratifying because it shows that Bjerrum's and Fuoss' treatments are valid for those systems which they intend to describe.

One final remark is in order. For solvents having medium dielectric constants the original approach of Bjerrum, as well as its refined forms as proposed by Fuoss (16) and later by Poirier and DeLap (17), give an exponential maximum $P(r)$ at $r = a$ followed by a relatively long and flat minimum. The steep part of $P(r)$, as shown in Figure 5, remains virtually unaffected by the modifications introduced into the original treatment of Bjerrum. The concentration of ion-pairs is determined by the integral $P(r)dr$ taken over the steep part of the curve, i.e., for r's close to a. The change in the distribution curve at larger values of r does not contribute to the final result. Therefore, all the treatments mentioned above lead to closely similar values for K_{assoc}. The differences in the shape of the distribution curves at larger values of r (Figure

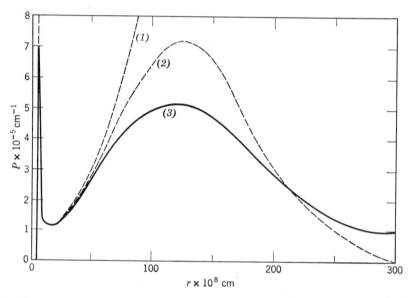

Fig. 5. Comparison of the distribution functions obtained for the same parameters by the methods of Bjerrum (*1*), of Fuoss (*2*), and of Poirier and DeLap (*3*).

5) affect the "long ion-pairs" but these are treated as free ions and, therefore, such a change of distribution does not affect either the concentration of "proper" pairs or the equilibrium constant.

The steep shape of the distribution curve in the vicinity of the cutoff point, i.e., for r only slightly greater than a, makes the value of the calculated K_{assoc} only slightly affected by the length of the critical distance, r_c, as long as its magnitude is not too small. This is shown by the data listed in Table II taken from the Poirier and DeLap (17) paper.

Our discussion demonstrates that the concept of free ions and ion-pairs is useful for diluted solutions of ionophores in media of relatively low dielectric constants. It becomes meaningless for concentrated solutions, e.g.,

TABLE II

Calculations of K_{assoc} for Different Dielectric Constants, D

D	K_{assoc}, calculated for $r_{crit} = 1.2$ Å	K_{assoc}, calculated for $r_{crit} = 2$ Å
2.38	5.7×10^{14}	5.8×10^{14}
2.90	1.0×10^{12}	1.0×10^{12}
4.42	3.8×10^{7}	5.1×10^{7}
8.5	6.2×10^{3}	9.8×10^{3}
12.1	3.6×10^{2}	5.6×10^{2}

for a molten salt. It loses also its usefulness in solvents of high dielectric constants, or at very low concentration of ions if Coulombic forces only are considered, because the two lobes of the distribution curve merge together. Figure 6 (taken from reference 19) shows how such a transition arises when the dielectric constant of the medium becomes sufficiently large, or the concentration of salt sufficiently low.

In view of all these difficulties, an alternative approach to the problem of ion-pairs may be desirable. The one outlined in the following section offers the advantage that other bonding forces, which may supplement the electrostatic attraction, can be introduced in the model. This possibility has been entirely neglected in the purely electrostatic treatments.

4. A THERMODYNAMIC APPROACH TO THE CONCEPT OF ION-PAIRS

Denison and Ramsey (20) seem to be the first in approaching the problem of ion-pair formation by thermodynamic methods. In their treatment only two situations were envisaged: (1) either the two oppositely charged ions are in physical contact, or (2) they are infinitely far from each other. Hence, the change in the electrostatic free energy of the system resulting from such a dissociation is given by

$$\Delta F_e = Ne^2/(r_1 + r_2)D$$

where r_1 and r_2 denote the radii of the ions which, for the sake of simplicity, are visualized as hard spheres immersed in a continuous medium having a dielectric constant D. The dissociation may be accomplished by the following sequence of steps:

1. Transfer of an ion-pair from its solution into evacuated space.
2. Separation of the ions in vacuum.
3. Immersion of the two isolated ions into solvent, while still keeping them far apart.

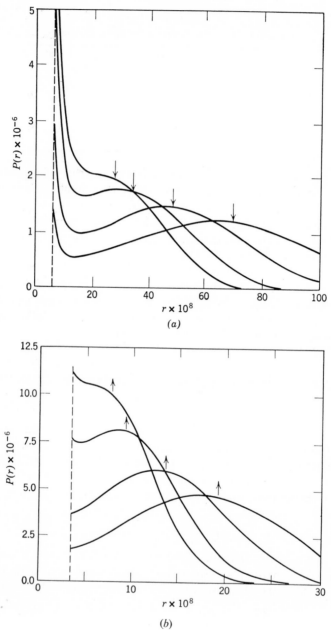

(a)

(b)

Fig. 6.   Transition in the shape of the distribution functions arising from the increase in the dielectric constant of the medium. The curves correspond to increasing concentration of ion-pairs. (a) $D = 20$; concentrations 0.5, 1.0, 2.0, and 3.0×10^{-3}. Genuine association. (b) $D = 78.6$; concentration 0.025, 0.05, 0.1, and 0.15. Pseudo-association (19).

In the first step, the ion-pair becomes desolvated and the free energy increases by $-\Delta F_{\text{sol ion-pair}}$. The free energy of the system increases again in the second step, namely, $\Delta F_v = Ne^2/(r_1 + r_2)$. In the third step, the gaseous ions become solvated and this decreases the free energy of the system by $\Delta F_{\text{sol free ions}}$. Hence the difference in the free energy of solvation of free, separated ions and that of an ion-pair is

$$\Delta F_{\text{sol free ions}} - \Delta F_{\text{sol ion-pair}} = -\{Ne^2/(r_1 + r_2)\}(1 - 1/D)$$

The free ions are more strongly solvated than the ion-pair, and the increase in the degree of solvation provides the driving force for the dissociation process. Therefore, the dissociation is much more extensive in well-solvating liquids than in the gas phase.

The above treatment considers only the *physical* free energy of solvation, i.e., it is concerned with the effects arising from the transfer of a charged sphere from vacuum into a continuous medium having dielectric constant D. This free energy of transfer, as shown by Born, is equal to $-(e^2/2r)(1 - 1/D)$, where r represents the radius of the investigated spherical ion. According to Ramsey's treatment, the free energy of physical solvation of a contact ion-pair composed of two ions of the same size is $-(e^2/2r)(1 - 1/D)$, i.e., it amounts to one-half of the total free energy of solvation of the two separated ions. The relative difference between the physical free energy of solvation of ion-pairs and that of the respective separated ions becomes progressively smaller as the dimensions of the ions diverge. For ions having radii r_1 and r_2, the association decreases the free energy by a factor $(1 + r_1/2r_2 + r_2/2r_1)^{-1}$. The effect of the chemical coordination of free ions, or ion-pairs, with solvent molecules is not considered. This phenomenon often contributes substantially to the free energy of dissociation and its importance will be discussed in a later section.

The thermodynamic formalism embodied in the equation $\Delta F_e = Ne^2/(r_1 + r_2)D$, permits us to calculate the effect of electrostatic interactions upon the entropy and enthalpy of dissociation. Thus,

$$\Delta S'_{\text{diss}} = \{Ne^2/(r_1 + r_2)D\}(\partial \ln D/\partial T)_p$$

and

$$\Delta H_{\text{diss}} = \{Ne^2/(r_1 + r_2)D\}(1 + \partial \ln D/\partial \ln T).$$

$\Delta S'_{\text{diss}}$ represents only the entropy decrease arising from the change in the degree of physical solvation, i.e., the entropy change of the surrounding medium. The fact that one particle (ion-pair) dissociates into two is not accounted for. The coefficient $(\partial \ln D/\partial \ln T)_p$ is always negative. For many liquids its absolute value exceeds unity (Table III), and hence $1 + \partial \ln D/\partial \ln T$ is usually also negative. This simple treatment accounts, therefore, for the exothermicity of the dissociation process observed for most solvent systems.

TABLE III
Temperature Dependence of the Dielectric Constants of Common Solvents

Solvent	Temperature range, °C	$\partial \ln D/\partial \ln T$	D at 0°C	Reference
Tetrahydrofuran	−70 to +25	−1.16	8.23	74
Tetrahydropyran	−40 to +25	−0.97	6.12	139
Dimethoxyethane	−70 to +25	−1.28	8.00	74
Diethyl ether	−70 to +25	−1.33	4.88	139
2-Methyltetrahydrofuran	−70 to +25	−1.125	6.92	139

The exothermicity of dissociation emphasizes again the importance of the physical solvation energy. The enthalpy of solvation is greater than the work needed to separate the charges in vacuum and, therefore, the solution of the separated, free ions is, from the energy point of view, more stable than that of the ion-pairs. On the other hand, solvation decreases the entropy of the system by immobilizing and polarizing the solvent molecules in the solvation shell and this factor opposes dissociation.

The formation of two particles (free ions) from one ion-pair increases the entropy of the system by a term ΔS_t, its value being determined mainly by a greater translational freedom of two ions when compared with that of the ion-pair. Hence, ΔF_{diss} should be given by

$$\Delta F_{\text{diss}} = Ne^2/(r_1 + r_2)D - T\Delta S_t$$

leading to

$$-\ln K_{\text{diss}} = e^2/(r_1 + r_2)DkT - \Delta S_t/R.$$

The term, $T\Delta S_t$, introduces concentration units; in fact, its omission left the original expression of Ramsey independent of units. This error was corrected in his later paper (21), namely, K_{diss} was given by the equation $-\ln K_{\text{diss}} = -\ln K_{\text{diss}}^0 + e^2/(r_1 + r_2)DkT$, where K_{diss}^0 denotes the dissociation constant of an "uncharged" ion-pair, i.e., of an associate of two fictitious species differing from the real ions only by the lack of charge. K_{diss}^0 accounts for the change in translational entropy as well as for other energy and entropy contributions, e.g., those resulting from the van der Waals interactions between the ions of a pair, etc.

The thermodynamic approach of Ramsey was elaborated further by Gilkerson (22), who pointed out that the simple treatment predicts, contrary to experimental findings, that ΔF_{diss} and K_{diss} should remain identical in a series of solvents having the same dielectric constant. Gilkerson's treatment starts with Kirkwood's zero approximation to the partition function of a particle in solution (23). Thus,

$$f = (2\pi mkT/h^2)^{3/2}gv_F\delta \exp\left(-E_0/RT\right),$$

where v_F is the available *free* volume per particle, g the partition function accounting for the internal degrees of freedom, and δ a constant slightly larger than unity. The dissociation equilibrium constant is, therefore,

$$K_{\text{diss}} = (2\pi\mu kT/h^2)^{3/2}\overline{gv\delta} \exp{(E_s/RT)} \exp{[-e^2/(r_1 + r_2)DkT]},$$

where $\mu = m_+ m_-/(m_+ + m_-)$ is the reduced mass of a pair,

$$\overline{gv\delta} = (g_+ g_- v_+ v_- \delta_+ \delta_-/g_\pm v_\pm \delta_\pm)$$

and E_s the difference of the specific interaction energies of ions and ion-pairs with the dipoles of the nearest solvent molecules. Gilkerson explicitly assumed that the energy needed to separate the charges in the solution is given by $e^2/(r_1 + r_2)D$ where D is the *macroscopic* dielectric constant of that solvent. This is a too drastic approximation for a treatment that attempts to account for the specific properties of solvents.

The influence of the solvent upon K_{diss} appears now in two terms: the free volume, $\bar{v} = v_+ v_-/v_\pm$, and the intereaction energy, E_s. The relative effect of the solvent's dipole moment upon the E_s value was calculated, and the obtained relation was confirmed by data derived from experiments performed in three solvents having approximately the same bulk dielectric constants but different dipole moments. Although, the usefulness of the latter calculations may be doubted, it should be stressed that Gilkerson's approach correctly emphasizes the importance of free volume and of specific solvation in calculating the equilibrium constants of ionic dissociations.

A different approach to the thermodynamic treatment of ions and ion-pairs has been proposed by Fuoss (24). His model consists of point-charge cations and spherical anions of radius a and volume v, both immersed in a continuum having a dielectric constant D. The ion-pair is defined as a species having a cation located *inside* the sphere representing an anion. The following experiment was then visualized.

To a solution containing Z cations and an equal number of anions present in a volume V, dZ cations and dZ anions are added. Let's assume that Z_1 cations are free and $Z_2 = Z - Z_1$ form pairs (i.e., each of them is located inside a sphere representing some anion). Now, out of the added dZ cations, dZ_1 remain free, whereas the remaining dZ_2 form pairs and, in view of the symmetry of the problem, dZ_2 of the added anions should also form ion-pairs. For a highly dilute solution, application of Boltzmann's method leads to the relation

$$dZ_2/dZ_1 = 2\{\exp{(-u/kT)}\}Z_1 v/(V - Zv),$$

where u is the potential energy of a cation inside an anion. Assuming that the anions behave like conducting spheres, one has to conclude that within their volume the potential remains constant. The potential ψ in the vicinity of each anion is given by the approximate form of the Poisson–Boltzmann equation, namely, $\Delta\psi = \kappa^2\psi$ for $r > a$ and $\Delta\psi = 0$ for $r < a$. Therefore, $u = -e^2/aD(1 + \kappa a)$, and integration of the equation gives dZ_2/dZ_1. The latter, coupled with the approximation $Zv \ll V$, leads then to

$$Z_2/V = (4\pi a^3 Z_1^2/3V^2) \exp [b/(1 + \kappa a)]$$

with b equal to $e^2/aDkT$. Denoting the total concentration of the salt by c and the fraction of the ions which are free by α, one finds $Z_1/V = \alpha c$ and $Z_2/V = (1 - \alpha)c$. Hence, for concentration given in units of moles/liter,

$$(1 - \alpha) = (4\pi a^3 N/3000)(\exp b)c\alpha^2 \exp [-b\kappa a/(1 + \kappa a)].$$

According to the Debye–Hückel theory, the second exponential gives the square of the activity coefficient, $f_\pm^2$, and thus,

$$K_a = (1 - \alpha)/(c\alpha^2 f_\pm^2) = (4\pi a^3 N/3000) \exp b.$$

The Fuoss treatment differs from that of Ramsey's in two details.

1. The former introduces the Debye activity coefficients which account for the ion–ion interaction, although this has been achieved in a rather artificial way.

2. The expression for the equilibrium constant involves a definite, non-exponential factor instead of the vaguely defined K_{diss}^0 which appears in the corrected Ramsey formula. However, the argument which makes dZ_2/dZ_1 proportional to $Z_1v/(V - Zv)$ is questionable. The spheres representing the enlarged anions form an *excluded* volume for the cations. Hence, one cannot accept the notion of an ion-pair represented by a cation located *inside* an enlarged anion, and consequently there is no justification to introduce the volume Z_1v when calculating the probability of ion-pair formation. It would be more rational to consider a spherical shell of thickness Δx surrounding each enlarged anion and to visualize that a point-shaped cation located inside such a shell represents an ion-pair. The success of Fuoss' equation, which accounts quantitatively for many experimental data, should be attributed to a plausible coincidence, namely,

$$(4\pi/3)\{(a + \Delta x)^3 - a^3\} \approx 4\pi a^2 \Delta x \approx 4\pi a^3/3$$

implying that $\Delta x \approx a/3$. This is reasonable because the extreme separation of ions in an ion-pair may be as large as $a/3$. This is a physically plausible distance, whereas the distance r_c, introduced in the original Bjerrum or Fuoss treatment, is physically unacceptable. Perhaps Δx represents the flat domain of the overall potential function within which the attractive forces (Coulombic and van der Waals) are nearly balanced by the repulsive forces operating in the system involving the two ions *and* the neighboring solvent molecules. The value of Δx should depend, therefore, on the nature of the solvent and on the size of its molecules. We shall see later that this indeed seems to be the case.

The idea of ion-pairs has led to many controversies and because of them we have discussed thoroughly the whole problem. The thermodynamic approach applies to highly dilute solutions. As stressed by Ramsey, in such a system the two oppositely charged ions exist *either* in contact *or* at such a large distance that their Coulombic interaction virtually vanishes. The intermediate states are highly improbable and this, in fact, justifies Ramsey's treatment. In concentrated solutions, the concept of ion-pair is ambiguous and inappropriate for the description of the system if the Coulombic attraction between oppositely charged ions is the only cause of pairing.

5. TERM INVOLVING CONCENTRATION UNITS

As pointed out in the preceding section, the expression for $\ln K_{\text{diss}}$ is expected to have the form

$$-RT \ln K_{\text{diss}} = Ne^2/aD - T\Delta S_t$$

the increase in the entropy ΔS_t being caused mainly by the gain in the translational degrees of freedom. In solution, $\Delta S_t \sim 10\text{--}15$ eu for the common units of moles/liter, if the rotational entropy of the dissociated products is greater than that of the associated ones. However, because an ion-pair represents a loose coupling of two species which freely rotate* even when associated, its dissociation decreases rather than increases the rotational entropy of the system. Hence, ΔS_t is expected to be substantially less than 10–15 eu.

The treatment of Gilkerson (22) introduces explicitly the $\exp(\Delta S_t/R)$ factor through the translational partition function. It is reasonable to assume that the partition functions for the internal degrees of freedom remain the same for ion-pairs and isolated ions, particularly when the ions are large and involve many atoms. Hence, the expression for K_{diss} is reduced to

$$K_{\text{diss}} = (2\pi\mu kT/h^2)^{3/2}(V_f/N)(h^2/8\pi^2\mu a^2 kT)\{\exp(-e^2/aDkT)\}\{\exp(E_s/RT)\}$$
$$= \{(\mu T)^{1/2}/0.12a^2\}(\text{free volume/total volume})\{\exp(-e^2/aDkT)\}\{\exp(E_s/RT)\}$$

in units of moles/liter. The third bracket of the unabridged expression represents the contribution of the free rotation of the ion-pair. Assuming that $T = 300°K$, $N\mu = 25$, and $a = 3$ Å, one finds $K_{\text{diss}} = 0.8$ (% of the free volume) $\{\exp(-e^2/aDkT)\}\{\exp(E_s/RT)\}$. For a free volume corresponding to 1% of the total volume, and in the absence of specific solvation, the preexponential factor is about 0.8, i.e., $\Delta S_t = -0.4$ eu. This value is too low

---

* Monoatomic ions have no rotational degree of freedom, but in solution such ions may be coordinated with solvent molecules; they possess then rotational entropy.

because the treatment neglects the vibrational partition function of the paired ions which may be quite appreciable.

The treatment proposed by Fuoss (24) gives the pre-exponential factor of ~ 15 for $a = 3$ Å and ~ 2 for $a = 6$ Å. In his equation the pre-exponential factor decreases more rapidly with increasing a than the corresponding factor in Gilkerson's expression. In fact, since an increase of a is usually associated with an increase of μ,* the Gilkerson expression is less affected by variations of a than it may appear from its first inspection. It is to be noted also that Fuoss' treatment does not involve the mass of the ions. This entity should, however, appear in the pre-exponential term.

The participation of free volume in the equation giving K_{diss} may be significant when the temperature dependence of K_{diss} is investigated. Denison and Ramsey (20) pointed out that the heat of dissociation arises from the temperature dependence of the dielectric constant of the solvent; their straightforward treatment gives

$$\Delta H_{diss} = \{Ne^2/(r_1 + r_2)D\}\{1 + \partial \ln D/\partial \ln T\}.$$

However, since the free volume increases with temperature, the expression $\Delta H_{diss} = (Ne^2/(r_1 + r_2)D)\{1 + \partial \ln D/\partial \ln T\} + RT\{\frac{1}{2} + \partial \ln V_f/\partial \ln T\}$ may be more correct. The added term decreases the predicted exothermicity of the dissociation.

6. CONTRIBUTION OF OTHER INTERACTIONS TO THE STABILITY OF ION-PAIRS

Although the main binding energy of ion-pairs arises from Coulombic interactions,† other forces may substantially contribute to their stability. For example, nonspherical ions may possess a dipole moment, as well as a charge. Dissociation of ion-pairs formed from such species was considered by Accascina, D'Aprano, and Fuoss (25) who included an appropriate term in the Ramsey equation, i.e.,

$$-\ln K_{diss} = -\ln K^0_{diss} + e^2/(r_1 + r_2)DkT + \mu e^2/d^2 DkT.$$

Here K^0_{diss} denotes the dissociation constant for the "uncharged" ion-pair, μ is the dipole moment of the unsymmetrical ion, and d is the distance between the center of the dipole and the center of the charge of the symmetrical counterion.

---

* This is particularly true for large organic ions where the increase of their size arises from the increase in the number of atoms forming their molecules.

† In aqueous solution, Coulombic interaction between a positive and a negative ion is not sufficient to secure the stability of the pair. In this medium, the contributions due to other interactions must be of paramount importance.

The contribution of the dispersion forces, which bind the two ions, is included in the term K_{diss}^0. For polarizable, colored ions, these interactions may be sufficiently strong to contribute significantly to the stability of the resulting ion-pair. For example, Grunwald (26) recently compared the dissociation of acids forming colored ions, viz., picric acid, with those yielding colorless anions. To account for the experimental data, it was necessary to assume that dispersion forces contribute as much as 4 kcal/mole to the energy of hydration of colored anions, and semiquantitative calculations confirmed these results. A similar problem was discussed by Noyes (27) who presented a treatment demanding a substantial contribution of nonelectrostatic forces to the free energy of hydration of inorganic ions (see also p. 244). The dispersion forces stabilize also the ion-pairs involving radical-anions derived from aromatic hydrocarbons, the respective binding energy being probably 2–3 kcal/mole.

In some systems, the destruction of the solvent structure caused by the presence of free ions contributes to the stabilization of ion-pairs. This factor is often important in aqueous solution, because liquid water possesses a well-defined and relatively stable structure. A recent example illustrating such an effect, which contributes to the stabilization of ion-pairs in methanol, was discussed by Kay et al. (28). Probably a similar effect is responsible for a larger degree of association of ions in D_2O than in H_2O (29). Another manifestation of this phenomenon may be found in the action of hydrophobic bonds advocated by Scheraga (30).

7. ACHIEVEMENTS AND SHORTCOMINGS OF THE "SPHERE IN CONTINUUM" MODEL

Until now we treated the problem of ion-pairs and free ions by considering an ion as a hard sphere immersed in a continuous medium. In spite of its oversimplification, this model of "sphere in continuum" is most successful, particularly when applied to large ions. It is obvious that the molecular, and therefore discontinuous, structure of the solvent might be disregarded when the ions are exceedingly large, but such an approximation becomes progressively less satisfactory as the dimensions of ions approach those of the solvent molecules. A more refined model must be used then; however, before discussing its improved versions, it may be desirable to summarize the substantial achievements of the "sphere in continuum" model.

_1._ The simple model of a "sphere in continuum" predicts a relation between the dissociation constant of ion-pairs and the dielectric constant of the solvent, viz.,

$$\ln K_{diss} = \ln K^0 - e^2/(r_1 + r_2)DkT.$$

It is tacitly assumed that K^0, r_1 and r_2 are not affected by the nature of the solvent, and hence a plot of $\ln K_{diss}$ vs. $1/D$ should be linear, its slope equal to $e^2/(r_1 + r_2)kT$. The distance, $r_1 + r_2$, separating the centers of the ions at contact may, therefore, be calculated from experimental data.

Numerous examples of such plots are reported in the literature, and for the sake of illustration a graph taken from a recent paper by Ramsey (31) is shown in Figure 7. A large change in the dielectric constant, which was varied

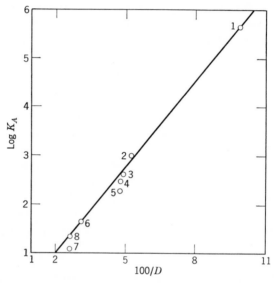

Fig. 7.   Linear relation between the log of the dissociation constant of an ion-pair and the reciprocal of the dielectric constant of solvents $\log K_A = - \log K_{diss}$ (31).

from 10 to 50, led to a 100,000-fold increase in the respective dissociation constant. Nevertheless, virtually all the experimental points are found on one straight line.

The results shown in Figure 7 refer to tetra-n-propyl ammonium bromide, which was deliberately chosen by Ramsey because its crystal structure is known. X-ray studies (32) showed that ($n = C_3H_7)_4N^+,Br^-$ crystallizes in the tetragonal space group 14, each bromide ion being surrounded by a tetrahedron of ammonium ions, and vice versa. The distance between the centers of the nearest, oppositely charged, ions was found to be 4.94 Å, whereas the separation of the ions in a pair was calculated from the slope of the line, $\ln K_{diss}$ vs. $1/D$, to be only 4.11 Å.

An ion in a crystal lattice is attracted by *all* its neighbors, whereas it remains under the influence of only one neighbor in a pair. Hence, one might expect a larger distance between the oppositely charged ions in a lattice than in a pair.* Observations derived from electron diffraction patterns (33) and microwave spectra (34) of the vapors of the halides of sodium, potassium, rubidium, and cesium support this conclusion. In each case, the interionic distance in a gaseous pair was found to be 10–17% smaller than in the respective crystal.

2. Studies of the conductance of $Bu_4N^+,B(Ph_4)^-$ in propylene carbonate and in mixtures of nitrobenzene and carbon tetrachloride (35) are perhaps the best examples of the success of the simple "sphere in continuum" model. The interionic distance, $r_1 + r_2$, was calculated from three sets of data: (a) from the equilibrium constant of the ion-pair association; this leads to $r_1 + r_2 = 7.0$ Å; (b) from the curvature of the phoreogram (the curve giving Λ as the function of $C^{1/2}$) leading to $r_1 + r_2 = 7.1$ Å; and finally (c) from the limiting conductance Λ_0 which in conjunction with Stokes' law gives $r_1 + r_2 = 7.2$ Å.

The limitations of the "sphere in continuum" model are clearly outlined in this discussion. The model accounts satisfactorily for the properties of electrolytes composed of large spherical ions in solvents of medium dielectric constants but with a low affinity for either ion. Deviations are expected when the investigated electrolyte is composed of small ions, such as Li^+ or Na^+, particularly when the molecules of solvent coordinate with the ions. Several approaches permit us to incorporate such effects into a theory based on the continuous nature of the solvent. (1) Allowance could be made for local dielectric saturations, local changes in viscosity, etc., in terms involving smooth variations of solvent properties with the distance from a reference ion. (2) "Discontinuous continuum" may be postulated, viz., it may be assumed that a layer of solvent in the immediate vicinity of the ion has properties distinctly different from its bulk. Both strata are treated then as continua. (3) A molecular approach may be developed in which the state of the nearest neighbors is described in terms of detailed intermolecular forces, e.g., specific coordination, etc.

Since the student of ionic polymerization is frequently concerned with such problems, we shall clarify them further under the following headings:

1. What is the correct local dielectric constant of the solvent?

2. What is the size of an ion in solution, and to what extent is it coordinated with solvent molecules?

---

* This is not a straightforward argument. Lattice energy per mole is greater than the interaction energy of 1 mole of isolated ion-pairs. Hence, one may deduce by similar, qualitative reasoning that a lattice is tighter than a hypothetical agglomerate pair. It is obvious, however, that ions of a pair should be deformed in the direction of its axis, whereas the ions in a crystal lattice reside in an approximately spherical field.

3. What is the structure of an ion-pair and how does it interact with the surrounding molecules?

However, before we start with this quest, let us review briefly the Walden rule and its implications.

8. STOKES RADII AND WALDEN RULE

One of the methods of determining $r_1 + r_2$, mentioned in the preceding section, calls for some comments. The limiting conductance, Λ_0, measures the mobility of both ions and, therefore, $\Lambda_0 = \lambda_0^+ + \lambda_0^-$, i.e., its value is given by the sum of contributions of the cations and anions to the charge transport. To separate Λ_0 into λ_0^+ and λ_0^-, the transference numbers of the ions have to be determined in the same solvent in which Λ_0 was measured. The only solvents, other than water, for which high precision transference numbers have been reported are: methanol (36), ethanol (37), nitromethane (38), and formamide (47). Using these data, Fuoss and Hirsch (35) calculated λ_0^+ and λ_0^- in methanol for NBu_4^+ and BPh_4^-, respectively, and showed that the mobilities of these ions differ only by 8%.

It is convenient to have a salt composed of a cation and an anion which contribute equally to its conductance. Triisoamylbutyl ammonium tetraphenyl boride nearly fulfils this requirement. The mobilities of $Bu(isoamyl)_3N^+$ and of $B(Ph)_4^-$ differ by 1% only. Recently, even a better standard has been synthesized (39), namely, tetraisoamyl ammonium tetraisoamyl boride.

Now, λ_0^+ and λ_0^- are related to Stokes radii of the corresponding ions, namely,

$$R_+ = 0.819/\lambda_0^+\eta \quad \text{and} \quad R_- = 0.819/\lambda_0^-\eta,$$

where η is the viscosity of the solvent and R_+ and R_- are the radii of the spheres which hydrodynamically behave like the corresponding ions. For sufficiently large ions it is expected that $R_+ = r_1$ and $R_- = r_2$, i.e., most probably such ions are not coordinated with the solvent and, therefore, they are not "coated" by its molecules either when free or when associated in pairs.

The Stokes relation accounts for the famous Walden rule which requires the product $\Lambda_0\eta$ of a salt to be constant and independent of the solvent. Obviously, the Stokes radii are a characteristic property of ions as long as their coordination with solvent molecules may be neglected, i.e., the Walden rule applies to large ions dissolved in solvents of low coordination power. However, even these restrictions are not sufficient to secure rigorous applicability of the rule. Precise experiments showed a definite trend of $\Lambda_0 \eta$ with the dielectric constant, D, of the solvent (40,41), namely, the reciprocal of $\Lambda_0 \eta$ seems to be linear with $1/D$. An ion in a polar solvent moves slower than an uncharged sphere of the same radius, and at least two phenomena contribute to this behavior: (1) a field generated by the ion compresses the surrounding solvent and therefore enhances its local viscosity; (2) the variable field of a moving ion induces orientation of solvent molecules, but due to the inevitable time lag this leads to a hindrance of its motion. The nonrelaxed orientation behind the ion pulls it back, and the process of orienting the solvent molecules in front of the ions adds a further resistance towards its forward motion. The relaxation effect was originally considered by Born whose treatment was refined by Hermans (42). In hydro-dynamic formalism this leads to an increase in the apparent Stokes radii of ions as shown by Boyd (43) and again by Zwanzig (44). The apparent Stokes radius, R, is given by the equation

$$R = R_\infty(1 + S/D)$$

where R_∞ represents the "true" Stokes radius observed in a medium of infinite dielectric constant. According to Boyd, the coefficient S is given by $2e^2\tau/27\pi\eta R_\infty^3$, whereas, the treatment of Zwanzig leads to

$$S = e^2\tau/9\pi\eta R_\infty^3(1 - D_\infty/D_0).$$

In these equations, τ denotes the dielectric relaxation time, and D_∞ and D_0 are the limiting dielectric constants referring to very high and very low frequencies.

Studies of Fabry and Fuoss (45) confirmed these relations. Within experimental uncertainties, the observed coefficient S was satisfactorily accounted for by either equation. Because R_∞ represents the "true" geometrical radius of an ion, its value should be compared with the respective r_1 or r_2 determined from other measurements.

A most striking example of the linear relation between R and $1/D$ has been reported by Coetzee and Cunningham (46). Using $N(isoamyl)_4^+, B(isoamyl)_4^-$ as a standard for which the transference numbers are most probably 0.5, they calculated the Stokes radii for NBu_4^+ in a series of solvents. Their results are shown in Figure 8 in the form of a plot of D vs. DR and lead to $R_\infty = 3.82$ Å.

Considerable deviations from the Walden rule may be expected in "organized" solvents such as water or methanol. Hydrogen bonding agglomerates these molecules and increases the viscosity of the respective liquid. The dissolved ions break the structure, disorganize locally the solvent, and decrease, therefore, its local viscosity. Consequently, the mobility of ions in a hydrogen-bonded liquid may be greater than expected from studies

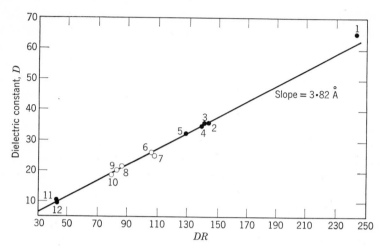

Fig. 8. Dependence of the Stokes radius on the dielectric constant of the solvent. (●) high precision results; (○) approximate (±2%) results (46).

performed in aprotic solvents. This increase in the Walden product was first reported by Frank (48) and confirmed by studies of Notley and Spiro (47) who investigated the conductance of alkali salts in formamide solutions.

9. THE DIELECTRIC CONSTANT OF SOLVENT

All the treatments previously discussed assumed that the solvent behaves as a continuum characterized by its bulk dielectric constant. However, the solvent molecules which are in the vicinity of ions, or in a space separating two relatively close ions, become polarized by the powerful electric fields (of the order of 10^6 V/cm) generated by the ions. Therefore, the local dielectric constant of the medium is modified; two examples will be given to illustrate this phenomenon.

1,2-Dichloroethane and ethylidene dichloride have virtually identical bulk dielectric constants, namely, 10.2 and 10.0, respectively. However, while ethylidene dichloride exists only in one conformation, dichloroethane may acquire a *trans-* or a *gauche*-conformation, the former being the most stable in the absence of an electric field. Therefore, an ordinary dichloroethane has a lower average dipole moment than the liquid exposed to an electric field which favors the *gauche*-conformation.

The dissociation of ion-pairs in dichloroethane is known to be about 10 times as extensive as in ethylidene dichloride (49), and this marked change of K_{diss} was explained (50) by assuming that the electric field of the ions increases the microscopic dielectric constant of dichloroethane. In the plot giving $\log K_{\text{diss}}$ as a linear function of the reciprocal of the dielectric constant, the point representing dichloroethane is displaced to the right (see Figure 7). From the degree of this displacement, the "effective" dielectric constant of dichloroethane was estimated to be 12.4.

The change in the conformation of dichloroethane, caused by the presence of the dissolved salts, was confirmed by other observations. Inami and Ramsey (51) examined the infrared spectra of dichloroethane in the absence and presence of salts and interpreted the observed spectra according to the method developed by Mizushima. The proportion of the gauche form was found to increase on salt addition. The increase in the fraction of the gauche form under the influence of an external electric field is demonstrated also by the negative dielectric saturation effect observed in dichloroethane by Piekara and Chelkowski (52).

A similar situation may be envisaged in dioxane solutions. Under normal conditions, a molecule of dioxane exists in a chair form in which the C—O—C dipoles oppose each other and, therefore, the bulk dielectric constant of this solvent is very low (2.4). However, the polar boat form is apparently favored in the vicinity of ions and this should increase the effective dielectric constant of dioxane (53).

The effects mentioned above are encountered only in some specific solvents; however, in every solvent we face the question of what is the correct value of the dielectric constant, D, which should be used in conjunction with the equation, $\Delta F_{\text{elec}}^0 = Ne^2/(r_1 + r_2)D$. The dilemma is clearly stated in a paper by Sadek and Fuoss (54). Let us separate the ions of a contact ion-pair. At the early stage of the process, the space between them remains empty because there is not yet enough room available to accommodate even a single molecule of solvent. Hence, the work performed at this stage of the dissociation is given by $(Ne^2/D')\{1/(r_1 + r_2) - 1/(r_1 + r_2 + \Delta)\}$ where D' should be substantially smaller than the bulk dielectric constant of the medium, and Δ denotes the extent of stretching needed to accommodate a molecule of solvent between the partially separated ions. The total work performed in separating the ions is, therefore,

$$\Delta F_{\text{elec}} = Ne^2/D'\{1/(r_1 + r_2) - 1/(r_1 + r_2 + \Delta)\} + Ne^2/D(r_1 + r_2 + \Delta)$$

instead of $\Delta F_{\text{elec}}^0 = Ne^2/(r_1 + r_2)D$. It is instructive to calculate the difference $\Delta F_{\text{elec}} - \Delta F_{\text{elec}}^0$ for two examples. Let us assume $D = 7.5$ (a value referring to tetrahydrofuran at room temperature) and $D' = 2$. Consider an ion pair having $r_1 + r_2 = 4$ Å and take $\Delta = 1$ Å. Under these assumptions, ΔF_{elec} is by 55.5% higher than ΔF_{elec}^0, reducing the effective average dielectric constant from 7.5 to 4.8 and decreasing the dissociation constant at room temperature by a factor of about 2×10^4. For $r_1 + r_2$ of 10 Å, a similar calculation leads to an increase of ΔF_{elec} by $\sim 25\%$ with respect to ΔF_{elec}^0; the effective dielectric constant decreases to 6 and the dissociation constant is reduced by a factor of 6.5 only. Hence, it is imperative to inquire what is the correct value of the effective dielectric constant whenever salts of small ions are investigated. In fact, the discrepancy between theory and experiment is probably underestimated in our examples, because the dielectric saturation considerably reduces the dielectric constant of the solvent located between adjacent ions.

The behavior of electrolytes in mixed solvents is even more complex. The composition of the mixture in the vicinity of ions, or ion-pairs, may deviate substantially from its average composition. Consequently, the local dielectric constant may greatly differ from that of the bulk of the mixtures and therefore the slope of the line giving $\ln K_{diss}$ as a function of $1/D$ becomes distorted. Usually, this leads to values of $r_1 + r_2$ which are too high, although the linear dependence of $\ln K_{diss}$ on $1/D$ is obeyed remarkably well over a wide range of compositions.

The complexity of mixed solvents is further illustrated by the following example (55). The addition of small quantities of benzene or hexane to methanol *increases* the dissociation of an ion-pair, although at higher proportion of hydrocarbon its value eventually decreases. This strange behavior is explained by the polymeric nature of methanol. The addition of another solvent causes its depolymerization (rupture of hydrogen bonds) and, therefore, the polarity of the mixture initially increases.

10. THE IONIC RADIUS

It is proper to remind the reader at the opening of this section of an important statement by Pauling: "The apparent ionic radius depends on the physical property under discussion and it differs for different properties." Hence, a single ionic radius may not be sufficient to describe an ion in different media, and if this is the case, which value is the most fundamental?

It has been customary to consider the crystal radii of ions, which are so valuable in describing ionic lattices, as a proper measure of dimensions of bare, nonsolvated ions in solutions. This attitude may be questioned. Indeed, it appears that the most fundamental dimension of an ion is its radius in the gaseous phase when its size is not perturbed by any external force. Such a radius, as has been pointed out by Stokes (56) should be considerably larger than the customary ionic radius.

To evaluate the van der Waals radii of gaseous ions, Stokes estimated first the atomic radii of the respective noble gases. These were calculated from the x-ray data for solid crystals of He, Ne, Ar, Kr, and Xe, and independently from the coefficients of the relevant Lenard–Jones potentials. The average of both estimates, which agree within $\sim 2\%$, were taken as a starting point of calculations in which one applies the quantum mechanical scaling principle for isoelectronic series, namely, $r'/r'' = (Z'' - S'')/(Z' - S')$. Here, Z' and Z'' are the relevant atomic numbers and S' and S'' the respective screening coefficients. Thus, the radii of alkali and alkali earth ions were obtained and they are listed in Table IV. Similar calculations led to the radii of gaseous halogen ions (see ref. 56).

TABLE IV

The van der Waals Radii of Gaseous Ions and Their Crystal Radii[a]

Ion	r_{vacuum}, Å	r_{cryst}, Å	Δr
Li^+	1.17	0.60	0.57
Na^+	1.352	0.95	0.40
K^+	1.671	1.33	0.34
Rb^+	1.801	1.48	0.32
Cs^+	1.997	1.69	0.31
Mg^{2+}	1.180	0.65	0.53
Ca^{2+}	1.480	0.99	0.49
Sr^{2+}	1.625	1.13	0.49
Ba^{2+}	1.802	1.35	0.45

[a] R. H. Stokes, *J. Am. Chem. Soc.*, **86**, 979 (1964).

The radii of gaseous alkali ions appear to be proportional to the respective metallic radii, the latter represent one-half of the separation distance between the nearest neighbors in crystals of alkali at $0°K$. The empirical relation gives (57) $r_{\text{metallic}} = 1.333 \, r_{\text{vacuum}}$, as shown in Figure 9, and this interesting result was justified by a theoretical consideration. It should be noted that such a simple relation does not hold for the crystal radii—the latter are also plotted in Figure 9—and the respective line shows that r_{cryst} values are linear, but not proportional, to r_{vacuum} values.

The lattice energy, E_0, of a solid metal is given by the sum of its heat of evaporation at $0°K$, L, and its ionization potential, I. For alkali metals, Stokes found (57) a most interesting relation between E_0 and r_{vacuum}, namely,

$$E_0 = L + I = 142/r_{\text{vacuum}} + 38,$$

as shown in Figure 10. The term of 38 kcal/mole was interpreted as a contribution of the self-energy of an electron moving in a metal lattice to its energy. In view of this interpretation, it is interesting to note that the photo-ionization threshold of alkali metals in liquid ammonia solution was found to be 34 kcal/mole (58). The self-energy of an electron at rest, if such a concept has any physical meaning at all, is zero, because its radius becomes infinitely large, as demanded by the Heisenberg uncertainty principle.

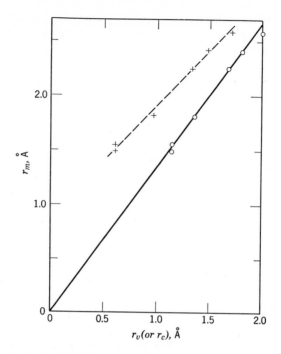

Fig. 9. Proportionality between the radii of alkali ions in vacuum (r_v) and their metallic radii (r_m). The crosses give the relation between crystal radii, r_c and r_m. The dashed line does not pass through the origin of coordinates (57).

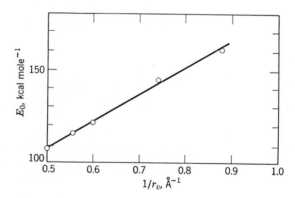

Fig. 10. Dependence of the lattice energy of an alkali metal on the radius of its ion in vacuum (52).

The radius of an ion *must* decrease on its immersion in a polar medium. This is a necessary consequence of the virial theorem. Immersion of an ion in a polar medium decreases the total energy of the system and, therefore, its kinetic energy must increase. The large gain in kinetic energy may be accommodated only by electrons, and an increase in their momenta requires a decrease of the radius of an electronic shell, i.e., it causes a contraction of the volume of an ion. A similar situation is encountered in the process of formation of a crystal lattice from gaseous ions. Not only their size, but even their shape becomes distorted as the lattice is formed; the originally spherical ions become polyhedrally shaped as demonstrated by the detailed electron contour map of sodium chloride (59).

The above observation leads to an extremely simple interpretation of the lattice energy of some ionic crystals (60). Formation of a crystal lattice by agglomeration of positive and negative ions of equal gaseous radii distorts their shape into polyhedra which contact each other and leave no empty space in the crystal. There is, however, no contraction of volume because the molar volume of salt is nearly equal to the sum of the molar volumes of the gaseous ions.* The lattice energy should, therefore, be equal to the total energy of gaseous ions, i.e.,

$$\text{lattice energy} = \sum \tfrac{1}{2}(e^2/r_{\text{vacuum}})$$

and, indeed, the available experimental data confirm this conclusion. Stokes' treatment led to a striking corollary, namely, that the lattice energy of such crystals is independent of their structure and therefore of the Madelung constant. This result is gratifying, since the assumption of $D = 1$, which underlies the calculation of the Madelung constant, appears to be artificial and unrealistic. In fact, some doubts were expressed on the validity of the method (61). Lattices formed from ions having unequal radii must contain some voids, and this affects their molar volume as well as their lattice energies (60).

The problem of the free energy of hydration of inorganic ions has been ingeniously treated by emphasizing the difference between the radii of gaseous and hydrated ions (56). In Born's approach, the physical free energy of solvation of an ion is taken as the difference of its self-energy in vacuum and in a polar solvent, and the equation

$$-\Delta F_{\text{solv}} = (NZe^2/2r)(1 - 1/D)$$

was derived by applying the "charging" procedure and assuming the same ionic radius, r, for the gaseous and the solvated ion. This approach failed utterly when applied to aqueous solutions. The first layer of water molecules

---

* Note that the ions are polyhedral in a crystal lattice, e.g., of cubical shape, but the isolated gaseous ions are spherical.

surrounding a cation is irrotationally bounded, and infrared and microwave studies showed that their dielectric constant is reduced to about 5. However, even if the appropriate correction is introduced, still an unacceptable value for $\Delta F_{hydration}$ is obtained. To reconcile the calculation with the experimental data it is necessary to postulate an impossibly low value of $D = 1.3$ for the bounded water. To justify such a low value Noyes (27) assumed that the molecules of water in the bounded layer are electronically saturated. This is most improbable. An alternative suggestion by Pitzer (62), who visualized a cavity surrounding an ion, also appears to be unreasonable.

The difficulties may be surmounted by postulating different radii for the gaseous and hydrated ions. The small difference in the enthalpies of solution of most ionic crystals in water indicates that the forces acting upon an hydrated ion have magnitudes comparable to those exerted by its crystal lattice (63). Therefore, the degree of compression is expected to be similar in both media, and hence, the radius of a bare ion in water should be approximately equal to its crystal radius. Consequently, the following model was proposed by Stokes (56). The appropriate crystal radius was attributed to each ion in aqueous solution. The positive alkali ions and Ba^{2+} were assumed to be surrounded by one layer of irrotationally bounded water molecules, the remaining bivalent and the trivalent cations were assumed to possess two such layers, and the negative ions none. The last assumption is justified by microwave studies (64) which show that water molecules adjacent to large halogen ions are free to rotate. The F^- ion is an exception—it is strongly hydrogen bonded to the adjacent water molecules. The average dielectric constant of the bonded water layer is taken to be a harmonic average of the irrotational water ($D = 5$) and the free water ($D = 78$), i.e., $D_{eff} = 9$. Hence

$$-\Delta F_{hydration} = \mu_{aq\ ion} - \mu_{gas\ ion} + \Delta\mu_S,$$

the last term representing a small correction resulting from nonelectrostatic interactions. Now,

$$\mu_{aq\ ion} = (NZ^2e^2/2)\{\Delta r/r_{cryst}(r_{cryst} + \Delta r)D_{eff} + 1/(r_{cryst} + \Delta r)D\},$$

where Δr is the thickness of the bounded water layer, whereas $\mu_{gas\ ion} = NZ^2e^2/2r_{gas\ ion}$. Using the values for $r_{gas\ ion}$ given in Table IV, Stokes got excellent agreement, within 2–3%, between the calculated and the observed free energies of hydration of inorganic salts.

Stokes' procedure involves one more additional term, namely, the work performed on moving the ions through the interfacial layer. However, this term, discussed by Randles (65), disappears when the hydration of a salt is computed, because the contributions of the positive and negative ions cancel out. This term remains, however, as an undetermined constant in the expression giving the $\Delta F_{hydration}$ of individual ions.

In conclusion, the radius of a bare ion in a polar solvent is smaller than its gaseous radius. The extent of its contraction depends on the dielectric constant of the surrounding medium—the ion becomes more compressed as its interaction energy with the solvent increases. The radius of a bare ion has to be distinguished from its hydrodynamic radius (Stokes radius). Under proper conditions, the ions may become coordinated with the solvent molecules and then the hydrodynamic size of such an ion increases appreciably. The approximate dimensions of such ions may be deduced from Stokes radii calculated from the respective λ_0^+ and λ_0^-. The association of ions into ion-pairs is expected to decrease their degree of coordination, and this contributes to the endothermicity of the association process.

11. COORDINATION OF IONS WITH SOLVATING MOLECULES

The coordination of an ion with polar solvent molecules, or with other solvating species, results mostly from charge–dipole interactions. The interaction energy between a point-charge, e, and a dipole of moment, μ, when separated by a distance r in a continuum having dielectric constant D, is $\mu e \cos \theta / Dr^2$. Here, θ denotes the angle between the direction of a dipole and the line joining the interacting species. To appreciate the degree of such an association, let us calculate the distance at which the energy of interaction of a monovalent ion with a dipole having a moment of 2 D is equal to kT at 25°C. This distance is 3.4 Å for $\theta = 0$ in a solvent having dielectric constant 20; it increases to about 5 Å when the dielectric constant of the medium decreases to 10, and it is still only 15 Å for species interacting in vacuum, i.e., for $D = 1$. In a realistic calculation of the interaction energy of an ion with neighboring solvent molecules, one should substitute the "local" dielectric constant for the bulk dielectric constant of the medium, and it is probable that this energy is not greater than 3 or 4 kcal/mole.

Numerous calculations of the solvation energies (particularly hydration energies) have been reported in the literature. Originally, macroscopic approaches were utilized in calculating the enthalpies of solvation. A microscopic approach, in which charge–dipole interactions were directly calculated, was initiated by the work of Bernal and Fowler (66), who greatly improved the treatment of such problems. For example, in their studies of hydration of inorganic salts the direction of the dipoles was assumed to coincide with the line joining the centers of the charge and of the dipole when solvation involves water molecules located in the vicinity of a cation. However, these two directions were assumed to be at an angle when anions are hydrated. The change of direction was justified by the necessity of having protons of water molecules located as closely as possible to the negative anion.

In many calculations, an ion and a polar molecule were visualized as rigid spheres possessing the respective charge, or dipole, at their centers. Such models cannot account for a difference between the solvation energies of a cation and an anion of the same size. The difference in orientation provides a parameter, which may discriminate between a positive and a negative ion, and thus the original calculation of Bernal and Fowler (66), as well as its refined version proposed by Eley and Evans (67), leads to a difference of about 10 kcal/mole in the enthalpy of hydration of cations and anions of the same radius.

The problem of orientation is particularly important when coordination involves large organic molecules, e.g., tetrahydrofuran (THF). This is clearly shown by the following, perhaps slightly oversimplified, models,

$$\oplus\ O\!\!\!+\!\!\!> \begin{array}{c} CH_2\!\!-\!\!CH_2 \\ | \\ CH_2\!\!-\!\!CH_2 \end{array}, \qquad \ominus\ \begin{array}{c} CH_2\!\!-\!\!CH_2 \\ | \\ CH_2\!\!-\!\!CH_2 \end{array}\!\!+\!\!> O.$$

The dipole of THF is located at the oxygen atom of the ether. It may, therefore, closely approach a cation, but it remains separated from the anion by the aliphatic CH_2 groups. This difference in distance greatly reduces the respective coordination energy, making it substantial for a cation and insignificant for an anion of the same radius.

An interesting approach that differentiates between the coordination of solvent molecules with cations and anions, has been recently developed by Buckingham (68). He considers the interaction of an ion (a point-charge, Ze) and a solvent molecule in which the charges are separated, viz., a fraction δe_i being located at the end of a vector $\bar{r}_i$. The electrostatic classic expression is expanded, giving thus the energy, u_e,

$$u_e = Ze \sum \frac{\delta e_i}{|\bar{R} + \bar{r}_i|}$$

as a power series of $\bar{R}$ ($\bar{R}$—the vector describing the location of the center of the solvent molecule in respect to the ion). This leads to

$$u_e = Ze \left\{ \frac{q}{R} - \frac{\mu \cos \varphi}{R^2} + \frac{1}{3}\Theta_{\alpha\alpha} \frac{3 \cos \theta_\alpha - 1}{R^3} + \cdots \right\}$$

where $q = \sum \delta e_i = 0$ for a neutral solvent molecule, μ is its dipole moment, and $\Theta_{\alpha\alpha}$ are the components of its quadrupole moment along the fundamental axis of the tensor. Higher order terms are neglected.

Introduction of a quadrupole moment differentiates automatically between the solvation of a cation and an anion. On reversing the sign of the charge of an ion, the sign of the dipole–charge interaction is also reversed, whereas the sign of the charge–quadrupole moment interaction remains unaffected. Hence, the total energy of interaction is given by the sum of $|u_{\text{dipole}}|$ and $|u_{\text{quadrupole}}|$ in one case, and by its difference in the other.

In his most extensive treatment of interaction energies, Buckingham also took into account the effect of polarizability, of dispersion forces, and of solvent–solvent interaction in the solvation shell. The contribution of the latter interaction depends on the structure of the shell, e.g., the dipole–dipole interaction energy between spherical solvent molecules is given by $(15/16)(\mu^2/R^3)\sqrt{6}$ for a tetrahedral configuration and $\{(18\sqrt{2} + 3)/4\}\mu^2/R^3$ for an octahedral arrangement.

One additional point should be mentioned. The actual interaction energy depends on the temperature of the investigated solution, because Brownian motion disrupts the energetically favored organization. Moreover, the difference between solvent–solvent interaction energies in the bulk of the liquid and in the solvation shell should be accounted for. Because the former interaction energy increases at lower temperatures, the exothermicity of solvation decreases with decreasing temperature of the solution. This effect is partially compensated by a simultaneous decrease in the entropy of solvation.

An interesting aspect of entropy of solvation has been raised recently by Conway and Verrall (124). They showed that the configurational entropy of mixing becomes nonideal when the ions and solvent molecules differ greatly in their size.

The solvation effects are not restricted to the liquid phase only. By using mass spectrometric methods applied to molecular flow conditions, Hogg, Haynes and Kebarle (69) succeeded to study heats and entropies of association of NH_4^+ ions with ammonia molecules into gaseous $NH_4^+(NH_3)_n$ aggregates. Thus, $\Delta H_{2,3} = -17.8$ kcal/mole and $\Delta H_{3,4} = -15.9$ kcal/mole, the subscripts referring to the processes: $NH_4^+(NH_3)_2 + NH_3 \rightleftarrows NH_4^+(NH_3)_3$ and $NH_4^+(NH_3)_3 + NH_3 \rightleftarrows NH_4^+(NH_3)_4$, respectively. These reactions involve a decrease in entropy of -38 and -40.5 eu respectively. The heat of coordination is substantially larger in the gas phase than in the liquid state, because in the former it is not necessary to disrupt a solvent–solvent structure, a process associated with endothermicity proportional to the heat of evaporation. Recently this work was extended; e.g., hydration of protons was studied (153).

12. DIMENSIONS OF SOLVATED IONS

In solution, a bare ion may be bounded to n solvent molecules. The precise meaning of n is somewhat ambiguous and depends on the type of measurements performed to obtain its magnitude. On the whole, it represents an average coordination number, and as such it need not be an integer. The solvated ions remain in equilibrium with the bulk of the solvent and, therefore, its molecules continuously exchange their positions moving from the solvation shells into the liquid's interior and vice versa. The exchange is expected to be rapid, although it may be slow if additional forces contribute to the binding of the solvation shell. An extreme example was provided by Hunt and Taube (70) who have shown, by using O^{18} labeling, that the exchange of water between $[Cr(H_2O)_6]^{3+}$ and its surroundings corresponds to a half-lifetime of about 40 hr. It is known that the Cr^{3+} ion forms coordinate links with its ligands, and this contributes substantially to the stability of the hydration shell. In spite of this, the binding in $[Cr(H_2O)_6]^{3+}$ does not appear to have any effect on the electrolytic behavior of that ion. For other trivalent ions, the exchange is much more rapid, its half-life being less than 3 min. In the absence of ligand type coordination, the exchange is even faster, usually the respective relaxation time is of the order 10^{-6} to 10^{-9} sec.

An interesting approach to the problem of solvent binding and of its exchange between the bulk of the liquid and the solvation shell was suggested by Samoilov (71,72). Denote by τ the average time of contact between two molecules of solvent placed in the bulk of the liquid and by τ_i the time spent by a molecule of solvent in the vicinity of an ion. The ratio

$$\tau_i/\tau = \exp(\Delta E/RT)$$

where ΔE represents the difference in the respective energies needed to separate by Δr a solvent molecule from an ion and to move apart, by the same distance, two adjacent solvent molecules. At this critical distance, the molecules may pass freely from one environment into the other. A procedure was developed (71) to calculate ΔE from the self-diffusion constant of the solvent and the temperature coefficient of an ion's mobility. For a series of ions in aqueous solutions the following values of ΔE were obtained,

Ion	Li^+	Na^+	K^+	Cs^+	Mg^{++}	Ca^{++}
ΔE, kcal/mole	0.73	0.25	−0.25	−0.33	2.61	0.45
τ_i/τ at 21.5°C	3.48	1.46	0.65	0.57	86.3	2.16

The magnitude of τ is independently estimated to be about 2×10^{-9} sec, and hence τ_i is less than 2×10^{-7} sec even for a strongly binding Mg^{++} ion.

It is most striking to find negative ΔE values for larger ions (K^+ and Cs^+). They indicate the *higher* mobility of water molecules in their vicinity than in the bulk of the liquid, and this phenomenon, named negative hydration, was observed in aqueous solutions of potassium chloride and iodide (73).

The number of solvent molecules bonded to an ion was frequently determined by a modification of the Hittorf method of evaluating transport numbers. An apparently inert nonelectrolyte was added to the conducting solution in anticipation that its molecules do not move during the course of an electrolysis. This assumption was shown to be erroneous for the aqueous systems studied in the past; the solute has to be polar to be soluble in water and then it contributes to the solvation of ions and, therefore, participates in their movement. Probably, this method might be more reliable in non-aqueous media. For example, a change in the concentration of radioactive benzene, added as a tracer to the tetrahydrofuran solution of an electrolyte, may determine reliably the amount of the transported solvent. However, this writer is not aware of any Hittorf-type experiments performed in non-aqueous media with the intention of determining the degree of solvation of an ion.

The size of the solvation shell may be estimated from the ion's mobility by applying for this purpose Stokes' law or some of its modifications (see Section 8). An example of such a study was reported recently by this writer and his co-workers (74) who investigated the conductance of salts of tetraphenyl boride in ethereal solvents over a wide temperature range (-70 to $+25°C$). For the sake of illustration, some of their results are shown in Figure 11 in the form of a Fuoss plot from which Λ_0 is calculated. Triisoamylbutyl ammonium tetraphenyl boride was included in the investigation because for this salt $\lambda_0^+ = \lambda_0^-$. From the observed Λ_0's of the alkali salts, the respective λ_0^+ of alkali ions were calculated by subtracting λ_0^- of the quaternary ammonium ion,

$$\lambda_0^- = \tfrac{1}{2}\Lambda_0(\mathrm{Bu,(isoamyl)_3N^+,BPh_4^-}).$$

The calculated Stokes radii for Na^+, Cs^+, and $Bu(isoamyl)_3N^+$ ions in tetrahydrofuran and in dimethoxyethane are given graphically in Figure 12. The data seem to be reliable within 15%. Their slight increase at lower temperatures could be genuine, reflecting an increase in the electrophoretic effect as Brownian motion becomes less violent. However, the observed changes are within the experimental errors and, therefore, their significance may be questioned. Comparison of the results as revealed by Figure 12 is most instructive. The Stokes radius of the $Bu(isoamyl)_3N^+$ ion is virtually the same in both solvents (about 4.2 Å), its value being only slightly smaller than that predicted from a molecular model or from molecular volumes. This indicates that this bulky ammonium ion is not coordinated with either solvent.

Let us digress at this juncture and point out that Stokes' law includes a coefficient of 6π, which applies to spheres having a radius substantially larger than that of solvent molecules. Robinson and Stokes (75) inquired whether this coefficient is applicable to noncoordinated ions. They estimated the radii of alkylated ammonium ions and compared them with the respective Stokes

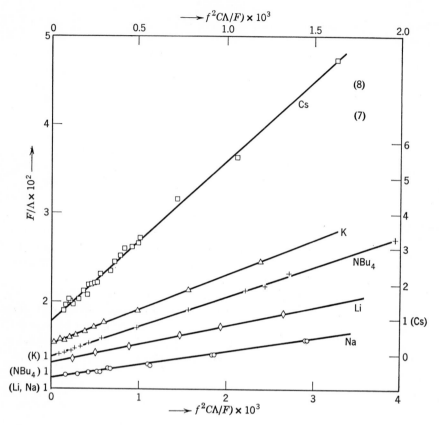

Fig. 11. Fuoss conductance plots for tetraphenyl-boride salts in tetrahydrofuran (76). f is the activity coefficient, F is the Fuoss correction function, and C is concentration.

TABLE V

Relation between Stokes Radii in Aqueous Solution (r_s) and Those Calculated from Molecular Volume or Models (r)[a]

Ion	r, Å	r_s, Å	r/r_s
$N(CH_3)_4^+$	3.47	2.04	1.70
$N(C_2H_5)_4^+$	4.00	2.81	1.42
$N(C_3H_7)_4^+$	4.52	3.92	1.15
$N(C_4H_9)_4^+$	4.94	4.71	1.05
$N(C_5H_{11})_4^+$	5.29	5.25	1.01

[a] R. A. Robinson and R. H. Stokes, *Electrolyte Solutions*, 2nd ed., Academic Press, New York, 1959.

radii determined from their mobilities in aqueous solutions. The results are shown in Table V and indicate that the Stokes relation ($r = \frac{1}{6}\pi\eta u$) applies for $r > 5$ Å but deviations are observed for smaller ions.

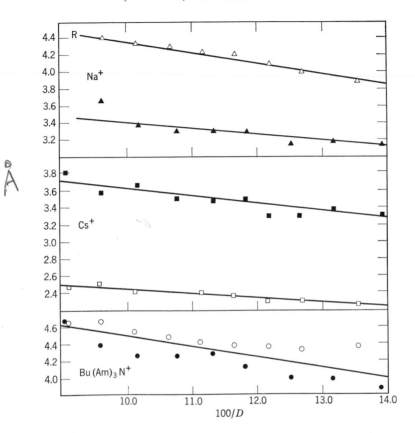

Fig. 12.   The Stokes radii of Na$^+$, Cs$^+$, and N(iso-C$_5$H$_{11}$)$_3$Bu$^+$ in tetrahydrofuran and in dimethoxyethane at different temperatures: (filled points) dimethoxyethane; (open points) tetrahydrofuran (74).

In tetrahydrofuran and in dimethoxyethane, the Stokes radii for Na$^+$ ions seem to be substantially larger than those expected for the bare ions (74). A value of about 4.0 Å was found in tetrahydrofuran, but strikingly in dimethoxyethane the radius was only 3.3 Å. The difference appears to be genuine and indicates that in tetrahydrofuran, the solvation shell is larger than in dimethoxyethane. Probably, the Na$^+$ ion is coordinated with four molecules of the former solvent but only with two of the latter because its molecule acts as a bidentate agent.

A reverse relation is found for the Cs$^+$ ion (74). Its Stokes radius is substantially larger in dimethoxyethane (about 3.4 Å) than the expected radius of the bare ion (1.7–1.8 Å). The increase is small, however, in tetrahydrofuran, the respective Stokes radius being only 2.2 Å. It appears, therefore, that this relatively large ion may still form a coordination shell with two molecules of the former solvent, each of them providing two oxygen atoms for coordination. However, the polarizing field at the Cs^+ surface is too weak to permit coordination with the monodentate tetrahydrofuran. This conclusion gains further support from the observed values of K_{diss} and ΔH_{diss} (see ref. 74).

The Stokes radius of the Li^+ in tetrahydrofuran at 25°C was claimed to be greater than that of the Na^+ (76). However, recent studies show that both ions have similar Stokes radii in tetrahydrofuran and in 2-methyl-tetrahydrofuran (139).

Neither tetrahydrofuran nor dimethoxyethane seem to be coordinated with negative ions. This is not surprising because reversing the orientation of the ether displaces the oxygen atoms (seats of the dipoles) from the inner layer of the solvation shell to its outside (see p. 246). Experimental evidence supporting this conclusion is provided by the work of Slates and Szwarc (77) and Chang, Slates, and Szwarc (78). Studies of the conductance of aromatic radical anions showed that their mobilities have the values expected on the basis of the diffusion coefficients of the respective neutral aromatic hydrocarbons. The latter obviously are not coordinated with the solvent molecules. The data given in refs. 77 and 78 indicate also that the mobilities of these radical anions are determined by that cross section which has the smallest area.

An ion may be solvated by a polar agent added to a poorly solvating medium. Such phenomena were reported by D'Aprano and Fuoss (79), and later by Fabry and Fuoss (80). Addition of a polar material to a solvent of a low dielectric constant has a dual effect. The long range interaction, which slows down the motion of an ion by virtue of a relaxation process (41,43,44), is diminished and this increases the conductance. However, if the added polar component coordinates with an ion, then its mobility decreases due to an increase in its size. Thus, this short range interaction decreases the conductance. In the studies of D'Aprano and Fuoss (79), acetonitrile ($\mu = 3.51$ D) was used as a solvent. Calculation based on Hirschfelder models shows that the interaction energy of $CH_3C{\equiv}N$ with the Br^- ion amounts to 1.5–2 kT only. On the addition of p-nitroaniline ($\mu = 6.32$), one finds a small but measurable decrease in Λ. The amount of the added p-nitroaniline was too small to appreciably change the dielectric constant of the medium and, therefore, the decrease in the conductance, $\Delta = \Lambda_{[aniline],\, C_1} - \Lambda_{[aniline],\, C_2}$ was attributed to the reaction

$$(MeCN) \cdot Br^- + NH_2C_6H_4NO_2 \rightleftharpoons (Br^- \cdot NH_2 \cdot C_6H_4NO_2) + MeCN \cdots K.$$

This leads to the relation

$$K \sim \Delta\Lambda/(C_2 - C_1)$$

and indeed the above quotient was found to be constant.

Fabry and Fuoss (80) employed in their studies a most interesting compound having an extremely high dipole moment, namely, triisopropanolamine borate,

its value was determined in dioxane (81) and found to be about 8 D. Studies of the conductance of salts of triisoamylbutyl ammonium in dioxane to which increasing amounts of triisopropanolamine borate were added showed that the Walden product, $\Lambda_0\eta$, first decreased and then increased as the amount of the dissolved borate became larger. This demonstrates that the effect of short range forces is eventually masked by the long range interaction which decreases the apparent Stokes radius as a consequence of an increasing dielectric constant of the medium.

A reverse effect was observed by Kraus and his co-workers (82,83) who studied the conductance of sodium, lithium, and silver perchlorates in pyridine. The mobility of these cations, viz., λ_0^+, increased on the addition of ammonia, indicating that the small and more nucleophilic ammonia molecules replaced the large pyridine molecules which were originally coordinated with the positive ions. This decreases the effective ionic radius of the solvated ion and increases its mobility.

13. STRUCTURE OF ION-PAIRS

The large dipole moment of ion-pairs causes them to interact strongly with other polar molecules, including those of polar solvents. The resulting interaction energy may be quite appreciable, e.g., for two point-dipoles, each having a moment of 5 D, the interaction energy at a distance of 5 Å in a medium of dielectric constant 2 is twice as high as kT at room temperature.

Two types of interaction should be distinguished: (*1*) those in which the intereacting polar molecule remains "outside" an ion-pair, and (*2*) those having the interacting molecule squeezed in between the two ions of a pair. The first type of interaction may lead to a parallel array of the dipoles (Figure 13*A*) or to an antiparallel alignment (Figure 13*B*). The former is more stable for oblate ellipsoids, whereas the latter is favored on energetic grounds by prolate ellipsoids.

The statistical mechanics treatment of this problem was presented by Fuoss (84) who showed that the parallel alignment is preferred when the ratio of the ellipsoid's axis along the dipole to that perpendicular to its direction is less than $2^{1/3}$. Otherwise, the antiparallel array becomes more stable. Hence, spherical, dipolar molecules associate in a parallel fashion.

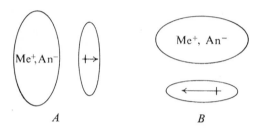

Fig. 13. The parallel and antiparallel arrangement of associated dipoles.

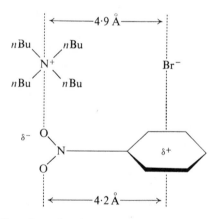

Fig. 14. Association of tetrabutyl ammonium bromide with nitrobenzene (53).

Two examples of "external" solvation of ion-pairs will be discussed. Studies of the conductance of tetra-n-butyl ammonium bromide in mixtures of methanol and nitrobenzene showed substantial deviations of the plot of $\log K_{assoc}$ vs. $1/D$ from the expected linearity (53). This was interpreted as a result of strong association between the ion-pair and nitrobenzene. The two partners fit each other, as shown in Figure 14, and this magnifies the effect. The proposed model of the associated pair gains support from NMR studies which show a small, but unmistakable, shift in the *para*- and *meta*-hydrogens but not in the *ortho*-hydrogens.

This shift is caused by the closeness of the Br^- ion to the *para-* and *meta-* hydrogens.

Bufalini and Stern (85) demonstrated that methanol solvates tetrabutyl ammonium bromide in benzene solution. Their deductions are based on the results of infrared studies of mixtures of benzene, methanol, and the undissociated salt—the O—H bond of the monomeric methanol is shifted in the presence of salt, apparently because of H-bonding to the anion.

The interaction leading to the separation of ions by a solvent molecule (or molecules) was clearly visualized by Winstein (86) and elaborated by Grunwald (87). The potential energy of two ions increases as they are pulled

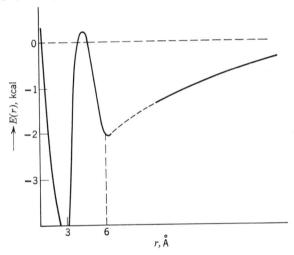

Fig. 15.   Dependence of an ion-pair energy on its separation distance (87).

away from their contact position. Initially, their attraction for each other is not attenuated by the solvent because the gap between them is too small to accommodate a solvent molecule (see also references 54, 55 and p. 239). However, at some critical distance a polar solvent molecule may be accommodated and then the energy of the system decreases as the solvent fills up the vacant space. Further stretching of the pair again increases the potential energy which asymptotically reaches its plateau value at infinite separation. These changes, depicted in Figure 15, suggest that two distinct species may exist: a contact ion-pair and a solvent-separated ion-pair.

The solvent-separated ion-pairs exist only in those media in which at least one of the ions, when free, is coordinated with solvent molecules. The association of such ions proceeds smoothly until the solvation shell touches the counterion. At this stage a solvent-separated pair is formed. Further approach requires expulsion of the solvation shell (a *discontinuous* process) and yields a contact-pair. In a continuous, structureless medium no such effect could be observed. A decrease of the interionic distance would then lead to a smooth and monotonic change in the properties of an ion-pair.

The concept of contact ion-pairs and solvent-separated ion-pairs was shown to be most fruitful in accounting for the kinetics of some processes involving ions and ion-pairs. For example, many solvolyses involving an asymmetric carbon atom and an intermediate carbonium ion may lead to inversion or retention of the configuration of the resulting product and they may, or may not, be accompanied by racemization of the starting material. Careful studies of the kinetics of all these processes, in conjunction with the effects exerted upon them by salts having a common counterion with the reagents, allowed Winstein (88) to show that contact ion-pairs, solvent-separated ion-pairs, and free ions participate in such reactions. The process is described in terms of the following mechanism:

$$A:X \rightleftharpoons A^+,X^- \rightleftharpoons A^+ \mid S \mid X^- \rightleftharpoons A^+ + X^-$$

contact ion-pair	solvent-separated ion-pair	free ions
$k_1 \downarrow \uparrow k_{-1}$	$k_2 \downarrow \uparrow k_{-2}$	$k_3 \downarrow \uparrow k_{-3}$
products	products	products

and the individual rate constants were determined eventually from the experimental data.

These ideas were confirmed by the numerous studies of Winstein's group, as well as by those of other workers, and their application to reactions of carbanions was pioneered by Cram (89).

The effect of solvation is revealed also by some changes observed in NMR spectra. Solvent influences the NMR spectrum of a solute and several mechanisms are responsible for the observed effects. (a) Anisotropic shielding of protons of solute by solvent molecules leads to shifts in the position of their lines. This effect was considered by Buckingham, Schaefer, and Schneider (146) and independently by Abraham (147). (b) The position of the absorption lines is affected by the dispersion forces through which the solute and solvent interact. This phenomenon was considered by Bothner-By (148). (c) Finally, the electric fields of ions or dipoles are expected to cause large effects by changing the charge distribution in the solute molecule, and this aspect of the problem was considered again by Buckingham (149).

Recently Nicholls and Szwarc (140) examined the NMR spectra of several ethers (tetrahydrofuran [THF], 2-methyl tetrahydrofuran [MeTHF], tetrahydropyrane [THP], and dioxane [DOX]) in the presence and in the absence of dissolved salts. Their concentration was high, and therefore the salts formed ion-pairs, the fraction of free ions or triple ions being negligible. It was found that $Na^+ClO_4^-$, $Li^+ClO_4^-$, and $Ag^+BF_4^-$ cause a downfield shift, their effectiveness increasing along the series $Na^+ < Li^+ < Ag^+$.

The results indicate therefore that the cation of a pair becomes associated with the oxygen atom of the ether molecule, and, thus, it induces deshielding of the neighboring protons. This conclusion was corroborated by the observation that the shift of the α protons is larger than that of the β or γ. On the other hand, a reverse effect was observed on the addition of sodium or lithium tetraphenyl boride. It will be shown later that these salts form solvent-separated pairs and hence the β, and to a lesser extent the α, protons of THF or MeTHF become influenced by the anisotropic action of the phenyl groups of BPh_4^- ions which cause an upfield shift.

An interesting example of solvation affecting NMR spectra was discussed by Fraenkel and Kim (150). These workers examined the NMR lines of *ortho-* and *meta*-protons of *para*-substituted anilines and their anilinium salts. Conversion of the base into a primary ammonium ion in nonpolar solvents only slightly affected the spectrum, whereas its conversion into a quaternary ammonium salt had a large effect. It was concluded that the inductive effect of the N^+ is reduced in the $-NH_3^+$ ion by its close association with the X^- ion. Such an association is less effective in the quaternary ammonium salt because the approach of X^- is hindered. Addition of methanol to chloroform solution of $-NH_3^+,X^-$ led to an upfield shift, the effect increasing with increasing concentration of methanol until the molar ratio $MeOH:\phi_s NH_3^+,X^-$ reached 4 and then remained constant. This indicates insertion of MeOH between the ions of the pair and formation of a stoichiometric solvation complex.

NMR studies also permit the observation of solvation of ion-pairs by suitable complexing agents. For example, Chan and Smid (128) confirmed the work of Nicholls and Szwarc (140) using, however, salts of fluorene carbanions. However, the shifts in the solvent spectrum disappeared when a powerful solvating agent, like glyme, was added. Moreover, the dissolved glyme showed different NMR spectra in the presence and absence of salt. These results clearly indicate that molecules of glyme replaced molecules of solvent in the solvation shell surrounding the ion-pair.

As has been pointed out earlier, the absorption spectrum of an ion-pair may differ from that of the respective free ion. A change in the absorption spectrum may also be expected on modifying the structure of an ion-pair. For example, Warhurst and his associates (90,91) observed that increased size of the counterion leads to a bathochromic shift in the spectra of negative radical-ions derived from aromatic hydrocarbons or ketones. For alkali salts, the increase in the absorption frequency, $\Delta\nu$, corresponding to λ_{max}, was found to be linear with $(r_c + const.)^{-1}$, r_c denoting the radius of the cation. These results are shown in Figure 16. A similar relation was observed for a series of salts of bivalent cations.

Warhurst rationalized his findings by assuming that Coulombic interaction between the cation and the negative radical-ion stabilizes more strongly the ground state than the excited state. The constant term in his relation corresponds, therefore, to the "radius" of the negative ion and the extrapolation to $r_c = \infty$ should give the absorption maximum of the free ion. A refined quantum mechanical treatment of this problem was presented by McClelland (92) for a series of aromatic ketyls, by Dieleman (93) for polyphenyls, and by Hush and Rowlands (94) for condensed polyaromatic hydrocarbons. A concise summary of these investigations is given in a recent review by McClelland (95).

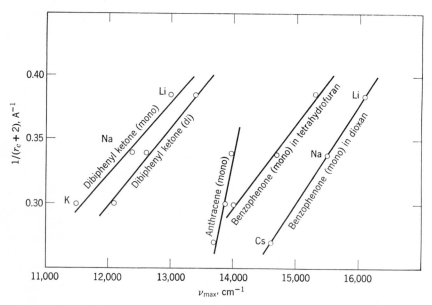

Fig. 16.   Linear dependence of ν_{max} (cm^{-1}) corresponding to maximum absorption of salt of aromatic radical-ions and of ketyles on $1/(r_c + 2)$ where r_c = crystal radius of the cation (91).

The effect caused by the cationic field depends on the polarization direction of the investigated transition with respect to the line determined by the centers of charges of the opposing ions. Hence, different transitions may be affected to a different degree. There are examples of spectra of organometallic compounds for which one transition is cation dependent whereas another is not (94).

Subsequent studies revealed that the absorption maxima are affected by solvents (96,97) and often are influenced by temperature (98). It has been suggested that the dissociation of an ion-pair into free ions is responsible for these phenomena (98). Alternatively, it has been proposed (99–103) that the increased polarity of the solvent leads to a greater dispersion of the cationic charge, and the effect should be then similar to that caused by an increase of the radius of the cation.

A correct explanation for at least some of these phenomena has been proposed by Hogen-Esch and Smid (104,105). They found that THF solutions of the alkali salts of the fluorenyl carbanion show two absorption peaks, one at 355 and the other at 373 mμ. Their relative intensities were not affected by dilution from 10^{-2} to 10^{-4} M, nor by addition of the tetraphenyl boride

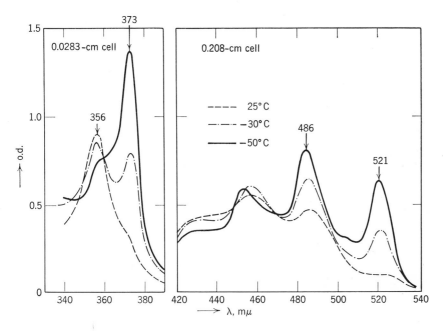

Fig. 17. Absorption spectrum of fluorenylsodium in tetrahydrofuran at different temperatures (105).

having a common counterion with the fluorenyl salt. However, the spectrum was drastically changed, as shown in Figure 17, as the temperature was lowered. These reversible changes of the absorption spectrum prove that two species coexist in the investigated solution, but neither of them is a free ion, or a triple ion resulting from the reaction $2F^-,Na^+ \rightleftarrows F^-,Na^+,F^- + Na^+$. This conclusion was verified by conductance studies—the proportion of free ions was found to be much too low for spectrophotometric detection. Hence, the solution apparently contains contact ion-pairs in equilibrium with the solvent-separated pairs. This study provides, therefore, the first spectroscopic evidence revealing the simultaneous existence of two types of ion-pairs (the extensive evidence based on the ESR investigation will be discussed in the last section of this chapter).

Only the 373 mμ peak is seen at low temperatures, indicating that it belongs to the solvent-separated pair. Further verification of this assignment comes from studies of the toluene solution. In this poorly solvating medium only the 355 mμ band appears—obviously being due to the contact ion-pair. The bathochromic shift associated with the conversion of a contact ion-pair into a solvent-separated pair is analogous to that arising from the increasing size of a cation in a contact-pair. This point is clearly supported by the data given in Table VI. The absorption maximum of the solvent-separated ion-pair is not affected by the nature of the counterion and is virtually identical with that of a free ion. In view of the large separation of the ions in a solvent-separated ion-pair, this has to be expected.

TABLE VI

Dependence of λ_{max} on the Radius of the Cation for 9-Fluorenyl Salts in THF at 25°C

Cation	r_c, Å	$\lambda_{max}(\mu)$
Li$^+$	0.60	349
Na$^+$	0.96	356
K$^+$	1.33	362
Cs$^+$	1.66	364
(NBu$_4$)$^+$	3.5	368
Solvent-separated pair[a]	~4.5	373
Free ion[b]	∞	374

[a] Independent of the nature of the counterion.
[b] Direct studies of a highly diluted solution.

The conversion of the contact ion-pairs of fluorenylsodium, F$^-$,Na$^+$, into solvent-separated pairs is exothermic:

$$F^-,Na^+ + n\text{THF} \rightleftarrows F^-(\text{THF})_n Na^+ \cdots K_i.$$

The equilibrium constant was determined spectrophotometrically (105) in the temperature range +25 to −70°C, and the results, presented in the form of a plot of log K_i vs. $1/T$, are shown in Figure 18. Thus, ΔH_i was found to be −7.6 kcal/mole, indicating a strong interaction between the solvent and the ion-pair. The respective ΔS was found to be −33 eu, this being due to immobilization of solvent molecules in the solvation shell.

The exothermicity of the process, contact ion-pair → solvent-separated-pair is not immediately apparent. One may be surprised that the replacement of a carbanion by a solvent molecule (or molecules), which becomes the neighbor(s) of the cation, decreases the total energy of the system. However, it is probable that the partial neutralization of both charges in the contact-pair prevents the cation from binding other solvent molecules on the "outside" of the pair. This coordinating power of the cation is restored by its partial separation from the anion. The large negative entropy of the process provides additional evidence supporting this idea.

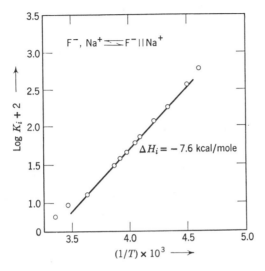

Fig. 18. Van't Hoff plot of log K_i (equilibrium constant for the transformation of contact ion-pair of fluorenylsodium into the solvent-separated pair)

$$F^-, Na^+ + nTHF \rightleftarrows F^-(nTHF)Na^+ \ (105).$$

As expected, the nature of the solvent affects the equilibrium constant, K_i. For example, at 25°C, and for the same counterion, a higher proportion of solvent-separated ion-pairs is observed in dimethoxyethane, which acts as a bidentate solvating agent, than in tetrahydrofuran. At room temperature, the latter solvent has a slightly higher dielectric constant (7.4) than the former (7.2). This clearly indicates that factors other than the dielectric constant are of paramount importance in determining the coordinating power of a solvent (see p. 238). In dimethylsulfoxide, a potent solvating agent, only solvent-separated ion-pairs are formed, but in spite of its relatively high dielectric constant the dissociation into free ions is moderate. Surprisingly, no solvent-separated ion-pairs are detected in dioxane. This observation is at variance with the findings of Grunwald (106), who reported that the salts of Li^+ and Na^+ ions crystallize from aqueous dioxane with dioxane of crystallization, i.e., this ether appears to be a more potent solvating agent than water.

Pursuing his studies of contact and solvent-separated ion-pairs, Smid (141) investigated the effect of other ethereal solvents upon the spectrum of fluorenyl salts. By this criterion, 3-methyl-tetrahydrofuran is a better solvent than the 2-isomer, both being less efficient than the unsubstituted ether. 2,5-Dimethoxy-tetrahydrofuran and 1,3-dioxolane turned out to be inferior solvents to the methyl-tetrahydrofurans, whereas 2-(CH_3OCH_2)-tetrahydrofuran is a most powerful ion-separating agent.

Some observations apparently inconsistent with the simple theories associating the bathochromic shift with the cation radius are now rationalized. For example, Streitwieser and Brauman (102) found a blue shift in the spectra of fluorenyl salts when Li^+ was replaced by Cs^+. These observations were made in cyclohexylamine and in dimethoxyethane (DME), and evidently in these media Li^+ salts form solvent-separated pairs while those of Cs^+ give contact ion-pairs (see Table VI).

The following question now arises: Is the concept of two types of ion-pairs applicable to all systems containing associated ions? There are, of course, systems in which only one type of ion-pair may exist. For example, for free ions which are not coordinated, or only weakly coordinated, with solvent molecules association gives contact ion-pairs only; no solvent-separated ion-pairs may exist in such solutions. On the other hand, for very strong coordination and for a sufficiently bulky counterion, the possible gain in Coulombic energy arising from the conversion of a solvent-separated ion-pair into a contact ion-pair may not be sufficient to permit the squeezing out of solvent molecules. In such systems, only solvent-separated ion-pairs exist. This point was discussed by Roberts and Szwarc (125).

These are rather obvious limitations which do not deny the virtual existence of two types of ion-pairs; they show only that the concentration of one form may become vanishingly small under some conditions. However, in many systems, the concept of two ion-pairs may be unattainable and the description of such systems cannot be achieved in these terms. To clarify the problem, let us return to the model proposed by Grunwald (87).

Grunwald's model is static in nature, the potential energy of the ion-pair is assumed to be uniquely determined by the interionic distance, r. The situation is, however, more complex. The ion-pair is embedded in a fluctuating environment of solvent molecules; it vibrates and the environment participates in this motion. The potential energy depends, therefore, on solvent because interaction with its molecules provides an *average* force field superimposed upon the interaction field of the pair. Such a field *changes with temperature* because temperature affects the average configuration of solvent molecules with respect to the pair, and hence the shape of the potential energy curve is also temperature dependent. This point has recently been stressed by Linder (126).

For deep and narrow potential wells, as assumed by Grunwald (see Figure 15) the concept of contact and solvent-separated ion-pairs as two thermodynamically distinct species, is justified. However, for shallow and wide potential wells, the distinction may become meaningless. The shape of the potential energy curve may gradually change with temperature, viz., it may possess a minimum at short interionic distances at higher temperatures, but it could be transformed into a well having a minimum at longer interionic distances at lower temperatures. This is schematically shown in Figure 19. In such systems, the pairs behave as contact ion-pairs at higher temperatures, as solvent-separated pairs at lower temperatures, and show intermediate properties at intermediate temperatures. Such pairs cannot be represented by two thermodynamically distinct species which coexist in equilibrium. Their properties, e.g., the absorption maxima or shifts in ESR line (see p. 289), may show a gradual change as the solution temperature, or the nature of the solvent, is

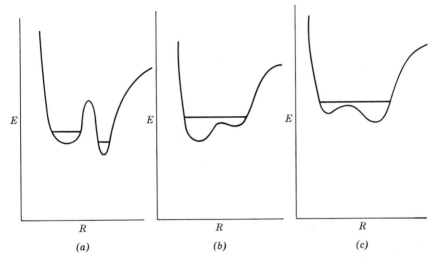

Fig. 19.   Schematic energy diagram for an ion-pair: (*a*) distinct two pairs, (*b*) merging of two pairs—low temperature favors contact, (*c*) merging of two pairs—low temperature favors separation. R is the distance between ions of a pair.

varied. It is probable that such a situation is often encountered in large radical-ions having a counterion that vibrates between two (or more) relatively well separated positions, e.g., in biphenyl⁻, terphenyl⁻, etc., and an example of such a continuous change has been recently reported by Nicholls and Szwarc (142).

The idea expressed in terms of a potential energy well has been presented in a different form (14,78) using the concept of a distribution function giving the probability of finding a pair separated by a distance *r*. This is shown in Figures 20*a*, *b*, and *c*. The concept of two thermodynamically distinct ion-pairs applies to systems having the distribution function represented by Figure 20*a*. The function in Figure 20*b* represents a system in which both forms of ion-pairs merge, and the function in Figure 20*c* illustrates a system in which the distinction between the solvent-separated pair and free ions is blurred.

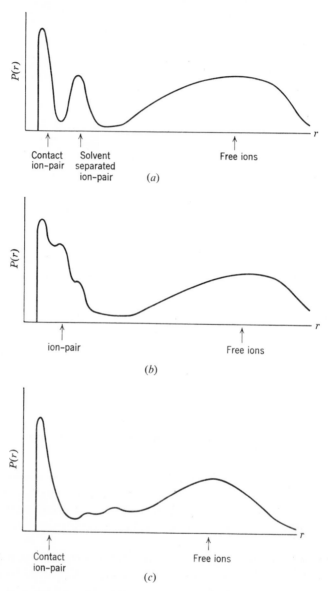

Fig. 20. Probability function representing the relation in an ion-pair system: (a) two distinct ion-pairs, (b) merging ion-pairs, (c) solvent-separated pair merging with free ions.

14. COMPLEXING OF ION-PAIRS WITH POLAR MOLECULES

In a poorly solvating medium, the contact ion-pairs may be transformed into coordinated ion-pairs by adding a suitable complexing agent. The equilibrium concentration of free ions is increased if the coordination of such an agent with the free ion is stronger than with the nonsolvated ion-pair; in the reverse case the concentration of free ions, solvated and nonsolvated, decreases. The above conclusion is valid whatever the structure of the coordinated ion-pair, i.e., whether it is solvated on its periphery or solvent-separated. The respective relations are clarified by the following calculations. Consider, e.g., a system

$$A^+,B^- \rightleftharpoons A^+ + B^- \cdots K_1,$$

in which two more equilibria are established on the addition of an agent S, viz.,

$$A^+,B^- + S \rightleftharpoons (A^+,B^-,S) \cdots K_2$$

$$A^+ + S \rightleftharpoons A^+S \cdots K_3$$

The total concentration of the free cations (or anions) is given by

$$[A^+] + [A^+,S] = [B^-] = XC$$

where C is the total concentration of the salt and X the fraction of ionized species. Application of the mass law leads then to the result giving the formal dissociation constant, K_D,

$$K_D = \frac{X^2 C}{1 - X} = K_1 \frac{1 + K_3[S]}{1 + K_2[S]}$$

where [S] denotes the concentration of the free (nonbonded) agent S. The initial concentration $[S]_0$ of the added coordinating agent S is related to [S] through the equation

$$[S]_0 = [S] + C\{(1 + 1/K_2[S])^{-1} + X[(1 + 1/K_3[S])^{-1} - (1 + 1/K_2[S])^{-1}]\}.$$

One may easily verify that X increases with S for $K_3 > K_2$, but it decreases if $K_3 < K_2$. In the first case, $(1 + 1/K_3[S])^{-1} > (1 + 1/K_2[S])^{-1}$, while a reverse inequality applies for $K_3 < K_2$. Hence, the square bracket in the equation giving $[S]_0$ is positive for $K_3 > K_2$ and negative for $K_3 < K_2$. Therefore, in either case $[S]_0$ increases with increasing [S], showing that X increases with [S] for $K_3 > K_2$ but decreases for $K_3 < K_2$. At a sufficiently large excess of S,

$$X^2 C/(1 - X) = K_1(K_3/K_2)$$

i.e., all the ion-pairs, as well as free A^+ ions, are coordinated with S. It is to be noted that $K_1(K_3/K_2)$ gives the equilibrium constant K_4 of the process

$$A^+,B^-,S \rightleftharpoons A^+S + B^- \cdots K_4.$$

The above described trend of X, arising from a variation in the concentration of the solvating agent, remains valid even if the coordination processes of ion-pairs and of the free A^+ ions involve higher, but equal, numbers of S molecules. Systems in which the ion coordinates with a greater number of S molecules than the ion-pair may be treated in a similar way by substituting a pseudo-constant $K_3' = K_3[S]^{m-n}$ for K_3 (m—the coordination number for the ion, n—for the ion-pair). It is obvious that the equality $K_3' > K_2$ must hold for a sufficiently large $[S]$, if $m > n$. Hence, at sufficiently high concentrations of such a coordinating agent, the dissociation into ions increases with S—a rather trivial result.

For systems in which the solvent is the coordinating agent, it may be convenient to incorporate its activity into the respective equilibria constants and replace K_2 and K_3 by K_2'' and K_3''. The overall equilibrium constant K_D which refers to the formal process: ion-pairs (in whatever form) $\rightleftarrows$ ions (in whatever form), acquires then the form:

$$K_D = K_1(1 + K_3'')/(1 + K_2'').$$

In the extreme case, $K_3'' \gg 1$ and $K_2'' \ll 1$, virtually all the free A^+ ions are solvent-coordinated and all the ion-pairs are not coordinated. The dissociation is given then by the equation:

$$A^+,B^- \rightleftarrows A^+(mS) + B^-, \ldots K_C$$

with $K_C = K_1 K_3''$. The dissociation constant, K_S, of solvent-coordinated ion pairs, ($K_3'' \gg 1$ and $K_2'' \gg 1$),

$$A^+(nS)B^- \rightleftarrows A^+(mS) + B^-, \ldots K_S$$

is given by the relation $K_S = K_1 K_3''/K_2''$.

In systems which are of the greatest interest to us, viz., salts of carbanions and radical-ions in ethereal solvents, coordination with an ion-pair involves mostly cations. A few examples may illustrate the techniques by which such phenomena may be observed and investigated.

Hogen-Esch and Smid (105) studied the changes taking place in the spectrum of lithium fluorenyl in dioxane solution caused by the addition of small quantities of strong solvating agents. It has been mentioned previously that in dioxane this salt forms contact ion-pairs which absorb at 346 mμ. However, the addition of dimethylsulfoxide (DMSO) dramatically changes the spectrum as shown in Figure 21. As the concentration of DMSO becomes larger, the optical density at $\lambda_{max} = 346$ mμ decreases and that at $\lambda_{max} = 373$ mμ increases accordingly. The simple stoichiometry of this process is demonstrated by the isosbestic point clearly seen in Figure 21. The amount of added DMSO is too small to appreciably modify the macroscopic properties of the solvent, and therefore virtually no dissociation of pairs into free ions takes place (the dielectric constant of dioxane is only 2.4). Hence, the reaction apparently converts the contact pairs into solvent-separated ones.

The ratio of optical densities at 346 and 373 mμ gives the molar proportions of the respective ion-pairs. A plot of the logarithm of this ratio against that of the DMSO concentration gives a straight line with a slope of 1.15, indicating that probably only one molecule of DMSO is needed for the formation of a solvent-separated pair.

Let us digress and note that there are systems in which two pairs coexist in equilibrium, but their absorption maxima are too close to each other and

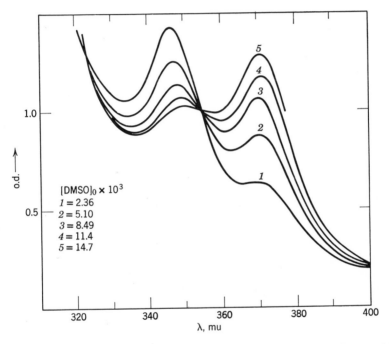

Fig. 21. Changes in the absorption spectrum of sodium fluorenyl in dioxane arising from the addition of increasing amounts of dimethylsulfoxide (DMSO) (105).

the spectral bands are too broad to produce distinct absorption peaks. In such systems, only one absorption band is seen, and a continuous change of conditions, leading from 100% of one pair to 100% of the other, causes only a gradual shift of the absorption maximum. Nevertheless, the existence of two species may be revealed by the study of the band width. The shift of the absorption maximum leads to a broadening of the band and eventually again to its sharpening. The narrow bands correspond to systems composed of one species only, whereas the maximum broadening takes place when the solution contains probably equal concentrations of both species. An example of such a study is found in the work of Waack et al. (133).

The spectrophotometric technique of Hogen-Esch and Smid was used by

Chan and Smid (128) in their studies of the relative solvating power of poly-glycol ethers $CH_3O(CH_2CH_2O)_xCH_3$. Their results are summarized in Table VII which is self-explanatory. It should be stressed that coordination involves

TABLE VII

Coordinating Power of $CH_3O(CH_2CH_2O)_xCH_3$[a] for Li^+,[b] Fluorenyl$^-$ in Dioxane

$$F^-,Li^+ + nS \rightleftarrows F^-,(nS),Li^+, \cdots, K$$

$$[F^-,Li^+] = 1.5 \times 10^{-3} M; T = 25°C$$

Ether = S	[S], moles/liter for $[F^-,Li^+] = [F^-,(nS)Li^+]$	n	K
$CH_3O \cdot (CH_2CH_2O) \cdot CH_3$	3.35	2.4	0.055
$CH_3O \cdot (CH_2CH_2O)_2 \cdot CH_3$	0.31	1	3.1
$CH_3O \cdot (CH_2CH_2O)_3 \cdot CH_3$	0.0089	1	130
$CH_3O \cdot (CH_2CH_2O)_4 \cdot CH_3$	0.0041	1	230
DMSO	0.0054	1	190

[a] $CH_3O(CH_2CH_2O)_xCH_3 = S$.
[b] $Li^+Fluorenyl^- = F^-Li^+$.

one molecule of ether, if $x > 1$, i.e., whenever more than two oxygen atoms are available in the coordinating molecule. However, for $x = 1$, i.e., for $CH_3O \cdot CH_2 \cdot CH_2 \cdot O \cdot CH_3$, two molecules are needed to convert the contact ion-pair into the solvent-separated one. The results of Chan and Smid parallel similar findings by Ugelstad et al. (129–132) who investigated the effects exerted by those ethers upon the reactivities of potassium *tert*-butoxide and sodium and potassium phenoxides. Other kinetic studies which shed light on these problems will be discussed in Chapter VII.

Slates and Szwarc (134) investigated the effect of polyethers upon the equilibrium established in tetrahydropyran between biphenyl and metallic sodium. The reaction is represented by the equation:

Biphenyl (solution) + metallic sodium $\rightleftarrows$ Biphenyl$^{\overline{\cdot}}$, Na$^+$ (solution) $\cdots K$

The ratio $[Biphenyl^{\overline{\cdot}},Na^+]/[Biphenyl]$ was determined spectrophotometrically over the temperature range -40 to $+25°C$. This ratio gives the respective equilibrium constant, K. Upon addition of the polyether, a new equilibrium is superimposed upon the previous one, viz.,

Biphenyl$^{\overline{\cdot}}$,Na$^+$ + polyether $\rightleftarrows$ Biphenyl$^-$,(polyether),Na$^+$ $\cdots K_3$

thus, increasing the optical density at λ_{max} of Biphenyl$^{\overline{\cdot}}$ (solvation does not affect λ_{max}). The results established that only one molecule of ether is involved in coordination and enabled calculation of the equilibrium constant K_3 and the heat of coordination, ΔH_s. In the course of these studies it was shown that two types of solvated ion-pairs may coexist in solution—one possessing the ether on its periphery, the other separated by the ether. This is the first example of isomerism involving a solvating agent. The equilibrium constant of the isomerization and the respective heat and entropy were determined.

15. ASSOCIATION OF IONS WITH ION-PAIRS

Free ions may associate with ion-pairs and form triple ions (107), e.g.,

$$A^+,B^- + A^+ \rightleftharpoons A^+,B^-,A^+, \ldots K_{A^+,B^-,A^+}$$

or

$$B^- + A^+,B^- \rightleftharpoons B^-,A^+,B^-, \ldots K_{B^-,A^+,B^-}$$

The formation of triple ions increases the conductance of a solution. For $K_{A^+,B^-,A^+} = K_{B^-,A^+,B^-}$, the concentration of free ions is only slightly decreased by the formation of triple ions, if the free ions and triple ions form only a small fraction of the original ion-pairs. The electric transport is, however, augmented by the contribution of triple ions. For $K_{A^+,B^-,A^+} \neq K_{B^-,A^+,B^-}$, the ratio $[A^+]/[B^-]$ deviates from unity. For example, it decreases, if $K_{A^+,B^-,A^+} > K_{B^-,A^+,B^-}$, because A^+ ions are preferentially removed from the system. However, since the product $[A^+][B^-]$ has to be constant the sum $[A^+] + [B^-]$ must increase. Thus, the conductance arising from the presence of the A^+ and B^- ions increases, provided the mobility of B^- is not much lower than that of A^+. Systems in which the two ions associate to a different degree with ion-pairs have been described by Wooster (108) and by Dole (109).

The formation of triple ions may have a pronounced effect upon ionic polymerization. Consider for example, the propagation of living polystyrene, $\sim\!\!S^-,Na^+$. In solvents such as THF, the apparent rate constant of propagation k_{ap} is given by the sum of two terms:

$$k_{ap} = (1 - \gamma)k_\pm + \gamma k_-$$

where $k_\pm$ and k_- are the rate constants of propagation of the ion-pair and free ion, respectively, and γ is the fraction of dissociated ion-pairs. Since $k_- \gg k_\pm$, dilution increases k_{ap}. Addition of a strongly ionized sodium salt, e.g., Na^+,BPh_4^-, decreases the k_{ap} because it depresses the degree of ionic dissociation of $\sim\!\!S^-,Na^+$. However, for a sufficiently high concentration of the salt the formation of triple ions may become appreciable, and if Na^+ ions associate more strongly than BPh_4^- ions, the concentration of the latter should decrease. Consequently, the concentration of the free $\sim\!\!S^-$ ions would then increase, thus enhancing the polymerization.

Living polymers possessing two ionizable ends, e.g.,

$$Cs^+,S\!\sim\!\!S^-,Cs^+$$

provide an interesting system in which the formation of triple ions may result from an intramolecular reaction (110), instead of the conventional inter-molecular process. Dissociation of an ionizable end of such a polymer produces a polystyryl ion linked by a chain to an $\sim\!\!S^-,Cs^+$ ion-pair, viz.:

$$Cs^+,{}^-S\!\!\sim\!\!S^-,Cs^+ \rightleftarrows Cs^+,{}^-S\!\!\sim\!\!S^- + Cs^+.$$

Even in the most dilute solutions such an ion has to remain in the close vicinity of the ion-pair located on the other end of the polymeric chain and, therefore, it may associate with its neighbor by closing the chain, i.e.:

$$Cs^+,{}^-S\!\!\sim\!\!S^- \overset{K_c}{\rightleftarrows} \underset{\sim\!\!\sim\!\!\sim\!\!\sim}{\sim\!\!S^-,Cs^+,S^-\!\!\sim}$$

The cyclization and the formation of a triple ion removes the free $\sim\!\!S^-$ ions from the system and upsets the ionization process:

$$\sim\!\!S^-,Cs^+ \rightleftarrows \sim\!\!S^- + Cs^+.$$

More living ends have to dissociate to replenish the consumed $\sim\!\!S^-$ ions, and consequently a new equilibrium is established in which the concentration of Cs^+ ions exceeds that of $\sim\!\!S^-$ ions. Thus, the solution of two living ends poly-mers contains less $\sim\!\!S^-$ ions and more Cs^+ ions than a solution of one living end polymer, $X\!\!\sim\!\!S^-,Cs^+$ at the same total concentration of living ends. This phenomenon manifests itself in two ways: (1) a solution of living polymer possessing two living ends is less reactive than that of the one living end species—because the concentration of the most reactive $\sim\!\!S^-$ ions is lower in the former than in the latter; (2) the conductance of the former solution is much higher than that of the latter because it contains more of the relatively mobile Cs^+ ions in addition to the triple ions which contribute also to the electric current.

At higher concentrations of electrolytes, or in media of lower dielectric constants, further associations may be observed. Thus, ion-pairs may assoc-iate to neutral pairs of ion-pairs, viz.,

$$2A^+B^- \rightleftarrows (A^+B^-)_2.$$

This reaction is equivalent to the associations

$$A^+,B^-,A^+ + B^- \rightleftarrows (A^+B^-)_2$$

and

$$B^-,A^+,B^- + A^+ \rightleftarrows (A^+B^-)_2$$

if the concentration of free ions is significant. Such associations, as well as still higher agglomerations, are frequently observed in solutions of lithium derivatives. Some of these phenomena will be discussed further in Chapter VIII.

Dimerization of ion-pairs becomes more significant as their dipole moment increases. For example, fluorenyllithium is well soluble in tetrahydrofuran; however, the addition of dimethylsulfoxide to its moderately concentrated solution leads to the precipitation of a crystalline adduct. The combined effect of increased dipole moment, caused by the separation of the ions, and of the associating power of dimethylsulfoxide leads to an extensive agglomeration and eventually to the formation of a new phase. A similar observation was reported by Waack et al. (133) who noted precipitation of some lithium salts from hexane solution on addition of small amounts of tetrahydrofuran.

Association of ions into ion-pairs and higher agglomerates is also observed in the gas phase. For example, study of the vapor pressure of numerous alkali halides (111) showed that dimers and trimers are present in the system. Mathematical treatment of these problems was developed by Blander (112) who applied his "dimensional" theory to those systems. Continuation of such investigations led to determination of the respective equilibrium constants. For example, for the system $2KBr \rightleftarrows (KBr)_2$ in the gas phase, the equilibrium constant was found to be given by the equation $\log K_{assoc} = -3.4 + 8400/T$ in units of liter mole^{-1} (113).

16. RATES OF ASSOCIATION OF IONS AND OF DISSOCIATION OF ION-PAIRS

Association of ions into ion-pairs is extremely rapid and usually the reaction is diffusion controlled. The first treatment of such a process was reported by Smoluchowski (114) who showed that the rate constant, k_d, of association of neutral colloidal spheres which stick together on every collision is given by:

$$k_d = (2N_0 kT/3\eta)(r_A + r_B)^2/r_A \cdot r_B$$

in units of liter moles^{-1} sec^{-1}. N_0 denotes Avogadro's number, η the viscosity of the medium, and r_A and r_B the radii of the combining spheres.

Interestingly, the rate constant of the diffusion-controlled reaction does not depend on the absolute values of r_A and r_B, i.e., on the dimensions of the combining particles, but only on the ratio of their radii, viz.,

$$(r_A + r_B)^2/r_A r_B = 2 + r_A/r_B + r_B/r_A.$$

It has a minimum value for $r_A = r_B = r$, and then

$$k_d = 8RT/3000\eta \text{ in liter moles}^{-1} \text{ sec}^{-1},$$

independent is of the magnitude of r.

The extension of Smoluchowski's work to ionic recombinations is due to Debye (115). The original treatment was supplemented by a Boltzmann factor which accounts for the electrostatic interaction energies between the charged spheres (ions). The respective rate constant involves, therefore, an additional factor, viz. $\delta/(\exp \delta - 1)$, where:

$$\delta = z_A z_B e^2 / DkTa.$$

The charges of the respective ions are z_A and z_B, a is the distance of their closest approach, and D is the dielectric constant of the medium. Thus, the combination of oppositely charged ions is somewhat faster than that of neutral particles.

The rate constant of the diffusion-controlled combination is usually of the order of 10^{10} liter moles^{-1} sec^{-1}. Therefore, for ion-pairs having a dissociation equilibrium constant of about 10^{-7} to 10^{-6} mole^{-1}, the dissociation rate constant may be as high as 10^3 to 10^5 sec^{-1}. Hence, the equilibria between ion-pairs and the free ions are rapidly established, the respective relaxation times being usually less than 10^{-3} sec. In most of the ionically polymerizing systems, the propagation step is much slower than ion-pair dissociation, and therefore the rates of the dissociation–association phenomena usually do not affect the kinetics of the investigated reaction, its course being determined by the equilibrium established between the growing species.

Exceptions are possible. The dissociation of ion-pairs into free ions is often exothermic, particularly for contact ion-pairs. This is caused by the substantial exothermicity of the solvation process which converts the non-coordinated ions into solvent coordinated species. Therefore, the association process may require activation energy, to permit the desolvation of ions, and this, in turn, may retard the conversion of solvent separated ion-pairs into contact ion-pairs. The rate of formation of solvent-separated pairs probably is not affected by these factors. Examples of such two-step associations have been reported recently, mainly by Eigen and his co-workers, and some results concerned with protolytic reactions in aqueous solutions have been reviewed by Eigen, Kruse, Maass, and DeMayer (116).

Studies of the association–dissociation reactions of ions are now in progress in several laboratories. The use of NMR, ESR, acoustic and ultrasonic waves, and of various "jump" methods (e.g., the temperature "jump" method) makes these investigations now feasible. Some studies involving ESR are reviewed in Section 21.

The rate of association of ion-pairs into higher agglomerates may be again diffusion-controlled, although not much is known about these processes. The dimerization of some organolithium compounds will be discussed in Chapter VIII. An interesting case of physical dimerization was discovered in the writer's laboratory (117,118). In tetrahydrofuran solution, the sodium salt of living polystyrene, $\sim\!\!S^-,Na^+$, was found to associate with the adduct of anthracene and living polystyrene $\sim\!\!S^-,A,Na^+$. The association is physical in nature, the optical spectrum of the resulting associate is given by the superposition of the spectra of the components. The existence of the associate is demonstrated by the lack of reactivity of associated $\sim\!\!\sim\!\!S^-,Na^+$ towards anthracene. The free $\sim\!\!\sim\!\!S^-,Na^+$ adds an equivalent amount of anthracene in a fraction of a second, whereas $\sim\!\!\sim\!\!S^-,Na^+$ associated with the anthracene adduct reacts extremely slowly; 20–30 min are needed to complete the reaction. The rate of association is given by a bimolecular rate constant of the order 2000–4000 liter moles^{-1} sec^{-1} (118), i.e., although the reaction is fast, it is very slow when compared with a diffusion-controlled process. It is probable that the reacting ion-pairs have to be desolvated prior to association, and this contributes towards the inertia of the process.

17. DETERMINATION OF CONDUCTANCE AND OF THE DISSOCIATION CONSTANT OF ION-PAIRS

In the last sections of this chapter we shall deal with some technical and experimental problems encountered in studies of electrolytic conductance. The theory and practice of ionic conductance are fully treated in several monographs (119) and in many original papers. The experimental problem is reduced to a method of determining the resistance of a conductivity cell containing a solution of the investigated electrolyte at some concentration and repeating this measurement for a series of decreasing concentrations at some fixed temperature. Usually, although not necessarily, the resistance is determined by an ac bridge operating at about 1000 cps, in order to avoid polarization of the electrodes. It is necessary to check whether the resistance is independent of frequency, and if this is not the case, a suitable extrapolation to an infinite frequency is recommended (for details see, e.g., ref. 75, pp. 94 and 95).

It is believed that the high resistance expected for cells containing solutions of electrolytes in low dielectric constant solvents may be conveniently measured by applying a high voltage, say 200 V, and determining the resulting current. The polarization of the electrodes is expected to contribute no more than ± 2 V and, thus, the resistance should be accurately determined within 1–2%. In this writer's laboratory it was observed that under these conditions electrodes become coated by some products, and consequently the resistance of the cell increases very considerably.

The conductivity—the reciprocal of the resistance—divided by the concentration of the electrolyte, c, gives the equivalent conductance, Λ, and for sufficiently dilute solutions the plot of $1/\Lambda$ vs. Λc is linear. Its intercept is equal to $1/\Lambda_0$, Λ_0 being the limiting conductance, and its slope equals $1/\Lambda_0^2 K_{diss}$, where K_{diss} refers to the equilibrium constant of the process

$$X^-,Y^+ \rightleftarrows X^- + Y^+.$$

In solvents of low dielectric constants, viz., less than 10, studies of conductance have to be extended to 10^{-6} or even $10^{-7}M$ solution in order to apply reliably the Ostwald mass law equation $1/\Lambda = 1/\Lambda_0 + \Lambda c/\Lambda_0^2 K_{diss}$. Even then, it is necessary to introduce some corrections to account for the interactions between the ions. The method developed by Fuoss leads to the equation:

$$F/\Lambda = 1/\Lambda_0 - (f^2/F)c\Lambda/\Lambda_0^2 K_{diss}$$

where

$$F = \{ \cdots \{1 - z[1 - z(1 - z)^{1/2}]^{1/2}\}^{1/2} \cdots \}^{1/2},$$

$$z = \frac{\alpha}{\Lambda_0^{3/2}} \sqrt{c\Lambda},$$

and

$$f^2 = 10^{-2}\beta(Fc\Lambda/\Lambda_0)^{1/2}.$$

Here, f is the activity coefficient of the ions, as deduced from the Debye–Hückel theory, α is the Onsager limiting slope given by

$$8.2 \times 10^5 \times \Lambda_0/(DT)^{3/2} + 82/\eta(DT)^{1/2},$$

$\beta = \{0.4343e^2/2DkT\}\{8\pi Ne^2/1000\ DkT\}^{1/2}$, and D and η denote the dielectric constant and the viscosity, respectively, of the solvent. Calculation of the variables F/Λ and $(f^2/F)\cdot c\Lambda$ is performed by computer. The improved equation of Fuoss gives a slightly steeper slope than the original Ostwald equation, and for dilute solution the increase amounted to about 10–20%. Empirically, it was found that this equation applies only to solutions for which $f^2 > 0.5$.

At higher concentrations and in solvents of low dielectric constants, the plots of F/Λ vs. $(f^2/F)c\Lambda$ deviate downward from the initially linear dependence, the trend being due to the formation of triple ions, e.g., $X^- + X^-,Y^+ \rightleftarrows X^-,Y^+,X^-$. Hence, if the experiments are not extended to a sufficiently low dilution, the extrapolation results in too low a value for Λ_0 and too high a value for K_{diss}.

The necessity of extending the conductance measurements to extremely high dilutions arises from two causes: (1) Coulombic interaction between ions slows down their mobilities and decreases the conductance; (2) the interaction of an ion with an ion-pair leads to the formation of a triple ion, which contributes to the conductance. Moreover, the formation of triple ions upsets the equilibrium, $X^-,Y^+ \rightleftarrows X^- + Y^+$, causing further dissociation of the ion-pairs. The latter two effects increase the conductance, and both become more pronounced as the dielectric constant of the solvent decreases. To appreciate the effect of decreasing dielectric constant, let us calculate the average distance, r, at which the interaction between two oppositely charged ions immersed in a medium of dielectric constant D is equal to kT at 25°C.

The results are collected in Table VIII. Dilutions at which interactions are comparable with kT are given in the last column of this table.

TABLE VIII

The Distance, r, and the Dilution at Which the Interaction Energy of the Two Oppositely Charged Ions is Equal to kT at 25°C

D	r, Å	Concentration, M
40	13.7	6.5×10^{-1}
30	18.2	2.8×10^{-1}
20	27.4	8.2×10^{-2}
15	36.7	3.4×10^{-2}
10	55	1.0×10^{-2}
7.5	73	4.3×10^{-3}
5.0	109	1.3×10^{-3}
4.0	137	6.5×10^{-4}
3.0	182	2.8×10^{-4}
2.0	274	8.2×10^{-5}

In practice, one finds no difficulties in determining reliably the cell resistance; however, an accurate determination of electrolyte concentration becomes difficult at high dilution. For example, the adsorption of the electrolyte on the walls of the cell becomes significant at $10^{-6}M$ solution. This makes the actual concentration of electrolyte lower than calculated from dilution. In view of that, $c\Lambda$ which is independent of c is experimentally more reliable than $1/\Lambda$ which does involve c ($\Lambda = 1/c \times$ resistance). At high dilution, the conductance of the pure solvent may not be negligible, and then its value has to be subtracted from the experimentally measured conductance of the investigated solution. This correction, which is relatively large for aqueous solutions, is usually small for most of the organic solvents. It becomes significant for salts dissociated to an extremely low degree.

Calibration of the conductance cell may be performed in the conventional way by determining the conductance of aqueous KCl. For cells of very low constant, Fuoss (120) recommends the use of $(NBu_4)^+(BPh_4)^-$ as a substandard. The cell constant is usually affected only slightly by temperature, and the problem of its temperature dependence is thoroughly discussed on pp. 97–99 of ref. 75. Some numerical examples quoted in this reference indicate that the cell constant changes only by 0.2% or less when the temperature is varied by 100°C. Experimental evidence for a very low temperature coefficient of a cell constant has been provided recently by Kay and his co-workers (143).

Studies of the conductance of living polymers and other organometallic compounds introduce some specific difficulties which arise from the extreme sensitivity of these compounds to moisture, O_2, CO_2, etc. Their presence not only destroys the investigated species changing, therefore, their concentration, but gives rise to products which may significantly affect the measured conductance. Hence, a technique had to be developed which permits dilution of the studied solution to concentrations of $10^{-6}M$ or less without introducing any damaging impurities into the investigated system. Moreover, the concentration of the organometallic compounds has to be determined *in situ*. The following procedure (121) solves all these problems.

The apparatus is shown in Figure 22. It consists of a sealed conductivity cell H and three optical cells D, E, and F having optical path lengths of 10 cm, 1 cm, and 0.2 cm, respectively, the last one being equipped with a 0.19-cm spacer. This permits an accurate determination of optical densities of solutions differing in their concentrations by a factor of 1000.

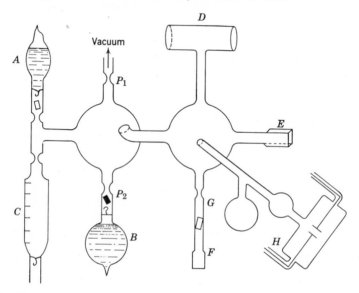

Fig. 22.   The apparatus used for studies of conductance of metalloorganic compounds.

A relatively concentrated solution of the investigated organometallic compound is sealed in ampoules A and B. The contents of the latter are eventually used to purge the apparatus and to destroy all the impurities adsorbed on its walls that are capable of reacting with the studied electrolyte.

The apparatus is evacuated on a high-vacuum line, flamed, and sealed off at construction P_1. The breakseal on ampoule B is crushed and then the walls of the whole unit are wetted with the stored purging solution of the organometallic compound. Thereafter, the rinsing solution is returned to ampoule B, by tilting the apparatus, and the walls are washed by condensing on them the solvent through external cooling. Eventually, all the solvent is frozen in ampoule B, which is then sealed off at constriction P_2. This procedure destroys all the damaging impurities and removes any electrolytes formed by the destruction of the organometallic compound.

The breakseal on ampoule A is now crushed and the investigated solution introduced into the purged unit. Part of the solution is transferred to the central bulb and diluted by condensing in it the remaining solvent. Its concentration is determined by its optical density measured in the 0.2-cm cell with the spacer. By tilting the unit, the solution is introduced then into the conductance cell and its resistance measured. Thereafter, the optical density is redetermined and thence, about two-thirds of the solution is transferred to C and the solvent distilled into the central bulb containing one-third of the original solution. The latter is, therefore, diluted to about one-third of its original concentration. Its exact concentration and conductance are then determined by the previously described procedure, and then again, about two-thirds of the investigated liquid is transferred to C and the remaining third diluted by the recondensed solvent. On repeating this procedure, one dilutes, stepwise, the investigated solution eventually down to $10^{-6} M$ concentration and determines its conductance and concentration at each stage of the dilution. Thus, the relation between $1/\Lambda$ and $c\Lambda$ may be established.

The method described above offers the following advantages. (1) The impurities and the products of their destruction are removed from the system. The latter, if left in the system, may significantly affect the conductance, particularly when the solution of the investigated electrolyte is highly diluted. (2) No impurities are introduced on dilution. In fact, the dilution is accomplished by removing the solute and not by adding the solvent. (3) Destruction or isomerization of the sensitive reagent is minimized because one proceeds from a more concentrated to a more diluted solution.

Usually, the conductance is determined at a fixed temperature by immersing the conductance cell in a thermostat. However, one may determine the conductance at a series of temperatures by immersing the cell in a thermostat kept, e.g., at $-70°C$ and then raise its temperature stepwise. After a new thermal equilibrium is established, the conductance is measured. This procedure gives the conductance at constant concentration as a function of temperature and such measurements are repeated for a series of concentrations. The results are corrected for the thermal contraction of the solvent and eventually utilized for constructing a series of curves, each giving the reciprocals of the cell resistance at constant concentration for different temperatures. For the sake of illustration, such curves are shown in Figure 23 and these are used for calculating $1/\Lambda$ as a function of $c\Lambda$ for a series of temperatures.

A few remarks about the accuracy of the method may not be out of place. Concentrations are always determined at a constant temperature, e.g., at 25°C, and therefore any changes of the extinction coefficient with temperature do not affect the results. However, it is necessary to establish the validity of Beer's law by an independent study. An internal check may be applied for some systems. For example, in studies of the conductance of the sodium salt of tetracene radical-anion, one may leave an excess of tetracene in the investigated solution. The experimental procedure, described above, dilutes equally the radical-anion and its parent hydrocarbon. The absorption peaks of both species do not overlap and, therefore, one may check the reliability of the dilution ratio by determining the optical densities of the respective hydrocarbon peak (77,78). The extinction coefficient of the hydrocarbon is not

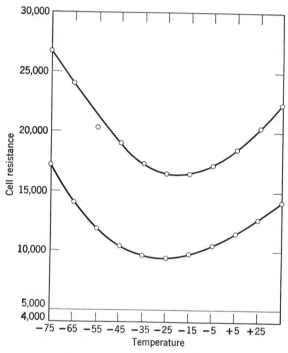

Fig. 23. Temperature dependence of the resistance of a solution of a metalloorganic compound at a constant weight fraction of the electrolyte. Each curve represents another concentration.

expected to vary with concentration, whereas the extinction coefficient of the salt might change on dilution because the absorption of the ion-pair and of the free ion need not necessarily be identical. Fortunately, in most cases the changes in the extinction coefficients of the salt caused by its dissociation are insignificant, although for extremely sharp absorption peaks this phenomenon may be troublesome.

It is important to know the temperature of the conducting solution. Therefore, the conductance cell is equipped with a thermocouple pocket touching the back of one of the electrodes. Thus, it is possible to check simultaneously the temperature of the conducting liquid and of the thermostatic bath. Furthermore, it is desirable to determine resistances at a constant concentration for a series of rising temperatures, and then again for a series of decreasing temperatures. In carefully performed experiments, the results are virtually the same for both sets of experiments.

It is advantageous to perform the experiments with larger volumes of concentrated solutions, say 50–100 ml. The sealing off of the ampoules generates minute amounts of moisture and perhaps some CO_2. This is insignificant when one deals with larger amounts of concentrated solution, but it may destroy a substantial fraction of the investigated electrolyte, if the solution is highly dilute.

Some organometallic compounds are inherently unstable in the solvents which one may desire to use for the studies. For such systems, no reliable data can be obtained. It is possible, however, to determine approximately the conductance by measuring the resistance and concentration at different times and extrapolating the data to zero time. It has been frequently observed that dilute solutions are less stable than the concentrated ones. Probably, side reactions which destroy the investigated salt are faster for the free ions than for the ion-pairs. Therefore, the sequence used in the procedure described above, viz., proceeding from a more concentrated solution to a more dilute, is advantageous.

18. SOME PROBLEMS CONCERNED WITH THE DETERMINATION OF Λ_0

In ethereal solvents such as tetrahydrofuran, tetrahydropyran, etc., which are of great interest to the student of anionic polymerization, the ionic dissociation constant, K_{diss}, of many organometallic compounds is extremely low, often less than $10^{-7}M$. In such cases, serious difficulties are encountered in the experimental determination of K_{diss} from the conductance data. The dissociation constant is given by the equation $K_{diss} = 1/\Lambda_0^2$ (slope of the respective Fuoss line) and, although the slope is experimentally reliable, the intercept, $1/\Lambda_0$, is too minute to be distinguished from zero if the line is steep, i.e., when K_{diss} is very small. This may be seen in Figure 24. The same difficulty arises, if any other method is applied in estimating Λ_0 from the experimental values of Λ.

For most of the systems, which will interest us, ionic dissociation is exothermic, and therefore K_{diss} increases at lower temperatures. Consequently, although the Fuoss line of the investigated compound may be too steep to permit a reliable extrapolation of Λ to Λ_0 at higher temperatures, say 20°C, the line becomes flatter and the extrapolation feasible at some sufficiently low temperature. The Walden product, $\eta\Lambda_0$, is only slightly affected by temperature (see p. 236), and therefore the extrapolation of $\eta\Lambda_0$ from lower temperatures to higher is reasonably reliable. Thus, the directly unavailable Λ_0 might be obtained by this method. Figure 25 illustrates this procedure for the case of the sodium salt of living polystyrene in THF. Data are taken from ref. 122; the Walden product is plotted against $1/T$ and the experimental points seem to form a straight line. This linear relationship is empirical; obviously for

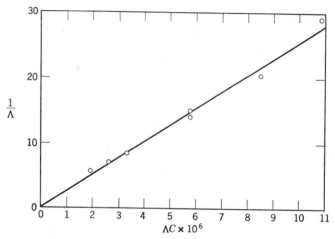

Fig. 24.   Plot of $1/\Lambda$ versus $c\Lambda$ for $\sim\!\!\mathrm{S}^-,\mathrm{Na}^+$ (living polystyrene) in THP at 25°C. Note that the intercept giving $1/\Lambda_0$ is indistinguishable from 0.

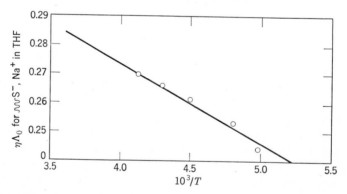

Fig. 25.   Plot of the Walden product of $\sim\!\!\mathrm{S}^-,\mathrm{Na}^+$ (living polystyrene) in THF against $1/T$. Used to extrapolate the low temperature data to high temperature in order to determine the respective Λ_0.

any slowly varying function a linear plot is expected within a limited range of variables.

An alternative method leading to the determination of Λ_0 is based on the equation, $\Lambda_0 = \lambda_0^+$ (cation) $+ \lambda_0^-$ (anion). Although the investigated electrolyte may be only scarcely dissociated, making the direct determination of its

Λ_0 unfeasible, another salt, sharing with the former a common ion, could be highly dissociated under the same conditions. Hence, Λ_0 of the latter can be determined directly from the respective conductance data. For example, the sodium salt of living polystyrene in THF at 25°C does not dissociate to a sufficient degree to permit the direct determination of its Λ_0 whereas, the evaluation of Λ_0 of the sodium tetraphenyl boride from its conductance study is feasible under the same conditions. Moreover, the same procedure gives Λ_0 of (isoamyl)$_3$(butyl)N$^+$,BPh$_4^-$, and it is known from relevant transference studies that both ions contribute equally to the conductance of the latter salt. Hence, by combining the results of both investigations, one may calculate the λ_0^+(Na$^+$) in THF at 25°C. It could be possible, at least in principle, to find some suitable electrolytes, the conductance of which would provide the value of λ_0^- of the living polystyrene ion, and then the Λ_0 of $\sim\!\!\sim$S$^-$,Na$^+$ may be calculated.

There is an alternative approach. The size and shape of large ions which do not coordinate with molecules of the solvent is negligibly affected by the environment. This, e.g., is true for negative ions in ethereal solvents, and particularly for polymeric ions such as the living polystyrene$^-$ ion. Its shape and size should be negligibly affected by the removal of an electron—a process transforming an $\sim\!\!\sim$S$^-$ ion into an $\sim\!\!\sim$S· free radical—or even by proton addition which converts $\sim\!\!\sim$S$^-$ into the dead $\sim\!\!\sim$SH polymer. Therefore, the self-diffusion constant of an $\sim\!\!\sim$S$^-$ ion should be virtually identical with the diffusion constant of the $\sim\!\!\sim$SH dead polymer having the same degree of polymerization. The diffusion constant, $\mathscr{D}$, and the mobility, u, of an ion are correlated by a fundamental relation, viz., $u = \mathscr{D}/kT$, where u and λ_0 represent the same entity but expressed in different units. Therefore, $\lambda_0^-(\sim\!\!\sim$S$^-$) may be calculated, if the diffusion constant of the respective dead polymer $\sim\!\!\sim$SH is experimentally determined. This approach was employed in studies of the conductance of $\sim\!\!\sim$S$^-$,Na$^+$ (122) and the results obtained by this method agreed well with those derived from extrapolation of the low temperature data.

Similar observations were reported by Slates and Szwarc (77) who investigated the conductance of the sodium salts of the naphthalene and anthracene radical-ions. The diffusion constants for naphthalene and anthracene have been reported (123) and the λ_0^-'s calculated from these data were only slightly larger than those derived from the conductance data. Obviously, the negative naphthalene$^-$ and anthracene$^-$ ions in THF are only insignificantly larger than the respective neutral hydrocarbons, demonstrating that their coordination with ethereal solvent is indeed negligible.

19. HEAT AND ENTROPY OF ION-PAIR DISSOCIATION

Data concerned with the heat of dissociation of ion-pairs into free ions were meager in the past. Most of the published studies dealt with the effect of solvents on the extent of dissociation at constant temperature, and the work published prior to 1956 was reviewed by Kraus (137,138). The scanty data which were available showed conclusively that the dissociation is exothermic as expected from the Denison and Ramsey treatment (20).

Systematic studies of the temperature dependence of K_{diss} have been recently undertaken in the writer's laboratory. The results illustrate the various aspects of the problems discussed in this chapter and, therefore, they deserve detailed discussion.

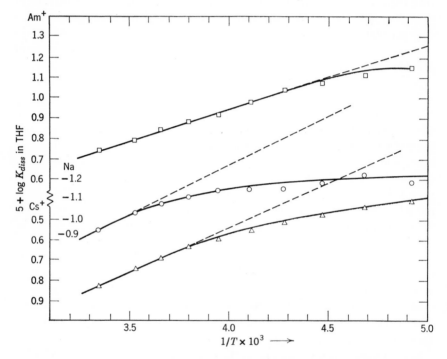

Fig. 26. Plot of $\log K_{diss}$ of $Bu(isoaml)_3N^+,BPh_4^-$ (○), Na^+,BPh_4^- (△), and Cs^+,BPh_4^- (□) versus $1/T$. Solvent THF.

Dissociation of three salts of tetraphenyl boride in tetrahydrofuran and in dimethoxyethane was investigated over a wide temperature range, viz., from $-70°$ to $+25°C$ (74). The results are presented in Figures 26 and 27 as plots of $\log K_{diss}$ versus $1/T$. All the curves seen in these figures are sloped upwards, indicating the exothermic nature of the dissociation, and their convexed curvature shows that the exothermicities decrease at lower temperatures. This indicates the increasing degree of solvation of ion-pairs at lower temperatures, and probably a change in their structure. The semi-quantitative treatment of the data is instructive, the results being summarized in Table IX.

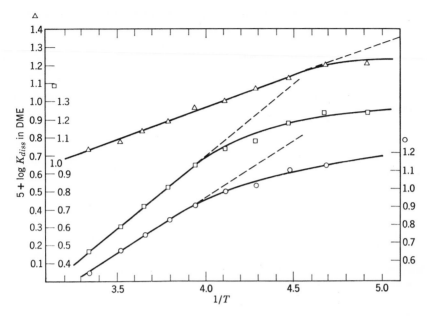

Fig. 27.   Plot of log K_{diss} of Bu(isoaml)$_3$N$^+$,BPh$_4^-$ (□), Na$^+$,BPh$_4^-$ (○), and Cs$^+$,BPh$_4^-$
(△) versus $1/T$. Solvent DME.

TABLE IX

The Thermodynamic Constants of Ionic Dissociation of Tetraphenylboride Salts in
DME and THF Calculated at 25°C

a, Å	System		$-\Delta H$, kcal/mole	$\Delta H_{DME}/\Delta H_{THF}$	Log K_{diss}
			Bu(Am)$_3$N$^+$,BPh$_4^-$		
7.8	THF	calcd.	0.95	1.75 calcd.	−4.25
		obsd.	1.3	1.75 obsd.	−4.25
	DME	calcd.	1.65		−4.30
		obsd.	2.3		−4.45
			Cs$^+$,BPh$_4^-$		
5.6	THF	calcd.	1.30		−5.65
		obsd.	1.55		−5.80
	DME	calcd.	2.3		−5.70
		obsd.	3.5		−4.55
			Na$^+$,BPh$_4^-$		
7.5	THF	calcd.	1.0	1.7 calcd.	−4.45
		obsd.	1.3	1.4 obsd.	−4.06
7.3	DME	calcd.	1.7		−4.70
		obsd.	1.8		−4.27

The theoretical dissociation constants, given in that table, were calculated from the Fuoss equation*

$$K_{diss} = (3000/4\pi a^3 N) \exp(-e^2/aDkT),$$

where $a = r_1 + r_2$ denotes the distance between the centers of the ions of the investigated pair. The data obtained in both solvents for triisoamylbutyl ammonium tetraphenyl boride may be satisfactorily accounted for by accepting $a = 7.8$ Å. This value is only slightly smaller than the sum of the respective Stokes radii, which were derived from the relevant λ_0^+ and λ_0^-. Since the Stokes radii agree well with those obtained from the molecular models, it seems that neither the free BPh_4^-, nor (isoamyl)$_3$BuN$^+$ ions are coordinated with the molecules of THF or DME. Thus, the fair agreement observed between the calculated and the experimental ΔH and K_{diss} suggests that the above salt forms contact ion-pairs which dissociate into bare, non-solvent-coordinated ions.

A similar agreement between the theory and experiment is found for sodium tetraphenyl boride if a relatively large value is chosen for a, viz., 7.5 Å for the salt dissolved in THF and about 7.3 Å for its DME solution. This implies that the relatively small Na$^+$ ion is coordinated with solvent molecules whether free or associated with its counterion into an ion-pair. The magnitudes of the respective λ_0^+ justify the proposed values of a. They show that the free Na$^+$ ion is surrounded by a coordination shell, which increases its radius to about 4 Å in THF and to about 3.2 Å in DME. The lower radius of the solvation shell in DME is reflected by the decrease in the respective a value. It seems, therefore, that sodium tetraphenyl boride forms in THF and in DME solvent-separated ion-pairs which dissociate into solvent-coordinated free Na$^+$ ions. Hence, both ions are always large, and their structures are not substantially affected by the dissociation or association processes. This accounts for the success of the theory.

A different situation is encountered for the solutions of the cesium salt. The dissociation in THF may be again satisfactorily accounted for by the simple treatment based on the "sphere-in-continuum" model. Numerical agreement with the experiment is achieved by postulating $a = 5.6$ Å—a value expected for a pair involving a bare Cs$^+$ ion. In fact, the Stokes radius of Cs$^+$ in THF appears to be only slightly greater than that of the non-coordinated Cs$^+$ ion, adding credulity to the descriptions of Cs$^+$,BPh$_4^-$ in THF as a contact ion-pair which dissociates into non-solvent-coordinated free Cs$^+$ ion. In this respect, Cs$^+$BPh$_4^-$ resembles (isoamyl)$_3$BuN$^+$,BPh$_4^-$, and the relatively large size of the Cs$^+$ ion permits us to satisfactorily apply the simple theory. On the other hand, this theory fails to account for the behavior

---

* The shortcomings of the Fuoss equation, as well as some objections raised against its derivation, were discussed on p. 230. Nevertheless, this equation is useful for the semi-quantitative treatment.

of Cs^+,BPh_4^- in DME. In this solvent the dissociation is much more extensive than in THF, and in addition $-\Delta H$ is larger in DME than in THF. To rationalize the experimental findings, one has to postulate that Cs^+,BPh_4^- forms a contact ion-pair even in DME, but on its dissociation the resulting free Cs^+ ion becomes coordinated with the solvent.

The striking difference in the behavior of contact ion-pairs and solvent-separated ion-pairs is well illustrated by the conductance studies of sodium salts of the radical-anion and dianion of tetraphenylethylene (125). The dependence of the dissociation constant on $1/T$ is shown in Figure 18, Chapter VI. The dissociation of the salt of the radical-anion, $T^{\bar{\cdot}}$, is given by the equation

$$T^{\bar{\cdot}},Na^+ \rightleftarrows T^{\bar{\cdot}} + Na^+,$$

that of the dianion by the equation

$$T^=,2Na^+ \rightleftarrows T^=,Na^+ + Na^+.$$

At 25°C the dissociation constant of the radical-ion salt is about 100 times as large as that of the salt of the dianion, although the heat of dissociation of the latter salt is about 6 kcal/mole more exothermic than that of the former, the respective $-\Delta H$'s being 7.2 and 1.3 kcal/mole.

It has been concluded that $T^{\bar{\cdot}},Na^+$ forms solvent-separated ion-pairs, whereas $T^=,2Na^+$ exists as contact ion-pairs, and both dissociate into THF-coordinated free Na^+ ions. This coordination contributes to the higher exothermicity of the dissociation of $T^=,2Na^+$.

It is interesting to inquire why $T^{\bar{\cdot}},Na^+$ forms a different ion-pair from $T^=,2Na^+$. The association of a solvent-coordinated ion with its counterion proceeds without any repulsion until a solvent-separated ion-pair is formed. Further approach requires expulsion of solvent molecules (or a molecule) which separates the ions, and this leads to some repulsion (54,55). Let the thickness of the solvation shell be Δ, the radii of the bare ions r_1 and r_2, and assume, for the sake of simplicity, that a constant dielectric constant D determines the Coulombic energy of the system. Hence, the gain in Coulombic energy produced by the solvent expulsion is

$$(e^2/D)\{1/(r_1 + r_2) - 1(r_1 + r_2 + \Delta)\}.$$

Roberts and Szwarc (125) pointed out that for a constant r_1 and Δ this gain is smaller as the radius of the counterion, r_2, becomes larger. The $T^{\bar{\cdot}}$ radical-ion is bulky, because the four phenyl groups, placed *cis* to each other, cannot be coplanar. However, the formation of the $T^=$ dianion destroys the C=C double bond, and this permits rotation around the C—C bond. Consequently, each pair of phenyl groups attached to one carbon atom may attain a coplanar configuration, the respective planes being mutually perpendicular. Therefore, in the dianion the sodium ion may closely approach the negative carbon atom, making r_2 small, but in the $T^{\bar{\cdot}}$ radical ion this is not the case and r_2 is large. This accounts for the difference in the type of solvation of these two ion-pairs.

The comparison of the behavior of Cs^+, BPh_4^- in THF and DME on the one hand, and of $T^{\bar{}}, Na^+$ and $T^=, 2Na^+$ in THF on the other, needs further stressing. The cesium salt forms contact ion-pairs in both solvents, but it dissociates into a non-solvent-coordinated free Cs^+ ion in THF, whereas a solvent-coordinated free cesium ion is formed in DME. Hence, an *increase*

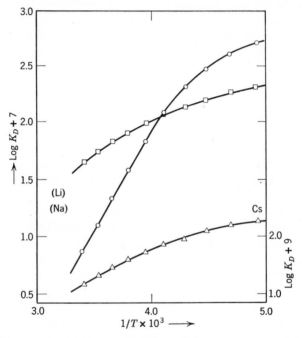

Fig. 28. Plot of log K_{diss} of salts of fluorenyl carbanion versus $1/T$ in THF. ($\square$) F^-, Li^+, ($\bigcirc$) F^-, Na^+, ($\triangle$) F^-, Cs^+.

in the dissociation constant is associated with an *increase* in the exothermicity of the process. In the case of $T^{\bar{}}, Na^+$ and $T^=, 2Na^+$, the dissociation produces in each system the solvent-coordinated free Na^+ ion. However, while the latter salt forms a contact ion-pair, the former exists as a solvent-separated pair. Consequently, a *higher* dissociation constant is associated with a *lower* exothermicity of the dissociation process.

The work of Hogen-Esch and Smid (144) adds much to our understanding of the problem of exothermicity of ion-pair dissociation. In the system sodiumfluorenyl–THF, the contact ion-pair is distinguished from the solvent-separated ion-pair by its different spectrum (see pp. 259 and 267). Hence, the equilibrium constant and the heat of the process,

contact ion-pair + solvent $\rightleftarrows$ solvent-separated pair, were determined from spectrophotometric data (see Fig. 18). The conductance of sodiumfluorenyl was studied over the temperature range −70 to +25°C, and the results led to the plot of log K_{diss} versus $1/T$ shown in Figure 28. At higher temperatures (around 25°C) the system is composed mainly of contact ion-pairs, and therefore the observed large heat refers to the dissociation of a contact ion-pair into a solvent-coordinated Na^+. At −70°C the equilibrium shifts to the right, and then virtually only solvent-separated ion-pairs are present in the solution. Hence, the heat of dissociation at that temperature is low and it refers to a solvent-separated pair dissociating into a solvent-coordinated Na^+ ion. The difference between these two values is equal to the spectrophotometrically determined heat of conversion of the contact into the solvent-separated pair. This confirms the proposed interpretation of the results.

It is interesting to note in the same figure the Van't Hoff plots of the Li^+ and Cs^+ salts. In both cases the heats of dissociation are low over the whole temperature range. No effects caused by the solvent coordination are observed in either system, because the lithium ion-pair is solvent separated, i.e., solvent coordinated to begin with, and yields a solvent-coordinated lithium ion, whereas the cesium salt forms a contact ion-pair which dissociates into a noncoordinated, bare cesium ion.

It is obvious from this discussion that studies of heats of dissociation are most valuable in elucidating the relations between solvent and the ion-pairs and free ions. Similar information has been recently gained from ESR studies of salts of radical-ions (154). These investigations are discussed in Section 21.

20. SOLVATION AND COORDINATION

It seems that a useful distinction should be made between two types of interactions which take place in the system: solvent (or other suitable agent) and free ions and ion-pairs. In fact, Bockris (145) in his review article suggested the term "primary solvation" to describe the comparatively firmly attached solvent molecules surrounding an ion (or ion-pair), whereas "secondary solvation" designates all the other interactions with solvent molecules which affect the behavior of the dissolved ions.

The term coordination may be equivalent to Bockris' "primary solvation." It refers to the strong interaction which produces a single kinetic and thermodynamic entity by uniting an ion (or ion-pair) and a set of solvent molecules. The stoichiometric relations are expected to be characteristic for a coordination, at least in the systems which exhibit a clear-cut behavior. Of course, it is possible to have two, or more, types of coordinated ions (or ion-pairs) coexisting in a particular system (134). On the other hand, any ion (or ion-pair) in a solvent must always be surrounded by neighboring solvent molecules, but if the interaction with them is sufficiently weak to prevent the formation of a new and unique entity, the resulting agglomerate will be designated as a solvated ion (or ion-pair). No stoichiometric relations are expected in that case, and it cannot be uniquely stated how many molecules are involved because the answer depends on the type of phenomenon which is investigated. It is apparent that a coordinated ion (or ion-pair) is still solvated by the next neighbors.

The proposed distinction is justified in view of the behavior of some systems discussed in this chapter. It is not rigorous, and many borderline cases may be quoted. It is hoped, however, that this nomenclature may lead to a distinction between physical interactions, in which the solvent may be treated as a continuum, and the proper chemical relations founded on its discreet molecular structure.

21. APPLICATION OF THE ESR TECHNIQUE IN STUDIES OF ION-PAIRS

The ESR spectra of radicals or radical-ions reveal hyperfine structure arising from the presence of magnetic nuclei in the investigated species. Each magnetic nucleus splits the ESR signal, and for nuclei of the same kind the splitting constants are proportional to the probability of finding the odd electron on each of them. For example, the ESR hyperfine structure of a free, nonassociated naphthalene⁻ radical-anion is caused by two sets of four equivalent protons, α and β. These split the signal into 25 lines; the splitting constant of α protons is greater than that of β because the odd electron is more likely to be on the α than on the β proton. The spectrum of the corresponding ion-pair, e.g., sodium naphthalenide, is more complex. The odd electron has a finite probability of being located on a ^{23}Na nucleus which has a spin of 3/2. Consequently, each of the original 25 lines is split further into four.

This additional splitting was first reported by Weissman (13), and its detection conclusively demonstrates the existence of ion-pairing. The converse is not necessarily true. An ion-pair could be formed although the relevant splitting may be undetected if it is too small. For example, in one of his earlier papers (13b) Weissman described two sets of lines discerned in the spectrum of sodium naphthalenide in tetrahydrofuran. One set showed the splitting due to ^{23}Na, while the other did not. The latter was attributed to the free naphthalene⁻ ion, but presently we are inclined to interpret it as the spectrum of solvent-separated ion-pairs. The relatively large distance parting the two ions of such a pair decreases the probability of finding the odd electron on the cation and, therefore, the respective splitting constant becomes too small for detection. On the whole, the splitting constants caused by cations are small, often not greater than 3 gauss. However, there are exceptions, e.g., a splitting constant as great as 10 gauss was reported for some cesium ketyle (157), and the spectrum of $CO_2^{\overline{\cdot}}$ radical-ions at 77°K is split by Na^+ by as much as 20 gauss (158).

The presence of two sets of lines in the ESR spectrum makes it possible to determine the relevant equilibrium constant. In his pioneering study Weissman (13b) interpreted the spectrum of sodium naphthalenide in tetrahydrofuran as evidence for the dissociation,

$$\text{Naphthalene}^{\overline{\cdot}},\text{Na}^+ \rightleftarrows \text{Naphthalene}^{\overline{\cdot}} + \text{Na}^+,$$

and accordingly calculated the equilibrium constant as $K = (\Gamma^2 + \Gamma)/C$, where Γ denotes the ratio of concentrations of the two species (i.e., the ratio of the intensities of the respective lines) and C is their total concentration. These constants were determined in three different solvents for a series of temperatures, and the results led to the relevant heats of dissociation. Because $C = $ const. for each series of experiments and $\Gamma^2 \ll \Gamma$, the calculated equilibrium constants are approximately proportional to those of the solvation process,

$$\text{contact-pair} \rightleftarrows \text{solvent-separated pair.}$$

This accounts for the fact that the reported values of ΔH are close to those of the solvation of contact ion-pairs into solvent-separated pairs.

The splitting constants attributed to the presence of ^{23}Na depend on temperature and on the nature of the solvent. First systematic studies of this phenomenon were reported again by Weissman (13b) who noted the increase of the splitting constant with rising temperature. Since then this subject was intensively studied by many workers, notably by Hirota (154), Symons (155), de Boer (156), and others.

The observed alkali metal splitting is the weighted average of the splittings characterizing the different states of the investigated species, provided the exchange between them is sufficiently rapid. Structurally different ion-pairs or different vibrational states of an ion-pair exemplify the pertinent subspecies. The time required to convert one type of ion-pair into another is usually much slower than the vibrational relaxation time (about 10^{-12} sec). Therefore the observed splitting constant, A_m, may be calculated as

$$A_m = \sum_{\substack{\text{different} \\ \text{pairs}}} (P_i A_i) \text{ with } A_i = \sum_{\substack{\text{different} \\ \text{vibrational} \\ \text{states}}} (P_{iv} A_v).$$

P_i and P_{iv} denote the relevant mole fraction; A_i, the splitting constants of the various ion-pairs; and A_{iv}, those of the respective, vibrationally different subspecies. The distinction is mainly a matter of convenience because unequivocal differentiation between structural and vibrational states is not always valid.

Temperature affects the values of P_i and P_{iv} and therefore it influences the observed splitting constant. For the sake of illustration, let us consider the case of sodium naphthalenide as discussed by Weissman (13b). Argument based on the symmetry consideration indicates that the $3s$ orbital of sodium and the half-occupied π orbital of the naphthalene$^-$ anion are orthogonal, provided the cation is located in a plane drawn through the 9 and 10 carbon atoms and perpendicular to the plane of naphthalene. Hence, if the displacement of the cation from this location is small, the splitting constant would be minute, but it should increase as the displacement becomes larger. Higher temperature increases the amplitude of vibration and, therefore, also the splitting constant.

Although the above explanation is applicable to some systems, it seems that most of the large changes in splitting constants should be attributed to the rapid equilibrium involving two (or more) types of ion-pairs. Each pair is solvated to a different degree, and therefore is characterized by a different average distance separating the ions of a pair. Denoting the splitting constants of the two kinds of ion-pairs by A_1 and A_2 and the equilibrium constant of the rapid process by K,

$$\text{Pair (1)} \rightleftharpoons \text{Pair (2)},$$

one finds the observed splitting constant A_m to be

$$A_m = (A_1 + KA_2)/(1 + K).$$

Usually the splitting constant of the more-solvated pair is smaller than that of the less-solvated (tighter) one. Moreover, for many of the investigated systems the solvation processes are *exo*thermic, and therefore, the observed splitting constants should decrease with decreasing temperature.

For equilibrium systems A_m shows a characteristic, nonlinear dependence on the reciprocal of temperature which was observed (154). In such a case the equilibrium constant K may be calculated for different temperatures, and from these data ΔH and ΔS of the investigated transformation may be determined. In some systems the heat is absorbed as the cation becomes coordinated with solvent molecules, i.e., the transformation is *endo*thermic. Potassium anthracenide in diethyl ether provides a pertinent example, and the splitting constant then increases, although only slightly, at lower temperatures. The stronger interaction of solvent–solvent compared to solvent–ion-pair accounts for such a trend.

Other changes observed in the ESR spectra also reveal the formation of ion-pairs, although in a less conclusive way. The most interesting is the change of the proton splitting constants of the radical-anion caused by the cation. This phenomenon was first reported by Reddoch (159) and a close correlation between the magnitude of such a perturbation and the alkali splitting constant was demonstrated by Hirota (154). The presence of cation slightly decreases the density of the odd electron on the anion, and subsequently the proton splitting constant is somewhat diminished. The pairing also leads to slight changes in the g values. The respective data were tabulated by Symons (155).

The perturbation of the electron density of the radical-anion by the neighboring cation may reveal sometimes the preferential location of the latter. A symmetrically located cation, or a cation that shows no preference for any specific location in the course of its motion, perturbs all the hyperfine lines evenly. However, the perturbation becomes unsymmetrical if at any time one location is favored more than the other (155b,155d,156b,160,161).

The most valuable, and unique, is the information provided by the ESR technique about the rates of the various transformations. In the case of a rapid equilibrium between two different ion-pairs the contribution of exchange to the line width of each hyperfine component is

$$1/T_2 = P_1^2 P_2^2 (\breve{\omega}_1 - \breve{\omega}_2)^2 \tau_1 (1 + K)$$

where $\breve{\omega}_1$ and $\breve{\omega}_2$ are the resonance frequencies of the respective pairs and τ_1 is the life time of the first pair. Thus, τ_1 may be determined from studies of the line width.

The line width depends on the Z component of the alkali nucleus spin, e.g., for Na^+ the lines corresponding to $I_m = 3/2$ or $-3/2$ are broader than those for $I_m = 1/2$ or $-1/2$. The reason for this behavior is revealed by the schematic drawing shown in Figure 29. For very slow exchange two sets of four lines appear in the spectrum (I). At extremely fast exchange the two sets merge into one (III). In the intermediate case the energy fluctuates within the limits shown in II. Thus, the extreme lines (3/2 and $-3/2$) become broader than the middle lines (1/2 and $-1/2$). Such a type of broadening provides, therefore, a conclusive proof for a fast equilibrium between two types of ion-pairs. Similar effects are observed when the cation hops between two locations (155,156,160,161) or moves from one side of the molecular plane to the other (164).

The rates of interconversion of ion-pairs were investigated by Crowley et al. (154c) and by Hirota (154d). In this process the alkali ion remains unaltered during the interconversion, whereas its identity is changed in the course of the dissociation–association process. Consequently, as the interconversion process becomes faster the initially broad lines eventually sharpen [Fig. 29 (III)]—the multiplet due to the cation being retained. In the dissociation–association process the lines broaden and eventually the alkali multiplet disappears when the exchange becomes sufficiently fast (163). However, even a fast dissociation–association process is revealed by ESR spectroscopy through the perturbation of the proton splitting, which is caused by the cation and by the subtle change in the g value (159).

ESR studies permit the study of some fine details of the exchange processes.

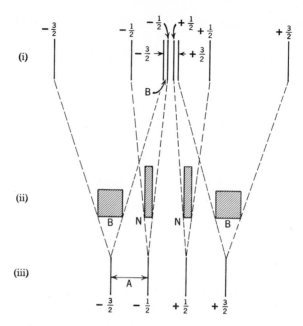

Fig. 29. Hypothetical spectra for two ion-pairs in equilibrium showing hyperfine coupling to a cation having $I_m = 3/2$ and hyperfine couplings constants A_1 and A_2; (i) slow, (ii) intermediate, and (iii) fast exchange.

For the sake of illustration we may quote the results of Adams and Atherton (162) who studied the reaction

$$(m\text{-dinitrobenzene})^{\bar{\cdot}},\text{Na}_\alpha^+ + \text{Na}_\beta^+ \rightleftharpoons \text{Na}_\alpha^+ + (m\text{-dinitrobenzene})^{\bar{\cdot}},\text{Na}_\beta^+.$$

They were able to show that the process involves the route

$$\text{Na}_\beta^+ + \text{NO}_2\text{—}\underset{\ominus}{\bigcirc}\text{—NO}_2, \text{Na}_\alpha^+ \rightarrow \text{Na}_\beta^+, \text{NO}_2\text{—}\underset{\ominus}{\bigcirc}\text{—NO}_2 + \text{Na}_\alpha^+,$$

i.e., the incoming cation goes to the uncomplexed nitro group. Such detailed studies of the mechanism are only possible through the application of magnetic resonance techniques.

References

(1) N. Bjerrum, *Kgtl. Danske Videnskab. Selskab.*, **4**, 26 (1906).
(2) R. M. Fuoss, *J. Chem. Educ.*, **32**, 527 (1955).
(3) E. Grison, K. Eriks, and J. L. de Vries, *Acta Cryst.*, **3**, 290 (1950).
(4) R. A. Ogg, *J. Chem. Phys.*, **15**, 337 (1947).
(5) K. Ziegler and H. Wollschitt, *Ann. Chem.*, **479**, 90 (1930).
(6) L. Pauling, *The Nature of the Chemical Bond*, 3rd ed., Cornell University Press, Ithaca, N.Y., 1960.

(7) C. B. Monk, *Electrolytic Dissociation*, Academic Press, New York, 1961, pp. 275–287.

(8) P. Debye and E. Hückel, *Z. Physik.*, **24**, 185 (1923).

(9) E. A. Guggenheim, *Discussions Faraday Soc.*, **24**, 53 (1957).

(10) F. Kohlrausch, L. Holborn, and H. Diesselhorst, *Wien. Ann.*, **64**, 417 (1898).

(11) L. Onsager, *Z. Physik.*, **28**, 277 (1927).

(12) N. Bjerrum, *Kgtl. Danske Videnskab. Selskab*, **7**, No. 9 (1926).

(13) (a) F. C. Adam and S. I. Weissman, *J. Am. Chem. Soc.*, **80**, 1518 (1958). (b) N. M. Atherton and S. I. Weissman, *ibid.*, **83**, 1330 (1961).

(14) R. M. Fuoss and L. Onsager, *J. Phys. Chem.*, **67**, 621 (1963).

(15) R. M. Fuoss and C. A. Kraus, *J. Am. Chem. Soc.*, **55**, 21 (1933).

(16) R. M. Fuoss, *Trans. Faraday Soc.*, **30**, 967 (1934).

(17) J. C. Poirier and J. H. DeLap, *J. Chem. Phys.*, **35**, 213 (1961).

(18) H. Reiss, *ibid.*, **25**, 400 (1956).

(19) R. M. Fuoss, *J. Am. Chem. Soc.*, **57**, 2604 (1935).

(20) J. T. Denison and J. B. Ramsey, *ibid.*, **77**, 2615 (1955).

(21) Y. H. Inami, H. K. Bodenseh, and J. B. Ramsey, *ibid.*, **83**, 4745 (1961).

(22) W. R. Gilkerson, *J. Chem. Phys.*, **25**, 1199 (1956).

(23) J. G. Kirkwood, *ibid.*, **18**, 380 (1950).

(24) R. M. Fuoss, *J. Am. Chem. Soc.*, **80**, 5059 (1958).

(25) F. Accascina, A. D'Aprano, and R. M. Fuoss, *ibid.*, **81**, 1058 (1959).

(26) E. Grunwald and E. Price, *ibid.*, **86**, 4517 (1964).

(27) R. M. Noyes, *ibid.*, **84**, 513 (1962).

(28) R. L. Kay, C. Zawoyski, and D. Fennell Evans, *J. Phys. Chem.*, **69**, 4208 (1965).

(29) R. L. Kay and D. Fennell Evans, *ibid.*, p. 4216.

(30) H. A. Scheraga, S. J. Leach, and R. A. Scott, *Discussion Faraday Soc.*, **40**, 268 (1965).

(31) H. K. Bodenseh and J. B. Ramsey, *J. Phys. Chem.*, **69**, 543 (1965).

(32) A. Zalkin, *Acta Cryst.*, **10**, 557 (1957).

(33) L. R. Maxwell, S. B. Hendricks, and V. M. Mosley, *Phys. Rev.*, **52**, 968 (1937).

(34) A. Honig, M. Mandel, M. L. Stich, and C. H. Townes, *ibid.*, **96**, 629 (1954).

(35) E. Hirsch and R. M. Fuoss, *J. Am. Chem. Soc.*, **82**, 1013, 1018 (1960).

(36) J. A. Davies, R. L. Kay, and A. R. Gordon, *J. Chem. Phys.*, **19**, 749 (1951).

(37) J. R. Graham and A. R. Gordon, *J. Am. Chem. Soc.*, **79**, 2350 (1957).

(38) S. Blum and H. I. Schiff, *J. Phys. Chem.*, **67**, 1220 (1963).

(39) J. F. Coetzee and G. P. Cunningham, *J. Am. Chem. Soc.*, **86**, 3403 (1964).

(40) R. M. Fuoss and C. A. Kraus, *ibid.*, **79**, 3304 (1957).

(41) R. M. Fuoss, *Proc. Natl. Acad. Sci. U.S.*, **45**, 807 (1959).

(42) J. J. Hermans, *Z. Physik*, **97**, 681 (1935).

(43) R. H. Boyd, *J. Chem. Phys.*, **35**, 1281 (1961).

(44) R. Zwanzig, *ibid.*, **38**, 1603 (1963).

(45) T. L. Fabry and R. M. Fuoss, *J. Phys. Chem.*, **68**, 971 (1964).

(46) J. F. Coetzee and G. P. Cunningham, *J. Am. Chem. Soc.*, **87**, 2529 (1965).

(47) J. M. Notley and M. Spiro, *J. Phys. Chem.*, **70**, 1502 (1966).

(48) H. S. Frank, *Proc. Roy. Soc. (London)*, Ser. A, **247**, 481 (1958).

(49) J. B. Ramsey and E. L. Colichman, *J. Am. Chem. Soc.*, **69**, 3041 (1947).

(50) Y. H. Inami, H. K. Bodenseh, and J. B. Ramsey, *ibid.*, **83**, 4745 (1961).

(51) Y. H. Inami and J. B. Ramsey, *J. Chem. Phys.*, **31**, 1297 (1959).

(52) A. Piekara and A. Chelkowski, *ibid.*, **25**, 794 (1956).

(53) J. B. Hyne, *J. Am. Chem. Soc.*, **85**, 304 (1963).

(54) H. Sadek and R. M. Fuoss, *ibid.*, **76**, 5905 (1954).

(55) H. Sadek and R. M. Fuoss, *ibid.*, p. 5897.

(56) R. H. Stokes, *ibid.*, **86**, 979 (1964).

(57) R. H. Stokes, *ibid.*, p. 2333.

(58) J. Häsing, *Ann. Physik.*, **37**, 509 (1940).

(59) H. Witte and E. Wölfel, *Z. Physik. Chem.*, **3**, 296 (1955).

(60) R. H. Stokes, *J. Am. Chem. Soc.*, **86**, 982 (1964).

(61) L. Onsager, *J. Phys. Chem.*, **43**, 189 (1939).

(62) W. M. Latimer, K. S. Pitzer, and C. M. Slansky, *J. Chem. Phys.*, **7**, 108 (1939).

(63) S. W. Benson and C. S. Copeland, *J. Phys. Chem.*, **67**, 1194 (1963).

(64) G. H. Haggis, J. B. Hasted, and T. J. Buchanan, *J. Chem. Phys.*, **20**, 1452 (1952).

(65) J. E. B. Randles, *Trans. Faraday Soc.*, **52**, 1573 (1956).

(66) J. D. Bernal and R. H. Fowler, *J. Chem. Phys.*, **1**, 515 (1933).

(67) D. D. Eley and M. G. Evans, *Trans. Faraday Soc.*, **34**, 1093 (1938).

(68) A. D. Buckingham, *Discussions Faraday Soc.*, **24**, 151 (1957).

(69) A. M. Hogg, R. M. Haynes, and P. Kebarle, *J. Am. Chem. Soc.*, **88**, 28 (1966).

(70) J. P. Hunt and H. Taube, *J. Chem. Phys.*, **18**, 757 (1950); **19**, 602 (1951).

(71) O. Ya. Samoilov, *Zh. Fiz. Khim.*, **29**, 1582 (1954).

(72) O. Ya. Samoilov, *Discussions Faraday Soc.*, **24**, 141 (1957).

(73) J. H. Wang, *J. Phys. Chem.*, **58**, 686 (1954).

(74) C. Carvajal, K. J. Tölle, J. Smid, and M. Szwarc, *J. Am. Chem. Soc.*, **87**, 5548 (1965).

(75) R. A. Robinson and R. H. Stokes, *Electrolyte Solutions*, 2nd ed., Academic Press, New York, 1959, pp. 124–125.

(76) D. N. Bhattacharyya, C. L. Lee, J. Smid, and M. Szwarc, *J. Phys. Chem.*, **69**, 608 (1965).

(77) R. V. Slates and M. Szwarc, *ibid.*, **69**, 4124 (1965).

(78) P. Chang, R. V. Slates, and M. Szwarc, *ibid.*, **70**, 3180 (1966).

(79) A. D'Aprano and R. M. Fuoss, *ibid.*, **67**, 1704, 1722 (1963).

(80) T. L. Fabry and R. M. Fuoss, *ibid.*, **68**, 907 (1964).

(81) H. C. Fu, T. Psarras, H. Weidmann, and H. K. Zimmerman, *Ann. Chem.*, **641**, 116 (1961).

(82) D. S. Burgess and C. A. Kraus, *J. Am. Chem. Soc.*, **70**, 706 (1948).

(83) C. J. Carignan and C. A. Kraus, *ibid.*, **71**, 2983 (1949).

(84) R. M. Fuoss, *ibid.*, **56**, 1031 (1934).

(85) J. Bufalini and K. H. Stern, *ibid.*, **83**, 4362 (1961).

(86) S. Winstein, E. Clippinger, A. H. Fainberg, and G. C. Robinson, *ibid.*, **76**, 2597 (1954).

(87) E. Grunwald, *Anal. Chem.*, **26**, 1696 (1954).

(88) S. Winstein and G. C. Robinson, *J. Am. Chem. Soc.*, **80**, 169 (1958), and many subsequent papers.

(89) D. J. Cram, K. R. Kopecky, F. Hauck, and A. Langemann, *ibid.*, **81**, 5754 (1959).

(90) H. V. Carter, B. J. McClelland, and E. Warhurst, *Trans. Faraday Soc.*, **56**, 455 (1960).

(91) A. Mathias and E. Warhurst, *ibid.*, **58**, 948 (1962).

(92) B. J. McClelland, *ibid.*, **57**, 1458 (1961).

(93) J. Dieleman, Thesis, University of Amsterdam, The Netherlands, 1962.

(94) N. S. Hush and J. R. Rowlands, *Mol. Phys.*, **6**, 201 (1963).

(95) B. J. McClelland, *Chem. Rev.*, **64**, 301 (1964).

(96) D. G. Powell and E. Warhurst, *Trans. Faraday Soc.*, **58**, 953 (1962).

(97) N. Hirota and S. I. Weissman, *J. Am. Chem. Soc.*, **86**, 2538 (1964).

(98) K. H. J. Buschow, J. Dieleman, and G. J. Hoijtink, *J. Chem. Phys.*, **42**, 1993 (1965).

(99) J. F. Garst, D. Walmsley, C. Hewitt, W. R. Richards, and E. R. Zabolotny, *J. Am. Chem. Soc.*, **86**, 412 (1964).

(100) J. F. Garst, R. A. Klein, D. Walmsley, and E. R. Zabolotny, *ibid.*, **87**, 4080 (1965).

(101) J. F. Garst and W. R. Richards, *ibid.*, p. 4084.

(102) A. Streitwieser and J. I. Brauman, *ibid.*, **85**, 2633 (1963).

(103) H. E. Zaugg and A. D. Schaefer, *ibid.*, **87**, 1857 (1965).

(104) T. E. Hogen-Esch and J. Smid, *ibid.*, p. 669.

(105) T. E. Hogen-Esch and J. Smid, *ibid.*, **88**, 307 (1966).

(106) E. Grunwald, C. Baughman, and G. Kohnstam, *ibid.*, **82**, 5801 (1960).

(107) R. M. Fuoss and C. A. Kraus, *ibid.*, **55**, 2387 (1933).

(108) C. B. Wooster, *ibid.*, **60**, 1609 (1938).

(109) M. Dole, *Trans. Electrochem. Soc.*, **77**, 385 (1940).

(110) D. N. Bhattacharyya, J. Smid, and M. Szwarc, *J. Am. Chem. Soc.*, **86**, 5024 (1964).

(111) S. Datz, W. T. Smith, and E. H. Taylor, *J. Chem. Phys.*, **34**, 558 (1961).

(112) M. Blander, *ibid.*, **41**, 170 (1964).

(113) K. Hagemark, M. Blander, and E. B. Luchsinger, *J. Phys. Chem.*, **70**, 276 (1966).

(114) M. Smoluchowski, *Z. Physik. Chem.*, **92**, 129 (1917).

(115) P. Debye, *Trans. Electrochem. Soc.*, **82**, 265 (1942).

(116) M. Eigen, W. Kruse, G. Maass, and L. De Mayer, in *Progress in Reaction Kinetics*, Vol. 2, Pergamon Press, New York, 1964, p. 285.

(117) R. Lipman, J. Jagur-Grodzinski, and M. Szwarc, *J. Am. Chem. Soc.*, **87**, 3005 (1965).

(118) J. Stearne, J. Smid, and M. Szwarc, *Trans. Faraday Soc.*, **62**, 672 (1966).

(119) R. M. Fuoss and F. Accascina, *Electrolytic Conductance*, Interscience, New York, 1959; see also ref. 70.

(120) J. J. Zwolenik and R. M. Fuoss, *J. Phys. Chem.*, **68**, 903 (1964).

(121) D. N. Bhattacharyya, C. L. Lee, J. Smid, and M. Szwarc, *ibid.*, **69**, 612 (1965).

(122) T. Shimomura, K. J. Tölle, J. Smid, and M. Szwarc, *ibid.*, **71**, 796 (1967).

(123) T. A. Miller, B. Prater, J. K. Lee, and R. N. Adams, *J. Am. Chem. Soc.*, **87**, 121 (1965).

(124) B. E. Conway and R. E. Verrall, *J. Phys. Chem.*, **70**, 1473 (1966).

(125) R. C. Roberts and M. Szwarc, *J. Am. Chem. Soc.*, **87**, 5542 (1965).

(126) B. Linder, *Discussions Faraday Soc.*, **40**, 164 (1965).

(127) M. Szwarc, *Makromol. Chem.*, **89**, 44 (1965).

(128) L. Chan and J. Smid, *J. Am. Chem. Soc.*, **89**, 4547 (1967).

(129) J. Ugelstad, O. A. Rokstad, and J. Skarstein, *Acta Chem. Scand.*, **17**, 208 (1963).

(130) J. Ugelstad, P. C. Mörk, and B. Jenssen, *ibid.*, **17**, 1455 (1963).

(131) J. Ugelstad and O. A. Rokstad, *ibid.*, **18**, 474 (1964).

(132) J. Ugelstad, A. Berge, and H. Listou, *ibid.*, **19**, 208 (1965).

(133) R. Waack, M. A. Doran, and P. E. Stevenson, *J. Am. Chem. Soc.*, **88**, 2109 (1966).

(134) R. V. Slates and M. Szwarc, *J. Am. Chem. Soc.*, **89**, 6043 (1967).

(135) A. I. Shatenstein, E. S. Petrov, and M. I. Belusova, *Organic Reactivity, Tartu State University*, **2**, 191 (1964).

(136) A. I. Shatenstein, E. S. Petrov, and E. A. Yakoleva, *J. Polymer Sci.*, in press.

(137) C. A. Kraus, *J. Phys. Chem.*, **58**, 673 (1954).

(138) C. A. Kraus, *ibid.*, **60**, 129 (1956).

(139) D. Nicholls, C. A. Sutphen, and M. Szwarc, *J. Phys. Chem.*, **72**, 1021 (1968).

(140) D. Nicholls and M. Szwarc, *J. Phys. Chem.*, **71**, 2727 (1967).

(141) L. L. Chan and J. Smid, to be published.

(142) D. Nicholls and M. Szwarc, *Proc. Roy. Soc.*, Ser. A, **301**, 223 (1967).

(143) D. L. Lydy, V. A. Mode, and J. G. Kay, *J. Phys. Chem.*, **69**, 87 (1965).

(144) T. E. Hogen-Esch and J. Smid, *J. Am. Chem. Soc.*, **88**, 318 (1966).

(145) J. O'M. Bockris, *Quart. Rev. (London)*, **3**, 173 (1949).

(146) A. D. Buckingham, T. Schaefer, and W. G. Schneider, *J. Chem. Phys.*, **32**, 1227 (1960).

(147) R. J. Abraham, *Mol. Phys.*, **4**, 369 (1961).

(148) A. A. Bothner-By, *J. Mol. Spectry.*, **5**, 52 (1960).

(149) A. D. Buckingham, *Can. J. Chem.*, **38**, 300 (1960).

(150) G. Fraenkel and J. P. Kim, *J. Am. Chem. Soc.*, **88**, 4203 (1966).

(151) J. Weiss, *J. Chem. Soc.*, **1942**, 245.

(152) L. Onsager, *J. Chem. Phys.*, **2**, 599 (1934).

(153) P. Kebarle, S. K. Searles, A. Zolla, J. Scarborough, and A. Arshad, *J. Am. Chem. Soc.*, **89**, 6393 (1967).

(154) (a) N. Hirota and R. Kreilick, *J. Am. Chem. Soc.*, **88**, 614 (1966). (b) N. Hirota, *J. Phys. Chem.*, **71**, 127 (1967). (c) A. H. Crowley, N. Hirota, and R. Kreilick, *J. Chem. Phys.*, **46**, 4815 (1967). (d) N. Hirota, to be published.

(155) (a) R. Catterall, M. C. R. Symons, and J. Tipping, *J. Chem. Soc. (London)*, **1966**, 4342. (b) T. E. Gough and M. C. R. Symons, *Trans. Faraday Soc.*, **62**, 271 (1966), (c) M. C. R. Symons, *J. Phys. Chem.*, **71**, 172 (1967). (d) T. A. Claxton, W. M. Fox, and M. C. R. Symons, *Trans. Faraday Soc.*, **63**, 2570 (1967).

(156) (a) E. de Boer, *Rec. Trav. Chim.*, **84**, 609 (1965). (b) E. de Boer and E. L. Mackor, *J. Am. Chem. Soc.*, **86**, 1513 (1964).

(157) B. J. Herold, A. F. N. Correia, and J. dos Santos Veiga, *J. Am. Chem. Soc.*, **87**, 2661 (1965).

(158) J. E. Bennett, B. Mile, and A. Thomas, *Trans. Faraday Soc.*, **61**, 2357 (1965).

(159) A. H. Reddoch, *J. Chem. Phys.*, **43**, 225 (1965); *ibid.*, 3411 (1965).

(160) N. M. Atherton and A. E. Goggins, *Trans. Faraday Soc.*, **61**, 1399 (1965); **62**, 1702 (1966).

(161) D. H. Chen, E. Warhurst, and A. M. Wild, *Trans. Faraday Soc.*, **63**, 2561 (1967).

(162) R. F. Adams and N. M. Atherton, *Trans. Faraday Soc.*, **64**, 7 (1968).

(163) N. Hirota and S. I. Weissman, *J. Am. Chem. Soc.*, **86**, 2537 (1964).

(164) M. Iwaizumi, M. Suzuki, T. Isobe, and H. Azumi, *Bull. Chem. Soc. Japan*, **40**, 1325 and 2754 (1967).

(165) W. F. Edgall, A. T. Watts, J. Lyford, and W. M. Risen, *J. Am. Chem. Soc.*, **88**, 1815 (1966).

Chapter VI

Radical-Ions, Solvated Electrons, and Electron Transfer Processes

1. RADICAL-IONS

A suitable molecule A may accept an extra electron and form a new species which we shall denote by $A\dot{\,}^-$. The negative sign appearing in this symbol emphasizes the presence of a negative charge characteristic of an ion; the dot indicates that the new species possesses an odd number of electrons and, therefore, a radical nature. Consequently, the term *radical-anion* seems to be appropriate for such an entity.

A reverse process, ionization through which a molecule A is deprived of one of its electrons, produces a positive radical-ion, and the symbol $A\dot{\,}^+$ is proposed for such an entity. Positive radical-ions (or *radical-cations*) are observed in mass spectrographs and then they are referred to as the parent ions of the analyzed gas. They are usually short-lived, because the vertical electron ejection process leaves them in vibrationally excited states. A more gentle removal of an electron from a neutral molecule, e.g., through a suitable chemical or electrochemical reaction, yields positive radical-ions in their ground state. Such positive radical-ions are intrinsically stable, although their binding energy may be negative if they can decompose into fragments which gain additional stability, e.g., by delocalization of their electrons.

A relatively stable negative radical-ion may be formed if the parent molecule possesses a sufficiently low-lying empty orbital. Otherwise, electron capture leads to dissociation. For example, H_2O or CH_3Br decomposes in the course of an electron transfer process into $H\cdot + OH^-$ or $CH_3\cdot + Br^-$, respectively, because the interactions of the pairs H—OH^- and CH_3—Br^- are repulsive.

Unsaturated compounds such as aromatic hydrocarbons, some substituted olefins, dienes, substituted acetylenes, ketones, etc., are particularly suitable materials from which stable negative, as well as positive, radical-ions may be formed. Molecules of such compounds are conveniently described in terms of π-orbitals, half of which are empty and, therefore, capable of accommodating additional electrons. Let us remark in passing that these empty orbitals are designated as antibonding ones. The term originates from the calculation which shows that an electron placed into such an orbital contributes less to the binding energy than an electron in a localized orbital associated with a particular atom. Nevertheless, energy levels of antibonding orbitals lie *below* the ionization potential of A⁻, i.e., the electron capture process is then exothermic and the relevant parent molecule has, therefore, a positive electron affinity.

Many radical-anions generated from unsaturated compounds are perfectly stable and, under favorable experimental conditions, may exist indefinitely. Even more stable, in respect to their decomposition, are the corresponding positive radical-ions formed from the parent molecule by removing an electron from the highest occupied π-orbital. However, the greater reactivity of radical-cations, as compared with the respective radical-anions, calls for more stringent precautions in their preparation. Their annihilation caused by interaction with the surrounding species is rather rapid because the potential energy barriers of various reactions in which they may participate are, on the whole, lower for the positive than for the negative radical-ions.

2. REVIEW OF EARLIER STUDIES OF RADICAL-IONS

It has been known for about one hundred years that aromatic hydrocarbons may react with alkali metals. For example, in 1867 Berthelot (1) described the formation of a black addition product on fusing metallic potassium with naphthalene in a closed tube. The first real comprehension of such processes, expressed, however, in terms which differ from our modern notation, should be attributed to Schlenk. As early as 1914 he described (2) the reaction of alkali metals with anthracene in ether solution, and reported the formation of two distinct compounds: a one-to-one adduct (sodium anthracene) and a two-to-one adduct (disodium anthracene). These two species were characterized by their visible spectra and by chemical analysis. Although the modern concepts of radicals and radical-ions were not yet developed in those days, Schlenk's description of sodium anthracene and of ketyls, which were extensively studied in his laboratory (28), is remarkably close to our present interpretation. Using the language of his day, Schlenk reported his findings in terms which closely correspond to our modern notation of electron transfer processes involving carbanions, radicals, and radical-ions.

In the following years, the radical nature of such adducts was explicitly stressed, e.g., in a paper by Willstätter et al. (3), and this interpretation was thoroughly developed, mainly by Schlenk and Bergmann (4). Some of their conclusions have already been discussed in Chapter I and, therefore, they need not be repeated at this time. Unfortunately, the emphasis laid on the radical nature of the alkali adduct distracted the attention of earlier workers from the ionic properties of these compounds. For example, sodium naphthalene was described by the formula

which implies that its molecule is electrically neutral. Had this been the case, one should not expect a significant role of solvent in the process of its formation. However, studies of Scott and his colleagues (5) demonstrated that in some specific solvents the reaction of metallic sodium with naphthalene produced the adduct, while in others no reaction was observed. For example, the characteristic green color of naphthalenide rapidly appeared when the interaction took place in tetrahydrofuran or dimethyl ether, but not in diethyl ether or benzene. Moreover, addition of benzene to the green solution of sodium naphthalene in tetrahydrofuran, followed by removal of the latter solvent through distillation, led to the reverse reaction—sodium naphthalene decomposed into naphthalene and sodium dust. All these observations indicated that sodium naphthalene, and therefore also the other analogous alkali adducts, had an ionic character. Indeed, Scott et al. (5) remarked that the green solution showed electric conductance, but they did not consider further the implications of their own observations.

The first objection to Willstätter's formulation of the alkali adducts was raised by Hückel and Bretschneider (6), who noted that the reduction by an alkaline earth metal such as calcium should closely resemble the reaction with sodium; and, indeed, both metals reduce naphthalene to 1,4-dihydronaphthalene in liquid ammonia. However, had the structure proposed by Willstätter been correct, the calcium adduct should be represented by the improbable structure

and should, therefore, exhibit different properties than the sodium adduct. This is contrary to observation. Consequently they proposed that the reduction involves an electron transfer:

producing a heteropolar C^-,Na^+ bond. Here, for the first time, the concept of an electron transfer process was explicitly expressed in interpreting this class of reactions.

The conclusive evidence which proved the radical-anion nature of the alkali adducts was furnished by the pioneering studies of Lipkin, Weissman, and colleagues (7), who reported and extensively discussed the paramagnetic properties of these compounds. Moreover, examination of their ESR spectra demonstrated that the extra-electron is delocalized and occupies the lowest antibonding π-orbital. Numerous studies published by various workers in the following years fully confirmed these conclusions.

3. FORMATION OF NEGATIVE AND POSITIVE RADICAL-IONS

Negative radical-ions may be formed in the gas phase by electron attachment, a subject fully discussed in Section 4. The attachment of solvated electrons with formation of the respective radical-ions may be studied by radiolysis (Sections 13 and 15) or by photolysis (Section 16). Radical-anions are also formed by reduction of suitable electron-acceptors by alkali and alkali-earth metals—a subject discussed in Section 12. Alternatively, the reduction may be performed electrochemically (see Sections 10 and 21); in this connection one should mention the interesting polarographic technique developed by Geske and Maki (179) which permits the formation of radical-ions in the cavity of the ESR spectrometer.

The formation of a radical-anion by electron transfer involving another radical-ion is discussed in Section 9, while the disproportionation producing radical-anions from dianions is dealt with in Section 17. Electron transfer from carbanions to acceptors forming radical-ions is reviewed in Section 20.

Finally, processes involving charge-transfer complexes are discussed in Section 22. These simultaneously form the positive and negative radical-ions. The positive radical-ions may also be formed electrolytically (Sections 10 and 21) and by radiolysis (Sections 13 and 15) or photolysis (Section 16). Oxidation by chemical means is discussed in Section 22; this involves the reaction of electron donors with various Lewis acids.

4. ELECTRON CAPTURE IN THE GAS PHASE AND ELECTRON AFFINITY

There are two ways in which stable negative ions may be formed in the gas phase following collision between an electron and a neutral molecule. Either the electron is captured and a negative parent radical-ion is formed, or the collision and the subsequent electron capture break up the molecule into a neutral fragment and a negative ion or ion-radical. It is the former case that interests us.

Electron capture by a neutral molecule may be regarded as a transition taking place between two electronic levels of the negative radical-ion. In the initial state, one of the electrons of the radical-ion occupies an unbound orbital, the potential energy curve of such a system being that of the neutral molecule in its ground state. The final state is, of course, the electronic ground state of the radical-anion. An electron is captured if its kinetic energy is equal to the potential energy of the negative radical-ion in its electronic ground

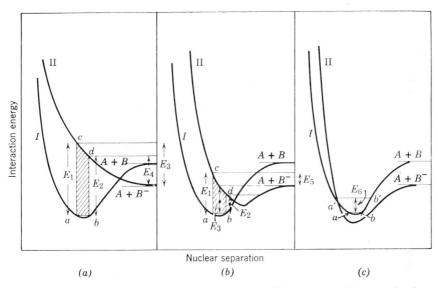

Fig. 1.   Potential energy curves showing three possible ways in which negative ions may be formed from a molecule AB by electron capture.

state, having, however, a configuration of the neutral molecule in its lowest vibrational state (a consequence of the Frank-Condon principle). The transition then takes place and produces a vibrationally excited radical-ion. The relevant relations are clarified by inspection of the potential energy curves of diatomic molecules shown in Figure 1.

In case (a) the potential energy of AB^- is repulsive and electron capture results in the dissociation of AB into $A \cdot + B^-$. The capture is efficient if the kinetic energy of the electron is greater than E_2 and smaller than E_1; electrons with energies higher than E_1 or lower than E_2 will be scattered but not captured.

In case (b) electrons having kinetic energies higher than E_2 but lower than E_3 form on capture AB^- radical-ions which may be stabilized by subsequent collision. Denoting by θ the relaxation time of spontaneous decomposition of a nonstabilized $AB^{\bar{}}$ into AB + electron, and by τ the average time between collisions, one finds that the probability of formation of a stable radical-ion arising from an electron capture process is $\breve{\omega} = \theta/(\theta + \tau)$. Since $1/\tau$ is proportional to the gas pressure, p, while θ is independent of it, $\breve{\omega} = p/(p + p')$ where p' is the pressure at which $\tau = \theta$.

Case (c) represents a situation in which only a vibrationally excited molecule AB has a high probability of electron capture. Of course, left to itself the radical-ion will lose its extra electron through a reverse process, but on collision with a gas molecule it may be stabilized like the radical-ion of case (b).

To recapitulate, diatomic molecules in the gas phase may capture electrons if their kinetic energies are within well-specified limits. Vibrationally excited negative radical-ions are produced if the potential energy curves are favorably interrelated. These ions lose their extra electron in about 10^{-8} sec, but they become stable if a favorable collision with a gas molecule has taken place during this time. Hence, at atmospheric pressure, when $\tau \sim 10^{-9}$ to 10^{-10} sec, stable radical-ions may be formed through an electron capture process involving suitable molecules.

Electron capture by polyatomic molecules resembles the process described for diatomic molecules. However, a polyatomic molecule has many degrees of freedom and a large number of available energy levels. Consequently, the restrictions imposed on the kinetic energy of the captured electron are greatly relaxed.

The ionization potential of a negative radical-ion is, by definition, identical with the electron affinity of its parent molecule. Hence, the latter may be determined by investigating thermal ionization of the radical-ions. Interesting studies based on this principle were reported by Becker, Wentworth, and associates. The method utilizes an electron capture detector designed by Lovelock (8) for gas-chromatographic analyses. Such a detector operates at a low electric field and utilizes low-energy electrons. The vapor to be investigated is mixed with a very large excess of a suitable carrier gas that flows through the device at atmospheric pressure. A mixture of argon containing 10% methane is most conveniently used for this purpose. The ionization is achieved by irradiating the gas with β-rays of tritium. The primary fast electrons are rapidly slowed down, their kinetic energy being reduced to the thermal level through inelastic collisions with the molecules of the carrier, and in the course of this process secondary electrons and positive ions, as well as neutral radicals, are produced. The stationary concentrations of the latter two species remain constant if the operating conditions are standardized. In particular, they are not affected by the presence of the investigated vapor because its partial pressure is negligibly small when compared with that of argon and methane. On the other hand, only the molecules of the investigated vapor are capable of capturing *thermal* electrons; neither methane nor argon can undergo an electron capture process. Therefore, the presence of the vapor reduces the concentration of thermal electrons in the plasma. The use of a chromatographic column ascertains the purity of the investigated compound.

By applying a short-duration positive potential to the collecting electrode of such a device one can measure the concentration of thermal electrons. In the argon–methane mixture the current drawn from the electrode was found to be proportional to the electron density in the plasma, its value being independent of the pulse duration (0.5–5 μsec) and of the applied voltage (10–80 V).

The following model (9) was proposed to account for the events occurring within the electron capture cell during the pulse sampling period. (*1*) The rate of formation of thermal electrons is assumed to be constant and not affected by the presence of the investigated vapor. (*2*) The pulse is assumed to remove most of the electrons and leave a large excess of positive ions, $\oplus$, and of radicals, R, in the plasma, i.e., $[\oplus] \gg [e^-]$ and $[R] \gg [e^-]$. (*3*) Thermal electrons are supposed to be lost by three reactions:

$$e^- + \oplus \rightarrow \text{products} \qquad k_N,$$

$$e^- + R \rightarrow R^- \qquad k_R,$$

$$e^- + A \underset{k_{-1}}{\overset{k_1}{\rightleftharpoons}} A^{\cdot -}.$$

Hence, in the absence of the investigated compound, A,

$$d[e^-]/dt = k_p R_\beta - \{k_N[\oplus] + k_R[R]\}[e^-],$$

and, in its presence,

$$d[e^-]/dt = k_p R_\beta - \{k_N[\oplus] + k_R[R]\}[e^-] - k_1[A][e^-] + k_{-1}[A^{\cdot -}].$$

Here $k_p R_\beta$ is the rate of formation of thermal electrons through the action of β-rays.

The stationary concentration of radical-ions, $[A^{\cdot -}]$, is determined by four processes: their formation, their thermal ionization and their destruction by the positive ions and by the radicals. Thus:

$$d[A^{\cdot -}]/dt = k_1[A][e^-] - k_{-1}[A^{\cdot -}] - \{k_{N_1}[\oplus] + k_{R_1}[R]\}[A^{\cdot -}].$$

Denoting the sums $\{k_N[\oplus] + k_R[R]\}$ and $\{k_{N_1}[\oplus] + k_{R_1}[R]\}$ by k_D and k_L, respectively, we may rewrite the above equations as:

$$d[e^-]/dt = k_p R_\beta - k_D[e^-] \quad \text{in the absence of A,}$$

$$d[e^-]/dt = k_p R_\beta - k_D[e^-] - k_1[A][e^-] + k_{-1}[A^{\cdot -}] \quad \text{in the presence of A,}$$

and

$$d[A^{\cdot -}]/dt = k_1[A][e^-] - k_{-1}[A^{\cdot -}] - k_L[A^{\cdot -}].$$

At a sufficiently long time ($t = \infty$), the simultaneous solution of these equations gives the stationary concentrations of electrons, $[e^-]_0$, in the absence of A, and their stationary concentration in the presence of A denoted by $[e^-]_A$. The electron capture coefficient, K, defined as

$$\{[e^-]_0 - [e^-]_A\}/[A][e^-]_A,$$

may be therefore calculated:

$$K = k_L k_1/k_D(k_L + k_{-1}).$$

The pseudo first order constants k_L and k_D refer to the diffusion-controlled reactions and, hence, their ratio is essentially temperature-independent, whereas k_{-1} increases exponentially with temperature. Thus, at sufficiently

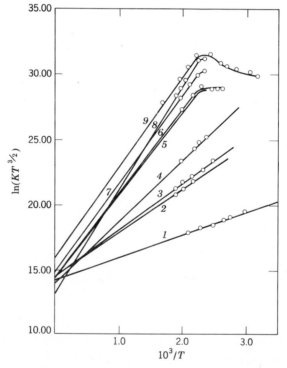

Fig. 2.   Plot of ln $(KT^{3/2})$ versus $1/T$, where K is the electron-capture coefficient for the gaseous reaction:

electron + aromatic hydrocarbon $\rightleftarrows$ (aromatic hydrocarbon)$^-$.

(1) naphthalene; (2) triphenylene; (3) phenanthrene; (4) chrysene; (5) benzo[c]phenanthrene; (6) anthracene; (7) pyrene; (8) benzanthracene; (9) azulene (9).

high temperatures $k_{-1} \gg k_L$ and then $K = (k_L/k_D)K_{eq}$, where K_{eq} is the equilibrium constant of the reaction

$$A + e \rightleftarrows A^{\overline{\cdot}}.$$

At low temperatures $k_{-1} \ll k_L$ and then $K = k_1/k_D$ becomes virtually temperature-independent because k_1 and k_D are rate constants of reactions controlled by diffusion. [A similar model which accounts for dissociative electron capture processes was developed recently by Wentworth, Becker, and Tung (123).]

At high temperatures $\ln K = \ln (k_L/k_D) + \ln (\nu/T^{3/2}) + \varepsilon/kT$, where ε is the electron affinity of A, and $\nu/T^{3/2} = f(A^{\overline{\cdot}})/f(A) \cdot f(e^-)$. The symbols $f(\ldots)$ denote the partition functions of $A^{\overline{\cdot}}$, A, and of the free electron, respectively. In the high-temperature region a plot of $\ln (KT^{3/2})$ versus $1/T$ should be linear, the slope being given by ε/k and the intercept by $\ln (k_L/k_D) + \ln f(A^{\overline{\cdot}})/f(A) + $ const. The ratio k_L/k_D is expected to be only slightly affected by the nature of the investigated hydrocarbon and the ratio $f(A^{\overline{\cdot}})/f(A)$ should be close to 2, the latter is the spin factor, whereas the other factors are nearly cancelled. In spite of the cancellation, the ratio of partition functions is not exactly constant, and it may vary within a factor of 1.5–2. Therefore, the intercepts of the lines obtained for various aromatic hydrocarbons are not exactly identical—their values may vary slightly, probably within ± 1.

The results obtained for some aromatic hydrocarbons (9) are shown in Figure 2; its inspection indicates only minor variations in the intercepts. Nevertheless, the assumption of a *constant* intercept does not seem to be justified, and it is doubtful whether the accuracy of the data could be genuinely improved by choosing a fixed value for the intercept (9). The electron affinities of 14 aromatic hydrocarbons calculated from the slopes of the respective lines are reported in references 9 and 10, and are listed in Table I. Only one value seems to be doubtful, that for electron affinity of anthracene; the magnitude of the error and its cause will be discussed later. Table I also includes the values of electron affinities of some aromatic aldehydes and ketones which were determined by the same technique (122).

The method described above applies to nondissociative electron capture processes. This limitation was emphatically stressed by Wentworth and Becker (11). Furthermore, the variation of the electron capture coefficients with temperature is a necessary condition for reliable determination of the electron affinities. Disregard of these restrictions may lead to erroneous values (12).

In the low-temperature region the electron capture coefficient, K, remains constant and its value then gives the ratio k_1/k_D. The pseudo-constant k_D may be estimated from the dependence of the current on the duration of the pulse. Thus, an approximate evaluation of k_1's is possible and their values,

TABLE I

Electron Affinities of Aromatic Hydrocarbons in the Gas Phase (9,10)

Compound	Electron affinity, eV	Intercept
Naphthalene	0.152 ± 0.016	14.21
Triphenylene	0.284 ± 0.020	14.38
Phenanthrene	0.308 ± 0.024	14.33
Chrysene	0.419 ± 0.036	13.98
Benzo[e]pyrene	0.486 ± 0.155	
Picene	0.490 ± 0.110	
Benzo[c]phenanthrene	0.542 ± 0.040	14.44
Anthracene	0.552 ± 0.061 (?)	14.44
Pyrene	0.579 ± 0.064	14.66
Dibenz[a,h]anthracene	0.676 ± 0.122	
Dibenz[a,j]anthracene	0.686 ± 0.155	
Benz[a]anthracene	0.696 ± 0.045	
Benzo[a]pyrene	0.829 ± 0.121	12.94
Azulene	0.587 ± 0.065	16.06

Electron Affinities of Some Aromatic Aldehydes and Ketones (122)

Acetophenone	0.334 ± 0.004	14.70
Benzaldehyde	0.42 ± 0.010	15.89
Naphthaldehyde-2	0.62 ± 0.040	14.46
Naphthaldehyde-1	0.745 ± 0.070	12.61
Phenanthrene aldehyde-9	0.655 ± 0.14	14.92

which are given in reference 9, were found to range from 3.5 to 25 × 10^{12} liter/mole sec. This nearly ten-fold variation of the rate constants of the diffusion-controlled processes again indicates that the assumption of a fixed intercept is questionable (see also Table XII in Section 13).

The results obtained for azulene are most interesting. The rate constant, k_1, of the electron capture process for this hydrocarbon appears to involve an activation energy of about 0.15 eV, whereas no activation energy was anticipated or observed for other hydrocarbons. This might indicate that only a vibrationally excited molecule of azulene may capture an electron, i.e., the relevant process corresponds to case (c) of Figure 1.

Although the technique developed by Becker, Wentworth, and Lovelock appears to be the most reliable and is admirably applicable to studies of electron affinities of polyatomic molecules in the gas phase, other experimental methods, nonetheless, deserve a brief discussion. In principle, one needs to determine the equilibrium constant of a system containing thermal electrons, molecules A, and their radical-ions, $A^{\overline{\cdot}}$. This calls for techniques by which one may analyze plasma for $A^{\overline{\cdot}}$ and e^-. Equilibrium between these species is attained when thermal electrons drift under the influence of a low and uniform potential gradient through an investigated gas toward an anode. The observed current gives a value proportional to the sum of $[e^-] + [A^-]$. Now, if an "electron filter" is placed in front of the anode, the electrons may be deflected and then only the heavy ions reach the electrode. Hence, the fraction of $[A^-]$ in the swarm may be determined. Early use of this technique is illustrated by Bradbury's investigation of electron affinity of oxygen molecules (13). This method was utilized in other studies, and demonstrated that CO, NH_3, CO_2, N_2O, H_2O, and H_2S have no affinity for electrons, whereas SO_2, NO, BF_3, Cl_2, Br_2, and I_2 have positive electron affinities. However, determination of electron affinities by this technique is rather unreliable.

Equilibria established in flames were investigated originally by Rolla and Piccardi (14). A fine wire inserted into the flame served as a cathode and a source of electrons. The distinction between electrons and negative ions was possible because their mobilities differ greatly. This technique led to the first estimates of electron affinities of I and Br atoms and subsequently to the determination of electron affinities of SO_2, SeO_2, and MoO_3 (14b,c,d). The method was refined by Sutton and Mayer (15), who used a device based on the "magnetron effect" (driving electrons into circular paths and preventing them from reaching the anode) in order to discriminate between ions and electrons. Subsequently, this device was applied with moderate success to other systems. Its further modification was described by Glockler and Calvin (16).

The magnetron technique was greatly improved by Page and his associates (17) and successfully used in studies of electron affinities of quinone and chloranil (18). Thus, a value of 32 ± 2 kcal/mole was derived for the electron affinity of quinone and 57 ± 6 kcal/mole for that of chloranil. Further developments (19) made possible the determination of the electron affinity of the OH radical. A value of 43.6 ± 3 kcal/mole was deduced (20), indicating that the previously reported high values are apparently unreliable.

Dissociation of alkali halides at high temperature ($1800°K$) gives, among others, M^+ and X^- ions. From their concentrations the equilibrium constant of the reaction

$$MX \rightleftarrows M^+ + X^-$$

was determined (21). These data permitted Mayer to calculate the electron affinity of X atoms, because the bond dissociation energy of MX and the ionization potential of M were known.

Injection of alkali metals, or their salts, into flames provides a convenient source of thermal electrons. Their concentration may be determined from attenuation of microwaves traversing the flame. In the presence of a suitable electron acceptor, A, the following equilibria are established:

$$Na \rightleftarrows Na^+ + e^- \qquad K_1$$

and

$$A + e^- \rightleftarrows A^- \qquad K_2.$$

Hence, K_2 may be determined if the concentrations of Na and A are known. This approach was extensively used by Sugden and his co-workers (116).

Electron affinities may be also deduced from some spectroscopic data. Studies of Person (22), who calculated the values for electron affinities of halogen molecules, provide a good example of such an investigation.

5. QUANTUM-MECHANICAL CALCULATION OF ELECTRON AFFINITIES AND THEIR CORRELATION WITH IONIZATION POTENTIALS

The simplest quantum-mechanical calculation of electron affinities of π-systems, e.g., of aromatic hydrocarbons, is based on Hückel's approach. In his treatment the electron affinity is given by the negative energy of the lowest unoccupied π-orbital, its value being $-(\alpha + \chi_{N+1} \cdot \beta)$. Unfortunately, this method is unreliable because the repulsion between electrons is not accounted for. Approximate calculations of electron affinities of alternant aromatic hydrocarbons and radicals, in which the self-consistent field approach was used, were reported by Hush and Pople (23). In their treatment, the energy of an orbital depends on whether other orbitals are occupied or empty. Similar calculations were subsequently performed by other workers, a different degree of sophistication being introduced into their approach. For the sake of illustration, the results of Hedges and Matsen (24), Hoyland and Goodman (25), and Becker and Chen (10) are compared in Table II. Inspection of these data shows that, on the whole, the theory is still unsatisfactory and the answers depend, to a great extent, on the choice of parameters used in computations. The calculations of Becker and Chen were based on the $\tilde{\omega}$-method (26) with $\tilde{\omega} = 3.75$, $\alpha_0 = 5.98$, and $\beta_0 = 1.23$. With this choice of parameters a very good agreement between calculated and observed values of electron affinities was obtained. However, the rather high value of $\tilde{\omega}$ may be questioned.

Electron affinities and ionization potentials of alternant hydrocarbons were first correlated by Hush and Pople (23), who concluded that their sum should be constant. In addition, Hedges and Matsen (24) stressed that the Hückel and the ASMOH theories suggest that the electron affinities and the corresponding ionization potentials should be symmetrical in respect to the work function of graphite. The experimental data now available seem to confirm these predictions.

TABLE II

Calculated Electron Affinities (in eV) of Aromatic Hydrocarbons

Hydrocarbon	Hedges and Matsen (24)	Hoyland and Goodman (25)	Becker and Chen (10)	Experi-mental
Naphthalene	−0.38	−0.21	0.17	0.15
Phenanthrene	−0.20	+0.25	0.31	0.31
Triphenylene	−0.28		0.29	0.28
Chrysene	0.04		0.49	0.42
Pyrene	0.68	0.55	0.55	0.58
Benzo[e]pyrene			0.54	0.49
Anthracene	0.49	0.61	0.55	0.55
Benzo[a]pyrene			0.71	0.83
Naphthacene	0.82			

According to Mulliken's definition, the electronegativity of an atom or of a molecule is the average of its ionization potential and electron affinity. Hence, the results of the calculations given above indicate that electronegativities of alternant aromatic hydrocarbons are constant, their most probable value being estimated at 4.1 or 4.2 eV. On this basis, the ionization potential of an aromatic hydrocarbon may be predicted if its electron affinity is known, or vice versa.

The semiempirical calculations of electron affinities of gaseous radicals have been reported by Gaines and Page (29). On the whole, their results agree well with the experimental data.

6. ELECTRON AFFINITIES IN SOLUTION

In solution, the addition of an electron to a suitable acceptor, A, differs in two respects from the reaction proceeding in the gas phase: (a) Dissipation of energy arising from the exothermicity of the process is ascertained by the interaction of the radical-ions with solvent molecules. (b) The heat of reaction is substantially larger than that in the gas phase because the solvation energy of the radical-ion substantially contributes to the exothermicity of the overall process. The heat of solvation may be calculated by the conventional Born approach:

$$\Delta H_{solv} = (e^2/2r)(1 - 1/D),$$

if the ions are assumed to be spherical (42). Certainly, this is a gross oversimplification. A more refined approach was proposed by Hush and Blackledge (27), who calculated the distribution of charge over all the atoms of a radical-ion, attributed then an effective radius, r_j, to each atom, and finally summed up all the Born-type interaction terms. Thus:

$$\Delta H_{solv} = \sum_j (q^2/2r_j)(1 - 1/D).$$

The results of their computations are listed in Table III. The heat of solvation decreases on increasing the area of the radical-ion; for sufficiently large radical-ions their values should be nearly constant. The calculations of Hush

TABLE III

Heats of Solvation of Aromatic Radical-Ions in Tetrahydrofuran

Parent hydrocarbon	ΔH, eV
Benzene	1.8
Naphthalene	1.1
Phenanthrene	0.8
Anthracene	0.8

exaggerate, however, the exothermicity of solvation because the "effective" dielectric constant is smaller than its bulk value. This point was discussed in the preceding chapter (see p. 238).

An interesting attempt to determine the solvation energy of aromatic radical-ions was reported by Prock et al. (269). They studied the photo-injection of electrons into liquids. These were produced by irradiating with visible light a negatively charged rhodium electrode immersed in a benzene solution of aromatic hydrocarbons such as naphthalene, phenanthrene, anthracene, and pyrene. The photocurrent ($\sim 10^{-11}$ A) far exceeded the dark current, and the decreased threshold, when compared with the photo-electric effect observed in the gas phase, is due to two factors: (*1*) electron affinity of the aromatic hydrocarbon which acquires an electron and becomes the current carrier; (*2*) solvation energy of the radical-ion. Indeed, the process may be represented by three steps,

Rh (solid) → Rh (solid)$^+$ + e$^-$ (gas) − work function ϕ,
e$^-$ (gas) + A (gas) → A$^{\dot{-}}$ (gas) − electron affinity, ε, in the gas phase,
A$^{\dot{-}}$ (gas) + A (sol.) → A$^{\dot{-}}$ (sol.) + A (gas) − solvation energy ΔE_{sol}.

Hence, the wave threshold, $h\nu$, is given by the equation $h\nu = \phi - \varepsilon + \Delta E_{sol}$ in which only the last term is unknown. Thus, the solvation energy is calculated from the experimental data.

Although the method is undoubtedly original and interesting, it remains to be seen how reliable it is. For example, the drop of potential between the electrode and the adjacent layer of liquid should be added to the left side of the equation, the omission of this term making $-\Delta E_{sol}$ too large. This may account for some surprising conclusions drawn by the authors.

The negative radical-ions are not specifically coordinated with molecules of ethereal solvents, whereas alkali ions such as Li^+ or Na^+ strongly interact with ethers (see, e.g., ref. 47). The lack of coordination of anions with ethers is evident from studies of mobilities of radical-anions (37). The respective self-diffusion constants, calculated from such data, are only slightly smaller than the diffusion constants of the parent hydrocarbons.

"Electron affinities" in solution were investigated by several workers who applied potentiometric, polarographic, and spectrophotometric methods in their studies. The details of these techniques and the significance of the results derived are discussed below.

7. POTENTIOMETRIC STUDIES OF ELECTRON AFFINITIES IN SOLUTION

The first reported potentiometric studies involving aromatic compounds were those of Bent and Keevil (31), who measured the potential established between a mercury amalgam electrode and a bright platinum electrode in contact with a saturated solution of, e.g., triphenylmethylsodium in ether. Although the potential was shown to obey the Nernst dilution equation, at least for triphenylmethylsodium, the results were not too conclusive. The solvation energies and the dissociation constants of ion-pairs were not known. In fact, Keevil and Bent (32) attempted to determine the dissociation constant of $C(Ph)_3^-, Na^+$ in diethyl ether, but their results are of doubtful value because the study was performed in too concentrated solutions. Subsequently, the dissociation constant was determined by Swift (50) and, on this basis, the electron affinity of the triphenylmethyl radical was calculated to be -48 ± 5 kcal/mole. The potentiometric technique was used in subsequent work to determine the electron affinities of benzophenone, fluorenone, bisbiphenylyl ketone, and of some aromatic hydrocarbons such as anthracene, stilbene, and tetraphenylethylene in diethyl ether solutions. Thus, the change in free energy for the process anthracene + sodium → sodium-anthracene was estimated as -12 to -13 kcal/mole.

A most successful approach to potentiometric studies was developed by Hoijtink and his associates (33). These workers measured the potential developed between two platinum electrodes, one placed in a standard solution of biphenyl and its radical-anion, and the other kept in contact with a solution of the investigated hydrocarbon and its radical anion. The procedure resembled a potentiometric titration of redox systems. The standard solution of the mixture, biphenyl (B) and biphenyl radical-anion ($B^{\overline{\cdot}}$), was gradually added to the solution of the investigated hydrocarbon, A. On addition of $B^{\overline{\cdot}}$, a virtually instantaneous electron transfer converted A into $A^{\overline{\cdot}}$ and, hence, the second electrode responds to the redox system $A–A^{\overline{\cdot}}$.

The original apparatus of Hoijtink was slightly modified in the writer's laboratory (34); its new version is shown in Figure 3. The buret containing the B + B$\bar{}$ solution was terminated by a 2-mm bore capillary, its lower tip being sealed and then punched with a needle to form six or seven tiny parallel capillaries. This arrangement considerably slowed down the diffusion of the liquid from reactor R to the upper platinum electrode, which was touching

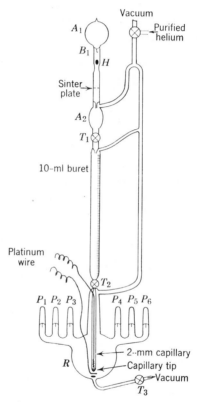

Fig. 3. Apparatus used in the potentiometric titration of aromatic hydrocarbons with sodium biphenyl (34): R, reactor; P_1–P_6, ampoules containing the hydrocarbon to be titrated; T_1, T_2, stopcocks; A_1, ampoule containing the B + B$\bar{}$, Na mixture; B_1, break seal; H, magnetic hammer; A_2, storage ampoule.

the sealed tip. The resistance of the unit, when filled with a $0.016M$ THF solution of sodium biphenyl, was about 10 megohms.

The potential between electrodes was measured by a valve voltmeter described by Scroggie (35). It functioned as a very stable impedance converter by means of which an input dc voltage in a high resistance circuit was converted into an identical output dc voltage generated in a very low

resistance circuit. The latter was then measured by any conventional voltmeter with a resistance exceeding 400 ohms.

All the operations were carried out in a helium atmosphere, the gas being
purified and freed from any traces of oxygen or moisture. The investigated
hydrocarbons were introduced from ampoules P (Figure 3) into the reactor
by crushing the respective breakseals. After completing the titration, the
titrated solution was sucked out, without exposing the electrodes to air, a
new solution was introduced, and another titration was performed. A typical
titration curve is shown in Figure 4.

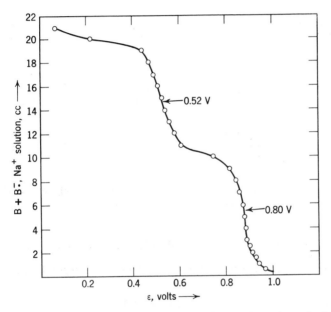

Fig. 4. A typical potentiometric titration curve for perylene titrated with sodium
biphenyl. The arrows denote the points corresponding to addition of 1/2 and 3/2 equivalents of sodium biphenyl. The respective potentials are ε_1 and ε_2.

Results of Hoijtink et al. (33) are given in Table IV where, for the sake of
comparison, the values reported by Chaudhuri, Jagur-Grodzinski, and
Szwarc (36) are also listed. In spite of some differences, both sets of data are
approximately linearly related. Several factors could contribute to the observed divergencies, and some of them will be considered now.

In his discussion of the potentiometric method, Hoijtink tacitly assumed
that the dimethoxyethane (DME) solutions of the investigated radical-ions
are virtually completely ionized, although he stressed the incomplete dissociation of $A^=,2Na^+$. Therefore, although he was concerned with the problem
of ion solvation, he neglected, in that early study, the effect of ion-pairing.

TABLE IV

Values of ε_0 for Various Hydrocarbons

Aromatic hydrocarbon	Original results of Hoijtink et al. in DME	Corrected results of Szwarc et al. in THF
1. Biphenyl	(0.0)	(0.0)
2. Naphthalene	0.09	0.043 ± 0.02
3. Triphenylene	0.19	0.132 ± 0.01[a]
4. Phenanthrene	0.17	0.142 ± 0.01
5. Benzo[e]pyrene	0.58	0.484 ± 0.02
6. Pyrene	0.60	0.529 ± 0.01
7. Benzo[a]anthracene	0.76	0.590 ± 0.02
8. 9,10-Dimethylanthracene		0.616 ± 0.01
9. Anthracene	0.78	0.642 ± 0.01
10. Benzo[a]pyrene	0.90	0.760 ± 0.02
11. Acenaphthylene	1.12	0.88 ± 0.03
12. Fluoranthrene	0.94	0.820 ± 0.02
13. Perylene	1.09	0.965 ± 0.01
14. Naphthacene	1.28	1.058 ± 0.02

[a] The original value given by Slates and Szwarc (37) was 0.128; after rechecking, the correct value was found to be 0.132.

The latter problem was considered by Jagur-Grodzinski et al. (34), who showed that the potential determined at the middle point of the titration curve is given by:

$$\varepsilon = \varepsilon_0 + 0.03 \log (K_B/K_A) + 0.03 \log ([A^{\overline{\cdot}},Na^+]_t/[B^{\overline{\cdot}},Na^+]_t).$$

Here ε_0 represents the correct standard potential corresponding to the ionic equilibrium

$$B^{\overline{\cdot}} + A \rightleftarrows A^{\overline{\cdot}} + B,$$

while K_B and K_A refer to the equilibrium constants of ion-pair dissociation:

$$B^{\overline{\cdot}},Na^+ \rightleftarrows B^{\overline{\cdot}} + Na^+$$

and

$$A^{\overline{\cdot}},Na^+ \rightleftarrows A^{\overline{\cdot}} + Na^+,$$

respectively. The last term of the equation giving ε accounts for the difference in the *total* concentration of $B^{\overline{\cdot}},Na^+$ in the buret and of $A^{\overline{\cdot}},Na^+$ in the reactor. At the middle point of titration the concentration of $A^{\overline{\cdot}},Na^+$ in the titrated solution is usually 3–4 times lower than that of $B^{\overline{\cdot}},Na^+$ in the buret, and hence the above correction term amounts only to less than 0.02 V. The correcting terms, $0.03 \log (K_B/K_A)$, usually are small. The required dissociation constants were determined (37) and thus, the proper ε_0 values could be calculated. The latter are listed in the last column of Table IV.

It has been shown recently (268) that sodium salts of radical-ions are virtually completely dissociated in hexamethyl phosphoramide (HMPA). Therefore, it is advantageous to carry out the potentiometric titrations in this solvent since the results give ε_0 directly. Moreover, some disturbing reactions which take place in the course of some titrations, e.g., of hetero-aromatics in tetrahydrofuran, are avoided in HMPA.

Fortunately, no corrections are necessary to account for the difference in dissociation of the various ion-pairs if the titration is carried out in dimethoxy ethane (the solvent used in Hoijtink's studies). Chang, Slates, and Szwarc (38) showed that in DME the relevant dissociation constants are closely similar for all the pertinent aromatic$^{\top}$,Na$^+$ ion-pairs. Therefore, the discrepancy between the two sets of data given in Table IV cannot be explained by this omission. The reliability of both sets of data will be discussed later, but first we wish to consider in some detail the dissociation of A$^{\top}$, alkali$^+$ ion-pairs into free ions.

8. DISSOCIATION OF SALTS OF RADICAL-ANIONS IN ETHEREAL SOLUTION

The dissociation of sodium salts of naphthalene$^{\top}$, biphenyl$^{\top}$, pyrene$^{\top}$, anthracene$^{\top}$, triphenylene$^{\top}$, tetracene$^{\top}$, and perylene$^{\top}$ in tetrahydrofuran (THF) has been studied conductometrically (37,38) over a wide temperature range. The results are shown graphically in Figure 5 as plots of log K_{diss} versus $1/T$. At room temperature the dissociation constant of the sodium salt of naphthalene$^{\top}$ differs by more than two powers of ten from those of perylene$^{\top}$ or naphthacene$^{\top}$. On the other hand, the dissociation constants are nearly identical at $-70°C$. All these dissociation processes are exothermic. At 25°C the exothermicity is high for naphthalene$^{\top}$,Na$^+$ (8.2 kcal/mole) but low for perylene$^{\top}$,Na$^+$ (2.2 kcal/mole). At the lowest temperature (about $-70°C$), the heat of dissociation of all these salts approaches 0. These findings indicate that small radical-anions, e.g., naphthalene$^{\top}$, form contact ion-pairs in THF at 25°C. Note the larger the radical-anion the weaker is its attraction for the counterion and, therefore, the more extensive the coordination of the Na$^+$ ion of the pair with THF molecules. The coordination of the counterion with the solvent becomes even more extensive at lower temperatures, i.e., the partial separation of the ions that takes place in the conversion of ion-pairs from a less solvated form to a more solvated is exothermic.

The dissociation of the sodium salts was also investigated in dimethoxy-ethane (DME) (38); the results indicate that this more powerfully solvating ether is coordinated with the sodium ion of all the investigated ion-pairs, even at 25°C. Consequently, the dissociation constants are relatively high, but the

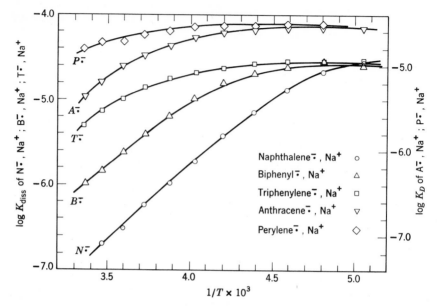

Fig. 5. Dissociation constants, K_{diss} of ion-pairs of aromatic radical-ions at different temperatures (range $+25$ down to $-70°\text{C}$) Solvent—tetrahydrofuran, sodium counterion. Note the striking difference between the heats of dissociation of naphthalene $\cdot$, Na^+ and perylene $\cdot$, Na^+. The lines marked A $\cdot$ and P $\cdot$ were shifted up by 0.2 units. This means all the lines nearly coincide at low temperature.

TABLE V

Dissociation Constants, Heats and Entropies of Dissociation of Sodium Salts
of Aromatic Radical Anions

	Biphenyl $\cdot$	Triphenylene $\cdot$	Perylene $\cdot$
In DME			
$10^6 K_{\text{diss}}$ at 25°C, M	4.6	5.6	6.0
$10^6 K_{\text{diss}}$ at -55°C, M	17.0	14.8	18.8
ΔH_{diss} at 20°C, kcal/mole	-2.1	-2.4	-2.5
ΔS_{diss} at 20°C, eu	-31.5	-32.5	-32.5
ΔH_{diss} at -55°C, kcal/mole	0	0	0
ΔS_{diss} at -55°C, eu	-22.0	-22.0	-21.0
In THF			
$10^6 K_{\text{diss}}$ at 25°C, M	1.0	5.2	15.5
$10^6 K_{\text{diss}}$ at -55°C, M	24.5	28.8	28.5
ΔH_{diss} at 20°C, kcal/mole	-7.3	-5.2	-2.2
ΔS_{diss} at 20°C, eu	-52	-42	-29
ΔH_{diss} at -55°C, kcal/mole	-1.6	-0.4	0.0
ΔS_{diss} at -55°C, eu	-28	-23	-21

exothermicities of these processes are rather low. This accounts for the relative independence of the dissociation constants of the nature of radical-ion, as is clearly demonstrated by the data given in Table V. Because it is instructive to compare the behavior of these ion-pairs in DME and THF the required heats and entropies of dissociation have also been included in the table. Attention should be paid to ΔS values. The solvation of free ions produced by the dissociation of the noncoordinated contact ion-pairs leads to a high negative ΔS of dissociation, e.g., its value for biphenyl^{-},Na^{+} in THF is -52 eu at 20°C. Much lower values, of the order -20 to -30 eu, are found for the dissociation of solvated and partially separated ion-pairs. This is a very general observation, as may be seen, for example, from the data presented in Table VII of reference 67.

It is also interesting to note that the dissociation of *solvent-separated* ion-pairs is higher in THF than in DME, other conditions being identical. This is due to the smaller size of the Na^{+} ion when coordinated with DME, its Stokes radius then being ~ 3.5 Å, as compared with ~ 4.2 Å (47) found for the ion coordinated with THF.

Our knowledge of salts of other alkali metals is less extensive. Data recently available (39) for lithium salts—contrary to a previous report (115)—indicate that the Stokes radii in THF are identical for the Li^{+} and Na^{+} ions. The evidence is provided by the observation that the limiting conductance, Λ_0, is the same for the sodium or lithium salts of anthracene^{-} or perylene^{-}. The lithium salts are, however, more solvated than the sodium. This is shown by the low heat of dissociation found for the lithium derivatives. Nevertheless, such lithium salts are often less dissociated than the respective *solvent-separated* sodium salts. This indicates that the lithium salts are solvated to a large extent on the periphery of the pair and, therefore, the tendency to separate the ions is less pronounced than that of the solvent-separated sodium salts. Thus the small lithium cation, even when solvated, may be more strongly attached to the anion than the solvent coordinated sodium ion. Moreover, the covalent character of the $\bar{\text{C}}$—Li^{+} bond may also contribute to this bias. The data collected in Table VI illustrate these relations. There is no doubt, however, that the free energies of formation of lithium salts from gaseous ions are greater than those of sodium salts (see Section 12).

Finally, a few words of caution may be pertinent. In a very interesting paper, Hoijtink et al. (40) discussed the dissociation of salts of aromatic radical-ions in tetrahydrofuran and in methyltetrahydrofuran. The conductance of each salt was investigated at *one* concentration ($10^{-4}M$) although over a wide temperature range, and the degree of dissociation was estimated from the temperature dependence of the conductance. This method may often be misleading. For example, Hoijtink considered a salt to be completely dissociated if the relevant conductance versus temperature curve showed no maximum

TABLE VI

Limiting Conductance of Lithium and Sodium Salts of Aromatic
Radical-Ions in THF (38,39)

Temp., °C	Perylene $\bar{\cdot}$		Anthracene $\bar{\cdot}$		Naphthalene $\bar{\cdot}$	
	Λ_0 (Li$^+$)	Λ_0 (Na$^+$)	Λ_0 (Li$^+$)	Λ_0 (Na$^+$)	Λ_0 (Li$^+$)	Λ_0 (Na$^+$)
25	107	106	121	120	129	128
5	86.4	85.6	97.5	95.2	104	102
−15	67.0	66.5	75.7	75.1	80.6	79.5
−35	48.9	49.5	56.4	55.8	60.0	59.0
−55	35.2	34.9	39.8	39.4	42.3	42.0
Dissociation Constants $\times 10^6$, M						
25	4.5	15.5	4.5	4.3	3.1	0.14
5	5.4	20.3	5.4	9.7	4.3	0.33
−15	6.1	25.3	6.3	16.8	6.0	1.07
−35	6.9	28.2	7.1	23.9	7.4	3.9
−55	7.0	28.5	7.4	28.7	9.0	13.1
Heat of Dissociation, ΔH, kcal/mole, at 25°C						
−1.4	−2.2	−1.9	−6.1		−2.8	−8.2

and monotonically decreased on lowering the temperature. This led to the conclusion that a $10^{-4}M$ solution of anthracene$\bar{\cdot}$,Na$^+$ at 25°C in THF is virtually completely dissociated, while, according to the data derived from Figure 5, its dissociation amounts only to about 8% under these conditions. Similarly, the lithium salt was considered to be completely dissociated at room temperature, whereas its dissociation is even slightly lower than that of the sodium salt. The dissociation of the lithium salt increases only insignificantly at −50°C.

Three factors are important in determining the temperature dependence of conductance at constant salt concentration: (1) "activation energy" of the solvent viscosity; (2) heat of dissociation of the investigated ion-pair; (3) effect of temperature on the Stokes radii of the ions. Whenever the "activation energy" of the solvent viscosity is higher than one-half of the exothermicity of dissociation of the investigated ion pair, the conductance diminishes monotonically with decreasing temperature. This, undoubtedly, is the case for the system anthracene$\bar{\cdot}$,Li$^+$ in THF because the heat of dissociation is only −1.9 kcal/mole (39), whereas the "activation energy" of THF viscosity is about 1.8 kcal/mole. Apparently a similar situation exists for the sodium salt (38).

The dissociation of salts of aromatic radical-ions was investigated by

Weissman and his co-workers with the ESR techniques. The results of his studies and their significance are discussed on p. 289.

9. EQUILIBRIA BETWEEN RADICAL-IONS AND THEIR PARENT MOLECULES

We are now in a position to consider the equilibria established between two radical-ions, $A^{\cdot-}$ and $B^{\cdot-}$, and their parent molecules A and B. Thus,

$$A^{\cdot-} + B \rightleftarrows B^{\cdot-} + A \qquad K_{AB,i}.$$

The equilibrium constant may be obtained from spectrophotometric studies because the spectra of $A^{\cdot-}$ and $B^{\cdot-}$, which extend to the visible or near-IR region, are often sufficiently different to permit their distinction and quantitative determination of concentrations of the respective radical-ions. Under most common experimental conditions, e.g., in tetrahydrofuran at concentrations of $10^{-4}M$ or higher, the radical-ions exist virtually as ion-pairs and, therefore, the spectrophotometrically observed equilibrium refers to the reaction:

$$A^{\cdot-},M^+ + B \rightleftarrows B^{\cdot-},M^+ + A \qquad K_{AB,p}.$$

There is a simple relation between $K_{AB,i}$ and $K_{AB,p}$:

$$K_{AB,p} = K_{AB,i}K_{diss,A^{\cdot-},M^+}/K_{diss,B^{\cdot-},M^+},$$

where $K_{diss,A^{\cdot-},M^+}$ and $K_{diss,B^{\cdot-},M^+}$ refer to the dissociation constants of $A^{\cdot-},M^+$ and $B^{\cdot-},M^+$ ion-pairs, respectively. Hence, if the latter constants are known, the determination of $K_{AB,p}$ leads to the value of $K_{AB,i}$. The equilibrium constant $K_{AB,i}$ gives the difference between the electron affinities of A and B in solution. The latter differs insignificantly from the $\varepsilon_A - \varepsilon_B$ determined in the gas phase (36) because the variations in the solvation energies of radical-anions in tetrahydrofuran, or more correctly the differences

$$\Delta H_{solv(A^{\cdot-})} - \Delta H_{solv(B^{\cdot-})} + \Delta H_{solv(B)} - \Delta H_{solv(A)}$$

are usually small.

The equilibria discussed above were investigated by Paul, Lipkin, and Weissman (41), whose pioneering work demonstrated the feasibility of the method and established the general pattern of electron affinities of the aromatic hydrocarbons. Unfortunately, various technical difficulties involved in such studies were not appreciated at that time. In addition, some confusion arose because it was not fully realized that the conventional method of preparation of aromatic radical-ions (reduction with alkali metals) may produce di-alkali as well as the monoalkali adducts. Consequently, the quantitative findings reported in this important investigation were somewhat in error. For example, the equilibrium constant of the reaction

$$\text{anthracene}^{\overline{\cdot}} + \text{naphthacene} \rightleftarrows \text{naphthacene}^{\overline{\cdot}} + \text{anthracene}$$

was found to be of the order of unity, while the potentiometric data (33,34,37) indicated an equilibrium constant of the order 10^6 to 10^7.

The studies of Weissman's group were, therefore, repeated in this writer's laboratory. The equilibria were investigated in THF at room temperature and the systems studied were selected on the basis of the following criteria:

(1) The difference in the reduction potentials of the investigated hydrocarbons should not exceed 0.15 V, i.e., the determined equilibrium constants should not be larger than 250. Otherwise, a reliable determination of the concentration of one of the radical-ions calls for an enormous excess of the other hydrocarbon. This, in turn, could lead to partial destruction of some radical-ions and, therefore, to erroneous results. Indeed, such a difficulty vitiated Weissman's study of the system

$$\text{naphthalene}^{\overline{\cdot}} + \text{anthracene} \rightleftarrows \text{anthracene}^{\overline{\cdot}} + \text{naphthalene}.$$

(2) The spectra of the investigated radical-ions should not overlap too closely. A successful spectrophotometric study obviously requires such a condition.

The following systems were eventually selected for the spectrophotometric studies: tetracene–perylene, anthracene–pyrene, 9,10-dimethylanthracene–pyrene, and triphenylene–naphthalene. The results are given in Table VII. The agreement between the differences of the observed potentials and those calculated from the respective equilibrium constants is remarkably good.

The equilibrium method is particularly valuable when small differences of electron affinities are to be determined. An alternative method, in which the same equilibrium constants are derived from kinetic studies, will be described later (p. 390).

TABLE VII

Equilibria $A^{\overline{\cdot}},Na^+ + B \rightleftarrows B^{\overline{\cdot}},Na^+ + A$ in THF at 25°C

Investigated pair	$K_{AB,p}$	$K_{AB,i}$	$\Delta\varepsilon V$ From $K_{AB,i}$	Observed[a]
Perylene (A)–tetracene (B)	52	34	0.092	0.093
Pyrene (A)–anthracene (B)	111	74.5	0.112	0.113
Pyrene (A)–9,10-dimethylanthracene (B)	91	29.5	0.088	0.087
Naphthalene (A)–triphenylene (B)	3	73	0.111	0.089

[a] Taken from reference 36.

10. POLAROGRAPHIC DETERMINATION OF ELECTRON AFFINITIES

The polarographic technique of determining electron affinities is closely related to the potentiometric method. Reduction on a dropping mercury electrode has to be performed in a medium of low proton activity to avoid rapid protonation of the primary product. Aqueous dioxane (75 and 96%), 2-methoxy-1-ethanol (Cellosolve), and dimethylformamide have been used as the most suitable solvents, with tetramethyl or tetrabutyl ammonium iodide as the supporting electrolyte. Depending on the reduced substrate, one or two polarographic waves were observed. The first is attributed to the formation of A^{-}, the second being associated with the formation of $A^{=}$.

This interesting field was opened through the pioneering work of Laitinen and Wawzonek (43–45), who initially investigated the polarographic reductions of styrene, α-methylstyrene, and 1,1-diphenylethylene and then extended their studies to the polarography of aromatic hydrocarbons. To account for the independence of the recorded potential on the pH of the solution, the reduction was assumed to proceed through the step

$$A + e \rightleftarrows A^{-} \qquad \text{(reversible and potential determining).}$$

This is followed by

$$A^{-} + e \rightarrow A^{=}$$

if a second wave is observed at higher potentials. The resulting product of the latter process is rapidly destroyed by protonation:

$$A^{=} + \text{proton} \rightarrow AH^{-} \text{ (irreversible)} \rightarrow AH_2 \text{ (irreversible).}$$

We should inquire whether the first step could be represented by an alternative equation,

$$A + X^{+} + e \rightleftarrows A^{-}, X^{+},$$

demanding that $X^{+} = NMe_4^{+}$ or NBu_4^{+} be involved in the energy-controlling step. Such a step may appear plausible in view of the presence of a supporting electrolyte, and it could be expected in media of low dielectric constant, e.g., in 96 or 75% dioxane. However, the half-wave potentials observed in 75% aqueous dioxane and in the 96% dioxane differed only by 0.02 V (46). The change in the proportion of water decreases the dielectric constant of the medium from about 13 to 3.5 and, therefore, the dissociation constant of the respective ion-pairs is greatly reduced. For example, the dissociation constant of tetrabutyl ammonium nitrate decreases by about six powers of ten as the amount of water in the solvent is lowered from 25 to 4%. Hence, had ion-pairing been involved in the actual electron transfer, a larger variation in the potentials should be expected.

The studies of Laitinen and Wawzonek were extended by Hoijtink and his co-workers (46,48) and independently by Bergman (49), who determined the half-wave reduction potentials of 78 aromatic hydrocarbons—a most comprehensive investigation of this subject. Some of his results are compared in Table VIII with those reported by the other two groups of investigators. Inspection of the table demonstrates a remarkable degree of agreement among all the reported findings. Even in solvents as different as acetonitrile

TABLE VIII

Polarographic Half-Wave Potentials, $\varepsilon_{1/2}$, for Aromatic Hydrocarbons and Some Related Hydrocarbons (in volts in reference to saturated calomel electrode)

Compound	Laitinen and Wawzonek[a]		Bergman[b]		Hoijtink[c]	
	$-\varepsilon_{1/2}$	$-\Delta\varepsilon_{1/2}$	$-\varepsilon_{1/2}$[d]	$-\Delta\varepsilon_{1/2}$	$-\varepsilon_{1/2}$	$-\Delta\varepsilon_{1/2}$
Biphenyl	2.70	+0.20	2.07	+0.09	2.70	+0.20
Naphthalene	2.50	0.00	1.98(1.99)[e]	0.00	2.50	0.00
Phenanthrene	2.45		1.93(1.93)[e]	−0.05	2.45	−0.05
Triphenylene			1.97	−0.01	2.50	0.00
Chrysene			1.80	−0.18		
Pyrene	2.11	−0.39	1.61(1.56)[e]	−0.37	2.13	−0.37
Anthracene	1.94	−0.56	1.46(1.41)[e]	−0.52	1.96(2.42)	−0.54
Naphthacene			1.13	−0.85	1.58(1.84)	−0.92
Perylene			1.25	−0.73	1.67	−0.83
Styrene	2.35	−0.15				
1,1-Diphenylethylene	2.26	−0.24			2.25	−0.25
Stilbene	2.14	−0.36			2.50	0.00
Tetraphenylethylene	2.05	−0.45			2.06	−0.44
α-Methylstyrene	2.54	+0.04				

All the data of Laitinen and Wawzonek and of Hoijtink were obtained in 75% dioxane + 25% water. Bergman's data were obtained in monomethyl ether of glycol. The data of Given (in parentheses) were observed in dimethylformamide (59b).

Extensive comparison of calculated and observed data in Cellosolve, 96% dioxane, and 75% dioxane are in Table I of Hoijtink's paper, Rec. Trav. Chim., **74**, 1525 (1955).

[a] H. A. Laitinen and S. Wawzonek, J. Am. Chem. Soc., **64**, 1765 (1942); S. Wawzonek and H. A. Laitinen, J. Am. Chem. Soc., **64**, 2365 (1942); S. Wawzonek and J. W. Fan, J. Am. Chem. Soc., **68**, 2541 (1946).

[b] I. Bergman, Trans. Faraday Soc., **50**, 829 (1954).

[c] G. J. Hoijtink and J. van Schooten, Rec. Trav. Chim., **71**, 1089 (1952); **72**, 691, 903 (1953). G. J. Hoijtink, J. van Schooten, E. de Boer, and W. I. Aalbersberg, ibid., **73**, 355 (1954). G. J. Hoijtink, ibid., **73**, 895 (1954); **74**, 1525 (1955).

[d] Expressed in reference to the mercury pool anode potential. To convert these values to those referred to the calomel electrode 0.52 V should be added to $-\varepsilon_{1/2}$.

[e] The data reported by Given (59b) in dimethylformamide. These are referred to the mercury pool anode potential.

(59a) or dimethylformamide (59b) the differences in measured potentials were not greater than 0.01 V. These observations indicate that the differences between the free energies of solvation of various aromatic radical-ions are small—certainly less than ± 0.1 eV or ± 2.5 kcal/mole.

The first polarographic reduction takes place extremely rapidly, whereas the protonation is relatively slow.* The measurements provide, therefore, the standard reduction potential of the investigated hydrocarbon. This conclusion was confirmed through ac polarographic studies (55) performed in dimethyl-formamide, a solvent used previously in dc polarography (59b). The correction term,

$$(RT/\mathscr{F}) \ln (D_A/D_{A^-}),$$

is negligible, because the diffusion constants, D_A and D_{A^-}, of the parent molecule and of its radical-ion are nearly identical (37). However, the second wave, associated with the formation of $A^=$, probably represents an irreversible process. Because of the enhanced reactivity of the dinegative ions toward protons (54,55), $A^=$ is rapidly removed, and hence the measured potential does not provide a thermodynamic value. Indeed, its magnitude is affected by the nature of solvent, type of the cation present in the system, and the concentrations of the reagents.

The polarographic data were used by Matsen (51) to calculate the absolute values of electron affinities in solution. The following sequence of reactions was considered:

A (gas) + e $\rightleftarrows$ A$^-$ (gas)   (ε_0, electron affinity),

A$^-$ (gas) + A (solution) $\rightleftarrows$ A$^-$ (solution) + A (gas)   overall solvation energy, ΔG_{solv},

e (in Hg) $\rightleftarrows$ e (gas)   work function of mercury, ϕ(Hg) = -4.54 eV,

Hg (liq) + Cl$^-$ (1 N,aq) $\rightleftarrows$ 1/2Hg$_2$Cl$_2$ + e (in Hg) $\big\{$ absolute potential of a saturated
Cl$^-$ (saturated Hg$_2$Cl$_2$) $\rightleftarrows$ Cl$^-$ (1 N,aq) $\big\{$ calomel electrode.

The absolute potential of a saturated calomel electrode was calculated by Latimer, Pitzer, and Slansky (52), who arrived at a value of -0.50 V. Thus:

$$\varepsilon_{1/2} = \varepsilon_0 + \Delta G'_{solv} - 5.07 \text{ eV}.$$

Alternatively, the above constant (-5.07 eV) correlating half-wave reduction potential with electron affinity in solution may be derived from the measured half-wave potential of the triphenylmethyl radical and its electron affinity as calculated by Swift (50).

Polarographic studies were extended to heteroaromatics such as pyridine, quinoline, etc. The subject has been reviewed by Kolthoff and Lingane (53),

* Rate constant of protonation of radical-ions by alcohols has been shown recently (101) to be of the order 10^4 M^{-1} sec^{-1} only; see e.g. Table XV, p. 347.

who pointed out that a catalytic hydrogen wave may be observed in addition to the normal type of reduction wave. The catalytic wave arises from the following reactions:

$$\text{(pyridine)} + H_2O \longrightarrow \text{(pyridinium, } N\text{–}H^+\text{)} + OH^-,$$

$$\text{(pyridinium, } N\text{–}H^+\text{)} + e \longrightarrow \text{(pyridinyl radical)} + H \quad \text{(catalytic polarographic wave)},$$

and finally

$$H + H \rightarrow H_2.$$

The net result is therefore

$$2H_2O + 2e^- \rightarrow H_2 + 2OH^-.$$

At pH < 6 the normal reduction wave is observed.

The catalytic wave should be avoided in an aprotic solvent. Indeed, Given (59b) reported an apparently unperturbed polarographic reduction of pyridine and quinoline in dimethylformamide: $\varepsilon_{1/2}$ was found to be 2.01 and 1.53, respectively, in reference to the mercury pool anode. These values seem, however, to be too low.* Indeed, if the difference between the reported half-wave potentials of naphthalene and quinoline ($+0.46$ V) are used for the quantum-mechanical calculation of L_N, the result is much too high, 1.1 instead of the usual value of about 0.5. Recently, extensive polarographic studies of heterocyclic aromatics have been reported (329).

Polarographic data inspired much theoretical work. In a series of papers (46,48,56) Hoijtink correlated the half-wave potentials with the energy of the lowest un occupied orbitals E. Thushe found

$$\varepsilon_{1/2} = \gamma E + \text{const.},$$

and Bergman's data, as well as his own, showed a good linear relation with E. Such a relation applies to alternant as well as to nonalternant hydrocarbons. Various approaches were used in the calculations, e.g., the original Hückel method gives $\gamma = -2.23$, Wheland's approximation (57) with overlap integral of 0.25 gives $\gamma = -1.97$, Longuet-Higgins' method (58) leads to $\gamma = -1.81$, etc. Correlation with the p-bands of the absorption spectra of the respective hydrocarbons were reported by Bergman (49), with methyl affinities (275) of aromatic hydrocarbons by Matsen (51), etc.

* Dimerization of the radical-ions of pyridine and quinoline could vitiate the results, see reference 268.

11. COMPARISON OF ELECTRON AFFINITIES DATA

Correlation of the polarographic data of Bergman (49) with the potentio-metric electron affinities (E.A.) determined in this writer's laboratory and corrected for the ion-pairing in THF solution is shown in Figure 6. The agreement is excellent, adding to the credulity of both sets of data. The slope of the line deviates by 16% from unity, indicating that solvation of the radical-ions by 2-methoxy-1-ethanol is more sensitive to the ion size than the interactions taking place in THF. Since $CH_3O.CH_2.CH_2OH$ may be hydrogen-bonded to the negative ion, and THF cannot, this conclusion seems to be plausible. The potentiometric data of Hoijtink (33), when used for such a correlation, show some scatter and lead to an even greater deviation of the slope from unity (about 30%).

Comparison of the electron affinities determined in the gas phase with those derived from potentiometric studies is also instructive. This is shown in Figure 7, and the experimental points fit, within experimental error, a 45° line, indicating that the solvation free energies of the investigated radical-ions are virtually constant in THF, also see ref. 269. The only serious deviation is shown by anthracene. The potentiometric data for anthracene were confirmed through studies of equilibria (34,37); it appears that the corresponding value determined in the gas phase is in error. Indeed, this value does not fit to other

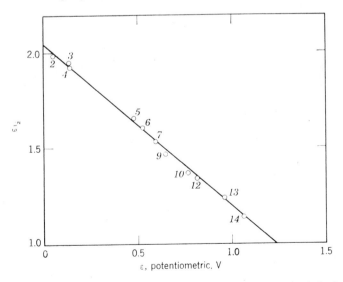

Fig. 6. Polarographic half-wave reduction potentials, $\varepsilon_{1/2}$, of aromatic hydrocarbons determined in 2-methoxy-1-ethanol compared with the respective potentiometric reduction potentials determined in tetrahydrofuran (both in V). Consult Table IV for the meaning of numbers.

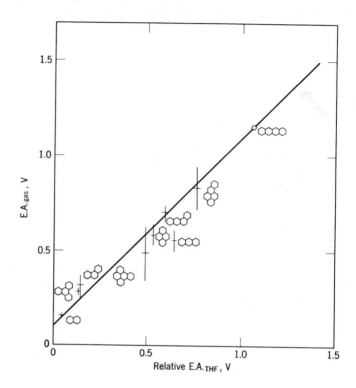

Fig. 7. Comparison of absolute electron affinities (E.A.) of aromatic hydrocarbons determined in the gas phase with the respective potentiometric relative reduction potentials determined in tetrahydrofuran. The line is drawn as the "best" line through experimental points having a fixed slope of unity. The length of the arms of each cross gives the experimental error of the respective determination.

correlations reported by Becker and Chen (10). For example, the point representing anthracene deviates from the line correlating the gaseous electron affinities with ionization potential or with methyl affinity of aromatic hydrocarbons (see Figure 4 of ref. 10). Correlation of gaseous electron affinities with Hoijtink's potentiometric data would lead to an improbable slope of the resulting line.

Electron affinity data obtained by the different techniques are collected and compared in Table IX. A remarkable agreement is in evidence among all the reported data, although the potentiometric values of Hoijtink are slightly too high. The reason for this deviation is intriguing. In Hoijtink's procedure, a sodium biphenyl solution was continually flowed through a narrow and long capillary as the potential was measured. This leads to an electrocapillary effect that increases the observed potential if the glass wall is negatively charged.

TABLE IX

Comparison ($\Delta\varepsilon$) of Electron Affinities Determined by Different Techniques

System	Perylene–naphthacene	Pyrene–anthracene	Naphthalene–anthracene
Polarographic			
Bergman	0.12	0.15	0.52
Given	—	0.15	0.58
Hoijtink	0.09	0.17	0.54
Potentiometric			
Hoijtink	0.19	0.17	0.69
Szwarc	0.09	0.11	0.60
Equilibrium, Szwarc	0.09	0.11	—

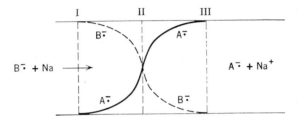

Fig. 8. Schematic representation of the concentration gradients in a liquid junction: (solid line) concentration of A⁻; (dashed line) concentration of B⁻.

Finally, the effect of the liquid junction calls for some discussion. The concentrations of the reagents across the boundary is represented by Figure 8. Diffusion of B⁻ between I and II leads to a potential which is, at least, partially balanced by the opposite potential caused by the diffusion of A⁻ from III to II. The decrease in the concentration of B⁻ taking place between II and III contributes negligibly to the potential of the liquid junction, because in this region the charge is transported mainly by A⁻. For the same reason the effect of the decreasing concentration of A⁻ between II and I is also insignificant. A thorough discussion of the liquid junction potential is given by McInnes (61).

12. EQUILIBRIA BETWEEN ALKALI METALS AND AROMATIC HYDROCARBONS

The reduction by alkali metals of aromatic hydrocarbons and some related compounds to the respective radical-anions has already been discussed.

Similar reductions may be achieved by alkaline-earth metals (62,63) and their amalgams. It was reported (63) that amalgams of Ca or Mg, but not the pure metals, may reduce aromatic hydrocarbons. This observation was accounted for by the lower photoelectric work function of the amalgam when compared with the pure metal (64). Undoubtedly, the work function of different metals is correlated with their reductive capacity, e.g., sodium or potassium (work function 2.3 and 2.2 V, respectively) easily reduce anthracene under conditions in which barium and magnesium (work function 2.7 and 3.7 V, respectively) are ineffective. Nevertheless, the proposed explanation for the amalgam activity seems unreasonable. Apparently, the lowering of the work function arises from the higher heat of reaction Ca^{++} (gas) + mercury (liq) $\rightarrow Ca^{++}$ (in mercury solution) when compared with the process Ca^{++} (gas) + calcium metal $\rightarrow Ca^{++}$ (in calcium lattice). Formation of calcium anthracene in solution requires not only emission of an electron but also removal of Ca^{++} ions, which apparently is unfavorable for the amalgam. It is probable, therefore, that the ease of the amalgam reduction and the inertness of the pure metal result from *kinetic* factors associated with this heterogeneous reaction, and the structure of the reacting surface apparently plays a decisive role in determining the rate of the process.

The equilibrium between an alkali metal, Alk, and an electron acceptor is represented by equation (I).

$$Alk + Acceptor \rightleftarrows Alk^+ + Acceptor^{-}. \tag{I}$$

The heat of this reaction, referred to gaseous ions, solid metallic alkali, and a crystalline acceptor is given by

$$\Delta H_I = \Delta H_{sub}(Alk) + I_p(Alk) + \Delta H_{sub}(Acceptor) - \varepsilon(Acceptor),$$

where ΔH_{sub} are the relevant heats of sublimation, $I_p(Alk)$ the ionization potential of the gaseous alkali atom, and ε (Acceptor) the electron affinity of the investigated species. The change in free energy is given by a similar equation,

$$\Delta G_I = \Delta G_{sub}(Alk) + I_p(Alk) + \Delta G_{sub}(Acceptor) - \varepsilon(Acceptor),$$

because the entropy change of the gaseous process

$$(Alk)_g + (Acceptor)_g \rightleftarrows (Alk^+)_g + (Acceptor^{-})_g$$

is probably insignificant. However, if the reaction takes place in solution, two additional processes have to be considered—solvation of the gaseous Alk^+ ions and of the negative $Acceptor^{-}$ radical-ions. Hence, the total heat of the reaction is decreased because $\Delta H_s(Alk^+)$ and $\Delta H_s(Acceptor^{-})$ are negative. In discussing the equilibrium, the terms $\Delta G_s(Alk^+)$ and $\Delta G_s(Acceptor^{-})$ have to be added to the equation giving the free energy change of the overall process taking place in solution. The problem becomes even more complex when ion-pairs, and not free dissociated ions, are formed in the reaction. Then, additional enthalpy or free energy terms referring to the ion-pair association must be introduced.

The solution equilibria between alkali metals and many acceptors often lie too far to the right to permit a reliable study of the equilibrium constant. This is particularly the case when the acceptor is a radical, e.g., triphenylmethyl. The difficulty may be avoided if mercury amalgam, instead of pure alkali metal, is used because the chemical potential of the alkali is then reduced. The necessary data giving the chemical potential of sodium in amalgam are available (30) and therefore the shift in the equilibrium may be calculated. Indeed, it is known that some metallo-organic compounds may be decomposed into their components if their solutions are stirred with mercury.

It has been known that the reaction of biphenyl with metallic sodium in ethereal solvents leads only to its partial conversion to sodium biphenyl (33) and the position of equilibrium is determined by the nature of solvent. The degree of conversion increases at lower temperatures, proving that the overall process is exothermic. In ethereal solvents and in not too dilute solutions the reaction yields ion-pairs, the proportion of free ions being negligible. Hence, the free energy of ion-pair solvation is of paramount importance in determining the position of such an equilibrium. Under these conditions the reaction is adequately represented by the equation

metallic sodium + biphenyl (solution) $\rightleftarrows$ Na$^+$, biphenyl$^-$ (solution),

and therefore the spectrophotometrically determined ratio

$$[Na^+, biphenyl^-]/[biphenyl] = K$$

is independent of the initial concentration of biphenyl.

The dependence of the equilibrium constant K on the nature of solvent and on temperature was investigated by Shatenshtein and his co-workers (65,66). Their results are shown in Table X, which also includes similar data obtained in this writer's laboratory. Plots of log K versus $1/T$ are shown in Figure 9, and the respective heats and entropies of reaction are listed in Table XI. It is obvious that solvation provides much of the driving force for the reaction and that the solvating power is strongly affected by steric factors. For example, at 25°C the equilibrium appears to be almost completely shifted to the right in 1,2-dimethoxyethane (DME); it yields only 7% sodium biphenyl in 1,2-methoxyethoxyethane, and the conversion is imperceptible in diethoxyethane. Interestingly, no conversion is observed in 1,1-dimethoxyethane; conversion reaches 100% in 1,2-dimethoxyethane, and only 6% in 1,3-dimethoxypropane. Also, it is significant that 1,2-dimethoxypropane is a much better solvating agent for Na$^+$ ion-pairs than 1,3-dimethoxypropane. While the equilibrium constant $K = [Na^+, B^-]/[B]$ has a value of 5.0 in

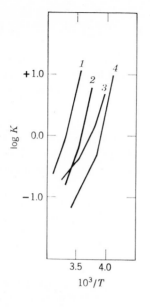

$\log K$

+1.0

0.0

−1.0

3.5   4.0

$10^3/T$

Fig. 9.   Equilibrium constants of the reaction
metallic sodium + biphenyl (solution) $\rightleftarrows$
sodium$^+$, biphenyl$^{\overline{}}$ (solution)
as functions of temperature for various solvents:
(1) dimethoxyethane–heptane, (2) methoxyethoxye-
thane–heptane (MEE-Hp), (3) tetrahydrofuran–
heptane, (4) diethoxyethane.

TABLE X

Equilibrium Constant for the Reaction
metallic sodium + biphenyl (solution) $\rightleftarrows$ Na$^+$, biphenyl$^{\overline{}}$, (solution)
$(K = [B^{\overline{}}, Na^+]/[B])$

Temp., °K	MEE	1,2-DMPr	THF	MeTHF[a]	DEE	THP	1,3-DMPr
318	0.07	0.07	0.08	—	—	—	—
313	0.12	0.09	0.10	—	—	—	—
303	0.28	0.20	0.20	—	—	—	—
293	0.75	0.49	0.36	0.02	0.07	—	—
283	2.55	1.40	0.66	0.036	0.11	—	—
273	7.0	5.0	1.50	0.055	0.19	0.06	—
263	—	—	2.90	0.11	0.39	0.10	0.12
258	—	—	—	0.14	0.61	—	0.20
253	—	—	—	0.20	1.25	0.17	0.34
248	—	—	—	0.29	2.25	—	0.61
243	—	—	—	0.45	8.7	0.29	1.20
238	—	—	—	0.64	—	—	2.75
233	—	—	—	1.18	—	0.48	—
228	—	—	—	—	—	0.75	—

MEE = 1,2-methoxyethoxyethane;   1,2-DMPr = 1,2-dimethoxypropane;   THF =
tetrahydrofuran;   MeTHF = 2-methyltetrahydrofuran;   DEE = 1,2-diethoxyethane;
THP = tetrahydropyrane; 1,3-DMPr = 1,3-dimethoxypropane.
[a] Data of Slates and Szwarc (67).

TABLE XI

Heat and Entropy Changes in the Reaction of Biphenyl with Metallic Sodium

Solvent[a]	K (273°K)	ΔH^b, kcal/mole	ΔS, eu
MEE	7.2 ± 0.8	−17.4 ± 0.5	−60 ± 2
1,2-DMPr	4.6 ± 0.2	−16.5 ± 0.7	−58 ± 3
THF	1.4 ± 0.1	−11.2 ± 0.5	−40 ± 2
DEE	0.2 ± 0.05	−9.6 (at the lowest temp. $\Delta H = -22$)	−38 ± 2 (at the lowest temp. $\Delta S = -86$)
1,3-DMPr	0.04	−15.5 ± 0.3	−63 ± 1
THP	0.06	−6.8 ± 0.2	−31 ± 1

[a] See Table X for explanation of abbreviations.

[b] These values should be taken with some reservation because the van't Hoff lines are slightly curved. The exothermicity becomes larger at lower temperatures. A striking example is provided by DEE, for which $-\Delta H = +9.6$ kcal/mole at about 20°C and increases to $+22$ kcal/mole at -30°C. Note that, small experimental errors in determination of K may lead to relatively large errors in ΔH.

the former solvent at 0°C, its value in the latter is only 0.04. Obviously, the cooperative coordination of both oxygens with the cation requires some specific spatial configuration which may be attained with the 1,2- but not with the 1,1- or 1,3-derivative.

Comparison of the equilibrium established between metallic lithium and biphenyl with that involving metallic sodium is most instructive. In diethoxyethane 45% of biphenyl is reduced by lithium, but only 15% by sodium. The difference is even larger in tetrahydropyrane, the conversions being 80 and 10%, respectively, for lithium and sodium (65). The heats of sublimation and the ionization potentials are greater for lithium ($\Delta H_{sub} = 37$ kcal/mole, $I = 5.36$ V) than for sodium ($\Delta H_{sub} = 26$ kcal/mole, $I = 5.12$ V). Therefore, the heat of solvation of the lithium ion-pair must be substantially larger than that of the sodium in order to account for the above observations. The solvation effects are, however, most specific because Shatenshtein found 4% of biphenyl to be reduced by sodium in dioxane but none by lithium. Further examples of the specificity of interaction between solvent and a cation are provided by some findings concerned with the reducing power of sodium and potassium. Sodium is a more powerful reducing agent than potassium in tetrahydrofuran or tetrahydropyrane, but the reverse order is found in 1,3-dimethoxypropane (66).

A word of caution: radical-anions become readily associated with other ion pairs in ethereal solvents. For example, shifts in the absorption maximum of the lithium diphenylketyl in tetrahydrofuran were observed when even small amounts of LiBr (less than $10^{-3}M$) were present (75). These findings are illustrated by Figure 10 taken from a paper by Powell and Warhurst (75), who interpreted their observations in terms of ion-pair agglomerations. Many more examples of such a behavior may be quoted (e.g., pp. 265, 269). It means that a new equilibrium is established whenever a suitable impurity, such as lithium or sodium alcoholate or hydroxide is present in the system, viz.,

$$\text{biphenyl}^{\bar{}}, \text{Na}^+ + \text{impurity} \rightleftharpoons \text{biphenyl}^{\bar{}}, \text{Na}^+, \text{impurity}.$$

Such an association shifts the main equilibrium to the right,

$$\text{sodium metal} + \text{biphenyl} \rightleftharpoons \text{biphenyl}^{\bar{}}, \text{Na}^+ \text{ (whether complexed or not)},$$

and consequently the spectrophotometrically determined concentration of sodium biphenyl increases in the presence of impurities. Such artifacts were

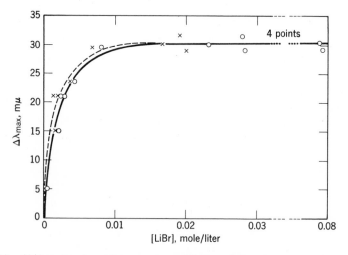

Fig. 10.   Shift in the absorption spectrum of lithium diphenylketyl in tetrahydrofuran arising from addition of lithium bromide (75).

observed in this writer's laboratory. They led to erroneous equilibrium constants, and to serious errors in the experimental ΔH of the reaction, because the slope of a van't Hoff line is most susceptible to small but systematic temperature-dependent errors in the respective equilibrium constants. A more detailed discussion of this problem is given in reference 117.

The complicating effect of impurities discussed in the preceding section may be advantageously exploited in studies of interesting equilibria of ion-pairs with various complexing agents (E), e.g.,

$$\text{biphenyl}^{\bar{}}, \text{Na}^+ \text{ (solution)} + n\text{E} \rightleftharpoons \text{biphenyl}^{\bar{}}, \text{Na}^+(\text{E})_n \qquad K_E.$$

The results of such studies were reported (67) for the systems sodium metal, biphenyl, and glyme-3 ($\text{CH}_3\text{O} \cdot \text{CH}_2 \cdot \text{CH}_2 \cdot \text{O} \cdot \text{CH}_2 \cdot \text{CH}_2 \cdot \text{OCH}_3$) or glyme-4

$(CH_3O \cdot CH_2 \cdot CH_2 \cdot O \cdot CH_2 \cdot CH_2 \cdot O \cdot CH_2 \cdot CH_2 \cdot OCH_3)$ in tetrahydropyrane and in 2-methyltetrahydrofuran. Denoting by K_E the equilibrium constant of the complex formation as described above, one may easily verify that

$$K_E(E)^n = (K_{ap} - K)/K,$$

where K_{ap} is the apparent equilibrium constant defined as the ratio

$$K_{ap} = \frac{\text{total concentration of } B^{\bar{}} \text{ in the presence of complexing agent}}{\text{concentration of the unconverted B}}$$

and K is the respective equilibrium constant found in the absence of the complexing agent:

$$K = [B_t^{\bar{}}]/[B] \qquad \text{in the absence of E.}$$

Here, B and $B_t^{\bar{}}$ denote biphenyl and any salt or complexed salt of the biphenyl$^{\bar{}}$ radical-ion as determined by the spectrophotometric method. The temperature dependence of K_{ap} and K for the system metallic sodium—biphenyl in tetrahydropyran solution and glyme-4 is shown in Figure 11 for various concentrations of glyme. From such graphs, plots of log $(K_{ap} - K)/K$ versus log [glyme] were constructed (see reference 67); these were found to be linear and have a slope of unity, proving that the complexing of glyme-4 (as well as of glyme-3) with sodium biphenyl ion-pair corresponds to a well-defined 1:1 stoichiometry, i.e., we are dealing with a *chemical* coordination and not with a physical association.

It was shown in Chapter 5 that some ion-pairs coexist in two forms— contact and solvent-separated, the absorption maximum of the latter being shifted to longer wavelength when compared with the former. A similar phenomenon is observed when a small amount of powerfully solvating agent is added to a solution of ion-pairs in a poorly solvating medium (118). The spectrum of sodium biphenyl, and of its 1:1 complex with glyme-3, absorbs at $\lambda_{max} = 400$ mμ within the whole range of the investigated temperatures, but, although at 25°C the absorption maximum for the complex glyme-4 still appears at 400 mμ, it is shifted to 407 mμ at -40°C. In fact, for every concentration of glyme-4 it was possible to find a temperature at which two peaks of equal optical density were seen, one at 400 mμ, the other at 407 mμ. It could be concluded that one of them corresponds to a non-complexed pair, whereas the other to a pair separated by the glyme. However, it was shown that the concentration of the former is lower than that calculated from the spectrophotometrically determined overall conversion and that of the latter is higher. Hence, the complex exists in two forms, one possesses the glyme on the outside of the ion-pair (like the complex with glyme-3) and absorbs at 400 mμ, while the other is separated by the glyme and has λ_{max} at 407 mμ. This is the first reported example of isomerism involving different positions of solvating agent in a solvated ion-pair.

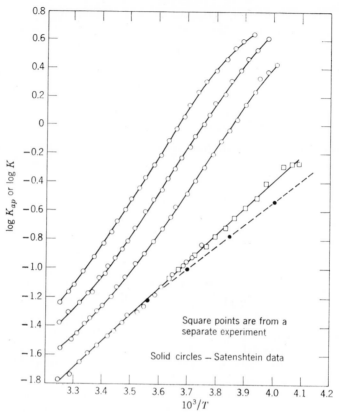

Fig. 11.   The lowest line gives the equilibrium constant of the reaction

metallic sodium + biphenyl (solution) $\rightleftarrows$ sodium$^+$, biphenyl$^-$ (solution)

in tetrahydropyrane at temperatures varying from 25° down to −40°C. Note the degree of agreement between Shatenshtein's data and those of Slates and Szwarc (67). The other three lines represent the effect of added glyme-4. The increase in total concentration of B$^-$ arises from formation of a complex between B$^-$, Na$^+$ and glyme.

13. DIRECT CAPTURE OF ELECTRONS IN γ-IRRADIATED SOLUTIONS AND GLASSES

It is well known that the primary action of high-energy γ-radiation, or x-rays, fast electrons, etc., causes ionization or electronic excitation of the irradiated molecules. Two theories of radiation chemistry inspired most of the work in this field. The earlier theory, developed by many workers (see, e.g., reference 276), proposes a heterogeneous dissipation of the radiation energy and creation of regions of high concentration of fragments—ions, electrons, radicals, and excited molecules. In these so-called spurs or hot spots most of the fragments combine or interact and, thus, they yield the observed products, while only a small fraction diffuses out and eventually achieves a homogeneous distribution throughout the irradiated material.

The second theory was put forward by Platzman (277). He suggested that the electrons formed in the primary ionization process become solvated and, therefore, "protected." Thus, many thermal electrons are eventually formed in such a process. Their capture by acceptors in the gas phase was discussed in Section 3. The same event may lead to electron capture in a liquid or glass. Two methods permit us to study such phenomena: (1) The irradiation may be performed at very low temperatures in an extremely viscous medium, essentially in a glass. This prevents or slows down the truly bimolecular combination of the positively and negatively charged species, and under such conditions their concentration may be increased to a level at which detection and investigation of their chemistry becomes feasible. (2) The irradiation is performed by a pulse of a very high intensity but extremely short duration. The events taking place after cessation of the pulse may be observed by suitable electronic techniques in periods as short as a few microseconds (see p. 345).

Radiolysis of organic glasses was extensively investigated by Hamill et al. Their first observations (68) dealt with systems involving frozen solvents, or their mixtures, such as 2-methyltetrahydrofuran, 3-methylpentane, isopentane, methylcyclohexane, diethyl ether, and ethanol. Samples were irradiated at $-196°C$, a typical dose rate being $\sim 10^{18}$ eV/liter min, and the dose was varied from $1-5 \times 10^{21}$ eV/liter. A broad absorption band was developed in irradiated polar glasses, its intensity increasing monotonically from 4000 Å up to the limits of applicability of the spectrophotometer, $\sim 13,000$ Å. This band was attributed to solvated electrons. Indeed, similar absorption spectra, reported in the past (69), had been assigned to solvated electrons. Modern techniques of pulse radiolysis (see Section 15) allow us to produce solvated electrons in a variety of media and to determine unequivocally their absorption spectra. For the sake of illustration the spectra of solvated electrons formed in aliphatic alcohols (144) are shown in Figure 12. The absorption observed in frozen glasses is usually broader because, due to the rigidity of the matrix, a variety of irregular "solvation shells" are formed and slightly different spectra are produced by electrons located in different shells (69).

Addition of naphthalene or biphenyl to the glass dramatically changed the absorption spectrum of the irradiated sample (68). Instead of the broad band, the characteristic absorption spectrum of naphthalene⁻ or biphenyl⁻ radical-anions appeared. It has been concluded, therefore, that the electrons formed in the matrix were trapped by the aromatic hydrocarbons, and thus the respective radical anions were produced. Extension of this work showed that the technique may yield radical-anions of benzophenone, nitrobenzene, and tetracyanoethylene (71), and in a later study (72) it was demonstrated that radical-anions may also be formed from phenanthrene, benzanthracene, triphenylene, *o*-terphenyl, pyrene, tri- and tetraphenylethylenes, cycloheptatriene, and hexamethylbenzene. On the basis of the known extinction coefficients of naphthalene $^{\overline{\cdot}}$ or biphenyl $^{\overline{\cdot}}$ radical-anions, the G value for the captured electrons was determined. In a glass produced by freezing 0.01 mole %

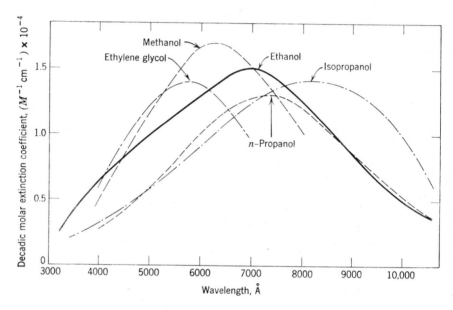

Fig. 12.   Spectra of solvated electrons formed in aliphatic alcohols at 23°C as determined by pulse radiolysis (144).

of naphthalene in 2-methyltetrahydrofuran (2-MeTHF) G was found to be 1.4. Its value reached the maximum of 4.0 for 1 mole % of naphthalene. The presence of carbon tetrachloride in the matrix suppressed the formation of negative radical-anions, e.g., the G value was reduced from 4 to 1.5 and eventually to 0 when 0.16 and 1.6 mole %, respectively, of carbon tetrachloride was mixed a with 1 mole % solution of naphthalene in methyltetrahydrofuran. This suggests that carbon tetrachloride competes with the aromatic hydrocarbons for the electrons; their capture by the former molecules causes their dissociation:

$$CCl_4 + e^- \rightarrow CCl_3 \cdot + Cl^-.$$

If two aromatic hydrocarbons are present in the glass, they are expected to compete for the electrons. Therefore the relative intensities of spectra of the respective radical-ions should depend on the mole ratio of the parent hydro-carbons. This, indeed, is the case, as shown by Figure 13. The recorded spectra refer to irradiated samples in which the concentration of one aromatic hydro-carbon was kept constant, while the proportion of the other increased. The decrease in the intensity of absorption due to the radical-anion of the former hydrocarbon and simultaneous increase in the absorption intensity of the other is seen clearly. Moreover, the isosbestic point, revealed by Figure 13,

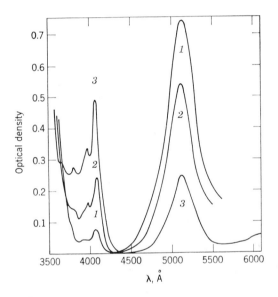

Fig. 13. Competition of two aromatic hydrocarbons for electrons formed by γ-irradiation in a frozen glass (72). Mole % triphenylene was 0.20, 0.10, and 0.04 for curves 1, 2, and 3, respectively; 0.158 mole % biphenyl.

proves that the total number of captured electrons remained constant. From such data the relative electron capture efficiencies of aromatic hydrocarbons were determined, and the relevant results are collected in Table XII. It was expected that the relative efficiencies would increase with the size of the mole-cule or with its polarizibility. This was not the case. However, Hamill reported a monotonic relation of efficiency with the electron affinities of the investigated aromatic hydrocarbon. This observation raises the question as to whether his results refer to an equilibrium established in the glass or are determined by the kinetics of the electron capture. It seems that the latter is the governing factor. The efficiencies determined by Hamill are only slightly affected by the nature of the aromatic substrate; the changes observed were not greater than a factor of 3. On the other hand, a large variation, by

TABLE XII

Relative Electron Capture Efficiency of Aromatic Hydrocarbons
in Organic Glasses (2-MeTHF) (72)

Hydrocarbon	Relative electron capture efficiency	Electron affinity in THF solution (relative to biphenyl) in V	Molar refraction
Biphenyl	1.00	0.0	52
Triphenylene	1.04	0.132	71
Phenanthrene	1.22	0.142	62
o-Terphenyl	1.40	—	74
Cycloheptatriene	1.94	—	30
Pyrene	1.95	0.529	67
Triphenylethylene	2.70	—	87
1,2-Benzanthracene	2.60	0.590	66
Tetraphenylethylene	2.76	—	118
Anthracene	2.76	0.642	65
Hexamethylbenzene	2.80	—	46

many powers of ten, should be expected if the equilibrium governs the outcome of the experiments. The relative electron capture efficiencies determined
in the glass are similar to those reported in the gas phase (see Sections
3 and 4, and reference 9). The gaseous electron capture coefficients are also
not much affected by the nature of the acceptor, and their values vary only
within a factor of 10. However, inspection of the latter data does not reveal
any regularities or correlations with electron affinities.

The relative electron capture cross-sections of aromatic hydrocarbons were
reinvestigated by Hamill and his co-workers (73,74) by means of an alternative technique—studying the competition for the electrons between an aromatic hydrocarbon and organic halides. This also permitted the determination
of the relative reactivities of halides toward electrons.

The matrix may also compete with the solute for the electrons. At a
sufficiently low concentration of an electron acceptor, the γ-irradiation of the
glass produces two spectra, that of the radical-ion and a broad band characteristic of the solvated electron. Subsequent irradiation with the selected
visible light bleaches the broad spectrum with concomitant increase in the
spectrum of the radical-anion. Absorption of the visible light ejects the solvated electrons from their shallow traps and permits their capture by the
aromatic hydrocarbon. Light absorbed by the radical-anions eventually
bleaches their spectrum. The ionization, or excitation, of radical-anions
facilitates the release of the captured electrons into the glass and leads to their
ultimate capture by the positive "holes." Of course, the latter had been formed

during γ-irradiation of a glass as a consequence of electron ejection from the matrix.

In hydrocarbon glasses the value of G for captured electrons is lower than that found in polar glasses. For example, in 3-methylpentane containing naphthalene G is only 0.15, but it increases to 0.4 on the addition of 1 mole % triethylamine. Apparently, the stability of the radical-anion and of the positive hole increases with their solvation, or alternatively the presence of amine facilitates electron transfer to the aromatic acceptor.

Finally, we may stress that the spectra of radical-anions formed in a frozen glass through γ-irradiation differ only slightly from those formed in the same medium by chemical reduction with sodium or potassium. On the whole, the agreement between both spectra is remarkably good. There is, however, a slight shift to the red in the spectrum developed in the glass. The fact that the γ-irradiation forms free ions, whereas chemical reduction produces ion-pairs, accounts for this observation (see also p. 259). In some systems the spectrum of a γ-irradiated glass shows a marked vibrational structure. For example, such a structure observed when glass containing anthracene was irradiated, is shown in Figure 14. For comparison the spectrum of anthracene⁻ in solution

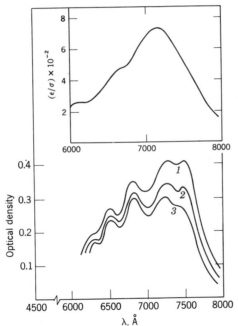

Fig. 14. Spectrum of anthracene⁻ radical-ion in a frozen glass at 70°K (below) and in solution at room temperature (formed by alkali-metal reduction). Note the vibrational structure in the lower spectrum and its absence in the upper one. ε/σ = extinction coefficient/frequency.

is included in the figure. The lack of structure in solution may be due to higher temperature of the sample, or to the presence of counterion, the soft vibration of which may broaden the spectrum.

An attempt was made to produce the radical-ion of styrene. Radiolysis of a glass containing styrene produced a spectrum with maxima at 4100 and 6000 Å (71) and these were attributed to styrene radical-anions. However, investigation of pulse radiolysis of styrene solution by Dainton and his associates (76) failed to reveal their existence. Only a single peak at 340 mμ, apparently that of living polystyrene, was observed in carefully purified organic solvents such as cyclohexane, dioxane, tetrahydrofuran, and benzene. The problem needs further study because Dainton's observations differ from those reported by Katayama (77) and by Keene, Land, and Swallow (78).

More conclusive evidence was obtained for the formation of radical-anions of butadiene and its derivatives. Radiolysis of frozen solutions of butadiene in methyltetrahydrofuran yields species that absorb at 388 and 570 mμ (79). Similar spectra were recorded on examining glasses containing butadiene derivatives, as shown by the data in Table XIII. However irradia-

TABLE XIII

Absorption Spectra of Radical-Anions Derived from Butadiene
and its Derivatives in γ-Irradiated MeTHF Glass

Hydrocarbon	λ_{max} (higher peak), mμ	λ_{max} (lower peak), mμ
1,3-Butadiene	388	570
Isoprene	390	565
cis-1,3-Pentadiene	388	563
2,4-Hexadiene	390	575
2,3-Dimethyl-1,3-butadiene	390	530
1,3-Cyclohexadiene	395	~ 575
1,3-Cyclooctadiene	380	~ 500

tion of glasses in which nonconjugated dienes, such as 1,4-pentadiene or 1,5-hexadiene, were dissolved resulted in a typical broad spectrum of solvent-trapped electrons.

Two methods of identification of the species produced from butadiene and its derivatives are available. The optical spectrum of 1,3-butadiene radical-anions was calculated on the basis of the simple Hückel molecular orbital approach, and the results led to λ_{max} of 382 and 473 mμ when $\beta = 2.62$ eV. The agreement with observation is reasonable, since it is known that this approach is much too crude to give more reliable values. Interestingly, the spectrum is only slightly affected by the presence of methyl substituents, as may be seen from Table XIII, indicating an insignificant influence of hyper-conjugation or induction on the spectra of the *negative* radical-ions. Alternatively, the ESR spectra of the glasses could be examined. The spectrum observed in a sample containing 1,3-butadiene revealed a quintuplet resembling the one reported by Levy and Myers (80) for the butadiene⁻ radical-anion generated electrolytically. However these workers obtained better resolution and could, therefore, determine the hyperfine splitting due to the four terminal hydrogens as well as the triplet due to the internal protons.

The irradiation carried out in methanol glass leads to obvious complications. Methanol glass stabilizes the positive "holes" by the reaction

$$CH_3OH^+ + CH_3OH \rightarrow CH_3OH_2^+ + CH_3O\cdot.$$

The released electrons are mobile and may be trapped by aromatic hydrocarbons or by some related compounds, e.g., by styrene. However, the resulting radical-anions eventually remove protons from neighboring methanol molecules to form the respective radicals (86). Thus, $C_6H_6^- + CH_3OH \rightarrow C_6H_7\cdot + CH_3O^-$, and the resulting cyclohexadienyl radical was identified by its optical (74) as well as by its ESR (86,87) spectrum. Of interest is a very sharp peak that appeared at about 320 mμ in the presence of styrene. This spectrum agrees well with that obtained in methyltetrahydrofuran glass in the presence of $Ph.CHBr.CH_3$ (74). In the latter case, the following reaction took place:

$$Ph.CHBr.CH_3 + e^- \rightarrow Ph.\dot{C}H.CH_3 + Br^-.$$

Hence, it appears that on irradiation of styrene in methanol an electron is captured and the respective radical-ion is initially formed. However, this species abstracts then a proton from a neighboring methanol molecule and produces the $Ph.CH.CH_3$ radical. The absolute rate constants of such proton-transfer reactions are discussed in Section 15.

Radical-anions may be formed in the course of γ-irradiation by an entirely different route. For example, irradiation of formate salts dissolved in potassium halide pellets (KCl, KBr, or KI) resulted in the formation of CO_2^- through rupture of the H—COO^- bond (81). The radical-anion was identified through its infrared and ESR spectra. This interesting technique was developed and explored by Hartman and Hisatsune (82).

CO_2^- radical-anions are also formed by an electron capture process (86). The γ-irradiation of methanol glass saturated with carbon dioxide produced a species identified through its ESR spectrum. This spectrum agreed with the one reported earlier (88) for CO_2^-. The CO_2^- spectrum is observed even if methanol contains benzene. Apparently, carbon dioxide competes efficiently with benzene for the electrons, and the proportion of products derived from benzene decreases significantly on addition of CO_2 (86).

14. FORMATION OF POSITIVE RADICAL-IONS BY γ-IRRADIATION

Let us discuss more thoroughly the actual process of electron formation and its trapping in organic glasses. Interaction of an ionizing agent with a molecule results in ejection of an electron (which may ionize adjacent molecules by virtue of its high kinetic energy) and leaves a positive residue in the glass. In frozen methyltetrahydrofuran, or more generally in glasses composed of ethers or some related compounds, the positive residue becomes stabilized by proton transfer. For example:

$$C_5H_{10}O^{+} + C_5H_{10}O \rightarrow C_5H_{10}\overset{+}{O}H + C_5H_9O \text{ (radical).}$$

Such a stabilization immobilizes (traps) the positive hole and makes it relatively unreactive, whereas the electron may move through a crystal by tunneling, its velocity being calculated as 1 Å in 10^{-14} sec (83). Eventually the electron is trapped by an acceptor or by a suitably distorted site in a glass. The latter, therefore, represents a shallow electron "trap" of low potential energy.

In some glasses, or microcrystalline solids, the electrons become stabilized by a process leading to immobile and inert negative centers, carbon tetrachloride or other organic halides being good examples (84). Thus, $CCl_4 + e^- \rightarrow CCl_3\cdot + Cl^-$, or more generally $RCl + e^- \rightarrow R\cdot + Cl^-$. What happens to the positive center? Such a center, e.g., CCl_4^{+} moves through the matrix by resonance involving the neighboring CCl_4 molecules:

$$CCl_4^{+} + CCl_4 \rightarrow CCl_4 + CCl_4^{+},$$

and this leads to its stabilization. We may remark, in passing, that no CCl_4^{+} ion is observed in the mass spectrum of CCl_4, even at relatively high pressures when the de-energization of an excited ion should be fast (85). Apparently either the gaseous CCl_4^{+} ion is in a repulsive state, or its potential energy minimum is very shallow.

Addition of substrates of sufficiently low ionization potential to glasses in which mobile positive holes are created leads to the formation of positive radical-cations. For example, addition of tetramethyl-p-phenylenediamine (TMPD) to frozen carbon tetrachloride dramatically changed the spectrum of this polycrystallinic solid when it was irradiated by γ-rays. Instead of the broad absorption at about 400 mμ, which is attributed to the positive holes, a characteristic spectrum of Wurster's blue (TMPD⁺) appeared (84). This is shown in Figure 15, where curve 4 refers to the irradiated pure CCl_4 and curve 1 to the solid in which TMPD was dissolved. Photobleaching of the irradiated sample decreased the intensity of the 400 mμ absorption and increased that due to TMPD‡ (shown by curve 2, Figure 15). UV light increased the mobility of the residual CCl_4^+ holes by enhancing their exchange with CCl_4 and, thus, some "diffused" to the amine and subsequently reacted with it:

$$CCl_4^{\ddagger} + TMPD \rightarrow CCl_4 + TMPD^{\ddagger}$$

Such "diffusion" is enhanced by heating the sample to 140° K and thereafter the radical-cation is stabilized by chilling the sample to 77° K. In an actual experiment all the CCl_4^+ holes disappeared and the intensity of the spectrum of the Wurster salt greatly increased (curve 3, Figure 15).

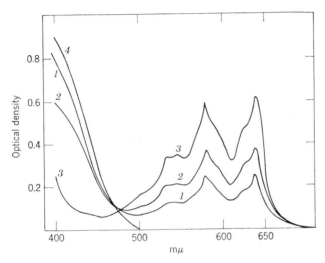

Fig. 15.   Absorption spectra of γ-irradiated TMPD (7 mM l^{-1}) in CCl_4 at 77°K: (*1*) irradiated to 3.6×10^{18} eV ml^{-1}; (*2*) as for *1*, then photobleached; (*3*) as for *1*, after warming to ~ 140°K and then chilling to 77°K; (*4*) pure CCl_4, same dosage (84).

The above technique was employed for the generation of many radical-cations, e.g., $Ph_3N^{\ddagger}$, $Ph_3P^{\ddagger}$, the radical-cations of tetramethylbenzidine etc. (84). All these species were produced by electron transfer from a donor to a positive hole, and not by direct ejection of an electron from the donor. The results shown in Figure 16 confirm this conclusion. The intensity of the spectrum of $Ph_3N^{\ddagger}$ radical-cation increased with concentration of the parent molecule, but eventually its value became constant and independent of the amine concentration. Such a situation arises when virtually all the positive holes formed in the bulk of the irradiated liquid are scavenged by the solute; this would not be the case had the radical-cations been formed by a direct ionization of the amine.

Irradiation of solutions of aromatic hydrocarbons in frozen carbon tetrachloride produced monopositive radical-ions of benzene, toluene, biphenyl, naphthalene, phenanthrene, anthracene, etc. (89). Of the 18 hydro-carbons investigated by this technique, anthracene was the only one for which the spectrum of the positive radical-ion had been reported previously (90). However, spectra of other aromatic, alternant radical-cations, not investi-gated by Hamill's technique, are known (90,91), viz., those of triphenylene, pyrene, tetracene, perylene, coronene, 3,4-benzpyrene, and quaterphenyl. These species were obtained either by photoionization of the parent hydro-carbons in rigid glasses or by using concentrated sulfuric acid at room tem-perature as an oxidizing agent.

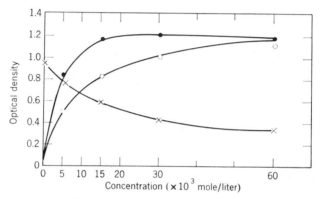

Fig. 16. Optical densities of $Ph_3N^{\ddot{+}}$ at 640 mμ (○) and of $CCl_4^{\ddot{+}}$ at 400 mμ (×) versus concentration of Ph_3N (84). Sample irradiated by γ-rays at 77°K 3.6 × 10¹⁸ eV ml.⁻¹ (●). After warming samples to 142°K, then returning to 77°K.

Finally, the γ-irradiation technique led to the formation of radical-cations derived from nonconjugated olefins (92). Isobutene, *cis*- and *trans*-butene-2, tetramethylethylene, and many other olefins were investigated. The spectra of the resulting radical-cations showed absorption maxima between 600 and 750 mμ.

How are hydrocarbon glasses affected by γ-irradiation? This problem was investigated by Gallivan and Hamill (93), who showed that positive holes are mobile in hydrocarbon matrices, and they migrate as far as the electrons, i.e., within a range of about 2 × 10⁴ Å. In fact, the mobility of positive holes is greater in hydrocarbon glasses than in glass produced from alco-hols or ethers. The latter trapped them forming radicals and stable cations. Gallivan and Hamill also presented some evidence indicating that both positive and negative radical-ions of biphenyl were formed when its solution in 3-methylpentane glass was irradiated. The simultaneous formation of positive and negative radical-ions was demonstrated for glasses formed from ketone (94), and their recombination led to chemiluminescence. For example, when pure frozen acetone was irradiated with γ-rays and then removed from liquid nitrogen it began to emit a bluish light which was visible for more than a minute.

15. PULSE RADIOLYSIS

Pulse radiolysis, as has been stated before, involves studies of transient species produced in an investigated sample by a pulse of ionizing radiation delivered in about one microsecond or less. Each pulse delivers energy of about 5–50×10^{17} eV per cubic centimeter of the sample. This energy is provided in the form of fast electrons accelerated to 1–15 million volts, or in the form of x-rays. The initial reaction involves solvent molecules that become ionized or electronically excited, and the fate of the resulting fragments may be followed, e.g., by a spectrophotometric technique. Addition of a chosen solute does not affect the primary process, although it alters the fate of the initial fragments. First studies of pulse radiolysis were reported in 1960 (95–97), and an excellent review of this subject was published in 1965 (98). Since the reader may find there many technical details and a complete list of the experimental results available at that time, we shall limit our discussion to several general remarks and then concentrate our attention on some specific topics pertinent to the subject of this book.

Much of the work in the field of pulse radiolysis is concerned with the generation of hydrated electrons in pure water and in aqueous solutions. The absorption spectrum of the hydrated electron has a maximum at 7200 ± 100 Å corresponding to an extinction coefficient of about 1.6×10^4. The most reliable extinction coefficient was determined by Rabani, Mulac, and Matheson (121), who converted the solvated electron into $C(NO_2)_3{}^-$ by reacting it with tetranitromethane. The extinction coefficient of the latter carbanion is known because the salts derived from the nitroform are stable. With equipment currently available, the products of pulse radiolysis may be detected at concentrations as low as $10^{-8}M$, i.e., optical densities as low as 0.002 are commonly measured (119) and sensitivity down to 0.00002 has been reported (120).

Although spectrophotometric methods are most frequently used for identification of the transients and for tracing their formation and decay, other techniques have also been utilized—for example, studies of electrical conduction (99) and of ESR spectra. The difficulty in conductance studies arises from the fact that it is not easy to maintain a constant voltage across the electrodes of the cell when its resistance changes rapidly. Moreover, the excess charge, due to the absorbed primary electrons, as well as the noise generated in the pulse, pose serious difficulties. To eliminate much of the noise, two oscilloscope tracings are always taken with reverse polarities of the cell electrodes. The conductivity portions of the two signals are then reversed while all the noise components remain unaffected.

Studies of pulse radiolysis render valuable kinetic data. It was possible to determine the absolute rate constants for capture of hydrated electrons by aromatic hydrocarbons, amines, quinones, etc. The results obtained by Hart, Gordon, and Thomas (100) for hydrated electrons are collected in Table XIV,

TABLE XIV

Absolute Rate Constants for Capture of a Solvated Electron

Substrate	Solvent	pH of solvent	k, liter mole^{-1} sec^{-1}
Tetracyanoethylene	Water	~ 7	1.5×10^{10}
1,3-Butadiene	Water	~ 7	8×10^9
Benzene	Water	~ 7	$< 7 \times 10^6$
Naphthalene	Water	~ 7	3.1×10^8
Nitrobenzene	Water	~ 7	3×10^{10}
Pyridine	Water	~ 7	1×10^9
Styrene	Water	~ 7	1.5×10^{10}
	Water	~ 13	1.1×10^{10}
Quinone	Water	~ 6.6	1.2×10^9
Biphenyl	Ethanol		4.3×10^9
Naphthalene	Ethanol		5.4×10^9
p-Terphenyl	Ethanol		7.2×10^9
Naphthacene	Ethanol		10.2×10^9

as are the results obtained in ethanol (101). Comparison of the data shows that the rate is insignificantly affected by the solvent. The last four lines in Table XIV show again that the rate constant of the electron-capture process is only slightly affected by the size of the aromatic system; these findings may be compared with those of Hamill listed in Table XII. The rate of capture seems to increase as one compares biphenyl and naphthalene with naphthacene, but not more than a factor of 2 or 3. The identity of the resulting radical-anions was established beyond any doubt by their absorption spectra (101–104).

In alcoholic solvents the aromatic radical-anions decay due to proton transfer from adjacent molecules (see also p. 386). The rate constant of the proton transfer from alcohols such as methanol, ethanol, n-propanol, and isopropanol to various aromatic radical-ions was investigated by Arai and Dorfman (101), and subsequent work led to the respective activation energies

TABLE XV

Rate Constants of Proton Transfer from Aliphatic Alcohols to
Aromatic Radical Anions at 25°C (101)

Radical-anion	Alcohol	k, liter mole^{-1} sec^{-1} × 10^{-4}
Biphenyl$^{\cdot}$	CH$_3$OH	7.
	C$_2$H$_5$OH	2.6
	n-C$_3$H$_7$OH	3.2
	iso-C$_3$H$_7$OH	0.5
Anthracene$^{\cdot}$	CH$_3$OH	8.
	C$_2$H$_5$OH	2.3
	n-C$_3$H$_7$OH	2.4
	iso-C$_3$H$_7$OH	0.4

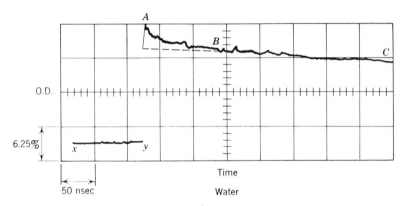

Fig. 17. Oscilloscope tracing of decay of solvated electrons in water. (A–B) (100 nsec), the rapid decay in spurs. (B–C) the slow, homogeneous decay of electrons which diffused into the bulk of the liquid. x–y base line, the pulse triggered at y.

(105). They found a good correlation between observed rates and acidity of the alcohols. The pertinent results are listed in Table XV.

Recent improvements of electronic techniques permitted the extension of pulse radiolysis to the nanosecond region (278). The results demonstrated directly, for the first time, the reality of spurs and made their quantitative studies feasible. This is shown in Figure 17 which gives the decay of the hydrated electron after a 12 nsec pulse. The rapid decay taking place within the time period AB is independent of the radiation intensity, whereas the slower decay BC does depend on the dose. The slower decay arises from the homogeneous bimolecular reaction of hydrated electrons in the bulk of the liquid; the initial rapid one is the heterogeneous process in the spur. This reaction is over in about 100 nsecs (10^{-7} sec).

The spur effect is quite small; in deaerated water the G value for the solvated electrons destroyed in spurs is 0.6, while the yield of those which diffused out is about 2.8. The hydrated electrons are destroyed in the spur by two reactions,

$$e_{aq}^- + OH \rightarrow OH^-$$

$$e_{aq}^- + H^+ \rightarrow H,$$

the OH radicals and protons being formed in the spur, viz.

$$H_2O \xrightarrow{h\nu} e_{aq}^- + H^+ + OH.$$

The evidence is provided by two observations: (a) the addition of sodium hydroxide in excess of $10^{-2}M$ reduces the spur effect by about 50%. Alkali removes H^+ and prevents the destruction of electrons through proton trapping. (b) The addition of $1M$ ethanol and $10^{-2}M$ alkali removes the spur effect completely because interaction with alcohol removes OH radicals. Note that this interpretation, if correct, gives an estimate for the rate constant of reaction $C_2H_5OH + OH \rightarrow$ products and $H^+ + OH^- \rightarrow H_2O$. The rate constant of the latter reaction was measured by two methods and its value is 1.4×10^{11} M^{-1} sec^{-1} (279).

Similar studies were performed in ethanol and isopropyl alcohol. The yield of electrons lost in spurs is given by $G = 0.5$; those reacting homogeneously correspond to $G = 1.0$.

16. SOLVATED ELECTRONS

Although application of pulse radiolysis advanced our knowledge of solvated electrons, this is not a new species. The blue solutions of alkali metals in liquid ammonia have been known for more than a century (145) although their exact nature is still in dispute. Extensive studies of their conductance, initiated by the pioneering investigations of Kraus (124), left no doubt that the dissolved metal dissociates into ions. At sufficiently low concentration the equivalent conductance increases with dilution and eventually reaches its limiting value, e.g., $\Lambda_0 = 1022$ for the sodium solution at $-33°C$. Since the limiting conductance of Na^+ ions in liquid ammonia is only about 130, the anion must be extremely mobile. These were the first observations which led to the concept of solvated electrons, i.e., it was postulated that

$$Na + solvent \rightleftarrows Na^+ \text{ (solvated)} + e^- \text{ (solvated)}.$$

The system alkali metal–liquid ammonia must be more complex than it appears, because at high concentration of metal the equivalent conductance *increases* with concentration (125). The problems associated with this increased conductance cannot be discussed here, and the reader who wishes to explore this subject is referred to the original literature (126,127a,129). It appears that a transition to the metallic state takes place in such systems; the valence electrons may move independently of the liquid since their orbitals may overlap and form conductance bands. Several alternative explanations have also been proposed, e.g., those involving a "microscopic Wien effect" caused by the field of the ions (128).

Let us return now to the conducting behavior of dilute alkali-metal solutions in liquid ammonia and in amines. Deviations of the equivalent conductance from the Debye–Hückel–Onsager law, which is applicable to dilute, ideal (i.e., completely dissociated) electrolyte solutions, may be explained in terms of ion-pairing. Indeed, Kraus interpreted his data in this way and visualized the pair in a vague sense as

$$\{M^+ \text{ (solvated)}, e^- \text{ (solvated)}\}$$

bonded through Coulombic interaction. Detailed analysis of the dependence of Λ/Λ_0 on concentration showed, however, that one equilibrium constant, K_1, is not sufficient to account for the observations:

$$M^+ \text{ (solvated)} + e^- \text{ (solvated)} \rightleftarrows M^+ \text{ (solvated)}, e^- \text{ (solvated)} \qquad K_1$$

but an agreement could be obtained if an additional dimerization was postulated; for example:

$$2(M^+, e^-) \rightleftarrows M_2.$$

An even better fit may be secured if three equilibria are postulated (130):

$$M \rightleftarrows M^+ + e^-,$$
$$M^- \rightleftarrows M + e^-,$$
$$2M \rightleftarrows M_2.$$

This is not surprising, because three adjustable coefficients are employed in calculations which account for the experimental data.

The existence of solvated M^+ and e^- ions in equilibrium with their pair is confirmed by several observations:

(*1*) Electrolysis of alkali-metal solutions leads to accumulation of the alkali ions (and of the blue colored species) in the cathode compartment, nothing being evolved or deposited on the anode.

(*2*) The Walden rule applies to alkali solutions in liquid ammonia and various amines.

(*3*) Salts having a common cation with the dissolved metal affect the conductance in a way accounted for by the mass law (132).

(*4*) The calculated equilibrium constant K_1 for alkali metal and for alkali salt solutions are of the same magnitude.

The M^+ ion formed from the dissolved metal does not differ from that produced on dissociation of other M^+, X^- salts; however, the nature of the solvated electron, of the pair M^+, e^- (often referred to as the monomer), and of the dimer need further elaboration (147).

Several plausible models were proposed for solvated electrons (127). In nonpolar solvents, e.g., in liquid rare gases, the electron seems to be located in a cavity (133–135). Electron–atom repulsion, if sufficiently strong, may stabilize the state of local fluid dilation and thus create the cavity within which the electron moves. Indeed, it is known that the electron–helium atom repulsion is large (136), and therefore this model seems to be applicable to this system. However, because the electron–argon interaction in liquid argon is attractive, a different model is needed to describe excess electron in this liquid. Preliminary data of Schnyders, Meyer and Rice (137), who determined the magnitude of the electron mobility (400 cm^2/V sec at 90°K) and of its temperature dependence, favor the free electron model in which the electron is treated as a quasi-free particle scattered by the atoms of the dense fluid.

In polar solvents the electron produces a polarization field that leads to its self-trapping. The binding energy is large, e.g., the heat of solution of an electron in liquid ammonia or in water is of the order 30–40 kcal/mole (138). The short-range electron–solvent repulsion leads, however, to the formation of a cavity in which the electron becomes localized—for instance in liquid ammonia the radius of the cavity is calculated to be about 3 Å. The long-range polarization effect makes such a cavity smaller than that formed in a nonpolar medium (e.g., in liquid helium). Another factor influencing the radius of the cavity arises from the surface energy term which is related to the surface tension of the liquid. The polarization of the liquid is also responsible for the leakage of electron charge density outside the cavity. For example, in liquid ammonia 30–40% of its charge leaks out.

The localization of the excess electron on a solvent molecule, such as ammonia, amines, or ethers, is unlikely because no low-energy antibonding orbitals are available in these species. If the resulting orbital for the electron is spread over many solvent molecules, the resulting system is referred to as a polaron and the solvent may then be treated as continuous dielectric. A concise review dealing with excess electrons in liquids has been published recently (327).

A substantial decrease in the density of alkali-metal solution relative to that of the components is a clear manifestation of the formation of cavities. The volume expansion accompanying dissolution of sodium in liquid ammonia was reported first by Kraus and Lucasse (139). The magnitude of the dilation is conveniently measured by

$$\Delta V = [V_{\text{solution}} - (V_{\text{NH}_3} + V_{\text{metal}})]/(\text{gram atoms of metal}),$$

ΔV being given as a function of metal concentration. The largest expansion was observed for lithium solutions; in fact, a saturated solution of lithium in ammonia is the lightest liquid known at room temperature, having a density of only 0.477 g/cc. This indicates creation of cavities with a volume of about 100 Å^3 each. Such a volume is normally occupied by three ammonia molecules.

Much of our knowledge of the structure of solvated electrons was derived from studies of their absorption spectra. The spectrum of a hydrated electron, which was determined by the pulse-radiolysis technique, has been mentioned. The technique of pulse radiolysis has also permitted determination of the spectra of solvated electrons in other solvents (141–144). The reported spectra resemble those found for the hydrated electron and for the "blue" ammonia solution, being broad and showing lack of structure even at $-78°C$. The oscillator strength varies from 0.6 to 0.9 and this implies that no intense bands exist at energies higher than 4 eV.

The mobile cavities which accommodate solvated electrons formed in solutions may be compared with localized electron traps formed in crystals doped with alkali metals. These traps represent vacancies in the crystal lattice which should be occupied by anions but which contain free electrons instead. ESR studies show that, although a trapped electron leaks out of the cavity and interacts with the neighboring cations, and to a lesser extent with the more remote ions, its highest density is in the trap. Such electrons are often referred to as F-centers.

Let us now consider the neutral but paramagnetic species referred to as M^+,e^- ion-pair or as the monomer. Several workers have attempted to describe its structure. The model proposed by Becker, Lindquist, and Alder (148) represents the associate as a pseudo-atom. The electron is supposed to circulate in an expanded orbital among the solvent molecules oriented by the field of the cation. The "monomer" is a distinct species different from an alkali atom. This is clearly demonstrated by the ESR spectra; the spin density on the alkali nucleus in a monomer is only 0.1% of that observed in gaseous alkali atoms.

At higher concentrations of alkali the paramagnetism of the solution decreases (164), indicating that some diamagnetic species are formed. This observation, as well as detailed conductance studies, led to the assumption of dimer formation. Becker, Lindquist, and Alder describe the dimer as diamagnetic, hydrogen-molecule-like species resulting from the association of two "expanded atom" alkali monomers. Such a species should have a very different spectrum from that of a solvated electron or a "monomer."

The above model was subsequently modified by Gold, Jolly, and Pitzer (146), who pointed out that the ammonia–alkali metal solutions obey Beer's law over a concentration range within which the solvated electrons associate into paramagnetic, nonconducting pairs, and even when the latter become dimerized into diamagnetic species. Since it is unlikely that the spectra of all these species are identical, it was concluded that the association yielding the "monomer" must be a "loose" one, like ion-pairing, and that the dimer is formed by further electrostatic aggregation leading to quadrupolar species held together by Coulombic forces. A similar suggestion was made by Symons and his co-workers (140).

The subject is, however, further confused by the ESR and NMR data. The Knight shifts in ammonia solution were measured by Hughes (149) and by Acrivos and Pitzer (150) and each group drew different conclusions from their own data. It appears that some other species may be present in such solutions. For example, Douthit and Dye (165) suggested that two monomers coexist in equilibrium with each other: the cation-centered species represented by an expanding atom model, and an ion-pair. A similar situation may be encountered when dimers are considered. Indeed, Becker, Lindquist, and Alder (148) pointed out that it is necessary to consider the existence of two different diamagnetic species, the "dimer" and a diamagnetic M^-, which was treated by Pitzer as a triple-ion. The structure of the latter species was recently reinterpreted by Golden, Guttman, and Tuttle (131) as a solvated (ammoniated) alkali anion, and a reasonable estimate of its solvation energy makes this suggestion plausible (154).

There is no doubt that some other species exist in various amine solutions. In these media Beer's law does not hold, and spectrophotometric studies showed that the "monomer" and "dimer" differ in their spectra (166). Moreover, ESR studies which provided the data about the structure of the spectrum, shape of the line, and its dependence on temperature and solvent (151–153) demand new models. The large hyperfine splitting due to the cation indicates a much higher electron density on the metal nucleus in amines than in liquid ammonia, which supports the idea of an expanded atom, at least for these solutions. In fact, no single model can yet account for all the data presently available.

Potassium and cesium dissolve in tetrahydrofuran and in dimethoxyethane to form the characteristic blue solutions (155). However, the dissolved species are diamagnetic (156,157) although still very reactive, e.g., they initiate rapidly anionic polymerization of styrene (158). It was suggested that the active entity is composed of two coupled electrons, $(e^-)_2$, but it is more probable that such solutions contain negative alkali ions produced by the reaction $2K \rightarrow K^+ + K^-$ (154).

A powerful aprotic solvent capable of dissolving alkali and alkali-earth metals was reported recently, namely, hexamethylphosphoramide. The formation of a blue solution on contact of alkali metals with this liquid was observed in 1961, but reports describing this system were not published until 1965 (159–162). Hexamethylphosphoramide dissolves lithium, sodium, and potassium. The blue solution is paramagnetic and stable at least for an hour at room temperature. It is extremely reactive and bursts into flame on contact with air. On standing, it turns red and becomes diamagnetic.

The ESR spectrum of the blue solution was investigated first by Fraenkel, Ellis, and Dix (159), who observed only a single line having no structure. Subsequent studies by Chen and Bersohn (163) revealed a rich spectrum showing 35 lines. This astonishing observation turned out to be an artifact. A trace of naphthalene present in the solvent led to the spectrum of naphthalene⁻ radical-ion.

In conclusion, the behavior of electron excess in the gas phase may be accounted for in terms of free electrons which are scattered by the molecules or radicals present in the system. The low density of matter justifies the model of plane wave-packet for the electron. In organized solids—crystals—the symmetry of the matrix leads to a different treatment of electron excess. The long-range order arising from translational symmetry, as well as the other symmetry conditions, permits the description of the excess of electrons in terms of collective coordinates. This leads to models of excitons. A liquid has high density, short range structural order, but not long range order. This complicates enormously the description of excess electrons in this phase. Depending on the molecular structure of the investigated liquid, attraction and repulsion of its molecules with an electron acquire different importance. This leads to alternative models of "holes" and "polarons," i.e., of solvated electrons.

Solvated electrons may be produced by a variety of techniques. In addition to radiolysis, which was discussed previously, photoionization led to interesting results. The pioneering work of Lewis and his students (169,170) demonstrated that electrons may be ejected from some phenols, amines, and dyes by irradiating their solutions in rigid glasses at low temperatures with UV or visible light. Observation of the irradiated glass demonstrated the presence of positive radical-ions; however, the solvated electrons were not detected at that time; they were identified later in the course of studies of the photolysis of lithium metal, as well as of N-lithiumcarbazole solutions in methylamine frozen into a glass (167). The characteristic bands of solvated electrons appeared in the near-infrared, and they disappeared on softening the glass—an obvious indication of the electron–positive ion recombination. It is also interesting to note that the spectrum of the solvated electron observed at 70°K differed somewhat from that recorded after the temperature of the glass was slightly raised. Linschitz (167) explained this phenomenon by postulating the formation of "poor" traps at 70°K. The low mobility of molecules of solvent, hindered by the rigidity of the medium, prevented them from attaining the most profitable orientation around the electron. At higher temperature the mobility became sufficiently high to permit the orientation of solvent molecules, and then a "proper" trap was formed.

Photoejection of electrons from radical-anions of aromatic hydrocarbons dissolved in frozen 2-methyltetrahydrofuran was investigated by Zandstra and Hoijtink (171). Although the spectrum of the solvated electron was not observed in this initial study, subsequent investigation (168) led to identification of absorption at $\nu_{max} = 7.7$ kK. This spectrum was independent of the nature of the generator (perylene$^-$, Li$^+$ or tetracene$^=$, 2Li$^+$) confirming its assignment to solvated electrons. Further proof was provided by the ESR spectra of the irradiated solution. The question was raised whether the electron is trapped in the bulk of solvent or around the cations. The answer was provided by investigating the competitive trapping of the ejected electrons by dissolved pyrene and the "traps" (whether in the bulk of the solvent or in the vicinity of cations). Such competitive studies showed that "traps" are more numerous than cations and, therefore, the ejected electrons had to be located in solvent cavities and not around the cations, Li$^+$ (172).

Solvated or hydrated electrons may be produced by photolysis of other negative ions (176,177). Recent studies involving photoionization of K$^+$,NH$_2^-$ led to better understanding of the kinetics of elementary processes that occur in this reaction (178).

An interesting method for the production of solvated electrons was suggested by Baxendale and Hughes (173):

$$H + OH^- \rightleftarrows e^- \text{ (solvated)} + H_2O,$$

and studies of Jortner and Rabani (174) confirmed its feasibility. A similar reaction has been investigated in liquid ammonia:

$$NH_2^- + \tfrac{1}{2}H_2 \rightleftarrows e^- \text{ (ammoniated)} + NH_3.$$

By using ESR for determining the concentration of solvated electrons the equilibrium constant of the above reaction was found to be about $10^{-6}M$ (175).

The reactions of solvated electrons were often confused with reactions of H atoms. Such mistakes were particularly common in studies of radiolysis of aqueous solutions. Discrimination between the reactions of solvated electrons and H atoms are possible if N$_2$O is used as the reagent. This oxide reacts extremely rapidly with solvated electrons, producing N$_2$ + O$^-$, but not at all with hydrogen atoms.

17. DISPROPORTIONATION OF RADICAL-ANIONS

Many electron acceptors may acquire a second electron and form dianions, e.g., anthracene$^=$, benzophenone$^=$, etc. In such systems the radical-anions may disproportionate. Denoting by A$^-$ the mononegative radical and by A$^=$ its dianion, one describes the disproportionation by

$$2A^- \rightleftarrows A + A^=, \qquad K_d.$$

Electron affinity of the parent molecule, A, is usually higher than that of $A^{\bar{}}$, because the Coulombic repulsion hinders the addition of a second electron. In terms of a simple Hückel MO theory, the repulsion energy, ΔE_{rep}, is given by the integral

$$\Delta E_{\text{rep}} = \int \int \bar{\phi}_{N+1}(1)\bar{\phi}_{N+1}(2)[e^2/r_{1,2}]\phi_{N+1}(1)\phi_{N+1}(2)d\tau_1 d\tau_2,$$

where ϕ_{N+1} is the wave function of the lowest, antibonding orbital. Calculations performed for aromatic alternant hydrocarbons (27) show that the magnitude of this repulsion is sufficiently large to make the disproportionation insignificant in the gas phase. However, a different situation is encountered in solution, where ionic species enjoy an extra stability arising from their solvation. Accepting the original Born approach and assuming that both $A^{\bar{}}$ and $A^{=}$ may be represented by identical spheres of radius, r, we find that the total free energy of solvation increases on disproportionation. Its gain, Δ_S, is given by $(e^2/r)(1 - 1/D)$, i.e., it is twice as large as the free energy of solvation of a single $A^{\bar{}}$ ion. Hence, the free energy of disproportionation of free radical-ions in solution is given by

$$\Delta G_{\text{d}} = \Delta E_{\text{rep}} - \Delta_S.$$

The above equation serves to calculate ΔG_{d} of aromatic radical-ions (27) if one neglects the change in entropy of disproportionation of gaseous ions, which, most probably, is small. The results of such computations are listed in Table XVI. Comparison with experiments shows again that the theory needs

TABLE XVI

Calculated Free Energy of Disproportionation (in eV) of Radical Anions in Solution (27)

Parent molecule	ΔE_{rep}	$2\Delta G_{\text{solv,A}^{\bar{}}}$	ΔG_{d}	
			Calculated[a]	Observed
Benzene	7.03	3.6	−3.4 (2.0)	
Naphthalene	5.64	2.2	−3.4 (1.9)	
trans-Stilbene	5.51	1.7	−3.8 (1.6)	−0.23[b]
Phenanthrene	5.17	1.6	−3.6	
Anthracene	5.10	1.6	−3.5 (1.7)	−0.58[c]
Tetraphenylethylene	4.76	1.2	−3.6	

[a] The values given in parentheses in this column arise from Lyons' (42) calculation of solvation energy.

[b] Hoijtink and van der Meij (106).

[c] Hoijtink et al. (33).

much refinement, although it correctly predicts that the repulsion and solvation energies simultaneously decrease with increasing size of the hydrocarbon. Such a compensating effect may account for the nearly constant difference between the first and the second reduction potential, as had been observed in the polarographic studies of aromatic hydrocarbons (33) and polyphenyls (106).

Let us stress again that the treatment discussed here applies only to *free* radical-ions in solution. However, in most solvents the radical-ions, and certainly the dianions, are paired with their respective counterions. In such a case, the disproportionation is represented by the equation:

$$2A^{\overline{\cdot}},M^+ \rightleftarrows A + A^=,2M^+,$$

where M^+ is the counterion. As may be easily proved, the free energy of disproportionation of ion-pairs is given by the difference between the first and second reduction potentials directly observed in potentiometric titrations performed in solvents favoring a substantial pairing, e.g., in tetrahydrofuran. The ion-pairing is expected to favor disproportionation, because binding of two counterions by the negative dianions is more than twice as powerful as that responsible for the linkage of $A^{\overline{\cdot}}$ with its partner. The situation is even more complex as shown by the temperature effects revealed in the spectra of radical-ions in methyltetrahydrofuran (107), and the most striking example is provided by the findings of de Boer (108). Lithium naphthalene in 2-methyltetrahydrofuran exists as the radical-ion at $-120°C$. The optical and ESR spectra conclusively identified this species. However, at higher temperatures the spectrum was drastically changed and the paramagnetism disappeared. Evidently, the radical-ions disproportionated, and indeed the spectrum of the resulting solution became identical with that reported for the naphthalene dianion (65,109). Since the disproportionation is favored by higher temperature, the reaction must be endothermic. Apparently the gain in energies of solvation and association is still not sufficient to overcome the energy loss caused by electron repulsion and, therefore, the disproportionation process is driven by entropy and not by energy.

The question arises as to why disproportionation increases the entropy of the system. Of course, conversion of free $A^{\overline{\cdot}}$ ion into $A^=$ dianion must *decrease* the entropy of the system, but the situation may be different for the ion-pairs. The naphthalene$^{\overline{\cdot}}$, Li^+ ion-pair probably forms a strong dipole that interacts powerfully with the solvent, orients its molecules and, therefore, decreases the entropy of the solution. Disproportionation produces a tight quadrupole which weakly interacts with the solvent. Hence, the solvent molecules become free, and the entropy of the system increases. This accounts for the findings. Let us add, however, that the association is exothermic, and thus it partially compensates for the increasing repulsion energy of the electron and for the *loss* of solvation energy.

The most thoroughly investigated disproportionation is that of the radical-ions of tetraphenylethylene; this species will be denoted here by $T^{\bar{\cdot}}$. It was known, since the early studies of Schlenk (2), that the reaction of tetraphenylethylene (T) with metallic sodium in ether or dioxane yields only the disodium salt ($T^=,2Na^+$). No radical-ions seemed to be formed even when an excess of the parent hydrocarbon was available. This observation was most puzzling, especially when the sensitive technique of ESR spectroscopy still failed to detect the radical-ions (110). However, recent investigations of Evans and his students (111a) proved that the ESR signal *may* be obtained if the reaction takes place in tetrahydrofuran, a finding soon confirmed by the independent studies of Garst and Cole (112). These results emphasized the important role of solvent; although the ESR spectrum was observed in tetrahydrofuran and dimethoxyethane, no signal was discerned in dioxane. Further studies of the reaction revealed that the intensity of the signal depends not only on the solvent but also on the nature of the counterion and on the temperature (111–113). Obviously, this is a complex system and the disproportionation therefore required further studies, the results of which are most illuminating and deserve detailed discussion.

It was remarked early in this section that disproportionation in ethereal solvents is represented by an equation such as

$$2T^{\bar{\cdot}},M^+ \rightleftarrows T+T^=,2M^+,$$

and thus the degree of disproportionation should not be affected by dilution. The disproportionation could be investigated by a spectrophotometric technique because the spectra of radical-ions and of dianions are different. Careful studies of this equilibrium in tetrahydrofuran revealed, however, that the apparent equilibrium constant, K_{ap}, defined by the ratio,

$$[T^= \text{ total}][T]/[T^{\bar{\cdot}} \text{ total}]^2,$$

decreases on dilution (113b,114). Here [$T^=$ total] denotes the concentration of *all* the dianions, in whatever form they are present, while [$T^{\bar{\cdot}}$ total] refers to the concentration of all the radical-ions. The effect of dilution on K_{ap} may be accounted for if the system is governed by three independent equilibria, each determined by its own equilibrium constant (114), e.g.:

$$2T^{\bar{\cdot}},Na^+ \rightleftarrows T + T^=,2Na^+ \qquad K_1, \qquad (1)$$

$$T^{\bar{\cdot}},Na^+ + T^{\bar{\cdot}} \rightleftarrows T + T^=,Na^+ \qquad K_2, \qquad (2)$$

$$2T^{\bar{\cdot}} \rightleftarrows T + T^= \qquad K_3.$$

The last equilibrium may be neglected because the concentration of the dianions, $T^=$, is expected to be vanishingly small. The equilibrium constants K_1 and K_2 are correlated through a simple relation:

$$K_1/K_2 = K_{\mathrm{diss,T^{\overline{\cdot}},Na^+}}/K_{\mathrm{diss,T^=,2Na^+}}.$$

The two constants on the right refer to the dissociation processes:

$$T^{\overline{\cdot}},Na^+ \rightleftharpoons T^{\overline{\cdot}} + Na^+ \qquad K_{\mathrm{diss,T^{\overline{\cdot}},Na^+}}$$

and

$$T^=,2Na^+ \rightleftharpoons T^=,Na^+ + Na^+ \qquad K_{\mathrm{diss,T^=,2Na^+}}.$$

The apparent disproportionation constant, K_{ap}, is then given by:

$$K_{ap} = K_1(1 + K_{\mathrm{diss,T^=,2Na^+}}/[Na^+])/(1 + K_{\mathrm{diss,T^{\overline{\cdot}},Na^+}}/[Na^+])^2$$

and, in accord with experimental findings, its value decreases on dilution, if

$$K_{\mathrm{diss,T^=,2Na^+}} < K_{\mathrm{diss,T^{\overline{\cdot}},Na^+}}.$$

Dissociation of both salts was investigated (114) and the results are shown in Figure 18, which reveals two facts: (1) The dissociation constant of the

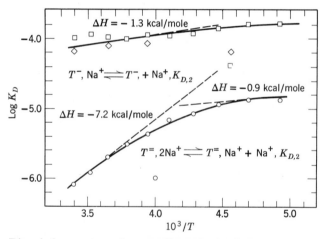

Fig. 18.   Dissociation constant of monosodium salt of radical-anion ($T^{\overline{\cdot}},Na^+$) derived from tetraphenylethylene (T) and of disodium salt of its dianion ($T^=,2Na^+$). Both spectrophotometric and conductance studies were used in determining the dissociation constants of the respective ionic agglomerates: ($\lozenge$) spectrophotometry, ($\square$ and $\bigcirc$) conductance.

$T^{\overline{\cdot}},Na^+$ ion-pair is relatively high, although the exothermicity of the process is low. Such a relation is characteristic of solvent-separated pairs. (2) The dissociation constant of $T^=,2Na^+$ is low, but this process is strongly exothermic. This behavior characterizes a contact pair that dissociates into solvated ions. It is possible now to estimate the term

$$(K_{\mathrm{diss,T^=,2Na^+}})[Na^+]^{-1}$$

and show that its value is much smaller than unity. Therefore, the approximation

$$K_{ap} = K_1/\{1 + (K_{diss,T^{\bar{\cdot}},Na^+})[Na^+]^{-1}\}^2$$

is valid, and this leads to the equation

$$1/K_{ap}^{1/2} = 1/K_1^{1/2} + (K_{diss,T^{\bar{\cdot}},Na^+}^{1/2}/K_1^{1/4})\{[T].[T^= total]\}^{-1/4}.$$

Plots of $1/K_{ap}^{1/2}$ versus $\{[T][T^= total]\}^{1/4}$ are shown in Figure 19, and from their intercepts and slopes the values for K_1 and $K_{diss,T^{\bar{\cdot}},Na^+}$ were derived. Thus,

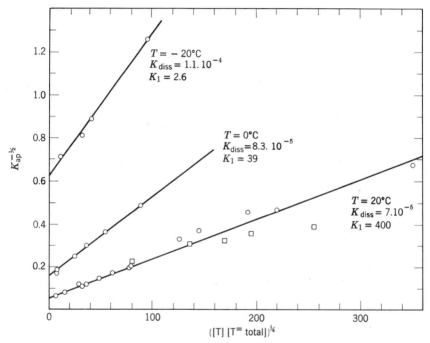

Fig. 19. Linear plots of $1/K_{ap}^{1/2}$ versus $\{[T][T^= total]\}^{-1/4}$ for the system tetraphenyl ethylene and its radical-ion and dianion: (○) Roberts and Szwarc, ref. 114; (□) data of Garst (113b).

K_1 and K_2 were computed and the final results, given in Table XVII, are shown graphically in Figure 20.

Let us assess the factors governing these disproportionations. Both reactions 1 and 2 are endothermic. Electron repulsion and *loss* of solvation energy of the cation contribute to this effect which is partially balanced by the greater binding energy of Na^+ to the dianion. On the other hand, the gain of entropy, due to desolvation of the ions or ion-pairs provides the driving force of this process. Note that the two $T^{\bar{\cdot}},Na^+$ (dipoles) orient solvent molecules

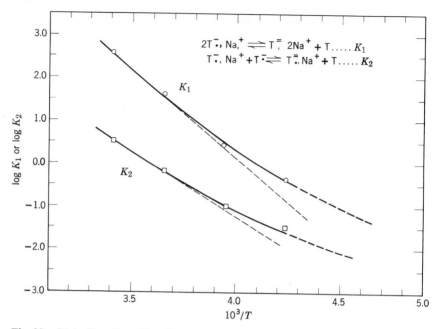

Fig. 20.   Plot of log K_1 and log K_2 versus $1/T$ for the disproportionation of radical-ions
of tetraphenylethylene.

TABLE XVII

Computation of K_1 and K_2

$$T^{\overline{\cdot}},Na^+ + T^{\overline{\cdot}},Na^+ \rightleftarrows T^=,2Na^+ + T \qquad K_1$$
$$T^{\overline{\cdot}},Na^+ + T^{\overline{\cdot}} \rightleftarrows T^=,Na^+ + T \qquad K_2$$

Temp., °C	K_1	K_2
20	400	3.3
0	39	0.72
−20	2.6	0.10
−37	0.44	0.038

At 20°C, $\Delta H_1 = 19 \pm 2$ kcal/mole, $\Delta S_1 = 75$ eu.
$\Delta H_2 = 13 \pm 2$ kcal/mole, $\Delta S_2 = 45$ eu.

more strongly than one $T^=,2Na^+$ (quadrupole), and a similar situation is
established in disproportionation 2.

The exothermicity of solvation of $T^{\overline{\cdot}},Na^+$ (solvent-separated ion-pair) is
indeed substantial. Consequently, disproportionation 1 is *more* endothermic
than disproportionation 2 (disproportionation causes desolvation), i.e.,

$\Delta H_1 > \Delta H_2$. In fact, this experimental result indicates that solvation by tetrahydrofuran of a $T^{\bar{\cdot}},Na^+$ ion-pair is even stronger than that of the free negative $T^{\bar{\cdot}}$ ion.

We may now compare our present interpretation of the disproportionation with that outlined at the beginning of this section. The early treatment was concerned solely with the solvation of anions. It appears, however, that the solvation of *cations* is more important, at least for some ion-pairs. The early approach stressed the contribution of the solvation energy (of anions) to the driving force of the disproportionation. Our present position is that the driving force arises from the gain in entropy of disproportionation caused by the *desolvation* of solvent-separated ion-pairs. On the basis of the early approach one could expect *more* disproportionation in a more powerfully solvating medium (implicitly in a better solvent for anions—a point not always appreciated). The findings of Evans (111) and of Garst (112,113) demonstrate the contrary—disproportionation is hindered in tetrahydrofuran, and even more so in dimethoxyethane or glyme, i.e., in media which powerfully solvate the cations. The process is favored, however, in ether or dioxane, i.e., in a solvent of low dielectric constant which interacts poorly with cations. All these observations are now rationalized.

A most peculiar situation is encountered in hexamethylphosphoramide (268,280). In this solvent the dissociation of $T^{\bar{\cdot}}$, Na^+ into free ions is virtually completed, whereas the dissociation $T^{=},Na^+$ into $T^{=} + Na^+$ is negligible. Hence, equilibrium (2)

$$T^{\bar{\cdot}} + T^{\bar{\cdot}},Na^+ \rightleftharpoons T^{=},Na^+ + T \qquad K_2,$$

determines the extent of disproportionation, i.e.,

$$[T^{=},Na^+][T]/[T^{\bar{\cdot}}][T^{\bar{\cdot}},Na^+] = K_2.$$

However, because $[T^{\bar{\cdot}},Na^+]$ is negligibly small so must be $[T^{=},Na^+]$, and therefore no detectable disproportionation takes place in this solvent.

The formation of $T^{\bar{\cdot}},M^+$ is favored by strong solvation of M^+ ions if $T^{\bar{\cdot}},M^+$ is a solvent-separated pair. Such an interaction is lost in the tight $T^{=},2M^+$ ion-agglomerate. Therefore, the concentration of $T^{\bar{\cdot}},M^+$ radical-ions should be high for $M^+ = Li^+$ and negligible for the salts of cesium. This again is confirmed by experiments.

Finally, let us ask why $T^{=},2Na^+$ forms contact ion-pairs (at least at not too low temperatures) while $T^{\bar{\cdot}},Na^+$ forms solvent-separated pairs. It seems that the geometry of the $T^{=}$ ion is different from that of the $T^{\bar{\cdot}}$ radical-ion. The latter resembles the parent hydrocarbon, tetraphenylethylene. Steric hindrance caused by placing phenyl groups in the *cis*-position prevents the $=CPh_2$ groups from attaining a coplanar conformation (181); the shape of the molecule is shown by Figure 21a. The same factors are responsible for the lack of coplanarity of the radical-ion $T^{\overline{\cdot}}$, and consequently the counterion cannot approach closely the negative charge, which resides mainly around the

$$\text{C}{=}\text{C}$$ bond.

The $\text{C}{=}\text{C}$ double bond is destroyed when the dianion, $T^{=}$, is formed. Consequently, the rotation around the $\text{C}{-}\text{C}$ axis becomes free and the dianion then attains the skewed conformation shown in Figure 21*b*. This conformation is favored by resonance, by the reduced repulsion between the extra electrons, and by the reduced steric hindrance, and it permits a close approach of the counterions to the coplanar negative centers of the

groups. It is obvious that the expulsion of the solvation shell surrounding a free cation, which takes place in the course of formation of a contact pair, is possible if the system gains a sufficiently large energy through the cation–anion interaction. Let us denote the thickness of the shell by Δx. The gain in Coulombic energy is given by the approximate expression

$$\text{const } \{(r_1 + r_2)^{-1} - (r_1 + r_2 + \Delta x)^{-1}\},$$

where $r_1 + r_2$ is the distance separating the ions of a contact pair. For a constant Δx the gain in energy increases as the distance $r_1 + r_2$ decreases. Hence, the approach to a bulky $T^{\overline{\cdot}}$ radical-ion ($r_1 + r_2$ large) does not provide a sufficient gain to favor the expulsion of solvent from around the cation. Consequently, the resulting product forms predominantly solvent-separated $T^{\overline{\cdot}},\text{Na}^+$ ion-pairs. The gain is large, however, in the pairing with $T^{=}$, because the cation may approach closely the planar fragments of the anion. Thus, $T^{=},2\text{Na}^+$ is a contact agglomerate.

The different behavior of contact and solvent-separated pairs accounts for the apparent anomaly stressed by Zabolotny and Garst (228). They noted that the degree of disproportionation of radical-ions derived from stilbene *increases* as the size of the cation decreases, i.e., the respective equilibrium constant for the lithium salt is greater than that for the potassium. This trend is the reverse of that observed in tetraphenylethylene systems. It is probable that the radical-anion of stilbene, as well as its dianion, forms a contact-pair even with Li^+ as the counterion, whereas a solvent-separated pair is formed when Li^+ or Na^+ becomes associated with the radical-anion of tetraphenylethylene. Thus, the gain arising from solvation contributes to the stability of the radical-anion in the latter system but not in the former.

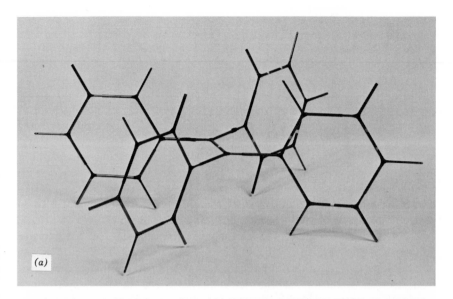

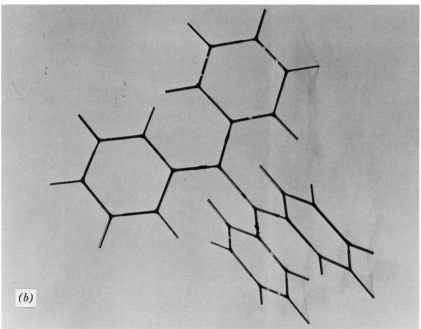

Fig. 21. Model of tetraphenylethylene (a) and of the dianion of tetraphenylethylene (b).

The above discussion clarifies the mechanisms responsible for the trends in the disproportionation constants caused by temperature, solvent, and nature of the cation. At ambient temperatures the disproportionation usually favors radical-anions if an excess of parent hydrocarbon is available. Tetraphenylethylene is an exception. The stability of $T^=,2Na^+$ results from its geometry which is different than that of the radical-ion. This was discussed in the preceding paragraphs; a variety of factors then contribute to the large shift of the pertinent equilibrium to the right. The concept of a different geometry of radical-ion and dianion is supported by the interesting findings of Garst et al. (210). They noted that the sodium salt of α-methylstilbene radical-anion disproportionates in 2-methyltetrahydrofuran at 25°C to a much greater extent than the salt of stilbene radical-anion, the equilibrium constants of disproportionation being $\geqslant 1000$ and 0.09, respectively. The extent of disproportionation of α-methylstilbene radical-anion is, under these conditions, comparable to that found for the tetraphenylethylene system ($K_1 \geqslant 1000$) and greater than that observed in the triphenylethylene system ($K_1 = 36$). As was pointed out by Garst, the alternative explanation invoking the reduced electron–electron repulsion in the tetraphenylethylene or triphenylethylene dianions, as compared with dianions of stilbene, cannot account for the behavior of α-methylstilbene radical-anions.

Further indication for the skewed geometry of the dianions of tri- and tetraphenylethylenes, and of some stilbene derivatives comes from studies of the triplets of stilbene. The geometry of such triplets may be similar to that of the corresponding dianions (210). Photochemical studies of Saltiel and Hammond (211) led them to postulate the existence of three triplets, cis, trans, and "phantom." It is probable that the last corresponds to the most stable skewed form.

The importance of geometric factors is seen in some other systems, e.g., cyclooctatetraene, which exists in a nonplanar tub conformation (D_{2d} symmetry group) and has a pronounced olefinic character. In tetrahydrofuran solution it reacts readily with alkali metal, giving a dianion that yields on hydrolysis cyclooctatriene and on carboxylation the corresponding acid (182). Potentiometric and polarographic studies (183) showed that the parent hydrocarbon is reversibly reduced in a two-electron process, i.e., the first reduction potential is *higher* than the second—a situation similar to that found for the tetraphenylethylene (34). ESR studies showed, nevertheless, that the respective radical-ion is present in the system, and its spectrum was analyzed (184). It consists of nine equally spaced lines having the intensities predicted for an octuplet. The hyperfine splitting constant was 3.21 gauss, the g value 2.0036. The spectrum was sharp when the reduction was performed in tetrahydrofuran with lithium, but broadened when potassium was used. Since the broadening was large for high conversion but insignificant for low conversion (low K/hydrocarbon ratio), it was concluded that the exchange is rapid between radical-ion and dianion, but slow between radical-ion and the parent hydrocarbon.

The study of C^{13} splitting (185), amounting to 1.28 gauss, demonstrated that the three valences radiating from each carbon atom of the ring have to be coplanar. This fact, in conjunction with other observations, several of which we shall discuss later, indicates that the radical-ion is planar and forms a regular octagon. The symmetry implies a charge density of 1/8 on each carbon atom, however the width of the spectrum Q is -25.7 gauss. According to the theory of π–σ interaction, originally developed by McConnell (186) and refined by Fraenkel et al. (187), the value of Q depends on the hybridization of the carbon atoms, and it should be -23.4 gauss for three equivalent sp_2 bonds (120° apart). This indeed was confirmed by examining the spectra of benzene radical-ion and CH_3 free radical. The same treatment leads, however, to the conclusion that $Q = -10$ gauss for a regular octagon (the bond angle 135°). Thus, the observed value is very different from that predicted. It was proposed (188) that this contradiction may be resolved by assuming "bent" bonds (189) still having 120° bond angles. Similar conclusions would apply to the cycloheptatrienyl radical, for which Q was found to be within 25.6 and 27.6 gauss (190,191).

The planar, regular octagonal conformation of the cyclooctatetraene radical-anion would require a Jahn-Teller distortion which should increase the g value over and above $g = 2.0023$ characteristic of a free electron. In addition, this effect should lead to a large line width and to a decrease in spin-lattice relaxation time (192a). This is not the case, and indeed re-examination of the problem (192b) indicates that, in the cyclooctatetraene radical-ion, in contradistinction to that derived from benzene, the Jahn–Teller effect may be insignificant.

The behavior of the cyclooctatetraene anion is reflected in the reactions of its dibenzo derivative. The parent hydrocarbon is again nonplanar and olefinic in character. Its radical-anion, formed by lithium reduction in tetrahydrofuran, was studied by Carrington et al. (229), who concluded from the analysis of its ESR spectrum that the odd electron is usually located on the two olefinic C=C bonds, the conjugation with the benzene rings being negligible (i.e., $\beta' \approx 0$). The spectrum was re-examined by Katz et al. (230), who produced this species by alkali-metal reduction (with lithium, sodium, or potassium) as well as by electrolytic reduction in dimethylformamide or dimethyl sulfoxide. Their analysis of the ESR spectrum shows that its structure may be accounted for by any value of β' from 0 to β, i.e., the conclusion $\beta' \approx 0$ apparently is not unique. Additional information about the nature of this species was obtained from polarographic studies. As in the cyclooctatetraene system, the exchange $A^{\overline{\cdot}} + A$ is slow, where $A^{\overline{\cdot}} + A^{=}$ is fast.

Still another system resembling cyclooactatetraene was investigated by Dauben and Rifi (231). They showed that tropylmethylether reacts with potassium in tetrahydrofuran, giving a deep blue anion $C_7H_7^-,K^+$, the existence of which had been questioned in the past. Simple Hückel molecular orbital treatment indicates that the system should have two degenerate antibonding orbitals occupied by one electron in the $C_7H_7\cdot$ radical and by two in the anion. Thus, energetically, the anion should be unstable. However, as predicted by the Jahn–Teller theorem, such a symmetrical cyclic π-system with a degenerate ground state should be stabilized by some perturbation which would remove its degeneracy. Indeed, a stabilization by 6 kcal/mole was anticipated by distortion of the bond length.

The blue species is diamagnetic; however, if the reduction is continued, a deep green radical-anion is formed (232). Its ESR spectrum was recorded at $-100°C$; it is an octet ($\alpha = 3.86$ gauss), further split into a heptet ($\alpha = 1.74$ gauss) by two magnetically equivalent Na^{23} ions. The latter splitting is excellent evidence for its dianion character.

This system therefore provides an interesting example of a diamagnetic monoanion and a paramagnetic dianion. The dianion should be subject to Jahn–Teller stabilization, but no evidence is available on this point.

Disproportionation of other radical-ions was less extensively studied. Radical-anions derived from aromatic ketones (ketyls) were discovered by Beckmann and Paul (227) and recognized as radicals by Schlenk and Weithel (28). An excellent review of the earlier literature of this subject may be found in the monograph by Hückel (212). Ketyls may perhaps dimerize to pinacone derivatives as well as disproportionate to dianions. The evidence for dimerization comes from synthetic studies, e.g., those of Bachman (213), who isolated quantitative amounts of the pinacone from products of hydrolysis of sodium benzophenone by dilute acetic acid. However, other preparative investigations led to different conclusions—for example, hydrolysis by water or alcohol yielded the monomeric alcohols (214).

Recently the equilibria established in the ketyl systems were investigated by physical methods, e.g., variation of optical density with concentration was studied by Warhurst and his students (215). The validity of Beer's law was established for a number of ketyls in dioxane in the concentration range 10^{-4} to $10^{-3}M$. It was concluded, therefore, that under these conditions association does not take place. (Alternatively this may indicate a complete association and negligible dissociation.) However, deviations from Beer's law were observed in ether (216). Extensive spectrophotometric and ESR studies of ketyl systems have been reported by Hirota and Weissman (217,218). The ESR spectra revealed the existence of two types of paramagnetic species, a monomeric and a dimeric, the latter being present, e.g., in dioxane solution of sodium benzophenone or sodium hexamethyl acetone (218). The monomeric species undergo rapid electron exchange with the respective ketones (217) and hence their ESR spectra collapse into four lines (due to Na^{23}) at sufficiently high concentration of ketone. Under these conditions the ESR spectrum of the dimer may be conveniently observed, since the electron transfer from dimer to ketone is extremely slow (see Table XX, p. 383).

The paramagnetic dimers are probably bonded through dipole–dipole interaction. Both sodium ions are magnetically identical, suggesting the following structure of the dimer:

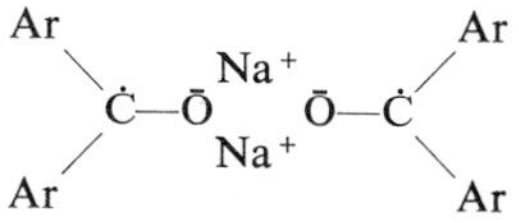

A closer examination of the ESR spectra of the Mg^{2+} salt permitted discrimination between several possible stereochemical structures of such dimers.

In addition to paramagnetic dimer, a diamagnetic dimer was observed. Equilibria between these two species were investigated by spectrophotometric techniques. The diagmagnetic species become abundant in nonpolar solvents and most probably these are the salts of pinacols. Hence, the equilibrium between a diamagnetic and paramagnetic dimer may be represented by the equation:

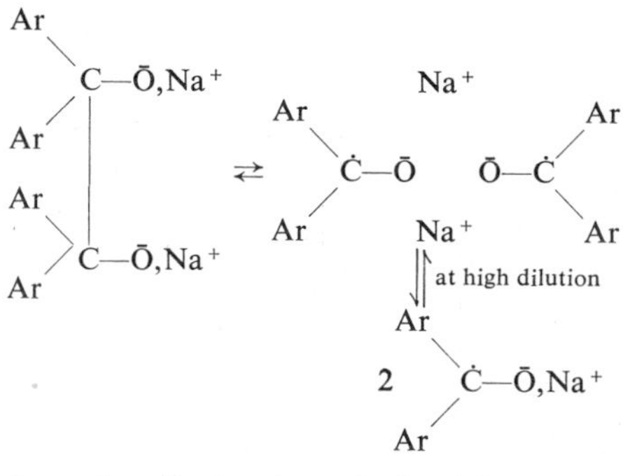

Disproportionation of radical-anions derived from diketones (benzil) was reported by Evans et al. (219). The equilibrium constant was determined to be about 0.6–0.7 for the sodium salt in tetrahydrofuran at 20°C. Apparently no dimerization takes place in this system.

18. DIMERIZATION OF RADICAL-IONS AND INITIATION OF ADDITION POLYMERIZATION

Radical anions derived from aromatic hydrocarbons (naphthalene, anthracene, etc.) do not dimerize. A substantial loss of resonance stability of the aromatic systems, which arises from the formation of a new covalent C—C bond, prevents such a process. Justification of this obvious conclusion, based on quantum-mechanics, was presented by McClelland (220). On the other hand, a rapid dimerization is expected when radical-ions are formed from vinyl or vinylidene monomers, and this step was proposed to account for the initiation of polymerization by electron transfer process (193).

When a molecule of a vinyl or vinylidene monomer acquires an extra electron from a suitable electron donor, the resulting radical-anion has only a fleeting existence. In solution it couples with another, similar radical-ion and forms a dimeric dicarbanion, or it may react with an ordinary molecule of the parent monomer and then form a dimeric radical-anion (193). These processes are illustrated by the following equations, styrene being used as an example:

$$2CH_2:CH(Ph)^{\overline{\cdot}} \rightleftarrows {}^-CH(Ph).CH_2.CH_2.CH(Ph)^- \qquad \text{dimeric dicarbanion,}$$
$$CH(Ph):CH_2^{\overline{\cdot}} + CH_2:CH(Ph) \rightleftarrows {}^-CH(Ph).CH_2.CH_2.CH(Ph)\cdot \qquad \text{dimeric radical-anion.}$$

Although the ionic species are denoted here by symbols representing the free ions, in most processes the reaction involves mainly ion-pairs.*

In the presence of an excess of monomer the dicarbanions initiate anionic polymerization, producing linear macromolecules endowed with two growing ends. The dimeric radical-anions may also initiate polymerization—anionic growth then ensues from the carbanion end, whereas the radical end initiates a radical-type propagation. The anionic growth is fast and usually proceeds without termination, giving rise to living polymers, whereas the radical propagation is rapidly terminated because free radicals destroy each other by dimerization or disproportionation or they aquire electrons from the initiating electron donors. Thus, the contribution of radical propagation to the overall polymerization is usually insignificant.

There are three distinct ways in which radical-anions derived from vinyl or vinylidene monomers may couple:

$$\overset{\ominus}{CXY}:CH_2 + CH_2:\overset{\ominus}{CXY} \rightleftarrows \overline{C}XY.CH_2.CH_2.\overline{C}XY$$

$$\overset{\ominus}{CXY}:CH_2 + \overset{\ominus}{CXY}:CH_2 \rightleftarrows \overline{C}XY.CH_2.CXY.\overline{C}H_2$$

$$CH_2:\overset{\ominus}{CXY} + \overset{\ominus}{CXY}:CH_2 \rightleftarrows \overline{C}H_2.CXY.CXY.\overline{C}H_2$$

The first mode of dimerization is greatly favored on thermodynamic grounds when X or Y (or both) are unsaturated groups, e.g., —COOR, —CN, —CH=CH₂, Ph, or some other aromatic or heteroaromatic moiety. At equi-

---

* Reaction of free ions may be studied in hexamethylphosphoramide because in this solvent the association of radical-ions with their counterions is negligible (220).

librium, if it is established, the first dimer should predominate to the virtual exclusion of the remaining two. One may question, however, whether the products of dimerization are determined by the equilibrium or by the kinetics of the coupling. Although a general answer to this problem is not known, it was found that in all cases in which dimers were isolated and identified only the thermodynamically stable species were produced. For example, the 1,4-dicarboxylic acids, derived from the first type of dimers, were quantitatively isolated as the only products of carboxylation of dicarbanions derived from styrene, α-methylstyrene (194), or 1,1-diphenylethylene (2).

The dimerization is, in principle, reversible. ESR study showed that the equilibrium concentration of radical-ions derived from vinyl or vinylidene monomers is vanishingly small and hence they cannot be detected by this technique. Apparently the equilibrium is shifted far to the right. For example, the concentration of 1,1-diphenylethylene radical-ions in equilibrium with the relevant dimeric dianions is less than $10^{-7} M$ when the dianions are at about $0.1 M$ concentration (195). Similar results were obtained for the dimeric dianions of α-methylstyrene.

Dimerization of radical-anions was studied only in a few systems. Matsuda, Jagur-Grodzinski, and Szwarc (198) followed the dimerization of 1,1-diphenylethylene radical-anions in tetrahydrofuran using the stop-flow technique. The reaction is extremely fast, the bimolecular rate constant being $1-2 \times 10^6$ liter mole^{-1} sec^{-1}. Even in hexamethyl phosphoramide, where the radical-ions are dissociated, the reaction is still very fast.* It was surprising, therefore, when Evans and Evans (247) reported that the dimerization in cyclohexane is slow. Their experiments were not confirmed by other workers because 1,1-diphenylethylene can *not* be reduced by metallic potassium in hydrocarbon solvents. Incidentally, the optical spectrum reported by Evans and Evans, and attributed to 1,1-diphenylethylene$^-$ is identical with that of Benzo-phenone$^-$ radical-anion.

A relatively slow dimerization is observed for radical-anions of tolane (PhC≡CPh). The bimolecular rate constant in tetrahydrofuran at 0°C was found to be 3–13 liter mole^{-1} sec^{-1} (256), depending on the nature of counterion. The results obtained in this writer's laboratory at 25°C lead to a rate constant of about 100 liter mole^{-1} sec^{-1}. However, in hexamethyl phosphoramide the dimerization does not take place at all (268); apparently the equilibrium is shifted far to the left when the radical-ions are not paired with their counterions.

The rate of dissociation of the dimeric dianions of α-methylstyrene into radical-ions was investigated by Asami and Szwarc (196). The dianions were prepared by quantitative reaction of α-methylstyrene with metallic potassium in tetrahydrofuran. An identical solution was then prepared using deuterated

---

* Recent studies (330) indicate that the free ions of $CH_2.CPh_2^-$ dimerize in hexamethyl phosphoramide with a rate constant of 2×10^4 M^{-1} sec.$^{-1}$.

α-methylstyrene, CD_2:$C(Ph)$.CD_3, instead of the ordinary monomer. Both solutions were stable. The dianions could be converted into the corresponding hydrocarbons by adding a drop of water, i.e.,

$$CH(Ph)(CH_3).CH_2.CH_2.CH(Ph)(CH_3) \text{ (mass 238)}$$

is then formed from the ordinary monomer and

$$CH(Ph)(CD_3).CD_2.CD_2.CH(Ph)(CD_3) \text{ (mass 248)}$$

is produced on protonation of the dianion resulting from the deuterated monomer.

To measure the rate of dissociation of the dimer, both solutions were mixed and the mixture was kept at constant temperature in a thermostat. Aliquots were withdrawn at various times; the dianions were protonated, isolated, and analyzed with a suitable mass spectrograph operating with 8 V acceleration potential. Under these conditions only the parent peaks appear in the mass spectrum. The aliquot withdrawn immediately after mixing the solution showed only the 238 and 248 peaks; no peak was observed at 243. However, after one day a peak appeared in the analyzed mixture having mass 243 and it grew more intense with time. This peak is due to the mixed dimer,

$$CH(CH_3)(Ph).CH_2.CD_2.CH(Ph)CD_3 \text{ (mass 243)}.$$

Its formation results from the following reactions:

$$^-\alpha MeS_H.\alpha MeS_H{}^- \underset{k_r}{\overset{k_{diss}}{\rightleftarrows}} 2\alpha MeS_H{}^{\overline{\cdot}}$$

$$+ \rightleftarrows {}^-\alpha MeS_H.\alpha MeS_D{}^-$$

$$^-\alpha MeS_D.\alpha MeS_D{}^- \underset{k_r'}{\overset{k_{diss}'}{\rightleftarrows}} 2\alpha MeS_D{}^{\overline{\cdot}}$$

where αMeS denotes α-methylstyrene and the subscripts H and D refer to the ordinary and deuterated compound. The isotope effect is probably negligible, i.e., $k_r = k_r'$ and $k_{diss} = k_{diss}'$. Therefore, for a 50:50 mixture the initial rate of formation of the mixed dimer is one-half the rate of dissociation. Thus, the rate constant of dissociation, $^-\alpha MeS.\alpha MeS^- \rightarrow 2\alpha MeS^{\overline{\cdot}}$ was found to be $6 \times 10^{-8} sec^{-1}$ at 25°C. Similar experiments were performed with the dimer of 1,1-diphenylethylene, but some experimental difficulties prevented reliable determination of the respective rate constant. It was shown, however, that its value is even lower than that found for the α-methylstyrene system.

Spach et al. (197) developed an alternative method of studying dissociation of the dimeric dicarbanions derived from 1,1-diphenylethylene. This hydrocarbon is a most reactive vinylidene "monomer" which, however, does not homopolymerize. The enormous steric strain, caused by the presence of two bulky phenyl groups on every second carbon atom of a hypothetical polymer, prevents the normal head-to-tail polymerization. Let us remark in passing that steric hindrance is larger when the bulky substituents are located on C atoms separated by another C than if they are attached to two adjacent atoms, a point easily appreciated from inspection of models.

The reasons outlined above explain why 1,1-diphenylethylene, D, does not react with its dimeric dianion, $^-DD^-$; viz., no reaction

$$^-DD^- + D \rightarrow {}^-DDD^-$$

is observed. However, if $^-DD^-$ is in equilibrium with its radical-ion, $D^{\overline{\cdot}}$, an exchange should be observed between $^-DD^-$ and D. Such an exchange was investigated in tetrahydrofuran (197)—radioactive D and nonradioactive $^-DD^-$, or vice versa, being used for labeling the reagents. Their separation, needed for counting, could be easily accomplished by carboxylation of the dianion. $^-DD^-$ yields, then, a substituted succinic acid which may be easily freed from the hydrocarbon. The following mechanism was envisaged:

$$^-DD^- \xrightarrow{k_1} 2D^{\overline{\cdot}} \qquad \text{slow, rate-determining step,} \qquad (1)$$

$$D^{\overline{\cdot}} + D^{14} \rightleftarrows D^{14\,\overline{\cdot}} + D \qquad \text{very fast,}$$

$$D^{14\,\overline{\cdot}} + D^{\overline{\cdot}} \rightarrow \text{radioactive } {}^-DD^- \qquad \text{again fast.} \qquad (2)$$

Hence, the rate of exchange, R_{ex}, should be first order in $^-DD^-$ and independent of D. However, experimental results showed a more complex kinetics, namely, in addition to the expected term $k_1[^-DD^-]$, another one, proportional to $[D][^-DD^-]^{1/2}$ appeared; i.e.,

$$R_{ex} = k_1[^-DD^-] + \text{const } [D][^-DD^-]^{1/2},$$

or $R_{ex}/[^-DD^-]$ was found to be a linear function of $[D]/[^-DD^-]^{1/2}$, as shown in Figure 22. Such a relation implies that the following route contributes to the exchange:

$$^-DD^- + D \rightleftarrows {}^-DD\cdot + D^{\overline{\cdot}} \qquad (3)$$

$$^-DD\cdot \underset{k_{-2}}{\overset{k_2}{\rightleftarrows}} D^{\overline{\cdot}} + D \qquad (4)$$

A rapid electron transfer between $D^{\overline{\cdot}}$ and D^{14} as well as between the radioactive $^-DD\cdot$ and the nonactive $^-DD^-$ leads then to the radioactive $^-DD^-$. Note that the radioactive $^-DD\cdot$ can be formed only by the combination of $D^{\overline{\cdot}}$ and D, either of which acquires radioactivity.

The formation of $^-DD\cdot$ [reaction (3)] *cannot* affect the equilibrium concentration of $D^{\overline{\cdot}}$, which is determined by dissociation (reaction 1). This process, however, provides an alternative route for the *production* and *consumption* of the radical-ions and therefore contributes to the rate of the ultimate exchange. Experimental results indicate that the formation of $^-DD\cdot$ from $^-DD^-$ and D is much faster than its decomposition into $D^{\overline{\cdot}} + D$ and, therefore, the rate of exchange arising from this route is governed by reaction (4). Since the system is in equilibrium, apart from the exchange, the investigated rate is given by

$$k_{-2}[D][D^{\dot{-}}] = k_{-2}K_1^{1/2}[D][^-DD^-]^{1/2},$$

i.e., the slope of the line shown in Figure 22 gives $k_{-2}K_1^{1/2}$.

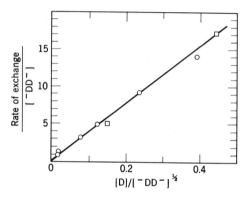

Fig. 22.   Rate of exchange between the dimeric dianion of 1,1-diphenylethylene ($^-DD^-$) and its radioactive monomer D (197); ($\bigcirc$) radioactive D; ($\square$) radioactive $^-DD^-$. The rate of exchange per $^-DD^-$ is plotted versus $[D]/[^-DD^-]^{1/2}$.

In the paper of Spach et al. (197), k_1 was reported to be 10^{-7} sec^{-1} at 25°C. However, it appears that the intercept is too small to be determined reliably, and the results should be interpreted by the inequality, $k_1 \leqslant 10^{-7}$ sec^{-1}. In fact, other evidence (195) suggests that $k_1 = 10^{-9}$ sec^{-1}. The rate constant of combination of $D^{\dot{-}}$ into $^-DD^-$ was investigated directly in a flow system (198) and its value seems to be 1–2×10^6 liter mole^{-1} sec^{-1}. Hence, the equilibrium constant, K_1, of dissociation of $^-DD^-$ into radical ions is about $10^{-15}M$. On this basis the value of k_{-2} was determined as about 800 liter mole^{-1} sec^{-1} (195).

Two comments are appropriate at this place. Judging from the results obtained in the 1,1-diphenylethylene system, the recombination of labile radical-ions, derived from monomers, is relatively slow when compared with recombination of ordinary radicals. The latter reaction is diffusion controlled, the respective rate constants being about 10^{10} liter mole^{-1} sec^{-1}. Apparently the delocalization of the odd electron and the need of rehybridization of the bonds is responsible for the inertia in radical-ion combination. The latter factor is manifested in the system, anthracene$^{\dot{-}}$ + styrene$^{\dot{-}}$. Recent study indicates that this mixed combination is even slower than combination of two $D^{\dot{-}}$, its rate constant being 10^4 liter mole^{-1} sec^{-1} (199). This is not surprising. The rehybridization of bonds in anthracene$^{\dot{-}}$, i.e., in a *stable* radical-ion, and the resulting loss of its resonance energy substantially increases the chemical inertia of combination. A similar value (195) for the combination constant was obtained for the system

$$\text{anthracene}^{\overline{\cdot}} + 1,1\text{-diphenylethylene}^{\overline{\cdot}} \rightleftarrows \text{}^{-}DA^{-}.$$

The above rate constants refer to the combination of ion-pairs. The combination of unpaired radical-ions is hindered by Coulombic repulsion and is often 100 or more times slower.

How much does a radical reaction contribute to polymerization initiated by radical-ions? The quantitative results obtained for the system

$$D^{\overline{\cdot}} + D^{\overline{\cdot}} \underset{\longleftarrow}{\overset{k_{-1}}{\longrightarrow}} {}^{-}DD^{-}$$

and

$$D^{\overline{\cdot}} + D \underset{\longleftarrow}{\overset{k_{-2}}{\longrightarrow}} {}^{-}DD\cdot$$

may serve as a starting point of our discussion. The rate constant k_{-2} appears to be about 800 liter mole^{-1} sec^{-1}, whereas $k_{-1} = 1\text{–}2 \times 10^6$ liter mole^{-1} sec^{-1}, i.e., the latter is at least 1000 times greater than the former. If a similar relation is valid for other monomer systems, the formation of dimeric radical-ions should be probable whenever the concentration of monomer exceeds that of its monomeric radical-ion by a factor of 1000 or more. The stationary concentration of radical-ions in polymerizing systems is difficult to assess, but its value must be lower at least by a factor of 1000 than that of the monomer if the degree of polymerization of the resulting product exceeds 1000. Hence, the dimeric radical-ions may play a role in systems yielding high molecular weight polymers if the polymerization is initiated by electron transfer reaction.

Participation of a radical-growth in polymerization reactions initiated by electron transfer might be revealed in a copolymerizing system composed of two monomers—one, say A, growing by anionic mechanism but unable to participate in a radical propagation, with the other, say B, being reactive in radical polymerizations but inert in anionic processes. If an electron transfer initiation yields in such a system a block polymer, viz.,

$$A.A\ldots A.B.B\ldots B, \text{ or } A.A\ldots A.B.B\ldots B.A.A\ldots A,$$

a clear manifestation of the participation of both the anionic and the radical ends in the growth process would be provided.

Such a copolymerizing system is not yet known. However, the system styrene–methyl methacrylate initiated by metallic lithium, investigated by Tobolsky and his students (200), could serve as a diagnostic tool in studies of these problems. A typical reaction mixture investigated by Tobolsky contained 20 ml of solvent (tetrahydrofuran or heptane) and 2 ml of each monomer. 50 mg of lithium dispersion was used to initiate the reaction, and the polymerization was quenched at the desired time by adding methanol. A reaction carried out at 20°C in the absence of solvent yielded a product containing up to 60% of styrene if the process was interrupted at 1% of conversion. As the polymerization proceeded, the proportion of styrene in the polymer rapidly decreased, being only 20% at 10% conversion. The resulting product had an intrinsic viscosity in benzene of about 0.2 at 25°C, indicating its high molecular weight nature. In tetrahydrofuran the reaction was much faster, and it was not feasible to stop the polymerization before 15% of the mixture had polymerized. The styrene content was less than 10%; its proportion remained constant up to about 40% conversion and then gradually increased. The reason for such an increase is obvious, as methyl methacrylate becomes consumed, styrene has to be incorporated into the product because lithium emulsion is still available for initiation.

Fractionation and extraction with suitable solvents indicated that the products did not form a mixture of homopolymers; neither were they composed of ordinary radical-type random copolymers. It was suggested, therefore, that the reaction yielded a block polymer (201). The electron transfer initiation produced monomeric radical-ions, and eventually dimeric radical-ions which initiated two types of growth. The anionic growth led to a block of homopolymethyl methacrylate, virtually free of styrene, whereas the chain grown from the radical end should form a block of 50:50 random copolymer of styrene and methyl methacrylate, as expected on the basis of the well-known study of Walling, Mayo et al. (202). In fact, under otherwise identical conditions, the polymerization initiated by butyllithium produced pure homopolymethyl methacrylate, indicating that the anionic polymerization with lithium counterion behaves conventionally. These findings were confirmed by independent studies of Pluymers and Smets (261).

Other results reported by Tobolsky et al. (200) seemed to fit this scheme well. In mixed solvents (tetrahydrofuran–heptane) the proportion of styrene in the polymer decreased with increasing concentration of tetrahydrofuran in the polymerizing medium. The ether accelerates anionic polymerization but it has a negligible effect on the radical growth. Hence, the observed result seemed to be accounted for. However, the intrinsic viscosity of the product was also decreased at higher ether concentration—a fact not considered by Tobolsky and contradicting his mechanism.

The scheme discussed by Tobolsky and elaborated mathematically in a paper with Hartley (203), as well as in an earlier paper by Stretch and Allen (204), assumes a conventional termination of radical-polymerization and lack of termination in anionic growth. Consequently, the higher the concentration of radical-ions the less significant is the contribution of the radical process. For example, the extremely rapid initiation by sodium naphthalene yields only polymethyl methacrylate (203). Strangely enough, polymerization initiated by metallic sodium yielded again only polymethyl methacrylate (205) and it was postulated therefore that the anionic growth or initiation is many times faster with sodium than with lithium counterion. This is questionable.

Some inconsistency becomes apparent when Tobolsky's data are carefully examined. The anionic growth is faster than the radical one, and therefore it should not be possible to incorporate more than 25% of styrene into the polymer if the hypothesis of the simultaneous anionic and radical propagation applies. Nevertheless, proportions of styrene as high as 60% were reported. A mathematical treatment developed by Tobolsky and Hartley (203) claimed to account for this difficulty. Denote by R_i the rate of simultaneous formation of the radical and anion ends through electron transfer and by $k_{p,r}$ and $k_{p,a}$ the respective rate constants of their propagation. The polymerizing system was assumed to be in a *stationary* state in respect to radical growth and, therefore, $R_i = k_t[R]^2$. Hence, the amount of the radical-grown polymer formed in time t is

$$P_r = k_{p,r}(R_i/k_t)^{1/2}M_0t$$

(we are interested in a low conversion, i.e., $[M] \sim ([M]_0)$). The amount of the polymer formed by the anionic propagation is

$$P_a = k_{p,a}\tfrac{1}{2}R_it^2M_0$$

because the anionic ends are assumed to be continually formed but not destroyed. Thus, the fraction of the radical-formed material,

$$P_r/(P_r + P_a) = \{1 + \tfrac{1}{2}(k_{p,a}k_t/k_{p,r}^2M_0)P_r\}^{-1}.$$

Its value therefore approaches unity when P_r and P_a tend to zero, i.e., the polymer produced in the very early stage of the process should result from radical growth only. Unfortunately, this conclusion is erroneous because the extrapolation to $P_r = 0$ invalidates the stationary-state assumption.

The attractive hypothesis of Tobolsky was challenged by other investigators. Overberger and Yamamoto (206), who repeated Tobolsky's experiments, examined the resulting products by the NMR technique. The high content of styrene in the initially formed material was confirmed but, contrary to expectation, the resulting polymers were shown to involve blocks of *homopolystyrene* instead of the predicted 50:50 random copolymer blocks of styrene and methyl methacrylate.

An alternative hypothesis was suggested therefore by Overberger (207,206). Styrene is assumed to be preferentially adsorbed on the surface of lithium particles and the initial polymerization is visualized to occur in the adsorbed layer. Therefore, a block of living homopolystyrene is produced initially, and only after this species is desorbed from lithium particles and enters the liquid phase does it react with methyl methacrylate. The latter reaction is irreversible because living polymers with methyl methacrylate end-groups cannot react with styrene (209). Hence, further polymerization produces blocks of homopolymethyl methacrylate. Such a model circumvents the objection raised by George and Tobolsky (208).

In conclusion, the participation of radical ends in polymerization initiated by the electron transfer process is not ruled out, but no convincing evidence for this phenomenon is as yet available. The behavior of the system styrene–methyl methacrylate initiated by lithium metal emulsion cannot be explained by such an hypothesis and therefore no evidence for simultaneous radical and anionic polymerization is provided by its study.

To complete our discussion of polymerization initiated by electron-transfer processes, we have to recall that some electron donors are capable of initiating polyaddition by a mechanism other than direct electron transfer to monomer. For example, sodium naphthalene initiates polymerization of ethylene oxide by the following sequence (271):

(I)

The proof of this mechanism is based on three observations:

(1) The presence of naphthalene moiety in the polymer;

(2) The appearance of transient red color at low temperature which is attributed to the ion (I);

(3) The conversion of only one-half the added sodium naphthalene into naphthalene.

This mechanism resembles the one proposed by Weissman et al. (41) to account for the carboxylation of naphthalene⁻ radical-ions. A similar mechanism was proposed for initiation of polymerization of cyclic siloxanes by sodium naphthalene (272).

Tobolsky and Hartley (273) tried to use the presence of initiator's moiety in the polymer as a diagnostic tool which would distinguish between the two possible modes of initiation by electron donors. However, such an observation, without further supporting evidence, is not sufficient to permit discrimination between the two alternatives (see appendix of reference 274).

We remarked at the beginning of this section that radical-anions derived from aromatic hydrocarbons do not dimerize—the formation of a new C—C bond is energetically unprofitable. The same reasons suggest that analogous radical-cations should not form dimers, and even less probable would be a dimerization in which a covalent bond is formed between a radical-ion and its parent aromatic hydrocarbon. Indeed, all these processes have not been observed. However, in the presence of an excess of aromatic hydrocarbon its radical-cation undergoes an interesting association. The ESR studies of such systems reveal the formation of new species (221,222). Their ESR spectra are similar to those of the respective radical-cations, but each line is doubled and the splitting is halved. It was concluded that association of A^+ and A (A being an aromatic hydrocarbon) yields a sandwich compound in which the two partners are arrayed in parallel, one on top the other, separated perhaps by a counterion and bonded by virtue of delocalization of the "positive hole." Obvious quantum-mechanical reasons make such an association improbable for the system $A^- + A$ and, indeed, this phenomenon was not observed in the anionic systems.

The quantum-mechanical treatment of bonding of A^+ and A is similar to that of charge-transfer complexes. Another type of dimer, resembling the A^+A, is found in photochemical studies when an excited molecule becomes associated with the nonexcited one (223). The formation of such an associate, known as an excimer (224), was manifested by a change in the fluorescence of the irradiated hydrocarbon, e.g., pyrene, the spectrum being affected by the concentration of the hydrocarbon. At low concentration the fluorescence of pyrene is violet and shows fine structure, but at concentrations above $10^{-4}M$ a broad, structureless band appears in the blue, its intensity being proportional to the square of concentration. The quantum-mechanical treatment of the excimer was discussed by Hoijtink (225), who represents its wave function by

$$E^- = A_1^* A_2 - A_1 A_2^*,$$

A being the ground state and A^* the second excited singlet wave function of pyrene. An interesting example of an intramolecular excimer, formed from [2.2]paracyclophane, was described by Ron and Schnepp (226).

Excimers are also formed in processes involving interaction between the negative, A^-, and positive A^+ radical-ions. This topic is discussed in Section 21.

Finally, dimerization is observed between dianions of aromatic hydrocarbons and vinyl monomers. For example:

Anthracene$^=$ + styrene $\longrightarrow$ [structure] $CH_2 \cdot CH(Ph)^-$;($^-AS^-$)

This reaction takes place directly (199) and not through the sequence of two processes:

$$\text{anthracene}^= + \text{styrene} \rightleftarrows \text{anthracene}^{\overline{\cdot}} + \text{styrene}^{\overline{\cdot}}$$
$$\text{anthracene}^{\overline{\cdot}} + \text{styrene}^{\overline{\cdot}} \rightarrow {}^-AS^-.$$

A similar dimer is formed directly when A$^=$ reacts with 1,1-diphenylethylene (D). However, the resulting product, $^-AD^-$, rapidly decomposes into A$^{\overline{\cdot}}$ + D$^{\overline{\cdot}}$ (195), i.e., at low concentrations the equilibrium favors the formation of the mixed dimer from A$^=$ and D but not from A$^{\overline{\cdot}}$ and D$^{\overline{\cdot}}$.

19. KINETICS OF ELECTRON TRANSFER REACTIONS BETWEEN RADICAL-IONS AND THEIR PARENT MOLECULES

Electron exchanges between radical-ions and their parent molecules are perhaps the simplest electron transfer processes. Their study was pioneered again by Weissman (233,234). The bimolecular rate constant of the reaction

$$\text{naphthalene}^{\overline{\cdot}}, M^+ + \text{naphthalene} \rightleftarrows \text{naphthalene} + \text{naphthalene}^{\overline{\cdot}}, M^+$$

was determined from the broadening of the ESR lines of naphthalene$^{\overline{\cdot}}$. The results are collected in Table XVIII and show the influence of counterions and solvents on rate of transfer. The rate was found to be especially high for the

TABLE XVIII

Influence of Counterions and Solvents on Rate of Transfer at 25°C (233)

$$\text{Naphthalene}^{\overline{\cdot}}, M^+ + \text{naphthalene} \xrightarrow{k} \text{naphthalene} + \text{naphthalene}^{\overline{\cdot}}, M^+$$

Counterion	Solvent[a]	k, liter mole^{-1} sec^{-1}
K$^+$	DME	$(7.6 \pm 3) \times 10^7$
K$^+$	THF	$(5.7 \pm 1) \times 10^7$
Na$^+$	DME	$\sim 10^9$
Na$^+$	THF	$\sim 10^7$
Li$^+$	THF	$(4.6 \pm 3) \times 10^8$

[a] DME = dimethoxyethane; THF = tetrahydrofuran.

sodium salt in dimethoxyethane (DME) and, to a lesser extent, for the lithium salt in tetrahydrofuran (THF). These observations have been accounted for by assuming participation of free naphthalene$\bar{\cdot}$ radical-ions in these two reactions, whereas ion-pairs were involved in the other processes. However, recent conductance studies (38,39) indicate that the sodium salt in DME and the lithium salt in THF form solvent-separated ion-pairs, the other solutions being composed of contact pairs. Hence, the results might be interpreted as evidence for rapid electron transfer from solvent-separated pairs which seems to be about 20 times as reactive as the contact pairs. Both exchanges proceed with low activation energy, probably not more than 2–3 kcal/mole. Recent evidence, confirming the participation of two (or more) ion-pairs in electron-transfer processes, is provided by Hirota (70). See also Section 21 of Chapter V.

The sodium salt in DME produces sharp ESR lines; the splitting caused by Na^{23} being negligible because the cation is relatively far away from the radical-ion. Addition of naphthalene broadens the signal (the consequence of exchange), but the lines sharpen again on addition of potassium iodide. Apparently the counterions exchange, i.e.,

$$N^{\bar{\cdot}},Na^+ + KI \rightleftarrows N^{\bar{\cdot}},K^+ + NaI$$

and electron transfer from the potassium salt is too slow to cause an appreciable broadening of the lines.

The technique described above was refined by Zandstra and Weissman (238). Subsequent studies of Atherton and Weissman (239) demonstrated that ESR spectroscopy enables one to discriminate between sodium naphthalene ion-pairs and the free naphthalene$\bar{\cdot}$ radical-ions (or perhaps between the contact and solvent-separated pairs of sodium naphthalene; see ref. 70). If both are present, the ESR spectrum of the investigated solution contains two sets of lines, the intensity of each being proportional to the concentration of the respective species.

On addition of naphthalene the lines of each set became broader (238), the broadening being again proportional to the concentration of naphthalene and independent of the concentration of naphthalene$\bar{\cdot}$. The degree of broadening was, however, different for each set. Using the method described in detail by Weissman (235), it was possible to determine from the broadening the rate constants of electron transfer from each species. Fortunately, the dissociation of pairs into free ions (or transformation of contact pairs into solvent-separated pairs, if we accept the alternative interpretation of the ESR signals), was found to be sufficiently slow in these solvents not to cause any perturbation of the respective spectra.

The pertinent results are summarized in Table XIX. The following points deserve comments. Rate constants of exchange are substantially lower in THP and MeTHF than in THF, and the respective "activation energies" are higher. However, reactivities of the free ions are not much different from those of ion-pairs, the THF system being an exception. In this solvent the activation energy for the free naphthalene$^{\bar{}}$ ion was reported to be much higher than that for the ion-pair (13 kcal/mole as compared with 5 kcal/mole)

TABLE XIX

Electron Transfer between Naphthalene$^{\bar{}}$ Ion-Pair (Na$^+$), or Its Free Ion and Naphthalene (238)

Solvent	Species	Temp., °C	$k \times 10^{-6}$, liter mole^{-1} sec^{-1}	E, kcal/mole
Tetrahydropyrane (THP)	Ion-pair	21	4.4	18
	Free ion	−23	9.7	19
2-Methyltetrahydrofuran (MeTHF)	Ion-pair	21	4.5	12
	Free ion	−23	2.8	13
Tetrahydrofuran (THF)	Ion-pair	50	34	5
	Free ion	13	38	13

Most of these activation energies seem to be too high.

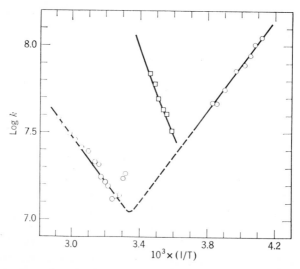

Fig. 23. Plot of exchange constant N$^{\bar{}}$ + N $\rightleftarrows$ N + N$^{\bar{}}$ (N = naphthalene) versus $1/T$ (238). The exchange is measured by ESR broadening. Circles refer to ion-pairs, squares to free ions. Units liter mole^{-1} sec^{-1}.

—a strange observation. In fact, the behavior of the sodium ion-pair in THF is most bizarre. As may be seen from Figure 23, the rate constant decreases with temperature but then increases again at lower temperatures. Several suggestions were made to account for such a temperature dependence, e.g., association of ion-pairs with naphthalene which could become appreciable at lower temperatures. Although this explanation was not rejected, it was not considered fully satisfactory.

Alternatively, one may visualize several vibrational subspecies, each contributing to the process, and postulate the highest rate constant for the lowest state. Denoting by k_0, k_1, etc. the *temperature-independent* rate constants of each subspecies, by K_1, K_2, etc. the ratios of concentrations [subspecies$_i$]/[subspecies$_0$], and finally by g_i and E_i the respective statistical weights and the separation energies, we find the observed k to be given by:

$$-k = (k_0 + \sum_i k_i K_i)/(1 + \sum_j K_j),$$

with $K_i = g_i \exp(-E_i/RT)$. Obviously, k *increases* with *decreasing* temperature if $k_0 > k_i$, but the function $k(T)$ has no minimum:

$$-dk/dT = \{\sum_i (k_0 - k_i)E_i K_i\}/RT^2(1 + \sum_j K_j)^2 > 0$$

Hence, this treatment, presented in reference 238 for 2 subspecies only, can account for the observations only if k_i is also temperature dependent. The change seen in Figure 23 is much too abrupt for a physical phenomenon of this type. It seems, therefore, that the problem needs additional investigation.

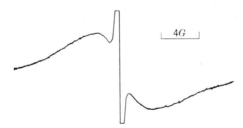

Fig. 24. ESR first derivative spectrum obtained when naphthalene (1.4M) was reduced with potassium in tetrahydrofuran (240). Only a small fraction of naphthalene is reduced and the presence of a large excess of naphthalene leads to a rapid exchange. The clipped center peak is due to the free ion and it indicates an off-scale reading.

Further information on this system comes from the interesting work of Chang and Johnson (240), who utilized the as yet little exploited "fast-exchange limit" technique. In the presence of a large excess of naphthalene the spectrum collapses into one very sharp line (for the free ion reaction) or into four lines (for the exchange involving sodium ion-pairs).* The collapsed spectrum is shown in Figure 24. The line is rigorously Lorentzian, thus

* This phenomenon was reported first by Adams and Weissman (234) who studied the exchange between sodium benzophenone radical-ion and benzophenone. It proves that the sodium ion retains its identity during exchanges.

confirming the fast-exchange limit. The second order rate constant of exchange can then be calculated from the equation derived by Piette and Anderson (241):

$$k = 2.04 \times 10^7 (\nabla / \Delta H [A]).$$

Here ∇ is the second moment, in gauss, of the spectrum of the investigated species in the absence of exchange, ΔH is the line width of the collapsed spectrum (separation between the extremes in the first derivative spectrum) corrected for modulation, natural line width, etc., and [A] is the concentration of the hydrocarbon. The method is useful if the two investigated species react with sufficiently different rates (241). Only one line was observed at high naphthalene concentration for N^{τ}, Na^+ and N^{τ}, K^+ in DME at temperatures ranging from -30 to $40°C$. However, two lines of appreciably different width were found for N^{τ}, K^+ in THF (see Figure 24) and a single line superimposed upon a broad uncollapsed spectrum was observed for N^{τ}, Na^+ in THF.

The faster rate was assigned to free ions. The smaller rate constants, attributed to ion-pairs, agreed well with those reported by Weissman (238) from the results of "slow-exchange limit." The major difference between Weissman's and Johnson's findings is found for the Na^+–THF system. The free ion line in the collapsed spectrum is prominent because of its narrowness.

Studies of exchange between N^{τ} and N were followed by similar investigations of electron transfer between the ketyl of benzophenone and its parent ketone (234). The ESR spectrum of the sodium ketyl in DME consists of more than 80 lines and it was conclusively shown that a splitting by the nuclear moment of Na^{23} does take place. The magnitude of this splitting is about 1 gauss, indicating that each molecule of ketyl retains its sodium counterion for at least 3×10^{-7} sec.

On addition of ketone, an electron transfer takes place. The question was raised whether the transfer takes place to benzophenone and then the newly formed free radical-ion becomes paired with *any* free cation, or whether the electron transfer is coupled with the transfer of Na^+ ion. These two alternatives may be represented schematically by:

$$Na^+, OC\Phi_2^{\tau} + \Phi_2 CO \rightarrow Na^+ + OC\Phi_2 + \Phi_2 CO^{\tau}, *$$
$$\text{any free } Na^+ + \Phi_2 CO^{\tau} \rightarrow \Phi_2 CO^{\tau}, Na^+ \tag{a}$$

and

$$\Phi_2 CO^{\tau}, Na^+ + OC\Phi_2 \rightleftarrows \Phi_2 CO + Na^+, OC\Phi_2^{\tau}. \tag{b}$$

For a rapid transfer (large concentration of benzophenone), the ESR spectrum should collapse into *one* line if the process takes place through route (a). On the other hand, four lines, each corresponding to one possible orientation of the nuclear spin, should be formed if the transfer proceeds by

---

* To comply with the principle of microscopic reversibility one has to assume that the reverse of the first step is slower than the rate of the second step.

mechanism (b). The latter spectrum was observed, indicating that we deal here with atom transfer rather than electron transfer. It is significant that the splitting due to Na^{23} is larger (1.1 gauss) in the collapsed spectrum than in the isolated ketyl (1.0 gauss). Hence, the electron density on the sodium nucleus *increases* in the course of the transfer process.

The second order rate constant was determined to be about 5×10^7 liter $mole^{-1} sec^{-1}$, i.e., for $2M$ concentration of benzophenone the mean life time of each ketyl molecule is less than 10^{-8} sec, whereas a sodium ion retains its association with radical-ion for more than 3×10^{-7} sec.

TABLE XX

Rate Constants, k, and Activation Energies, E, for the Electron Transfer Reaction: ketyl + ketone → exchange

Ketyl	Solvent[a]	Temp., °C	$k \times 10^{-8}$, liter mole^{-1} sec^{-1}	E, kcal/mole
Na$^+$, xanthone$^{\bar{}}$	DME	37	4.6	5.1
	THF	25	4.5	4.3
	THP	7	4.7	—
	MeTHF	0	4.3	—
Rb$^+$, xanthone$^{\bar{}}$	DME	25	2.5	—
Na$^+$, benzophenone$^{\bar{}}$	DME	25	1.1	6.3
	THF	12	1.1	6
K$^+$, benzophenone$^{\bar{}}$	THF	25	2.5	—
Rb$^+$, benzophenone$^{\bar{}}$	DME	25	1.6	—
Mg^{2+} (benzophenone$^{\bar{}}$)$_2$	DME	24	$<10^{-2}$	—
Ca^{2+} (benzophenone$^{\bar{}}$)$_2$	DME	24	$<10^{-2}$	—
(Na$^+$, benzophenone$^{\bar{}}$)$_2$	MeTHF	24	$<10^{-2}$	—
	Dioxane	24	$<10^{-2}$	—
Free xanthone$^{\bar{}}$ ion	Acetonitrile	25	10	—

[a] DME = dimethoxyethane; THF = tetrahydrofuran; THP = tetrahydropyrane; MeTHF = 2-methyltetrahydrofuran.

More comprehensive kinetic studies of electron exchange between ketyls and ketones were reported by Hirota and Weissman (217). Their results, summarized in Table XX, show that the transfer is faster in tetrahydrofuran than in tetrahydropyrane, and even faster in dimethoxyethane. The transfer is extremely slow for the dimers. Apparently the transition state has a sandwich-type structure, as shown by equation (b), and for steric reasons such a structure is unatainable for a dimer. It may be that the electron transfer requires dissociation of the dimer and the dissociation process is then the rate-determining step.

The activation energies are relatively high, i.e., the A factors are of the order 10^{12} to 10^{13} liter mole^{-1} sec^{-1}. Apparently some desolvation of the ion-pair takes place in the transition state.

Electron exchange processes between parent molecules and their radical-anions formed electrolytically in dimethylformamide were investigated by Adams and his co-workers (282). Under these conditions the radical-ions are *not* associated with counterions. The results obtained for the systems anthracene, *p*-benzoquinone, duroquinone, and 1,4-naphthaquinone led to the following bimolecular rate constants of exchange: 5×10^8, 4×10^8, 0.6×10^8, and 4×10^8, respectively (in units of M^{-1} sec^{-1}). These studies were extended to substituted nitrobenzenes and led to similar results. It was concluded that the exchange occurs via molecular orbital overlap, i.e., an electron is not transferred over long distance in the solution. This statement is consistent with the current opinion on processes involving radical-ions not associated with counterions.

The exchange of an electron between potassium 2,2'-bipyridyl radical-anion and its parent molecule was investigated by Reynold (252). His intention was to determine whether "complex" formation takes place during the electron transfer when the investigated species is a potent complexing agent. No peculiarity was observed in this system. The bimolecular rate constant of exchange was found to be $\sim 4 \times 10^6$ liter mole^{-1} sec^{-1} at 20°C, the activation energy of transfer 10 kcal/mole, and the entropy of activation about $+5$ eu.

The exchange between a radical and its carbanion involving an electron transfer process was investigated by Jones and Weissman (236),

$$(p\text{-}NO_2.C_6H_4)_3C\cdot + Alk^+,\overline{C}(C_6H_4\text{-}p\text{-}NO_2)_3 \rightarrow \text{exchange}.$$

Again, broadening of the ESR lines of the neutral radical permitted determination of the respective bimolecular rate constants. Their values were in the range of 10^8 to 10^9 liter mole^{-1} sec^{-1} at 25°C, as in other systems, but the activation energies were only of the order 1–2 kcal/mole. The rate constants were affected by the solvent and counterion, indicating participation of ion-pairs. The exchange was found to be 2–4 times faster in acetonitrile than in ethereal solvents and the free ions appears to be only slightly more reactive than ion-pairs.

A very interesting problem was posed by Zandstra and Weissman (237). Consider a system composed of radical-ions and its parent molecules in which electron transfer takes place. Denote by t the average time spent by the odd electron on a molecule. The molecules exist in N nuclear-spin states, of which g_k states correspond to the same energy level and give rise to the same line k in the ESR spectrum. Electron transfer leads, therefore, to broadening of this line if the transfer takes place from a molecule belonging to group g_k to another that does not belong to this group. Hence, the characteristic time determining the broadening is given by t_k where

$$1/t_k = (1/t)(N - g_k)/N.$$

Obviously, t_k is always greater than t and it becomes infinity if there is only one spin state (in such a case the molecular magnetic field within which the electron moves is not changed by the electron transfer).

The above equation is based on the assumption that the probability of transfer is independent of the spin state of the acceptor. Is this assumption valid?

The treatment given above indicates that the addition of parent molecule broadens different lines in the ESR spectrum to a different extent, because g_k is, on the whole, different for different lines. Moreover, this approach permits one to calculate the broadening ratios of, say, line k and k'. The result of calculation may be verified experimentally and the agreement would then confirm the basic assumption of constant probability of transfer. Studies of Zandstra and Weissman (237) verified this relation for several lines of the ESR spectrum of potassium naphthalene in the presence of naphthalene (237) and showed, therefore, that the probability of electron transfer is indeed independent of the spin state of the acceptor.

Although the ESR technique is versatile and conveniently adaptable for electron-transfer studies, other methods should not be forgotten. For example, a recent investigation of Doran and Waack (270) of *cis-trans* isomerization induced by radical-anion formation provides an interesting method for studying the kinetics of some electron transfer processes. These workers demonstrated that *cis*-stilbene is isomerized into the *trans*-compound when a small fraction of olefin is converted into stilbene $\bar{\cdot}$. The reaction was followed spectrophotometrically and it was shown that it does not take place in the absence of radical-ion. Presumably, the process is governed by the following reaction:

$$cis\text{-stilbene} + \text{stilbene}\,\bar{\cdot} \rightleftarrows \text{stilbene}\,\bar{\cdot} + trans\text{-stilbene},$$

and therefore the rate of the electron transfer determines the rate of isomerization. It was not elucidated whether, and to what extent if any, the dianion participates in the reaction. A similar behavior was observed in the 1,2,3,4-tetraphenylbutadiene system in which three geometric isomers exist.

Electron transfer reactions of the type $A\bar{\cdot} + B \rightarrow B\bar{\cdot} + A$ were also investigated. The ESR technique was employed (242), but only qualitative results were obtained. These reactions were successfully investigated by means of pulse radiolysis (243). The following principle is applied. Solvated electrons, formed by a pulse in isopropyl alcohol, are scavenged by aromatic hydrocarbons. Since the rate of electron capture is nearly independent of the electron affinity of the scavenger (see p. 338), the hydrocarbon A, which is present in large excess, initially captures all the electrons. The scavenging is completed in about 0.5 μsec. The system also contains another hydrocarbon, B, of high electron affinity but present at low concentration. Thus, the reaction

$$A^{-} + B \xrightarrow{k_{AB}} B^{-} + A$$

follows the initial electron capture, and its progress is monitored by the usual spectrophotometric method. Biphenyl and p-, m-, and o-terphenyls were used as the scavengers that produced the respective electron-donors, whereas naphthalene, phenanthrene, p-terphenyl, pyrene, and anthracene were the electron-acceptors. The kinetics is slightly complicated by protonation, and hence

$$-d[A^{-}]/dt = k_{AB}[A^{-}][B] + k_{2A}[A^{-}][PrOH] + k_{3A}[A^{-}][PrOH_{2}^{+}]$$

and

$$d[B^{-}]/dt = k_{AB}[A^{-}][B] - k_{2B}[B^{-}][PrOH].$$

The rate constants k_{2A} and k_{2B} refer to the protonation of A^{-} and B^{-}, respectively, by isopropyl alcohol, and these were determined in an independent study (see p. 346); k_{3A} is the rate constant of protonation by the $PrOH_{2}^{+}$ acid but the term $k_{3A}[A^{-}][PrOH_{2}^{+}]$ turned out to be very small. Solution of these differential equations led to the results given in Table XXI.

TABLE XXI

Electron Transfer Rate Constants for Aromatic
Molecules in Isopropanol at 25°C (243)

Donor anion radical	Acceptor molecule	Absolute rate constant, liter mole^{-1} sec^{-1} × 10^{-9}
Diphenylide	Naphthalene	0.26 ± 0.08
	Phenanthrene	0.6 ± 0.3
	p-Terphenyl	3.2 ± 0.7
	Pyrene	5.0 ± 1.8
	Anthracene	6.4 ± 2.0
p-Terphenylide	Pyrene	3.6 ± 1.1
	Anthracene	5.5 ± 0.9
m-Terphenylide	Pyrene	3.5 ± 1.2
o-Terphenylide	Pyrene	4.0 ± 1.8

It is obvious that the investigated processes involve free radical-ions not paired with counterions. In this system $PrOH_{2}^{+}$ form the positive counterions, and their interaction with radical-anions leads to protonation and not to ion-pair formation. It is also interesting to note that these reactions, although very fast and favorable energetically, are not always diffusion-controlled. This is certainly the case for the slower process (e.g., diphenylide with naphthalene or phenanthrene), i.e., in those reactions not every encounter leads to electron-transfer.

Let us compare the electron transfer reactions

$$A^{\overline{\cdot}} + A \rightleftarrows A + A^{\overline{\cdot}}$$

$$B^{\overline{\cdot}} + B \rightleftarrows B + B^{\overline{\cdot}}$$

and

$$A^{\overline{\cdot}} + B \rightleftarrows A + B^{\overline{\cdot}},$$

and denote the respective rate constants by k_{AA}, k_{BB}, and k_{AB} and the equilibrium constant of the last reaction by K_{AB}. Marcus has shown (295) that

$$k_{AB} = (k_{AA} \cdot k_{BB} \cdot K_{AB} \cdot f)^{1/2},$$

where f is a unique function of K_{AB}, such that $f \equiv 1$ when $K_{AB} = 1$.

In an ingenious experiment Bruning and Weissman (294) tested this theory. They investigated, by ESR technique, the exchange between the optically pure radical-ion (α-naphthyl)$\cdot$(Ph)$\cdot$CH$\cdot$CH$_3$ and its parent hydrocarbon, as well as the exchange in the racemic mixture. The first system provides the rate constant $k_{dd} = k_{ll}$, i.e.,

$$A(d)^{\overline{\cdot}} + A(d) \rightleftarrows A(d) + A(d)^{\overline{\cdot}},$$

whereas from the rate constant of exchange in the racemic system, one may calculate the rate constant k_{dl},

$$A(d)^{\overline{\cdot}} + A(l) \rightleftarrows A(d) + A(l)^{\overline{\cdot}}.$$

As may be verified, $k_{dd} = k_{ll} = k_{dl}$ if Marcus' equation applies. The experimental results obtained for the potassium salts at 25°C were as follows: in DME $k_{dd} = (0.68 \pm 0.07) \times 10^8 \ M^{-1} \sec^{-1}$; $k_{dl} = (1.1 \pm 0.3) \times 10^8 \ M^{-1} \sec^{-1}$; in THF $k_{dd} = (0.19 \pm 0.04) \times 10^8 \ M^{-1} \sec^{-1}$; $k_{dl} = (0.2 \pm 0.1) \times 10^8 \ M^{-1} \sec^{-1}$. It seems, therefore, that Marcus' rule applies, although Weissman felt that a small, but distinct, deviation is observed in dimethoxyethane.

The kinetics of electron transfer to vinyl monomers was studied and it was shown that the rate of the overall reaction is determined by dimerization of the resulting monomeric radical-anion and not by the transfer (195,198, 244). For example, polymerization of styrene may be initiated by sodium naphthalene. The initial rate of disappearance of the radical-ion seems to be given by the equation

$$(-dN^{\overline{\cdot}}/dt)_0 = \text{const} \ (N^{\overline{\cdot}}/N)^2.$$

Hence, the equilibrium $N^{\overline{\cdot}} + $ styrene $\rightleftarrows N + $ styrene$^{\overline{\cdot}}$ is rapidly established and is followed then by a relatively slow dimerization of styrene$^{\overline{\cdot}}$ radical-ions.

Relatively slow electron transfer processes were reported by Evans et al. (245), who studied the reactions of sodium chrysene and sodium picene with 1,1,3,3-tetraphenylbutene-1. The reaction is claimed to be first order in the olefin and in the radical-ion, and the respective rate constants were found to be 0.013 liter mole^{-1} sec^{-1} for chrysene$^{\bar{}}$,Na$^+$ and 10^{-5} liter mole^{-1} sec^{-1} for picene$^{\bar{}}$,Na$^+$, at 40°C. It was shown previously (248) that butyllithium does not react with 1,1,3,3-tetraphenylbutene-1, because steric hindrance prevents the approach of the Bu$^-$ moiety to the C=C bond. The reaction with electron donors presumably involves the transfer of electrons through the phenyl groups of the olefin and, consequently, it becomes sterically possible in spite of the shielding. The reliability of these results is questionable. The rates are much too slow for genuine electron-transfer processes. Judging from the data given by Evans (266), the relevant equilibrium constants cannot be smaller than 0.01 and, hence, the endothermicity of the reaction cannot account for the results. It seems that the observed process is due to some other reaction, the nature of which was not recognized by the authors.

Donors derived from aromatic hydrocarbons of lower electron affinity, e.g., naphthalene or phenanthrene, react too fast to permit a kinetic study (a conventional mixing batch technique was used). Those derived from hydrocarbons of much higher electron affinity, e.g., pyrene or anthracene, do not react at all—the equilibrium being obviously unfavorable.

An analogous study was performed with tetraphenylethylene (246), i.e., with another hindered hydrocarbon which does not react with butyllithium. Nevertheless, the electron transfer reactions which did take place with tetraphenylethylene were too fast for conventional kinetic investigation. Also, the reverse reactions, such as the transfer from the dianion of tetraphenylethylene to anthracene, were found to be very fast indeed.

Finally, we should consider some intramolecular electron transfer processes. This problem was treated by McConnell (249), who discussed radical-ions having the general structure

$$\left\{ \bigcirc\!\!-(CH_2)_n-\!\!\bigcirc \right\}^{\bar{}}$$

ESR spectra of such molecules were observed by Weissman (250) and by Voevodskii et al. (251) for $n = 1$ or 2. They found the rate of electron transfer from one ring to another to be fast as compared to the σ-proton hyperfine splitting ($\sim 10^7$ sec^{-1}).

When an electron moves through a molecule or a solid, there is a tendency for the nuclear motions to become correlated with the electronic motion. This correlation represents a breakdown of Born–Oppenheimer approximation, and such a tendency of the nuclear motion to follow the electronic motion leads to "self-trapping" of the electron (formation of a polaron).

The self-trapping greatly reduces the rate of transfer, which decreases exponentially with n. Calculation suggests that the increase by one CH_2 group in the length of the polymethylene chain reduces the rate by, roughly, a factor of 10. No experimental data are yet available to test these predictions.

20. ELECTRON TRANSFER REACTIONS FROM CARBANIONS TO AROMATIC HYDROCARBONS

In the presence of a suitable electron-acceptor, a carbanion may be oxidized by the electron transfer process. For example:

$$Ph_3C^-,Na^+ + A \text{ (anthracene)} \rightarrow Ph_3C\cdot + A^{\bar{\cdot}},Na^+.$$

Electron transfer from Ph_3C^-,Na^+ to benzophenone was described at the beginning of this century by Schlenk (262), and a similar reaction involving cyclooctatetraene was reported by Wittig (263). Powerful electron acceptors, such as aromatic, nitro compounds, may acquire an electron even from a poor donor, e.g., alkoxides (264) or thiolates (265).

Organolithium compounds (e.g., butyllithium) may add to olefins or some aromatic hydrocarbons, a new C—C covalent bond being formed in such a reaction. However, if bulky groups shield the C=C bond, the addition may be prevented by steric hindrance. For example, BuLi does not add to tetraphenylethylene or 1,1,3,3-tetraphenylbutene-1 (248). Apparently the electron affinities of these olefins are not sufficiently high to permit the electron transfer process either, although electron transfer is observed with naphthalene or phenanthrene radical-anions (266). Electron transfer becomes possible, however, with 1,2,3,4-tetraphenylbutadiene (267). Formation of the respective radical-anions was demonstrated by ESR studies as well as by spectrometric observations. The structure of the organolithium compound has a pronounced effect upon the rate of the electron transfer process. The radical-ion is formed almost instantaneously with benzyllithium in tetrahydrofuran, somewhat slower with n-butyllithium, and very slowly with methyl- or phenyllithium.

Kinetic studies of electron transfer reactions involving carbanions are meager. A thorough kinetic study of the system

$$Ph_2\bar{C}.CH_2.CH_2.\bar{C}Ph_2(^-DD^-) + A \rightleftharpoons Ph_2\bar{C}.CH_2.CH_2.\dot{C}Ph_2(^-DD\cdot) + A^{\bar{\cdot}} \quad K_A,$$

were reported by Szwarc and his associates (195,253). In this equation A denotes anthracene and $A^{\bar{\cdot}}$ its radical-anion. Although, for the sake of simplicity, the free ion notation is used, the investigated reaction involves ion-pairs. The dimeric radical-ion, $^-DD\cdot$, decomposes into monomeric radical-ion $D^{\bar{\cdot}}$ and 1,1-diphenylethylene D,

$$^-DD\cdot \xrightarrow{k_{-2}} D^{\bar{\cdot}} + D,$$

and the rapid electron transfer, $D^{\cdot} + A \rightarrow A^{\cdot} + D$, completes the process. The system attains its stationary state in a few seconds, and thereafter the rate of the overall reaction, determined by the disappearance of $^-DD^-$ or by the formation of $A^{\cdot}$, is given by the equation:

$$d[A^{\cdot}]/dt = k_{-2}K_A[^-DD^-][A]/[A^{\cdot}].$$

The validity of this scheme was established by testing the integrated form of the kinetic equation and by investigating the retardation of the overall process caused by the initial addition of $A^{\cdot}$.

The kinetics of decomposition of $^-DD^-$ retains its mathematical form whatever electron acceptor is added, provided that its electron affinity is sufficiently high. This conclusion was established by Gill, Jagur–Grodzinski, and Szwarc (254), who studied the processes involving 9,10-dimethylanthracene or pyrene instead of anthracene. Each of these investigations led to a value for the combined rate constants, i.e., $k_{-2}K_A$ was calculated from the studies of the anthracene system, while $k_{-2}K_{DMA}$ and $k_{-2}K_\pi$ were derived from the results concerned with the 9,10-dimethylanthracene (DMA) and pyrene (π) systems. The constants K_{DMA} and K_π refer to the equilibria

$$^-DD^- + DMA \rightleftarrows {}^-DD\cdot + DMA^{\cdot} \qquad K_{DMA},$$

and

$$^-DD^- + \pi \rightleftarrows {}^-DD\cdot + \pi^{\cdot} \qquad K_\pi.$$

It is easy to verify that the ratios K_{DMA}/K_A and K_π/K_A are the equilibrium constants of electron transfer processes

$$A^{\cdot} + DMA \rightleftarrows A + DMA^{\cdot} \qquad K_{DMA}/K_A,$$

and

$$A^{\cdot} + \pi \rightleftarrows A + \pi^{\cdot} \qquad K_\pi/K_A.$$

Hence, such kinetic studies permit us to calculate the differences of electron affinities of acceptors, which were independently determined by other direct methods (see, e.g., pp. 319 and 325). The agreement between the kinetic and potentiometric data, or those derived from direct studies of equilibria, is satisfactory, and this verifies the proposed kinetic scheme.

Does electron transfer take place directly or through an intermediate adduct? In the $^-DD^-$ system no intermediate could be seen indicating either direct transfer or extremely short-lived intermediate. However, analogous studies involving a dimeric dianion of α-methylstyrene,

$$\bar{C}(CH_3)(Ph).CH_2.CH_2\bar{C}(CH_3)(Ph) = {}^-\alpha\alpha^-,$$

conclusively showed that an intermediate adduct is formed (255). The dimer, $^-\alpha\alpha^-$ (potassium counterion, tetrahydrofuran solvent) absorbs at $\lambda_{max} = 341$ mμ. On addition of anthracene this absorption peak disappears and a new one is formed at $\lambda_{max} = 451$ mμ. The addition is very rapid and virtually quantitative, its bimolecular rate constant being about 4000–5000 liter mole^{-1}

sec^{-1} at ambient temperature (257), and therefore the reaction is virtually over in 0.2 sec. The optical density of the new absorption peak was determined by a flow technique, and the plot of OD (451 mμ) (measured 0.2 sec after mixing the reagents) versus [anthracene]/[$^-\alpha\alpha^-$] is given in Figure 25 for a constant initial concentration of the dimer. It is obvious that each carbanion end of the dimer is capable of reacting with anthracene and therefore the plot shown in Figure 25 levels off for an [A]/[$^-\alpha\alpha^-$] ratio of 2. In the adduct the carbanion is bonded through a covalent C—C bond with the carbon-9

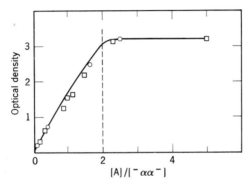

Fig. 25.   Change in optical density at 451 mμ on addition of anthracene (A) to dimeric α-methylstyrene dianion ($^-\alpha\alpha^-$) (255). Concentration of the dimer ($^-\alpha\alpha^-$) is constant.

atom of the aromatic. This addition resembles the one taking place between anthracene and living polystyrene (258,259) or ethyl- or butyllithium (260). However, while the latter adducts are stable and do not decompose, the adduct formed with $^-\alpha\alpha^-$ decomposes within a few minutes, yielding anthracene radical-anion. The initial rate of this decomposition was measured at constant concentration of the dimer and variable concentration of anthracene and the results are presented in Figure 26 as plots of $(d[A^-]/dt)_0$ versus [A]/[$^-\alpha\alpha^-$]. As may be seen from the figure, the initial rate increases with increasing [A]/[$^-\alpha\alpha^-$] ratio, it reaches maximum for [A]/[$^-\alpha\alpha^-$] = 1 and then decreases, eventually becoming very low for [A]/[$^-\alpha\alpha^-$] = 2. This most unusual behavior was explained by postulating a rapid decomposition of an α-methylstyrene dimeric dianion associated with *one* molecule of anthracene. The dianion possessing *two* molecules of anthracene, one on each of its ends, was found to be relatively stable, measurable decomposition being observed only after about one hour. Let us denote by p the ratio [A]/2[$^-\alpha\alpha^-$]. If the probability of addition of anthracene to one end of the dimer is independent of the fate of the other, then in a mixture of $^-\alpha\alpha^-$ and A a fraction $2p(1-p)$ of dimers should be associated with one molecule of anthracene, a fraction p^2 with two and $(1-p)^2$ with none. This conclusion was confirmed by gas-chromatographic analyses of protonated products.

The assumption of rapid decomposition of the monoanthracenated dimer, $^-\alpha\alpha A^-$, leads to proportionality between the initial rate of decomposition and $p(1 - p)$. Hence, the rate increases with p until for $p = 1/2$, i.e., $[A]/[^-\alpha\alpha^-] = 1$, it reaches it maximum value, and then decreases virtually to zero.

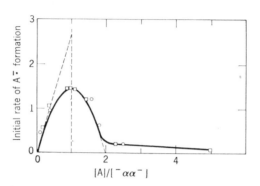

Fig. 26.   Initial rate of decomposition of anthracene-$^-\alpha\alpha^-$ adduct at various initial ratios of $[A]/[^-\alpha\alpha^-]$ (255). Measured by the rate of formation of $A^{\overline{\cdot}}$ radical-ion. Concentration of dimer ($^-\alpha\alpha^-$) is constant.

The kinetics of the decomposition was thoroughly investigated. It was shown that the following mechanism accounts for the facts:

$$^-\alpha\alpha A^- \rightleftharpoons ^-\alpha\alpha\cdot + A^{\overline{\cdot}} \qquad \text{rapid, equilibrium constant } K_\alpha,$$

$$^-\alpha\alpha\cdot \xrightarrow{k'_2} \alpha^{\overline{\cdot}} + \alpha \qquad \text{slow rate-determining,}$$

$$^-\alpha\alpha\cdot + \alpha^{\overline{\cdot}} \rightarrow ^-\alpha\alpha\alpha^- \qquad \text{again rapid.}$$

The symbol α denotes a molecule of α-methylstyrene and $\alpha^{\overline{\cdot}}$ its radical anion.

One may wonder now why $^-\alpha\alpha A^-$ decomposes rapidly while the decomposition of $^-A\alpha\alpha A^-$ is slow. It seems that the repulsion between the two negative ends of $^-\alpha\alpha A^-$ contributes to the decomposition and provides enough driving force to shift the equilibrium K_α sufficiently to the right. This repulsion is greatly reduced in $^-A\alpha\alpha A^-$, and consequently the equilibrium concentration of $^-A\alpha\alpha\cdot$ may be much lower than that of $^-\alpha\alpha\cdot$. Moreover, decomposition of $^-\alpha\alpha\cdot$ into $\alpha^{\overline{\cdot}} + \alpha$ may be faster than that of $^-A\alpha\alpha\cdot$ into $^-A\alpha\cdot + \alpha$ or $A^{\overline{\cdot}} + 2\alpha$. It is known that the adduct of anthracene and α-methylstyrene tetramer dianion, $^-\alpha\alpha\alpha\alpha^-$, decomposes only slowly, yielding $A^{\overline{\cdot}}$. No maximum is observed when the initial rate of this decomposition is determined as a function of the $[A]/[^-\alpha\alpha\alpha\alpha^-]$ ratio.

In the system $^-\alpha\alpha^- + \pi$ (pyrene) again $^-\alpha\alpha\pi^-$ decomposes rapidly while $^-\pi\alpha\alpha\pi^-$ is relatively stable (331). Hence, $(d[\pi^{\overline{\cdot}}]/dt)_0$, measured for $[^-\alpha\alpha^-]_0 = $ const as a function of $[\pi]_0/[^-\alpha\alpha^-]_0$ passes again through a maximum.

21. ELECTROLYTIC GENERATION OF RADICAL-IONS

In the electrolytic process the electrode acts as a donor (cathode) or as an acceptor (anode). The substrate in the vicinity of a cathode becomes reduced, whereas oxidation takes place in the vicinity of an anode. An electrode process involving one-electron-transfer may convert a suitable substrate into a radical-ion, and such reactions occur, e.g., in polarographic systems which were discussed in Section 10. Adaptation of polarographic techniques for ESR studies was first reported by Geske and Maki (179), and since then these methods have become widely used for generation of radical-ions in the cavity of ESR spectrometer.

Electrochemical generation of radical-ions is advantageous and often superior to the conventional chemical reduction (e.g., by alkali metals) or oxidation by electron acceptors. It allows the use of a far wider range of solvents, is more controllable because the potential may be adjusted at will, and permits the variation of the counterions by judicious choice of supporting electrolyte. The various handling procedures employed in such investigations are now fairly standard; their details are described, e.g., by Adams (281).

The study of electrode processes permits determination of k_{el}, viz. the rate constant of a heterogeneous electron-transfer at an electrode surface. The theoretical treatment of such reactions (283) led to an important correlation between k_{el} and the homogeneous rate constant of the respective electron exchange, k_{ex}, namely, $k_{ex} \approx 10^3 \times k_{el}{}^2$. Numerous relaxation methods (see, e.g., ref. 284) led to the determination of k_{el}; however, their utility is somewhat obviated by experimental and theoretical difficulties. The steady-state direct current voltametry has proven to be more reliable (285). For example, this technique, in conjunction with rotated-disk polarography, was adopted by Adams and his students (282) in their investigation of exchanges between radical-anions and their parent molecules. Unfortunately, the data showed a poor agreement with Marcus' theory (283), and a possible explanation of this failure was discussed in their paper.

An interesting electrochemical redox process, leading through electron-transfer to chemiluminescence, was reported by Hoijtink (286) and then investigated in greater detail by Hercules (287), Visco and Chandross (288), and Santhanam and Bard (289). Positive and negative radical-ions may be formed successively on the *same* electrode if an alternating square-wave potential is applied (287). These species diffuse then into the surrounding liquid where they annihilate each other. Consider such a reaction which is represented on p. 394 with the aid of diagrams giving the energy levels of bonding and anti-bonding molecular orbitals of an alternate hydrocarbon.

anti-
bonding ──X──   ──────   ──────   ──X──

          ──── ── + ──── ──   →   ──── ── + ──── ── fluores-
                                                                cence
bonding ──X─X──   ──X⊖──   ──X─X──   ──X⊖──
        radical-anion   radical-cation   ordinary molecule excited molecule
                                           (ground state)

A collision between radical-anion and radical-cation leads to electron transfer; however, since the process is adiabatic the transferred electron initially moves into the antibonding orbital of the acceptor. This yields an electronically excited molecule which eventually returns to its ground state with emission of fluorescent light. The collision of the radical-ions may also yield excited triplets which luminesce then through triplet–triplet annihilation.

Closer examination of this phenomenon shows that the system is rather complex. For example, some radical-cations may be reduced directly on the electrode and yield radical-anions; some of the excited molecules may undergo spontaneous nonradiative decay which reduces the chemiluminescence, and finally some quenching by the parent species may also take place. Furthermore, it was shown by Chandross et al. (290) that the cation and anion may dimerize, i.e., form an excimer which then radiates,

$$A^{\overline{\cdot}} + A^{\dot{+}} \rightleftarrows A_2^* \overset{h\nu}{\to} 2A.$$

The diffusive character of these processes implies that the intensity of the resulting luminescence is a function of current density and of the distance from the electrode. Mathematical treatment of this problem was developed by Feldberg (291) who showed how the pertinent rate constants may be computed from the experimental data which give the spatial distribution of light intensity for various current densities.

Variation of luminescence intensity with the applied voltage and frequency (sinusoid wave) was measured (288). Intensity increased linearly with the voltage up to a point (~ 15 V) and then leveled off. At frequencies below 100 cps more than 95% of light was modulated; its intensity decreased as the frequency was raised. The luminescence was observed with 9,10-diphenyl anthracene, anthracene, perylene, rubrene, quaterphenyl, and 1,1,4,4-tetraphenyl butadiene. Dimethyl formamide was the most frequently used solvent.

Further studies showed that luminescence may also be observed with donors other than the corresponding radical-anions, e.g., electroluminescence of rubrene was induced by dimethyl formamide, water, n-butyl amine, or triethyl amine, all of which may serve as donors (292). Recently it was shown that a direct capture of solvated electrons, which were produced by an electrolytic process, also leads to light emission (293).

Electroluminescence phenomena of a similar nature were reported by other investigators [e.g., Maricle, Maurer and their colleagues (296,297)]. An interesting problem was raised: Is a direct electrode-generation of excited molecules possible? Theoretical work of Marcus (303) makes this suggestion unlikely and further arguments against this hypothesis were raised recently by Chandross and Visco (304).

Electrochemically generated radical-ions may initiate anionic polymerization. Such processes were investigated by Funt and his students. The early work (298–300) demonstrated that anionic polymerization may take place under such conditions; however, termination processes were not excluded and consequently only dead polymers were produced. Later Yamazaki et al. (301) showed that under properly chosen conditions living polymers are formed, and their findings were soon confirmed by Funt et al. (302). Similar results were obtained at low temperature in liquid ammonia with 4-vinyl pyridine as the monomer (323). A comprehensive review of electrolytically initiated polymerization was published by Funt (325).

22. CHARGE TRANSFER COMPLEXES AND THEIR SIGNIFICANCE AS INITIATORS OF IONIC POLYMERIZATION

Charge-transfer complexes are well recognized entities, their structure being best explained in terms of Mulliken's theory (305). The association of a donor D with an acceptor A gives a complex described in quantum mechanics language by

$$\psi = a\psi_1(D, A) + b\psi_2(D^+, A^-)$$

with $|a| > |b|$, whereas the excited state is described by ψ',

$$\psi' = b\psi_1(D, A) - a\psi_2(D^+, A^-).$$

The excitation from state ψ to ψ' leads to the characteristic spectra of charge-transfer complexes. Its maximum absorption is affected by solvent, and qualitatively these changes were explained by assuming that the polar form is more stabilized through solvation than the ground state (ψ). Such stabilization may lead to formation of ion-pairs D^+, A^-. For example, the solid charge-transfer complex of tetramethylphenylene diamine (TMPD) with chloranile dissociates in acetonitrile into $TMPD^+$ and chloranile$^-$, both radical-ions being identified by their characteristic absorption spectra (306). Such a dissociation does not take place in nonpolar solvents. These types of investigations were extended to other complexes by Foster (e.g., 307) and independently by Briegleb (308).

Similar phenomena may be induced by photoactivation. For example, photoaddition of maleic anhydride to benzene apparently involves photoactivation of a charge-transfer complex (309). The photoinduced ESR signals of charge-transfer complexes provide perhaps the most direct evidence for such a transformation of a nonpolar form into a polar one (310,311).

Studies of ESR spectra of charge transfer complexes revealed several interesting features. A complete transfer of an electron from a donor to an acceptor produces two independent paramagnetic species, and hence the ESR spectrum should be given by the superposition of those of $D^{\dot{+}}$ and $A^{\dot{-}}$, if the interaction between these species is weak. When the interaction is strong the energy levels become split, the lower corresponding to a dia-magnetic singlet and the higher to a paramagnetic triplet; the latter level may then be thermally populated if the energy gap is small. Examples of such phenomena were described in the literature (e.g., 312–316).

Formation of radical-ions from charge-transfer complexes suggests that such systems may initiate ionic polymerization, particularly if the monomer is the component of the complex. For example, polymerization of N-vinyl carbazole was initiated by electron acceptors (317,318) and it was proposed that this process is caused by an electron transfer. Nevertheless, the details of initiation were not discussed, although many features of the reaction were recorded (319,320). Several other polymerizations were subsequently attri-buted to electron-transfer initiation (321,322) but no attempt was made to verify the mechanism.

Extensive kinetic study of vinylcarbazole polymerization initiated by tetra-nitromethane in nitrobenzene was reported by Pac and Plesch (324). These authors concluded that the reaction involves electron transfer and the forma-tion of radical-cations is responsible for the initiation which leads to conven-tional cationic polymerization. The radical end was assumed to be inactivated by tetranitromethane which often acts as a radical trap. Further investigation of this polymerization and studies of some related systems were completed recently in this writer's laboratory. The results confirmed the cationic character of the propagation; however, they indicated that the initiation does not involve radical-cations, but it is due to a direct transfer of NO_2^+ cation leading to the formation of carbonium-ions and the $C(NO_2)_3^-$ counterions (see ref. 328).

References

(1) M. Berthelot, *Ann. Chim.*, **12**, 155 (1867).

(2) W. Schlenk, J. Appenrodt, A. Michael, and A. Thal, *Chem. Ber.*, **47**, 473 (1914).

(3) R. Willstätter, F. Seitz, and E. Bumm, *Chem. Ber.*, **61**, 871 (1928).

(4) W. Schlenk and E. Bergmann, *Ann.*, **463**, 1 (1928); **464**, 1 (1928).

(5) (a) N. D. Scott, J. F. Walker, and V. L. Hansley, *J. Am. Chem. Soc.*, **58**, 2442 (1936). (b) J. F. Walker and N. D. Scott, *J. Am. Chem. Soc.*, **60**, 951 (1938).

(6) W. Hückel and H. Bretschneider, *Ann.*, **540**, 157 (1939).

(7) (a) D. Lipkin, D. E. Paul, J. Townsend, and S. I. Weissman, *Science*, **117**, 534 (1953). (b) S. I. Weissman, J. Townsend, D. E. Paul, and G. E. Pake, *J. Chem. Phys.*, **21**, 2227 (1953).

(8) (a) J. E. Lovelock, *Nature*, **189**, 729 (1961). (b) J. E. Lovelock, *Anal. Chem.*, **33**, 162 (1961); **35**, 474 (1963). (c) J. E. Lovelock and N. L. Gregory, in *Gas Chromatography* (N. Brenner, ed.), Academic Press, New York, 1962, p. 219. (d) J. E. Lovelock, P. G. Simmonds, and W. J. A. Van den Heuvel, *Nature*, **197**, 249 (1963).

(9) W. E. Wentworth, E. Chen, and J. E. Lovelock, *J. Phys. Chem.*, **70**, 445 (1966).

(10) R. S. Becker and E. Chen, *J. Chem. Phys.*, **45**, 2403 (1966).

(11) W. E. Wentworth and R. S. Becker, *J. Am. Chem. Soc.*, **84**, 4263 (1962).

(12) G. Briegleb, *Angew. Chem.*, **76**, 326 (1964).

(13) N. E. Bradbury, *Phys. Rev.*, **44**, 883 (1933).

(14) (a) L. Rolla and G. Piccardi, *Atti Accad. Nazl. Lincei*, **2**, 29, 128, 173 (1925). (b) G. Piccardi, *ibid.*, **3**, 413, 566 (1926). (c) G. Piccardi, *Z. Physik*, **43**, 899 (1927). (d) L. Rolla and G. Piccardi, *Atti Accad. Nazl. Lincei*, **5**, 546 (1927).

(15) P. P. Sutton and J. E. Mayer, *J. Chem. Phys.*, **2**, 145 (1934); **3**, 20 (1935).

(16) G. Glockler and M. Calvin, *J. Chem. Phys.*, **3**, 771 (1935).

(17) F. M. Page, *Trans. Faraday Soc.*, **56**, 1742 (1960).

(18) A. L. Farragher and F. M. Page, *Trans. Faraday Soc.*, **62**, 3072 (1966).

(19) A. L. Farragher, Ph.D. Thesis, University of Arton, Birmingham, England, 1966.

(20) J. Kay and F. M. Page, *Trans. Faraday Soc.*, **62**, 3081 (1966).

(21) J. E. Mayer, *Z. Physik.*, **61**, 798 (1930).

(22) W. B. Person, *J. Chem. Phys.*, **38**, 109 (1963).

(23) N. S. Hush and J. A. Pople, *Trans. Faraday Soc.*, **51**, 600 (1955).

(24) R. M. Hedges and F. A. Matsen, *J. Chem. Phys.*, **28**, 950 (1958).

(25) J. R. Hoyland and L. Goodman, *J. Chem. Phys.*, **36**, 12 (1962).

(26) A. Streitwieser, *J. Am. Chem. Soc.*, **82**, 4123 (1960).

(27) N. S. Hush and J. Blackledge, *J. Chem. Phys.*, **23**, 514 (1955).

(28) (a) W. Schlenk and T. Weithel, *Chem. Ber.*, **44**, 1182 (1911). (b) W. Schlenk and A. Thal, *Chem. Ber.*, **46**, 2840 (1913); **47**, 473 (1914).

(29) A. F. Gaines and F. M. Page, *Trans. Faraday Soc.*, **62**, 3086 (1966).

(30) J. H. Hildebrand and H. E. Bent, *J. Am. Chem. Soc.*, **49**, 3011 (1927).

(31) (a) H. E. Bent and N. B. Keevil, *J. Am. Chem. Soc.*, **58**, 1228, 1367 (1936). (b) N. B. Keevil, *J. Am. Chem. Soc.*, **59**, 2104 (1937).

(32) N. B. Keevil and H. E. Bent, *J. Am. Chem. Soc.*, **60**, 193 (1938).

(33) G. J. Hoijtink, E. de Boer, P. H. van der Meij, and W. P. Weijland, *Rec. Trav. Chim.*, **75**, 487 (1956).

(34) J. Jagur-Grodzinski, M. Feld, S. L. Yang, and M. Szwarc, *J. Phys. Chem.*, **69**, 628 (1965).

(35) M. G. Scroggie, *Wireless World*, **58**, 14–18 (1952).

(36) J. Chaudhuri, J. Jagur-Grodzinski, and M. Szwarc, *J. Phys. Chem.*, **71**, 3063 (1967).

(37) R. V. Slates and M. Szwarc, *J. Phys. Chem.*, **69**, 4124 (1965).

(38) P. Chang, R. V. Slates, and M. Szwarc, *J. Phys. Chem.*, **70**, 3180 (1966).

(39) D. Nicholls, C. Sutphen, and M. Szwarc, *J. Phys. Chem.*, **72**, 1021 (1968).

(40) K. H. J. Buschow, J. Dieleman, and G. J. Hoijtink, *J. Chem. Phys.*, **42**, 1993 (1965).

(41) D. E. Paul, D. Lipkin, and S. I. Weissman, *J. Am. Chem. Soc.*, **78**, 116 (1956).

(42) L. E. Lyons, *Nature*, **166**, 193 (1950).

(43) H. A. Laitinen and S. Wawzonek, *J. Am. Chem. Soc.*, **64**, 1765 (1942).

(44) S. Wawzonek and H. A. Laitinen, *J. Am. Chem. Soc.*, **64**, 2365 (1942).

(45) S. Wawzonek and J. W. Fan, *J. Am. Chem. Soc.*, **68**, 2541 (1946).

(46) G. J. Hoijtink, J. van Schooten, E. de Boer, and W. I. Aalbersberg, *Rec. Trav. Chim.*, **73**, 355 (1954).

(47) C. Carvajal, K. J. Tölle, J. Smid, and M. Szwarc, *J. Am. Chem. Soc.*, **87**, 5548 (1965).
(48) G. J. Hoijtink and J. van Schooten, *Rec. Trav. Chim.*, **71**, 1089 (1952); **72**, 691, 903 (1953).
(49) I. Bergman, *Trans. Faraday Soc.*, **50**, 829 (1954); **52**, 690 (1956).
(50) E. Swift, *J. Am. Chem. Soc.*, **60**, 1403 (1938).
(51) F. A. Matsen, *J. Chem. Phys.*, **24**, 602 (1956).
(52) W. M. Latimer, K. S. Pitzer, and C. M. Slansky, *J. Chem. Phys.*, **7**, 108 (1939).
(53) I. M. Kolthoff and J. L. Lingane, *Polarography*, 2nd ed., Vol. 2, Interscience, New York, 1952, p. 809.
(54) A. C. Aten and G. J. Hoijtink, *Z. Physik. Chem. (Frankfurt)*, **21**, 192 (1959).
(55) A. C. Aten, C. Büthker, and G. J. Hoijtink, *Trans. Faraday Soc.*, **55**, 324 (1959).
(56) G. J. Hoijtink, *Rec. Trav. Chim.*, **73**, 895 (1954).
(57) G. W. Wheland, *J. Am. Chem. Soc.*, **63**, 2025 (1941).
(58) H. C. Longuet-Higgins, *J. Chem. Phys.*, **18**, 265 (1950).
(59) (a) S. Wawzonek, E. W. Blaha, R. Berkey, and M. E. Runner, *J. Electrochem. Soc.*, **102**, 235 (1955). (b) P. H. Given, *J. Chem. Soc.*, **1958**, 2684.
(60) E. de Boer, *Advan. Organometal. Chem.*, **2**, 115 (1964).
(61) D. A. McInnes, *The Principles of Electrochemistry*, Reinhold, New York, 1939.
(62) W. E. Bachmann, *J. Am. Chem. Soc.*, **55**, 1179 (1933).
(63) A. Mathias and E. Warhurst, *Trans. Faraday Soc.*, **58**, 948 (1962).
(64) Quoted by E. de Boer in his review on p. 123 of reference 60.
(65) A. I. Shatenshtein, E. S. Petrov, and M. I. Belousova, *Organic Reactivity*, **1**, 191 (1964) (Tartu State University, Estonia, U.S.S.R).
(66) A. I. Shatenshtein, E. S. Petrov, and E. A. Yakoleva, International Symposium on Macromolecular Chemistry, Prague, 1965, *J. Polymer. Sci. C*, **16**, Part 3, 1729 (1967).
(67) R. V. Slates and M. Szwarc, *J. Am. Chem. Soc.*, **89**, 6043 (1967).
(68) P. S. Rao, J. R. Nash, J. P. Guarino, M. R. Ronayne, and W. H. Hamill, *J. Am. Chem. Soc.*, **84**, 500 (1962).
(69) H. Linschitz, M. G. Berry, and D. Schweitzer, *J. Am. Chem. Soc.*, **76**, 5833 (1954).
(70) N. Hizota, *J. Phys. Chem.*, in press.
(71) (a) M. R. Ronayne, J. P. Guarino, and W. H. Hamill, *J. Am. Chem. Soc.*, **84**, 4230 (1962). (b) See also J. P. Guarino, M. R. Ronayne, and W. H. Hamill, *Radiation Res.*, **17**, 379 (1962).
(72) N. Christodouleas and W. H. Hamill, *J. Am. Chem. Soc.*, **86**, 5413 (1964).
(73) E. P. Bertin and W. H. Hamill, *J. Am. Chem. Soc.*, **86**, 1301 (1964).
(74) W. H. Hamill, J. P. Guarino, M. R. Ronayne, and J. A. Ward, *Discussions Faraday Soc.*, **36**, 169 (1963).
(75) D. G. Powell and E. Warhurst, *Trans. Faraday Soc.*, **58**, 953 (1962).
(76) K. Chambers, E. Collinson, F. S. Dainton, and W. Seddon, *Chem. Commun.*, **1966**, 498.
(77) M. Katayama, *Bull. Chem. Soc. Japan*, **38**, 2208 (1965).
(78) J. P. Keene, E. J. Land, and A. J. Swallow, *J. Am. Chem. Soc.*, **87**, 5284 (1965).
(79) T. Shida and W. H. Hamill, *J. Am. Chem. Soc.*, **88**, 5371 (1966).
(80) D. H. Levy and R. J. Myers, *J. Chem. Phys.*, **41**, 1062 (1964); **44**, 4177 (1966).
(81) K. O. Hartman and I. C. Hisatsune, *J. Chem. Phys.*, **44**, 1913 (1966).
(82) K. O. Hartman and I. C. Hisatsune, *J. Phys. Chem.*, **69**, 583 (1965).
(83) W. L. McCubbin and I. D. C. Gurney, *J. Chem. Phys.*, **43**, 983 (1965).
(84) T. Shida and W. H. Hamill, *J. Chem. Phys.*, **44**, 2369 (1966).

(85) L. P. Blanchard and P. LeGoff, *Can. J. Chem.*, **35**, 89 (1957).
(86) T. Shida and W. H. Hamill, *J. Am. Chem. Soc.*, **88**, 3689 (1966).
(87) J. A. Leone and W. S. Koski, *J. Am. Chem. Soc.*, **88**, 224 (1966).
(88) P. M. Johnson and A. C. Albrecht, *J. Chem. Phys.*, **44**, 1845 (1966).
(89) T. Shida and W. H. Hamill, *J. Chem. Phys.*, **44**, 2375 (1966).
(90) W. I. Aalbersberg, G. J. Hoijtink, E. L. Mackor, and W. P. Weijland, *J. Chem. Soc.*, **1959**, 3049, 3055.
(91) P. Bennema, G. J. Hoijtink, J. H. Lupinski, L. J. Oosterhoff, P. Selier, and J. D. W. van Voorst, *Mol. Phys.*, **2**, 431 (1959).
(92) T. Shida and W. H. Hamill, *J. Am. Chem. Soc.*, **88**, 5376 (1966).
(93) J. B. Gallivan and W. H. Hamill, *J. Chem. Phys.*, **44**, 2378 (1966).
(94) T. Shida and W. H. Hamill, *J. Am. Chem. Soc.*, **88**, 3683 (1966).
(95) M. S. Matheson and L. M. Dorfman, *J. Chem. Phys.*, **32**, 1870 (1960).
(96) R. L. McCarthy and A. McLachlan, *Trans. Faraday Soc.*, **56**, 1187 (1960).
(97) J. P. Keene, *Nature*, **188**, 843 (1960).
(98) L. M. Dorfman and M. S. Matheson, *Prog. Reaction Kinetics*, **3**, 237 (1965).
(99) J. W. Boag, *Am. J. Roentgenology*, **90**, 896 (1963).
(100) E. J. Hart, S. Gordon, and J. K. Thomas, *J. Phys. Chem.*, **68**, 1271 (1964).
(101) S. Arai and L. M. Dorfman, *J. Chem. Phys.*, **41**, 2190 (1964).
(102) G. E. Adams, J. H. Baxendale, and J. W. Boag, *Proc. Roy. Soc. (London)*, Ser. A, **277**, 549 (1964).
(103) I. A. Taub, M. C. Sauer, and L. M. Dorfman, *Discussions Faraday Soc.*, **36**, 206 (1963).
(104) I. A. Taub, D. A. Harter, M. C. Sauer, and L. M. Dorfman, *J. Chem. Phys.*, **41**, 979 (1964).
(105) S. Arai, E. L. Tremba, J. R. Brandon, and L. M. Dorfman, *Can. J. Chem.*, **45**, 1119 (1967).
(106) G. J. Hoijtink and P. H. van der Meij, *Z. Physik. Chem. (Frankfurt)*, **20**, 1 (1959).
(107) J. Dieleman, Thesis, Univ. of Amsterdam, 1962.
(108) E. de Boer, unpublished results.
(109) K. H. J. Buschow and G. J. Hoijtink, *J. Chem. Phys.*, **40**, 2501 (1964).
(110) D. W. Ovenall and D. H. Whiffen, *Chem. Soc. (London) Spec. Publ.*, **12**, 139 (1958).
(111) (a) A. G. Evans, J. C. Evans, E. D. Owen, and B. J. Tabner, *Proc. Chem. Soc.*, p. 226 (1962). (b) J. E. Bennett, A. G. Evans, J. C. Evans, E. D. Owen, and B. J. Tabner, *J. Chem. Soc.*, **1963**, 3954. (c) A. G. Evans and B. J. Tabner, *ibid.*, **1963**, 4613.
(112) J. F. Garst and R. S. Cole, *J. Am. Chem. Soc.*, **84**, 4352 (1962).
(113) (a) J. F. Garst, E. R. Zabolotny, and R. S. Cole, *J. Am. Chem. Soc.*, **86**, 2257 (1964). (b) J. F. Garst and E. R. Zabolotny, *ibid.*, **87**, 495 (1965).
(114) R. C. Roberts and M. Szwarc, *J. Am. Chem. Soc.*, **87**, 5542 (1965).
(115) D. N. Bhattacharyya, C. L. Lee, J. Smid, and M. Szwarc, *J. Phys. Chem.*, **69**, 608 (1965).
(116) H. Smith and T. M. Sudgen, *Proc. Roy. Soc. A*, **211**, 31, 58 (1952).
(117) R. V. Slates, Thesis, Syracuse University, 1967.
(118) L. L. Chan and J. Smid, *J. Am. Chem. Soc.*, **89**, 4547 (1967).
(119) J. P. Keene, *J. Sci. Instr.*, **41**, 493 (1964).
(120) I. A. Taub, Proc. Intern. Symposium, *Pulse Radiolysis*, April, 1965, Academic Press, London–New York, 1965.
(121) J. Rabani, W. A. Mulac, and M. S. Matheson, *J. Phys. Chem.*, **69**, 53 (1965).
(122) W. E. Wentworth and E. Chen, *J. Phys. Chem.*, **71**, 1929 (1967).

(123) W. E. Wentworth, R. S. Becker, and R. Tung, *J. Phys. Chem.*, **71**, 1652 (1967).

(124) C. A. Kraus, *J. Am. Chem. Soc.*, **30**, 1323 (1908).

(125) D. S. Berns, in "Solvated Electron," *Advan. Chem. Ser.*, **50**, 82 (1965).

(126) G. Lepoutre and M. J. Sienko, Eds., *Metal-Ammonia Solutions*, Benjamin, New York, 1964.

(127) J. Jortner and S. A. Rice, in "Solvated Electron," *Advan. Chem. Ser.*, **50**, 7 (1965); (a) *ibid.*, p. 23.

(128) L. C. Kenausis, E. C. Evers, and C. A. Kraus, *Proc. Natl. Acad. Sci. U.S.*, **48**, 121 (1962).

(129) J. C. Thompson, in "Solvated Electron," *Advan. Chem. Ser.*, **50**, 96 (1965).

(130) E. Arnold and A. Patterson, *J. Chem. Phys.*, **41**, 3098 (1964).

(131) S. Golden, C. Guttman, and T. R. Tuttle, *J. Am. Chem. Soc.*, **87**, 135 (1965).

(132) D. S. Berns, G. Lepoutre, E. A. Bockelman, and A. Patterson, *J. Chem. Phys.*, **35**, 1820 (1961).

(133) R. A. Ferrell, *Phys. Rev.*, **108**, 167 (1957).

(134) C. G. Kuper, *Phys. Rev.*, **122**, 1007 (1961).

(135) W. T. Sommer, *Phys. Rev. Letters*, **12**, 271 (1964).

(136) N. R. Kestner, J. Jortner, M. H. Cohen, and S. A. Rice, *Phys. Rev.*, **140**, A56 (1965).

(137) H. Schnyders, L. Meyer, and S. A. Rice, *Phys. Rev.*, **150**, 127 (1966).

(138) J. H. Baxendale, *Radiation Res.*, *Suppl.* **4**, 139 (1964).

(139) C. A. Kraus and W. W. Lucasse, *J. Am. Chem. Soc.*, **43**, 2529 (1921).

(140) M. C. R. Symons, M. J. Blandamer, R. Catterall, and L. Shields, *J. Chem. Soc.*, **1964**, 4357.

(141) J. H. Baxendale, E. M. Fielden, and J. P. Keene, *Science*, **148**, 637 (1965).

(142) J. Eloranta and H. Linschitz, *J. Chem. Phys.*, **38**, 2214 (1963).

(143) M. Anbar and E. J. Hart, *J. Phys. Chem.*, **69**, 1244 (1965).

(144) L. M. Dorfman, in "Solvated Electron," *Advan. Chem. Ser.*, **50**, 36 (1965).

(145) W. Weyl, *Pogg. Ann.*, **121**, 601 (1864).

(146) M. Gold, W. L. Jolly, and K. S. Pitzer, *J. Am. Chem. Soc.*, **84**, 2264 (1962).

(147) E. Arnold and A. Patterson, *J. Chem. Phys.*, **41**, 3089 (1964).

(148) E. Becker, R. H. Lindquist, and B. J. Alder, *J. Chem. Phys.*, **25**, 971 (1956).

(149) T. R. Hughes, *J. Chem. Phys.*, **38**, 202 (1963).

(150) J. V. Acrivos and K. S. Pitzer, *J. Phys. Chem.*, **66**, 1693 (1962).

(151) K. Bar-Eli and T. R. Tuttle, *J. Chem. Phys.*, **40**, 2508 (1964).

(152) J. L. Dye and L. R. Dalton, *J. Phys. Chem.*, **71**, 184 (1967).

(153) D. E. O'Reilly and T. Tsang, *J. Chem. Phys.*, **42**, 3333 (1965).

(154) T. R. Tuttle, C. Guttman, and S. Golden, *J. Chem. Phys.*, **45**, 2206 (1966).

(155) J. L. Down, J. Lewis, B. Moore, and G. Wilkinson, *J. Chem. Soc.*, **1959**, 3767.

(156) F. Cafasso and B. R. Sundheim, *J. Chem. Phys.*, **31**, 809 (1959).

(157) F. S. Dainton, D. M. Wiles, and A. N. Wright, *J. Chem. Soc.*, **1960**, 4283.

(158) F. S. Dainton, D. M. Wiles, and A. N. Wright, *J. Polymer Sci.*, **45**, 111 (1960).

(159) G. Fraenkel, S. H. Ellis, and D. T. Dix, *J. Am. Chem. Soc.*, **87**, 1406 (1965).

(160) H. Normant and M. Larchevegne, *Compt. Rend.*, **260**, 5062 (1965).

(161) H. Normant, J. Normant, T. Cuvigny, and B. Angelo, *Bull. Soc. Chim.* (*France*), **1965**, 1561.

(162) H. Normant, *Bull. Soc. Chim.* (*France*), **1966**, 3362.

(163) H.-L. J. Chen and M. Bersohn, *J. Am. Chem. Soc.*, **88**, 2663 (1966).

(164) C. A. Hutchison and R. C. Pastor, *J. Chem. Phys.*, **21**, 1959 (1953).

(165) R. C. Douthit and J. L. Dye, *J. Am. Chem. Soc.*, **82**, 4472 (1960).

(166) M. Ottolenghi, K. Bar-Eli, H. Linschitz, and T. R. Tuttle, *J. Chem. Phys.*, **40**, 3729 (1964).

(167) H. Linschitz, M. G. Berry, and D. Schweitzer, *J. Am. Chem. Soc.*, **76**, 5833 (1954).

(168) J. D. W. van Voorst, and G. J. Hoijtink, *J. Chem. Phys.*, **42**, 3995 (1965).

(169) G. N. Lewis and D. Lipkin, *J. Am. Chem. Soc.*, **64**, 2801 (1942).

(170) G. N. Lewis and J. Bigeleisen, *J. Am. Chem. Soc.*, **65**, 520 (1943).

(171) P. J. Zandstra and G. J. Hoijtink, *Mol. Phys.*, **3**, 371 (1960).

(172) G. J. Hoijtink and J. D. W. van Voorst, *J. Chem. Phys.*, **45**, 3918 (1966).

(173) J. H. Baxendale and G. Hughes, *Z. Physik. Chem. (Frankfurt)*, **14**, 323 (1958).

(174) J. Jortner and J. Rabani, *J. Am. Chem. Soc.*, **83**, 4868 (1961).

(175) E. J. Kirschke and W. L. Jolly, *Science*, **147**, 45 (1965).

(176) G. Dobson and L. I. Grossweiner, *Trans. Faraday Soc.*, **61**, 708 (1965).

(177) M. S. Matheson, W. A. Mulac, and J. Rabani, *J. Phys. Chem.*, **67**, 2613 (1963).

(178) M. Ottolenghi and H. Linschitz, in "Solvated Electron," *Advan. Chem. Ser.*, **50**, 149 (1965).

(179) D. H. Geske and A. Maki, *J. Am. Chem. Soc.*, **82**, 2671 (1960).

(180) E. Clementi, A. D. McLean, D. L. Raimondi, and M. Yoshimine, *Phys. Rev.*, **133**, A1274 (1964).

(181) G. Favini and M. Simonetta, *Theoret. Chim. Acta*, **1**, 294 (1963).

(182) W. Reppe, O. Schlichting, K. Klager, and T. Toepel, *Ann.*, **560**, 1 (1948).

(183) T. J. Katz, W. H. Reinmuth, and D. E. Smith, *J. Am. Chem. Soc.*, **84**, 802 (1962).

(184) T. J. Katz and H. L. Strauss, *J. Chem. Phys.*, **32**, 1873 (1960).

(185) H. L. Strauss and G. K. Fraenkel, *J. Chem. Phys.*, **35**, 1738 (1961).

(186) H. M. McConnell, *J. Chem. Phys.*, **24**, 632, 764 (1956).

(187) I. Bernal, P. H. Rieger, and G. K. Fraenkel, *J. Chem. Phys.*, **37**, 1489 (1962).

(188) H. L. Strauss, T. J. Katz, and G. K. Fraenkel, *J. Am. Chem. Soc.*, **85**, 2360 (1963).

(189) C. A. Coulson and W. E. Moffitt, *Phil. Mag.*, **40**, 1 (1949).

(190) D. E. Wood and H. M. McConnell, *J. Chem. Phys.*, **37**, 1150 (1962).

(191) J. dos Santos-Veiga, *Mol. Phys.*, **5**, 639 (1962).

(192) (a) H. M. McConnell and A. D. McLachlan, *J. Chem. Phys.*, **34**, 1 (1961). (b) L. C. Snyder and A. D. McLachlan, *J. Chem. Phys.*, **36**, 1159 (1962).

(193) (a) M. Szwarc, M. Levy, and R. Milkovich, *J. Am. Chem. Soc.*, **78**, 2656 (1956). (b) M. Szwarc, *Nature*, **178**, 1168 (1956).

(194) J. L. R. Williams, T. M. Laakso, and W. J. Dulmage, *J. Org. Chem.*, **23**, 638 (1958).

(195) J. Jagur-Grodzinski and M. Szwarc, *Proc. Roy. Soc. (London)*, *Ser. A*, **288**, 224 (1965).

(196) R. Asami and M. Szwarc, *J. Am. Chem. Soc.*, **84**, 2269 (1962).

(197) G. Spach, H. Monteiro, M. Levy, and M. Szwarc, *Trans. Faraday Soc.*, **58**, 1809 (1962).

(198) M. Matsuda, J. Jagur-Grodzinski, and M. Szwarc, *Proc. Roy. Soc. (London)*, *Ser. A*, **288**, 212 (1965).

(199) S. C. Chadha, J. Jagur-Grodzinski, and M. Szwarc, *Trans. Faraday Soc.*, **63**, 2994 (1967).

(200) K. F. O'Driscoll, R. J. Boudreau and A. V. Tobolsky, *J. Polymer Sci.*, **31**, 115 (1958).

(201) K. F. O'Driscoll and A. V. Tobolsky, *J. Polymer Sci.*, **31**, 123 (1958).

(202) C. Walling, E. R. Briggs, W. Cummings, and F. R. Mayo, *J. Am. Chem. Soc.*, **72**, 48 (1950).

(203) A. V. Tobolsky and D. B. Hartley, *J. Polymer Sci. A*, **1**, 15 (1963).

(204) C. Stretch and G. Allen, *Polymer*, **2**, 151 (1961).

(205) K. F. O'Driscoll and A. V. Tobolsky, *J. Polymer Sci.*, **37**, 363 (1959).
(206) C. G. Overberger and N. Yamamoto, *J. Polymer Sci. A-1*, **4**, 3101 (1966).
(207) J. E. Mulvaney, C. G. Overberger, and A. Schiller, *Fortschr. Hochpolymer.-Forsch.*, **3**, 106 (1964).
(208) D. B. George and A. V. Tobolsky, *J. Polymer Sci. B*, **2**, 1 (1964).
(209) R. K. Graham, D. L. Dunkelberger, and W. E. Goode, *J. Am. Chem. Soc.*, **82**, 400 (1960).
(210) J. F. Garst, J. G. Pacifici, and E. R. Zabolotny, *J. Am. Chem. Soc.*, **88**, 3872 (1966).
(211) J. Saltiel and G. S. Hammond, *J. Am. Chem. Soc.*, **85**, 2515, 2516 (1963). See also G. S. Hammond et al., *ibid.*, **86**, 3197 (1964).
(212) W. Hückel, *Theoretische Grundlage der organischen Chemie*, Akadem. Verlag, 1934.
(213) W. E. Bachmann, *J. Am. Chem. Soc.*, **55**, 1179 (1933).
(214) W. Schlenk and E. Bergmann, *Ann.*, **464**, 1 (1928).
(215) H. V. Carter, B. J. McClelland, and E. Warhurst, *Trans. Faraday Soc.*, **56**, 455 (1960).
(216) H. E. Bent and A. J. Harrison, *J. Am. Chem. Soc.*, **66**, 969 (1944).
(217) N. Hirota and S. I. Weissman, *J. Am. Chem. Soc.*, **86**, 2537, 2538 (1964).
(218) N. Hirota and S. I. Weissman, *ibid.*, **82**, 4424 (1960).
(219) A. G. Evans, J. C. Evans, and E. H. Godden, *Trans. Faraday Soc.*, **63**, 136 (1967).
(220) B. J. McClelland, *Chem. Rev.*, **64**, 301 (1964); see p. 304.
(221) I. C. Lewis and L. S. Singer, *J. Chem. Phys.*, **43**, 2712 (1965).
(222) O. W. Haworth and G. K. Fraenkel, *J. Am. Chem. Soc.*, **88**, 4514 (1966).
(223) T. Förster and K. Kasper, *Z. Physik. Chem.*, **1**, 275 (1954), and *Z. Elektrochem.*, **59**, 976 (1955).
(224) B. Stevens and E. Hutton, *Nature*, **186**, 1045 (1960).
(225) G. J. Hoijtink, *Z. Elektrochem.*, **64**, 156 (1960).
(226) A. Ron and O. Schnepp, *J. Chem. Phys.*, **44**, 19 (1966).
(227) E. Beckman and T. Paul, *Ann.*, **266**, 1 (1891).
(228) E. R. Zabolotny and J. F. Garst, *J. Am. Chem. Soc.*, **86**, 1645 (1964).
(229) A. Carrington, H. C. Longuet-Higgins, and P. F. Todd, *Mol. Phys.*, **8**, 45 (1964).
(230) T. J. Katz, M. Yoshida, and L. C. Siew, *J. Am. Chem. Soc.*, **87**, 4516 (1965).
(231) H. J. Dauben and M. R. Rifi, *J. Am. Chem. Soc.*, **85**, 3041 (1963).
(232) N. L. Bauld and M. S. Brown, *ibid.*, **87**, 4390 (1965).
(233) R. L. Ward and S. I. Weissman, *ibid.*, **79**, 2086 (1957).
(234) F. C. Adam and S. I. Weissman, *ibid.*, **80**, 1518 (1958).
(235) S. I. Weissman, *Z. Elektrochem.*, **64**, 47 (1960).
(236) M. T. Jones and S. I. Weissman, *J. Am. Chem. Soc.*, **84**, 4269 (1962).
(237) P. J. Zandstra and S. I. Weissman, *J. Chem. Phys.*, **35**, 757 (1961).
(238) P. J. Zandstra and S. I. Weissman, *J. Am. Chem. Soc.*, **84**, 4408 (1962).
(239) N. M. Atherton and S. I. Weissman, *ibid.*, **83**, 1330 (1961).
(240) R. Chang and C. S. Johnson, *ibid.*, **88**, 2338 (1966).
(241) L. H. Piette and W. A. Anderson, *J. Chem. Phys.*, **30**, 899 (1959).
(242) J. M. Fritsch, T. P. Layloff, and R. N. Adams, *J. Am. Chem. Soc.*, **87**, 1724 (1965).
(243) S. Arai, D. A. Grev, and L. M. Dorfman, *J. Chem. Phys.*, **46**, 2572 (1967).
(244) M. Levy and M. Szwarc, *J. Am. Chem. Soc.*, **82**, 521 (1960).
(245) A. G. Evans and J. C. Evans, *J. Chem. Soc.*, **1963**, 6036.
(246) A. G. Evans and B. J. Tabner, *J. Chem. Soc.*, **1963**, 5560.
(247) A. G. Evans and J. C. Evans, *Trans. Faraday Soc.*, **61**, 1202 (1965).
(248) A. G. Evans and D. B. George, *J. Chem. Soc.*, **1961**, 4653.

(249) H. M. McConnell, *J. Chem. Phys.*, **35**, 508 (1961).

(250) S. I. Weissman, *J. Am. Chem. Soc.*, **80**, 6462 (1958).

(251) V. V. Voevodskii, S. P. Solodovnikov, and V. M. Chibrikin, *Dokl. Akad. Nauk SSSR*, **129**, 1082 (1959).

(252) W. L. Reynold, *J. Phys. Chem.*, **67**, 2866 (1963).

(253) J. Jagur, M. Levy, M. Feld, and M. Szwarc, *Trans. Faraday Soc.*, **58**, 2168 (1962).

(254) D. Gill, J. Jagur-Grodzinski, and M. Szwarc, *Trans. Faraday Soc.*, **60**, 1424 (1964).

(255) J. Jagur-Grodzinski and M. Szwarc, *ibid.*, **59**, 2305 (1963).

(256) D. Dadley and A. G. Evans, *J. Chem. Soc. (B)*, **1967**, 418.

(257) (a) R. Lipman, J. Jagur-Grodzinski, and M. Szwarc, *J. Am. Chem. Soc.*, **87**, 3005 (1965). (b) J. Stearne, J. Smid, and M. Szwarc, *Trans. Faraday Soc.*, **62**, 672 (1966).

(258) S. N. Khanna, M. Levy, and M. Szwarc, *Trans. Faraday Soc.*, **58**, 747 (1962).

(259) A. A. Arest-Yakubovich, A. R. Gantmakher, and S. S. Medvedev, *Dokl. Akad. Nauk SSSR*, **139**, 1331 (1961).

(260) D. Nicholls and M. Szwarc, *Proc. Roy. Soc. (London), Ser. A*, **301**, 223, 231 (1967).

(261) S. Pluymers and G. Smets, *Makromol. Chem.*, **88**, 29 (1965).

(262) W. Schlenk and R. Ochs, *Ber.*, **49**, 608 (1916).

(263) G. Wittig and D. Wittenberg, *Ann. Chem.*, **606**, 1 (1957).

(264) G. A. Russell and E. G. Janzen, *J. Am. Chem. Soc.*, **84**, 4153 (1962).

(265) F. J. Smentowski, *J. Am. Chem. Soc.*, **85**, 3036 (1963).

(266) A. G. Evans, J. C. Evans, and B. J. Tabner, *Proc. Chem. Soc.*, **1962**, 338.

(267) R. Waack and M. A. Doran, *J. Organometal. Chem.*, **3**, 92 (1965).

(268) E. Franta, J. Chaudhuri, A. Cserhegyi, J. Jagur-Grodzinski, and M. Szwarc, *J. Am. Chem. Soc.*, **89**, 7129 (1967).

(269) A. Prock, M. Djibelian, and S. Sullivan, *J. Phys. Chem.*, **71**, 3378 (1967).

(270) M. A. Doran and R. Waack, *J. Organometal. Chem.*, **3**, 94 (1965).

(271) D. H. Richards and M. Szwarc, *Trans. Faraday Soc.*, **55**, 1644 (1959).

(272) M. Morton, A. Rembaum, and E. E. Bostick, *J. Polymer Sci.*, **32**, 530 (1958).

(273) A. V. Tobolsky and D. B. Hartley, *J. Am. Chem. Soc.*, **84**, 1391 (1962).

(274) F. Bahsteter, J. Smid, and M. Szwarc, *J. Am. Chem. Soc.*, **85**, 3909 (1963).

(275) J. H. Binks and M. Szwarc, *J. Chem. Phys.*, **30**, 1494 (1959) also "Theoretic Organic Chemistry," *Chem. Soc. Spec. Publ.*, 262 (1959).

(276) A. H. Samuel and J. L. Magee, *J. Chem. Phys.*, **21**, 1080 (1953).

(277) R. L. Platzman, *U.S. At. Energy Comm.*, **305**, 34 (1953).

(278) J. W. Hunt and J. K. Thomas, *Rad. Res.*, to be published.

(279) M. Eigen, W. Kruse, G. Maass, and L. de Maeyer, *Progress in Reaction Kinetics*, Vol. 2, Pergamon Press, London, 1964, p. 285.

(280) A. Bhadani, J. Jagur-Grodzinski, and M. Szwarc, to be published.

(281) R. N. Adams, *J. Electroanal. Chemistry*, **8**, 151 (1964).

(282) P. A. Malachesky, T. A. Miller, T. Layloff, and R. N. Adams, *Proc. Sym. Exchange Reactions*, Upton, New York, 1965, p. 157.

(283) R. A. Marcus, *J. Phys. Chem.*, **67**, 853 (1963).

(284) B. E. Conway, *Theory and Principles of Electrode Processes*, Ronald, New York, 1965.

(285) J. Jordan and R. A. Javick, *Electrochim. Acta*, **6**, 23 (1962).

(286) G. J. Hoijtink, unpublished results.

(287) D. M. Hercules, *Science*, **145**, 808 (1964).

(288) R. E. Visco and E. A. Chandross, *J. Am. Chem. Soc.*, **86**, 5350 (1964).

(289) K. S. V. Santhanam and A. J. Bard, *ibid.*, **87**, 139 (1965).

(290) E. A. Chandross, J. W. Longworth, and R. E. Visco, *J. Am. Chem. Soc.*, **87**, 3259 (1965).
(291) S. W. Feldberg, *ibid.*, **88**, 390 (1966).
(292) D. M. Hercules, R. C. Lansbury, and D. K. Roe, *ibid.*, **88**, 4578 (1966).
(293) E. A. Chandross, private communication.
(294) W. Bruning and S. I. Weissman, *J. Am. Chem. Soc.*, **88**, 373 (1966).
(295) R. A. Marcus, *J. Phys. Chem.*, **67**, 853 (1963); *Ann. Rev. Phys. Chem.*, **15**, 155 (1964).
(296) D. L. Maricle and A. H. Maurer, *J. Am. Chem. Soc.*, **89**, 188 (1967).
(297) A. Zweig, D. L. Maricle, J. S. Brinen, and A. H. Maurer, *J. Am. Chem. Soc.*, **89**, 473 (1967).
(298) B. L. Funt and K. C. Yu, *J. Polymer Sci.*, **62**, 359 (1962).
(299) B. L. Funt and F. D. Williams, *J. Polymer Sci. A*, **2**, 865 (1964).
(300) (a) B. L. Funt and S. W. Laurent, *Can. J. Chem.*, **42**, 2728 (1964); (b) B. L. Funt and S. N. Bhadani, *ibid.*, **42**, 2733 (1964); *J. Polymer Sci. A*, **3**, 4191 (1965).
(301) N. Yamazaki, S. Nakahama, and S. Kambara, *J. Polymer Sci. B*, **3**, 57 (1965).
(302) B. L. Funt, D. Richardson, and S. N. Bhadani, *Can. J. Chem.*, **44**, 711 (1966).
(303) R. A. Marcus, *J. Chem. Phys.*, **43**, 2654 (1965).
(304) E. A. Chandross and R. E. Visco, to be published.
(305) R. S. Mulliken, *J. Phys. Chem.*, **56**, 801 (1952).
(306) H. Kainer and A. Uberle, *Chem. Ber.*, **88**, 1147 (1955).
(307) R. Foster and T. J. Thomson, *Trans. Faraday Soc.*, **58**, 860 (1962).
(308) W. Liptay, G. Briegleb, and K. Schindler, *Z. Elektrochem.*, **66**, 331 (1962).
(309) D. Bryce-Smith and J. E. Lodge, *J. Chem. Soc. (London)*, **1962**, 2675.
(310) C. Lagercrantz and M. Yhland, *Acta Chem. Scand.*, **16**, 508, 1043, 1799 (1962).
(311) R. L. Ward, *J. Chem. Phys.*, **38**, 2588 ; **39**, 852 (1963).
(312) D. Bijl, H. Kainer, and A. C. Rose-Innes, *J. Chem. Phys.*, **30**, 765 (1959).
(313) L. S. Singer and L. Kommandeur, *J. Chem. Phys.*, **34**, 133 (1961).
(314) M. Bose and M. M. Labes, *J. Am. Chem. Soc.*, **83**, 4505 (1961).
(315) D. B. Chesnut and W. D. Phillips, *J. Chem. Phys.*, **35**, 1002 (1961).
(316) A. Ottenberg, C. J. Hoffman, and J. Osiecki, *J. Chem. Phys.*, **38**, 1898 (1963).
(317) (a) H. Scott, G. A. Miller, and M. M. Labes, *Tetrahedron Letters*, **1963**, 1073; (b) H. Scott and M. M. Labes, *J. Polymer Sci. B*, **1**, 413 (1963).
(318) L. P. Ellinger, *Polymer*, **5**, 559 (1964).
(319) J. W. Breitenbach and O. F. Olaj, *J. Polymer Sci. B*, **2**, 685 (1964).
(320) L. P. Ellinger, *Polymer*, **6**, 549 (1965).
(321) K. Takakura, K. Hayashi, and S. Okamura, *J. Polymer Sci. B*, **3**, 565 (1965).
(322) K. Nomori, M. Hatano, and S. Kambara, *J. Polymer Sci. B*, **4**, 261 (1966).
(323) D. Laurin and G. Paravano, *J. Polymer Sci. B*, **4**, 797 (1966).
(324) J. Pac and P. H. Plesch, *Polymer (London)*, **8**, 237 and 252 (1967).
(325) B. L. Funt, *Macromolecular Reviews*, Vol. 1, A. Peterlin, M. Goodman, S. Okamura, B. H. Zimm, and H. F. Mark, Eds., Interscience, New York, 1967.
(326) L. S. Marcoux, J. M. Fritsch, and R. N. Adams, *J. Am. Chem. Soc.*, **89**, 5766 (1967).
(327) S. A. Rice, *Account Chem. Res.*, **1**, 81 (1968).
(328) S. Penczek, J. Jagur-Grodzinski, and M. Szwarc, *J. Am. Chem. Soc.*, **90**, 2174 (1968).
(329) B. J. Tabner and J. R. Yandle, *J. Chem. Soc. A*, **1968**, 381.
(330) T. Staples, J. Jaghr-Grodzinski, and M. Szwarc, to be published.
(331) S. Chadha, J. Jaghr-Grodzinski, and M. Szwarc, *Trans. Faraday Soc.*, in press.

Chapter VII

Propagation of Anionic Polymerization

1. PROPAGATION OF SIMPLE ANIONIC POLYMERIZATION

A variety of growing species may contribute to ionic propagation, and kinetic studies of anionic polymerization provided ample evidence for their existence. The use of living polymers enormously simplified such investigations. In a living polymer system termination is excluded by a proper choice of experimental conditions. Moreover, living polymers may be prepared prior to kinetic experiments, and therefore the initiation process is also eliminated. Hence, only propagation takes place on the addition of monomer to a solution of living polymers.

The concentration of the growing macromolecules is predetermined by the experimenter, and since it remains constant during the whole course of reaction the propagation is kinetically a first order reaction, i.e.,

$$-d[\text{M}]/dt = \text{const. [M]}.$$

In this equation M represents the monomer and the constant is equal to $k_p[\text{LP}]$, where k_p is the absolute rate constant of propagation and [LP] the concentration of living polymers. The latter is determined through independent methods, and thus the absolute rate constant of propagation is directly given by the kinetic data.

One of the first applications of this stratagem was described by Allen, Gee, and Stretch (1), who determined the absolute rate constant of styrene propagation in dioxane. Polymerization was initiated with a known amount of sodium naphthalene, and the propagation was followed in a dilatometer. Since the conversion of sodium naphthalene into sodium salt of living polystyrene is virtually instantaneous and quantitative, the concentration of living polymers was given by the known concentration of the initiator. Each kinetic run rigorously followed first order law, and the pseudo first order constants were found to be proportional to the concentration of living polymers. Thus, the propagation in this system involves apparently one species only, presumably the relevant $\sim\text{S}^-,\text{Na}^+$ ion-pair. The bimolecular propagation constant k_p was found to be about 4 liter mole^{-1} sec^{-1} at 25°C, and from its temperature dependence, investigated over a rather narrow range from 15 to 50°C, the activation energy was calculated as 9 ± 3 kcal/mole leading to a frequency factor of about 10^7 liter mole^{-1} sec^{-1}.

The studies of Gee's group were repeated and extended to other salts of living polystyrene by Bhattacharyya, Smid, and Szwarc (2). The decrease in styrene concentration was followed spectrophotometrically and the concentration of living polymers was determined in the course of each run by measuring the optical density at their absorption maximum at about 340 mμ. The second order character of the reaction was fully confirmed. No systematic trend was observed in the respective rate constants in spite of 24-fold variation in the concentration of living polystyrene. This may be verified by inspecting Table I.

TABLE I

Polymerization of Sodium Salt of Living Polystyrene in Dioxane at 25°C

[LP], $M \times 10^3$	[Styrene]$_0$, $M \times 10^3$	% conversion	$k_p = k_\oplus$, liter mole^{-1} sec^{-1}
0.50	7.5	28–68	3.2
3.02	63.5	15–55	3.4
4.95	89.5	20–73	3.4
9.4	93.8	44–89	3.3
12.0	41.6	76–97	3.8

These experiments were followed for periods varying from 17 to 1500 sec, depending on the product $k_p[LP]$. $k_\oplus$ is the propagation constant of the ion-pair.

The dependence of the propagation constant on the nature of the counterion is most interesting. As shown in Table II, which also includes the data reported by Dainton, Ivin, and their students (3), the reactivity of ion-pair substantially increases with size of the counterion, making $\sim\!\!\text{S}^-,\text{Li}^+$ the least, and $\sim\!\!\text{S}^-,\text{Cs}^+$ the most, reactive pair. The significance of this observation will be discussed later.

TABLE II

Homopropagation of Living Polystyrene in Dioxane at 25°C

	$k_\oplus$, liter mole^{-1} sec^{-1}, as reported by various investigators		
Counterion	Bhattacharyya et al. (2)	Gee et al. (1)	Dainton et al. (3)
Li$^+$	0.9		
Na$^+$	3.4	3–5	6.5
K$^+$	20		28
Rb$^+$	21.5		34
Cs$^+$	24.5		15

Investigation of anionic polymerization of styrene in dioxane is complicated by a phenomenon not mentioned in earlier reports. For no apparent reason, an irreversible transformation of living polystyrene often occurs in this solvent. The characteristic absorption peak of $\sim$S$^-$ carbanions observed at $\lambda_{max} = 340$ mμ diminishes and a new peak appears in the 500 mμ region. A similar phenomenon was observed in tetrahydrofuran, although less frequently and usually only in aged solutions. Its study was reported elsewhere (29) and the results of that investigation are discussed in Chapter XII. It should be stressed, however, that the kinetic data are reliable only when such transformation is prevented.

One more point deserves some comments. The dielectric constant of dioxane is low, about 2.4 at 25°C. Consequently, the degree of dissociation of ion-pairs into free ions is expected to be vanishingly small, and, indeed, no measurable conductance was detected in dioxane solution of living polystyrene (2). The low dielectric constant of the medium may, however, favor association of ion-pairs into unreactive quadrupoles or into some higher agglomerates. Such phenomena are encountered in hydrocarbon solutions of living polymers (see pp. 495–507 of Chapter VIII) and they profoundly influence the kinetics of polymerization.

The data collected in Table I show that the observed propagation constants of the anionic polymerization of styrene in dioxane are independent of the concentration of living polymers, and this apparently indicates lack of any association. However, the sodium salt of polystyryl anions prepared by Allen et al. (1), Bhattacharyya et al. (2), or Dainton et al. (3) all possessed *two* active ends per chain. Therefore, the association, if any, could be *intra-* and not *inter-*molecular, and thence its degree would not be affected by dilution. To clarify this problem two samples of living sodium polystyryls were prepared, one endowed with one growing end and the other with two. Their reactivities in dioxane were compared (4) and the results definitely proved that both types of polymers have identical reactivity—that is, no appreciable association of active ends takes place in this solvent in spite of its low dielectric constant.

2. IONS AND ION-PAIRS

The simplicity of the kinetics of propagation observed in dioxane should be contrasted with the more complex behavior of living polystyrene in tetrahydrofuran. The first kinetic study of this system was reported by Geacintov, Smid, and Szwarc (5), who developed a capillary-flow technique applicable for such studies. Their results revealed a small but undeniable trend in the observed propagation constant, clearly seen from inspection of Figure 1. The decrease in the concentration of sodium polystyrene from about 10^{-2}

to 10^{-3} M led to a 30% increase in the respective k_p, indicating that some reactive species are formed on dilution. Comparison of the viscosities of the living and dead polymers proved that association of active ends does not take place in tetrahydrofuran, leaving, therefore, the hypothesis of ionic dissociation as an alternative explanation for the observed phenomenon. Unfortunately,

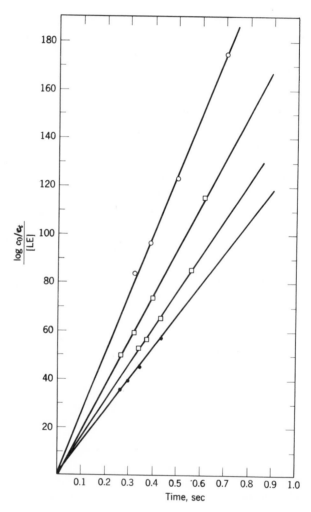

Fig. 1. Effect of living polystyrene concentration, [LE], on apparent bimolecular propagation constant, k_p, defined by the equation $-d[S]/dt = k_p[LE][S]$. In tetrahydrofuran, with Na$^+$ as counterion, and at 25°C: ($\bigcirc$) [LE] = 1.1 $\times$ $10^{-3}M$, ($\square$) 7.6 $\times$ $10^{-3}M$, ($\blacksquare$) 7.4 $\times$ $10^{-3}M$ (temp. $-7°$), ($\bullet$) 14.5 $\times$ $10^{-3}M$. Note that k_p values are given by the slopes of lines shown in the figure and that k_p *decreases* with increasing [LE].

some technical difficulties confused the issue and prevented the authors from drawing this conclusion. In fact, studies of Worsfold and Bywater (6), reported a year earlier, showed that the electrolytic dissociation constant of sodium polystyryl in tetrahydrofuran is $1.5 \times 10^{-7} M$ at 25°C. Thus, under experimental conditions of Geacintov et al. (5) about 1% of ion-pairs was dissociated.

The omission of Geacintov et al. (5) was soon rectified by Bhattacharyya, Lee, Smid, and Szwarc (7). Using an improved spectrophotometric technique, which allowed them to investigate the kinetics of propagation even for as low a concentration of living polymers as $10^{-5} M$, they demonstrated a linear relation between the apparent bimolecular propagation constant, k_p, and the reciprocal of the square root of living ends concentration, $[LE]^{-1/2}$. These relations are illustrated in Figure 2 for the Li^+, Na^+, K^+, Rb^+, and Cs^+ salts of living polystyrene in tetrahydrofuran at 25°C. They were soon confirmed by independent, although less extensive, studies of Schulz et al. (8), who limited their investigations to the sodium salt. Although the experimental technique utilized by the latter group was different from that developed by the Syracuse team, the quantitative agreement between both sets of data is remarkable. This may be seen by inspecting Figure 4; see also reference 9.

The above linear relation suggests that free ions and ion-pairs participate in the propagation process. For the equilibrium

$$\sim\!\!S^-, M^+ \rightleftarrows \sim\!\!S^- + M^+ \qquad K_{\text{diss}},$$

the fraction, f, of the free polystyryl ions, $\sim\!\!S^-$, is given by the relation

$$f^2/(1 - f) = K_{\text{diss}}/[LE].$$

The electrolytic equilibrium is not disturbed by propagation because this reaction does not change the concentration of living polymers.* Hence, the observed rate constant, k_p, is given by the sum $(1 - f)k_\oplus + f k_\ominus$, where $k_\oplus$ and $k_\ominus$ are the propagation constants of the ion-pair and the free ion, respectively. For $f \ll 1$, the approximation $f = (K_{\text{diss}})^{1/2}[LE]^{-1/2}$ is valid, and then

$$k_p = k_\oplus + (k_\ominus - k_\oplus)(K_{\text{diss}})^{1/2}[LE]^{-1/2}.$$

This relation is confirmed by the experimental findings presented in Figure 2. Since the slopes shown in that figure are positive, $k_\ominus$ has to be greater than $k_\oplus$. In fact, it will be shown later that for the polymerization of styrene proceeding at 25°C in tetrahydrofuran, as well as in many other ethereal systems, $k_\ominus \gg k_\oplus$. Then, further simplification is permissible, giving,

$$k_p = k_\oplus + k_\ominus(K_{\text{diss}})^{1/2}[LE]^{-1/2}.$$

---

* Propagation increases the length of the polymer molecule but the variation of molecular weight, if not too low, does not affect the dissociation constant of the ion-pairs.

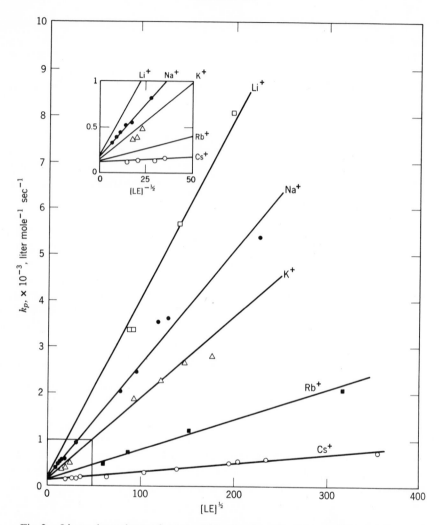

Fig. 2.   Linear dependence of apparent bimolecular rate constant of living polystyrene propagation on reciprocal of square root of living polymer concentration, with tetrahydrofuran as solvent, at 25°C (7). Different lines refer to different counterions—Li⁺, Na⁺, K⁺, Rb⁺, and Cs⁺.

It is important to note that the relation $k_\ominus > k_\oplus$ is by no means trivial, a point stressed by this writer (10). If the polarization of the approaching monomer is the main factor affecting the rate of propagation, a free ion which generates a more powerful electric field than an ion-pair should be more reactive. However, the opposite situation might be encountered if a push–pull mechanism operates. The monomer becomes, then, simultaneously polarized by the living anion, acting on one of its ends, and by the cation attracting the other. Such a concerted interaction perhaps could be more powerful than polarization due to a single charge.*

Conductance studies of salts of living polystyrene in tetrahydrofuran (THF) led to the dissociation constants listed in the second column of Table III (7b,11). For the sodium salt the value obtained by the Syracuse group

TABLE III

Dissociation Constants of $\sim S^-, M^+$ Ion-Pairs in THF at 25°C

$$\sim S^-, M^+ \rightleftarrows \sim S^- + M^+ \qquad K_{diss}$$

	$K_{diss} \times 10^7$, M	
Counterion	Conductance	Kinetics
Li$^+$ (one active end)	1.9[a]	2.2[a]
Na$^+$ (one active end)	1.5[b] (extrap.)	
Na$^+$ (two active ends)	1.5,[a] 1.5,[b] 1.5[c]	1.5,[a] 0.8[d]
K$^+$ (two active ends)	0.7[a]	0.8[a]
Rb$^+$ (two active ends)		0.11[a]
Cs$^+$ (one active end)	0.028[a]	0.021[a]
Cs$^+$ (two active ends)	0.165[a,e]	0.0046[a,e]

[a] Bhattacharyya, Lee, Smid, and Szwarc (7b).

[b] Shimomura, Tölle, Smid, and Szwarc (11).

[c] Worsfold and Bywater (6).

[d] Hostalka and Schulz (8b). This result was disputed. See refs. 8 and 13 and the reply in ref. 12.

[e] The reasons will be given later as to why different values of the apparent K_{diss} of cesium salt of living polystyrene are observed when the polymer possesses two active ends.

agrees well with that reported by Worsfold and Bywater (6). The variations in K_{diss}, arising from changes in the nature of the counterion, closely resemble those found for some other salts of alkali metals in tetrahydrofuran (14). Having the dissociation constants of alkali salts of living polystyrene, we may calculate the absolute propagation constant of the free $\sim S^-$ anions from the slopes of the lines shown in Figure 2. Thus, $k_\ominus$ was found to be 65,000 liter mole^{-1} sec^{-1} at 25°C.

It was assumed that the increase in the apparent propagation constant, k_p, arising from dilution of the polymerized solution is caused by the increase in proportion of the free $\sim S^-$ ions. The effect of added salts, sharing a common cation with the living polystyrene, confirms this hypothesis. Such an addition suppresses the ionic dissociation of the living polymer and, therefore, it should retard the propagation. The alkali tetraphenyl borides are conveniently suited for this purpose, because their degree of ionization in tetrahydrofuran is much higher than that of the respective living polystyrenes (14). Therefore,

* The push-pull mechanism might require less activation energy than the mechanism operating in the free anion propagation. However, the entropy of activation is expected to be more negative in the former case.

in the presence of an excess of such a salt, virtually all the free cations are derived from its dissociation, and the fraction, f, of the free polystyryl anions is then determined by the concentration of the cations, $[M^+]$, i.e.,

$$f = K_{diss}/[M^+].$$

The concentration of M^+ may be calculated if the dissociation constant of the tetraphenyl boride (14) and its concentration in the polymerized solution are

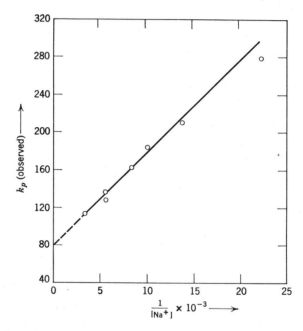

Fig. 3. Linear dependence of the apparent bimolecular rate constant of living polystyrene on reciprocal of $[Na^+]$ in THF at 25°C. Sodium ions were introduced by dissolving sodium tetraphenyl boride.

known. The observed rate constant of polymerization in the presence of salt should be given, therefore, by the equation

$$k_p = (1 - f)k_\oplus + fk_\ominus = k_\oplus + (k_\ominus - k_\oplus)K_{diss}/[M^+],$$

i.e., a plot of k_p versus $1/[M^+]$ would be linear, its intercept giving again $k_\oplus$ but the slope being $(k_\ominus - k_\oplus)K_{diss}$ instead of $(k_\ominus - k_\oplus)(K_{diss})^{1/2}$ as required for the previously described experiments. These predictions were fully confirmed by studies of Bhattacharyya et al. (7) and Shimomura et al. (11), and for the sake of illustration a plot of the observed k_p versus $1/[Na^+]$ is shown in Figure 3.

The kinetic data obtained in the presence and absence of the added salt permit direct calculation of K_{diss} of the living polymer and of the absolute propagation constant of its free ion. For example, the slope of the line shown in Figure 3 gives $k_\ominus K_{diss}$ for the sodium salt of living polystyrene in tetrahydrofuran. The slope of the corresponding line, shown in Figure 2 and derived from the experiments performed in the absence of added salt, provides $k_\ominus(K_{diss})^{1/2}$. The omission of $k_\oplus$ in both these cases is justified *a posteriori* by showing that $k_\ominus \gg k_\oplus$. By squaring the ratio of these slopes we obtain K_{diss}, and the square of the latter divided by the former gives $k_\ominus$. The values of dissociation constants of alkali salts of living polystyrene obtained by this method are listed in the last column of Table III, and agree well with those derived from the conductance study.

A variant of this approach was described by Schulz (8b). By applying the mass law to a solution of living polymer containing an added salt which shares with it a common cation, the following relation is derived:

$$K_{diss} = \{\tfrac{1}{2}\kappa K_{B^-,Me^+}[-1 + \sqrt{1 + 4[B^-,Me^+]/K_{B^-,Me^+}(1 - \kappa^2[LE])}]\}^2.$$

K_{diss} is the dissociation constant of the living polymer, K_{B^-,Me^+} that of the added salt, [LE] and $[B^-,Me^+]$ the concentrations of the respective species, and κ the experimentally acquired variable defined as $(k_p - k_\oplus)\alpha$. Here k_p is the observed rate constant of propagation in the presence of the added salt, while $k_\oplus$ and α are the intercept and slope, respectively, of the line obtained by plotting the observed propagation constant in the *absence* of salt versus $1/[LP]^{1/2}$. Note that $\kappa = f(K_{diss})^{1/2}$, where f is the fraction of dissociated living polymers.

The above algebraic procedure is equivalent, in principle, to the simpler graphic method outlined previously. It has two drawbacks: it heavily depends on the exact value of $k_\oplus$, and it does not clearly show the reliability and self-consistency of the available data. Schulz tested this approach for the system tetrahydrofuran solution of living polystyrene containing sodium tetraphenyl boride (8b). Since the experimental conditions were most unfortunately chosen (9,13), the calculated K_{diss} was susceptible even to minute errors in k_p, as was pointed out by Tölle et al. (9), and consequently erroneous values for the dissociation constant were obtained. Nevertheless, Schulz's data agree with those reported by the Syracuse group if treated graphically, as may be seen from inspection of Figure 4.

A purely technical point may be raised. The propagation constant of ion-pairs, $k_\oplus$, is given by the intercept of an appropriate experimental line. In most systems the reactivity of free living anions exceeds by a large factor that of ion-pairs. Thus, the lines giving the dependence of k_p upon $1/[LP]^{1/2}$ are steep and, in the required experimental scale, the intercepts are often indistinguishable from 0. Experiments performed in the presence of added salts which suppress the dissociation yield flatter lines, and then the intercepts become more reliable (compare Figs. 2 and 3). In fact, the rate constant of ion-pairs may be determined by studying the kinetics of propagation in the presence of a large excess of the salt (3,8b,15). Although such a procedure may create some problems (see p. 440), on the whole it gives reliable results.

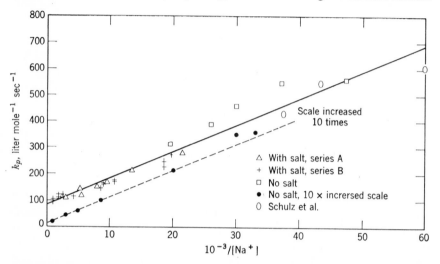

Fig. 4. Combined results of kinetic studies of living polystyrene polymerization in tetrahydrofuran at 25°C in the presence and absence of added sodium tetraphenyl boride (11). Note that all the points fit a common line if K_{diss} of $\sim\!S^-,Na^+$ is taken to be 1.5 × $10^{-7}M$. In order to extend the graph to very high dilutions and, therefore, very high values of k_p, the respective points, marked by full circles, were presented in a 10 times smaller scale. The broken line is, therefore, the extension of the solid line. The results obtained by three groups of workers are concordant (see refs. 9,12, and 13).

The kinetic data obtained in the absence and presence of added salt may be compared by plotting k_p as a linear function of $1/[M^+]$, if the dissociation constant of living polymer is known. Such a plot is illustrated by Figure 4. In a most general case the concentration of the free cations, $[M^+]$, is given by the solution of three simultaneous equations:

$$[M^+][A^-] = K_{salt}\{[salt] - [A^-]\},$$
$$[M^+][LP^-] = K_{diss}\{[LP]_T - [LP^-]\},$$
$$[M^+] = [A^-] + [LP^-],$$

where A^- and LP^- denote the anions derived from the added salt and from the living polymer, respectively, $[LP]_T$ represents the total concentration of the polymer, and K_{salt} is the dissociation constant of the added salt. Usually $[LP^-] \ll [LP]_T$, and then the fraction, f, of free living polymers is given by

$K_{diss}/[M^+]$. If this is not the case, f must be calculated from the more awkward equation

$$f/(1 - f) = K_{diss}/[M^+].$$

Finally, we may inquire whether the treatment outlined above needs refinement to account for the deviations of the pertinent activity coefficients from unity. Such corrections were introduced by Schulz (8b), who used the Debye–Hückel equation for their calculations. The problem of activity coefficients was treated also by Shimomura et al. (17), who computed them from the experimental activity coefficients determined for chlorides or bromides in aqueous solution, using a scaling factor arising from different dielectric constants of the medium employed and water. The latter approach gives higher values for the activity coefficients than does the Debye–Hückel equation. In most cases, the activity coefficients are close to unity and therefore their omission is permissible. However, for values lower than 0.9, corrections may be desirable. The activity coefficients appear in the calculation of the equilibrium concentration of the free ions and therefore affect the value of f—the fraction of the free living anions, whereas the kinetic equation $k_p = (1 - f)k_\oplus + fk_\ominus$ is not affected directly. As the activity coefficient decreases, f becomes larger, and therefore the observed value of k_p may increase more steeply than expected. It is not yet clear whether or not other factors, e.g., the formation of triple-ions, are offsetting the Debye–Hückel type of deviations from ideality. Unfortunately, the kinetic techniques presently available are not sufficiently refined to justify further study of this problem.

The kinetic studies of anionic polymerization of living polystyrene in tetrahydrofuran led to the following results. (1) The reactivity of ion-pairs decreases along the series $Li^+ > Na^+ > K^+ \sim Rb^+ > Cs^+$. For example, the respective propagation constants at 25°C are about 150 for $\sim\!S^-,Li^+$, 80–90 for $\sim\!S^-,Na^+$, about 60–50 for $\sim\!S^-,K^+$ and $\sim\!S^-,Rb^+$, and 20–25 for $\sim\!S^-,Cs^+$, all in units of liter mole^{-1} sec^{-1}. (2) The decrease in the propagation constants of ion-pairs is paralleled by a decrease in their dissociation constants, as may be seen from inspection of Table III. (3) The free polystyryl anions, $\sim\!S^-$, are enormously more reactive than the ion-pairs. Their rate constant of propagation, $k_\ominus$, is 65,000 liter mole^{-1} sec^{-1} at 25°C, i.e., about 800 times as high as the propagation constant of $\sim\!S^-,Na^+$ ion-pairs.

Because the free polystyryl anions are so much more reactive than their ion-pairs, their contribution to polymerization may be substantial even if their proportion is low. For example, in a $10^{-3}M$ solution of sodium salt of living polystyrene the free $\sim\!S^-$ anions form only 1% of all the growing polymers. Nevertheless, they contribute 89% to the polymerization, whereas the remaining 99% of living polymers contribute only 11% to the reaction.

This situation causes some broadening of the molecular weight distribution of the resulting product (see Section 7 of Chapter II), although the rapid dissociation of the ion-pairs and association of the free ions (see p. 453) makes the effect relatively insignificant.

Comparison of the reactivities of ion-pairs in tetrahydrofuran and in dioxane is most intriguing. In the latter medium the lithium pair is the least reactive, $k_\oplus = 0.9$ liter mole^{-1} sec^{-1}, whereas it is the most reactive pair in the former solvent, $k_\oplus \sim 150$ liter mole^{-1} sec^{-1}. On the other hand, the cesium salt is the least reactive ion-pair in tetrahydrofuran but the most reactive in dioxane, although the respective absolute rate constants of propagation of $\sim\!\!S^-,Cs^+$ are approximately the same in both media, viz., $k_\oplus \sim 20$–25 liter mole^{-1} sec^{-1}. To appreciate the significance of these facts, we have to learn more about the effect of temperature on anionic polymerization.

3. EFFECT OF TEMPERATURE ON ANIONIC PROPAGATION. HEAT OF DISSOCIATION OF SALTS OF LIVING POLYSTYRENE

In the course of early kinetic studies of the anionic polymerization of styrene in tetrahydrofuran it was noted that temperature has only a slight effect upon the rate of propagation. These observations led Dainton et al. (18) to the conclusion that the activation energy of anionic propagation is abnormally low, and their qualitative findings were confirmed by quantitative studies of Geacintov et al. (5). The latter determined the absolute rate constants of propagation of sodium salt of living polystyrene at $-7, 0$, and $+25°C$ and calculated the activation energy of this process to be 1 kcal/mole only, corresponding to a large negative entropy of activation. Such data contrasted sharply with those reported by Gee et al. (1) for the polymerization proceeding in dioxane for which the activation energy was determined at about 9 kcal/mole and the entropy of activation was "normal."

The apparent rate constant, k_p, determined in each individual kinetic run is a complex entity. Its value is determined by *two* processes characterized by the rate constants $k_\oplus$ and $k_\ominus$, and by the extent to which each contributes to the propagation. Therefore, as was pointed out by Bhattacharyya et al. (7b), calculation of activation energy from the temperature dependence of k_p is misleading. The degree of dissociation of ion-pairs in ethereal solvents (and in many other media) *increases* with *decreasing* temperature. In the living polystyrene system such dissociation produces the extremely reactive $\sim\!\!S^-$ free ions and, hence, the contribution of the more reactive species to propagation is higher at lower temperatures. This factor makes the apparent activation energy of anionic propagation in tetrahydrofuran abnormally low; under suitable conditions, it may lead even to a "negative activation energy."

To handle the problem properly we have to determine the heat of dissociation of ion-pairs into free ions and the activation energies of the individual elementary steps, i.e., those associated with the propagation by free ions and by ion-pairs. The heat of dissociation of $\sim\!\!S^-,M^+$ into the free ions is determined by the van't Hoff method. Plots of log K_{diss} versus $1/T$ are shown in Figure 5 for the sodium and cesium salts, respectively, in tetrahydrofuran. Two points emerge from its inspection:

(1) The dissociation constant of the sodium salt increases by a factor of about 300 as the temperature decreases from 25 to $-70°C$. The formation of the free $\sim\!\!S^-$ and Na^+ ions is therefore appreciably exothermic, the relevant $-\Delta H$ being about 9 kcal/mole. On the other hand, the dissociation of the cesium salt is only slightly exothermic, $-\Delta H = 2$ kcal/mole.

(2) The degree of dissociation of the sodium salt is much higher than that of the cesium salt. For example, the equilibrium constants are $1.5 \times 10^{-7}M$ and $0.03-0.05 \times 10^{-7}M$, respectively, at 25°C.

Let us consider the significance of these observations. It is known (14) that the free sodium cation strongly interacts with molecules of tetrahydrofuran and, consequently, becomes coordinated with them if dissolved in that ether. The cesium ion remains virtually bare in tetrahydrofuran, i.e., no tight coordination shell is formed around it in that medium. These relations are revealed by studies of the respective Stokes radii (14). The intrinsically small sodium ion is bulkier in tetrahydrofuran (Stokes radius ~ 4.2 Å) than the basically large cesium cation (Stokes radius ~ 2.2 Å only). The degree of solvation of cations is modified on formation of ion-pairs. There is evidence that both the sodium and the cesium salts of living polystyrene form contact ion-pairs in tetrahydrofuran, i.e., the cations are nonsolvated, or at least only peripherally solvated in the undissociated salt. The coordination of the free sodium cation with the ether, which takes place in the course of dissociation, therefore provides an additional driving force, facilitating the ionization of the pair and increasing the exothermicity of this reaction. Hence, in spite of its tighter structure, the sodium polystyrene is more ionized in tetrahydrofuran, and its dissociation is more exothermic than that of the cesium salt.

In a more powerfully solvating medium even the bulky cesium ion may become coordinated with molecules of the solvent. Such a situation is encountered in dimethoxyethane (14). Indeed, the dissociation of the cesium

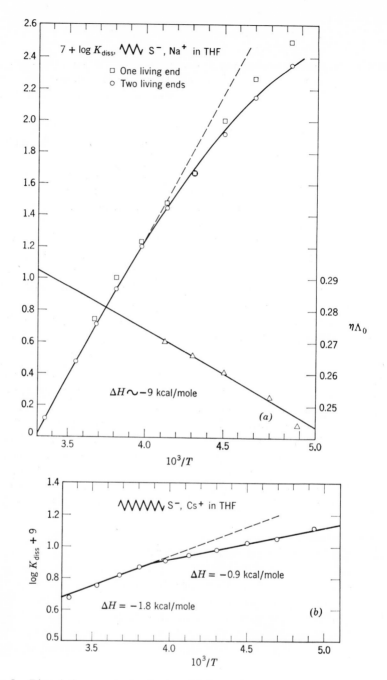

Fig. 5. Dissociation constants of salts of living polystyrene in tetrahydrofuran for temperatures ranging from -70 to $+25°C$ (11). (a) Sodium salt. Note the *high* heat of dissociation. The triangles give the Walden product $\eta\Lambda_0$ at different temperatures. (b) Cesium salt. Note the *low* heat of dissociation.

salts of living polystyrene is much higher in that solvent than in tetrahydrofuran—K_{diss} is 1×10^{-7} M in the former while only $\sim 0.04 \times 10^{-7}$ M in the latter solvent at 25°C. The exothermicity of dissociation is also larger, i.e., $-\Delta H = 5$ kcal/mole in dimethoxyethane (17) as compared with $-\Delta H$ less than 2 kcal/mole found in tetrahydrofuran.

4. PROPAGATION OF POLYMERIZATION BY FREE POLYSTYRYL ANIONS

Complete kinetic studies which provide information about the rates and activation energies of propagation by free polystyryl anions are limited at present to three solvents—tetrahydrofuran (11), dimethoxyethane (17), and tetrahydropyran (20). The pertinent data are collected in Table IV. Less comprehensive data limited to one temperature only are available for 2-methyltetrahydrofuran (19) and for mixtures of tetrahydrofuran with dioxane (21) and benzene (22).

The temperature dependence of the propagation constant, $k_\ominus$, of the free polystyryl anions is depicted by the Arrhenius plots shown in Figures 6 and 7. The deviation from linearity, observed at the lowest temperatures and revealed by Figure 6, is intriguing. Whether this behavior is genuine or an artifact of

TABLE IV
Propagation Constants of Free Polystyrene Anion

Solvent	$k_\ominus \times 10^{-4}$ liter mole^{-1} sec^{-1}				$E,$ kcal/mole	log A, liter mole^{-1} sec^{-1}
	25°C	0°C	−30°C	−60°C		
Tetrahydrofuran,[a] study of the						
Na$^+$ salt	6.5	1.6	0.39	0.14⎫	5.9	9.05
Cs$^+$ salt	6.3	2.2	0.62	0.11⎬		
Dimethoxyethane,[b] study of the						
Cs$^+$ salt	∼4	1.6–2.2	0.6–0.7	0.1	∼5	8.3
Tetrahydropyrane,[c]						
Li$^+$ salt	5.9–7.3	2.8	1.4	0.6	5.1	8.9
	(23°C)		(−20°C)	(−40°C)		

[a] Shimomura, Tölle, Smid, and Szwarc (11).
[b] Shimomura, Smid, and Szwarc (17).
[c] Fisher, Smid, and Szwarc (20).

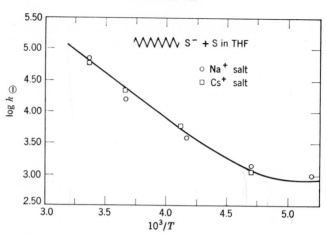

Fig. 6. Arrhenius plot of absolute rate constants of propagation of free polystyryl anion in tetrahydrofuran obtained from: (○) studies of the sodium salt and (□) studies of the cesium salt.

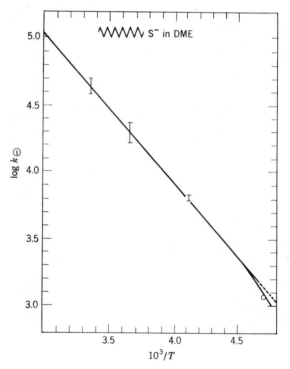

Fig. 7. Arrhenius plot of absolute rates of propagation of the free polystyryl anion in dimethoxyethane.

the experimental technique is not known; the propagation observed in dimethoxyethane or in tetrahydropyran does not reveal such effects.

TABLE V

Absolute Rate Constants of Propagation by Free Polystyryl Anions in Different Solvents at 25°C

Solvent	$k_\ominus$, liter mole^{-1} sec^{-1}	Ref.
Tetrahydrofuran	63,000–65,000	(11)
Dimethoxyethane	~40,000	(17)
2-Methyltetrahydrofuran	27,000–30,000	(19)
Tetrahydropyrane	59,000–72,000	(20)
Tetrahydrofuran–dioxane mixtures	~60,000	(21)
Tetrahydrofuran–benzene mixtures	40,000–70,000	(22)

The data collected in Table V show an insignificant influence of solvent upon the rate of propagation, and those collected in Table IV demonstrate that the activation energy and the entropy of activation are also virtually identical in tetrahydrofuran, tetrahydropyran, and dimethoxyethane. Apparently the free anions interact weakly with ethers and do not become coordinated with their molecules.* This conclusion is supported by other observations. For example, the mobilities of aromatic radical-anions in those solvents are only slightly smaller than those of their parent hydrocarbons (23) demonstrating the absence of tight solvation shells around these ions.

The propagation constant of the free polystyryl anion is much higher than that of the free polystyryl radical, the latter being only 27 liter mole^{-1} sec^{-1} at 25°C (24). Two factors contribute to this result: (1) The activation energy of the free $\sim\!\!\!S^-$ ion propagation (5–6 kcal/mole) is lower than that of the free $\sim\!\!\!S\cdot$ radical (7.8 kcal/mole). (2) The A factor is larger for the anionic than for the radical reaction, the pertinent values being 10^8 to 10^9 liter mole^{-1} sec^{-1} for the former as compared with $\sim\!10^7$ liter mole^{-1} sec^{-1} for the latter.

The lower activation energy could be expected. The approaching monomer is polarized by the free anion. This generates a relatively long-range attractive force and reduces, therefore, the repulsion responsible for the potential energy barrier. The larger A factor also seems plausible. The negative charge of the anion becomes dispersed in the transition state and, consequently, the degree of order in the surrounding liquid is diminished, thereby increasing the entropy of activation. Of course, such an effect does not operate in the free-radical polymerization. It is also possible that the incipient C—C bond

---

* Of course, the environment should affect the reactivity of free $\sim\!\!\!S^-$ ion if it interacts with the growing centers. Unfortunately, no polymerization induced by free ions was investigated in media which strongly interact with carbanions.

is longer in the anionic process than in the radical addition. The freedom of the reacting monomer in the transition state of ionic polymerization would then be greater than that expected in the radical process.

It is instructive to compare the reactivity of the $\sim\sim S^-$ anion with that of the free $\sim\sim S^+$ cation. Propagation of styrene polymerization induced by free carbonium ions was investigated in γ-irradiated systems (25). The available data indicate that the $k_\oplus$ is of the order of 10^6–10^8 liter mole^{-1} sec^{-1}, i.e., larger by several orders of magnitude than the corresponding $k_\ominus$. Since carbonium ions are expected to be intrinsically more reactive than carbanions, these results are not surprising.

Investigations of polystyrene systems are supplemented by similar studies of α-methylstyrene polymerization. Worsfold and Bywater (26) were the first to follow the kinetics of propagation of living poly-α-methylstyrene in tetrahydrofuran, although they did not discriminate between free ions and ion-pairs. The differentiation was achieved later by Dainton, Ivin, and their students (27), who determined dilatometrically the apparent rate constants of propagation at various dilutions and evaluated the relevant dissociation constant of ion-pairs by conductometric technique. They found the dissociation constant of the lithium salt to be $5.6 \times 10^{-9}M$ in tetrahydropyran at $-30°C$ and that of the sodium salt to be $3.8 \times 10^{-7}M$ in tetrahydrofuran at $-31°$. The charge density on the carbanion of poly-α-methylstyrene should be higher than that on the C^- atom of the living polystyrene, because the methyl group is an electron-donating substituent. Hence, the dissociation constant of $\sim\sim S^-,Na^+$ (27×10^{-7} in tetrahydrofuran at $-30°C$) should indeed be higher than that found for $\sim\sim\alpha\text{-MeS}^-,Na^+$. Nevertheless, the heats of dissociation are comparable—8 and 7.5 kcal/mole, respectively. The kinetic results pertaining to living poly-α-methylstyrene system are collected in Table VI and graphically presented in Figure 8. It was concluded again that the reactivity of free growing anion is negligibly affected by the nature of the

TABLE VI

Anionic Polymerization of α-Methylstyrene Induced by Free Anions (27)

Counterion	Solvent	Temp., °C	$10^2 \, k_\ominus K_{diss}^{1/2}$, (mole/liter)$^{-1/2}$ sec^{-1}	$K_{diss} \times 10^9$, M	$k_\ominus$, liter mole^{-1} sec^{-1}
Li$^+$	Tetrahydropyran	-3	2.5	5.6	340 ± 120
		-11.5	1.6	6.8	190 ± 40
		-16.2	0.75	7.4	87 ± 20
Na$^+$	Tetrahydrofuran	-30	3.0	380	48 ± 6
		-59	1.2	2600	7.7 ± 0.8
K$^+$	Tetrahydrofuran	-59	0.24	100	7.7 ± 0.4

ethereal solvent and, in fact, the data derived from studies of tetrahydrofuran and tetrahydropyran systems were presented on a common Arrhenius line. Thus, the activation energy was calculated to be 7.2 kcal/mole corresponding to the log A of 8.2.

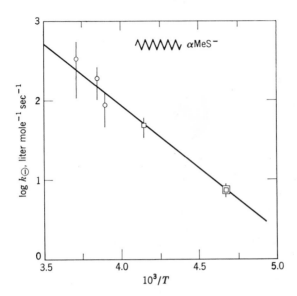

Fig. 8.   Arrhenius plot for propagation of the free anion of poly-α-methylstyrene in tetrahydrofuran (□) and tetrahydropyrane (○) (27). Vertical lines indicate the extent of experimental uncertainty.

A comparison of activation energies of styrene and α-methylstyrene polymerization shows that the latter is higher than the former only by 2 kcal/mole (28). This small increase contrasts with the substantial decrease in the respective heats of polymerization: $-\Delta H$ is larger by about 8–9 kcal/mole for styrene polymerization than for α-methylstyrene. The low heat of polymerization of the latter monomer reflects the considerable steric strain developed in the molecule of poly-α-methylstyrene. However, as pointed out by Alfrey and Ebelke (30) such a strain should be smaller in the transition state of propagation and the above results confirm their prediction.

An interesting question was raised by Vesely (130), namely, does the propagation involve one step only, viz.,

$$\sim\sim M^- + M \rightarrow \sim\sim MM^-,$$

or are two steps required, namely an electron transfer,

$$\sim\sim M^- + M \rightarrow \sim\sim M\cdot + M^{\overline{\cdot}},$$

producing a radical-ion followed by the combination of the partners,

$$\sim M \cdot + M^{\bar{\cdot}} \rightarrow \sim MM^-.$$

Addition of styrene (S) to the dianion of anthracene illustrates a process in which a two-step mechanism might be most plausible, namely,

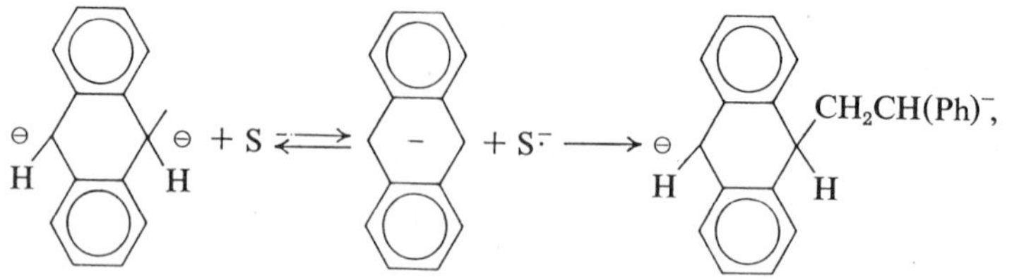

because the separation of electrons and the formation of stable radical-ion ($A^{\bar{\cdot}}$) should facilitate this route. However, even in this most favorable case it was shown (131) that the addition occurs in one step only. Consequently, this writer believes that the propagation of anionic polymerization does not involve electron transfer prior to the bond formation.

5. PROPAGATION OF ANIONIC POLYMERIZATION BY ION-PAIRS

Anionic polymerization propagated by ion-pairs is much more complex than the reaction initiated by free ions. A free living anion has no companion which could affect its reactivity, whereas in an ion-pair the anion is chaperoned by a counterion. Hence, the rate of polymerization induced by ion-pairs may be modified by varying the nature of cation—a problem we must treat in greater detail. Moreover, while the free-ion polymerization is only slightly influenced by solvent, its change may enormously affect the reactivity of ion-pairs. For example, the propagation constant of sodium polystyrene is about 3 liter $mole^{-1} sec^{-1}$ in dioxane, 80 in tetrahydrofuran, and about 3600 in dimethoxyethane at 25°C. The problem is further complicated by the existence of different types of ion-pairs, e.g., contact and solvent-separated pairs may coexist in the same solution (90).

An ion-pair probably becomes partially separated in the transition state of propagation. The following diagram might depict the course of such a reaction:

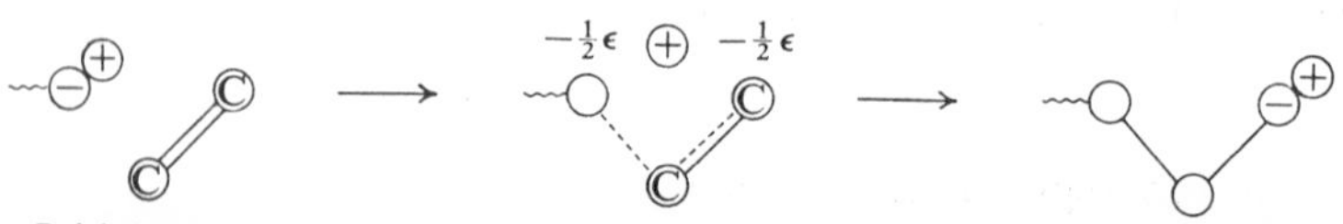

Initial state   Transition state   Final state

In a poorly solvating medium the partial dissociation is not facilitated by solvation, or rather coordination, of the cation with solvent molecules, and then the increasing Coulombic interaction between ions decreases the reactivity of a pair. In such a case, the lithium salt is expected to be the least reactive, while the reactivities of other alkali salts should increase monotonically with the radii of the pertinent cations making the cesium salt the most reactive. This relatively simple situation apparently is encountered in the anionic propagation of styrene, and perhaps also α-methylstyrene, in dioxane, as implied by the data collected in Table VII.

TABLE VII

Propagation by Living Polystyrene Salts in Various Solvents at 25°C

| Solvent | $k_{\oplus}$, liter mole^{-1} sec^{-1} | | | | | |
	Li$^+$	Na$^+$	K$^+$	Rb$^+$	Cs$^+$	Ref.
	Living Polystyrene Salts					
Dioxane	0.9	3.4	20	21.5	24.6	(2)
		6.5	28	34	15	(3)
Benzene (nonassociated)			47	24	18	(34)
Tetrahydropyrane	< 10^a	14,12^a	73	83	53	(3,16, 132)
		10				(15)
2-Methyltetrahydro- furan	57	11	7.5		22	(19)
Tetrahydrofuran	∼160	80	∼100(?)	∼50	25	(7)
Dimethoxyethane		3600			∼150	(17)
	Living Poly-α-Methylstyrene Salts					
Dioxane		0.02	0.1	0.06		(3)
Tetrahydropyran	2.6	0.047	0.25	0.35	0.26	(31)
Tetrahydrofuran		0.04				(26)

^{a} Aging the solution inhibits the reaction making $k_{\oplus}$ smaller.

As seen from Table VII, the patterns are more complex in other media. There are several factors which may account for such behavior. For example, one may argue that no partial dissociation is needed in the transition state of propagation if the cation is sufficiently large. The following idealized diagram clarifies this idea:

Partial dissociation in the transition state—the counterion is small

No partial dissociation in the transition state—the counterion is large

The coordination of the cation with the solvent is the most important factor which affects the pattern of reactivities. This may facilitate the partial dissociation of the pair and speed up propagation. Since the coordination with small ions is more pronounced, the lithium or sodium ion-pairs may, after all, turn out to be more reactive than expected from the "tightness" of the respective ion-pairs.

Does the propagation involve one type of ion-pair or two or more different pairs? The temperature dependence of propagation induced by ion-pairs provides some insight into this question. The rate constant of polymerization of cesium living polystyrene in tetrahydrofuran (11) decreases with falling

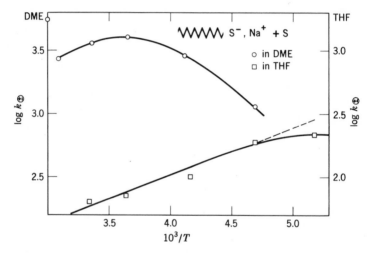

Fig. 9. Arrhenius plot of propagation constant $k_{\oplus}$ of $\sim\!S^-$,Na$^+$ ion-pairs in tetrahydrofuran and in dimethoxyethane (11,17). Note the "negative activation energy" in the former solvent and the maximum in the plot giving the results obtained in the latter medium.

temperature and the Arrhenius activation energy was found to be about 5.7 kcal/mole with $\log A_{\oplus} = 5.35$ liter mole^{-1} sec^{-1}. This is a "normal" behavior. The low A factor may indicate a higher degree of solvent orientation around the partially dissociated pair formed in the transition state of propagation. In fact, dissociation of cesium living polystyrene in tetrahydrofuran decreases the entropy of the system by 44 eu. Hence, the additional negative entropy of activation of about 14 eu is quite plausible.

Sodium living polystyrene behaves differently. Figure 9 shows a plot of $\log k_{\oplus}$ for $\sim\!S^-$,Na$^+$ in tetrahydrofuran and in dimethoxyethane. In tetrahydrofuran the rate constant of this polymerization *increases* with *decreasing* temperature, thus leading to the formal "negative activation energy" of about

-1.5 kcal/mole. The potential energy barrier for any elementary process cannot be negative and, therefore, this peculiar temperature dependence indicates some complexity of the reaction. At least three possible explanations may account for this phenomenon.

(*1*) A reaction $A + B \rightarrow C$ may proceed by two steps: a fast reversible complex formation producing minute amounts of (AB),

$$A + B \rightleftharpoons (AB) \qquad K_a,$$

followed by a rate-determining unimolecular rearrangement,

$$(AB) \rightarrow C \qquad k_r.$$

The observed overall rate constant k is equal then to $K_a k_r$ and the experimental activation energy is given by $\Delta H_a + E_r$. If the association is exothermic and the absolute value of the negative ΔH_a greater than the activation energy E_r, of the rearrangement, then the apparent activation energy of the overall process becomes negative. This hypothesis will be explored in the following section, but it seems that it does not apply to the system discussed here. All attempts to demonstrate the existence of a complex $\sim S^-, Na^+$; S have failed (133).

(*2*) The rate of reaction in solution, and especially of ionic reaction, depends on solvent. In the language of transition state theory one explains the solvent effect in terms of solvation of the initial and transition states. For example, if solvent I solvates the transition state better than solvent II, the reaction in the former requires less activation energy than in the latter and, therefore, it proceeds faster. Of course, the entropy factors may modify the final results although usually they do not reverse the order of reactivities. We are therefore not surprised to find, sometimes, an enormous difference in the rates of a reaction in two different solvents. Now, the same solvent at two different temperatures represents, after all, two different media. For example, tetrahydrofuran at $+25$ and $-70°C$ has different densities, viscosities, dielectric constants, etc. One may say, with some justification, that it is only the label on the bottle that remains the same, but the content is different at different temperatures. It is possible, therefore, that the potential energy barrier for a reaction may be higher at $25°$ and substantially lower at $-60°$ and thus as the temperature falls the reaction may proceed faster. Again, it appears that this is not the correct explanation of the negative activation energy observed in the propagation of sodium living polystyrene, nevertheless such a situation might be encountered in some systems.

(*3*) Finally, one of the reagents of a bimolecular reaction may exist in two forms, say a mixture of A' and A'' is represented formally as a single reagent A participating in the reaction $A + B \rightarrow C$. However, the actual course of such a reaction is determined by two elementary processes proceeding simultaneously:

$$A' + B \rightarrow C \text{ (or perhaps } C') \qquad k',$$

$$A'' + B \rightarrow C \text{ (or perhaps } C'') \qquad k'',$$

and then the formal rate constant $k = fk' + (1 - f)k''$. In this equation f is the fraction of the systems A which are present in the form of A' and $1 - f$ of those being in the form of A''. Let us assume that the equilibrium,

$$A' \rightleftarrows A'' \qquad K_A = (1 - f)/f,$$

is maintained at each temperature. If A'' is the more reactive species and the process $A' \rightarrow A''$ is exothermic, the observed apparent activation energy may be negative. For example, say $1 - f \ll 1$ and $k'' \gg k'$, then $k \approx K_A k''$ and the observed activation energy $E_{ap} = \Delta H_A + E''$. For a negative ΔH_A (an exothermic transformation) such that $|\Delta H_A| > E''$, the rate of the reaction increases with decreasing temperature because the gain in the proportion of the more reactive species more than counteracts the decelerating effect of the genuine activation energy. In a more general situation, still subject to the conditions $k'' \gg k'$, the observed rate constant, k, is given by

$$k = k'' K_A/(1 + K_A),$$

and the observed "activation energy," E_{ap}, is

$$E_{ap} = E'' + \Delta H_A/(1 + K_A).$$

The Arrhenius plot obtained for such a system is curved. The "activation energy" is negative at higher temperature range if $-\Delta H_A > E''$ and $K_A \ll 1$. At lower range it may become positive because nearly all the systems form A'' at some sufficiently low temperature and then further cooling cannot contribute much to the concentration of these highly reactive species. Of course, in this low temperature region E_{ap} eventually approaches E'', since K_A becomes very large.

The Arrhenius curve discussed here exhibits a maximum at a temperature at which $K_A + 1 = -\Delta H_A/E''$. Determination of T_{max} is valuable, since it provides additional information from which the constants characterizing the system can be calculated.

At sufficiently high temperatures the fraction of the reactive A'' species becomes vanishingly small, and then $(1 - f)k'' \ll k'$. In this region the reaction is relatively slow because only the less reactive A' species contribute to its progress and the observed rate constant becomes then identical with k'. Of course, further increase of temperature leads now to an *increase* in the observed rate constant, E_{ap} approaching eventually the value of E'. Therefore, for a system in which the whole temperature range is accessible to kinetic studies, the Arrhenius plot would be S-shaped as shown in Figure 10.

It seems that the "negative activation energy" observed in polymerization of sodium living polystyrene in tetrahydrofuran could be explained in the terms outlined above. The $\sim$S$^-$,Na$^+$ ion-pairs are of the contact type at temperatures higher than, say, $-40°$C, although at lower temperatures the proportion of solvent-separated pairs becomes appreciable. Two observations lend support to this statement: (*1*) The exothermicity of dissociation of $\sim$S$^-$,Na$^+$ is 8–9 kcal/mole if the system is investigated in tetrahydrofuran above $-40°$C. Its value decreases on further cooling, being reduced to about 5–6 kcal/mole at $-70°$. Since the heat of dissociation of the solvent-separated

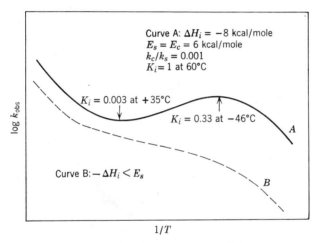

Fig. 10. Hypothetical Arrhenius plot for propagation involving contact and solvent-separated ion-pairs. The symbols ΔH_i $E_s E_c$, k_s, k_c, and K_i refer to contact and solvent-separated ion-pairs.

pairs is close to zero, one concludes that about 25% of the salt forms the separated pairs at $-70°$C. (*2*) The absorption spectrum of ion-pairs is shifted to a longer wavelength as the contact pair is transformed into a solvent-separated one (32). The $\sim$S$^-$,Na$^+$ absorbs at $\lambda_{max} = 340 \pm 2$ mμ in the region -40 up to 25°C. Lowering the temperature below $-40°$C leads, however, to a bathochromic shift, λ_{max} being 347 ± 2 mμ at $-80°$C. Significantly, such a temperature-dependent shift is not observed in the spectrum of the cesium salt in tetrahydrofuran, λ_{max} remaining approximately constant at about 344–346 mμ.

Although most of the $\sim$S$^-$,Na$^+$ pairs form contact species in tetrahydrofuran at 25°C, it is the minute fraction of highly reactive solvent-separated pairs that is responsible for the observed rate of polymerization. Following our previous argument, we represent the observed "negative activation energy" of -1.5 kcal/mole as the sum $\Delta H_i + E_s$, where ΔH_i is the heat of solvation of contact pairs as they become transformed into solvent-separated ones, and E_s is the activation energy of polymerization induced by the latter. The exothermicity of solvation is usually by 1 kcal/mole lower than the exothermicity of dissociation of the contact pairs and, hence, $\Delta H_i \leqslant -7.5$ kcal/mole. Therefore E_s must be smaller than 6 kcal/mole. However, the activation energy of the propagation induced by free ions *is* 6 kcal/mole and it is highly improbable that the solvent-separated pair propagates with *lower* activation energy than the free ion. Hence the activation energy of propagation induced by these pairs must be approximately equal to that of the free-ion propagation, i.e., $E_s \approx E_\ominus = 6$ kcal/mole. The proportion of solvent-separated ion-pairs at $-70°C$ was shown to be about 0.25. Since $\Delta H_i = -7.5$ kcal/mole, this fraction is reduced by about a factor of 500 at 25°C. Assuming a negligible contribution of contact pairs to the observed polymerization, one finds the rate constant of propagation induced by solvent-separated pairs to be about 20,000 liter mole^{-1} sec^{-1} at 25°, i.e., the reactivity of these species is comparable to that of the free ions. [A slightly higher value, 30,000 liter mole^{-1} sec^{-1}, was reported in reference 11.]

The latter conclusion may be checked. The activation energy of propagation induced by a solvent-separated pair is 6 kcal/mole. The observed $k_\oplus$ should increase, therefore, by about a factor of 4 as the temperature of polymerization is lowered from 25 to $-70°C$. This is precisely the case, $k_\oplus$ being 80–90 liter mole^{-1} sec^{-1} at 25°C, increasing to about 300 at $-70°C$.

It is known (32,33) that the fraction of solvent-separated sodium pairs is higher in dimethoxyethane than in tetrahydrofuran. Kinetic studies of anionic polymerization of styrene in dimethoxyethane (17) confirm this claim. The observed rate constant $k_\oplus$ in that solvent is exceedingly high, e.g., 3600 liter mole^{-1} sec^{-1} at 25°C. Moreover, the Arrhenius plot of $\log k_\oplus$ versus $1/T$ is now given by a curve, shown also in Figure 9, with a maximum at $T \sim 0°C$. Our previous discussion implies that the maximum should be expected when the fraction of solvent-separated ion-pairs becomes high, the condition for its appearance being expressed mathematically by the equation $K_S + 1 = -\Delta H_S/E_{\text{sep}}$. Accepting the reasonable value of -7 kcal/mole for ΔH_S and 5 kcal/mole for E_{sep}, one finds $K_S = 0.4$ at $T = 0°C$. The contribution of contact pairs to the rate of propagation probably is negligible and, therefore, the observed $k_\oplus \times 4100$ liter mole^{-1} sec^{-1} leads to the propagation constant of solvent-separated pairs of about 14,000 liter mole^{-1} sec^{-1} at 0°. Extrapolation leads again to a value of about 20,000 at 25°C.

In conclusion, the kinetic studies of styrene polymerization in tetrahydrofuran (11) and in dimethoxyethane (17) lead to three important ramifications:

(*1*) Solvent-separated ion-pairs may contribute to the propagation and in some media their share in the polymerization is dominant. (*2*) The reactivities of solvent-separated pairs exceed by several powers of ten that of the contact pairs. Hence, their role is important even if their proportion is small. (*3*) The reactivities and activation energies of solvent-separated pairs are similar to that of free ions. Both seem to be only slightly affected by the nature of ethereal solvent in which the reaction takes place. The data presently available do not yet give the answer to the question of how the counterion affects the reactivity of a solvent-separated pair. This writer believes that a change of counterion has only a minor effect upon the rate of propagation induced by a solvent-separated pair, whereas the rate and the character of propagation induced by a contact pair may be significantly, and often enormously, affected by the nature of counterion.

It has been concluded above that the reactivity of solvent-separated ion-pairs is only slightly affected by the nature of ethereal solvent. Contrary views are expressed by Schulz et al. (132). Their study of kinetics of polymerization of sodium polystyryl in tetrahydropyrane led to the rate constant of solvent-separated pairs of about 50 M^{-1} sec^{-1} at 20°C. However, the numerous assumptions introduced by these workers cast doubt about their conclusion.

Several factors determine the reactivity of contact-ion-pairs: the tightness of the ion pair, the extent of its dissociation and the degree of solvation in the transition state. The contribution of the reactive solvent-separated pairs present in the solution should be taken into account when deriving the data. Since the variation of the counterion affects each of these factors differently, a complicated pattern of reactivities may be observed. In Table VIII are listed the Arrhenius parameters reported for the polymerization of styrene and α-methylstyrene in tetrahydropyrane (27,31). These data illustrate well the complex role of the opposing factors. For example, the very low A factors are indicative of a substantial contribution of the solvent-separated ion-pairs to the propagation. Nevertheless, some patterns do emerge, and the broad generalizations based on them are probably justified, although some quantitative statements may be altered later when our knowledge of this subject will be more extensive. In this connection let us keep in mind that experimentation in this field is exceedingly difficult, and the data given in the tables need not be final. Our own experience, as well as that of other workers, shows that various unwarranted factors may greatly affect the observed results, the impurities arising from the decomposition of living polymers often being the main cause of confusion.

Studies in nonethereal solvents are less extensive, although polymerizations proceeding in hydrocarbons and involving lithium counterions were thoroughly investigated. These reactions differ in some respects from others and therefore will be treated separately in Chapter VIII.

TABLE VIII
Arrhenius Parameters for Ion-Pair Propagation in Tetrahydropyrane (27,31)

Counterion	$k_\oplus$ at 25°C, liter mole^{-1} sec^{-1}	$E_\oplus$, kcal/mole	log $A_\oplus$, M^{-1} sec^{-1}	Temperature range, °C
		Styrene		
Li$^+$	—	—	—	
Na$^+$	14	6.7[a]	6.0[a]	
K$^+$	73	5.3	5.8	
Rb$^+$	83	4.9	5.5	
Cs$^+$	53	4.9	5.3	
		α-Methylstyrene		
Li$^+$	2.6	0.	0.4	-16 to -3
Na$^+$	0.05	5.6	2.8	-12 to $+9$
K$^+$	0.25	6.4	4.1	-20 to $+5$
Rb$^+$	0.35	7.0	4.7	-20 to $+5$
Cs$^+$	0.26	7.3	4.8	-20 to $+5$

[a] Schulz et al. (132) claim the activation energy of contact ion-pair, $E_\ominus$, to be between 7.5 and 8.3 kcal/mole corresponding to log $A_\oplus$ between 6.6 and 7.2.

The investigation of polystyryl polymerization in benzene with sodium, potassium, rubidium, and cesium counterions has been reported (34). The cesium salt seems to be nonassociated; its propagation obeys the first order law for concentrations ranging from 10^{-4} to $10^{-3}M$. The rate constant, $k_\oplus$, was determined to be 14.5 liter mole^{-1} sec^{-1} at 20°C, and the activation energy, calculated from a rather narrow temperature range of 10 to 30°C, is 7.3 kcal/mole. This leads to a relatively low A factor of $10^{6.5}$ liter mole^{-1} sec^{-1}.

Only two experiments were performed with the rubidium salt. Assuming no association at $10^{-4}M$ concentration, the propagation constant was found to be 19 liter mole^{-1} sec^{-1}.

The kinetics of propagation of the potassium salt gives a variable order. The plot of log $\{(1/M)(-dM/dt)\}$, M being the monomer concentration, versus log M is curved, as shown in Figure 11. Judging from the curvature the reaction is first order at the concentration of living polymers of $10^{-4}M$, but obeys half-order law at the concentration of $10^{-3}M$. It is assumed that the polymers may associate into dimers, and only the nonassociated polymers propagate. On this basis the experimental curve may be fitted to a theoretical one for the system,

$$(\sim\!\!S^-,K^+)_2 \rightleftarrows 2\sim\!\!S^-,K^+$$

if the dissociation constant is 6×10^{-4} M and the propagation constant is 38 liter mole^{-1} sec^{-1} at 20°C.

From the temperature dependence of the overall rate, the heat of dissociation was calculated to be 7.2 kcal/mole and the activation energy of the nonassociated species was found to be 7.6 kcal/mole. These values are reasonable—e.g., the entropy of dissociation is then 11 eu, as should be expected for such a process.

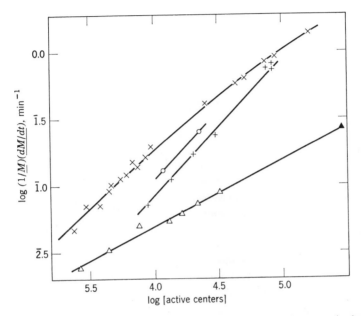

Fig. 11. Pseudo-first-order propagation constant of living polystsrene in benzen e solution plotted versus living polymer concentration on a log-log scale (34). The slope gives the kinetic order of the reaction in respect to living polymers. (×) Polystyryl potassium at 20.3°C, the line was calculated by assuming $K = 6 \times 10^{-4}$ and $k_p = 38.1$ mole^{-1} sec^{-1}. (○) Polystyryl rubidium at 19.5°C. (+) Polystyryl cesium at 20.3°C. (△) Polystyryl sodium at 30.3°C, the points being displaced downward by 0.3 log unit. (▲) A single point reported by Smid and Szwarc and corrected to 30°C. Note that a still lower formal rate constant is obtained for Li$^+$salt; see p. 497.

The kinetic evidence shows a high degree of association of the sodium salt in the whole concentration range investigated. The plot of $\log \{(1/M)(-dM/dt)\}$ versus $\log M$, shown in Figure 11, is linear and has a slope of 1/2. The single result reported previously by Szwarc and Smid (28) fits well the observed relation which covers nearly a 1000-fold variation in the concentration.

The sodium polystyryl was prepared by an electron transfer reaction in tetrahydrofuran. The solvent was distilled off *in vacuo* and pure benzene introduced; the polymer was then freeze-dried and eventually redissolved in fresh benzene. Nevertheless, the analysis showed presence of one molecule of THF for each molecule of the living polymer. Hence, this kinetic study was performed with a monoetherate of $\sim\!\!\!\sim\!S^-,Na^+$ and not with the pure non-coordinated salt. Other salts were prepared directly in benzene, and therefore pure nonsolvated salts were obtained.

Roovers and Bywater (34) conclude that association of living polymers in benzene increases with decreasing size of the cation, and that the reactivities of the nonassociated ion-pairs in benzene are not greatly different from those found in dioxane. However, an interesting question may be raised. All these polymers were prepared by electron transfer initiation and therefore they were endowed with two active ends. Hence, the association should be *intra-* and not *inter*-molecular, consequently the concentration dependence is puzzling. Further studies of this problem may be desirable.

6. PROPAGATION IN MIXED SOLVENTS

Two systematic studies of the kinetics of anionic polymerization in mixed solvents have been reported. Van Beylen et al. (21) studied the polymerization of lithium polystyrene in mixtures of tetrahydrofuran and dioxane, while Worsfold and Bywater (22) described a similar study in mixtures of tetrahydrofuran and benzene. In the latter case, the log of the propagation constants of

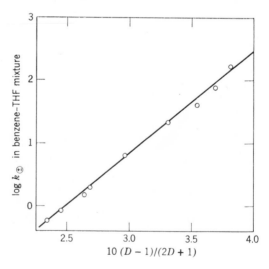

Fig. 12. Propagation constant of lithium salt of living polystyrene in tetrahydrofuran–benzene mixture plotted versus the Kirkwood function $(D - 1)/(2D + 1)$, D being the dielectric constant of the mixture (22).

the $\sim\!\!\sim\!\!S^-,Li^+$ ion-pairs varied linearly with the Kirkwood function, $(D - 1)/(2D + 1)$, where D is the dielectric constant of the medium. This is shown in Figure 12. The situation appears to be more complex in tetrahydrofuran–dioxane, as shown in Figure 13—here the plot of log $k_\oplus$ versus $(D - 1)/(2D + 1)$ or versus $1/D$ is definitely curved. The least curvature is obtained in the plot of log $k_\oplus$ versus mole fractions of dioxane, as shown in

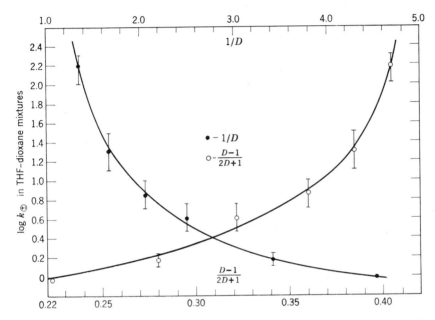

Fig. 13. Propagation constant of lithium salt of living polystyrene in tetrahydrofuran–dioxane mixture plotted versus the Kirkwood function $(D - 1)/(2D + 1)$ and versus the reciprocal of the dielectric constant D of the mixture (21).

Figure 14. Surprisingly, the experimental line is concave, indicating that the addition of a small amount of dioxane to tetrahydrofuran greatly reduces the reactivity of the ion-pair, whereas a comparable addition of tetrahydrofuran to dioxane has a relatively small effect upon the rate of polymerization. It is believed that tetrahydrofuran interacts more powerfully with ion-pairs than does dioxane, e.g., fluorenylsodium or -lithium forms only contact pairs in dioxane, but a substantial fraction of solvent-separated pairs is present in tetrahydrofuran (32). Therefore, one could anticipate a partial demixing of the mixture in the vicinity of ion-pairs, making the local concentration of tetrahydrofuran higher. The kinetic results seem to contradict this conclusion and remain unexplained at the present time. Density, viscosity, or the dielectric constant of such mixtures monotonically vary with their composition, indicating that no complexes are formed in this system. It seems, therefore, that additional studies of mixed solvent may be needed.

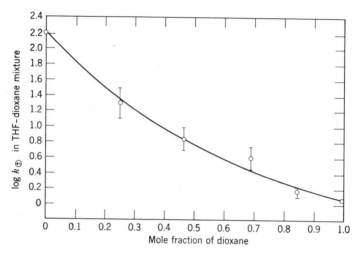

Fig. 14. Propagation constant of lithium salt of living polystyrene in tetrahydrofuran-dioxane mixture plotted versus mole fraction of dioxane in the mixture (22). Note the the concave shape of the curve, which indicates a large retarding effect of small amounts of dioxane added to tetrahydrofuran and a relatively low enhancing effect of small amounts of tetrahydrofuran added to dioxane.

7. MONOMER–GROWING POLYMER COMPLEX

Complexing of monomer with growing polymer was considered on several occasions. For example, Fontana and Kidder (35) concluded from their kinetic studies of propylene polymerization initiated by the adduct $AlBr_3 +$ HBr that the propagation involves two steps:

$$\sim\!\!M^+,(AlBr_4^-) + M \rightleftarrows \sim\!\!M^+,(AlBr_4^-),M \qquad K,$$

and

$$\sim\!\!M^+,(AlBr_4^-),M \rightarrow \sim\!\!M\cdot M^+,(AlBr_4^-) \qquad k_p.$$

The uptake of the monomer at the steady state stage, i.e., after all the available catalyst and promoter were converted into growing ends, was found to obey the equation

$$-d[M]/dt = k_p K[C][M]/\{1 + K[M]\},$$

in which [C] denotes the *total* concentration of the growing ends present in either the complexed or the noncomplexed form. This relation is distinct from that expected for a simple bimolecular kinetics and it follows from the proposed mechanism if a *substantial* fraction of growing ends forms complexes. The dissociation of the ion pairs, $\sim\sim M^+,(AlBr_4^-)$ or $\sim\sim M^+,(AlBr_4^-),M$, into the respective ions probably is not significant, since the reaction was studied in hydrocarbon solvent.

An alternative explanation of Fontana and Kidder's results was proposed by Mayo and Walling (36), who suggested that the contact ion-pairs $\sim\sim M^+,(AlBr_4)^-$ exist in equilibrium with a small fraction of solvent-separated species and only the latter are able to propagate the polymerization. A solvent-separated pair is assumed to collapse into an unreactive contact pair whenever a monomer addition takes place. The monomer-induced collapse of the solvent-separated pair is assumed to supplement the conventional spontaneous collapse governed by the unimolecular rate constant, k_r. This is an interesting idea—it implies that monomer may replace a molecule of solvent which separates the ions, and the formation of a new C^+ center leads then to the postulated result. Following this thought, it was assumed that the spontaneous solvation of ion-pair, governed by the pseudo-unimolecular rate constant, k_s, is the rate-determining step of polymerization. Thus, the observed rate constant of propagation would be

$$k_s k_p[M]/(k_r + k_p[M]),$$

as required by the observation. The authors implicitly postulate that the monomer catalyzes the collapse but not the formation of solvent-separated pairs. Since the propagation increases only the length of the polymer tail, it should not affect the equilibrium established between contact and solvent-separated pairs. Hence, if this reaction catalyzes the collapse of ion-pairs it must also enhance the rate of their formation—contrary to the implicit assumption, and therefore the proposed mechanism is not consistent.

Formation of a complex between a vinyl (or diene) monomer and a living polymer involving lithium counterion was proposed by Korotkov et al. (37,38) and by Medvedev et al. (39,40). The mechanism, exemplified by the equations

$$\sim\sim M^-,Li^+ + M \rightleftarrows \sim\sim M^-,Li^+,M \qquad K,$$
$$\sim\sim M^-,Li^+,M \rightarrow \sim\sim M.M^-,Li^+ \qquad k_r,$$

leads to a simple bimolecular kinetics, if the complex forms only a small fraction of growing ends. Therefore, conventional kinetic studies cannot prove, or disprove, such an hypothesis. A plausible evidence may be provided by some physical (e.g., spectroscopic) identification of the complex. In the absence of such an evidence other arguments may be helpful, e.g., unusual temperature dependence of the reaction, or some peculiarities in copolymerization. The problems concerned with temperature dependence were discussed in a previous section (see p. 426), while the peculiarity of copolymerization will be dealt with in Chapter IX. This writer believes that genuine examples of such a reaction will eventually be discovered, but it is doubtful whether polymerization of living polymers possessing lithium counterions may be explained in these terms.

Recently the idea of complex was invoked (41) with the intention of accounting for the observations interpreted in terms of ion-pairs and free ions. The monomer is assumed to form a complex with a living polymer that requires a second monomer to propagate, i.e.,

$$\sim\!\!M^-,Cat^+,M\ +\ M \rightarrow \sim\!\!M^-,Cat^+ \overset{M}{\underset{M}{\cdots}} \rightarrow \sim\!\!M.M^-,Cat^+\ +\ M.$$

Thus, a collision of the complex with a monomer catalyzes its destruction, simultaneously inducing the monomer addition. Furthermore, a similar process may be induced by a noncomplexed living polymer:

$$\sim\!\!M^-,Cat^+,M\ +\ \sim\!\!M^-,Cat^+ \rightarrow \sim\!\!M.M^-,Cat^+\ +\ \sim\!\!M^-,Cat^+.$$

This mechanism leads to an awkward kinetics which, according to the authors, is in a better agreement with the observations than the mechanism postulating the presence of free ions and ion-pairs. Unfortunately, Figini's analyses of the mathematical treatment revealed its inconsistencies (42). For example, the above mechanism leads to the apparent propagation constant independent of initiator concentration, contrary to experimental findings.

Another variant of the "complex" mechanism was advocated recently by Schulz, Figini, and Löhr (43). They assumed two modes of propagation:

$$\sim\!\!M^-,Cat^+\ +\ M \rightarrow \sim\!\!M.M^-,Cat^+ \qquad k_1$$

simultaneously with

$$\sim\!\!M^-,Cat^+\ +\ M \rightleftarrows \sim\!\!M^-,Cat^+,M$$
$$\sim\!\!M^-,Cat^+,M\ +\ M \rightarrow \sim\!\!M.M^-,Cat^+,M \qquad Kk_2' = k_2.$$

Thus, the observed rate of polymerization would be given by

$$(k_1[M]\ +\ k_2[M]^2)[\sim\!\!M^-,Cat^+],$$

the first term prevailing at low monomer concentration, the squared term becoming important at high concentrations of monomer. This idea was tested by Shimomura et al. (44), who demonstrated that the propagation is first order in monomer even at its highest feasible concentration. Moreover, the observed propagation constant at high monomer concentration, viz. 0.1–$0.2M$, was identical to that found at the low concentration, $0.001M$. This is illustrated by Figure 15.

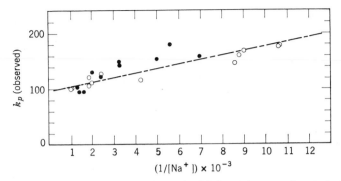

Fig. 15. Propagation constant of sodium salt of living polystyrene in tetrahydrofuran at 25°C plotted versus $1/[Na^+]$ (44). Full circles refer to runs performed at high concentration of the monomer (about $0.1-0.2M$). Open circles refer to those in which its concentration was low (about $0.001M$). The results show that monomer concentration does not affect the observed rate constant.

Complex formation between growing polymer and its monomer was often postulated in mechanisms accounting for the Ziegler–Natta catalyses (e.g. 134,135). It is highly probable that such complexes are formed in these systems and that they play an important role in determining the stereospecificity of the reaction (see also Section 14).

Finally, one more hypothetical case may be considered. It will be shown later that the free counterion may associate with a suitable complexing agent, and such a reaction shifts the equilibrium between the ion-pairs and free ions, increasing the concentration of the anions. Thus, in the system

$$\sim\!\!M^-,Cat^+ \rightleftarrows \sim\!\!M^- + Cat^+,$$

a complex formation between Cat^+ and monomer should produce higher concentration of free $\sim\!\!M^-$ ions. Since these are more reactive than ion-pairs, the rate of propagation would increase faster with monomer concentration than is expected from a simple bimolecular mechanism.

Formation of π-complexes of olefins, including styrene, with silver ions is well known (45). However, attempts to demonstrate the existence of such a complex in the systems Li^+ or Na^+ and styrene were unsuccessful (46). Further search for such systems may be desirable.

Erussalimskii et al. (137) discussed the formation of a monomer-initiator complex which either decomposes spontaneously and irreversibly into inactive species or is converted by another molecule of monomer into a growing species.

8. OTHER COMPLEXES OF LIVING POLYMERS

Living polymers may become complexed in a variety of ways. In the preceding section we were concerned with complexing involving monomers.

Dimerization of living polymers, i.e., complexing with itself, was briefly mentioned at the end of Section 6, and will be further discussed in Chapter VIII. Let us consider, now, other types of complexing which often may be of considerable significance.

Studies of Fuoss and Kraus (47) revealed that an ion-pair may become associated with a free ion and form a triple-ion. Association of a free cation or anion with an ion pair may, therefore, take place in a system involving living polymers. An interesting example of *intra*molecular formation of triple-ions was described by Bhattacharyya et al. (7b,48). Living cesium polystyryl may be prepared with one or with both ends active. In tetrahydrofuran the conductance and the kinetic studies of the polymer endowed with one growing end led to an identical value for the relevant dissociation constant, $K_{diss} \sim 2.1-2.8 \times 10^{-9} M$. On the other hand, contradictory results were obtained from similar studies of the polymers endowed with two growing ends. The conductance method led to $K_{diss} = 1.65 \times 10^{-8} M$, a value which is much higher than that determined for the polymer possessing one growing end only, whereas K_{diss} calculated from the results of kinetic studies was found to be $4.6 \times 10^{-10} M$, i.e., much too low. The apparent contradictions are reconciled if one considers the possibility of *intra*molecular association of a free $\sim\!\!S^-$ end with the nondissociated pair forming the other end of the polymer. Thus, the following equilibria should be established:

$$Cs^+, {}^-S\!\!\sim\!\!S^-, Cs^+ \; \rightleftharpoons \; Cs^+, {}^-S\!\!\sim\!\!S^- + Cs^+, \quad K_{diss}$$

$$Cs^+, {}^-S\!\!\sim\!\!S^- \; \rightleftharpoons \; \overbrace{\sim\!\!S^-, Cs^+, S^-\!\!\sim} , \quad K_{cyc}$$

Here K_{diss} represents the genuine dissociation constant evaluated from studies of the polymers endowed with one growing end only. K_{cyc} is the equilibrium constant of cyclization caused by the formation of triple-ions. Analysis of the above scheme shows that the conductance of the solution is higher than expected on the basis of simple ionic dissociation. Cyclization removes the free $\sim\!\!S^-$ ions and shifts the ionic dissociation to the right. Hence, the concentration of charged species increases and, consequently, one derives from the conductance data a too high value for the dissociation constant.

The triple $\overbrace{\sim\!\!S^-, Cs^+, S^-\!\!\sim}$

ions are apparently less reactive than the free $\sim\!\!S^-$ ions. Cyclization increases the concentration of Cs^+ ions and thus substantially decreases the concentration of the most reactive $\sim\!\!S^-$ ions. The propagation is therefore slower than expected and consequently the conventional interpretation of the kinetic data gives a much too low value for K_{diss}. Actually, the equilibrium constant, K_{cyc}, and the propagation constant of triple-ions may be evaluated by proper treatment of all the available data. Thus, the reactivity of the triple-ion was calculated to be 2000 liter mole^{-1} sec^{-1}, i.e., these species are much more reactive than the ion-pairs but substantially less so than the free ions. K_{cyc} was calculated to be about 5.5 for a polymer having a degree of polymerization of about 25.

Dependence of K_{cyc} upon the molecular weight of the polymer permits testing the proposed scheme. The deviations observed in the conductance and reactivity studies should diminish as the polymers become longer. This was

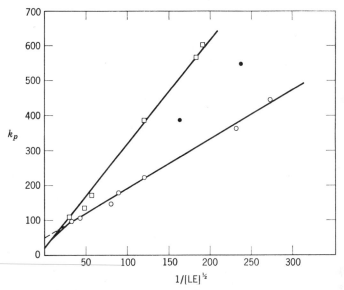

Fig. 16. Rate constant of propagation of cesium salt of living polystyrene possessing two active ends (○) and one active end only (□) (7b). The sample of the investigated living polymer possessing two active ends had a $DP \approx 25$. The two black circles refer to experiments performed with a polymer possessing two active ends of $DP \approx 500$.

shown to be the case—for example, the observed propagation constant of cesium polystyryl endowed with two growing ends is plotted in Figure 16 versus $1/[LE]^{1/2}$. Two samples were examined, one having a degree of polymerization of 25, the other of about 500. As required by the theory, a steeper slope was obtained for the latter.

The phenomenon of intramolecular triple-ion formation is probably a general one; however, it is conspicuously manifested in the case of cesium salts because the cesium ions, and ion-pairs, are poorly solvated by tetrahydrofuran. It would be interesting to extend these studies to other solvents, since careful examination of these phenomena may provide valuable information on the degree of flexibility of polymers in different solvents.

Living polymers may become associated with external agents, and such reactions may increase or decrease their reactivity. For example, Huisgen and Mack (49) found that the rate of benzyne formation from bromobenzene through the action of lithium piperidine was slowed down by the addition of lithium bromide. Obviously, complexing of lithium piperidine with lithium bromide reduced the reactivity of the former. A similar phenomenon was observed when phenyllithium was used instead of lithium piperidine.

A thorough study of the effects of salts on the reactivity of phenyllithium was reported by Waack and Doran (50). The lithium derivative was used to

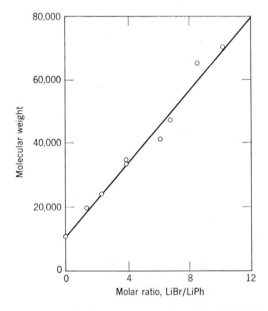

Fig. 17. Molecular weight of polymer versus molar ratio LiBr/LiPh (50) in benzene at 20°C. [LiPh] = 0.26M. Initial styrene to LiPh ratio was 11.0.

initiate anionic polymerization of styrene in tetrahydrofuran. The conditions were such that the initiation was competing with propagation for the monomer and therefore the molecular weight of the product reflected the relative reactivity of the initiator. The results demonstrated that LiCl, LiBr, and LiI are approximately equally effective in reducing the reactivity of phenyllithium, whereas lithium fluoride and sodium chloride had no effect. The molecular weight of the resulting polystyrene increased linearly with the increasing mole ratio of LiBr/LiPh, as shown in Figure 17. Analysis of the data permitted the establishment of the 1:1 stoichiometry of the complex and determination of its equilibrium constant as 13.5 $\pm$ 1.5 liter mole^{-1} at 20°C.

The degree of complexing depends on the nature of the organolithium compound, e.g., the association of LiBr with benzyllithium was found to be less pronounced than with phenyllithium.

Qualitative observations demonstrated that alkoxides resulting from destruction of living polymers associate with the remaining active ends, the association being particularly notable in tetrahydropyran and methyltetrahydrofuran. Such an association, does not affect the spectrum of the living polymer, although it greatly reduces its reactivity. Much of the uncertainties found in the kinetic data pertaining to propagation of living polymers arises from this phenomenon.

Association of living polymers with polyether increases their reactivity. For example, the reactivity of sodium polystyryl in tetrahydropyran is greatly enhanced by the addition of small amounts of glyme, e.g., $CH_3(OC_2H_4)_3OCH_3$ or $CH_3(OC_2H_4)_4OCH_3$. The kinetics of such a polymerization was studied by Shinohara et al. (51), who demonstrated that the observed effects should be attributed to two factors:

(1) Glyme becomes complexed with the growing end converting it from a contact into a glyme-separated pair. This enhances the propagation, since the glyme-separated pair is much more reactive than the contact pair.

(2) Coordination of glyme with the free cation converts it into a new species, $Cat^+(glyme)_n$, and this process shifts the equilibrium

$$\sim\!\!\sim\!M^-,Cat^+ \rightleftarrows \sim\!\!\sim\!M^- + Cat^+$$

to the right. Thus, the concentration of the highly reactive free $\sim\!\!\sim\!M^-$ ions increases, causing further enhancement of the reactivity of living polymers.

In the presence of a constant concentration of glyme the observed propagation constant, k_{obs}, is a linear function of the reciprocal of the square root of living polymer concentration, as it is in the absence of glyme, but the intercepts and slopes depend now on the concentration of glyme. This is illustrated by Figure 18, which demonstrates that both—the intercepts and the slopes—increase with increasing concentration of the coordinating agent. If only one molecule of glyme is involved in the complex with living polymer (ion-pair) or with a free sodium ion, then

$$k_{obs} = k_\oplus + K_1 k'_\oplus[glyme] + k_\ominus(K_{diss})^{1/2}\{1 + K_2[glyme]\}^{1/2},$$

where $k_\oplus$ is the propagation constant of the ordinary ion-pair, $k'_\oplus$ that of an ion-pair coordinated with glyme, and $k_\ominus$ denotes the propagation constant of the free $\sim\!\!\sim\!S^-$ ion.

The symbols K_1 and K_2 are the equilibrium constants of the complexing

$$\sim\!\!\sim\!S^-,Na^+ + glyme \rightleftarrows \sim\!\!\sim\!S^-,(glyme),Na^+ \qquad K_1$$

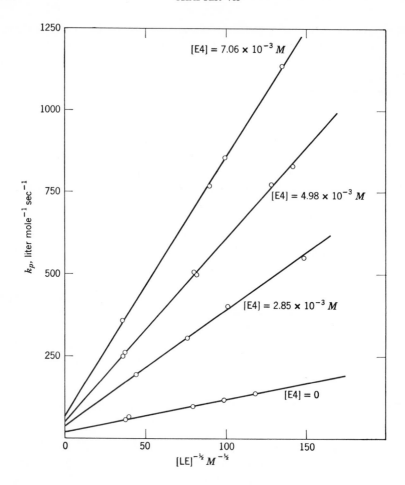

Fig. 18*a*. Apparent propagation constant, k_p, of sodium polystyryl in tetrahydro-pyran at 25°C in the presence of various concentrations of glyme-4 (E4) (51). Note the increase in the slopes and intercepts of the lines giving k_p as the function of $[LE]^{-1/2}$.

and

$$\text{Na}^+ + \text{glyme} \rightleftarrows \text{Na}^+(\text{glyme}) \qquad K_2,$$

whereas K_{diss} is the dissociation constant of the ordinary $\sim\!\!\!\text{S}^-,\text{Na}^+$ ion-pair into free ions. However, if free sodium ion forms predominantly complexes with two molecules of glyme, viz.,

$$\text{Na}^+ + 2 \text{ glyme} \rightleftarrows \text{Na}^+(\text{glyme})_2 \qquad K_2',$$

then the observed propagation constant is given by

$$k_{\text{obs}} = k_\oplus + K_1 k_\oplus'[\text{glyme}] + k_\ominus(K_{\text{diss}})^{1/2}\{1 + K_2'[\text{glyme}]^2\}^{1/2}.$$

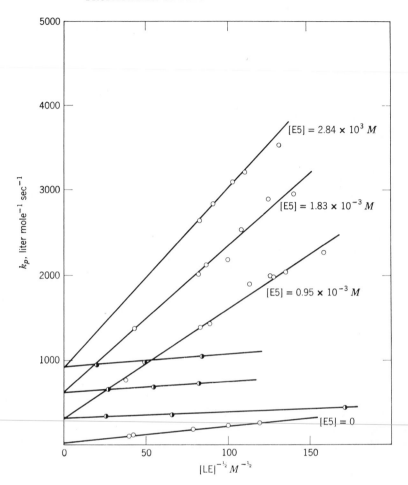

Fig. 18b. Apparent propagation constant, k_p, of sodium polystyryl in tetrahydropyran at 25°C in the presence of various concentrations of glyme-5 (E5) (51). Note the increase in the slopes and intercepts of the lines giving k_p as the function of $[LE]^{-1/2}$. The half-shadowed circles represent experiments performed with an excess of Na^+,BPh_4^-.

The validity of both equations hinges on the assumption that the bulk of living polymers forms the ordinary ion-pairs. It was shown that the first relation is applicable to the system $\sim\!\!\!S^-,Na^+$ and glyme-5 $(CH_3O[C_2H_4O]_4CH_3)$, whereas the second is obeyed in the system involving glyme-4 $(CH_3O[C_2H_4O]_3CH_3)$. Thus, for both systems the plots of intercepts versus concentration of glyme are rigorously linear, the slopes of such lines giving the respective constants $K_1k'_\oplus$, viz., 8×10^3 liter² mole⁻² sec⁻¹ for glyme-4 and 300×10^3 liter² mole⁻² sec⁻¹ for glyme-5. Unfortunately, it was not yet possible to separate the products into individual constants. The plot of the square of the slopes taken from Figure 18b are also linear with respect to concentration of glyme-5. Thus, the association constant of Na^+ with glyme-5 into a 1:1 complex is found to be 1.6×10^5 liter mole^{-1}. For glyme-4 the analogous plot is linear if square of concentration is used as the abscissa. The equilibrium constant of association for Na^+ with glyme-4 into a 1:2 complex is then found to be 1.2×10^5 liter^2 mole^{-2}.

To complete the survey of the complexing phenomena, let us review the reaction of living polystyrene with anthracene. Anthracene reacts rapidly with living polystyrene; the reaction leads to the formation of a covalent bond between the carbanion of the polymer and the carbon 9 of the aromatic hydrocarbon. Thus, the process may be treated as an example of a copolymerization forming a new anion, extremely inert, which does not propagate further. Formation of such an anion was observed by Medvedev's group (52) and independently by Khanna et al. (53). It was found that the adduct is labile and its rate of decomposition into its original components was determined by investigating the exchange between radioactive anthracene and the nonactive adduct (54). Thus, the rate constant of the decomposition was found to be 3×10^{-3} sec^{-1} at 28°C, and the activation energy of the process appears to be 10 kcal/mole.

The kinetics of the addition of anthracene was investigated by a flow and stop-flow technique (55,56). The results were most intriguing. The preliminary studies (55) demonstrated that, in the presence of an excess of anthracene, an extremely rapid addition converts nearly quantitatively *all* the growing ends into the anthracene adducts if each macromolecule possesses only one growing end. However, only one-half of the growing ends were converted into the anthracenated adducts during the fast reaction, i.e., within about two or three seconds, if each polymeric molecule was endowed with *two* growing ends. Further conversion proceeded with a rate approximately 1000 times slower than the rate of the fast reaction. This surprising observation suggested that an anthracenated end, symbolized by $\sim\!\!\sim\!S\,.\,A^-,Na^+$ (or K^+), may rapidly associate with a free $\sim\!\!\sim\!S^-\,Na^+$ end to form an inert *intra*molecular complex which then very slowly reacts with anthracene. The degree of polymerization of the investigated polymers was about 25, and their concentration in the experiments about $10^{-4}M$. Hence, the two ends of each macromolecule were, on the average, only 15 Å apart, whereas the average separation of the ends of different polymers exceeded 200 Å. Therefore, the *intra*molecular complexing should be much faster than the *inter*molecular combination, and thus it could efficiently compete with the bimolecular addition of anthracene.

More comprehensive studies (56) fully confirmed these ideas. The spectrum of a mixture of anthracenated and nonanthracenated polymers, each endowed with one active end only, was identical with that obtained by superposition of the spectra of the components. This indicates that the association is of a physical nature, the binding arising from dipole–dipole or dipole–induced dipole interactions, or from dispersion forces due to the high polarizibility of the reacting species. The evidence for an association is undeniable because addition of anthracene to such a mixture resulted only in a very slow reaction provided anthracenated polymers were in excess.

Kinetic studies of anthracene addition to living polystyrene demonstrated

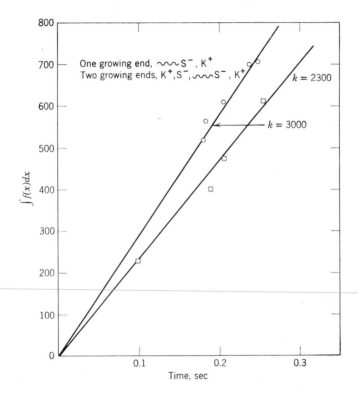

Fig. 19. The bimolecular integral $\int f(x)\,dx$ plotted versus time. Anthracene addition to potassium salt of living polystyrene endowed with one growing end ($\bigcirc$) and two growing ends ($\square$) (56):

$$f(x) = ([A]_0 - [LP]_0)^{-1} \ln \{[LP]_0([A]_0 - x)/[A]_0([LP]_0 - x)\}.$$

Note that the concentration of living *polymers* and not of *growing ends* is used in the calculation of $f(x)$.

that the reaction is bimolecular. This is demonstrated by the plots shown in Figures 19 and 20. The bimolecular rate constant, k_{ad},

$$\sim S^-,K^+ + A \rightarrow \sim S.A^-,K^+ \qquad k_{ad},$$

was found to be about 2000–3000 liter mole^{-1} sec^{-1} at 25°C, the same value

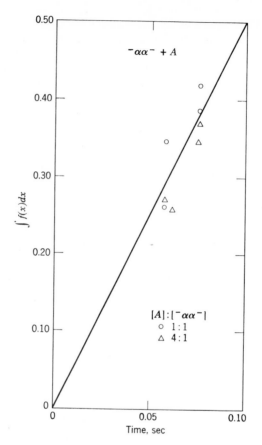

Fig. 20. The bimolecular integral $\int f(x)\, dx$ plotted versus time. Anthracene addition to the dimeric dianion of α-methylstyrene,

$$K^+,\overline{C}(CH_3)(Ph).CH_2.CH_2.\overline{C}(CH_3)(Ph),K^+ = {}^-\alpha\alpha^-,$$
$$f(x) = ([A]_0 - [LE]_0)^{-1} \ln \{[LE]_0([A]_0 - x)/[A]_0([LE]_0 - x)\}.$$

Note that the concentration of *growing ends* is used in the calculation of $f(x)$ (56).

being found whether the polymers are endowed with one or two growing ends, provided that the concentration of macromolecules, and not of the growing ends, is used in the kinetic calculations. Similar studies were performed with the dianion of α-methylstyrene dimer. Although this oligomer also possesses two growing ends, no intramolecular association takes place and the reaction proceeds to completion. This is not surprising, because the ring closure is impossible for the dimer due to steric hindrance.

To determine the rate constant of association, k_a,

$$\sim\sim S^-,K^+ + \sim\sim S.A^-,K^+ \to \text{complex} \qquad k_a,$$

the kinetics of anthracene addition was investigated in a stop-flow system in which a solution of living polystyrene, possessing one end only, was mixed with a similar solution containing, however, an excess of anthracene. On mixing, two reactions ensued: (*1*) The excess of anthracene introduced with the second solution reacted with the free $\sim$S$^-$,K$^+$ present in the first, converting it into an anthracenated polymer. (*2*) The anthracenated polymers, originally present in the second solution, became associated with living polymers introduced with the first solution, and this association prevented the latter from rapid reaction with anthracene. Hence, the rapid anthracene addition was partially inhibited by the presence of anthracenated polymers, and analysis of the kinetics of these competitive reactions permitted the evaluation of the ratio k_a/k_{ad}. Since k_{ad} was previously determined, k_a could be calculated and was found to be about 2000 liter mole^{-1} sec^{-1}.

Several interesting conclusions may be drawn from these results. First, one may inquire why $\sim$S$^-$,K$^+$ associates with $\sim$SA$^-$,K$^+$ and not with itself. Apparently the binding is greatly strengthened by the high polarizibility of the anthracenated partner ($\sim$SA$^-$,K$^+$). The equilibrium constant of the complex dissociation,

$$\text{complex} \rightleftarrows \sim\text{S}^-,\text{K}^+ + \sim\text{S}.\text{A}^-,\text{K}^+,$$

may be calculated. In the presence of about $10^{-3}M$ concentration of anthracene the complex is 1000 times less reactive than the free $\sim$S$^-$,K$^+$. It was shown (56) that the addition of a second anthracene molecule to the complex proceeds with a rate independent of the anthracene concentration, implying that the complex dissociation determines the rate of this addition. Hence, the ratio of the forward to backward reaction, i.e., the equilibrium constant of complex formation, is about 10^6 liter mole^{-1}. On the other hand, no association was observed in a $10^{-3}M$ solution of $\sim$S$^-$,K$^+$, indicating that the homodimerization, if any, has an equilibrium constant lower than 10^{-2} liter mole^{-1}. These figures show how strikingly the association is affected by the structure of the interacting components.

Another interesting comment may be drawn from the fact that the association of $\sim$S$^-$,K$^+$ with $\sim$S.A$^-$,K$^+$, although physical in nature, proceeds with a bimolecular rate constant of less than 10^4 liter mole^{-1} sec^{-1}. Such a reaction would be expected to be diffusion-controlled and, therefore, its rate constant should exceed 10^9 liter mole^{-1} sec^{-1}. Apparently, the process involves some activation energy which arises from the need of desolvation of the reactants in the transition state.

Anthracene is not the only hydrocarbon that associates with living polystyrene. Formation of some unidentified complexes with other aromatic hydrocarbons was reported by Levy and Cohen–Bosidan (57). Further studies of these phenomena should be interesting.

The interaction of aromatic hydrocarbons with living polymers complicates studies of relative rates of initiation of anionic polymerization by aromatic radical ions. The resulting aromatic hydrocarbon may convert a living polymer into a dormant one, and therefore no simple relation can be established between the rates of initiation and propagation. The latter depends now on the nature of the aromatic compound with which the growing end is associated. This difficulty invalidates the calculations based on the assumption of a constant rate of propagation (58).

9. WIEN EFFECT IN IONIC POLYMERIZATION

Dissociation of ion-pairs into free ions is favored by an external electrostatic field. The phenomenon, known as the second Wien effect, increases the conductance of solutions of weak electrolytes—the potential gradient decreases the fraction of undissociated ion-pairs or ionogens, converting them into conducting free ions. In ionically polymerized systems, free ions grow faster than ion-pairs and therefore an electric field should enhance the rate of reaction.

Comprehensive studies of effects which an electric field exerts upon ionic polymerization were reported by Sakurada, Ise, and their students. Their first paper (59a) is concerned with cationic polymerization of p-methoxystyrene induced by iodine in ethylene dichloride. The reaction was carried out in a glass reactor equipped with two platinum electrodes placed 10 cm apart. The process was investigated in the absence and in the presence of electric field under otherwise identical conditions. As may be seen from Figure 21, the initial rate of polymerization increases with monomer concentration, at least in the lower concentration range, being however always by about 40% faster when the field is applied (1.5 kV/cm) than in its absence. Even more convincing are the results presented in Figure 22, which demonstrate linear dependence of rate of polymerization on strength of the applied field.

Application of high-potential gradient may lead some temperature rise caused by Joule effects and one may question whether higher temperature is responsible for the observed enhancement of polymerization. It was shown, however, that the investigated polymerization has a negative activation energy of about -2.5 kcal/mole and therefore any electric heating should slow it down. It is possible that the high gradient leads to electrolysis and formation of reactive ions or radicals. This objection was refuted by demonstrating that no polymerization whatsoever occurs in the presence of an electric field but in the absence of iodine. One may argue, then, that the presence of iodine is essential for the electrolysis. This objection is difficult to refute, and the argument against it was rather speculative.

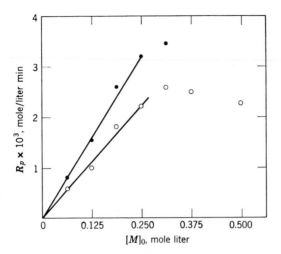

Fig. 21. Cationic polymerization of p-methoxystyrene initiated by $1.0 \times 10^{-4}M$ I_2 in CH_2Cl_2 (59a). Rate of polymerization plotted versus monomer concentration. (●) polymerization performed in electric field of 1.5 kV/cm. (○) polymerization performed in the absence of electric field.

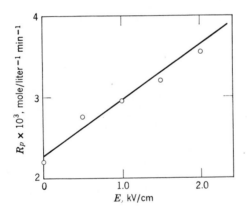

Fig. 22. Initial rate of polymerization of $0.22M$ p-methoxystyrene initiated by $1.0 \times 10^{-4}M$ I_2 in CH_2Cl_2 as a function of electric field (59a).

However, if an electric field enhances the polymerization by means other than through Wien's effect, then these may be perceived even in a radical polymerization. This is not the case. The rate of styrene polymerization and the molecular weight of the resulting product formed in a process induced by benzoyl peroxide were not affected even by a field as powerful as 20 kV/cm. The same observations were reported for other radical polymerizations. It must be concluded, therefore, that ionic polymerization induced by iodine is, at least partially, propagated by free carbonium ions which are more reactive than ion-pairs and that the degree of ionization may be influenced by an external electric field.

Extension of these studies to polymerizations initiated by high-energy radiation is less satisfactory (60). The reproducibility of the pertinent experiments is poor; nevertheless, it was established that an electric field enhances the polymerization and increases the number of macromolecules formed. It is believed that at low temperature the polymerization induced by high-energy radiation is ionic in nature. A fraction of oppositely charged ions, formed by the impact of high-energy photons or fast moving electrons, escape immediate recombination and these initiate ionic polymerization propagated by free ions. The experiments of Sakurada and Ise were performed with styrene in methylene chloride at $-78°C$ and, hence, the polymerization proceeds through a cationic mechanism. The theoretical work of Onsager (61,62) indicated that the probability of combination of two oppositely charged ions, initially separated by some fixed distance, decreases under the influence of external field. It seems, therefore, that in radiation-induced polymerization the yield of the initiating ions increases under the influence of potential gradient; this increases the rate of the reaction and the *number* of polymeric molecules formed.

There is a basic difference in the field effect upon polymerization induced by iodine and by high-energy radiation. In the first case we deal with a proper Wien effect—increase in the fraction of dissociated ion-pairs. In the second, the effect comes from the opposite cause—combination of free ions is hindered, thus increasing the yield of initiator and reducing the rate of termination. Therefore, in the latter case the electric field had a marked effect upon the molecular weight of the product.

The study of field effect has since been extended to anionic polymerization (63). Polymerization of styrene initiated by lithium salt of low molecular weight living polystyrene was chosen for these investigations. The reaction was studied in benzene–tetrahydrofuran mixtures in the presence and absence of electric field (5 kV/cm). Spectrophotometric observations proved that the concentration of living ends remained constant during the reaction in either case. Hence, it is obvious that the electric field did not induce electroinitiation (89). The results are summarized in Table IX. It is interesting to note that no effect was observed in a mixture containing 10% of tetrahydrofuran, a perfectly plausible result, because the concentration of free ions, and therefore also their contribution to polymerization, is imperceptible under these conditions. The effect observed at higher concentrations of tetrahydrofuran became more pronounced implying a monotonic relation between dielectric constant of the medium, and the Wien effect.

TABLE IX

Effect of Electric Field on Anionic Polymerization of Styrene in
Benzene–Tetrahydrofuran Mixtures

Volume fraction of THF, %	Dielectric constant	$[LE] \times 10^5$, M	Field strength, kV/cm	k_p, liter mole^{-1} sec^{-1}	
				Obs.	Lit.[a]
10	2.69	7.6	0	2.1	2.1
			5	2.1	—
30	3.58	1.9 ·	0	14.1	15
			5	20.4	—
40	4.05	3.9	0	22.0	28
			5	38	—
44.5	4.27	1.0	0	—	34
			5	110	—

[a] Worsfold and Bywater (22).

10. EXCHANGE BETWEEN IONS AND ION-PAIRS

The equilibria established among the various species which participate in ionic polymerization were discussed in previous sections. A complete description of such systems requires knowledge of the rates with which these equilibria are established—for example, we should know how rapidly ion-pairs dissociate into free ions or how rapidly contact pairs are transformed into solvent-separated ions. In addition to their intrinsic interest, these problems are important to polymer chemists because they influence the molecular weight distribution of the final product.

In Chapter II we dealt with the question of molecular weight distribution obtained under various conditions of polymerization. In the absence of complicating factors, propagation of living polymers leads to a narrow, Poisson-type, molecular weight distribution if only one species participates in the reaction. The presence of two or more species, each propagating with different rates, broadens the distribution, and the degree of broadening depends on the rates of their exchange. For example, if free ions and ion-pairs are the only carriers of polymerization the deviation of molecular weight distribution from the Poisson type is determined by the rate of dissociation of ion-pairs into free ions. Hence, the rate constant of such a dissociation may be calculated from the experimentally observed deviations of the distribution from the Poisson type.

This approach was developed by Schulz and his co-workers (64–66). Their treatment led to the equation

$$k_d = \{(2 - x)/2x\}\{\alpha c^{1/2}/(U - P_n^{-1})(1 + k_\oplus c^{1/2}/\alpha)\},$$

where k_d is the rate constant of ion-pair dissociation, x the fraction of monomer conversion, c the concentration of living polymers, $k_{\oplus}$ the rate constant of ion-pair growth, and $\alpha = k_{\ominus}(K_{\text{diss}})^{1/2}$, the product of the growth constant of the free ions and the square root of the dissociation constant of ion-pairs. The last coefficient, α, is obtained by plotting the observed rate constant of propagation versus the reciprocal of the square root of living polymer concentration. The other required data—$\overline{DP}_n$, the number average degree of polymerization, and the nonuniformity coefficient $U(= \overline{DP}_w/\overline{DP}_n - 1)$, were obtained through careful fractionation of the product on a modified Williams-type column (67). Thus, the rate constant of sodium polystyryl dissociation in tetrahydrofuran at 25°C was determined to be 75 sec^{-1}. Since the dissociation constant is 1.5×10^{-7} liter mole^{-1}, the bimolecular rate constant of association of ions seems to be 5×10^8 liter mole^{-1} sec^{-1}. This value is by two or three powers of ten lower than that calculated from Debye's equation, perhaps indicating the slow collapse of solvent-separated ion-pairs into contact pairs. These studies were extended recently by Figini (136) who investigated anionic polymerization in methyl-cumyl ether.

It was pointed out (68) that an alternative mechanism of exchange may operate in these systems:

$$\text{\small$\sim$}S^- + \text{\small$\sim$}S^-,Na^+ \rightleftarrows \text{\small$\sim$}S^-,Na^+ + \text{\small$\sim$}S^-.$$

The mathematical treatment outlined above does not distinguish between these two alternatives. However, the calculated constant, k_d, should be independent of living polymer concentration if the dissociation–association is responsible for the broadening, whereas its apparent value becomes concentration-dependent when the other mechanism operates.

The exchange between contact and solvent-separated pairs provides another mechanism causing the broadening of molecular weight distribution. Judging from Schulz's results, the polystyrene obtained in tetrahydrofuran in the presence of an excess of sodium tetraphenyl boride had a Poisson distribution. Since the contribution of free carbanions is then eliminated, the polymerization arises only from the growth of ion-pairs. It has been shown (11) that a minute fraction of highly reactive solvent-separated ion-pairs exists in this solvent at 25°C and their rate of growth is probably 10,000 times faster than that of contact pairs. Therefore, the sharpness of molecular weight distribution implies that the exchange between the contact and solvent-separated pairs is very rapid or that their proportion in the solution is too low to influence the observed distribution. The first explanation seems to be more plausible.

11. EFFECT OF MONOMER STRUCTURE ON RATE OF PROPAGATION

The propagation constant of homopolymerization depends in a complex manner upon the structure of the polymerized monomer because its change affects simultaneously, although differently, its own reactivity and that of the growing chain. More instructive, and useful for theoretical interpretation, are the data obtained from copolymerization studies in which the nature of one partner is kept constant while the structure of the other is systematically varied. Such investigations are discussed in Chapter IX. Nevertheless, since we have to face the problem of how the homopropagation constant of anionic polymerization is affected by the structure of monomer, we shall review this subject here, albeit briefly.

The kinetics of anionic homopropagation of many vinyl monomers has been investigated in tetrahydrofuran with sodium as the counterion; the

TABLE X

Anionic Homopolymerization Rate Constants of Vinyl Monomers in THF at 25°C
(counterion Na^+,[active end] $\approx 3 \times 10^{-3}M$,[monomer] $\approx 10^{-2}M$)

Monomer	k_p, liter mole^{-1} sec^{-1}	Ref.	Monomer	k_p, liter mole^{-1} sec^{-1}	Ref.
Vinylmesitylene	0.9	(69)	Styrene	950	(5,11)
α-Methylstyrene	2.5	(26,27)	1-Vinylnaphthalene	~ 500	(71)
p-Methoxystyrene	52	(70)	2-Vinylnaphthalene	~ 300	(71)
o-Methylstyrene	170	(70)	2-Vinylpyridine	7300	(72)
p-$tert$-Butylstyrene	220	(70)	4-Vinylpyridine	3500	(72)
p-Methylstyrene	210	(70)	9-Vinylanthracene	0.2	(73)

relevant data are collected in Table X. The observed variations of the propagation constants are large, but several difficulties hinder their interpretation. The basic complexity of the problem, which was stressed above, is augmented by other factors that make this task even more difficult. Ionic polymerization is simultaneously propagated by several distinct species, e.g., by free ions and ion-pairs, whereas the reported constants refer to the overall process. For example, the low propagation constant of p-methoxystyrene polymerization is due, at least partially, to the relatively low dissociation of the respective living polymer into free ions. The participation of free ions in this process is clearly demonstrated by the dependence of the observed k_p on the concentration of growing ends (70).

It is instructive to compare at this point the rates of polymerization of 2-vinylpyridine, 4-vinylpyridine, and styrene. The presence of nitrogen atom in the vinylpyridines decreases the negative charge density on the vinyl group, so these monomers are expected to be more reactive than styrene (see p. 542). The increase of the observed propagation constant with decreasing concentration of growing ends again demonstrates that the highly reactive free ions participate in the polymerization, and this leads to higher a k_p than would otherwise be expected. It was shown (74) that living poly-2-vinylpyridine is much less dissociated in tetrahydrofuran than living polystyrene; the dissociation constants are about 10^{-9} and 10^{-7} liter mole^{-1}, respectively, for the sodium salts at 25°C. Hence, the reactivity of the free polyvinylpyridene anion is even greater than suggested by the data given in Table X. It is interesting to note that living poly-4-vinylpyridine is more dissociated than the living polymer derived from the 2-isomer (74), although a reverse conclusion could be drawn on the basis of the relevant propagation constants observed at $3 \times 10^{-3} M$ concentration of the growing ends. Apparently the propagation of the respective free ion derived from 4-vinylpyridine is much slower than that of the ion of the living poly-2-vinylpyridine.

The low dissociation constants of living poly-2-vinylpyridine deserves further comments. It seems that the coordination of Na$^+$ ion with the nitrogen atoms of the neighboring pyridine moieties stabilizes the ion-pair and hinders its dissociation,

This hypothesis was discussed by Sigwalt (74); it is in accord with the general observation that amines may coordinate with ion-pairs. The proposed intramolecular association probably has a bearing on the stereochemistry of this polymerization. Indeed, it is significant that organomagnesium compounds initiate stereospecific polymerization of 2-vinylpyridine; they yield however, atactic polymers on reacting with 4-vinylpyridine (75). Further studies of these problems would be interesting.

The results listed in Table X clearly reflect the influence of steric hindrance on the rate of polymerization. The extremely low propagation constant of α-methylstyrene undoubtedly is due to the strain developed in the resulting polymer, and this factor is responsible for the relatively high rate constant of depropagation of living poly-α-methylstyrene, a process which cannot be neglected in kinetic studies of this polymerization. The subject of simultaneous propagation and depropagation is fully discussed in Chapter III.

The slow polymerization constant of vinyl mesitylene again is traced to steric hindrance, which prevents coplanarity of vinyl and phenyl groups. We shall return to this subject in Chapter IX when the copolymerization of vinylmesitylene with living polystyrene will be discussed.

Unusual kinetic features are exhibited by the homopolymerization of methyl methacrylate. The reaction initiated by fluorenyllithium in toluene–ether or toluene–tetrahydrofuran, mixtures leads eventually to propagation, which is first order in respect to monomer, but the bimolecular rate constant depends on its initial concentration. This behavior resembles that observed by Erussalimskii et al. (137) in polymerization of acrylonitrile. Some peculiarities of such systems are discussed in Chapter VIII and the mechanism of these reactions is further elaborated in Chapter XI.

12. EFFECT OF CHAIN LENGTH ON RATE OF ITS GROWTH

The lack of termination, characteristic of the living polymer system makes it possible to determine the absolute rate constant of propagation for each individual step of the growth process, $P_n^* + M \rightarrow P_{n+1}^*$. It was pointed out in Chapter II that the living polymer technique permits us to produce living polymers of a narrow molecular weight distribution, or synthesize well-defined living oligomers such as dimers, trimers, or tetramers. A solution of such polymers, or oligomers, may be mixed with an equivalent amount of monomer, and then, on the average, only one molecule of monomer is added to each growing chain. Hence, the kinetics of an individual step, $P_n^* + M \rightarrow P_{n+1}^*$, may be investigated, and from the rate of monomer consumption the propagation constant, $k_{p,n}$, may be determined. This constant should be distinguished from the *average* rate constant $k_p = \langle k_{p,n} \rangle$, which is obtained from conventional kinetic studies of polymerization.

Unpublished results of the Syracuse group have shown that for all the investigated systems, $k_{p,n}$ is a constant independent of n if the degree of polymerization is greater than 2 or 3.

13. SOME PECULIARITIES IN ANIONIC PROPAGATION

Anionic polymerization of vinylmesitylene was investigated in tetrahydrofuran (counterion Na^+) at temperatures ranging from $+25$ down to $-70°C$ (69). Studies of conductivity and the retardation of polymerization by sodium tetraphenyl boride demonstrate that the free ions, as well as ion-pairs, propagate this reaction, and at 25° the free ions seem to be 1000 times as reactive as ion-pairs.

Living polyvinylmesitylene (VM) in tetrahydrofuran is relatively stable at room temperature. Its spectrum shows a maximum at 360 mμ, but a slight increase in optical density at 435 mμ is noted after 24 hours. The homopolymerization is slow; the basically pseudo-unimolecular reaction is gradually retarded, the retardation being accompanied by an increase in optical density at 435 mμ. This new peak remains virtually constant after completion of the reaction. All these facts led to the conclusion that the monomer reacts in a dual way: (*a*) conventionally, by being added to the growing polymer, and (*b*) as a proton transfer agent, viz.:

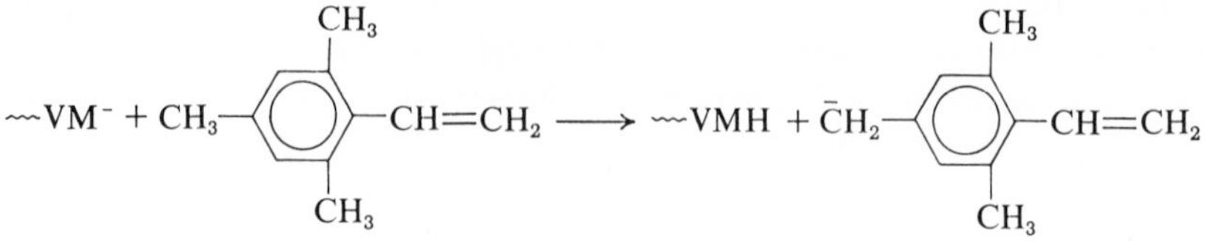

The resulting primary ion is much less reactive than the secondary ion:

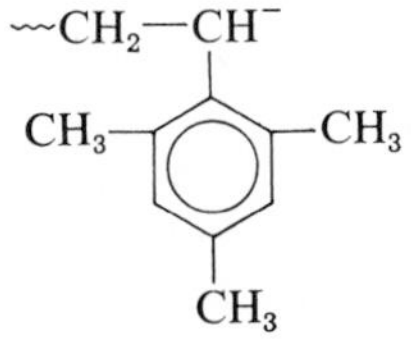

which carries the propagation (see also ref. 11), and therefore its formation leads to the retardation of polymerization. The retardation is manifested by the deviation of the curve, giving log (M_0/M_t) as a function of time, from the expected straight line. This is shown in Figure 23—lowering the temperature of polymerization slows down the proton transfer more than the propagation. A similar effect is observed on addition of sodium tetraphenyl boride, i.e., the free ions are more effective in proton transfer than the ion-pairs. The intramolecular proton transfer is also possible, although it is much slower. It is manifested by the slight shift in the absorption spectrum of the aged solution (from λ_{max} 350 to 360 mμ), by the decrease in its conductivity, and by the greatly reduced reactivity of living polymers on storage of their solution. The reaction is represented by the equation

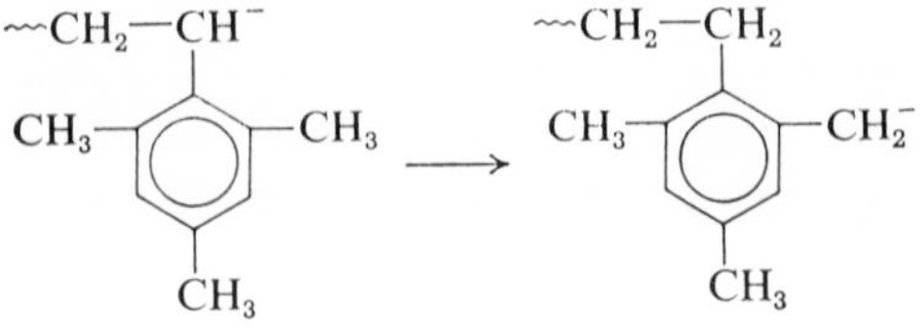

and it converts a secondary ion into a primary one. The latter is less reactive and less dissociated than the former.

The retardation of vinylmesitylene polymerization resembles phenomena

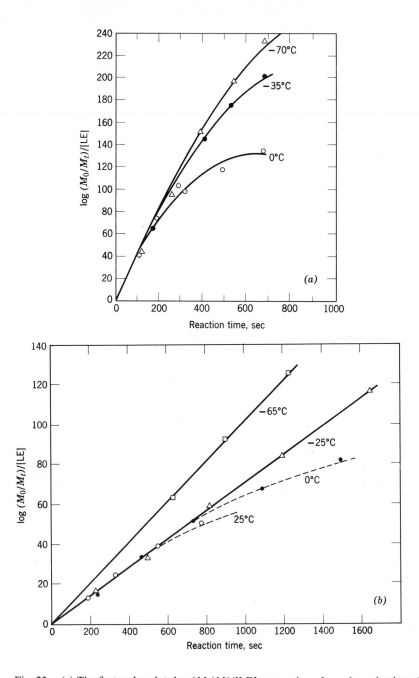

Fig. 23. (a) The first order plot, log (M_0/M_t)/[LE] versus time, for polymerization of sodium salt of living vinylmesitylene in tetrahydrofuran at different temperatures (69). The retardation of polymerization becomes more pronounced at higher temperatures. (b) Effect of sodium tetraphenyl boride on polymerization of living vinylmesitylene in tetrahydrofuran (69). Note, the presence of the salt reduces the deviation from the first order behavior.

reported in radical polymerization, e.g., the "degradative" polymerization of allyl chloride (76) or the self-inhibition of polymerization of vinyl benzoate (77). The latter is caused by the competition between two modes of monomer addition—either the vinyl group or the benzene ring of the benzoate group is involved in the reaction. Addition to the vinyl group yields a reactive radical which propagates polymerization, while an inert radical is formed in the other mode of addition, the latter being too stable to continue the chain reaction.

A somewhat similar situation is encountered in the polymerization of 9-vinylanthracene. This monomer is intrinsically reactive (78), although steric hindrance prevents again coplanarity of the vinyl group and the aromatic moiety. Nevertheless, it has been impossible to obtain a high molecular weight product, whatever the technique used in its polymerization. The propagation is slow, e.g., bulk polymerization initiated by di-*tert*-butyl peroxide at 100°C proceeded with a rate corresponding to a first order constant of $6 \times 10^{-5} \sec^{-1}$ only (79). Anionic polymerization was investigated by Eisenberg and Rembaum (73) and independently by Michel and Baker (80); the latter also studied the cationic polymerization of 9-vinylanthracene (81). Initiation by naphthalene or biphenyl radical-ions, or by butyllithium, produced oligomers, their degree of polymerization ranging from 4 to 12. Neither the molecular weight nor the yield of the product seems to be affected by the temperature of the reaction, although the rate of polymerization was greatly enhanced by heating. In fact, it is claimed that the polymerization initiated by butyllithium has an exceptionally high activation energy of 17 kcal/mole, although the significance of this value is not yet clear. Under properly chosen conditions it was possible to achieve a 100% conversion of monomer to polymer when Na^+, K^+, or Li^+ was the counterion. The polymerization involving Cs^+ counterion always resulted in a low polymer yield.

In the anionic polymerization the concentration of the growing species appeared to be constant in each run. The consumption of vinyl anthracene was given by a first order law, but the low molecular weight of the product indicates an efficient chain-transfer to the monomer. The latter conclusion is supported by the following observations. The apparent propagation constant, k_p, was found to be independent of initiator concentration—for example, when the reaction was initiated in tetrahydrofuran at 20°C with sodium naphthalene, the concentration of which varied from 5×10^{-3} to $2 \times 10^{-2} M$, the observed k_p was constant and equal to 0.2 liter mole^{-1} sec^{-1}. The molecular weight of the product also remained constant throughout the whole process, being independent of conversion. Moreover, if additional monomer was supplied to the polymerized system, more polymer was produced without affecting its molecular weight.

The observed bimolecular constant characterizing the anionic polymerization was interpreted as the propagation constant of living polymers (73). This is questionable because the reaction involves chain-transfer. Unfortunately, no clear-cut experiments have been performed to elucidate the mechanism of that chain-transfer process. Does it occur through transfer of an atom, or of some other moiety, from the monomer to the growing species (or vice versa)? This appears to be the most common transfer mechanism in radical-polymerization, but it is unlikely in the anionic process involving 9-vinyl-anthracene. It seems that this polymerization leads to the formation of dormant polymers which eventually initiate a new chain and, if that is the case, the reported constant, k_p, is not the genuine propagation constant because only a fraction of initially growing molecules participate in the reaction at any time. Such a situation may account for the apparent high activation energy of the propagation. By raising the temperature one increases not only the rate constant of propagation but perhaps also the fraction of growing polymers.

The dormant polymers may be produced by several mechanisms. For example, a "wrong" addition of monomer may form an unreactive species. In fact, such a reaction, which leads to termination, was suggested for the polymerization of Leuch's anhydrides to polyamino acids (82). It was shown that anthracene (A) reacts with living polymers (52,53), the addition taking place at carbon-9, forming a carbanion on carbon-10. Such a reaction may occur in the 9-vinylanthracene (9-VA) system,

producing a very stable, and therefore unreactive, carbanion. Spectroscopic evidence for the presence of such a unit in the oligomer was presented by Michel (81) and by Rembaum and Eisenberg (83). These authors proposed that the 1,6-addition, leads to its formation, but the above suggested alternative may be equally plausible. The dormant polymer may eventually transfer its electron to a monomer and the resulting monomeric radical-ion may initiate a new chain:

$$\sim CH_2 \underset{H}{\overset{}{\bigcirc\bigcirc\bigcirc}} =CH—CH_2^- + \text{(9-VA)} \rightleftarrows \sim CH_2 \underset{H}{\overset{}{\bigcirc\bigcirc\bigcirc}} =CH—CH_2\cdot + \text{(9-VA)}^-$$

Such electron transfer processes were thoroughly investigated and are well documented (84–88). Combination of the resulting radicals may then complete the process. According to this scheme, the rate of polymerization is determined by the above equilibrium followed by the re-initiation and propagation. The system discussed here is most interesting for further studies, but systematic and imaginative experiments are needed to clarify the mechanism of this polymerization.

14. STEREOSPECIFICITY OF IONIC PROPAGATION

Two types of linkages, namely isotactic and syndiotactic, join the segments of polymeric chains built from vinyl or unsymmetrical vinylidene monomers. This fact was recognized for a long time, although the terms isotactic and syndiotactic were coined less than 15 years ago. The polymers are defined as being stereoregular if virtually all their segments are joined through one type of linkage only; they are classified as atactic if both linkages appear along their chains, their distribution being somewhat random. Several unsuccessful attempts to synthesize stereoregular polymers were recorded in the earlier literature, but it was not until 1955 that Natta and his associates (91) reported the first, fully successful preparation of isotactic polypropylene. Propylene was polymerized with some heterogeneous catalysts commonly referred to as the Ziegler–Natta initiators, and the regular structure of the resulting polymer was conclusively proved by thorough x-ray examination.

The heterogeneous nature of these polymerization systems greatly complicates their studies. Indeed, the detailed mechanism of the stereoregulating power of these initiators is still poorly understood, although many hypotheses were presented to account for their action. A comprehensive review of this subject was published by Bawn and Ledwith (92).

Extensive investigations of various stereoregular polymerizations followed Natta's discovery. In the course of these studies it was shown that homogeneous anionic polymerization of acrylic and methacrylic esters is capable of yielding stereoregular polymers (93). These systems are more amenable to mechanistic studies and, therefore, several groups of workers undertook detailed investigations of such reactions hoping eventually to elucidate the course of stereo-regular propagation.

The mechanistic studies were greatly facilitated by the important work of Bovey and Tiers (94) who developed an NMR technique that permits one to recognize the isotactic or syndiotactic linkages in polymeric chains, as well as to distinguish all three possible sequences of consecutive segments (triads). These triads are classified as isotactic, syndiotactic, or heterotactic depending on whether the sequence involves two isotactic or two syndiotactic linkages in a row, or an isotactic linkage followed by a syndiotactic one, or vice-versa.

Bovey and Tiers pointed out that the two methylenic hydrogens of polymethyl methacrylate are magnetically equivalent in a syndiotactic linkage, viz.,

$$
\begin{array}{c c c}
X & H & Y \\
| & | & | \\
P-C-C-C-P, \\
| & | & | \\
Y & H & X \\
\end{array}
$$

and, therefore, they produce only one line in the relevant NMR spectrum. However, in an isotactic linkage their immediate magnetic environment is different, even if a free rotation around the neighboring C—C bonds is allowed. This becomes obvious after inspecting the following diagram.

Consequently, the methylenic protons of an isotactic linkage should produce an AB pattern in the NMR spectrum and this indeed was observed. Furthermore, three different magnetic environments are expected for the α-methyl protons, depending on whether the methyl group is located in the middle of an isotactic, a syndiotactic, or a heterotactic triad. Hence, the presence of each triad should be revealed by the characteristic chemical shift of the respective protons, and this again was demonstrated. Thus the CH_3 protons in an isotactic triad absorb at $\tau = 8.78$ ppm, those in a syndiotactic one at $\tau = 9.09$ ppm, and the protons in a heterotactic triad produce a line at $\tau = 8.95$ ppm. The relative intensities of the various lines give the abundance of the various triads in the investigated polymer. These concepts and their ramifications were recently reviewed by Bovey (95,96).

As NMR technique improved, further differentiation became possible. The influence of the more distant groups could be observed, and these effects, first predicted by Bovey (97), were eventually demonstrated by Yoshino (98). His team identified all the ten different chemical shifts corresponding to the six different types of tetrads of poly(α-β-d_2-vinyl chloride). Undoubtedly, further refinements of the techniques and of the instrumentation will permit a still more sophisticated differentiation of the various sequences, thus providing a deeper insight into the structure of polymeric chains. The abundance of various sequences is governed by some statistical laws. For example, Coleman and Fox (99) derived the required functional relations between the abundance of the various diads (single linkages) and triads. Extension of this work to tetrads and pentads was reported by Frisch, Mallows, and Bovey (100). Some of these relations are independent of the mechanism of addition and therefore provide checks for the assignment of the observed lines. Others depend on the stochastic laws which govern the structure of the chain, and hence the observed abundance of the various sequences permits one to establish the laws governing a particular mechanism of polymerization. It should be stressed that some mechanisms of anionic polymerization lead to nonstochastic sequences of segments in the resulting polymers. This indeed was established by detailed studies of their NMR spectra (99,118). Similar problems are encountered in some ionic copolymerizations (see Chapter IX, p. 555).

For the sake of illustration, some identity relations are quoted. These are, of course, always valid whatever the mechanism of polymerization, and they apply even to non-Markoff chains. Denoting the mole fraction of isotactic diads (linkages) by m, those of syndiotactic ones by r, and the mole fraction of the isotactic, syndiotactic, and heterotactic triads by i, s, and h, respectively, one finds

$$m + r = 1,$$
$$i + h + s = 1,$$
$$m = i + \tfrac{1}{2}h \qquad r = s + \tfrac{1}{2}h,$$
$$i = (mmm) + \tfrac{1}{2}(mmr) \qquad s = (rrr) + \tfrac{1}{2}(rrm),$$
$$h = (mmr) + 2(rmr) = (rrm) + 2(mrm),$$
$$(mmr) + 2(rmr) = (rrm) + 2(mrm).$$

Here the symbols in brackets denote the six respective tetrads; the sequence of their linkages is self-explanatory.

When a specific mechanism operates further relations are expected. A classic example is provided by Bovey and Tiers (94) who showed graphically the relation between i, s, and h expressed in terms of one parameter only. Such a relation must be maintained if the polymerization obeys the Bernoulie statistics, i.e., when the structure of the last linkage uniquely determines the probability of the following placement.

The NMR spectra of polymers may become more complex if two or more conformations are favored and when the rotations converting one conformer into the other are sufficiently slow. Examples of such phenomena are reported by Yoshino et al. (103).

To account for stereoregularity of propagation we have to comprehend the interactions through which the stereostructure of the newly formed segment is determined by the geometry of the preceding units. The latter problem is different for two distinct types of addition. In some systems the stereostructure of an added segment is finally determined in the course of the reaction. For example, consider the addition of a $CH_2\!\!=\!\!CXY$ monomer to a polymer endowed with a $\sim\!CH_2$ Cat reactive end group in which the C—Cat

$$\sim\!CH_2\diagdown_{\substack{C\\ X^{\cdots}\diagup\diagup\searrow Y}}\diagup Cat$$

bond is strong and does not dissociate easily. The addition involves insertion into the C—Cat bond, e.g.,

$$\sim\!CH_2 \diagdown \underset{\substack{X \diagdown Y}}{C} \diagup Cat + CH_2\!\!=\!\!CXY \longrightarrow \sim\!CH_2 \diagdown \underset{\substack{X \diagdown Y}}{C} \diagup CH_2 \diagdown \underset{\substack{X \diagdown Y}}{C} \diagup Cat ,$$

and the new segment emerges with a well-defined geometry which is not altered in the course of the subsequent growth of the polymer. The growth is stereo-regular and obeys Bernoulie statistics if the geometry of the *last* unit determines that of the newly added one. A similar situation exists in a propagation described by the equation[*]

$$\sim\!CH_2 \diagdown \underset{\substack{X \diagdown Y}}{C} \diagup CH_2,Cat + CXY\!\!=\!\!CH_2 \longrightarrow \sim\!CH_2 \diagdown \underset{\substack{X \diagdown Y}}{C} \diagup CH_2 \diagdown \underset{\substack{X \diagdown Y}}{C} \diagup CH_2,Cat$$

Here again the spatial configuration of the last segment is permanently determined, and on its addition the new unit acquires a geometry which cannot be modified by the subsequent steps of polymerization.

The situation is different if the end group of the growing polymer has no established parity, e.g., when it has a planar, trigonal structure. Such a geometry may characterize some free carbonium ions, free carbanions, or free radicals. Alternatively, one faces the same problem when ion-pairs undergo a rapid Walden inversion as the result of their easy dissociation or partial dissociation. In such systems the parity of the last unit becomes established only in the course of the addition of a monomer, and the

[*] This type of addition is probably encountered in Ziegler–Natta polymerization of propylene.

propagation is stereoregular if the geometry of the *penultimate* unit determines the spatial arrangement around the *last* unit at a time when the latter becomes linked up to a newly added monomer.

Three types of interactions may account for stereoregularity of propagation: steric factors may force the new unit into spatial arrangement determined by bulky groups of the preceding segments, the polar factors could favor one conformation more than the other; and the simultaneous coordination of the last unit and of the incoming monomer, or of the penultimate unit and the last unit, with a suitable counterion may determine the stereochemistry of the addition. The last interaction takes place only in ionic or coordination polymerization; it does not pertain to a classic radical polymerization.

The effect of the bulky groups present in the monomer upon the stereoregularity of the resulting polymer was noted in the early studies of polymerization of acrylates and methacrylates (93b,104). Comprehensive studies of this subject were reported recently by Tsuruta (105,106).

The direct polar effects were considered by Fordham (107) who discussed radical polymerization of vinyl chloride. However, the most important seem to be the effects caused by the coordination of polar groups with counterions. These operate only in ionic or coordination-type polymerization and their significance was discussed by Bawn and Ledwith (92), Furukawa (108), Okamura (109), and others. These ideas were conceived and tested for small molecules by Cram (110,111) who visualized a rigid six-membered ring in the transition state of an isotactic placement. His treatment considers also the size of the substituents and their influence on the stereochemistry (138).

The importance of coordination explains the striking effects of solvents upon the degree of stereospecificity of polymers. For example, Fox et al. (93a) reported that anionic polymerization of methyl methacrylate (Li^+ counterions) yields isotactic polymer if it is carried out in hydrocarbon, a syndiotactic polymer when the reaction proceeds in solvents having relatively high dielectric constants, and stereoblocks when run in mixed solvents. More detailed studies of Kern et al. (112) show that the degree of syndiotacticity increases as the Lewis base strength of solvents becomes higher. The effects are even more pronounced in isotactic polymerizations; they are observed at concentrations of the added ethers as low as $10^{-2}M$ (113). These phenomena cannot be accounted for in terms of the variable dielectric constant of the medium because the bulk dielectric constant is hardly affected by the additives. They must be ascribed to a specific solvation (coordination) of ion-pairs (92, 114,115). Apparently, contact ion-pairs favor isotactic polymerization, while the pairs separated by solvating agents tend to produce syndiotactic linkages.

Anionic polymerizations initiated by Grignard compounds are particularly affected by solvent and additives. Extensive studies of the stereoregularities of the resulting polymers were reported by Nishioka (119), and recently this subject was reinvestigated by Allen (120). The additives, like ethers, affect not only the stereoregularity of the product, and the molecular weight distribution of the product, but also the rate of polymerization. The rate increases on addition of the coordinating agent; it reaches maximum at some concentration of the agent and then decreases.

The existence of two different propagating species in anionic polymerization of methyl methacrylate is evident from recent results of Bywater (116,117) who observed two absorption peaks in the UV spectrum of living polymethyl methacrylate and noted that their relative intensities were solvent dependent. This implies that each absorption peak represents a distinct form of ion-pairs, their relative amounts depending on solvent and temperature. Another important evidence for the participation of two types of growing ends in the polymerization is obtained from detailed NMR studies of polymers produced in hydrocarbon solvents in the presence of ethers. The sequences of isotactic and syndiotactic linkages are *not* given by any Markoff chain (99,118), indicating again the existence of two growing species which are in dynamic equilibrium with each other. Apparently one deals with an etherated and nonetherated form of living polymers, each of them showing preference for a different type of placement.

Further insight into the mechanism of polyaddition may be obtained from studies of the modes of opening the C=C double bond. Consider for the sake of illustration *cis*- and *trans*-1-d_1-propene:

Isotactic polymerization of such monomers may produce diisotactic chains of the *erythro*- or *threo*-type, viz.,

Indeed, Natta et al. (121) observed that *cis*- and *trans*-d_1-propene yield different polymers distinguished through their characteristic polarized infrared spectra. The assignment of configuration of these polymers was achieved by Miyazawa and Ideguchi (101). They showed that the *cis*-isomer produces an *erythro*-polymer, whereas a *threo*-polymer is formed from the *trans*-olefin. These results were interpreted as evidence for a *cis*-opening of a double bond in the course of polyaddition induced by the Ziegler–Natta catalyst. The same conclusion was previously reached by Natta et al. (102) who studied the diisotactic polymerization of *trans*-propenyl isobutyl ether. Note that the representation of the polymer chain through a planar zigzag naively implies that the *trans*-opening takes place.

Subsequently, Yoshino et al. (122) demonstrated that anionic polymerization of deuterated methyl acrylate initiated with lithium aluminium hydride produces diisotactic polymer. On the basis of their NMR analyses one may conclude that, contrary to the previous findings, a *trans*-addition takes place in this polyaddition. Even more surprising are the results of Schuerch et al. (123) who demonstrated that Grignard-initiated polymerization of α-*cis*-β-d_2-acrylate led to an isotactic polymer with respect to α-carbon, while the orientation around the β-carbon corresponds to stereo-blocks. These findings show that the concept of *cis*- and *trans*-opening should be re-examined.

In a polyaddition, *cis*- and *trans*-openings have no precise meaning in terms of the structure of transition state. The concept is unequivocal in a four-center reaction, viz.,

$$
\begin{array}{c}
\text{X—Y} \\
A\diagdown \vdots \quad \vdots \diagup C \\
\text{C—C} \\
\diagup \qquad \diagdown \\
B \qquad\quad D.
\end{array}
$$

However, if a two-step process takes place, then the mode of approach of the monomer to the growing polymer uniquely determines the configuration around the β-carbon (assuming the addition takes place at this position). The incorporation of the C=C bond is completed when the next monomer is added, and if the rotation around the β-α-bond is permissible, the final outcome may appear as if it were either a *cis*- or a *trans*-opening of the C=C bond.

To clarify these ideas, let us consider the addition of the monomer,

$$
\begin{array}{c}
\text{topside} \\
H \qquad\quad X \\
\beta \quad \diagdown C{=}C \diagup \quad \alpha \\
\diagup \qquad\quad \diagdown \\
D \qquad\qquad H \\
\text{bottomside}
\end{array}
$$

to an anion,

$$
\begin{array}{c}
H \qquad D \\
\diagdown \quad \diagdown \diagup \\
\text{\textasciitilde}C \qquad C \qquad \diagup \\
\diagdown \quad \diagup (2) \diagdown \\
C \qquad\quad C{-}Y \\
\diagup \qquad\qquad \diagdown \\
X \ (3) \ Y \ \ (1) \ X.
\end{array}
$$

Assume that some kind of interaction between carbons (1) and (3) forces the orientation around the last carbon to be isotactic in respect to (3) when a new monomer becomes bonded to (1) and that the monomer must approach the anion from above as indicated by the arrow. However, the monomer may reach the anion with its "topside" or "bottomside." Each of these alternative cases establishes a well-defined configuration around the β-carbon, but such a configuration is isotactic in respect to carbon (2) for one alternative, while it becomes syndiotactic for the other. Either addition may lead to the desirable configuration around the α-carbon by proper rotation around the β–α axis. Such a rotation may take place *after* the $C(1)$—$C(\beta)$ bond was formed. Thus, a series of monomer additions may produce the isotactic structure around α-carbons, while the configuration around β-carbons may remain random.

These ideas were developed in a most interesting paper by Fowells, Schuerch, Bovey, and Hood (124). Further consideration of the problem suggests that the following questions may be posed:

(*a*) Is the structure of the growing carbanion rigid or is the rotation around the β–α axis free? The rigidity of the structure *does not* preclude a rotation in the transition state. Such a rotation could be imposed by the system and it could lead to a fixed geometry of the new carbanion defined in respect to the preceding unit. This geometry may then be maintained in the process of addition of the next monomer.

(*b*) If the geometry of the carbanion is rigidly maintained, one may ask whether the approach of the monomer is restricted to one side of the carbanion or whether both sides are available for the reaction. In the former case, which is the "reactive side"?

(*c*) Does the rotation occur in the transition state? This problem is related to question *a*.

(*d*) Assuming a rigid geometry of a carbanion and availability of only one side for the reaction, we may ask whether the approaching monomer is free to reach the carbanion with either of its sides (topside or bottomside), or whether the system favors one type of approach more than the other.

Detailed discussion of these possibilities reveals that several distinct mechanisms may yield polymers of the same structure. One should inquire, therefore, what configuration is imposed on the product by each particular mechanism. For the sake of illustration, three types of mechanisms will be considered.

(*1*) Living polyacrylate forms a contact ion-pair, say with a Li^+ counterion. The Li^+ cation is bonded to oxygen atoms of the carbonyl groups of the last and of the penultimate units. Thus, the geometry of the last unit is related to the configuration of the penultimate unit. Let the approaching monomer be first bonded to the Li$^+$. Its spatial relation with respect to the preceding segments becomes, therefore, uniquely determined, and by the same token its β-carbon approaches the growing anion in a unique way. The new C—C bond is then formed, and eventually the linkage between Li$^+$ and the former penultimate unit is ruptured. This sequence of events is depicted in Figure 24. The above mechanism ascertains a unique structure around the α- as well as β-atoms of the polymer, and the resulting additions may be formally represented by *cis*-opening of the C=C bond.

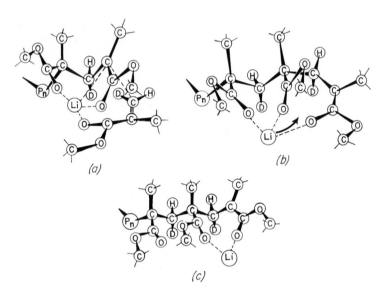

Fig. 24. (*a*) An isotactic-like approach of the monomer to the chelated contact ion-pair. (*b*) The new C—C bond has been formed with the methylene D on the same side of the zigzag as the ester function. (*c*) The Li$^+$ moves up to the new anion, with concurrent rotation of the new penultimate ester group, forming the same chelated structure as in *a*.

(2) The Li$^+$ of the ion-pair is still bonded to the two carbonyl groups as in case *1*. However, due to the presence of some solvating agent, say tetrahydrofuran, the other coordinating sites of the Li$^+$ become saturated by the ether and therefore are not available for the approaching monomer. Nevertheless, the spatial structure of the carbanion leaves only one of its sites available for the reaction. Now, two situations may be visualized. The monomer, which must reach the carbanion from "above" (see Figure 25), is bound to approach it from one side and not from the other, e.g., with its downside. This restriction may be imposed by steric or polar factors and the resulting configuration is shown in Figure 25a. *After* formation of the C—C bond, a rotation around the β–α axis brings the α-carbon into a proper

position in respect to Li$^+$; the cation, therefore, becomes bonded to the carbonyl group of the new last unit and the link with the former penultimate unit is ruptured simultaneously. The resulting polymer again has well-defined configurations around α- and β-carbons, both being rigidly maintained along

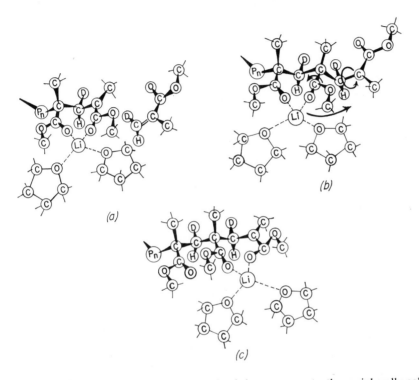

Fig. 25. (a) A syndiotactic-like approach of the monomer to the peripherally solvated contact ion-pair. There is no coordination of the monomer carbonyl with the counterion, and nonbonded interactions force the approach to be syndiotactic-like. (b) The new C—C bond has been formed. (c) The Li$^+$ and its peripheral solvent shell move up to the terminal unit, with concurrent rotation of the ester function. As the new anion resides largely on the carbonyl, there is a simultaneous rotation about the new α,β-bond to reduce charge separation. This results in an *erythro-meso* placement, the methylene D being on the opposite side of the zigzag from the ester groups and the α-carbon now in an incipient isotactic configuration.

the chain. However, the addition appears formally as a *trans*-opening of the C=C bond. This sequence of events is shown in Figure 25.

(*3*) If the restriction imposed upon the approaching monomer is lifted, i.e., when its topside and bottomside have comparable chances of being added to the growing anion, then the mechanism outlined in *2* leads to stereo-regularity of α-carbons, while the configurations around β-carbons become random. This case has not yet been observed.

It is hoped that these examples clarify some of the problems encountered in stereo-regular polymerization. It is important to realize that the above ideas take cognizance of the structure of the ionic species which propagate the polymerization. This rationalizes the effects of counterions, solvents, solvating additives, etc. We see again that the structure of ion-pair is of the utmost importance in determining the course of ionic polymerizations.

Stereo-regularity of other ionic polymerizations has been investigated. However, no other system has been as thoroughly explored as the acrylates and methacrylates. The stereospecific polymerization of styrene was reported by Sinn et al. (125), Kern et al. (126), and Williams et al. (127). The significance of the last study is discussed in Chapter XI (see p. 633). Stereospecific polymerization of N,N-disubstituted acrylamides was investigated by Butler et al. (128) and stereospecific polymerization of t-butylvinyl ketones by Overberger and Schiller (129). The stereospecific polymerization of dienes is discussed in Chapter VIII (see p. 513).

References

(1) (a) G. Allen, G. Gee, and C. Stretch, *J. Polymer. Sci.*, **48**, 189 (1960). (b) C. Stretch and G. Allen, *Polymer*, **2**, 151 (1961).

(2) D. N. Bhattacharyya, J. Smid, and M. Szwarc, *J. Phys. Chem.*, **69**, 624 (1965).

(3) F. S. Dainton, G. C. East, G. A. Harpell, N. R. Hurworth, K. J. Ivin, R. T. LaFlair, R. H. Pallen, and K. M. Hui, *Makromol. Chem.*, **89**, 257 (1965).

(4) M. Shinohara, unpublished results of Syracuse group.

(5) C. Geacintov, J. Smid, and M. Szwarc, *J. Am. Chem. Soc.*, **83**, 1253 (1961); **84**, 2508 (1962).

(6) D. J. Worsfold and S. Bywater, *J. Chem. Soc. (London)*, **1960**, 5234.

(7) D. N. Bhattacharyya, C. L. Lee, J. Smid, and M. Szwarc: (a) *Polymer*, **5**, 54 (1964); (b) *J. Phys. Chem.*, **69**, 612 (1965).

(8) (a) H. Hostalka, R. V. Figini, and G. V. Schulz, *Makromol. Chem.*, **71**, 198 (1964); (b) H. Hostalka and G. V. Schulz, *Z. Physik. Chem. (Frankfurt)*, **45**, 286 (1965).

(9) K. J. Tölle, J. Smid, and M. Szwarc, *J. Polymer Sci. B*, **3**, 1037 (1965).

(10) M. Szwarc, *Makromol. Chem.*, **35A**, 123 (1960).

(11) T. Shimomura, K. J. Tölle, J. Smid, and M. Szwarc, *J. Am. Chem. Soc.*, **89**, 796 (1967).

(12) H. Hostalka and G. V. Schulz, *J. Polymer Sci. B*, **3**, 1043 (1965).

(13) M. Szwarc, *Pure Appl. Chem.*, **12**, 127 (1966).

(14) C. Carvajal, K. J. Tölle, J. Smid, and M. Szwarc, *J. Am. Chem. Soc.*, **87**, 5548 (1965).

(15) W. K. R. Barnikol and G. V. Schulz, *Makromol. Chem.*, **68**, 211 (1963); **86**, 298 (1965).

(16) K. J. Ivin, unpublished data.

(17) T. Shimomura, J. Smid, and M. Szwarc, *J. Am. Chem. Soc.*, **89**, 5743 (1967).

(18) F. S. Dainton, D. M. Wiles, and A. N. Wright, *J. Polymer Sci.*, **45**, 111 (1960).

(19) M. Fisher, M. van Beylen, J. Smid, and M. Szwarc, to be published.

(20) M. Fisher, J. Smid, and M. Szwarc, to be published.

(21) M. van Beylen, D. N. Bhattacharyya, J. Smid, and M. Szwarc, *J. Phys. Chem.*, **70**, 157 (1966).

(22) D. J. Worsfold and S. Bywater, *J. Phys. Chem.*, **70**, 162 (1966).

(23) R. V. Slates and M. Szwarc, *J. Phys. Chem.*, **69**, 4124 (1965).

(24) M. S. Matheson, E. E. Auer, E. B. Bevilacqua, and E. J. Hart, *J. Am. Chem. Soc.*, **73**, 1700 (1951).

(25) K. Ueno, F. Williams, K. Hayashi, and S. Okamura, *Trans. Faraday Soc.*, **63**, 1478 (1967). F. Williams, K. Hayashi, K. Ueno, H. Hayashi, and S. Okamura, *ibid.*, **63**, 1501 (1967).

(26) D. J. Worsfold and S. Bywater, *Can. J. Chem.*, **36**, 1141 (1958).

(27) J. Comyn, F. S. Dainton, G. A. Harpell, K. M. Hui, and K. J. Ivin, *J. Polymer Sci. B*, **5**, 965 (1967).

(28) M. Szwarc and J. Smid, *Progr. Reaction Kinetics*, **2**, 250 (1964).

(29) G. Spach, M. Levy, and M. Szwarc, *J. Chem. Soc. (London)*, **1962**, 355.

(30) T. Alfrey and W. H. Ebelke, *J. Am. Chem. Soc.*, **71**, 3235 (1949).

(31) F. S. Dainton, K. H. Hui, and K. J. Ivin.

(32) T. E. Hogen-Esch and J. Smid, *J. Am. Chem. Soc.*, **88**, 307 (1966).

(33) P. Chang, R. V. Slates, and M. Szwarc, *J. Phys. Chem.*, **70**, 3180 (1966).

(34) J. E. L. Roovers and S. Bywater, *Trans. Faraday Soc.*, **62**, 701 (1966).

(35) C. M. Fontana and G. A. Kidder, *J. Am. Chem. Soc.*, **70**, 3745 (1948).

(36) F. R. Mayo and C. Walling, *ibid.*, **71**, 3845 (1949).

(37) G. V. Rakova and A. A. Korotkov, *Dokl. Akad. Nauk SSSR*, **119**, 982 (1958).

(38) S. W. Bresler, A. A. Korotkov, M. J. Mosevitskii, and Y. Y. Poddubnyi, *Zh. Tekhn. Fiz.*, **28**, 114 (1958).

(39) Yu. L. Spirin, A. R. Gantmakher, and S. S. Medvedev, *Vysokomolekul. Soedin.*, **1**, 1258 (1959).

(40) Yu. L. Spirin, A. A. Arest-Yakubovich, D. K. Polyakov, A. R. Gantmakher, and S. S. Medvedev, *J. Polymer Sci.*, **58**, 1181 (1962).

(41) A. A. Korotkov and A. F. Podolsky, *J. Polymer Sci. B*, **3**, 901 (1965).

(42) R. V. Figini, *J, Polymer Sci. B*, **4**, 223 (1966).

(43) G. V. Schulz, R. V. Figini, and G. Löhr, *Makromol. Chem.*, **96**, 283 (1966).

(44) T. Shimomura, J. Smid, and M. Szwarc, *Makromol. Chem.*, **108**, 288 (1967).

(45) S. Winstein, E. Clippinger, A. H. Fainberg, and G. C. Robinson, *J. Am. Chem. Soc.*, **76**, 2597 (1954).

(46) S. Chadha and M. Szwarc, unpublished results.

(47) R. M. Fuoss and C. A. Kraus, *J. Am. Chem. Soc.*, **55**, 21, 476, 1019, 2387 (1933).

(48) D. N. Bhattacharyya, J. Smid, and M. Szwarc, *J. Am. Chem. Soc.*, **86**, 5024 (1964).

(49) R. Huisgen and W. Mack, *Chem. Ber.*, **93**, 332 (1960).

(50) R. Waack and M. A. Doran, *Chem. Ind. (London)*, **1964**, 496.

(51) M. Shinohara, J. Smid, and M. Szwarc, *J. Am. Chem. Soc.*, **90**, 2175 (1968).

(52) A. A. Arest-Yakubovich, A. R. Gantmakher, and S. S. Medvedev, *Dokl. Akad. Nauk SSSR*, **139**, 1351 (1961).

(53) S. N. Khanna, M. Levy, and M. Szwarc, *Trans. Faraday Soc.*, **58**, 747 (1962).

(54) R. Asami, S. Khanna, M. Levy, and M. Szwarc, *Trans. Faraday Soc.*, **58**, 1821 (1962).

(55) R. Lipman, J. Jagur-Grodzinski, and M. Szwarc, *J. Am. Chem. Soc.*, **87**, 3005 (1965).

(56) J. Stearne, J. Smid, and M. Szwarc, *Trans. Faraday Soc.*, **62**, 672 (1966).

(57) M. Levy and F. Cohen-Bosidan, *Polymer*, **1**, 517 (1960).

(58) H. Kawazura, *Makromol. Chem.*, **59**, 201 (1963).

(59) (a) I. Sakurada, N. Ise, and T. Ashida, *Makromol. Chem.*, **95**, 1 (1966); (b) I. Sakurada, N. Ise, Y. Tanaka, and Y. Hayashi, *J. Polymer Sci. A-1*, **4**, 2801 (1966).
(60) I. Sakurada, N. Ise, and S. Kawabata, *Makromol. Chem.*, **97**, 17 (1966).
(61) L. Onsager, *J. Chem. Phys.*, **2**, 599 (1934).
(62) L. Onsager, *Phys. Rev.*, **54**, 554 (1938).
(63) I. Sakurada, N. Ise, H. Hirohara, and T. Makino, *J. Phys. Chem.*, **71**, 3711 (1967).
(64) G. Löhr and G. V. Schulz, *Makromol. Chem.*, **77**, 240 (1964).
(65) R. V. Figini, G. Löhr, and G. V. Schulz, *J. Polymer Sci. B*, **3**, 985 (1965).
(66) R. V. Figini, H. Hostalka, K. Hurm, G. Löhr, and G. V. Schulz, *Z. Physik. Chem. (Frankfurt)*, **45**, 269 (1965).
(67) G. V. Schulz, K. C. Berger and A. G. R. Scholz, *Z. Phys. Chem.*, **69**, 856 (1965).
(68) M. Szwarc, *Polymer Preprints*, **8**, (1), 22 (1967).
(69) D. N. Bhattacharyya, J. Smid, and M. Szwarc, *J. Polymer Sci. A*, **3**, 3099 (1965).
(70) M. Shima, D. N. Bhattacharyya, J. Smid, and M. Szwarc, *J. Am. Chem. Soc.*, **85**, 1306 (1963).
(71) F. Bahsteter, J. Smid, and M. Szwarc, *J. Am. Chem. Soc.*, **85**, 3909 (1963).
(72) C. L. Lee, J. Smid, and M. Szwarc, *Trans. Faraday Soc.*, **59**, 1192 (1963).
(73) A. Eisenberg and A. Rembaum, *J. Polymer Sci. B*, **2**, 157 (1964).
(74) M. Tardi, D. Rougé, and P. Sigwalt, *European Polymer J.*, **3**, 85 (1967).
(75) G. Natta, G. Mazzanti, P. Longi, and G. Dall'Asta, *J. Polymer Sci.*, **51**, 487 (1961).
(76) P. D. Bartlett and R. Artschul, *J. Am. Chem. Soc.*, **67**, 812, 816 (1945).
(77) (a) E. D. Morrison, E. H. Gleason, and V. Stannett, *J. Polymer Sci.*, **36**, 267 (1959). (b) M. Litt and V. Stannett, *Makromol. Chem.*, **37**, 19 (1960).
(78) F. Carrock and M. Szwarc, *J. Am. Chem. Soc.*, **81**, 4138 (1959).
(79) D. Katz, *J. Polymer Sci. A*, **1**, 1635 (1963).
(80) R. H. Michel and W. P. Baker, *J. Polymer Sci. B*, **2**, 163 (1964).
(81) R. H. Michel, *J. Polymer Sci. A*, **2**, 2533 (1964).
(82) M. Sela and A. Berger, *J. Am. Chem. Soc.*, **75**, 6350 (1953) ; **77**, 1893 (1955).
(83) A. Rembaum and A. Eisenberg, *Macromolecular Reviews*, Vol. 1, A. Peterlin, M. Goodman, S. Okamura, B. H. Zimm, and H. F. Mark, Eds., Interscience, New York, 1967, p. 57.
(84) (a) J. Jagur, M. Levy, M. Feld, and M. Szwarc, *Trans. Faraday Soc.*, **58**, 2168 (1962). (b) J. Jagur-Grodzinski and M. Szwarc, *Proc. Roy. Soc. London A*, **288**, 224 (1965).
(85) J. Jagur-Grodzinski and M. Szwarc, *Trans. Faraday Soc.*, **59**, 2305 (1963).
(86) D. Gill, J. Jagur-Grodzinski, and M. Szwarc, *Trans. Faraday Soc.*, **60**, 1424 (1964).
(87) G. A. Russell, E. G. Janzen, and E. T. Strom, *J. Am. Chem. Soc.*, **86**, 1807 (1964).
(88) A. A. Arest-Yakubovich, *Vysokomolekul. Soedin.*, **6**, 290 (1965).
(89) See, e.g., B. L. Funt and F. D. Williams, *J. Polymer Sci. A*, **2**, 865 (1964).
(90) M. Szwarc, *Proc. N. Y. Acad. Sci.*, in press.
(91) (a) G. Natta, *J. Polymer Sci.*, **16**, 143 (1955). (b) G. Natta, P. Pino, P. Corradini, F. Danusso, E. Mantica, G. Mazzanti, and G. Moraglio, *J. Am. Chem. Soc.*, **77**, 1708 (1955). (c) G. Natta and P. Corradini, *Ric. Sci. Suppl. A*, **25**, 695 (1955).
(92) C. E. H. Bawn and A. Ledwith, *Quarterly Rev. (London)*, **16**, 361 (1962).
(93) (a) T. G Fox et al., *J. Am. Chem. Soc.*, **80**, 1768 (1958). (b) B. S. Garrett et al., *ibid.*, **81**, 1007 (1959). (c) J. D. Stroupe and R. E. Hughes, *ibid.*, **80**, 2341 (1958).
(94) F. A. Bovey and G. V. D. Tiers, *J. Polymer Sci.*, **44**, 173 (1960).
(95) F. A. Bovey, *Advan. Polymer Sci.*, **3**, 149 (1963).
(96) F. A. Bovey, *Pure Appl. Chem.*, **12**, 525 (1966).
(97) F. A. Bovey, E. W. Anderson, and D. C. Douglass, *J. Chem. Phys.*, **39**, 1199 (1963).
(98) T. Yoshino and J. Komiyama, *J. Polymer Sci. B*, **3**, 311 (1965).
(99) B. D. Coleman and T. G Fox, *J. Chem. Phys.*, **38**, 1065 (1963); *J. Polymer Sci. A*, **1**, 3183 (1963).

(100) H. L. Frisch, C. L. Mallows, and F. A. Bovey, *J. Chem. Phys.*, **45**, 1565 (1966).
(101) T. Miyazawa and Y. Ideguchi, *J. Polymer Sci. B*, **1**, 389 (1963).
(102) G. Natta, M. Farina and M. Peraldo, *Chim. Ind. (Milan)*, **42**, 255 (1960).
(103) T. Yoshino, Y. Kikuchi, and J. Komiyama, *J. Phys. Chem.*, **70**, 1059 (1966).
(104) M. L. Miller and C. E. Rauhut, *J. Am. Chem. Soc.*, **80**, 4115 (1958); *J. Polymer Sci.*, **38**, 63 (1959).
(105) T. Tsuruta, T. Makimoto, and H. Kanai, *Makromol. Chem.*, **1**, 31 (1966).
(106) T. Makimoto, T. Tsuruta, and J. Furukawa, *Makromol. Chem.*, **50**, 116 (1961).
(107) J. W. L. Fordham, *J. Polymer Sci.*, **39**, 321 (1959).
(108) J. Furukawa, *Polymer (London)*, **3**, 487 (1962).
(109) Y. Oshumi, T. Higashimura, and S. Okamura, *J. Polymer Sci. A*, **3**, 3729 (1965).
(110) D. J. Cram and K. R. Kopecky, *J. Am. Chem. Soc.*, **81**, 2748 (1959).
(111) D. J. Cram, *J. Chem. Ed.*, **37**, 317 (1960).
(112) D. Braun, M. Herner, U. Johnsen, and W. Kern, *Makromol. Chem.*, **51**, 15 (1962).
(113) C. F. Ryan and P. C. Fleischer, *J. Phys. Chem.*, **69**, 3384 (1965).
(114) K. Butler, P. R. Thomas, and G. J. Tyler, *J. Polymer Sci.*, **48**, 357 (1960).
(115) C. Schuerch, *Ann. Rev. Phys. Chem.*, **13**, 195 (1962).
(116) S. Bywater, P. E. Black, and D. M. Wiles, *Can. J. Chem.*, **44**, 695 (1966).
(117) D. M. Wiles and S. Bywater, *J. Polymer Sci. B*, **2**, 1175 (1964).
(118) B. D. Coleman and T. G Fox, in *International Symposium on Macromolecular Chemistry, Paris 1963*, M. Magat, Ed. (*J. Polymer Sci. C*, **4**),Interscience, New York, 1964, p. 345.
(119) A. Nishioka, H. Watanabe, K. Abe, and Y. Sono, *J. Polymer Sci.*, **48**, 241 (1960).
(120) P. E. M. Allen and A. G. Moody, *Makromol. Chem.*, **81**, 234 (1965); **83**, 220 (1965).
(121) (a) G. Natta, M. Farina, and M. Peraldo, *Atti Accad. Nazl. Lincei*, **25**, 424 (1958). (b) M. Peraldo and M. Farina, *Chim. Ind. (Milan)*, **42**, 1349 (1960).
(122) (a) T. Yoshino, J. Komiyama, and M. Shinomiya, *J. Am. Chem. Soc.*, **86**, 4482 (1964). (b) T. Yoshino, M. Shinomiya, and J. Komiyama, *ibid.*, **87**, 387 (1965).
(123) C. Schuerch, W. Fowells, A. Yamada, F. A. Bovey, F. P. Hood, and E. W. Anderson, *ibid.*, **86**, 4481 (1964).
(124) W. Fowells, C. Schuerch, F. A. Bovey, and F. P. Hood, *ibid.*, **89**, 1396 (1967).
(125) H. Sinn, C. Lundborg, and K. Korschner, *Angew. Chem.*, **70**, 744 (1958).
(126) W. Kern, D. Braun, and M. Herner, *Makromol. Chem.*, **28**, 66 (1958); **36**, 232 (1960).
(127) J. L. R. Williams, T. M. Laakso, and W. J. Dulmage, *J. Org. Chem.*, **23**, 638 (1958)
(128) K. Butler, P. R. Thomas, and G. J. Tyler, *J. Polymer Sci.*, **48**, 357 (1960).
(129) C. G. Overberger and A. M. Schiller, in *First Biannual American Chemical Society Polymer Symposium*, H. W. Starkweather, Jr., Ed. (*J. Polymer Sci. C*, **1**),Interscience, New York, 1963, p. 325.
(130) K. Vesely, *Pure Appl. Chem.*, **4**, 407 (1962).
(131) S. C. Chadha, J. Jagur-Grodzinski, and M. Szwarc, *Trans. Faraday Soc.*, **63**, 2994 (1967).
(132) L. Böhm, W. K. R. Barnikol, and G. V. Schulz, *Makromol. Chem.*, **110**, 222 (1967).
(133) Unpublished data of the Syracuse group.
(134) F. Patat and H. Sinn, *Angew. Chem.*, **70**, 496 (1958).
(135) F. J. Karol and W. L. Carrick, *J. Am. Chem. Soc.*, **83**, 2654 (1961).
(136) R. V. Figini, *Makromol. Chem.*, **107**, 170 (1967).
(137) B. L. Erussalimskii, I. V. Kulevskaya, and V. V. Masurek, *International Symposium on Macromolecular Chemistry, Prague 1965*, O. Wichterle and B. Sedláček, Chairmen (*J. Polymer Sci. C.* **16** Part 3),Interscience, New York, p. 1355.
(138) T. J. Leitereg and D. J. Cram, *J. Am. Chem. Soc.*, **90**, 4011, 4019 (1968).

Chapter VIII

Anionic Polymerization in Systems Involving Lithium Counterions

1. SOME COMMENTS ABOUT ORGANOLITHIUM COMPOUNDS

Organolithium derivatives may be classified as members of a larger class of electron-deficient compounds. Electron deficiency is defined as the structure in which the number of nearest neighbor atom–atom connections, i.e., of chemical bonds, exceeds the number of the available pairs of valence electrons (1). The criteria of what constitutes a valid atom–atom connection (a bond) is not always unambiguous, although generally a bond order–bond length relation may serve as a satisfactory diagnostic evidence for its existence.

One of the most characteristic features of alkyllithium compounds, and of many other organolithium derivatives, is their high degree of association into tightly bonded agglomerates such as tetramers or hexamers. For example, colligative properties of *n*-butyllithium in hydrocarbon solvents indicate that this compound exists, even in the most dilute solution, as hexamer (2,3). A similar degree of association was claimed for ethyllithium (4,5) whereas *tert*-butyllithium seems to be tetrameric in hexane or benzene (6).

The high strength of the bonds which link the monomeric units into an agglomerate is manifested by their exceedingly low vapor pressure. For example, the vapor pressure of butyllithium is only 4×10^{-4} mm Hg at 60°C (7) and the heat of its vaporization exceeds 30 kcal/mole. Dissociation of the agglomerates appears to be still negligible even in the gas phase, although the evidence derived from the mass spectra of ethyllithium (8) indicates that tetramers, in addition to hexamers, are present in its vapor. Further evidence for the stability of the agglomerates is provided by NMR studies (see Section 2) which indicate a relatively slow intermolecular exchange of alkyl groups; e.g., in a mixture of methyl- and ethyllithium dissolved in hydrocarbons the exchange time is about 0.1 sec, or longer, at room temperature (14).

Studies of the solid, crystalline organolithium compounds were most instructive and helpful in elucidating the details of their structure. Dietrich (9), who obtained a three-dimensional Fourier projection of ethyllithium from

476

single-crystal data, concluded that four lithium atoms form a distorted tetra-hedral, the Li—Li bonds being 2.42, 2.60, and 2.63 Å long. Methyllithium probably forms a regular tetrahedral, the length of the Li—Li bonds being 2.56 Å (10). The short Li—Li distance, as compared with that found in the Li_2 molecule (2.67 Å), and the sharpness of the Li—C—Li bond angles are indicative of electron-deficient bridging. The tightness of the lithium core is further manifested by the appearance of $Li_6R_5^+$ and $Li_4R_3^+$ ions in the mass spectrum of ethyllithium vapor. These, together with Li_2R^+, are the most abundant species formed by the impact of electrons accelerated by a 75 V potential (8).

All these facts suggest that multicenter bonding sites are formed in organo-lithium agglomerates (6,11). Apparently each lithium atom is associated with three alkyl carbon atoms, the former contributing three orbitals, whereas each carbon atom provides a single sp^3 orbital. Their overlap yields a molecu-lar orbital embracing four centers, which accommodates four valence elec-trons (10). The relevant energy level diagram for an alkyllithium tetramer is shown in Figure 1 and, on the basis of the available thermochemical data, the average bond energy for each bonding electron pair is calculated to be 82 kcal/mole (12).

Organolithium compounds involving resonance-stabilized groups, such as benzyl, allyl, diphenylmethyl, or their derivatives, are found to be dimeric in hydrocarbon solvents (see, e.g., ref. 13). However, the bonding character in these dimers is entirely different from that linking alkyllithium derivatives and it arises from the dipole–dipole interaction.

The problem is even more fundamental. Although the precise nature of the C—Li bond in the aliphatic lithium compounds is still unknown, there is little doubt that basically this is a covalent linkage with some ionic contribu-tion which might be relatively high. Of course, such a description is pertinent to monomeric R—Li, which unfortunately is not observed directly, and it becomes questionable when a multicenter bond is discussed. However, when the negative charge becomes delocalized through a resonance-stabilized moiety, the bond becomes ionic and then the model of an ion-pair may serve better for its description. This is evident when the optical spectra of the pertinent lithium derivatives are examined. These compounds are colored and absorb in the same region as other alkali salts. For example, the spectra of living polystyrene are essentially identical for the cesium, potassium, sodium, and *lithium* salts in tetrahydrofuran, and the spectrum of the latter salt remains basically the same even in benzene solution where virtually all the salt exists in dimeric form (15). Hence, this dimerization, contrary to the agglomeration observed in alkane derivatives, arises from electrostatic interactions leading to the formation of a quadropole from two dipoles. A similar association is observed for the sodium salt in benzene (16).

The peculiar characteristics of the lithium compounds that distinguish them from other alkali derivatives are not surprising. The elements of the first row in the periodic table greatly differ from other members of the respective columns. The small size of the lithium ion is certainly a factor, although not the only one, which leads to its peculiarities—for example, the greater degree

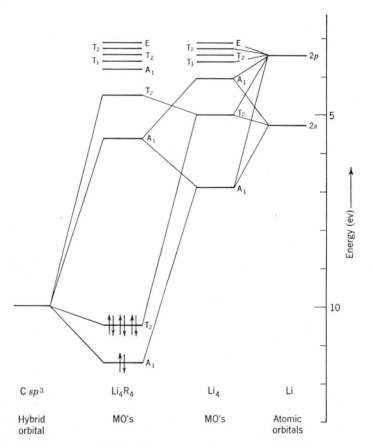

Fig. 1.   Energy level diagram for alkyllithium tetramer (12).

of association of ion-pairs involving lithium cations may be accounted for by its small radius. Let us also emphasize that the character of the C—Li bond in aliphatic compounds seems to be different from that established in other alkali derivatives. The latter form infusible and insoluble solids, and it seems that their structure is described in terms of ionic lattices (25). This conclusion may be questioned, however, because recent x-ray studies of methyl sodium (26) indicate its tetrameric structure. There is no doubt, however, that organolithium crystals are built from individual, although complex, molecules and not from ions.

2. FRAGMENTATION OF ORGANOLITHIUM AGGLOMERATES

Organolithium agglomerates seem to be in equilibrium with minute amounts of some fragments which apparently are responsible for many reactions involving the lithium derivatives. The nature and the reactivities of these fragments are, therefore, of the greatest interest to kineticists who study processes in which lithium compounds participate.

The exchange of alkyl groups between alkyllithiums exemplifies such a process. The ^{7}Li NMR spectrum is affected by the nature of the investigated organolithium compound. For example, the resonance signal of methyllithium is shifted by 16 cps from that of ethyllithium, both spectra showing only one reasonably sharp line. The spectrum of their mixture in hydrocarbons reveals these two lines in addition to new ones attributed to the mixed product (14). The fact that all these lines remain sharp indicates a slow exchange, with a correlation time longer than 0.1 sec.

A different situation was observed in mixtures dissolved in ether (23). At 20°C the absorption of ^{7}Li produced only one line; its position varied with the composition of the mixture, moving from the frequency characteristic for the one component to that observed for the other. Obviously, a rapid intermolecular exchange of alkyl groups takes place under these conditions.

The line broadens as the solution is cooled and, eventually, at −80°C it splits into several components, as illustrated by Figure 2. Examination of the spectra reveals the existence of four distinct types of ^{7}Li, each having its own characteristic environment. Apparently ^{7}Li may be surrounded in this system by three methyl groups, by two methyls and one ethyl, by one methyl and two ethyls, or finally by three ethyl groups. Such a hypothesis referred to as the principle of "local environment," is compatible with the electronic structure postulating overlap between an sp^2 orbital of Li and the sp^3 orbitals of neighboring carbon atoms; each lithium atom then possesses three alkyl neighbors. Assuming a random mixture of alkyllithium units, one may calculate from the composition of the solution the intensity of each line in the NMR spectrum. The agreement with observation is good, as may be seen from Figure 2. The chemical shifts of the extreme lines provide further evidence favoring the hypothesis. These lines should be identical with those observed in the solution of the individual pure compounds—as is, indeed, the case.

One additional point deserves comments. Colligative properties of methyllithium in ether (27) prove this compound to be tetrameric in that solvent.

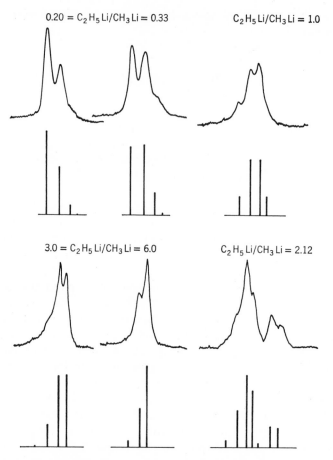

Fig. 2. ⁷Li resonance spectra of mixtures of ethyllithium and methyllithium in ether at −80° (23). The spectrum at lower right represents a solution containing lithium ethoxide in addition to methyl and ethyllithium. The calculated spectra, based on the "local environment" hypothesis, are shown beneath the observed spectra.

Such a conclusion is at variance with the earlier results of Wittig et al. (3), who used the ebullioscopic method for molecular weight determination and concluded that ethyllithium is trimeric in boiling ether. It is possible, however, that the lithium compound dissociates under these conditions (23) and the equilibrium

$$(LiR)_4 \rightleftarrows 2(LiR)_2$$

may account for the results. If methyl- and ethyllithium are indeed tetrameric in ether, then the relatively sharp lines of the mixed products indicate either a very slow or a very fast *intra*molecular exchange of alkyl groups. In the latter case, *five*, and not *four*, different ^{7}Li signals should be observed. Unfortunately, the lines reported in the literature (23) are not sufficiently well resolved to permit unequivocal discrimination between the alternatives.

Two types of mechanisms may account for the exchange:

(*a*) Association followed by dissociation, e.g.,

$$(Li_4R'_4) + (Li_4R''_4) \rightleftarrows complex \rightleftarrows 2 \text{ mixed dimers,}$$

(*b*) Dissociation followed by association, e.g.,

$$(Li_4R'_4) \rightleftarrows 2(Li_2R'_2),$$
$$(Li_4R''_4) \rightleftarrows 2(Li_2R''_2),$$
$$(Li_2R'_2) + (Li_2R''_2) \rightleftarrows (Li_4R'_2R''_2).$$

The latter mechanism is more plausible. Dissociation into dimers rather than trimers was favored by Brown (12) on the basis of his mass spectrographic studies. These revealed the existence of fragments having even numbers of lithium atoms, whereas fragments with an odd number of lithiums were not observed.

A much more powerful argument, favoring the dissociation of agglomerates into dimers, was developed by Hartwell and Brown (24), who investigated the exchange between *t*-butyllithium and lithiomethyl trimethyl silane. These compounds were chosen for two reasons: both form tetramers, even in hydrocarbon solvents (6,28,29); moreover, it had been expected that their ultimate distribution in equilibrated mixture would be nearly random because the $(CH_3)_3C\cdot$ group is not very different in size from $(CH_3)_3Si\cdot CH_2$. The ^{7}Li resonance of *t*-butyllithium appears at -0.89 ppm, whereas the absorption of ^{7}Li in the silane derivative takes place at -1.96 ppm, aqueous ^{7}LiBr serving as the reference. Hence, the application of the NMR technique was feasible in these studies (see also ref. 116).

The approach to equilibrium is slow in cyclopentane; the half-lifetime of the exchange is about 6 hr at 28°C (24). The first product of the reaction was $(Li_4R'_2R''_2)$, where $R' = (CH_3)C$ and $R'' = (CH_3)_3Si\cdot CH_2$, and then the tetramers $Li_4R'R''_3$ and $Li_4R'_3R''$ were formed, but the former was always at higher concentration than the latter. These facts, as we shall see, are most significant and illuminating.

Consider an equimolar mixture of $(Li_4R'_4)$ and $(Li_4R''_4)$. The dissociative equilibria, tetramer $\rightleftarrows 2$ dimers, imply that the dimer $(Li_4R'_2R''_2)$ *must* be the first product of mixing; the initial rate of its formation is then determined by the rate of dissociation of the tetramer that dissociates more slowly. On the other hand, $(Li_4R'_3R'')$ and $(Li_4R'R''_3)$ would be the first products had the dissociation produced monomeric and trimeric fragments. Mechanism (*a*) is also eliminated by the fact that $[Li_4R'R''_3] > [Li_4R'_3R'']$—both dimers should be formed initially in equal amounts if (*a*) were operative.

The proposed mechanism (tetramers dissociating into dimers) implies that the tetramers ($Li_4R'R_3''$) and ($Li_4R_3'R''$) are formed by the symmetric dissociation of the mixed tetramer ($Li_4R_2'R_2''$), i.e.,

$$(Li_4R_2'R_2'') \rightleftarrows 2(Li_2R'R''),$$

followed by the combination

$$(Li_2R'R'') + (Li_2R_2'') \rightleftarrows (Li_4R'R_3'')$$

or

$$(Li_2R'R'') + (Li_2R_2') \rightleftarrows (Li_4R_3'R'').$$

Assuming that all the combinations are equally fast, we conclude that the tetramer that dissociates to a greater extent should produce a higher proportion of the unsymmetrical product. Hence, the results of Hartwell and Brown indicate that $Li_4(Me_3SiCH_2)_4$ is more dissociated than $Li_4(t\text{-Bu})_4$.

The exchange studies discussed above were repeated using toluene as a solvent (24). The reaction followed the same course as the exchange studied in cyclopentane, but it was 10^3 to 10^4 times faster. This observation appears to be most interesting; it indicates a higher degree of dissociation of the tetramers in aromatic than in aliphatic hydrocarbons. Apparently the dimers are stabilized in aromatic solvents through the interaction with π-electrons.

Although the proposed scheme of exchange is very plausible, it need not be unique. For example, the following route,

$$Li_2R_2' + Li_4R_4'' \rightleftarrows Li_4R_2'R_2'' + Li_2R_2'',$$

may also contribute to the process, provided that the above reaction is faster than the combination of dimers. Such a mechanism, if operative, could involve hexamers as intermediates and it may become important when the equilibrium concentration of dimers is exceedingly low.

The exchange between tetrameric methyllithium and the dimeric benzyllithium was also investigated (40). Apparently only the mixed aggregate having a 1:1 composition was formed.

Exchange between organic and inorganic lithium compounds is also feasible. Waack et al. (31) investigated the reaction between tetrameric methyllithium and lithium bromide, the latter is aggregated in tetrahydrofuran or ether (32–34). The NMR resolution was very good and the results are shown in Figure 3. The reaction is very fast at $0°C$ but becomes slow at $-100°C$. A definite, although broad, signal with an intermediate chemical shift is evident in the spectrum recorded at $-100°C$. The variation of the ratio of methyllithium to lithium bromide apparently does not affect the position of the broad signal, indicating that the mixed agglomerate has a fixed composition corresponding, most probably, to a 1:1 stoichiometry. Indeed, a solid 1:1

complex of methyllithium and lithium bromide has been reported in the literature (33).

An interesting feature of the system deserves comment. On decreasing the temperature of the solution one finds that the methyllithium signal becomes sharper before any appreciable sharpening of the lithium bromide signal takes place. This again implies that a direct exchange,

$$\text{Li}_4\text{Me}_4 + (\text{LiBr})_n \rightleftarrows \text{mixed aggregates},$$

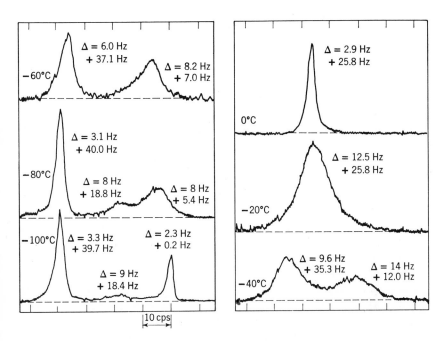

Fig. 3. Nuclear magnetic spectrum of ^{7}Li for mixtures of CH$_3$Li (0.62 mole fraction) and LiBr (0.38 mole fraction) in THF at various temperatures (31). Total [Li] = 0.57M.

does not occur, and the reaction proceeds, therefore, only through the dissociation. Apparently, at low temperature, the dissociation of Li$_4$Me$_4$ is very slow, while the rates of dissociation of lithium bromide and of the mixed aggregate are still appreciable.

Facile exchange was observed between lithium bromide and phenyl- or benzyllithium, inferring that this process is common to all types of organolithium species. The exchange takes place also between lithium chloride and alkoxide and various organic lithium compounds.

Studies of Brown and colleagues provide a powerful argument for the existence of dimers which are responsible for the exchanges discussed above.

Nevertheless, this evidence does not imply that dimers are involved in other reactions of lithium compounds. For example, dimers may be present at a higher equilibrium concentration than monomers, and being reactive in the exchange they may hide any contribution arising from the reaction of monomers. However, dimers may not react with compounds containing C=C bonds, whereas the monomers could be very reactive. In such a case, only the addition of monomers would be observed.

Finally, let us consider the possibility of ionic dissociation, e.g.,

$$(Li_4R_4) \rightleftarrows (Li_4R_3)^+R^- \rightleftarrows (Li_4R_3)^+ + R^-.$$

Such reactions might be important and may be perhaps responsible for the nucleophilic reactivity of organolithium compounds. In mixed aggregates involving alkoxides, or other lithium salts, the ionization

$$(Li_4R_3,OR') \rightleftarrows (Li_4R_3)^+,R'O^-$$

should be more probable than

$$(Li_4R_3,OR') \rightleftarrows (Li_4R_2,OR)^+,R^-$$

because the $R'O^-$ anion is more stable than carbanion R^-. This may account for the inhibiting effect of alkoxides, and of other lithium salts, on the nucleophilic reactivity of alkyllithiums.

3. COMPLEXING OF LITHIUM COMPOUNDS WITH LEWIS BASES

The electron-deficient nature of lithium agglomerates facilitates their association with a variety of Lewis (n-type) bases. Such an association may, or may not, lead to the fragmentation of the original lithium compound. For example, complexing of triethylamine or lithium ethoxide with hexameric ethyllithium in hydrocarbon solvents takes place without fragmentation of the original agglomerate (17,18), although fragmentation occurs when the base is present in large excess.

It was proposed that an agglomerate has some vacant "faces" with which the base becomes associated, e.g., it is claimed that hexameric ethyllithium has two such "faces" and, therefore, a complex with two molecules of amine is formed (17). According to this hypothesis, further complexing is prevented by the lack of suitable sites but, because new sites are formed through fragmentation of the agglomerate, the uptake of the complexing agent may be increased by the dissociation of the original species. For many systems the degree of fragmentation in donor solvents was determined by classic methods based on the colligative properties of the resulting solutions, but because such methods were questioned (23) other approaches had to be explored.

Infrared studies shed much light on the problem of fragmentation (19,20, 35). For example, alkyllithium absorbs at 500 cm^{-1}, and this band shows a large isotope effect when ^{6}Li replaces ^{7}Li. It is likely, therefore, that the band arises from a C—Li bending-stretching vibration. Its position is affected by solvent and its disappearance indicates a complete dissociation of the agglomerates into monomeric units. The band is observed even in ether solution proving that only a partial fragmentation, and not complete disruption, takes place under these conditions.

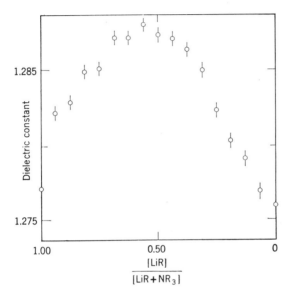

Fig. 4.   Dielectric constant vs. [triethylamine]/[ethyllithium] ratio in dilute benzene solution (17), where R = ethyl.

The dielectric constant of a solution of alkyllithium mixed with a base provides another diagnostic tool revealing the change in the structure of a lithium compound. Variation of the dielectric constant of a solution of triethylamine and ethyllithium (17) is shown in Figure 4. The total concentration of the solutes (Σ moles per unit volume) was kept constant, but the ratio

$$[(EtLi)_6]/\{[(EtLi)_6] + [amine]\}$$

was varied. At the ratio 1/2 a maximum is clearly seen in the graph, and this implies that a polar 1 : 1 complex is formed from the only slightly polar components. A similar technique was employed by Eastham and his colleagues (21), who measured the electric capacity of a cell containing alkyllithium dissolved in hexane to which increasing amounts of diethyl ether were added. The results of such a "titration" are shown in Figure 5. The complex is much more polar than the components, and the breaks seen in curves B and C indicate that all the lithium compound is complexed at this stage of titration. Thus, the stoichiometry of the complex was found to be two LiR for each Et_2O. Similar studies were performed using tetrahydrofuran and triethylamine as the respective Lewis bases. Eastham concluded therefore that the process of fragmentation of alkyllithiums results in the formation of dimers

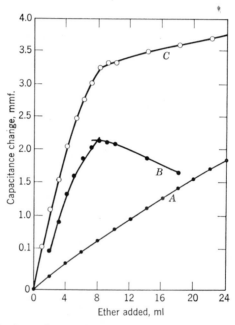

Fig. 5.   Curve A: change in capacitance of a cell containing 90 ml hexane solution to which increments of ether were added. Curve C: same except that 0.144 equivalent of butyllithium was present. Curve B: difference between C and A. Titrations were carried out at 25° (ether = 9.5M) (ref. 21).

(22). This conclusion need not be correct. The data provide us only with the stoichiometry of the complex and give no information about its state of aggregation. However, the stoichiometry of the complex may be accounted for in terms of electron deficiency of the lithium agglomerate. The dimer has a deficiency of two electrons and these are supplied by one lone electron pair of an ether or an amine.

An alternative method, which also leads to the determination of the stoichiometry of the ether–organolithium complexes, was described by Waack et al. (36). Saturated vapor, maintained in equilibrium with a benzene solution of the investigated lithium compound to which a variable amount of ether (e.g., tetrahydrofuran) is added, was analyzed by gas chromatography. To simplify the analysis the solution contained a small, but constant, amount of hexane. The ratio tetrahydrofuran/hexane was determined for each concentration of the ether, the results being shown in Figure 6. The sharp break

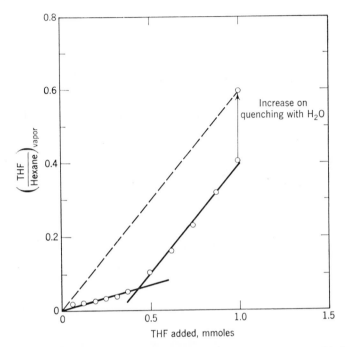

Fig. 6.   Observed ratio of tetrahydrofuran/hexane in vapor vs. THF added to 10 ml of benzene solution containing 0.19 mmole of RLi and 0.44 mmole of hexane (internal standard) (27). In the absence of RLi, the THF/hexane ratio in the vapor is linear over the range of 0–5 mmoles THF added. This is indicated by the broken line. R = $C_5H_{11}C(Ph)_2^-$.

in the curve indicates that the complexing is strong, and the position of the kink gives the stoichiometry of the complex, viz., one tetrahydrofuran for two lithium units.

4. ADDITION OF ORGANOLITHIUM COMPOUNDS TO C=C DOUBLE BONDS

The addition reaction

$$RLi + C{=}C \rightarrow R{\cdot}C{-}C^-,Li^+$$

takes place in the initiation and propagation steps of anionic polymerization involving Li$^+$ counterions. One could expect the rate of the addition to be a monotonic function of the basicity of the respective R$^-$ carbanion. However, the agglomeration of RLi, which is encountered in many systems involving organolithium compounds, complicates the matter. The agglomeration makes the kinetics of the addition complex and leads to fractional orders of the reaction. Consequently, a simple correlation between the reactivity and the basicity of R$^-$ is not observed. For example, the initiation of styrene or of a diene polymerization in benzene is much faster with a secondary alkyllithium than with a primary lithium compound (37,38) although a primary carbanion is more reactive than a secondary one. The large increase in the reactivity of *sec*-butyllithium, as compared with the *n*-butyllithium, is at least partially due to the higher degree of dissociation of the agglomerate (*sec*-BuLi)$_n$ than of (*n*-BuLi)$_n$.* The initiation is even faster if 1,1-diphenylhexyllithium (39), intrinsically an unreactive compound, is used instead of *sec*-BuLi, because its association is weak. On the other hand, the association is very strong for the planar vinyllithium or phenyllithium and, consequently, these compounds initiate very slowly the polymerization of styrene in benzene or ether (41,42), although the initiation is rapid in tetrahydrofuran.

Extensive studies of correlation between the structure of an organolithium compound and its rate of addition to a standard monomer (styrene) were reported recently by Waack and Doran (114). The reaction was studied in tetrahydrofuran, i.e., in a solvent which partially dissociates the aggregates. The interpretation of the data is complicated due to the reasons outlined above, namely, to what extent the change in structure of the reagent affects its ease (or equilibrium) of dissociation into the reactive fragment and how far it affects the reactivity of the latter. The combined effect of both factors leads to a series of increasing reactivities, namely,

Triphenyl methyl < vinyl < methyl < phenyl < *p*-tolyl ~ allyl <
    benzyl ~ crotyl < α-methyl ~ benzyl ~ *n*-butyl ~ ethyl.

The rate of initiation may be conveniently studied by the spectrophotometric technique. For example, alkyllithiums do not absorb light at 334 mμ, where living polystyrene has a strong absorption band. Hence, the rate of initiation may be determined by the rate of increase of optical density at this wavelength. This approach was utilized by Worsfold and Bywater (43).

The distinction between initiation and propagation is important. The pioneering studies of the kinetics of polymerization initiated by butyllithium were concerned with the rate of the overall process (44,45). The progress of polymerization was measured by the rate of consumption of the monomer, but such data are difficult to interpret. The curves giving the monomer uptake were sigmoidal, particularly so at high initiator concen-

* *Tertiary*-butyllithium is less reactive than *secondary*.

trations. The rates of polymerization, if determined by the maximum of
$-d$(monomer)$/dt$ (44), are particularly ambiguous, but even the "initial"
rates, obtained from the apparently linear portion of the first order plot of
monomer concentration (45) are still complex functions of initiation and
propagation. Moreover, it is difficult to determine the "linear" portion of
such a curve without some bias.

In spite of all these difficulties, Tobolsky (44) and Welch (45) correctly
deduced that the *monomeric* butyllithium is the active initiator, and that most
of the growing polymers are inactive due to their association with the unreacted butyllithium. In fact, Welch (45) found the rate of polymerization to
be independent of initiator concentration, provided its value exceeds some

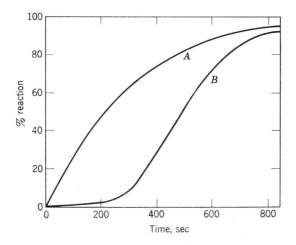

Fig. 7. Percentage of conversion of alkyllithium, measured by the increase of optical
density at $\lambda_{max} = 334$ mμ, plotted versus time (54). (*A*) Reaction of $1.09 \times 10^{-3}M$ *sec*-
butyllithium with $5.33 \times 10^{-4}M$ styrene in benzene solution at 30°; (*B*) reaction of
$1.34 \times 10^{-3}M$ *sec*-butyllithium with $8.67 \times 10^{-2}M$ styrene in cyclohexane solution
at 40°.

critical magnitude. This observation was confirmed by Minoux and Marchal
(91). It was concluded, therefore, that the degree of association of the growing
polymers with BuLi must be high and, therefore, the concentration of
nonassociated species is only slightly affected by the variation of the total
concentration of the initiator. Under such conditions only a fraction of
butyllithium is consumed, even when all the monomer has been polymerized.

Direct spectrophotometric studies of the initiation (43) showed that the
reaction of butyllithium with styrene in benzene solution is given by a 1/6
order, an observation soon confirmed by other workers (47,48). Typical
curves giving the increase of concentration of growing polymers with time
are shown in Figure 7 (curve *A*). Such a behavior was observed in several

systems, provided benzene, toluene, or some other *aromatic* hydrocarbon was the solvent. There is no induction period and the initial slope of the conversion curve is proportional to some fractional power of concentration of the initiator. For example, the rate of initiation by the secondary butyllithium is given by a 1/4 order (46,54) and that by *n*-butyllithium follows a 1/6 order (43). In those systems the reported fractional order, is, therefore, compatible with the degree of association of the alkyllithium, viz., 6-fold for *n*-butyllithium (2,3) and 4-fold for *tert*-butyllithium (6). Hence, the mechanism

$$(RLi)_n \rightleftarrows nRLi$$

$$RLi + monomer \rightarrow growing\ polymer$$

appears to be very plausible.

Such schemes become even more acceptable if some other reactions involving the pertinent lithium compound show a similar kinetic behavior. For example, the addition of butyllithium to 1,1-diphenylethylene (49) and the metalation of fluorene by butyllithium (55) were thoroughly studied. In both systems the rate of reaction was shown to be proportional to the 1/6 power of the butyllithium concentration. The kinetic order of the addition of *t*-butyllithium to 1,1-diphenylethylene was found to be 1/4 (66), this being again in agreement with the observed tetrameric aggregation of this compound. Reactions of other organolithium compounds were extensively studied by Waack and his colleagues (50), who determined their orders from plots of log (rate) vs. log (concentration).

Several factors may complicate these investigations. The active species are assumed to be in equilibrium with the aggregates. This assumption implies that their rate of formation is much faster than their rate of consumption through addition to a C=C double bond. However, even in the systems in which equilibrium is established the concentration of active species *may* be affected by the mechanism leading to their formation (51). To clarify the significance of this statement, let us consider a hypothetical example

$$(LiBu)_6 \rightleftarrows (LiBu)_4 + (LiBu)_2, \tag{A}$$

$$(LiBu)_2 \rightleftarrows 2LiBu, \tag{B}$$

$$LiBu + olefin \rightarrow adduct. \tag{C}$$

The rate of reaction (C) is assumed to be much slower than the rates of the backward reactions (A) and (B) and, therefore, the equilibrium should be established. Nevertheless, if either of the reactions (D) or (E)

$$(LiBu)_4 \rightleftarrows 2(LiBu)_2, \tag{D}$$

$$2(LiBu)_4 \rightleftarrows (LiBu)_6 + (LiBu)_2, \tag{E}$$

were excluded, the equilibrium concentration of LiBu would *not* be given by the 1/6 power of concentration of the hexameric BuLi but by the equation

$$[LiBu] = \{[(LiBu)_6]/[(LiBu)_4]\}^{1/2} \cdot const.$$

As a result of the addition (C) the tetrameric butyllithium accumulates in the system and consequently the overall addition process becomes autoinhibited. On the other hand, if the dissociation of $(LiBu)_4$ into $2(LiBu)_2$ does take place (a most probable event), and if its rate is faster than that of (C), then the rate of the addition is proportional to the $1/6$ power of concentration of the hexameric butyllithium.

This result may be generalized. The rate of addition should show $1/6$ order in respect to total LiBu if the equilibrium concentrations of *all* the intermediate agglomerates is maintained and *not perturbed* by reaction (C). Such conditions are satisfied when the mechanism of formation of the intermediate agglomerates includes some cross-reactions such as (D) or (E). Of course, we assume implicitly that the concentration of all the fragments is very low as compared with that of $(BuLi)_6$.

In some systems the rate of formation of the active species may be too slow to maintain their equilibrium concentration and, thence, the rate of the addition is determined by the slowest step in the series of consecutive reactions which lead to the formation of this intermediate. For example, let us assume a fast dissociation (D),

$$(LiR)_4 \rightleftarrows 2(LiR)_2, \tag{D}$$

but an extremely slow decomposition (B),

$$(LiR)_2 \rightleftarrows 2LiR. \tag{B}$$

The rate of addition, $k_C(LiR)(olefin)$ is then given by

$$k_{+B}K_D^{1/2}[(LiR)_4]^{1/2},$$

if $k_C[olefin]$ is much greater than $k_{-B}[LiR]$. It is probable that such a situation is encountered when the addition takes place in aliphatic hydrocarbons.* Indeed, the studies of Hartwell and Brown (24) showed a substantial decrease in the rate of exchange, at least by a factor of 1000, when lithium alkyls reacted in cyclopentane instead of toluene. This observation implies a substantially slower rate of dissociation of tetramers into dimers in cyclopentane when compared with their dissociation in toluene. The same trend may be expected for the dissociation of dimers into monomers.

At sufficiently low concentration of the initiator one expects a transition from the kinetics in which the equilibrium is established to that in which the equilibrium is strongly perturbed. Consider, for the sake of argument, the system

$$(LiR)_6 \rightleftarrows 6LiR,$$
$$LiR + olefin \rightarrow adduct.$$

An n-fold decrease in the concentration of total alkyllithium decreases the concentration of the monomeric species only by a factor $n^{1/6}$. Hence, the rate

---

* This situation does not apply to the reaction in cyclohexane (private communication from Dr. Bywater).

of addition decreases slightly, whereas the rate of dissociation is greatly affected. Consequently, the supply of monomeric species may not be sufficiently rapid to maintain the equilibrium, and then the rate of reaction becomes dependent on the rate of dissociation. Kinetically, the reaction behaves as if its order has increased at lower concentrations of the initiator. This phenomenon may be responsible for some curious observations of Lundborg and Sinn (53), who investigated the polymerization of isoprene initiated by butyllithium. At concentrations of initiator exceeding $10^{-2}M$ the initiation was proportional to the 1/6 power of BuLi concentration, but its order increased to 1/3 at a concentration of BuLi of about $10^{-5}M$.*

The simple schemes imply that only one species is active in the addition. This need not always be the case. For example, Waack et al. (42) suggested that in tetrahydrofuran both monomers and dimers of phenyllithium participate in the metalation of triphenylmethane. In such a case, no simple order may be attributed to the reaction, its rate being given by the sum

$$\{k_m 2f + k_d(1 - f)\} \text{ [total PhLi]} \cdot \text{[Ph}_3\text{CH]},$$

where f is the fraction of the dimers which are dissociated and k_m and k_d are the rate constants of metalation by the monomers and dimers, respectively. For a small f, the approximate expression

$$\text{rate} = \text{const}_1 \text{[total PhLi]}^{1/2} + \text{const}_2 \text{[total PhLi]}$$

is valid and, within a not too wide range of concentration, the reaction shows then an apparent fractional order larger than 1/2. In fact, Waack (42) reported an apparent order of 0.64 for the concentration range $5 \times 10^{-3}M$ to $2M$. Studies of the addition of phenyllithium to 1,1-diphenylethylene led to the 0.66 order for the addition of [PhLi] if its concentration is varied in the range 1×10^{-2} to $5 \times 10^{-1}M$.

Finally, let us consider the effect of the addition product on the rate of addition. The product of the reaction is again an organolithium derivative and consequently it may become associated either with molecules of its own kind or with those of the organolithium initiator. These reactions may affect the equilibrium, or the stationary state, which governs the concentration of the active species. For example, consider again the system

$$(\text{LiBu})_6 \rightleftarrows 6\text{BuLi},$$

$$\text{BuLi} + \text{C}{=}\text{C} \rightarrow \text{Bu} \cdot \text{C}{-}\text{C}^-, \text{Li}^+.$$

Assume that under the experimental conditions prevailing in a polymerization a direct, or indirect, dissociation of the hexamer is extremely slow. Then, the ordinary initiation, which is observed in benzene solution, may be immeasurably slow. However, the carbanions eventually formed may interact with $(\text{LiBu})_6$, and a hypothetical reaction such as

$$\sim\text{C}^-, \text{Li}^+ + (\text{LiBu})_6 \rightarrow (\sim\text{C}^-, \text{Li}^+, \text{LiBu}) + \text{LiBu} + (\text{LiBu})_4,$$

* However, the reaction was claimed to be first order in monomer even under these conditions.

or some other similar processes, may provide then new routes leading to the formation of active LiBu. Alternatively, the associates themselves, e.g.,

$$(\sim C^-, Li^+, BuLi),$$

may act as efficient initiators, e.g.,

$$(\sim C^-, Li^+, BuLi) + 2C=C \rightarrow 2\sim C^-, Li^+.$$

Such a set of reactions, or any conceptually similar scheme, would lead to autoacceleration and then the initiation process is represented by the sigmoidal curve *B* of Figure 7.

Bywater and Worsfold (54) forcefully stressed the basic difference between the alkyllithium addition which takes place in aliphatic hydrocarbons as compared with the reaction observed in aromatic hydrocarbons. By applying the spectrophotometric technique, they examined the polymerization of styrene and isoprene initiated by *sec*-butyllithium in benzene and in cyclohexane. Both monomers behaved similarly in each solvent, but the behavior in the two solvents was strikingly different. In benzene the initiation proceeds without any induction period, a fact pointed out earlier. In cyclohexane there is an induction period followed by autocatalytic initiation; in fact the sigmoidal curve *B* of Figure 7 describes the progress of this reaction. Eventually the rate declines as the initiator becomes exhausted. The same behavior was previously reported (43,57) for the systems styrene, butadiene, or isoprene initiated by *n*-butyllithium in cyclohexane, and more recently for the initiation of isoprene polymerization by *n*-butyllithium in benzene (72). The latter observation may be questioned; it is possible that the sigmoidal curve observed by Urwin et al. (72) arises from some impurities initially present in their system.

In conclusion, it is likely that the scheme outlined in the preceding paragraph applies to polymerizations initiated in aliphatic hydrocarbon solvents. It is also possible that some impurities, initially present in the system, are responsible for the induction periods and for the sigmoidal shape of the observed curves. However, the same stringent purification procedures, which led to nonsigmoidal curves when the reaction was performed in benzene, failed to eliminate the induction period when the reaction was carried out in aliphatic solvents. On the contrary, deliberate addition of trace amounts of air to a reaction of alkyllithium with isoprene in cyclohexane completely eliminated the slow reaction. Also, less stringent drying of butadiene produces the same effect (see p. 78 of ref. 84). These facts indicate that the induction period represents a genuine behavior of these systems and not an artifact caused by impurities.

It is probable that other kinetic patterns will be encountered as more experimental data become available. For the sake of discussion let us consider the initiation of butadiene polymerization by lithium polystyrene in benzene. Polystyryllithium is dimeric in this solvent (see Section 5), i.e., virtually all the living polymers are present in the form of $(\sim\!\!S^-,Li^+)_2$. The state of aggregation of living lithium polybutadiene is not well known; in aliphatic hydrocarbons the polymer seems to form hexameric species (57,58) and it is probable that a powerful association into large aggregates (tetramers?) takes place in benzene. We know nothing about mixed aggregates, but let us assume that they are much more strongly associated than $(\sim\!\!S^-,Li^+)_2$. In such a case, the initiation

$$\sim\!\!S^-,Li^+ + \text{butadiene (BD)} \rightarrow \sim\!\!S,BD^-,Li^+$$

would be followed by the formation of mixed aggregates and therefore by a rapid decrease in the concentration of monomeric $\sim\!\!S^-,Li^+$. This in turn should be reflected in the kinetics, which might then reveal the characteristic features of an autoinhibited reaction. Indeed, such behavior was observed by Harris (59), who studied this process by a stop-flow technique.

To complete our discussion let us consider one more example—the addition of 1,1-diphenylethylene to living polystyryllithium in benzene. Lithium polystyryl, as has been mentioned repeatedly, exists in benzene in dimeric form. Let us assume that the resulting $\sim\!\!CH_2C(Ph)_2^-,Li^+$, abbreviated to $\sim\!\!D^-,Li^+$, forms in this medium homodimers as well as mixed dimers with the living polystyrene. Hence, the following equilibria should be considered:

$$2\sim\!\!S^-,Li \rightleftarrows (\sim\!\!S^-,Li^+)_2, \qquad \tfrac{1}{2}K_1,$$

$$\sim\!\!S^-,Li^+ + \sim\!\!D^-,Li^+ \rightleftarrows \begin{pmatrix} \sim\!\!S^-,Li^+ \\ \sim\!\!D^-,Li^+ \end{pmatrix}, \qquad K_{12},$$

$$2\sim\!\!D^-,Li^+ \rightleftarrows (\sim\!\!D^-,Li^+)_2, \qquad \tfrac{1}{2}K_2.$$

For brevity we shall denote by u and v the respective monomeric species and by x, y, and z the dimeric ones. The process is represented by the reaction

$$\sim\!\!S^-,Li^+ + CH_2{:}C(Ph)_2,(D) \xrightarrow{k_p} \sim\!\!S,D^-,Li^+;$$

its progress may be investigated spectrophotometrically by following the disappearance of $\sim\!\!S^-,Li^+$ or the appearance of $\sim\!\!D^-,Li^+$.

Let us denote u/v by α and (total $\sim\!\!S^-$)/(total $\sim\!\!D^-$) by β and assume that u and v are much smaller than x, y, or z. Hence,

$$\beta = (2x + y)/(2z + y)$$

and, since the reaction does not involve termination,

$$2x + 2y + 2z = 2C = \text{const.}$$

Finally, denote by f the ratio $u/(u + v) = \alpha/(1 + \alpha)$ and by f' the ratio $(2x + y)/2C = \beta/(1 + \beta)$. Provided that all the equilibria are maintained during the course of reaction, the following relations are fulfilled:

$$\tfrac{1}{2}K_1u^2 = x, \qquad K_{12}uv = y, \quad \text{and} \quad \tfrac{1}{2}K_2v^2 = z$$

and

$$K_1 u^2 + 2K_{12} uv + K_2 v^2 = 2C$$

where K's are the respective dissociation constants and C is the total concentration of the agglomerates (dimers).

For $[D]_0 \gg 2C_0$ the solution of this equation gives:

$$-2Cdf'/dt = -d(2x + y)/dt = k_p[D]_0\sqrt{2C/K_1}\cdot G$$

where G is a function of f' which involves K_1, K_2, and K_{12} as coefficients. In a special case, when $K_{12}^2 = K_1 K_2$ (a plausible situation), the equation is reduced to the form

$$-2Cdf'/dt = -d(2x + y)/dt = (k_p[D]_0/K_1^{1/2})\sqrt{2C}\cdot f',$$

i.e., the reaction is *internally* first order, although *externally* it shows a 1/2 order behavior in respect to *total* concentration of living polymers. The apparent first order rate constant gives, therefore, $k_p/K_1^{1/2}$. Extensive studies of this process by Laita (60) proved indeed that the addition of 1,1-diphenylethylene to lithium polystyrene in benzene is internally first order in living polystyrene, at least up to 90% conversion, and showed that externally the process is given by a 1/2 order dependence on the living polymer concentration, as demanded by the above mathematical treatment. Hence, for this system the relation $K_{12}^2 = K_1 \cdot K_2$ is obeyed, i.e., the dissociation constant of the mixed dimer (apart from the statistical factor of 2) is given by the geometric average of the dissociation constants of the respective homodimers.

Summing up, we may say that initiation by alkyllithiums in hydrocarbon solvents (aliphatic and aromatic) is caused by some reactive fragment, or fragments, formed from the original lithium agglomerates. It is probable that the true initiating species are the monomeric lithium alkyls, but this conclusion, although very plausible, is not established finally. The available data are derived from kinetic studies, but experimental and conceptual difficulties complicate these investigations. The understanding of the role of mixed agglomerates is imperative for further progress. Fortunately, some systems are relatively simple. For example, it is probable that the simple scheme $(RLi)_n \rightleftarrows nRLi$ followed by $RLi + C{=}C \rightarrow R\cdot C\cdot C^-,Li^+$ applies, at least during the early stages of the process, whenever the kinetic data lead to an order of reaction which is compatible with the state of aggregation of the initiator.

5. PROPAGATION OF ANIONIC POLYMERIZATION INVOLVING LITHIUM COUNTERIONS

In polymerization involving lithium counterions the propagation is much simpler than the initiation because the reaction does not perturb the equilibrium established between the aggregates and the fragments formed by their dissociation. The most thoroughly investigated system is that of lithium polystyrene, and therefore we shall start our discussion by considering its behavior.

The first studies of lithium polystyrene propagation not complicated by initiation were reported by Worsfold and Bywater (43). The polymerization was initiated in benzene by n-butyllithium, but the propagation was studied about one hour after the onset of the reaction, i.e., after all the initiator was consumed. The completion of initiation was established by spectrophotometric technique. The optical density at 334 mμ remained constant during the whole period following the initiation stage, proving that the concentration of living polymers did not increase in that time. The absence of the initiator may be checked by an alternative technique, e.g., by analyzing a sample of the reacting solution to which a small amount of water or methanol has been added. The presence of any residual butyllithium is manifested by the formation of butane. The "seeding" technique recommended by some workers (13,62,63) is questionable; owing to the peculiar characteristics of this polymerization rather large amounts of monomer are required to consume all the initiator. This problem has been fully discussed in Chapter II (Section 6), where it has been pointed out that mere waiting is not sufficient to complete the initiation.

The kinetics of propagation proves that the reaction is 1/2 order in living polystyrene. The original study (43) covered concentrations ranging from 10^{-3} down to $10^{-5}M$, and the values of the reported constants were confirmed by the subsequent studies of Morton and his colleagues (13,65). Similar data were reported by Spirin, Gantmakher, and Medvedev (58,64), who investigated this polymerization in toluene; and recent studies of Johnson and Worsfold (57) showed that even in cyclohexane the propagation is still 1/2 order in living polystyrene. Inspection of Figure 8 permits the reader to compare the reported data and appreciate the degree of agreement.

The 1/2 order kinetics of propagation was explained (43) by the following mechanism:

$$(\sim\!S^-,Li^+)_2 \rightleftarrows 2\sim\!S^-,Li^+, \qquad K_{diss}$$
$$\sim\!S^-,Li^+ + S \rightarrow \sim\!SS^-,Li^+. \qquad k_p$$

Virtually all the living polystyrene is dimeric, only a minute fraction being dissociated into active monomeric species. According to this hypothesis, the formal first order constant, $-d \ln [S]/dt$, is given by the product

$$(1/\sqrt{2})k_p K_{diss}^{1/2}[\sim\!S^-,Li^+]^{1/2}.$$

Similar studies performed in cyclohexane (57) showed that the overall rate constant $k_p K_{diss}^{1/2}$ of the propagation in cyclohexane is smaller than that observed in benzene by about a factor of 3. It is probable that the change of solvent leads to a 9-fold decrease in K_{diss}; indeed, the dissociation is expected to be more extensive in benzene than in cyclohexane. However, the result may also indicate an appropriate decrease of k_p.

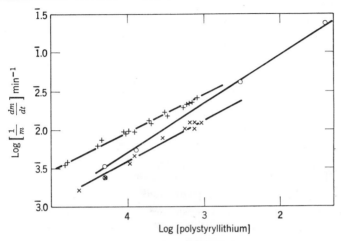

Fig. 8. Variation of rate constant of propagation with total concentration of poly-styryllithium. Results of (+) Worsfold and Bywater in benzene (43), (○) Spirin et al. in toluene (58), (×) Johnson and Worsfold in cyclohexane (57).

The data of Worsfold and Bywater (43):

T, °C	10	15	20	25	30.3
$10^2(1/\sqrt{2})k_p K_{diss}^{1/2}$ ($M^{1/2}$ sec^{-1})	0.26	0.44	0.65	0.94	1.55

permit us to calculate $E_p + \frac{1}{2}\Delta H_{diss}$. The Arrhenius plot is linear, and from its slope a value of 14 kcal/mole was obtained for $E_p + \frac{1}{2}\Delta H_{diss}$.

It is interesting to compare these data with those referring to the initiation of polymerization by n-BuLi in benzene. The latter give $k_i(K'_{diss})^{1/6}$, where k_i is the bimolecular rate constant of the addition of *monomeric* BuLi to styrene, and K'_{diss} is the equilibrium constant of the dissociation,

$$(\text{BuLi})_6 \rightleftharpoons 6\text{BuLi}.$$

The values of $k_i(K'_{diss})^{1/6}$ are very small: $2.5 \times 10^{-6}\ M^{-1/6}$ sec^{-1} at 10°C and $25 \times 10^{-6}\ M^{-1/6}$ sec^{-1} at 30°C. On this basis, $E_i + 1/6\Delta H'_{diss}$ is calculated to be 18 kcal/mole. As expected, the energy of association per monomeric unit appears to be higher for butyllithium than for polystyryllithium.

The slowness of the initiation, and its low order, often prevent the complete utilization of the initiator. For example, for $[\text{BuLi}]_0 \geqslant 0.1M$ and $[\text{S}]_0 = 1M$ a substantial fraction of the initiator is left at the end of polymerization. The propagation in the presence of BuLi is complex. Continuous creation of new growing centers may often lead to a rate expression involving the square of monomer concentration (44), and the formation of mixed aggregates complicates the kinetics further. This writer suspects that the mixed aggregates may exert a large influence on the propagation, although some evidence to the contrary has been provided by Worsfold and Bywater (43). In fact, it would be most interesting to study the propagation of living lithium polystyrene in benzene in the presence of subsequently added butyl- or ethyllithium. It should be relatively easy to account for the effect arising from the formation of the new growing species and, consequently, one could reliably examine the possible effect of mixed aggregates upon the course of the reaction. The viscosity of the solution of living lithium polyisoprenyl decreases drastically when butyllithium is added. This clearly indicates that the following reaction takes place:

$$(\text{living polymers})_n + m(\text{BuLi}) \rightleftarrows j(\text{living polymers})_k(\text{BuLi})_y,$$

where $n > k$. A similar association is expected in the lithium polystyryl system.

All the studies of lithium polystyrene that have been reported in the literature utilized living polymers endowed with only one active end per chain. Hence, their dimerization is *inter*molecular. On the other hand, living lithium polystyryls endowed with *two* growing ends per chain should undergo an *intra*molecular association. This type of association should lead to first order kinetics of propagation instead of the usual 1/2 order. Moreover, the equilibrium constant of the dissociation

is expected to increase with increasing degree of polymerization. Hence, the longer the polymer the faster is its growth. Attempts to prepare such polymers were made in this writer's laboratory but unfortunately no satisfactory samples have been obtained so far.

The kinetic studies suggest that living lithium polystyrene exists in benzene and in cyclohexane in dimeric form. Have we any other evidence to confirm this conclusion? The association of living polymers or polymers possessing functional end groups was investigated by Brody, Richards, and Szwarc (61). If the active groups are responsible for the association, then the viscosity of living polymer solution should decrease when such groups are destroyed. Hence, the high ratio of viscosities of the investigated solution, determined before and after "killing," is diagnostic of the association. This approach was extensively explored by Morton and his students (13,62). Solutions of living lithium polystyrene in benzene were prepared at sufficiently high concentration to ascertain the 3.4 relation between their molecular weight $(\overline{M}_w)$

and the viscosity (η) of their solution, viz.,

$$\eta = K(\overline{M}_w)^{3.4}.$$

Suppose that the living polymers are aggregated to a degree n, the aggregation arising from the interaction between the —S$^-$,Li$^+$ groups. Addition of a drop of water or methanol converts the —S$^-$,Li$^+$ groups into the inert —SH ends and destroys, therefore, any aggregation. For $n = 2$, the viscosity of such a solution, and hence also the time of its flow in a viscometer, should decrease by a factor $2^{3.4}$ on "killing" the polymers.*

As an illustration, the results reported by Morton and Fetters (62) are collected in Table I, which shows that the degree of association, n, may be

TABLE I

Determination of Degree of Association of Lithium Polystyrene in Benzene Solution at 25°C (62)

Flow time, sec		[RLi]$_0$ × 10^3,	[S],	
Living polymers	Dead polymers	M	M	N^a
6050	525	5.84	5.00	2.03
4630	400	2.66	3.90	2.03
1030	98.1	1.39	2.75	1.99
825	81.1	1.23	2.80	1.98
5770	560	0.23	1.75	1.99

^{a} $N = \{\overline{M}_w$ (living polymer)$\}/\{\overline{M}_w$ (dead polymer)$\}$. For values of N close to 2.000, the degree of association is $n \approx N$.

determined with a reliability of about 10% if favorable experimental conditions have been chosen. However, the errors may be substantially larger in other systems. For example, viscometric technique led Sinn et al. (67) to conclude that polyisoprenyllithium is tetrameric in heptane. This result was confirmed by light-scattering measurements performed in cyclohexane (68), while Morton's group (65) claimed for lithium polyisoprenyl n values in hexane which varied from 1.97 to 2.05. The presence of small amounts of impurities, which may "kill" living polymers or lead to their dissociation through formation of mixed aggregates, provides the most plausible explanation for these differences. In fact, highly refined techniques have to be developed to avoid such difficulties.

The dimeric nature of living lithium polystyrene in cyclohexane was confirmed by a light-scattering technique (57). Again, the molecular weight

* The calculation has to be modified if $n > 2$. In such a case, *star-shape* polymers, instead of *longer* polymers, are formed by the association. These calculations refer to monodispersed polymers.

$(\overline{M}_w)$ of the living and dead polymers were determined and the data then led to $n = 2$.

Kinetic studies of Worsfold and Bywater (43) provide us with the values of the composite constants, $k_p K_{\text{diss}}^{1/2}$. Can we determine the absolute values of k_p and K_{diss}? An interesting effort aimed at this goal was reported by Morton et al. (62,63), who hoped that the viscometric determination of the degree of association, n, would be sufficiently reliable to yield a meaningful deviation from its limiting value of 2; they then expected the fraction, x, of the disso- ciated dimers to be evaluated. In the ideal case $n = 2/(1 + x)$, provided *all* the impurities, excess of initiator, etc., are most rigorously excluded. This is rarely the case. Hence, although the approach is theoretically sound, its practical application is virtually impossible. The expected deviations of n from their limiting value of 2.00 are extremely small, often less than 2%, while the experimental uncertainty, shown, e.g., by the data collected in Table I, approaches 10%. For example, a value of $n = 1.94$ was claimed for a solution kept at 40°C, and therefore dissociated to a relatively high degree, about 3%. An error of 2% in n suffices to change the calculated K_{diss} by at least 400%. Moreover, the method depends heavily on the exact choice of the exponent, viz., 3.4. Its value may, however, be as high as 3.5 or as low as 3.3, and such a deviation, by ± 0.1, introduces an error of 2% in n and, there- fore, of 400% or more in K_{diss}. In view of this analysis the values of K_{diss}, derived from minute deviations of n from its limiting value of 2, cannot be treated seriously. Unfortunately, these data have been persistently reported, and are often quoted in the literature. Since K_{diss} is questionable, ΔH_{diss} lacks any substance, even more so because the experiments were performed in a very narrow range of temperature (30–50°C).

In conclusion, the viscometric technique is attractive; it is simple and the experimentation is easy. If properly applied, it may give the correct gross values for the degree of association of living polymers. The method is most useful in qualitative studies, e.g., when the dissociating or associating action of some added agents is investigated (61). Another good example of its utility is provided by Morton's studies of "capping" phenomena (69). The viscosity of living lithium polystyrene was measured, and thereafter a small amount of another monomer, M, was added. In this way the $\leadsto S^-,Li^+$ groups were converted into $\leadsto M^-,Li^+$ ends, without affecting to any significant extent the molecular weight of the original polymer. The change in the solu- tion viscosity, if any, indicates whether the degree of aggregation of the new living polymer is higher, lower, or the same as that of living polystyrene.* However, the method *cannot* provide any reliable information about minute

---

* In performing the "capping" experiments one has to establish that the addition of the monomer is *faster* than homopolymerization of its living polymer. This definitely is the case in the addition of butadiene or isoprene to living lithium polystyryl (see Chapter IX, p. 530).

variations of n, and therefore it is not practical to determine K_{diss} by this technique.

Let us return to our original problem: what are the absolute values of k_p and K_{diss} for the system lithium polystyrene in benzene? It seems that a reasonable estimate may be made on the basis of well-founded observations. Inspection of Figure 8 indicates that the square-root law is valid down to about a $10^{-5}M$ concentration of living polystyrene. It appears, therefore, that no more than 20% of living polystyrene is dissociated at this concentration and, hence, $K_{diss} \leqslant 10^{-6}M$. Since $(2)k_p K_{diss}^{1/2}$ is about 10^{-2} liter$^{1/2}$ mole$^{-1/2}$ sec^{-1} (at $\sim 25°C$), we conclude that $k_p \geqslant 10$ liter mole^{-1} sec^{-1}. It is improbable that k_p is much larger than that* because the propagation constant of $\sim\!\!\sim$S$^-$,Li$^+$ is only 1 liter mole^{-1} sec^{-1} in dioxane and less than 10 liter mole^{-1} sec^{-1} in tetrahydropyran (see Table VII of Chapter VII). The growing ends form contact ion-pairs in those solvents and $\sim\!\!\sim$S$^-$,Li$^+$ in benzene probably has the same structure. Therefore, $K_{diss} = 10^{-6}M$ provides the most probable estimate for the dissociation constant at 25°C and, hence, ΔF of dissociation is about 8.4 kcal mole^{-1} (a reasonable value of free energy needed to separate two dipoles). The dissociation should increase the entropy of the system by about 10–12 eu. On this basis, the value of ΔH_{diss} is calculated to be about 12 kcal mole^{-1}. Since the data of Worsfold and Bywater (43) give $E_p + \frac{1}{2}\Delta H_{diss} = 14$ kcal mole^{-1}, we find E_p to be about 8 kcal mole^{-1}, corresponding to A factor of 4×10^6 liter mole^{-1} sec^{-1}. The activation energy of 8 kcal mole^{-1} for the propagation of the monomeric living lithium polystyrene in benzene is close to the value of 9 kcal mole^{-1} reported for the propagation of living sodium polystyryl in dioxane (70)—an encouraging agreement.

Why is the dimeric lithium polystyryl $(\sim\!\!\sim$S$^-$,Li$^+)_2$ at least 100 times less reactive than the monomeric species? Consider the schematic equation:

and compare it with the one representing the propagation by a single ion-pair

---

* Bywater estimates $k_p \approx 50$ M^{-1} sec^{-1} (private communication).

(see p. 424). It is obvious from such inspection that the rearrangement needed for the addition of monomer to a pair of ion-pairs is more extensive than that taking place in the addition to a single ion-pair. Furthermore, the potential energy barrier probably is higher. Steric hindrance, caused by the overcrowding, may also contribute to the height of the potential energy barrier and this would reduce the reaction rate even more.

The kinetic data available for other polymeric systems are less extensive. Kinetics of propagation of lithium polyisoprenyl was investigated by several groups. In aliphatic hydrocarbons the propagation proceeds according to the 1/4 order law, as may be seen from the data shown in Figure 9. The results of

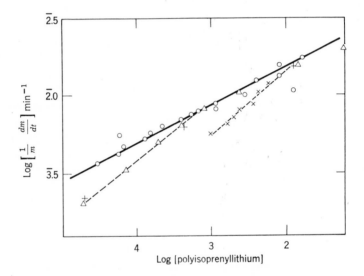

Fig. 9.   Variation of rate of propagation with total concentration of polyisoprenyl-lithium. Results of ($\bigcirc$) Worsfold and Bywater in cyclohexane (68), ($\times$) Morton et al. in hexane (63), ($+$) Spirin et al. in heptane (58), and ($\triangle$) Sinn and Patat in heptane (71).

Spirin et al. (58), Sinn and Patat (71), and Worsfold and Bywater (68) are mutually concordant, especially at higher concentrations of living polymers. The first two groups investigated the reaction in heptane, while Worsfold and Bywater studied it in cyclohexane. At the lowest concentration there are deviations, probably caused by some loss of living species. The earlier results of Morton et al. (63) deviate from those of the other workers but their later publication (69) led to the agreement with others.

The kinetic data suggest that although the bulk of lithium polyisoprenyl is tetrameric in aliphatic solvents, the propagation is carried out by monomeric species. This model is supported by molecular weight studies; both the viscometric technique (67) and studies of light scattering (68) confirmed a 4-fold aggregation of the living polymer. The data of Morton (63) differ from those of other workers, a 2-fold association being claimed.

Lithium polyisoprenyl seems to be dimeric in benzene* (65,72). This behavior may reflect the higher solvating power of benzene when compared with hexane or heptane. The ability of benzene to break the higher agglomerates should be attributed to the basic nature of π-electrons. Similar behavior of ethers and amines was discussed in Section 3 of this chapter.

On the basis of their kinetic observations, O'Driscoll et al. (74) claimed that butyllithium is trimeric in the presence of a large excess of styrene. They

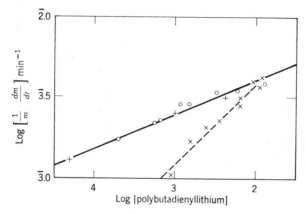

Fig. 10. Variation of rate of propagation with total concentration of polybutadienyllithium. Results of (○) Johnson and Worsfold in cyclohexane (57), (+) Spirin et al. in heptane (58,73), (×) Morton et al. in hexane (63).

believe that $(BuLi)_n$ becomes associated with the aromatic compounds; the peculiar retardation of the polymerization caused by added durene (75) was cited as an additional evidence for their hypothesis. Closer inspection of the data, however, casts some doubt on the reality of these phenomena, although the idea of complex formation between a cation and a monomer is most attractive (see Section 7 of Chapter VII).

The propagation of polybutadiene was investigated in heptane by Spirin et al. (58,73) and in cyclohexane by Johnson and Worsfold (57). The excellent agreement between their findings is obvious from Figure 10. The slope of the unbroken line shown in the figure leads to a rate equation involving the 1/6 power of lithium polybutadienyl concentration. The results of Morton (13,65) again deviate from those of the other workers; the order of the reaction with the living polymer seems to be between 1/2 and 1/3. In their most recent publication (69) a lower order (1/4) is claimed. These findings, as well as

* This conclusion is questioned by Bywater (private communication).

those which refer to the kinetics of lithium polyisoprenyl propagation, are inconsistent with the viscometric studies reported in the same paper (69). The viscometric results, obtained by the "capping" technique supposedly demonstrate a dimeric aggregation of lithium polyisoprenyl and lithium polybutadienyl and are listed in Table II. The approach is elegant and straightforward, nevertheless the final conclusions are puzzling.

TABLE II

Association of Living Polystyrene to which a Small Amount of Monomer M was Added (69)

Counterion Li^+, at 30°C

			Flow time, sec		
System	Solvent	$[RLi]_0$ × 10^3, M	Living polymers	Dead polymers	N^a
Polystyrene + isoprene	Benzene	2.4	1924	187.5	1.98
	Cyclohexane (40°C)	2.5	1571	158.4	1.96
Polystyrene + butadiene	Benzene	0.60	398.8	38.4	1.99
	Hexane	1.6	1171	119.0	1.97
Polyisoprene	Hexane	0.94	169	16.6	1.98

[a] $N = \{\overline{M}_w \text{ (living polymer)}\}/\{\overline{M}_w \text{ (dead polymer)}\}$.

Comparison of the viscosities of living polystyrene with that of the "capped" polymer would provide the most desirable test to prove, or disprove, the constancy of the degree of association. Unfortunately, the published data give only the flow times of the living and dead "capped" polymers. Nonetheless, the experiment should provide an unequivocal answer even in this form, and, therefore, the results listed in Table II are surprising. They raise the question of whether the kinetic studies or the viscometric data provide more reliable information about the degree of association of living polymers possessing Li^+ counterions. It seems that the latter should be questioned because the technical difficulties probably were not overcome. For example, the viscometric technique was applied in studies of equilibria.

$$(\text{polymer-Li})_2 + (\text{oligomer-Li})_2 \underset{}{\overset{K_1}{\rightleftharpoons}} 2(\text{polymer-Li,oligomer-Li}).$$

The results are given in Table III and lead to $K_1 \approx 1$. However, in view of random mixing K_1 should be 4.

Accepting the state of aggregation of the living polymers given by the kinetic data, we conclude that lithium polybutadienyl is more strongly associated than lithium polyisoprenyl. It seems that the living polymers behave similarly to lithium alkyls: n-butyllithium is hexameric whereas

TABLE III

Equilibrium Constant K_1 (69) of the Reaction (in Hexane, at 30°C) of
(polymer-Li)$_2$ + (oligomer-Li)$_2$ ⇌ 2(polymer-Li,oligomer-Li)

[polymer-Li] $\times 10^3$, M	[oligomer-Li] $\times 10^3$, M	N^a	K_1^b	
1.1	6.8	1.34	0.88	(0.85)^c
1.0	4.6	1.40	0.90	

[a] $N = \{\overline{M}_w \text{ (living polymer)}\}/\{\overline{M}_w \text{ (dead polymer)}\}$.

[b] $K_1 = [\text{(oligomer-Li,polymer-Li)}]^2/[\text{(polymer-Li)}_2][\text{(oligomer-Li)}_2]$. Fraction x of dimeric (polymer-Li)$_2$ which reacted with the oligomers and produced mixed dimers is given by the equation $x = 2 - N$.

[c] Calculated by this writer on the basis of the given N.

sec- or tert-butyllithiums are tetrameric. Apparently steric effects operate in both systems.

Three attempts were made to determine the absolute rate constants of polymerization of lithium polyisoprenyl. Sinn et al. (53,67,76) investigated the kinetics of polymerization of isoprene down to extremely low concentrations ($5 \times 10^{-7}M$). Elaborate precautions were taken to ascertain the absence of impurities; however this point may be still questioned. The dependence of the rate of polymerization on the concentration of the living polymer was investigated by concentrating the extremely dilute solutions, and not by diluting the concentrated aliquots. The order of the reaction was found to increase at concentrations below $10^{-4}M$, and at the concentration $5 \times 10^{-6}M$ the rate was first order in lithium polyisoprenyl. One may deduce, therefore, that the dissociation of agglomerates is completed at this concentration and the observed rate constant gives the required absolute constant, k_p. Thus, k_p in heptane at 20°C was claimed to be 0.65 liter mole^{-1} sec^{-1}. Unfortunately, the reaction order in living polymers increased as the concentration was still lowered, and at $C = 5 \times 10^{-7}M$ the reaction appeared to follow the second order law. A rather complex mechanism was proposed to account for the changes in the reaction order (76). Formation of a complex between monomeric lithium polyisoprenyl and the monomer (isoprene) was postulated, and such a complex was assumed to react then with another molecule of the monomer. It seems more likely that the observed changes of order are caused by the presence of some lithium salts formed by destruction of the initiator. Bywater demonstrated that in an all-glass high-vacuum system pretreated with butyllithium, the concentration of reactive impurities is still about $3 \times 10^{-6}M$, even after the most thorough purification of the reactants and solvent. Experiments carried out at initiator concentrations lower than $3 \times 10^{-5} M$ cannot, therefore, be accurate, even under the most favorable conditions. The presence of salts depresses the propagation as well as other reactions of organolithium compounds (16,78,79); apparently mixed unreactive aggregates are formed (see also Chapter VII, Section 8).

The type of mechanism proposed by Sinn is frequently encountered in the literature concerned with polymerization induced by lithium compounds as well as in other ionic polymerizations (e.g., refs. 80–82). In most cases it was invoked to explain observations that later were shown to arise from some artifacts. Although the idea of monomer association with the counterion is most attractive, the evidence in its favor should be critically examined. General remarks on this subject are given in Section 7 of Chapter VII.

Does monomer dissociate the agglomerates? This is an interesting question. If the degree of dissociation is large it could be studied by the viscometric technique, viz., measuring the viscosity of living polymers before and after monomer addition. The polymerization is slow and therefore should not interfere with such an experiment.

An ingenious attempt to determine the absolute rate constant of propagation of lithium polyisoprenyl in cyclohexane was reported by François, Sinn, and Parrod (83). Conventional dilatometric studies of the propagation of living polymers led to a 1/2 order rate expression, a doubtful result in view of the latest findings (58,68,69,71). Thereafter, the reaction was investigated in the presence of increasing amounts of ethyllithium and this led to retardation. The authors then assumed the following scheme:

$$2(\sim\text{Isp}^-,\text{Li}^+) \rightleftarrows (\sim\text{Isp}^-,\text{Li}^+)_2, \qquad K_1,$$

$$\sim\text{Isp}^-,\text{Li}^+ + \text{Isp} \rightarrow \text{propagation}, \qquad k_p,$$

$$(\sim\text{Isp}^-,\text{Li}^+) + \text{EtLi} \rightleftarrows \text{mixed dimer}, \qquad K_2.$$

The scheme is questionable because it not only assumes a dimeric form for living polyisoprene, while its aggregation appears to be higher, but it also postulates for ethyllithium only a monomeric form or the form of a mixed dimer, the hexameric aggregation being ignored. The mathematical treatment of the proposed mechanism may be substantially simplified by introducing a plausible approximation,

$$8K_1 C \ll (1 + K_2[\text{EtLi}])^2,$$

C being the total concentration of all the organolithium compounds. It follows then that the reciprocal of the rate is given by

$$1/\text{rate} = 1/k_p C + (K_2/k_p)([\text{EtLi}]/C).$$

The plot of the reciprocal of the rate vs. $[\text{EtLi}]/C$ is shown in Figure 11; it is indeed linear, as demanded by the treatment, and from its intercept and slope the values of k_p and K_2 were obtained. Moreover, K_1 may be calculated after k_p has been determined. The reported value of k_p is about 2 liter mole^{-1} sec^{-1}. From the experimental point of view this may be disputed because the intercept is virtually indistinguished from 0.*

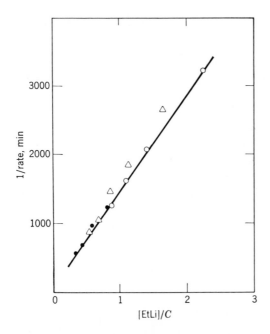

Fig. 11.   Polymerization of isoprene in the presence of ethyllithium. Reciprocal of rate plotted vs. the ratio [EtLi]/[polyisoprenyllithium] (83).

The details of this study have been presented to illustrate the problems encountered by the student concerned with the literature of this subject. Although the assumptions are questionable, the experimental results conformed with mathematical predictions.

The determination of the absolute rate constants of propagation is undoubtedly extremely difficult. Theoretical and experimental problems are enormous due to a high, and perhaps variable, degree of association of lithium compounds in nonpolar solvents, and to their capacity for forming mixed aggregates with impurities. Much more work is needed in this field.

---

* Note also that the experiments cannot be extended to [EtLi]/$C = 0$ because the approximation $8K_1C \ll (1 + K_2[\text{EtLi}])^2$ is then not valid ($K_1C \gg 1$ under feasible experimental conditions).

6. EFFECT OF LEWIS ACIDS AND BASES ON PROPAGATION

Polar materials exert a large influence upon the rate of polymerization initiated by lithium compounds in hydrocarbon solvents (44,45,81), as well as on other reactions of organolithium compounds (93). The effects are often dramatic, even when the concentration of the additive is comparable to that of the initiator. Both the initiation and propagation are affected, e.g., Kropacheva and her colleagues (86) showed that tetrahydrofuran markedly accelerates initiation by alkyllithium. Indeed, the effect upon initiation

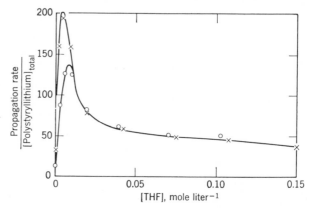

Fig. 12.   Effect of tetrahydrofuran on rate of propagation in the polymerization of styrene in benzene (77). Concentration of polystyryllithium: ($\bigcirc$) $\sim 1.1 \times 10^{-3}M$, ($\times$) $\sim 1.4 \times 10^{-4}M$.

appears to be of paramount importance, and the initiation process in the presence of tetrahydrofuran becomes too fast for conventional kinetic study (77).

The effect of tetrahydrofuran on propagation is less pronounced and often rather complex. Studies of styrene propagation in benzene showed a large increase in the apparent propagation constant when small amounts of tetrahydrofuran were added, but the rate decreased on further addition (77) as shown in Figure 12. The reaction obeys the 1/2 order law on the ascending part of the curve, but when the ratio $[THF]/[BuLi]_0$ exceeds 20 or 30 the propagation becomes nearly independent of the ether concentration and shows a first order character.

Tetrahydrofuran apparently dissociates dimeric lithium polystyryl and converts the living polymers into solvated monomeric species. A detailed kinetic analysis indicates the presence of four different species in this system:

inert dimers $(\sim S^-,Li^+)_2$, reactive monomers $(\sim S^-,Li)$, reactive mono-etherates, $(\sim S^-,Li^+,THF)$, and less reactive dietherates $(\sim S^-,Li^+,2THF)$, all coexisting in equilibrium with each other. The results permit determining composite constants based on the following scheme:

$$(\sim S^-,Li^+)_2 \rightleftarrows 2(\sim S^-,Li^+), \qquad\qquad K_1,$$

$$(\sim S^-,Li^+) + THF \rightleftarrows (\sim S^-,Li^+,THF), \qquad K_2,$$

$$(\sim S^-,Li^+,THF) + THF \rightleftarrows (\sim S^-,Li^+,2THF), \qquad K_3,$$

$$(\sim S^-,Li^+) + S \rightarrow propagation, \qquad\qquad k_{p1},$$

$$(\sim S^-,Li^+,THF) + S \rightarrow propagation, \qquad k_{p2},$$

$$(\sim S^-,Li^+,2THF) + S \rightarrow propagation, \qquad k_{p3}.$$

Thus, at 25°C, $k_{p1}K_1^{1/2} = 0.01$, $k_{p2}K_1^{1/2}K_2 = 25.3$, and $k_{p2}/K_3 = 0.9$, all the data given in mole/liter and sec units. The increase in the rate observed in the presence of small amounts of ether need not be interpreted as an indication for higher reactivity of monoetherates when compared with monomeric $\sim S^-,Li^+$. This result basically reflects the fact that the fraction of all reactive species increases on addition of tetrahydrofuran.

In the first order region of the propagation, all the living polymers are converted into dietherates. The kinetic data then give directly the absolute rate constant of propagation for the dietherate: $k_{p3} = 35$ liter mole^{-1} sec^{-1} at 30°C. The dietherate in benzene is, therefore, much less reactive than the solvent-separated ion-pair; the propagation constant of the latter is about 20,000 to 30,000 liter mole^{-1} sec^{-1} (see Chapter VII). Apparently the dietherate forms an "outside" coordinated contact ion-pair, a type described by Slates and Szwarc (87).

It is puzzling to find the monoetherate more reactive than the dietherate. The coordination with ether involves cation and not anion; hence, a direct reaction with carbanion should not be hindered by further solvation. On the other hand, if the monomer becomes coordinated with the cation *prior* to its addition to the carbanion, then the addition of a second molecule of ether may hinder the propagation. We may account tentatively for this result by the following symbolic equation in which the arrow represents a molecule of the ether (THF):

Insertion of the ether, i.e., formation of the separated pair in the activated complex, may facilitate the addition represented by the four-center transition state. The dietherate

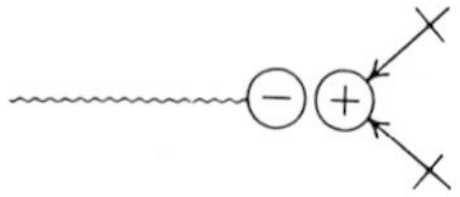

is assumed to be sterically blocked. Thus, the monoetherate may be even more reactive than the nonsolvated $\sim$S$^-$,Li$^+$.

Tetrahydrofuran is less efficient in accelerating the polymerization of lithium polyisoprenyl in cyclohexane. At about $10^{-3}M$ concentration of the living polymers, the rate of propagation was doubled on addition of $0.1M$ tetrahydrofuran (68), and the order of the reaction increased from 1/4 to 1/2. Thus, a 100-fold excess of the ether is still not sufficient to dissociate all the tetrameric lithium polyisoprenyls; under the same conditions, a 20-fold excess suffices to dissociate all the dimeric lithium polystyryl.

One of the most interesting coordination agents is tetramethylethylenediamine (TMEDA). This base becomes powerfully coordinated with butyllithium forming a five-membered ring chelate:

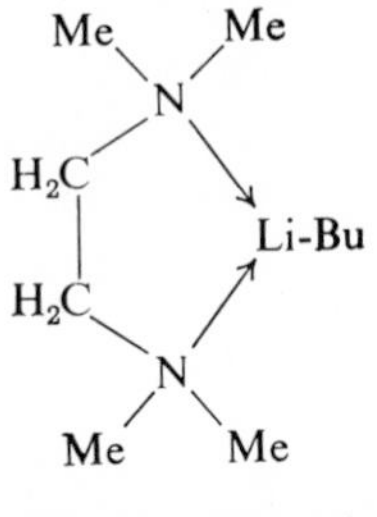

The complex is monomeric, it appears as a pale yellow liquid that is miscible in all proportions with hydrocarbons. Its high reactivity was first reported by Eberhardt and Butte (88) and later by Langer (89). The remarkable reactivity of the complex prevents its storage in any other solvent but aliphatic hydrocarbon.

The unusual properties of the complex are demonstrated by the fact that merely bubbling hydrogen gas through dilute solution of TMEDA·BuLi at 25°C results in a rapid precipitation of LiH. In contrast, one needs 100 atm pressure and prolonged exposure (60 hr) to reduce uncomplexed butyllithium. Metalation of benzene provides another example. On mixing the components in benzene at 25° one achieves 70% conversion to PhLi·TMEDA in about an hour. Again, free butyllithium does not react even at 100°. This procedure is general, and thus quantitative formation of new organolithium·TMEDA complexes may be afforded. The products may often be isolated in crystalline form. In fact, even TMEDA may be metalated, and therefore the complex slowly decomposes on storage.

For a polymer chemist, the most interesting point is the ability of the complex to initiate polymerization of ethylene to a high molecular weight polyethylene; in fact, under suitable conditions samples of 140,000 molecular weight were obtained (89). Polymerization initiated by ordinary butyllithium led only to waxy materials ($\overline{M}_w \approx 17,000$) even at a pressure of 15,000 psi (90).

Polymerization of ethylene by the complex in the presence of a suitable hydrocarbon telogen, such as benzene, toluene, or xylene, yields telomers (88,89). In the case of propylene acting as a telogen long linear α-olefins are produced.

The increase in reactivity of the Li—C bond upon solvation of the lithium has been attributed to an increase in its ionic character (56). It seems that the complexing to TMEDA produced a genuine ion-pair and the extraordinary reactivity of such alkyllithiums manifests the high reactivity of a partially dissociated alkyl carbanion. It is interesting to inquire whether carbanions are formed under other conditions. For example, one may speculate about the feasibility of ionic dissociations such as $(RLi)_2 \rightleftarrows (RLi_2)^+, R^-$. This type of polarization was visualized by Brown et al. (23).

Reverse effects are observed when the association involves Lewis acids, e.g., diethylzinc or triethylaluminium (92). These become associated with the growing anions and retard the propagation. For example, the presence of trialkylaluminium at concentrations only slightly higher than that of living polystyrene is sufficient to prevent its polymerization in benzene, but a 10-fold excess of diethylzinc is needed to achieve the same result. The respective equilibrium constants of association were determined from the degree of inhibition of the polymerization (92).

The inactivation of living polymers by alkyllithium should be viewed in the same way. These are Lewis acids, and become complexed with the growing anions.

7. POLYMERIZATION OF POLAR MONOMERS IN NONPOLAR SOLVENTS

The polymerization of methyl methacrylate initiated in toluene by organolithium compounds was investigated most intensively. The stereospecific effects observed in this process have been discussed in Section 14 of Chapter VII, and some peculiarities, encountered in the propagation and attributed to the "helical" growth, are treated in Chapter XI. Butyllithium reacts with methyl methacrylate in two distinct ways. It may attack the C=C double bond (ordinary initiation of vinyl polymerization), or it may react with the alkyl carboxylate (97). The latter reaction is extremely fast at room temperature and consumes two molecules of RLi for each carboxylate group (94). In fact, the original initiator disappears almost instantaneously on mixing the reactants (95,96).

Polymerization in toluene at $-30°C$ was investigated by Wiles and Bywater (98). The reaction is obviously most complex. An appreciable fraction of monomers consumed in the early stages of the polymerization yielded oligomers having a molecular weight of about 800. After 25% of the monomer was polymerized, the rate increased and roughly, a first order reaction followed, i.e., at this stage of the process the concentration of growing ends became stationary. On addition of further monomer to a polymerized batch, the reaction was spontaneously reinitiated and proceeded with the rate observed in the steady period of the previous polymerization. No low molecular weight oligomers were formed during the reinitiated polymerization. Apparently the "initiator" operating in the later stages of the first polymerization, or in the course of reinitiated reaction, is different from butyllithium. Moreover, it is likely that some alkoxides are associated with growing ends and modify their reaction (84).

Much information was gained by fractionating the products formed at various conversions and on multiple monomer addition. The bimodal molecular weight distribution is formed in the early stage of the reaction and it persists throughout the polymerization. Moreover, the longer chains seem to grow faster.

Lithium methoxide is definitely formed in the process. Its amount was determined by converting it with acetic acid to methanol (98,99). The percentage of BuLi converted into methoxide is given in Figure 13 as a function of monomer conversion. About 50% of butyllithium forms lithium methoxide in the period during which the first few percents of polymerization take place. Strikingly, the formation of lithium methoxide is inhibited if 1,1-diphenylhexyllithium is used as the initiator. Rempp et al. (100) have reported that addition of 1,1-diphenyl ethylene to living polystyrene prevents the reaction with the carboxylate group and only the C=C bond of the monomer is involved then in the process. This explains the finding of Bywater (117).

The presence of methoxide provides a possible explanation of the bimodal molecular weight distribution (see Chapter XI for the alternative explanation)—i.e., two types of growing ends participate in the reaction, one associated with methoxide and the other not. The exchange between them has to be slow or the molecular weight would not be affected. The polymer formed through initiation by 1,1-diphenylhexyllithium does not show bimodal distribution. This observation supports the contention that the complexing with $LiOCH_3$ is responsible for the unusual distribution—only a minute amount of methoxide is formed under these conditions (see Fig. 13).

Polymerization in the presence of added ether or tetrahydrofuran (0.1–15%) is even more complex. Because the data presently available are not sufficient to unravel all the puzzling features of this reaction we shall not discuss further the literature dealing with this subject.

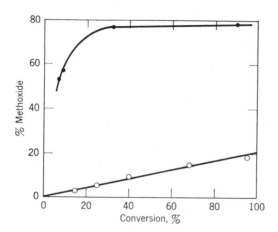

Fig. 13. Formation of lithium methoxide in the polymerization of methylmethacrylate (0.125M) in toluene at −30° (98). Initiator: (○) 3.2 × 10⁻³M diphenylhexyllithium, (●) 3.85 × 10⁻³M butyllithium.

The literature concerned with the anionic polymerization of other polar monomers is voluminous but not instructive. Most of the papers provide information about the yield, molecular weight and properties of the produced materials, which were isolated under conditions not always fully specified. Hardly any kinetic or mechanistic study is reported. For a review of this subject the reader may consult reference 84 (p. 88).

8. STEREOSPECIFIC POLYMERIZATION OF DIENES

In the field of dienes, such as butadiene or isoprene, stereospecificity is shown by the preference of a particular type of linkage. There are three basically different modes of linkage: the *trans*-1,4, the *cis*-1,4, and the 1,2 (or 3,4) linkages. The latter, if configurationally stereospecific, may lead to iso- or syndio-tactic 1,2 polymers (101). The great interest in diene polymerization initiated by metallic lithium, or lithium alkyls, arose from the discovery of Stavely et al. (102), who demonstrated that isoprene polymerized in hydrocarbon solvents by lithium emulsion yields all-*cis*-1,4-polyisoprene. In this respect, lithium is exceptional; polymerization initiated by other alkali metals is predominantly 3,4, and the 1,4 linkage has mainly a *trans*-configuration. The pertinent results are given in Table IV.

TABLE IV

Microstructure of Polyisoprene Polymerized in Hydrocarbon Solvents

Counterion	1,4-*cis*	1,4-*trans*	3,4-	1,2-	Ref.
Li	91	—	9	—	(103)
	93	—	7	—	(104)
	94	—	6	—	(105)
	94	—	6	—	(106)
	80	15	5	—	(68)
Na	—	12	77	11	(103)
	—	46	45	9	(104)
	3	40	50	7	(105)
	—	43	51	8	(106)
	29	29	42	—	(68)
K	—	55	36	9	(104)
	20	40	32	7	(105)
	—	52	40	8	(106)

The large differences in the results of various workers reflect the uncertainties of various methods used in the analysis. The reliability of these methods does, indeed, need careful checking.

The all-*cis*-1,4 structure is obtained whether metallic lithium (heterogeneous initiator) or lithium alkyls (homogeneous initiator) are employed in the polymerization, but it is imperative to carry out the reaction in a hydrocarbon medium and to exclude any polar impurities. In polar solvents, or even in the presence of small amounts of polar materials, e.g., ethers, the polymerization yields mainly the 3,4 linkage (107,108). The ability of various ethers, sulfides, etc. to modify the microstructure has been investigated by many workers (e.g., 110). Tetrahydrofuran appears to be one of the most potent agents modifying the microstructure of polyisoprene (68).

With butadiene, polymerized by lithium or its derivatives in hydrocarbons, the reaction yields a large proportion of the 1,4 structure but of a mixed *cis* and *trans* content (106). Again, the proportion of 1,2 linkage increases in this polymerization when polar substances are present (85,109).

The ionic character of the C—Li bond seems to have a decisive influence upon the microstructure of the polydiene (111). One may ask, however, whether the solvating agent gradually modifies the property of a C—Li bond, or whether two or more types of C—Li bonds exist in dynamic equilibrium with each other and the solvating agent modifies the equilibrium concentration of these forms. It is also important to know whether the configuration is fixed in the last unit of the growing chain, or becomes fixed when a next monomer is added and the unit in question becomes a penultimate segment.

If the last unit forms an allyl ion, the middle $\overset{2}{C}=\overset{3}{C}$ double bond is not yet fully developed and then a rotation along the C(2)–C(3) axis, although strongly hindered (112), may still take place. On the other hand, if a truly covalent C—Li bond is formed, then the 2,3 double bond is finally established in the last unit and the configuration is then fixed. In the latter case, the required configuration may be imposed on the molecule of the monomer prior to its addition. It is plausible to treat a diene as a didentate agent, and assume that Li becomes coordinated with it:

The *intra*molecular complexing with the other C=C bond explains the results of Bywater and Worsfold (115) who found the dissociation of lithium polyisoprenyl in tetrahydrofuran to be 100 times lower than that of lithium polystyryl. Various versions of this model have been considered by several workers (105,111,113) and they apply to both butadiene and isoprene. Why, therefore, does isoprene, and not butadiene, give the all-1,4-*cis* structure? Two factors may be responsible for the difference in the behavior of these dienes: (a) A steric factor which favors the *cis* form more in isoprene than in butadiene. This factor may prevent the frequent rotation which might take place in the last unit of polybutadiene but not in polyisoprene. (b) An inductive factor which should make the C—Li bond tighter in polyisoprene than in polybutadiene and therefore make the chelated structure more rigid.

The spectrum of lithium polyisoprene or lithium polybutadiene in hydrocarbon is virtually identical with that observed in polar solvents. This observation casts some doubt on the hypothesis implying a covalent C—Li bond.

The explanation of trans-1,4 addition was suggested by Szwarc (113). These ideas are explained by the models shown in Figure 14. They attempt to rationalize the effect of the size of the counterion upon the stereochemistry of the transition state, however other factors still have to be accounted for.

The unusual properties of organolithium compounds are certainly associated with the structure of the C—Li bond which differs in several respects from that involving other alkali metals. It is hoped that the problem of its structure will be clarified by future studies.

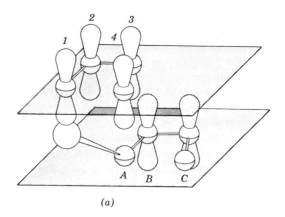

(a)

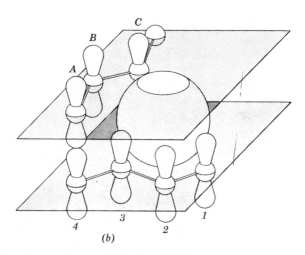

(b)

Fig. 14.   Schematic representation (113) of the diene addition to growing polymer:
(a) covalent C—Li bond; (b) alkyl⁻ large counterion⁺.

References

(1) R. E. Rundle, *J. Phys. Chem.*, **61**, 45 (1957).

(2) D. Margerison and J. P. Newport, *Trans. Faraday Soc.*, **59**, 2058 (1963).

(3) G. Wittig, F. J. Meyer, and G. Lange, *Ann.*, **571**, 167 (1951).

(4) F. Hein and H. Schramm, *Z. Physik. Chem.*, **A151**, 234 (1930).

(5) T. L. Brown and M. T. Rogers, *J. Am. Chem. Soc.*, **79**, 1859 (1957).

(6) M. Weiner, G. Vogel, and R. West, *Inorg. Chem.*, **1**, 654 (1962).

(7) E. Warhurst, *Discussions Faraday Soc.*, **2**, 239 (1947).

(8) J. Berkowitz, D. A. Bafus, and T. L. Brown, *J. Phys. Chem.*, **65**, 1380 (1961).

(9) H. Dietrich, *Acta Cryst.*, **16**, 681 (1963).

(10) E. Weiss and E. A. C. Lucken, *J. Organomet. Chem.*, **2**, 197

(11) T. L. Brown, D. W. Dickerhoof, and D. A. Bafus, *J. Am. C* (1962).

(12) T. L. Brown, *Advan. Organomet. Chem.*, **3**, 365 (1965).

(13) M. Morton, E. E. Bostick, and R. Livigni, *Rubber Plastic Age*,

(14) T. L. Brown and J. A. Ladd, *J. Organomet. Chem.*, **2**, 373 (1964)

(15) A. F. Johnson, D. J. Worsfold, and S. Bywater, *Can. J. Chem.*, **4** (1964).

(16) J. E. L. Roovers and S. Bywater, *Trans. Faraday Soc.*, **62**, 701 (1966).

(17) T. L. Brown, R. L. Gerteis, D. A. Bafus, and J. A. Ladd, *J. Am. Chem. Soc.*, **86**, 2135 (1964).

(18) T. L. Brown, J. A. Ladd, and G. N. Newman, *J. Organomet. Chem.*, **3**, 1 (1965).

(19) R. West and W. Glaze, *J. Am. Chem. Soc.*, **83**, 3580 (1961).

(20) A. N. Rodinov, T. V. Talalaeva, D. N. Shigorin, G. N. Tyumofeyuk, and K. A. Kocheshkov, *Dokl. Akad. Nauk SSSR*, **151**, 1131 (1963).

(21) F. A. Settle, M. Haggerty, and J. F. Eastham, *J. Am. Chem. Soc.*, **86**, 2076 (1964).

(22) Z. K. Cheema, G. W. Gibson, and J. F. Eastham, *J. Am. Chem. Soc.*, **85**, 3517 (1963).

(23) L. M. Seitz and T. L. Brown, *J. Am. Chem. Soc.*, **88**, 2174 (1966).

(24) G. E. Hartwell and T. L. Brown, *J. Am. Chem. Soc.*, **88**, 4625 (1966).

(25) G. E. Coates, "Organo-Metallic Compounds," 2nd ed., Wiley, New York, 1960.

(26) E. Weiss, G. Sanermann, and H. Plerz; Reported at the 3rd International Symposium on Organometalic Chemistry, Munich, 1967.

(27) P. West and R. Waack, *J. Am. Chem. Soc.*, **89**, 4395 (1967).

(28) G. E. Hartwell and T. L. Brown, *Inorg. Chem.*, **3**, 1656 (1964).

(29) R. H. Baney and R. J. Krager, *ibid.*, **3**, 1657 (1964).

(30) L. M. Seitz and T. L. Brown, *J. Am. Chem. Soc.*, **88**, 2174 (1966).

(31) R. Waack, M. A. Doran, and E. B. Baker, to be published.

(32) P. West and R. Waack, *J. Am. Chem. Soc.*, **89**, 4395 (1967).

(33) T. V. Talalaeva, A. N. Rodionov, and K. A. Kocheshkov, *Dokl. Akad. Nauk SSSR*, **140**, 847 (1961); **154**, 174 (1964).

(34) M. Chabauel, *J. Chim. Phys.*, **63**, 1143 (1966).

(35) A. N. Rodionov, D. N. Shigorin, T. V. Talalaeva, and K. A. Kocheshkov, *Dokl. Akad. Nauk SSSR*, **143**, 137 (1962).

(36) R. Waack, M. A. Doran, and P. E. Stevenson, *J. Am. Chem. Soc.*, **88**, 2109 (1966).

(37) I. Kuntz, *J. Polymer. Sci. A*, **2**, 2827 (1964).

(38) G. Holden and R. Milkovich, U.S. Pat. 3,231,635 (1966).

(39) D. O. Harris, J. Smid, and M. Szwarc, to be published.

(40) R. Waack, unpublished results.

(41) R. Waack and M. A. Doran, *Chem. Ind. (London)*, **1961**, 1165.

(42) R. Waack, P. West, and M. A. Doran, *ibid.*, **1966**, 1035.

(43) D. J. Worsfold and S. Bywater, *Can. J. Chem.*, **38**, 1891 (1960).

(44) K. F. O'Driscoll and A. V. Tobolsky, *J. Polymer Sci.*, **35**, 259 (1959).

(45) F. J. Welch, *J. Am. Chem. Soc.*, **81**, 1345 (1959).

(46) H. L. Hsieh, *J. Polymer Sci. A*, **3**, 163 (1965).

(47) R. C. P. Cubbon and D. Margerison, *Proc. Roy. Soc. (London), Ser. A*, **268**, 260 (1962).

(48) H. Sinn and F. Patat, *Angew. Chem. (Intern. Ed. English)*, **3**, 93 (1964).

(49) A. G. Evans and D. B. George, *J. Chem. Soc.*, **1961**, 4653.

(50) R. Waack and P. E. Stevens; *J. Am. Chem. Soc.*, **87**, 1183 (1965).
(51) M. Szwarc and J. Smid, in "Progress in Reaction Kinetics," Vol. 2, Pergamon Press, New York, 1964, p. 219.
(52) T. L. Brown, *J. Organometal. Chem.*, **5**, 191 (1966).
(53) C. Lundborg and H. Sinn, *Makromol. Chem.*, **41**, 242 (1960).
(54) S. Bywater and D. J. Worsfold, *J. Organomet. Chem.*, **10**, 1 (1967).
(55) A. G. Evans and N. H. Rees, *J. Chem. Soc.*, **1963**, 6039.
(56) A. V. Tobolsky and C. E. Rogers, *J. Polymer Sci.*, **38**, 205 (1959).
(57) A. F. Johnson and D. J. Worsfold, *J. Polymer Sci. A*, **3**, 449 (1965).
(58) Yu. L. Spirin, A. R. Gantmakher, and S. S. Medvedev, *Dokl. Akad. Nauk SSSR*, **146**, 368 (1962).
(59) D. O. Harris, preliminary studies performed in Syracuse.
(60) Z. Laita, unpublished results from Syracuse.
(61) H. Brody, D. H. Richards, and M. Szwarc, *Chem. Ind. (London)*, **1958**, 1473.
(62) M. Morton and L. J. Fetters, *J. Polymer Sci. A*, **2**, 3311 (1964).
(63) M. Morton, E. E. Bostick, R. A. Livigni, and L. J. Fetters, *J. Polymer Sci. A*, **1**, 1735 (1963).
(64) Yu. L. Spirin, D. K. Polyakov, A. R. Gantmakher, and S. S. Medvedev, *Dokl. Akad. Nauk SSSR*, **139**, 899 (1961).
(65) M. Morton, L. J. Fetters, and E. E. Bostick, in *First Biannual American Chemical Society Polymer Symposium*, H. W. Starkweather, Jr., Ed. (*J. Polymer Sci. C*, **1**), Interscience, New York, 1963, p. 311.
(66) R. A. H. Casling, A. G. Evans, and N. H. Rees, *J. Chem. Soc. (B)*, **1966**, 519.
(67) H. Sinn, C. Lundborg, and O. T. Onsager, *Makromol. Chem.*, **70**, 222 (1964).
(68) D. J. Worsfold and S. Bywater, *Can. J. Chem.*, **42**, 2884 (1964).
(69) M. Morton, R. A. Pett, and J. F. Fetters; presented at the I.U.P.A.C. Meeting, Tokyo, 1966.
(70) (a) G. Allen, G. Gee, and C. Stretch, *J. Polymer Sci.*, **48**, 189 (1960). (b) C. Stretch and G. Allen, *Polymer (London)*, **2**, 151 (1961).
(71) H. Sinn and F. Patat, *Angew. Chem.*, **75**, 805 (1963).
(72) D. N. Cramond, P. S. Lawry, and J. R. Urwin, *European Polymer J.*, **2**, 107 (1966).
(73) Yu. L. Spirin, A. A. Arest-Yakubovich, D. K. Polyakov, A. R. Gantmakher, and S. S. Medvedev, *J. Polymer Sci.*, **58**, 1181 (1962).
(74) K. F. O'Driscoll, E. N. Ricchezza, and J. E. Clark, *J. Polymer Sci. A*, **3**, 3241 (1965).
(75) K. F. O'Driscoll and R. Patsiga, *J. Polymer Sci. A*, **3**, 1037 (1965).
(76) H. Sinn and O. T. Onsager, *Makromol. Chem.*, **55**, 167 (1962).
(77) D. J. Worsfold and S. Bywater, *Can. J. Chem.*, **40**, 1564 (1962).
(78) R. Waack and M. A. Doran, *Chem. Ind. (London)*, **1964**, 496.
(79) R. Huisgen and W. Mack, *Chem. Ber.*, **93**, 332 (1960).
(80) A. A. Korotkov, N. N. Chesnokova, and L. B. Trukhmanova, *Vysokomolekul. Soedin.*, **1**, 46 (1959).
(81) A. A. Korotkov and A. F. Podolsky, *J. Polymer Sci. B*, **3**, 901 (1965); see, however, R. V. Figini, *J. Polymer Sci. B*, **4**, 223 (1966) for comments.
(82) M. Szwarc and J. Smid, in "Progress in Reaction Kinetics," Vol. 2, Pergamon Press, New York, 1964; p. 243.
(83) B. François, V. Sinn, and J. Parrod, in *International Symposium on Macromolecular Chemistry, Paris, 1963*, M. Magat, Ed. (*J. Polymer Sci. C*, **4**), Interscience, New York, 1964, p. 375.
(84) S. Bywater, *Fortschr. Hochpolymer.-Forsch.*, **4**, 66 (1965).

(85) A. A. Korotkov, *Angew. Chem.*, **70**, 85 (1958).
(86) E. N. Kropacheva, B. A. Dolgoplosk, and E. M. Kuznetsova, *Dokl. Akad. Nauk SSSR*, **130**, 1253 (1960).
(87) R. V. Slates and M. Szwarc, *J. Am. Chem. Soc.*, **89**, 6043 (1967).
(88) G. G. Eberhardt and W. A. Butte, *J. Org. Chem.*, **29**, 2928 (1964).
(89) A. W. Langer, *Trans. N. Y. Acad. Sci.*, **27**, 741 (1965).
(90) W. E. Hanford, J. R. Roland, and H. S. Young, U.S. Pat. 2,377,779 (1945).
(91) J. Minoux and J. Marchal, *Compt. Rend.*, **250**, 3650 (1960).
(92) F. J. Welch, *J. Am. Chem. Soc.*, **82**, 6000 (1960).
(93) R. Waack and P. West, *J. Organomet. Chem.*, **5**, 188 (1966).
(94) Z. Laita and M. Szwarc, *J. Polymer Sci. B*, **6**, 197 (1968).
(95) B. J. Cottam, D. M. Wiles, and S. Bywater, *Can. J. Chem.*, **41**, 1905 (1963).
(96) D. L. Glusker, E. Stiles, and B. Yoncoskie, *J. Polymer Sci.*, **49**, 297 (1961).
(97) A. A. Korotkov, S. P. Mitsengendler, and V. N. Krasulina, *J. Polymer Sci.*, **53**, 217 (1961).
(98) D. M. Wiles and S. Bywater, *Chem. Ind. (London)*, **1963** 1209.
(99) D. M. Wiles and S. Bywater, *J. Phys. Chem.*, **68**, 1983 (1964).
(100) D. Freyss, P. Rempp, and H. Benoit, *Makromol. Chem.*, **87**, 271 (1965).
(101) C. E. H. Bawn and A. Ledwith, *Quart. Rev. (London)*, **16**, 361 (1962).
(102) F. W. Stavely et al., *Ind. Eng. Chem.*, **48**, 778 (1956).
(103) H. Morita and A. V. Tobolsky, *J. Am. Chem. Soc.*, **79**, 5853 (1957).
(104) A. V. Tobolsky and R. J. Boudreau, *J. Polymer Sci.*, **51**, 553 (1961).
(105) R. S. Stearns and L. E. Forman, *J. Polymer Sci.*, **41**, 381 (1959).
(106) F. C. Foster and J. L. Binder, *Advan. Chem. Ser.*, **17**, 7 (1957).
(107) H. Hsieh and A. V. Tobolsky, *J. Polymer Sci.*, **25**, 245 (1957).
(108) H. Hsieh, D. J. Kelley, and A. V. Tobolsky, *J. Polymer Sci.*, **26**, 240 (1957).
(109) V. A. Kropacheva, B. A. Dolgoplosk, and N. I. Nikolaev, *Dokl. Akad. Nauk SSSR*, **115**, 516 (1957).
(110) A. V. Tobolsky and C. E. Rogers, *J. Polymer Sci.*, **40**, 73 (1959).
(111) Yu. L. Spirin, A. R. Gantmakher, and S. S. Medvedev, *Dokl. Akad. Nauk SSSR*, **1**, 1258 (1959).
(112) C. Walling, "Free Radicals in Solution," Wiley, New York, 1957.
(113) M. Szwarc, *J. Polymer Sci.*, **40**, 583 (1959).
(114) R. Waack and M. A. Doran, *J. Org. Chem.*, **32**, 3395 (1967).
(115) S. Bywater and D. J. Worsfold, *Can. J. Chem.*, **45**, 1821 (1967).
(116) T. L. Brown, *Accounts of Chem. Res.*, **1**, 23 (1968).
(117) D. M. Wiles and S. Bywater, *Trans. Faraday Soc.*, **62**, 150 (1966).

Chapter IX

Kinetics of Anionic Copolymerization

1. GENERAL PROBLEMS OF ANIONIC COPOLYMERIZATION

A mixture of two or more monomers may copolymerize anionically if the anion derived from one monomer reacts and forms a covalent bond with the other monomer, and vice versa. This condition is not fulfilled for any two monomers. For example, the polystyryl anion reacts with methyl methacrylate and, thus, the methyl methacrylate anion becomes attached at the end of the polymer. However, no styrene addition to the polymethyl methacrylate anions is ever observed (1). On the other hand, both cross-additions take place in the styrene–butadiene system.

The addition of numerous anions to four typical monomers—acrylonitrile, methyl methacrylate, styrene, and butadiene—was examined by Wooding and Higginson (2), who concluded that such a process may be compared with a general acid–base reaction: the more basic the anion the more active it is (see Table I). However, correlation of the basicity of an ion with its ability to add a monomer is not always justified (3). The acid–base reaction refers to a proton transfer process, $A{-}H + B^- \rightleftarrows A^- + H{-}B$; its driving force is governed by the bond dissociation energies of $A{-}H$ and $B{-}H$, by the electron affinities of the $A\cdot$ and $B\cdot$ radicals, and by the relevant solvation energies. In the monomer addition reaction a bond $\sim\!\!\!\sim M_1{-}M_2^-$ is formed and its dissociation energy is influenced, e.g., by the energy gained in the formation of a $C{=}C$ double bond. Thus, not all of the factors that govern one process need be relevant for the other.

Let us consider now a series of additions, $\sim\!\!\!\sim A^- + M_i \rightarrow \sim\!\!\!\sim A{-}M_i^-$, for a set of monomers $M_1, M_2 \cdots M_i$ reacting with a fixed anion $\sim\!\!\!\sim A^-$. The monomers may be classified as being more or less "acidic," the readiness of the addition reaction serving as a useful criterion of acidity. Such classification may be valid for a series of closely related monomers. It becomes ambiguous when their structures diverge too much because the order of "acidities" of monomers of greatly different structures may depend on the choice of the anion.

Finally, it should be stressed that the anionic addition may involve a free

TABLE I

Addition of Anions to Four Typical Monomers (2)

Anion derived from acid	pK in ether	Acrylonitrile	Methyl methacrylate	Styrene	Butadiene
In ether solution at 20°C					
Methanol	17	+	−	−	−
Ethanol	18	+	−	−	−
Acetophenone	19	+	−	−	−
Triphenylcarbinol	19	+	−	−	−
Indene	21	+	+	−	−
Phenylacetylene	21	+	+	−	−
Diphenylamine	23	+	+	−	−
Fluorene	25	+	+	−	−
Aniline	27	+	−	−	−
p-Methoxyaniline	27	+	−	−	−
Xanthene	29	+	+	+	+
Triphenylmethane	33	+	+	+	+
In liquid ammonia solution at −33.5°C					
Ethanol	18	+	−		
Acetophenone	19	+	−		
Triphenylcarbinol	19	+	−		
Indene	21	+	−		
Phenylacetylene	21	+	−	−	
Diphenylamine	23	+	+	−	
Fluorene	25	+	−	−	
Acetylene	26	+	+	−	
Aniline	27	+	+	+	
Xanthene	29	+	+	+	
Triphenylmethane	33	+	+	+	
Ammonia	36	+	+	+	−

⟿A⁻ ion or its ion-pair and different orders of "acidities" may be found for these two species. Moreover, the nature of the counterion may profoundly affect the relative reactivities of ion-pairs. The solvent also plays an important role in the reaction. For instance, it affects the equilibria between free ions and ion pairs, it modifies the structure of ion-pairs, etc.

Early investigations of anionic copolymerization led to some confusion and errors; the system styrene–methyl methacrylate provides an example. Walling et al. (4) reported that the copolymerization of an equimolar mixture of styrene and methyl methacrylate yields a product containing only 1% of styrene if the reaction is initiated by metallic sodium or potassium. Landler (5) studied the same polymerization in liquid ammonia and concluded that the respective reactivity ratios are 0.123 and 6.4, i.e., the copolymers produced in his experiments contained a relatively high proportion of styrene. However, subsequent investigations of Graham et al. (1) conclusively show that styrene behaves like an inert solvent in the anionic polymerization of methyl methacrylate. For example, a mixture of these two monomers polymerized with sodium or lithium fluorenyl produces pure polymethyl methacrylate, even in the presence of a large excess of styrene. The fluorenyl salts react with methyl methacrylate only, and not with styrene; hence, the latter monomer is not involved in the initiation. On the other hand, a small amount of styrene is incorporated into the copolymer if butyllithium is employed as the initiator. This finding is significant; it indicates that errors may arise in the interpretation of results if initiation is confused with propagation. Indeed, in contradistinction to sodium or lithium fluorenyl, butyllithium reacts with *both* monomers.

In many studies the reactivity ratios of copolymerization were obtained by the conventional method, viz., by comparing the composition of the copolymer with that of the feed. Such technique is acceptable if the following conditions are fulfilled:

(1) The resulting copolymer should have a high molecular weight. The composition of the first few units may substantially differ from that characterizing the remaining portion of the chain, and this would distort the experimental findings if a material of low molecular weight is examined. For example, Landler's results probably were affected by such a distortion of the copolymer composition.

(2) The effects caused by initiation must be clearly distinguished from those arising from propagation. The resulting errors are again large for low molecular weight copolymers.

Both these remarks pertain, of course, to any copolymerization.

In some ionic "copolymerizations" the product results from two homopropagations, each proceeding independently of the other. The cationic polymerization of vinylcarbazole–oxetane mixture provides a recent example of such a system (6). In terms of the conventional scheme of copolymerization (7) this situation arises whenever k_{11} and k_{22} are relatively large but k_{12} and k_{21} vanish. The common initiator may induce both polymerizations and, in the simplest case, the relative rates of initiation are proportional to the ratio of the monomer concentrations. For living polymers the composition of the mixture, which could be mistaken for a copolymer, then becomes proportional to $(M_1/M_2)^2$, where M_1 and M_2 denote the concentrations of the respective monomers in the feed. Such a situation was discussed by O'Driscoll (8).

For polymerizing systems involving termination, the composition of the product depends on the kinetics of this step. For example, if termination is caused by a "wrong monomer addition" (see Chapter XII), its rate is proportional to the concentration of the relevant monomer. Therefore, the composition of the "copolymer" formed in such a system is proportional to the ratio (M_1/M_2). On the other hand, had a reaction of monomer M_2 with a growing M_1-chain induced its termination, and vice versa, the composition of the resulting mixture of polymers would be proportional then to $(M_1/M_2)^3$.

Obviously, the conventional scheme of copolymerization cannot account for the composition of a polymeric blend produced by two independent homopropagations, and the results obtained from such experiments therefore cannot be used to derive the reactivity ratios.

In heterogeneous systems, further complications may be encountered. For example, Overberger and Yamamoto (9) studied the polymerization of mixtures of styrene and methyl methacrylate initiated by an emulsion of metallic lithium. They postulated a preferential adsorption of styrene on the surface of lithium particles which results in homopolymerization of this monomer. The living polystyrene eventually becomes detached from the surface, diffuses into the bulk of the solution, and then reacts with methyl methacrylate. Thus a block polymer is formed. Such a product was indeed isolated and identified and, of course, its composition was not given by the copolymerization scheme. A somewhat similar suggestion was made by Korotkov et al. (10), who investigated the polymerization of the same pair of monomers initiated by solid methyllithium.

An alternative explanation for this phenomenon was proposed by Laita and Szwarc (11). They visualized the polymerization as occurring in the pores of the heterogeneous catalyst. Under these conditions, all the methylmethacrylate present in a pore is rapidly consumed by the large local excess of catalyst. Because the lithium compound reacts extremely fast with acrylates, only low molecular weight products are formed under these conditions. Styrene reacts slowly with the initiator and, therefore, its polymerization to a high molecular weight product may take place after all the methyl methacrylate has been consumed. Eventually, the living polystyrene formed in the pore diffuses out into the bulk of the liquid and then it forms a block polymer.

In mixed solvents a partial demixing may occur in the vicinity of ions and ion-pairs. Thus, ionic polymerization taking place in a mixture of two monomers may proceed in a local environment surrounding the growing centers which differs in its composition from that of the bulk liquid. The composition of the resulting copolymer deviates, then, from that anticipated on the basis of a conventional copolymerization scheme, even if the classic mechanism rigorously applies to the investigated system. This explanation was invoked by Korotkov et al. (12), who discussed the peculiarity of the styrene–butadiene polymerization initiated by butyllithium (see p. 528).

2. KINETICS OF ANIONIC COPOLYMERIZATION

The classic treatment of radical copolymerization (7) considers two types of growing polymers, one terminated by radical $\sim\!\!M_1^\cdot$, the other by $\sim\!\!M_2^\cdot$. The monomers, M_1 and M_2, then compete for the two growing species and the composition of the product is determined by the two reactivity ratios r_1 and r_2. Anionic systems are more complex. In those media in which association or dissociation of ion-pairs does not take place, only two growing species, $\sim\!\!M_1^-,X^+$ and $\sim\!\!M_2^-,X^+$, participate in the propagation. In such a case the classic scheme of copolymerization may still apply. A similar simplicity is expected in hypothetical systems in which all the growing polymers are virtually dissociated into free ions. However, when the ion-pairs become associated or cross-associated, forming inactive (dormant) agglomerates, the situation becomes more complex. The composition of the resulting copolymer is still determined by the two reactivity ratios, provided the rates of association–dissociation are fast in respect to the rate of polymerization. In such systems only the two types of the nonassociated ion-pairs propagate, and their stationary concentration is determined by the conventional equation, $k_{12}P_1M_2 = k_{21}P_2M_1$ (where P represents the actively growing polymer). The stationary state is not perturbed by the agglomeration if the equilibrium is rapidly established. However, the kinetics of the reaction is affected by the formation of the dormant polymers which grow eventually.

The same approach is valid for systems composed of ion-pairs, which are unreactive, and of free ions which are the only species participating in the polymerization. The concentration of the free ions, and therefore the rate of polymerization, depends on the dissociation constants of the respective ion-pairs:

$$\sim\!\!M_1^-,X^+ \;\rightleftarrows\; \sim\!\!M_1^- + X^+, \qquad K_{diss,1},$$

and

$$\sim\!\!M_2^-,X^+ \;\rightleftarrows\; \sim\!\!M_2^- + X^+, \qquad K_{diss,2}.$$

These equilibria are coupled by the presence of a common counterion, X^+, and this complicates the mathematical treatment of the kinetics of polymerization. Nevertheless, the composition of the resulting product is again determined only by the two reactivity ratios, r_1 and r_2, provided the dissociation of ion-pairs and the combination of the free ions with the counterions are much faster than polymerization.

The classic approach fails when ion-pairs and free ions significantly contribute to polymerization. In such systems the propagation involves four growing species, $\sim\!\!M_1^-,X^+$, $\sim\!\!M_1^-$, $\sim\!\!M_2^-,X^+$, and $\sim\!\!M_2^-$, and the reaction is described by eight rate constants, $k_{11\pm}$, k_{11-}, $k_{12\pm}$, k_{12-}, etc. Whenever the dissociation–association processes are fast, the composition of the growing species is determined by the modified condition of copolymerization:

$$\{k_{12\,\pm}\,[\sim\!M_1^-,X^+] + k_{12\,-}\,[\sim\!M_1^-]\}[M_2]$$
$$= \{k_{21\,\pm}\,[\sim\!M_2^-,X^+] + k_{21\,-}\,[\sim\!M_2^-]\}[M_1].$$

This equation, in conjunction with the two equilibria,

$$[\sim\!M_1^-]/[\sim\!M_1^-,X^+] = K_{\text{diss},1}[X^+],$$
$$[\sim\!M_2^-]/[\sim\!M_2^-,X^+] = K_{\text{diss},2}[X^+],$$

and the stoichiometric conditions

$$[\sim\!M_1^-] + [\sim\!M_2^-] = [X^+]$$

and

$$[\sim\!M_1^-] + [\sim\!M_1^-,X^+] + [\sim\!M_2^-] + [\sim\!M_2^-,X^+] = [LP],$$

uniquely determine the mole fractions f_1 and f_2 of polymers 1 and 2, respectively, and the fractions of dissociation, α and β, of each polymer. In the absence of termination, the concentration of all living polymers [LP] remains constant during the whole course of copolymerization.

For α and β much smaller than 1, the pertinent equation may be simplified:

$$f_1\{k_{12\,\pm} + (k_{12\,-} - k_{12\,\pm})\alpha\}[M_2] = f_2\{k_{21\,\pm} + (k_{21\,-} - k_{21\,\pm})\beta\}[M_1],$$

$$K_{\text{diss},1}[LP](f_1\alpha + f_2\beta) = \alpha, \qquad K_{\text{diss},2}[LP](f_1\alpha + f_2\beta) = \beta,$$

and these, in conjunction with the condition $f_1 + f_2 = 1$, uniquely determine the variables f_1, f_2, α, and β for any $[M_1]$, $[M_2]$, and [LP]. However, α and β still depend on $[M_1]$ and $[M_2]$ at constant [LP] and hence the "observed" cross-propagation constants,

$$k_{12,\text{obs}} = \{(1 - \alpha)k_{12\,\pm} + \alpha k_{12\,-}\}$$

and

$$k_{21,\text{obs}} = \{(1 - \beta)k_{21\,\pm} + \beta k_{21\,-}\},$$

vary with the composition of feed at constant [LP], and with [LP] at constant composition of the feed. Obviously, in such systems the reactivity ratios, r_1 and r_2, are not constant, e.g.,

$$r_1 = \{(1 - \alpha)k_{11\,\pm} + \alpha k_{11\,-}\}/\{(1 - \alpha)k_{12\,\pm} + \alpha k_{12\,-}\}.$$

They depend on α and β, respectively, and their values are affected, therefore, by the composition of the feed and the concentration of the initiator. Only if a coincidence makes the ratios $k_{11\,\pm}/k_{12\,\pm} = k_{11\,-}/k_{12\,-}$ and $k_{22\,\pm}/k_{21\,\pm} = k_{22\,-}/k_{21\,-}$, do r_1 and r_2 become constant and independent of the composition of the feed and the concentration of the initiator.

In conclusion it must be re-emphasized that the calculation of the reactivity ratios is reliable if the polymerizing systems involve only two types of growing species. The concept is still valid even when the growing species are in equilibrium with some dormant polymers, provided the equilibria are rapidly established. On the other hand, the stationary concentration of the reacting end groups is given by the equation,

$$k_{12}[\sim\!M_1][M_2] + k'_{diss}[dormant\text{-}2] - k'_{assoc}[\sim\!M_2]^n$$
$$= k_{21}[\sim\!M_2][M_1] + k''_{diss}[dormant\text{-}1] - k''_{assoc}[\sim\!M_1]^m$$

whenever the equilibrium is established slowly. The dormant species are assumed to form n- and m-meric agglomerates, respectively, which directly dissociate into monomeric species. The equations become even more complex if the kinetics of the dissociation–association processes are more complex or when cross-association takes place. Similar complications arise in those polymerizations in which initiation takes place simultaneously with propagation.

3. CLASSIC STUDIES OF ANIONIC COPOLYMERIZATION

In view of the preceding discussion it is obvious that the classic approach to anionic copolymerization is justified only for a few systems. Further difficulties in its application are caused by some characteristic features of living polymers. In radical polymerization the molecular weight of the product is virtually independent of conversion. It is possible, therefore, to choose the conditions which permit us to obtain a high molecular weight product at low conversion. These conditions are ideal for determining r_1 and r_2 from the composition of the polymer and of the feed. In systems involving living polymers, low conversion is often associated with low degree of polymerization and, hence, the experimentation is more difficult and the outcome could be unreliable.

Interesting results, obtained by classic approach, were reported by Tobolsky and his associates (17,23), who investigated anionic copolymerizations of styrene with its para-substituted derivatives (p-methyl- and p-methoxy). The reaction was initiated by butyllithium in tetrahydrofuran at 0°C. The reactivity ratios were determined for three pairs of monomers, styrene–p-methylstyrene, styrene–p-methoxystyrene, and p-methylstyrene–p-methoxystyrene, For styrene–p-methylstyrene the constants are $r_1 = 2.0$ and $r_2 = 0.4$. The results reported demonstrate the nucleophilic character of the reaction, i.e., the reactivities of the monomers toward lithium polystyryl decrease along the series styrene > p-methylstyrene > p-methoxystyrene. The products $r_1 r_2 = 1$. The system styrene–p-methylstyrene was investigated in benzene by O'Driscoll and Patsiga (18), who arrived at closely similar values for the respective reactivity ratios, viz., $r_1 = 2.5$ and $r_2 = 0.4$. Since free ions are responsible for most of the polymerization taking place in tetrahydrofuran, while ion-pairs (probably of contact type) form the growing species present in benzene, one might conclude that the relative reactivities of free ions and ion-pairs toward styrene and p-methylstyrene are closely similar. This conclusion should be checked and it would be desirable to extend these observations to other pairs of similar monomers.

The presence of polar solvents affects the reactivity ratios making styrene relatively more reactive than its methyl or methoxy derivative (23). These observations were confirmed by Dawans and Smets (24), who initiated the copolymerization with phenylmagnesium bromide, and by Overberger et al. (25), who employed the butyllithium as the initiator.

Much work was done on the system styrene–dienes. Kelley and Tobolsky (14) investigated the copolymerization of isoprene with styrene; the reaction was initiated by metallic sodium, by lithium, and by butyllithium in polar and in nonpolar solvents. Their results, which are collected in Table II,

TABLE II

Per Cent of Styrene in Isoprene–Styrene Copolymer Initiated by Metallic Sodium, Lithium, and Butyllithium (14)
(results are extrapolated to 0% conversion)

Solvent	Li	BuLi	Na
Benzene	15	18	66
Bulk	15	17	66
Et$_3$N	59	60	77
Et$_2$O	68	68	75
THF	80	80	80

demonstrate the large effect on the composition of the copolymer exerted by the counterion and solvent. The copolymer is rich in styrene, and has virtually the same composition whether lithium or sodium forms the counterion, provided the reaction takes place in polar solvent. On the other hand, lithium yields a polymer rich in isoprene, while sodium still produces a copolymer rich in styrene, when the reaction occurs in hydrocarbons. Kelley and Tobolsky conclude that the composition of the copolymer is determined by the degree of ionic character of the carbon–metal bond, a hypothesis further elaborated by Tobolsky and Rogers (15). In fact, the composition of the copolymer was correlated with the proportion of the cis-form found in homopolyisoprene produced by the same catalyst system (15)—an increase in the ionic character of the bond supposedly decreases the relative reactivity of isoprene and the amount of the *cis*-form in homopolyisoprene.

Although such a correlation is indisputable, the proposed explanation may be questioned. Examination of the UV spectra of living polystyrenes associated with various counterions and dissolved in polar and in nonpolar solvents shows that the C—Li or C—Na bonds are always ionic if the relevant carbanions are conjugated with groups such as phenyl, C=C double bonds, or some similar unsaturated entities (26). Hence, an alternative explanation for the strange behavior of lithiated polymers is needed.

The copolymerization involving Li^+ counterion and proceeding in benzene is strongly influenced by a small amount of polar substances added to the reacting mixture. Addition of 4% of diethyl ether increases the styrene content in the copolymer almost to the value observed in the pure ether. Tetrahydrofuran is even more efficient. The copolymer composition is reported to reach the limiting value, characteristic of pure tetrahydrofuran, when only one molecule of the ether is available for each living polymer (19). It appears that the composition of the copolymer is determined by the nature of solvation shell formed around the growing ion-pair and is not greatly dependent on the bulk properties of the medium.

The system styrene–butadiene polymerized by lithium alkyls in hydrocarbon solvents attracted much attention. The most interesting observations were reported by Korotkov and his colleagues (12). Under the above conditions the homopolymerization of styrene is faster than that of butadiene if the reaction is initiated by butyllithium in benzene. Nevertheless, in an equimolar mixture of both monomers the reaction starts slowly and only butadiene polymerizes in the early stages of the process. Thereafter, when most of the butadiene has been consumed, the reaction speeds up and a substantial portion of styrene then appears in the product. According to Korotkov's interpretation, the growing ion-pairs are preferentially solvated by butadiene and hence only this monomer is involved in the initial polymerization.

Surely, the ability of molecules to solvate ions or ion-pairs is not related to their rate of polymerization—in fact, a good solvating agent need not be a monomer. However, had a monomer been a solvating agent, it may have polymerized preferentially by virtue of its high local "concentration" in the vicinity of the growing centers. One may argue, therefore, that butadiene is a better solvating agent than styrene; however, being a more sluggish monomer it would lead to an initially slow polymerization. After most of the butadiene has been consumed, styrene may reach the growing centers and then the reaction accelerates because the latter monomer is more reactive than the former.

The copolymerization of styrene with isoprene resembles that of styrene with butadiene. According to Medvedev and his associates (21), who initiated the reaction with ethyllithium in toluene, the polymerization is suddenly enhanced when the diene is virtually consumed. This observation was confirmed by Bawn (22), who noted also an abrupt change in the color of solution at this stage of the reaction which coincided with the sudden acceleration of polymerization. Obviously, as long as some isoprene was present the solution contained only lithium polyisoprenyl end groups, with virtual exclusion of the polystyryl groups. This imposed the pale yellow color of $\sim$(isoprenyl$^-$,Li$^+$) upon the reacting mixture. After exhaustion of the diene, the red color of $\sim$(styryl$^-$,Li$^+$) appeared as these end groups became abundant. The same observation was subsequently reported by Worsfold (27) who investigated the system by spectrophotometric methods.

A dramatic change in the behavior of these systems is observed when the reaction is carried out in tetrahydrofuran at $-78°C$ using Cs$^+$ as the counterion. Under these conditions the red color of $\sim$styryl$^-$,Cs$^+$ is noted first and eventually it changes into the pale yellow of the dienyl anion (33). Here is good evidence that the composition of copolymer may be affected by the nature of the counterion.

Although the reported facts were fully confirmed by the observations of Kuntz (19), a new interpretation of the results was proposed by O'Driscoll and Kuntz (13). In view of the findings of Worsfold and Bywater (16) it was assumed that virtually all the lithium polystyryls are present in dimeric, inactive form $(P_1)_2$ and only a minute fraction is actively growing. The same was arbitrarily assumed for lithium polybutadienyl, P_2. Thus,

$$[P_1] = \{K_1[(P_1)_2]\}^{1/2} \quad \text{and} \quad [P_2] = \{K_2[(P_2)_2]\}^{1/2},$$

and the reaction is then described in terms of differential equations,

$$-d[M_1]/dt = k'_{11}[M_1][(P_1)_2]^{1/2} + k'_{21}[M_1][(P_2)_2]^{1/2},$$
$$-d[M_2]/dt = k'_{12}[M_2][(P_1)_2]^{1/2} + k'_{22}[M_2][(P_2)_2]^{1/2},$$

and

$$k'_{21}[M_1][(P_2)_2]^{1/2} = k'_{12}[M_2][(P_1)_2]^{1/2},$$

with the obvious condition

$$2[(P_1)_2] + 2[(P_2)_2] = [LP] = \text{const.}$$

Superficially, the differential equations of O'Driscoll and Kuntz appear to be different from the conventional equations of the classic treatment of copolymerization; in fact, they are identical. The unusual results reported by Korotkov et al. (12) and by Kuntz (19) were subsequently rationalized by assuming $k'_{12} \gg k'_{11}$ and $k'_{22} \gg k'_{21}$. Although the constants k'_{11}, k'_{12}, k'_{21}, and k'_{22} are complex, e.g.,

$$k'_{11} = k_{11}K_1^{1/2} \quad \text{and} \quad k'_{12} = k_{12}K_1^{1/2},$$

the reactivity ratios r_1 and r_2 have their usual meaning:

$$r_1 = k'_{11}/k'_{12} = k_{11}/k_{12} \quad \text{and} \quad r_2 = k'_{22}/k'_{21} = k_{22}/k_{21}.$$

Only these two constants appear in the final results, which give the composition of the polymer as a function of conversion. Judicious choice of the r_1 and r_2 values led to the results shown in Figure 1, in which a fair agreement is revealed between calculations and observations.

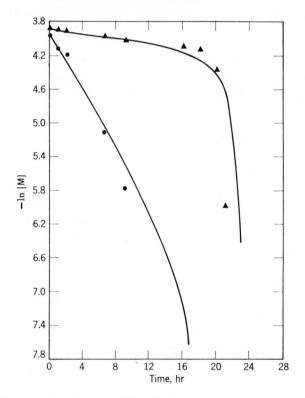

Fig. 1.   Consumption of styrene (▲) and of butadiene (●) in copolymerization of styrene–butadiene initiated by $0.017M$ butyllithium in heptane at 30°C (13). The solid curves are calculated on the basis of O'Driscoll and Kuntz's mechanism.

The cross-association has not been considered. This additional equilibrium,

$$(P_1)_2 + (P_2)_2 \rightleftarrows 2(P_1P_2), \qquad K_c,$$

does not affect the differential equations; it modifies only the stoichiometric condition, viz., the latter acquires the form

$$2[(P_1)_2] + 2[(P_2)_2] + 2[(P_1P_2)] = [LP] = \text{const.},$$

subject to the condition $[(P_1P_2)]^2 = K_c[(P_1)_2][(P_2)_2]$.

   In effect, this phenomenon does not change the main features of the process but increases only the fraction of polymers stored in the dormant form.

Consequently, although the rate of copolymerization is slowed down, the functional dependence of the composition of the remaining feed, or of the polymer formed, on the fraction of conversion is unaltered.

The success of O'Driscoll and Kuntz's explanation does not finally disprove Korotkov's idea which, in principle, is sound. In fact, the concept of preferential solvation was invoked by many workers (25,28–31), although the evidence for it should be treated cautiously. We therefore need some criteria that conclusively differentiate between this hypothesis and the explanation proposed by O'Driscoll and Kuntz. Two types of experiments are significant. (1) The preferential solvation may lead to a saturation phenomenon, i.e., at sufficiently high concentration of the monomer its rate of addition may reach a limiting value. Further increase in the concentration of that monomer, which supposedly accumulates around the growing ends, should no longer affect the rate. (2) An even more conclusive test is based on studies of the absolute rate of the reaction $\sim\sim M_1^-,X^+ + M_2 \rightarrow \sim\sim M_1 M_2^-,X^+$. For constant concentrations of $\sim\sim M_1^-,X^+$ and M_2 the observed rate should be higher in the *absence* of M_1 than in its *presence*, if the solvation hypothesis is valid. These tests were performed for the systems styrene–butadiene and styrene–isoprene in cyclohexane (see Section 5 below) and both studies showed that the preferential solvation (or the preferential pre-equilibrium) is not responsible for the observed results.

The treatment of O'Driscoll and Kuntz provides a formal account of Korotkov's observations but it raises the question as to why $k'_{12} \gg k'_{11}$ (i.e., $k_{12} \gg k_{11}$) and $k'_{22} \gg k'_{21}$ (i.e., $k_{22} \gg k_{21}$). The physical phenomena which are responsible for this behavior deserve further studies. It is probable that the addition of diene to a nonsolvated lithium ion facilitates the reaction. This effect is lost when the Li^+ becomes solvated by ether or amine. Apparently the larger ions, e.g., Na^+ or Cs^+, do not favor the formation of such a "π-complex" and, therefore, show no preference for the dienes.

The peculiarities revealed by Korotkov's studies are not restricted to anionic polymerization. It has been pointed out (20) that a similar situation is encountered in the radical-copolymerization of styrene and vinyl acetate.

The case of styrene–butadiene copolymerization brings to focus a feature of anionic copolymerization that complicates the classic method of determination of r_1 and r_2 based on composition studies. In anionic, or more generally in ionic, polymerizations the reactivities of monomers toward either type of growing chains differ to such an extent that the reaction tends to yield a block polymer rather than a proper copolymer. Although the large difference in reactivities of two monomers does not violate the principles of copolymerization, the analytical techniques and proper sampling becomes extremely difficult and consequently the calculated values of r_1 and r_2 are often questionable.

Anionic copolymerizations of several monomers other than dienes were studied by classic methods. The system methacrylonitrile–methyl methacrylate was investigated by Foster (35). Both anions react faster with the first monomer, a fact reflecting the higher electronegativity of —CN than of —CO·OCH$_3$. Therefore, the anionic copolymerization of this pair yields a product resembling a block polymer, while in radical polymerization these monomers tend to alternate. An even greater difference in reactivities was found for the pair acrylonitrile–methyl methacrylate. In fact, the latter monomer is hardly copolymerized in the presence of the former (36). Vinylbutyl sulfone was found to be more reactive than methacrylonitrile but less than acrylonitrile (36). The following series of reactivities was therefore proposed:

acrylonitrile > vinylbutyl sulfone > methacrylonitrile > methyl methacrylate > styrene.

The system acrylonitrile–styrene was investigated in isooctane and in ether at $-12°C$ (37). The reaction was initiated by butyllithium and r_1 and r_2 were reported to be 14 and 0.2, respectively. These values seem to be questionable.

4. ABSOLUTE RATE CONSTANTS OF COPOLYMERIZATION

The classic studies of copolymerization yield only the reactivity ratios, r_1 and r_2. Utilization of living polymers permits direct determination of the absolute rate constants k_{11}, k_{12}, k_{21}, and k_{22}. Thus, a far more penetrating attack on the problem of copolymerization becomes feasible.

Two experimental methods are available for such studies:

(1) If the optical spectrum of anion $\sim\sim M_1^-$ differs from that of $\sim\sim M_2^-$, the reaction may be followed spectrophotometrically. Thus, living polymer $\sim\sim M_1^-,X^+$ is mixed with monomer M_2 and the disappearance of the $\sim\sim M_1^-,X^+$ ions, or the appearance of the $\sim\sim M_2^-,X^+$, may be observed. Ordinary mixing techniques suffice if the reaction is slow (38); a stop-flow technique must be applied if the cross-over is fast.

(2) The above technique cannot be applied if the spectra of the relevant ions are similar. An alternative, and more general approach is then useful. Excess of living polymer $\sim\sim M^-,X^+$ is mixed with monomer M_2, and the reaction is quenched at various times. The concentration of the residual M_2 monomer is then determined. Thus, the initial rate of the reaction,

$$\sim\sim M_1^-,X^+ + M_2 \xrightarrow{k_{12}} \sim\sim M_1 M_2^-,X^+,$$

is calculated and since $[\sim\sim M_1^-,X^+]_0$ and $[M_2]_0$ are known the computation of k_{12} is feasible.

Several technical difficulties may be encountered in such studies. The reaction may be very fast and can be over in less than a second. In such a case a capillary flow technique is most useful (see pp. 188 and 191). One should also consider more carefully the overall process. The cross-propagation, determined by the rate constant k_{12}, is followed by the homopropagation

$$\sim\!M_1M_2^-,X^+ + M_2 \xrightarrow{k_{22}} \sim\!M_1M_2M_2^-,X^+, \text{ etc.}$$

Hence, the consumption of the monomer is given by the differential equation (39,40):

$$-d[M_2]/dt = k_{12}[\sim\!M_1^-,X^+][M_2] + k_{22}([\sim\!M_1^-,X^+]_0 - [\sim\!M_1^-,X^+])[M_2].$$

A general solution of this equation is complicated; however, a few special cases may be considered:

(A) $k_{12} \approx k_{22}$. Then

$$-d \ln [M_2]/dt = k_{12}\{[\sim\!M_1^-,X^+]_0 - \alpha([\sim\!M_1^-,X^+]_0 - [\sim\!M_1^-X^+])\},$$

where $\alpha = k_{22}/k_{12}$. For $\alpha \ll 1$, and particularly at the early stage of the reaction when $([\sim\!M_1^-,X^+]_0 - [\sim\!M_1^-,X^+])/[\sim\!M_1^-,X^+]_0 \ll 1$, the right-hand bracket is virtually constant and equal to $[\sim\!M_1^-,X^+]_0$. Hence, the plot of ln $[M_2]_0/[M_2]$ versus time gives a straight line through the origin, the slope of which is equal to $k_{12}[\sim\!M_1^-,X^+]_0$. In practice ($\alpha \neq 0$) the line is slightly curved, but it is easy to determine its initial slope, which gives k_{12}.

(B) $k_{12} \gg k_{22}$. In this case, the homopropagation may be virtually neglected and the reaction behaves as if it were of second order, i.e.,

$$-d[M_2]/dt \approx k_{12}\{[\sim\!M_1^-,X^+]_0 - ([M_2]_0 - [M_2])\}[M_2].$$

The integrated result leads to a linear plot of

$$(a - b)^{-1} \ln \{b(a - x)/a(b - x)\} \text{ versus time,}$$

the resulting straight line passing through the origin with slope of k_{12}. In this equation a and b denote the initial concentrations of living polymers ($\sim\!M_1^-,X^+$) and of the monomer (M_2), respectively, while x represents the amount of polymerized M_2 per unit volume of the solution. For $k_{22} \neq 0$, but small, such a line becomes slightly curved; nevertheless, its initial slope is again easily determined and it gives k_{12}.

(C) $k_{12} \ll k_{22}$. Such a reaction exhibits an autocatalytic behavior and at first approximation the amount of monomer polymerized in the early stages of the process increases proportionally with square of time of reaction, the proportionality constant being related to $k_{12}k_{22}$. Thus, k_{12} may be calculated if k_{22} is known. The determination of the absolute rate constant of anionic homopropagation is relatively straightforward; this problem has been discussed in Chapter VII and therefore k_{12} may be calculated from the available data.

To determine which of the first two approximations fits a particular system better, the results are plotted twice in accordance with each kinetic law. Obviously the "true" value of k_{12} must be larger than or equal to that derived from the first-order plot and smaller than that obtained from the second-order plot. These two limits converge as the fraction of conversion decreases. Moreover, the approximation which represents better the system under investigation gives a plot that deviates less from linearity than the other one. Suffice it to say that for many systems good linear plots are obtained by employing either the first-order or the second-order treatment.

In the above discussion we postulated the existence of two centers only, $\sim M_1^-,X^+$ and $\sim M_1 M_2^-,X^+$ (or $\sim M_1 M_2 \cdots M_2^- X^+$). The problem has to be modified if more than two types of living polymers coexist in the reacting mixture. For the sake of illustration let us consider the system

$$\sim M_1^-,X^+ \xrightleftharpoons{K_1} \sim M_1^- + X^+$$

and

$$\sim M_1 M_2^-,X^+ \xrightleftharpoons{K_2} \sim M_1 M_2^- + X^+$$

in which the monomers are mainly polymerized by the free ions. The reaction

$$\sim M_1^-,X^+ + M_2 \rightarrow \sim M_1 M_2^-,X^+$$

decreases the concentration of $\sim M_1^-,X^+$ ion-pairs and therefore increases their degree of dissociation provided $K_1 \geqslant K_2$. Hence, the apparent rate constant k_{12} increases with conversion and the previously discussed plots (of the first or of second order—depending on the system) become concave. Indeed, some of the plots produced in this writer's laboratory revealed such a curvature. A convex curve may be expected if $K_2 \gg K_1$.

5. SPECTROPHOTOMETRIC DETERMINATION OF CROSS-PROPAGATION CONSTANTS

The addition of styrene to lithium polybutadienyl in benzene was investigated by Morton and Ells (38). Butyllithium was employed in the preparation of living polybutadiene and eventually styrene was added to the resulting polymer. The reaction was followed spectrophotometrically at 436 mμ. This choice of wavelength cannot be recommended because the λ_{max} of living polystyrene (Li$^+$) is at 335 mμ. Plots of $\log\{(OD)_t/(OD)_0\}$ versus time were linear, as shown in Figure 2. The apparent bimolecular rate constants, $k''_{B,S}$ were calculated from their slopes, and these are listed in Table III. Their values vary linearly with the square root of the concentration of lithium polybutadienyl, as shown in Figure 3. This implies that polybutadienyl dimerizes in benzene and only a minute fraction of the nonassociated chains

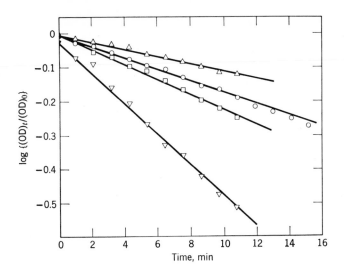

Fig. 2. First-order plot for reaction of styrene with lithium polybutadienyl in benzene at 29°C (38):

Point	[Styrene], M	[Lithium polybutadienyl], $M \times 10^3$
△	0.13	1.69
○	0.14	1.07
□	0.11	0.63
▽	0.26	0.60

TABLE III

Addition of Styrene to Polybutadienyllithium in Benzene at 29°C (38)

$[\sim\text{Bu}^-,\text{Li}^+]$, $M \times 10^3$	[Styrene], M	$k''_{B,S} \times 10^3$, liter mole^{-1} sec^{-1}
0.53	0.13	7.3
0.60	0.26	6.7
0.63	0.11	7.5
1.07	0.14	4.6
1.69	0.13	3.05

The rate constants listed were calculated on the assumption that rate of cross-propagation is proportional to concentration of lithium polybutadienyl. This polymer is associated in benzene and, if the nonassociated species only participate in the reaction, the rate should be proportional to $[\sim\text{Bu}^-,\text{Li}^+]^{1/n}$, where n is the degree of association of living polymer.

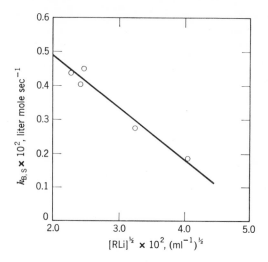

Fig. 3.   Linear dependence of cross-propagation constant, $\sim$butadiene$^-$, Li$^+$— styrene, on square root of concentration of living polymer in benzene at 29°C (38).

participate in the reaction. The association should cause some curvature of the plots shown in Figure 2, but this is not evident from the data.

Morton and Ells also investigated the addition of butadiene to lithium polystyryl in benzene. They found this reaction to be extremely fast, in fact too rapid to permit the use of the conventional mixing technique. Hence, the apparent bimolecular rate constants were only estimated, viz., $k''_{S,B} \approx 1$ liter mole^{-1} sec^{-1} at [$\sim$S$^-$,Li$^+$] $\approx 10^{-3}M$.

These findings are important because they explain the peculiar behavior of the styrene–butadiene system and confirm the scheme proposed by O'Driscoll and Kuntz (13).

Recently the addition of butadiene to lithium polystyryl was reinvestigated by Harris (41), who employed a stop-flow technique. The rate of addition was found to be 100 times faster than reported by the earlier workers. The apparent bimolecular rate constants showed a strictly linear dependence on the square root of [$\sim$S$^-$,Li$^+$], even when the half-lifetime of the initial reaction was as short as 0.1 sec. It was concluded, therefore, that the unimolecular rate constant of the dissociation, ($\sim$S$^-$,Li$^+$)$_2 \rightleftarrows 2\sim$S$^-$,Li$^+$, taking place in benzene at 25°C is larger than 10 sec^{-1}.

Extensive studies of the copolymerization of styrene and butadiene in cyclohexane were reported by Johnson and Worsfold (32). In this medium the addition of butadiene to living polystyrene seems to be instantaneous. The differential equations of copolymerization are reduced, therefore, to

$$-d[Bu]/dt = k_{22}\{[\sim\!Bu^-,Li^+] + k_{12}[\sim\!S^-,Li^+]\}[Bu]$$
$$= k_{22}[\sim\!Bu^-,Li^+][Bu] + k_{21}[\sim\!Bu^-,Li^+][S]$$

and

$$-d[S]/dt = k_{21}[\sim\!Bu^-,Li^+][S],$$

i.e.,

$$-d[monomer]/dt = k_{22}[\sim\!Bu^-,Li^+][Bu] + 2k_{21}[\sim\!Bu^-,Li^+][S].$$

Hence, the overall rate of copolymerization is expected to be only slightly higher than that of homopropagation because $k_{21}/k_{22} \ll 1$. The latter ratio is obtained directly from the composition of the copolymer because

$$\{[Bu]/[S]\}_{in\ copolymer} = (k_{22}/k_{21})\{[Bu]/[S]\}_{in\ feed} + 1.$$

The agreement between k_{22}/k_{21} determined in this way and the value derived from direct spectrophotometric studies proves that the presence of butadiene does not affect the rate constant of styrene addition to living lithium polybutadienyl. This conclusion was confirmed by spectrophotometric studies of styrene addition in the presence and in the absence of butadiene. The final results led to $r_1 < 0.04$ and $r_2 = 26$, both determined at 40°C.

Investigations of the system styrene–isoprene initiated by *sec*-butyllithium in cyclohexane were reported by Worsfold (27). Since the secondary lithium compound reacts faster than *n*-butyllithium, the initiation had been completed at the time when the propagation was investigated. Both cross-propagations were followed by measuring the optical density at 328 mμ (the absorption maximum of living polystyrene). Addition of styrene or isoprene to living polystyrene showed the square root dependence on $[\sim\!S^-,Li^+]$, while the rates of reaction of living polyisoprenyl with either monomer were proportional to about the 1/4 power of $[\sim\!isoprenyl,Li^+]$. These relations are demonstrated by Figures 4 and 5. The absolute rate constants of all the steps are given in Table IV.

TABLE IV

Rate Constants of Copolymerization of Styrene and Isoprene in Cyclohexane at 47°C (27) (counterion Li⁺)

$k'_{SS} = 2.3 \times 10^{-2}$ (mole liter^{-1})$^{1/2}$ sec^{-1}	$k'_{SI} = 51 \times 10^{-2}$ (mole liter^{-1})$^{1/2}$ sec^{-1}
$k'_{IS} = 2.4 \times 10^{-4}$ (mole liter^{-1})$^{1/4}$ sec^{-1}	$k'_{II} = 34 \times 10^{-4}$ (mole liter^{-1})$^{1/4}$ sec^{-1}

$$k'_{SS} = k_{SS}(K_S)^{1/2}, \quad k_{II} = k'_{II}(K_I)^{1/4}, \quad etc.$$

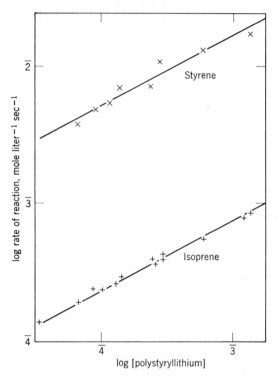

Fig. 4.   Addition of styrene and isoprene to living lithium polystyryl in cyclohexane at 60°C (27). The slopes of the lines give the order of the reaction in respect to [$\sim$S$^-$,Li$^+$]. Note that the lines are parallel to each other; the slopes are 1/2, i.e., living lithium polystyrene is dimeric in cyclohexane.

The absence of living polystyrene in a 1:1 mixture of monomers was demonstrated spectrophotometrically. The rate of consumption of isoprene (determined by vapor phase chromatography) was not affected by the presence of substantial amounts of styrene (note the two squares in Figure 5). The rates of styrene addition to living polyisoprene were found to be slightly higher, and not lower, when isoprene was present in the polymerized mixture (note the triangles in Figure 5). The higher rates observed in the presence of styrene are possibly due to the buildup of small amounts of polystyryllithium. The presence of less than 0.1% of such chain ends could account for the observations if their association is maintained at the normal level.

The results of these studies, as well as those of Johnson and Worsfold, disprove the hypothesis of solvation of growing ends by the diene. It remains to be seen whether systems that show such behavior may be discovered.

A spectrophotometric stop-flow technique was employed in studies of a few additional systems, viz., addition of butadiene to living sodium polystyryl

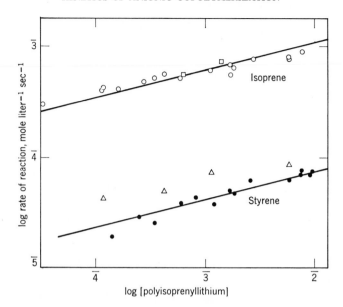

Fig. 5. Addition of styrene and isoprene to lithium polyisoprenyl in cyclohexane at 60°C (27). The slopes of the lines give the order of the reaction in respect to [∼isoprenyl⁻, Li⁺]. Note that the lines are parallel to each other; the slopes are 1/4, i.e., lithium polyisoprenyl appears to be tetrameric in cyclohexane. (□) In presence of styrene; the rate of isoprene addition is not affected by the presence of styrene. (△) In presence of isoprene; the rate of styrene addition is slightly *enhanced*, and not *inhibited*, by the presence of isoprene.

in tetrahydrofuran (34), addition of styrene to living poly-1-vinylnaphthalene (42), and addition of α-methylstyrene to living poly-1-vinylnaphthalene (43). The results of these investigations are discussed in the following sections.

6. COPOLYMERIZATION STUDIES IN TETRAHYDROFURAN

Absolute rate constants of many cross-propagations were determined for anionic polymerizations proceeding in tetrahydrofuran. The results led to some generalizations which will be reviewed in this section.

The addition of butadiene, isoprene, and 2,3-dimethylbutadiene to living sodium polystyryl in tetrahydrofuran (34) revealed a large inductive effect of methyl groups. The following values were obtained for the bimolecular rate constant of addition at 25°C:

Butadiene, $k_{12} = 32.7 \pm 1.0$ liter mole⁻¹ sec⁻¹ at $[\sim S^-, Na^+] = 2\text{–}3 \times 10^{-3} M$,

Isoprene, $k_{12} = 17.0$ liter mole⁻¹ sec⁻¹ at $[\sim S^-, Na^+] = 3\text{–}6 \times 10^{-3} M$,

2,3-Dimethyl-
butadiene, $k_{12} = 0.4\text{–}0.5$ mole liter⁻¹ sec⁻¹ at $[\sim S^-, Na^+] = 3\text{–}15 \times 10^{-3} M$.

The first system was investigated by the stop-flow technique; the stirred-flow technique (see p. 207) was employed in the second system and a batch technique in the third. Increase in the concentration of living polystyrene apparently leads to a decrease in k_{12} and therefore the observed rate constants are composite, being given by $k_{12 \pm} + k_{12 -} (1.5 \times 10^{-7}/[\sim S^-, Na^+])^{1/2}$. The first term representing the contribution of ion-pairs, is probably small; if it may be neglected, then the rate constant of the butadiene addition to the free living polystyryl anion is calculated to be about 100 times greater than the reported values of the k_{12}'s. This remark is general and applies to the other systems discussed in this section.

Leaving aside the problem of the reactivity of the free $\sim S^-$ ions and the $\sim S^-, Na^+$ ion-pairs, we may note that isoprene is half as reactive as butadiene while 2,3-dimethylbutadiene is about 70 times less reactive. It seems that a methyl group attached to a C=C double bond reduces its reactivity 70 times. Hence, one of the ends of isoprene is unreactive while the other retains its normal reactivity. This accounts for the factor of 2 found for the ratio k_{12}(butadiene)/k_{12}(isoprene). In 2,3-dimethylbutadiene the reactivity of both ends is drastically reduced, making this monomer inert. It would be interesting, therefore, to investigate how much of the addition to isoprene takes place on each end of the molecule. Also studies of reactivities of 1,3-pentadiene and 2,4-hexadiene are desirable.

Studies of copolymerization of α-methylstyrene and β-methylstyrene with living sodium polystyryl provide further information about the effect of a methyl group attached to a C=C double bond upon its reactivity. The results show that both monomers are approximately 40 to 50 times less reactive than styrene—the k_{12}'s are 27 and 18 liter mole^{-1} sec^{-1} for the α- and β-isomers, respectively, while $k_{11} \approx 950$ liter mole^{-1} sec^{-1}. Those data refer to the polymerization performed in tetrahydrofuran at 25°C when $[\sim S^-, Na^+] = 3 \times 10^{-3} M$. The lower reactivity of the β-isomer may reflect some steric hindrance caused by a methyl group attached to the reaction center.

For a few systems involving styrene and some substituted styrenes, all four copolymerization rate constants were determined. The pertinent data are collected in Table V. The pattern of reactivities in the system styrene–α-methylstyrene (44) is affected by two factors: (*a*) the steric hindrance that considerably slows down the addition of α-methylstyrene to α-methylstyrene anion and, to a much lesser extent, the addition of styrene to α-methylstyrene anion, or vice versa; (*b*) the inductive effect of the methyl group that enhances the addition of styrene to poly-α-methylstyrene anion and, by the same token, impedes the reverse addition. It is also interesting to note that $r_1 r_2$ calculated from these data is 0.1, clearly showing that the assumption $r_1 r_2 = 1$ cannot be valid for all pairs of monomers. This relation definitely fails when steric factors are important.

TABLE V

Four Propagation Constants of Anionic Copolymerization
in Tetrahydrofuran at 25°C
(counterion Na^+, concentration of living polymers about $3 \times 10^{-3}M$)

Pair	k_{11}	k_{12}	k_{21}	k_{22}	r_1	r_2	r_1r_2
	liter mole^{-1} sec^{-1}						
Styrene–α-methyl-styrene	950	27	780	2.5	35	0.03	0.1
Styrene–p-methyl-styrene	950	180	1150	210	5.3	0.18	0.95
Styrene–p-methoxy-styrene	950	50	1100	50	19	0.045	0.85
Styrene–vinylmesi-tylene	950	0.9	77	0.3	1000	0.004	$\sim$4

It has been pointed out (45) that steric hindrance caused by a bulky substituent located at the α-carbon of a vinyl group affects the rate of addition more than it affects the equilibrium constant. This hypothesis is confirmed by the experimental data available now (see p. 423).

In the system styrene–p-methylstyrene only the inductive effect operates; the steric hindrance is avoided. The reactivity ratios can be computed from the individual rate constants, viz., $r_1 = 5.3$ and $r_2 = 0.18$. These may be compared with Tobolsky and Boudreau's findings (17), $r_1 = 2$ and $r_2 = 0.4$. Although the agreement is not perfect, it is fair in view of the differences in the systems and in the methods of investigation. One should note that in this case $r_1r_2 \approx 1$, a result to be expected for a pair of closely similar monomers.

The system styrene–p-methoxystyrene behaves similarly to styrene–p-methylstyrene. Here again the inductive effect, caused by the p-methoxy group, is responsible for the observed pattern. However, since the electron-donating power of this group is greater than that of methyl, the cross-propagation rate constant, k_{12}, is lower in this system than the corresponding constant for p-methylstyrene. The product r_1r_2 once more is found to be close to unity.

The inductive effect of the methyl group in the α-position is greater than in the *para*-position. This statement cannot be tested in anionic copolymerization because the steric and inductive effects operate in the same direction. However, this point becomes obvious when cationic systems are studied. In spite of steric hindrance α-methylstyrene is *more* reactive than p-methylstyrene (62).

Steric hindrance is responsible for the extremely low reactivity of vinyl-mesitylene (52). The hindrance prevents coplanarity of the vinyl group with the phenyl ring, and this greatly reduces the activity of the monomer (52). Extensive studies of the relation between monomer structure and its reactivity were reported by Shima et al. (46). The results are collected in Table VI.

TABLE VI

Rate Constants, k_{12}, of Addition of Styrene Derivatives
to Sodium Polystyryl in Tetrahydrofuran at 25°C (46)
($[\sim\!\!S^-,Na^+] \approx 3 \times 10^{-3}M$)

Monomer	k_{12}, liter mole^{-1} sec^{-1}	Monomer	k_{12}, liter mole^{-1} sec^{-1}
Vinylmesitylene	0.9	p-Vinylbiphenyl	1,660
β-Methylstyrene	18	p-Fluorostyrene	1,800
α-Methylstyrene	27	1,1-Diphenylethylene	2,500
p-Methoxystyrene	50	1-Vinylnaphthalene	8,000
p-t-Butylstyrene	110	2-Vinylnaphthalene	8,600
2,4-Dimethylstyrene	160	p-Chlorostyrene	23,000
p-Methylstyrene	180	2-Vinylpyridine	> 30,000
o-Methylstyrene	320	4-Vinylpyridine	> 30,000
Styrene	950		

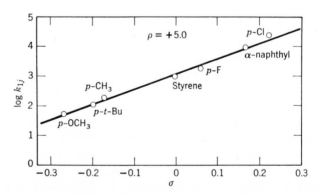

Fig. 6. Hammett σ–ρ plot for addition of p-substituted styrenes to living sodium polystyryl in tetrahydrofuran at 25°C (46).

The nucleophilic character of this addition is undeniable, and this point is clearly demonstrated by the σ–ρ Hammett plot shown in Figure 6. The high ρ value of about 5.0 should be contrasted with the relatively low ρ value of 0.5 determined for the analogous series of radical reactions (47). This is not surprising; the k_{12} values listed in Table VI refer to the addition involving predominantly the free ⌇styryl⁻ ions. It would be interesting to extend such studies to systems involving ion-pairs. Preliminary work of Yasuoka (48), who studied the addition of substituted styrenes to sodium polystyryl in dioxane, demonstrated a fair σ–ρ Hammett relation, surprisingly leading again to $\rho \approx 5$. In this solvent the reaction is caused by ion-pairs. Repetition of these studies and their extension are highly desirable.

The effect of substituents upon the rate of copolymerization initiated by a Ziegler-Natta type of catalyst was investigated by Natta and his associates

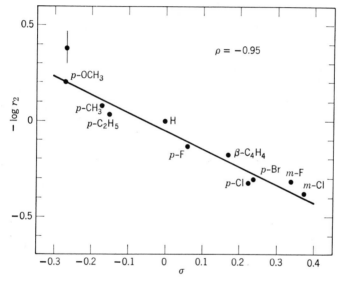

Fig. 7. Hammett σ–ρ plot for copolymerization of substituted styrenes with styrene (49). The polymerization was initiated by Ziegler-Natta heterogeneous catalyst. Note the negative ρ.

(49). The results fit well a Hammett σ–ρ plot, as shown in Figure 7, but with a *negative* ρ ($= -0.9$). This unexpected result apparently indicates that the polymerized monomer first becomes coordinated with the positive cation and then is inserted into the C^-—metal$^+$ bond. It remains to be seen whether such a trend in reactivities could be observed in homogeneous anionic polymerization.

Structure affects the reactivity of carbanions, as may be seen from the data collected in Table VII. The observed trends are complicated by the fact that the free ions carry out most of the reaction. The variation of structure affects the dissociation constant of ion-pairs as well as the reactivity of the relevant carbanion, and both factors are reflected in the data reported. Nevertheless, it is safe to conclude that structure affects only slightly the reactivity of closely similar carbanions, although it greatly modifies the reactivity of the respective monomers. Inductive effects appear to be less significant, but steric effects are still large.

TABLE VII

Reactivities of Carbanions toward Styrene in Tetrahydrofuran at 25°C
(counterion Na^+, concentration of living polymers about $3 \times 10^{-3}M$)

Polyanion	k_{21}, liter mole^{-1} sec^{-1}	Polyanion	k_{21}, liter mole^{-1} sec^{-1}
$\sim\!\!\sim$α-Methylstyrene$^-$	780	$\sim\!\!\sim$Vinylmesitylene$^-$	77
$\sim\!\!\sim$Styrene$^-$	950	$\sim\!\!\sim$1-Vinylnaphthalene$^-$	30
$\sim\!\!\sim$p-Methylstyrene$^-$	1150	$\sim\!\!\sim$1,1-Diphenylethylene$^-$	0.7
$\sim\!\!\sim$p-Methoxystyrene$^-$	1100		

7. THE PENULTIMATE EFFECTS

The simple scheme of polymerization assumes that the reactivity of a growing chain is determined by its last unit, although it was realized that penultimate units may also exert some effect upon rate of polymerization (50). However, in only a few radical polymerizations were such effects observed (51). It is difficult to reveal the penultimate effect if the necessary data are derived from the variations in the composition of the copolymer arising from changes in the composition of the feed. On the other hand, application of living polymers permits the direct study of such an effect. One needs to prepare a series of living polymers such as $\sim\!\!\sim M_1 \cdot M_1^-, X^+$, $\sim\!\!\sim M_2 \cdot M_1^-, X^+$, etc., and to determine the absolute rate constant of addition of some monomer to each of them. This may be achieved with the previously described techniques. Two examples will be discussed.

(1) The addition of vinylpyridine to living polystyrene, or its derivatives, proceeds instantaneously in tetrahydrofuran. Hence, mixing (in a flow system) of living polystyrene with a slight excess of 2-vinylpyridine yields a polymer having $\sim\!\!\sim$styrene(vinylpyridine)$^-$,Na^+ end groups (53). The addition of vinylpyridine to such polymers and to the ordinary living homopolyvinylpyridine was investigated. The results are given in Table VIII. Substitution of vinylpyridine in the penultimate unit by styrene, or by its derivative, accelerated the reaction by more than a factor of 4.

We may inquire whether the rate of addition to the homopolyvinylpyridine is "normal," while it is enhanced by the substitution of styrene for vinylpyridine in the penultimate unit, or whether the reaction is hindered by the

TABLE VIII

Penultimate Effects at 25°C in Addition of Vinylpyridine
to Living Polymers Having $\sim\sim M_i$(vinylpyridine)$^-$,Na$^+$
End Groups (53)
(solvent, tetrahydrofuran)

The penultimate monomer M_i	$k,$ liter mole^{-1} sec^{-1}
Vinylpyridine	7,300
Styrene	31,000
p-Methylstyrene	30,000
p-Methoxystyrene	30,000
α-Methylstyrene	25,000

presence of vinylpyridine in the penultimate units, being "normal" with the other polymers. The answer depends on what is considered "normal" and, therefore, it is somewhat arbitrary. Nevertheless, it seems that the presence of vinylpyridine in the penultimate units may exert a peculiar effect on the course of the addition. The interaction of the nitrogen atom of this group with the counterion, or the withdrawal of electrons from the growing carbanion by the adjacent pyridine ring, may reduce the reactivity of the living polymer. In this sense the reactivity of the polymers possessing styrene (or its derivative) in the penultimate group may be "normal." It is perhaps significant that the rates of addition are virtually identical whether styrene, p-methylstyrene, or even α-methylstyrene is placed in the unit, next to the last, while the rate is reduced so much by the presence of vinylpyridine.

Do the segments located still farther away from the growing end of a polymer exert any notable effect upon the rate of polymerization? This question was answered by Lee et al. (53), who investigated the rate of addition of vinylpyridine to polymers having the structure:

$$\sim\sim\text{styrene (vinylpyridine)}_n\text{(vinylpyridine)}^-,\text{Na}^+$$

for $n = 1, 2, 3,$ and 5. Superficially it appeared that the effect decays slowly because the results led to the values (in liter mole^{-1} sec^{-1}) for k of 30,600 for $n = 1.1$, 11,000 for $n = 2.04$, 8,020 for $n = 3.2$, and 7,000 for $n \geqslant 5$. However, it was shown that one may account for these findings by postulating the simultaneous presence of two types of end groups, viz., $\sim\sim$S(VPy$^-$),Na$^+$ and $\sim\sim$S(VPy)$_n$(VPy$^-$),Na$^+$, the reactivity of the latter being independent of n for $n \geqslant 1$. The proportion of these two groups in a polymer is determined by two competing reactions

$$\sim\sim\text{S(VPy}^-),\text{Na}^+ + \text{VPy} \xrightarrow{k'} \sim\sim\text{S(VPy)(VPy}^-),\text{Na}^+,$$

$$\sim\sim\text{S(VPy)}_n(\text{VPy}^-),\text{Na}^+ + \text{VPy} \xrightarrow{k''} \sim\sim\text{S(VPy)}_{n+1}(\text{VPy}^-),\text{Na}^+,$$

and the solution of the appropriate differential equations gives

$$n = (1 - k''/k')(1 - x) - (k''/k') \ln x$$

where x denotes the mole fraction of $\sim\!\!S(VPy^-),Na^+$ units present in a polymer obtained by adding n moles of vinylpyridine to a mole of $\sim\!\!S(VPy^-),Na^+$. The observed rate constants of vinylpyridine addition to

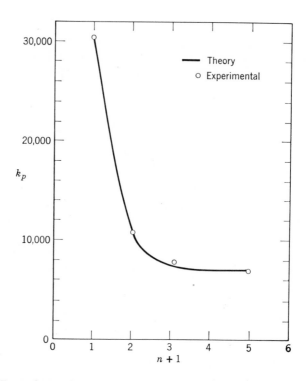

Fig. 8.   Dependence of formal rate constant of the addition on n

$\sim\!\!$styrene(vinylpyridine)$_n$(vinylpyridine)$^-$,Na$^+$ + vinylpyridine

at 25°C in tetrahydrofuran (53). The solid line was calculated on the basis of the derived kinetic scheme.

such a polymer are $k = xk' + (1 - x)k''$. The experimental results, quoted above, were plotted, therefore, in Figure 8 against $n + 1$, and the solid line was calculated by assuming $k' = 30,000$ liter mole^{-1} sec^{-1} and $k'' = 7000$ liter mole^{-1} sec^{-1}. The agreement between theory and experiment is satisfactory.

(2) Copolymerization of styrene with living poly-1-vinylnaphthalene showed most interesting features (42). The addition of two or three equivalents

of styrene to living polyvinylnaphthalene led to the disappearance of the spectrum of

$$\sim\!CH_2\!\cdot\!CH^-, Na^+$$

($\lambda_{max} = 558$ mμ), but the expected spectrum of living polystyrene ($\lambda_{max} = 340$ mμ) did not appear. Instead a new absorption peak developed at $\lambda_{max} = 440$ mμ. These spectral changes are shown in Figure 9. However, the characteristic spectrum of living polystyrene appeared permanently when a large excess of styrene, 20-fold or more, was added. It seems that the reaction involves three steps:

$$\text{(1)}$$

$$\text{(2)}$$

$$\text{(3)}$$

The product produced by addition of the first molecule of styrene gives a benzyl-type anion, which forms a bond with the preceding napththalene moiety. Such a species resembles the adduct of anthracene and living polystyrene (54); it forms a dormant polymer (see Chapter XII) and it very slowly adds a second styrene unit. The addition of the second molecule of styrene destroys the complexing with naphthalene and so the resulting polymer

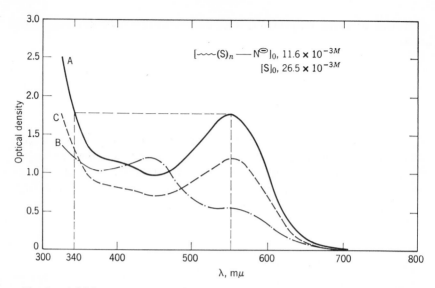

Fig. 9. Addition of styrene to living polyvinylnaphthalene in tetrahydrofuran at 25°C (42): (*A*) spectrum of living poly-1-vinylnaphthalene, (*B*) spectrum observed immediately after mixing the reagents, (*C*) spectrum observed 24 hours after reagents were mixed.

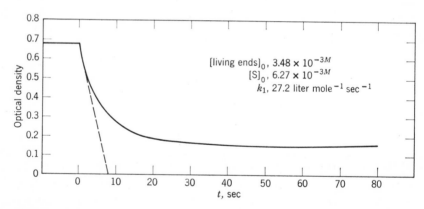

Fig. 10. Change in optical density at 558 mμ observed in a stop-flow technique for reaction (1)

behaves like an ordinary living polystyrene; it rapidly grows, adding more styrene.

The kinetics of the first step was investigated by a stop-flow technique. The optical density at 558 mμ was followed and a typical result is shown in Figure 10. Thus, k_1 was determined to be 30 liter mole^{-1} sec^{-1}. The last stage of the reaction represents an ordinary homopropagation of living polystyrene; its rate constant is known, viz., $k_3 \approx 900$ liter mole^{-1} sec^{-1}. In order to determine the rate constant of the second step, the kinetics of the overall reaction was studied in a stirred-flow reactor. The pertinent balance equations are

$$(k_1 X + k_2 C + k_3 y)[S]t = [S]_0 - [S],$$
$$k_2 X[S]t = X_0 - X,$$
$$(k_1 X - k_2 C)[S]t = C,$$
$$k_2 C[S]t = y,$$

where X denotes the concentration of living polymers terminated by a vinylnaphthalene end group, C is the concentration of the intramolecular complex formed between the styryl group and the preceding naphthalene moiety, y is the concentration of the ordinary living polystyrene produced when at least two styrene units are linked up with living polyvinylnaphthalene, and [S] and t are the concentration of styrene and the residence time, respectively. All the concentrations are those maintained in the reactor at its stationary state, whereas the subscript zero refers to the concentrations of the reagents in the feed flowing into the reactor. The balance equations (see p. 207) arise from the conditions that the amount of the reagent (or the intermediate) consumed (or formed) in the reactor is given by the difference between its amount supplied through the feed and withdrawn in the outgoing flow. This set of equations leads to the relations*:

$$([S]_0 - [S] - C - 2y)/C[S]^2 = k_2 k_3 t^2$$

and

$$C = k_1[S]t(X_0 - y)/\{1 + (k_1 + k_2)[S]t\}.$$

Thus, by plotting the left side of the first equation versus the square of the residence time one should obtain a straight line with a slope equal to $k_2 k_3$. The value of the left side of the equation can be calculated by applying a series of successive approximations. The results of six experiments in which $[S]_0$ and X_0 were varied led to a constant value of $k_2 k_3$. For the sake of illustration, three of such plots are shown in Figure 11. Thus, k_2 was found to be 2 liter mole^{-1} sec^{-1}.

The proposed kinetic scheme may be checked. The addition of some finite amount of styrene to the complex cannot convert all of it into living polystyrene, even when all the monomer is consumed. The addition of the first

* Each experiment gives [S] for chosen and known values of $[S]_0$, X_0, and t. These data are used for the calculation of C and y.

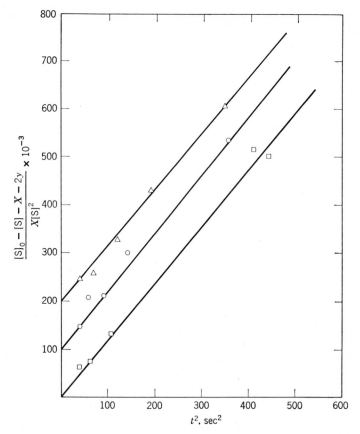

Fig. 11. Kinetics of styrene addition to living polyvinylnaphthalene in tetrahydrofuran at 25°C (42). The slopes give $k_2 k_3$. The successive lines are shifted upward by 100 and 200 units, respectively.

Point	X_0	$[S]_0$
□	$2.7 \times 10^{-3} M$	$4.5 \times 10^{-3} M$
○	$2.6 \times 10^{-3} M$	$11.5 \times 10^{-3} M$
△	$3.5 \times 10^{-3} M$	$54 \times 10^{-3} M$

molecule of the monomer and the conventional propagation compete for styrene. The mathematics of such competing reactions was discussed in Chapter II (see p. 50), where it was shown that after all the monomer is polymerized the fraction, f, of the complex converted into living polystyrene is uniquely determined by $[S]_0$, X_0, and k_3/k_2. Because k_3 is known, the experimental data give k_2. These studies led to $k_2 = 1.3$ liter mole^{-1} sec^{-1}, in fair agreement with the previous determination.

The latter method, in which the ultimate concentration of the complex is determined, is less reliable than the stirred-flow method because of some technical difficulties concerned with the mixing of the reagents and thermostating the reactor. Nevertheless, such an approach has been successfully used by Ureta et al. (55) to determine the rate of addition of styrene to carbanions formed from 1,1-diphenylethylene.

The kinetics of styrene addition to living polyvinylnaphthalene illustrates an extreme example of penultimate unit effect caused by the formation of a bond between the active end of a growing polymeric molecule and the preceding segment of that polymer. Let us consider the thermodynamics of this reaction. Dormant polymers are formed on addition of anthracene to living polystyrene (see p. 446). It was shown that such an addition is reversible because the binding causes the formation of a weak covalent C—C bond. The equilibrium constant is high and therefore only a small fraction of living (growing) polymers is present in the solution. Dormant polymers are not formed on addition of naphthalene. Binding of anthracene is favored by its relatively high electron affinity, whereas the electron affinity of naphthalene is lower by 0.5 V and therefore the equilibrium constant of the addition, if any, is shifted far to the left. Apparently the gain in energy is not sufficient to overcome the loss of entropy arising from the binding of a molecule of free naphthalene. The *intra*molecular addition of the benzyl-type carbanion to the preceding naphthalene unit creates a different situation. Figure 12 shows a model of such an end group and demonstrates that a strainless C—C covalent bond may be formed by the process linking C$^-$ with the carbon-8 of the naphthalene moiety. This reaction is associated with lesser decrease in entropy because the large loss of translational entropy is avoided. Consequently, although *inter*molecular association of ⌇S$^-$ carbanions with naphthalene is unfavorable, the *intra*molecular cyclization is possible. Such a cyclization is prevented by steric strain when the next styrene unit is added, and it becomes, entropywise, unprofitable for a longer chain.

How does the complex grow? Two mechanisms may be visualized: (*a*) Styrene adds directly to the complex; the opening of the ring takes place simultaneously with addition of the second styrene unit. In this case k_2 is a proper bimolecular rate constant; it is low because of the nature of the reaction. (*b*) Alternatively, an equilibrium may be established between the cyclic and the open form, viz.,

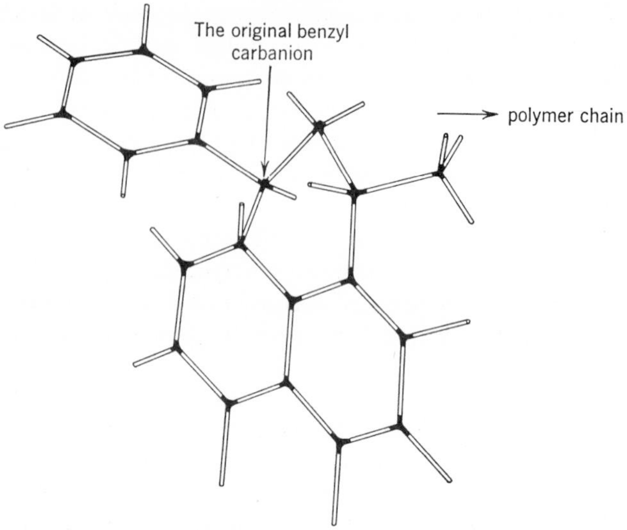

Fig. 12.   Model of the polymer end group

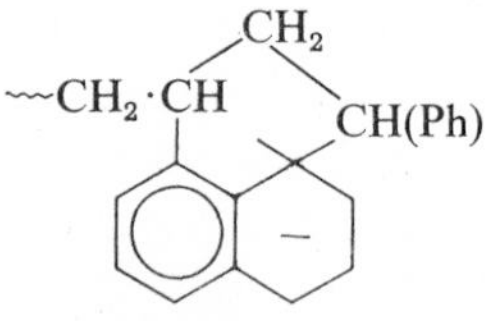

$$\sim\!CH_2\!\cdot\!CH \underset{\underset{\bigcirc}{}}{\overset{\overset{CH_2}{\diagdown}}{\diagup}} CH(Ph)$$

which shows that the cyclization is sterically feasible.

and then the open form reacts like "normal" living polystyrene. In the latter case k_2 is a composite constant, $k_2 = K_c k_3$, where K_c is the equilibrium constant of cyclization. Taking $k_3 = 900$, we find $K_c = 2 \times 10^{-3}$. In view of the results reported by Khanna et al. (54) the second mechanism appears to be more plausible.

8. REVERSIBLE COPOLYMERIZATION

The addition of styrene to living polyvinylnaphthalene shows another interesting feature. The complex formed on addition of a small excess of styrene is unstable. Its spectrum disappears within 24 hours and the spectrum of living polyvinylnaphthalene reappears. This phenomenon is documented by Figure 9. The subsequent addition of a small excess of styrene led again to complex and, once more, the spectrum of living polyvinylnaphthalene reappeared after 24 hours. Such a cycle was repeated four times; each time living polyvinylnaphthalene was regenerated, although each time at decreasing concentration. The study of this phenomenon was performed with polystyrene possessing only one terminal naphthalene unit and, therefore, the reaction

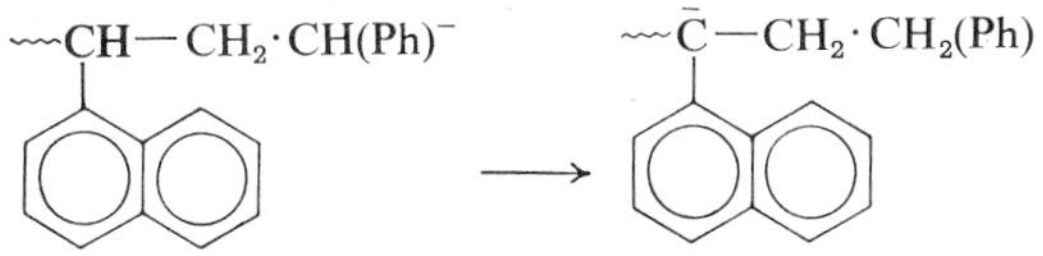

could perhaps account for the first cycle but not for the subsequent results. Hence, it was concluded that the formation of the complex is reversible and the equilibrium concentration of styrene should be determined by the reaction:

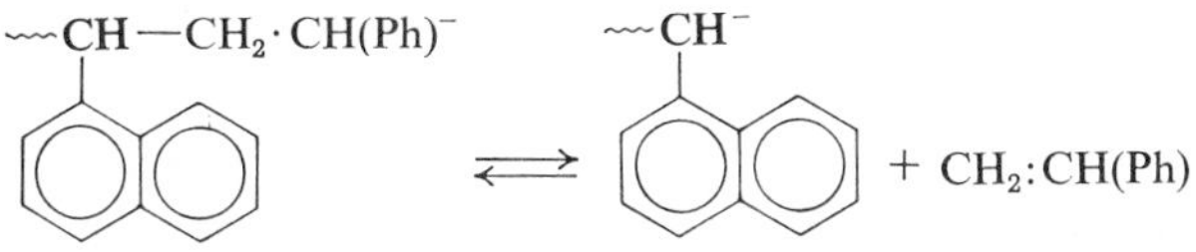

However, the reacting mixture *must* contain a small proportion of the ordinary living polystyrene resulting from addition of the excess of styrene to those units that have succeeded in adding two or more styrene units in a row. Hence, another equilibrium is established:

$$\sim\sim\{CH_2 \cdot CH(Ph)\}_n CH_2 \cdot CH(Ph)^-$$
$$\rightleftharpoons \{CH_2 \cdot CH(Ph)\}_{n-1} \cdot CH_2 \cdot CH(Ph)^- + CH_2 : CH(Ph)$$

The two equilibria are coupled and apparently the equilibrium of the overall process

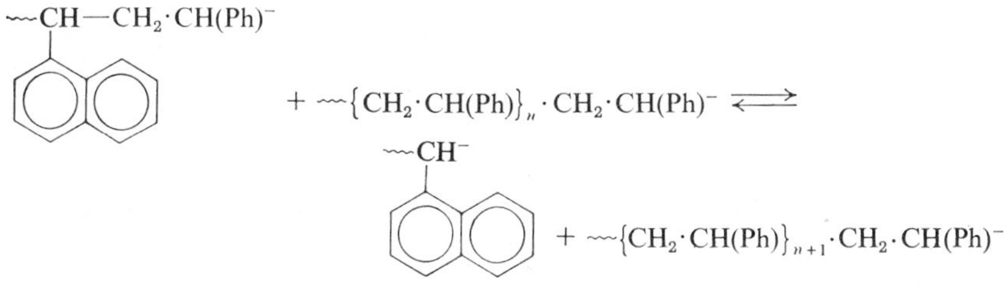

favors the right side. Consequently, the rich become richer and the poor become poorer. The polymers possessing only one styrene unit lose them; those possessing many gain.

The proposed scheme may be tested. The addition of a monomer, propagation of which is thermodynamically unfavorable, should lead to the formation of a stable complex. Such a complex must not decompose even if the excess of monomer is low because the relevant overall equilibrium now favors the left side. Such a system was investigated by Stearne et al. (43). They added α-methylstyrene to living polyvinylnaphthalene. Although the monomer was in excess, its concentration was kept below the equilibrium concentration for

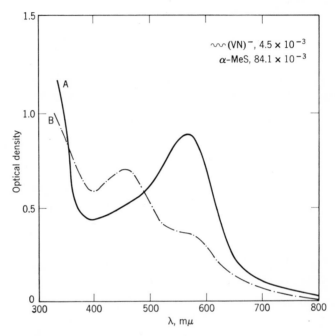

Fig. 13. Spectral changes taking place in the addition of α-methylstyrene, α-MeS, to living polyvinylnaphthalene, ($\sim$VN)$^-$, in tetrahydrofuran (43): (A) original spectrum of poly-1-vinylnaphthalene, (B) spectrum observed immediately after mixing the reagents. This spectrum was not changed even after 2 or 3 days.

the α-methylstyrene–living poly-α-methylstyrene system. The formation of the complex is demonstrated by the appearance of its spectrum, as shown in Figure 13. The system remained stable over a period of several days, and the reappearance of the absorption spectrum of living polyvinylnaphthalene was not observed.

The equilibrium

was studied in a stirred-flow reactor (see p. 207). The results give the rates of the forward and backward reactions, and the equilibrium constant (43). Thus,

$$k_{-12} = 0.4 \text{ liter mole}^{-1} \text{ sec}^{-1},$$

and

$$k_{12} = 0.024 \text{ sec}^{-1},$$

$$K_{12} = 16.7 \text{ liter mole}^{-1}.$$

Steric strain caused by the presence of the α-methyl group makes the addition of α-methylstyrene to living polyvinylnaphthalene 75 times slower than the addition of styrene. The same effect was observed upon addition of these two monomers to living polystyrene but, since phenyl is smaller than naphthyl, the ratio of the respective rate constants is then less than 40.

The equilibrium study was repeated in a static system, radioactive α-methylstyrene being used to facilitate the analytical problems. The results led to $K_{12} = 18$ liter mole^{-1}, in perfect agreement with those derived from the stirred-flow experiments.

The addition of styrene to living polymer terminated by carbanion derived from 1,1-diphenylethylene was found also to be reversible (55), viz.,

$$\sim\!CH_2\cdot C(Ph)_2^-,Na^+ + CH_2:CHPh \underset{k'_{-12}}{\overset{k'_{12}}{\rightleftharpoons}} \sim\!CH_2\cdot CH(Ph)_2\cdot CH_2\cdot CH(Ph)^-,Na^+;$$

a stirred-flow technique was again employed in this investigation. The problem is complicated by the subsequent, virtually irreversible, polymerization of the resulting living polystyrene. The respective balance equations were solved and k'_{12} was found to be 0.5–0.7 liter mole^{-1} sec^{-1} while $k'_{-12} = 13$ sec^{-1}. Thus, K'_{12} is only 5×10^{-2} liter mole^{-1}. These values reflect the considerable steric strain created by this addition.

9. APPLICATION OF NMR TO STUDIES OF COPOLYMERIZATION

Pioneering studies of Bovey and Tiers (56) demonstrated that the NMR technique may provide information about the sequences of segments formed in a polymeric chain (see p. 463). Such data are most interesting in investigations of copolymers because the statistics of growth can be then determined. Monomer distribution in copolymers formed by radical-propagation is usually determined by the statistics of the zero-order Markovian process (57). This is the classic type of copolymerization, and NMR studies of copolymers of ethylene–vinyl chloride and ethylene–vinyl acetate confirm the expected pattern of segments (58). In anionic copolymerization, different statistics may be expected; the penultimate unit effect may be pronounced (see Section 7 of this chapter), or the growing ends may exist in different forms which are in dynamic equilibrium with each other. In the latter case, the distribution is given by non-Markovian statistics (59,60). Recently copolymerization of propylene oxide and maleic anhydride was investigated by the NMR technique by Kern and Schaefer (61), who prepared the copolymers by employing seven different types of homogeneous catalysts. They showed that some catalysts, such as $ZnCl_2$, $CdCl_2$, or $SnCl_4$, tend to produce sequences of three propylene oxide segments in a row, while $SbCl_5$ does not exhibit such a tendency.

Many more NMR studies of copolymers are expected in the future. It is hoped that the results will shed much light on the mechanisms of these reactions and will be particularly illuminating in studies of ionic copolymerization.

References

(1) R. K. Graham, D. L. Dunkelberger, and W. E. Goode, *J. Am. Chem. Soc.*, **82**, 400 (1960).
(2) N. S. Wooding and W. C. E. Higginson, *J. Chem. Soc. (London)*, **1952**, 774.
(3) M. Szwarc, *Makromol. Chem.*, **35**, 132 (1960).
(4) C. Walling, E. R. Briggs, W. Cummings, and F. R. Mayo, *J. Am. Chem. Soc.*, **72**, 48 (1950).
(5) (a) Y. Landler, *Compt. Rend.*, **230**, 539 (1950). (b) Y. Landler, *J. Polymer Sci.*, **8**, 63 (1952).
(6) R. Gumbs, S. Penczek, J. Jagur-Grodzinski, and M. Szwarc, to be published.
(7) (a) T. Alfrey and G. Goldfinger, *J. Chem. Phys.*, **12**, 205, 322 (1944). (b) F. R. Mayo and F. M. Lewis, *J. Am. Chem. Soc.*, **66**, 1594 (1944). (c) F. T. Wall, *ibid.*, **66**, 2050 (1944).
(8) K. F. O'Driscoll, *J. Polymer Sci.*, **57**, 721 (1962).
(9) C. G. Overberger and N. Yamamoto, *J. Polymer Sci.*, A, **1**, 4, 3101 (1966).
(10) A. A. Korotkov, S. P. Mitzengendler, and L. L. Dantzig, *J. Polymer Sci.* B, **4**, 809 (1966).
(11) Z. Laita and M. Szwarc, *J. Polymer Sci.*, B, **6**, 197 (1968).
(12) (a) G. V. Rakova and A. A. Korotkov, *Dokl. Akad. Nauk SSSR*, **119**, 982 (1958); (b) A. A. Korotkov and N. N. Chesnokova, *Vysokomolekul. Soedin.*, **2**, 365 (1960); (c) A. A. Korotkov and G. V. Rakova, *ibid.*, **3**, 1482 (1961).
(13) K. F. O'Driscoll and I. Kuntz, *J. Polymer Sci.*, **61**, 19 (1962).
(14) D. J. Kelley and A. V. Tobolsky, *J. Am. Chem. Soc.*, **81**, 1597 (1959).
(15) A. V. Tobolsky and C. E. Rogers, *J. Polymer Sci.*, **38**, 205 (1959).
(16) D. J. Worsfold and S. Bywater, *Can. J. Chem.*, **38**, 1891 (1960).
(17) A. V. Tobolsky and R. J. Boudreau, *J. Polymer Sci.*, **51**, S53 (1961).
(18) K. F. O'Driscoll and R. Patsiga, *J. Polymer Sci.*, A, **3**, 1037 (1965).
(19) I. Kuntz, *J. Polymer Sci.*, **54**, 569 (1961).
(20) M. Szwarc and J. Smid, *in* "Progress in Reaction Kinetics," Vol. 2, Pergamon Press, New York, 1964; p. 219, see also p. 268.
(21) Y. L. Spirin, D. K. Polyakov, A. R. Gantmakher, and S. S. Medvedev, *J. Polymer Sci.*, **53**, 233 (1961).
(22) C. E. H. Bawn, *Rubber Plastic Age*, **42**, 267 (1961).
(23) B. D. Phillips, T. L. Hanlon, and A. V. Tobolsky, *J. Polymer Sci.* A, **2**, 4231 (1964).

(24) F. Dawans and G. Smets, *Makromol. Chem.*, **59**, 163 (1963).
(25) C. G. Overberger, T. M. Chapman, and T. Wojnarowski, *J. Polymer Sci. A*, **3**, 2865 (1965).
(26) A. F. Johnson, D. J. Worsfold, and S. Bywater, *Can. J. Chem.*, **42**, 1255 (1964).
(27) D. J. Worsfold, *J. Polymer Sci. A1*, **5**, 2783 (1967).
(28) C. S. Marvel and J. F. Dunphy, *J. Org. Chem.*, **25**, 2209 (1960).
(29) S. Okamura and T. Higashimura, *Kobunshi Kagaku*, **17**, 635 (1960).
(30) S. Okamura, T. Higashimura, and T. Takada, *ibid.*, **18**, 839 (1961).
(31) C. G. Overberger and V. G. Kamath, *J. Am. Chem. Soc.*, **85**, 446 (1963).
(32) A. F. Johnson and D. J. Worsfold, *Makromol. Chem.*, in press.
(33) A. Rembaum, F. R. Ells, R. C. Morrow, and A. V. Tobolsky, *J. Polymer Sci.*, **61**, 155 (1962).
(34) M. Shima, J. Smid, and M. Szwarc, *J. Polymer Sci. B*, **2**, 735 (1964).
(35) F. C. Foster, *J. Am. Chem. Soc.*, **72**, 1370 (1950).
(36) F. C. Foster, *ibid.*, **74**, 2299 (1952).
(37) N. L. Zutty and F. J. Welch, *J. Polymer Sci.*, **43**, 445 (1960).
(38) M. Morton and F. R. Ells, *ibid.*, **61**, 25 (1962).
(39) J. Smid and M. Szwarc, *ibid.*, **61**, 31 (1962).
(40) D. N. Bhattacharyya, C. L. Lee, J. Smid, and M. Szwarc, *J. Am. Chem. Soc.*, **85**, 533 (1963).
(41) D. Harris, unpublished work, University of Syracuse group, 1967.
(42) F. Bahsteter, J. Smid, and M. Szwarc, *J. Am. Chem. Soc.*, **85**, 3909 (1963).
(43) J. Stearne, J. Smid, and M. Szwarc, *Trans. Faraday Soc.*, **60**, 2054 (1964).
(44) D. N. Bhattacharyya, C. L. Lee, J. Smid, and M. Szwarc, *J. Am. Chem. Soc.*, **85**, 533 (1963).
(45) T. Alfrey and W. H. Ebelke, *ibid.*, **71**, 3235 (1949).
(46) M. Shima, D. N. Bhattacharyya, J. Smid, and M. Szwarc, *ibid.*, **85**, 1306 (1963).
(47) C. Walling, E. R. Briggs, K. B. Wolfstirn, and F. R. Mayo, *ibid.*, **70**, 1537 (1948); see also C. Walling, "Free Radicals in Solution," Wiley, New York, 1957, p. 137.
(48) K. Yasuoka, unpublished work, University of Syracuse group, 1965.
(49) G. Natta, F. Danusso, and D. Sianesi, *Makromol. Chem.*, **30**, 238 (1959).
(50) E. Merz, T. Alfrey, and G. Goldfinger, *J. Polymer Sci.*, **1**, 75 (1946).
(51) W. G. Barb, *ibid.*, **11**, 117 (1953); **18**, 310 (1955).
(52) F. Carrock and M. Szwarc, *J. Am. Chem. Soc.*, **81**, 4138 (1959).
(53) C. L. Lee, J. Smid, and M. Szwarc, *Trans. Faraday Soc.*, **59**, 1192 (1963).
(54) S. N. Khanna, M. Levy, and M. Szwarc, *ibid.*, **58**, 747 (1962).
(55) E. Ureta, J. Smid, and M. Szwarc, *J. Polymer Sci. A1*, **4**, 2219 (1966).
(56) F. A. Bovey and G. V. D. Tiers, *J. Polymer Sci.*, **44**, 173 (1960).
(57) F. P. Price, *J. Chem. Phys.*, **36**, 209 (1962).
(58) J. Schaefer, *J. Phys. Chem.*, **70**, 1975 (1966).
(59) B. D. Coleman and T. G Fox, *J. Chem. Phys.*, **38**, 1065 (1963).
(60) L. Peller, *ibid.*, **43**, 2355 (1965).
(61) R. J. Kern and J. Schaefer, *J. Am. Chem. Soc.*, **89**, 6 (1967).
(62) C. G. Overberger et al., *J. Am. Chem. Soc.*, **74**, 4848 (1952).

Chapter X

The Kinetics and Mechanism of *N*-Carboxy-α-Amino Acid Anhydride (NCA) Polymerization to Poly-Amino Acids

1. LEUCH'S ANHYDRIDES (NCA)

Polymerization of *N*-carboxy-α-amino acid anhydrides (NCA) to poly-peptides represents an important and interesting process which yields polymers not easily prepared by other techniques. The propagation step involves anions or growing ends of polymers which exhibit nucleophilic character, and therefore inclusion of this subject in this monograph is fully justified.

NCA polymerization had attracted the attention of many investigators* and has been extensively studied during the last 20 years. Propagation proceeds through a chain polyaddition, i.e., growth is determined by the sequence of steps $P_n + M \rightarrow P_{n+1}$ rather than $P_n + P_m \rightarrow P_{n+m}$ (see, however, p. 601). The overall reaction involves initiation and propagation, and often it includes some termination.

N-Carboxy-α-amino acid anhydrides react in a variety of ways and, therefore, it is difficult to discuss their polymerization in a conventional manner, i.e., by separately considering the initiation, propagation, and termination steps. It is believed that the reader may comprehend more easily the nature of the polymerization if the material is presented in terms of the chemical behavior of the monomer toward various reagents. Such a course will be followed, therefore, in the present chapter.

The *N*-carboxy-α-amino acid anhydrides, referred to as Leuch's anhydrides, or briefly as NCA's, are well-defined, colorless, crystalline substances of sharp melting points when pure. The synthesis of the simplest member of this class of compounds, viz. 1,3-oxazolidine-2,5-dione, was described by Leuchs (1) in 1906. In subsequent papers (2) he reported the preparation of some of its derivatives, namely, the *N*-carboxy anhydrides of *N*-phenylglycine, *C*-phenyl-glycine, phenylalanine, and leucine. Numerous derivatives of 1,3-oxazoli-

---

* For recent review of the subject see also ref. 91.

dine-2,5-dione have been synthesized since, and an impressive list of about 100 Leuchs' anhydrides, together with references on their preparation and melting points, is given in a recent review by Katchalski and Sela (3).

$$
\begin{array}{ll}
(4) & (5) \\
\mathrm{CH_2\!-\!CO} \\
\quad\big| \qquad\quad \diagdown \\
\quad\qquad\qquad\mathrm{O} \quad (1) \\
\quad\big| \qquad\quad \diagup \\
\mathrm{NH\!-\!CO} \\
(3) & (2)
\end{array}
$$

The Leuchs' anhydrides are extremely reactive compounds, and most of them may be stored only if kept below -20 or $-30°C$ under strictly anhydrous conditions. At $10°C$ they dissolve in water with little decomposition,* but eventually they hydrolyze into carbamic acids

$$
\begin{array}{lll}
\mathrm{RCH\!-\!CO} & & \mathrm{R\cdot CH\cdot COOH} \\
\quad\big| \qquad \diagdown & & \qquad\big| \\
\qquad\qquad \mathrm{O} + \mathrm{H_2O} \rightarrow & \\
\quad\big| \qquad \diagup & & \qquad\big| \\
\mathrm{NH\!-\!CO} & & \mathrm{NH\cdot COOH}
\end{array}
$$

which readily decarboxylate into the respective amino acids. The kinetics of the hydrolysis and the decarboxylation of NCA's in aqueous solution were investigated by Bartlett and Jones (4). The reaction shows a first order character, the respective rate constants being 10^{-3} to 10^{-2} sec^{-1} at $25°C$. However, this process is catalyzed by acids and bases.

On exposure to moisture, the crystalline anhydrides undergo an auto-accelerated polymerization with evolution of carbon dioxide. This solid-state reaction proceeds without any observable change in the form of the crystal, and its effect on the x-ray pattern has been investigated briefly by Miller, Fankuchen, and Mark (5). Later, it was reported by Okamura (63) that the solid state polymerizations of NCA's derived from γ-benzyl glutamate and from alanine proceed faster with the racemic material than with the pure enanthiomorph. This striking observation spurred Morawetz and his associates (64) to investigate more thoroughly these reactions. Their studies included NCA's derived from γ-benzyl glutamate, γ-methyl glutamate, and ε-carbobenzoxylysine. Both the pure L-compounds and the racemic mixtures were prepared, and their solid-state polymerizations were investigated at constant temperature and humidity. In addition, powdered x-ray diffraction photographs of the respective NCA's were taken in order to elucidate the crystal structure of the racemic compounds. It was shown that the racemic crystals are *not* mixtures of the D- and L-enanthiomorphs, a conclusion supported by the fact that the melting points of γ-methyl D,L-glutamate NCA and

---

* This reaction is greatly affected by the pH of the solution. For example, extensive polymerization occurs at pH = 7.4 (see ref. 24).

ε-carbobenzoxy-D,L-lysine NCA are higher than those of the respective L-anhydrides.

Morawetz et al. (64) characterized the kinetics of these polymerizations by their induction periods and the maximum rates. They noted that for glutamic acid NCA's, the maximum rates are quite insensitive to the partial pressure of water, although increased humidity shortens the induction period. The NCA of ε-carbobenzoxy-L-lysine differs somewhat from the other NCA's

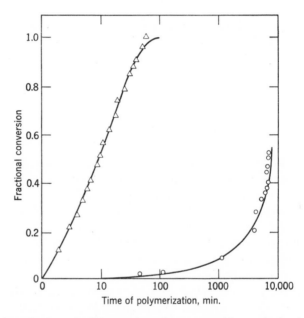

Fig. 1. Polymerization of solid racemic and pure enanthiomorphs in the solid state. (△) γ-methyl-L-glutamate NCA; (○) γ-methyl-D,L-glutamate NCA.

in its kinetic behavior. There is no induction period and the rate of polymerization is accelerated considerably by increased humidity. The observation of Okamura was confirmed that in a humid atmosphere the racemic γ-benzyl D,L-glutamate NCA reacts faster than the L-enanthiomorph; however, in a dry environment the former has a longer induction period and a lower reaction rate. On the other hand, the racemic NCA derivatives of γ-methyl glutamate and of ε-carbobenzoxylysine were found unreactive, whereas the pure enanthiormorphs polymerize rapidly. This is illustrated in Figure 1.

It would be expected that the fast polymerization of racemic γ-benzyl D,L-glutamate should lead to an alternating D,L-polypeptide. Unfortunately, this is not the case. Physical studies of the copolymer seem to indicate a random structure and it is unlikely that the product forms a mixture of D- and L-homopolymers because such a mixture is known to form a highly soluble complex (65). It seems, therefore, that the racemic copolymer is composed of blocks of varying lengths of D- and L-units.

On melting the crystals, a rapid decarboxylation takes place with the formation of a low molecular weight polypeptide. This spontaneous polymerization was first noticed by Leuchs (1,2), and was subsequently thoroughly studied by Katchalski and his co-workers (6) who carried out the reaction in vacuum at elevated temperatures. Under such conditions, conversion of the anhydride to polypeptide appears to be quantitative.

The mechanism of high temperature bulk polymerization of NCA's is still obscure. It was suggested that the reaction involves isomerization of the anhydride to isocyanate, viz.,

$$\begin{array}{ccc} \text{R} \cdot \text{CH} \cdot \text{CO} & & \text{R} \cdot \text{CH} \cdot \text{COOH} \\ | & \diagdown & | \\ | & \text{O} \rightleftarrows & | \\ | & \diagup & | \\ \text{NH} \cdot \text{CO} & & \text{N}{=}\text{C}{=}\text{O} \end{array}$$

because the reverse step takes place in the Curtius synthesis of NCA. The polymerization of isocyanates to linear anhydrides

$$\begin{array}{ccccc} \text{HO} \cdot \text{CO} \cdot \text{CH} \cdot \text{R} & & \text{N}{=}\text{C}{=}\text{O} & \text{HO} \cdot \text{CO} \cdot \text{CH} \cdot \text{R} & \text{N}{=}\text{C}{=}\text{O} \\ | & + & | & \rightarrow & | & | \\ \text{N}{=}\text{C}{=}\text{O} & & \text{HO} \cdot \text{CO} \cdot \text{CH} \cdot \text{R} & \text{NH} \cdot \text{CO} \cdot \text{O} \cdot \text{CO} \cdot \text{CH} \cdot \text{R} \end{array}$$

may be facile* and the subsequent rapid decarboxylation of the latter is known to give a peptide bond, i.e.,

$$\text{HO} \cdot \text{CO} \cdot \text{CH(R)} \cdot \text{NH} \cdot \text{CO} \cdot \text{O} \cdot \text{CO} \cdot \text{CH(R)} \cdot \text{N}{=}\text{C}{=}\text{O} \rightarrow$$
$$\text{HO} \cdot \text{CO} \cdot \text{CH(R)} \cdot \text{NH} \cdot \text{CO} \cdot \text{CH(R)} \cdot \text{N}{=}\text{C}{=}\text{O} + \text{CO}_2.$$

These decarboxylations were extensively studied by Dieckmann and Breest (7) who elucidated their mechanism.

The proposed mechanism demands the presence of a terminal isocyanate group in the resulting polymer. However, such a group may be easily lost, e.g., by reaction with moisture, and therefore it is not surprising that it was undetected in the final product. The isocyanate mechanism resembles the one proposed once by Kopple (8) to account for the solution polymerization of NCA.

The thermal decarboxylation of NCA's involves the 2-carbonyl group of the 1,3-oxazolidine-2,5-dione ring. This was proved (9) by labeling one of the CO groups of glycine NCA with C^{13} and examining the C^{13} content of the carbon dioxide evolved during thermal polymerization.

---

* The reaction $\text{R} \cdot \text{N}{=}\text{C}{=}\text{O} + \text{HO} \cdot \text{CO} \cdot \text{R}_1 \rightarrow \text{R} \cdot \text{NH} \cdot \text{CO} \cdot \text{O} \cdot \text{CO} \cdot \text{R}_1$ has to be catalyzed by bases. It is not known whether the presence of adjacent carboxyl groups facilitates this process.

The reaction of Leuch's anhydrides with primary and secondary amines proceeds in two ways, viz.,

$$
\begin{array}{c}
\text{R} \cdot \text{CH}\!-\!\text{CO} \\
\big| \qquad\quad \diagdown \\
\qquad\qquad \text{O} + \text{HNR}_1\text{R}_2 \\
\big| \qquad\quad \diagup \\
\text{NH}\!-\!\text{CO}
\end{array}
\quad
\begin{array}{c}
\text{NH}_2 \cdot \text{CH(R)} \cdot \text{CO} \cdot \text{NR}_1\text{R}_2 + \text{CO}_2 \\
\\
\\
\text{R}_1\text{R}_2\text{N} \cdot \text{CO} \cdot \text{NH} \cdot \text{CH(R)} \cdot \text{COOH.}
\end{array}
$$

The detailed mechanism of these two additions is the subject of the following section.

2. REACTIONS OF LEUCHS' ANHYDRIDES WITH PRIMARY AND SECONDARY AMINES

The reaction of NCA's with amines was reported by Fuchs (10) in 1922 and by Wessely (11) in 1925. From the resulting mixture they isolated amides formed by an interaction such as:

$$
\begin{array}{c}
\text{CH}_2\!-\!\text{CO} \\
\big| \qquad\quad \diagdown \\
\qquad\qquad \text{O} + \text{NH}_2\text{R} \rightarrow \\
\big| \qquad\quad \diagup \\
\text{Ph} \cdot \text{N}\!-\!\!-\!\text{CO}
\end{array}
\quad
\begin{array}{c}
\text{CH}_2 \cdot \text{CO} \cdot \text{NHR} \\
\big| \qquad\qquad\quad + \text{CO}_2 \\
\\
\text{Ph} \cdot \text{NH}
\end{array}
$$

and, as shown by subsequent work, these substances are the main products of the addition, if the amine is in large excess. For example, 90% of the phenylalanine amide was obtained when phenylalanine NCA was reacted with an excess of amine in ethyl acetate (12). Improvements of this procedure eventually led to a quantitative preparation (13) of dimethylamides of glycine, D,L-alanine, D,L-phenylalanine, and sarcosine from their respective NCA's.

In an excess of Leuch's anhydride, the amine formed from NCA by the action of the initiating amine reacts again with a molecule of NCA and produces a dimer possessing a terminal amine group. Repetition of these reactions leads therefore, to amine-terminated polypeptides. Hence, the "normal" or "simple" amine-propagated polymerization of Leuch's anhydrides is described by the overall reaction

$$\sim\!\text{CO} \cdot \text{CH(R)} \cdot \text{NH}_2 + (\text{NCA}) \rightarrow \sim\!\text{CO} \cdot \text{CH(R)} \cdot \text{NH} \cdot \text{CO} \cdot \text{CH(R)}\text{NH}_2 + \text{CO}_2.$$

Waley and Watson (14) were the first to report on a kinetic study of this reaction. They initiated the polymerization of sarcosine NCA by a preformed low molecular weight polysarcosine possessing a terminal amine group. This type of initiation, as the writers pointed out, simplifies the kinetics of the reaction, because the process is reduced to the propagation step only, i.e., one deals with the growth of a living polymer fed by a monomer. The rate of propagation was determined by the increase in carbon dioxide pressure at constant volume. This method leads, however, to some complications, viz., as the CO_2 pressure rises, the extent of carboxylation of the amine to the respective substituted carbamic acid becomes larger. The consequences of this reaction will be considered later (see p. 567).

The terminal amine groups are not destroyed by the reaction as shown by their titration at the onset of the polymerization and after its completion. This, therefore, proves that termination is absent or, at least,[13] negligible under the conditions maintained in the experiments of Waley and Watson.

The absence of termination and of chain transfer, viz., the existence of living polymers, is also demonstrated by the molecular weight of the resulting polymer, its number average degree of polymerization being given then by the simple relation

$$\overline{DP}_n = p + \text{(monomer supplied)/(initiator)}_0,$$

in which p denotes the degree of polymerization of the initiating oligomer. Furthermore, in the absence of termination and for an initiation which is not slower than the propagation step, the resulting product should have a narrow, Poisson-type, molecular weight distribution. This conclusion was verified by Waley and Watson (14). Polysarcosine prepared by their technique was found to be monodispersed, and this observation was subsequently confirmed by Fessler and Ogston (15) and by Pope et al. (16). Similar observations were reported for other polypeptides prepared from their respective NCA's by the action of primary amines, e.g., for poly-γ-benzyl L-glutamate resulting from *n*-hexylamine initiation in dimethylformamide (17). Finally, the preparation of block polymers provided additional proof for the lack of termination. For example, such a polymer was obtained (13) in the polymerization of D,L-phenylalanine NCA initiated by a polypeptide formed from sarcosine NCA. Numerous examples of other block polymers prepared by this technique have been reported.

A detailed examination of the kinetics of the amine-initiated polymerization of sarcosine led Waley and Watson (14) to postulate a reversible addition of amines to NCA, viz.,

$$
\begin{array}{c}
CH_2\text{—}CO \\
| \qquad\quad \diagdown \\
\quad\qquad O + HNR_1R_2 \rightleftarrows \\
| \qquad\quad \diagup \\
CH_3N\text{——}CO
\end{array}
\qquad
\begin{array}{c}
\qquad NR_1R_2 \\
\qquad | \\
CH_2\text{—}C\text{—}OH \\
| \qquad\quad \diagdown \\
\qquad\qquad O \\
| \qquad\quad \diagup \\
CH_3N\text{——}CO
\end{array}
$$

The intermediate adduct may be represented, and perhaps more realistically, as a resonating zwitterion (18)

$$
\begin{array}{c}
R_1R_2NH^+ \\
| \\
CH_2\text{—}C\text{—}O \\
| \qquad\qquad O \;\;\ominus \\
CH_3N\text{——}C\text{⋯}O
\end{array}
$$

because this form accounts better for the low entropy of activation of the addition process (see p. 569). The decomposition of such an intermediate into a relatively unstable carbamic acid, or its zwitterion

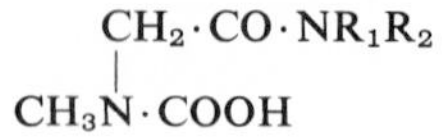

$$CH_2 \cdot CO \cdot NR_1R_2$$
$$|$$
$$CH_3N \cdot COOH$$

or

$$CH_2 \cdot CO \cdot NH^+R_1R_2$$
$$|$$
$$CH_3N \cdot COO^-$$

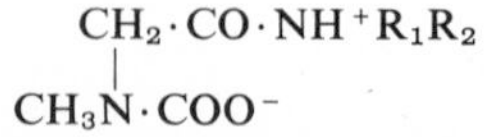

was demonstrated by Bailey (19) who isolated and identified these compounds. As shown by him and by others (see, e.g., ref. 31) these acids decarboxylate rapidly into the amides of the respective amino acids.

The detailed step sequence in the reaction of primary and secondary amines with NCA's may now be represented by the following scheme (20):

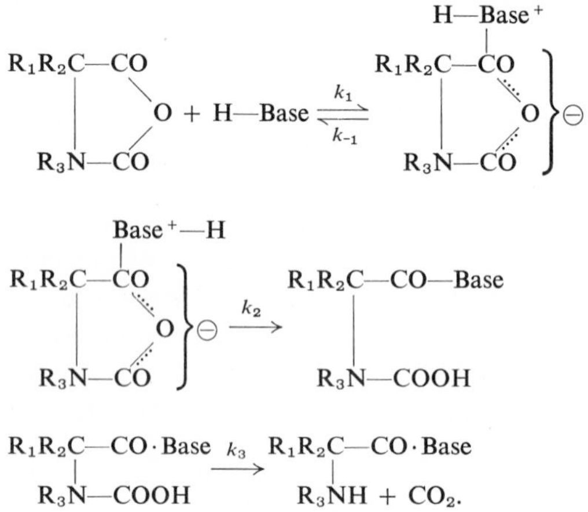

$$
R_1R_2C{-}CO \atop R_3N{-}CO
\diagdown O + H{-}Base \underset{k_{-1}}{\overset{k_1}{\rightleftharpoons}}
\left[{R_1R_2C{-}CO \atop R_3N{-}CO} \diagup O \right]^{\ominus} \quad H{-}Base^+ \tag{a}
$$

$$
\left[{R_1R_2C{-}CO \atop R_3N{-}CO} \diagup O \right]^{\ominus} \; Base^+{-}H \xrightarrow{k_2} {R_1R_2C{-}CO{-}Base \atop R_3N{-}COOH} \tag{b}
$$

$$
{R_1R_2C{-}CO \cdot Base \atop R_3N{-}COOH} \xrightarrow{k_3} {R_1R_2C{-}CO \cdot Base \atop R_3NH + CO_2.} \tag{c}
$$

Isotope labeling again proved (21) that decarboxylation involves only carbon-2 of the oxazolidine-2,5-dione (see also ref. 8) and studies of the kinetic isotope effect (22) demonstrated that this step is not rate-determining in the overall process. Therefore, the rate of the overall reaction is determined either by (a), i.e., the formation of the complex, or (b), viz., by opening of the ring between atoms 1 and 5. This conclusion applies also to the "normal" amine-propagated NCA polymerization.

Although, the most frequent interaction of amines with NCA's involves the 5-CO group, the reaction with the 2-CO group is not entirely excluded (23). This much less frequent event leads to the formation of an α-ureido acid

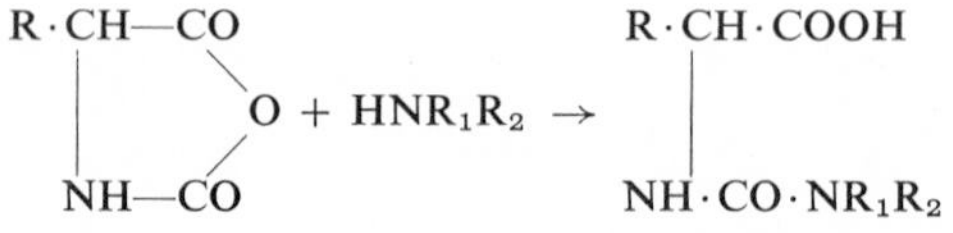

$$
{R \cdot CH{-}CO \atop NH{-}CO} \diagup O + HNR_1R_2 \rightarrow {R \cdot CH \cdot COOH \atop NH \cdot CO \cdot NR_1R_2}
$$

which cannot propagate the polymerization of NCA's and therefore, it terminates the growth of polymeric chains (23). It should be noted that such a termination takes place in *the course* of polymerization as a result of a wrong mode of monomer addition; it becomes impossible *after* completion of the polymerization, viz., a living polypeptide cannot be spontaneously terminated by this process.

As shown by the data collected in Table I, the extent of α-ureido acids formation depends on the nature of the attacking amine (24): for more basic

TABLE I

Yield of α-Ureido Acids in the Reaction of Amines with NCA's[a]

Amine	Glycine NCA	Sarcosine NCA, per cent	Phenylalanine NCA
Diethyl	100	—	80
tert-Butyl	90	0	60
Isopropyl	45	0	35
Ethyl	55	—	10
Dimethyl	Low	Low	2.5
Phenyl	—	—	0

[a] E. Katchalski and M. Sela, *Advan. Protein Chem.*, **13**, 249 (1958).

amines it seems to be more pronounced. This process requires a higher activation energy than the addition to the 5-CO group, and therefore, such a termination becomes more significant at elevated temperatures, e.g., in the thermal bulk polymerization of molten NCA's which yield polypeptides almost devoid of free α-amino groups.

Other termination steps are also possible. For example, in polypeptides formed from γ-benzyl L-glutamate NCA, Hanby et al. (25) demonstrated the presence of a stable, terminal pyrrolidone ring. The reaction proceeds as follows:

$$\sim CO \cdot CH - NH_2 \qquad \qquad \sim CO \cdot CH - NH$$
$$\quad | \qquad\qquad\qquad\qquad\qquad\qquad | \qquad |$$
$$\quad CH_2 \quad CO \cdot O \cdot CH_2 \cdot Ph \rightarrow \quad CH_2 \quad CO + Ph \cdot CH_2 \cdot OH$$
$$\qquad \diagdown \diagup \qquad\qquad\qquad\qquad\qquad \diagdown \diagup$$
$$\qquad CH_2 \qquad\qquad\qquad\qquad\qquad\qquad CH_2$$

and it may take place *during* or *after* completion of the polymerization. For example, the amine-initiated polymerization of γ-benzyl L-glutamate NCA (M/I ≈ 5) carried out in dioxane at room temperature was completed in about one-half hour. The product contained initially terminal amino groups but most of them disappeared within several days. Similar observations were reported by Doty et al. (26).

The early work of Wessely suggested the occurrence of some side reactions which led to the formation of diketopiperazines (11) or hydantoins (27). These reactions may put an upper limit to the degree of polymerization which can be obtained in polyadditions of NCA's but it is more probable that these products arise from an unsuccessful initiation (see p. 602) rather than from a termination. Their participation in the process decreases the yield of carbon dioxide evolved during polymerization.

An interesting termination which may account for the excess of carboxylic groups over amine end groups in the resulting polypeptide was suggested by Sluyterman and Labruyere (28b). The reaction seems to apply to the water-initiated polymerization of glycine and alanine NCA's carried out at elevated temperatures and leads to terminal hydantoin rings, viz.,

$$\sim NH \cdot CO \cdot CH(R) \cdot NH_2 + \underset{\underset{O\text{————}CO}{|\qquad\qquad|}}{OC\text{—}CH(R)\text{—}NH} \rightarrow$$

$$\sim NH \cdot CO \cdot CH(R)NH \cdot CO \cdot CH(R)NH \cdot COOH \rightarrow$$

$$\sim NH \cdot CO \cdot CH(R) \cdot N \underset{\diagdown}{\overset{\diagup}{\underset{CO\text{—}NH}{\overset{CO\text{—}CH(R)}{\big|}}}} + H_2O.$$

This mechanism elaborates the one proposed by Sela and Berger (23).

3. INITIATION OF "NORMAL" NCA's POLYMERIZATION BY OTHER AGENTS

In his first paper, Leuchs (1) reported that on mixing a small amount of water with glycine NCA, carbon dioxide was evolved and a resin was formed. This observation was generalized by him and by others, and for a while polymerization of NCA's in moist solvents was considered as an attractive method for preparing polypeptides. In fact, in 1947 Woodward and Schramm (28a) reported on a polymerization of L-leucine and D,L-phenylalanine NCA's in moist benzene which supposedly yielded polypeptides having molecular weights exceeding 1,000,000. Unfortunately, this claim turned out to be erroneous. The molecular weight of the resulting copolymer was determined by osmometry in benzene solution, and in this solvent extensive agglomeration of the investigated polyaminoacids takes place. End group analysis by Coleman and Farthing (29) and by other investigators (30) proved that the degree of polymerization of such a polypeptide did not exceed 100.

Reaction with water produces an amino acid which eventually initiates (28) the apparently "normal" amine-propagated polymerization of NCA. The initiation and hydrolysis are substantially slower than the propagation, e.g., at least a 600-fold water excess is required to hydrolyze NCA without inducing any appreciable polymerization. Hence, polymerization in the presence of water behaves kinetically as an autocatalytic reaction. This process is most useful for the preparation of polyamino acids possessing free terminal α-amino and ω̆-carboxyl groups. The action of alcohols and ammonia probably takes an analogous course—the produced amine-terminated derivatives act as initiators of NCA polymerization.

4. DETAILED KINETIC STUDIES OF "NORMAL" AMINE-PROPAGATED POLYMERIZATION OF NCA's

The pioneering work of Waley and Watson (14) was soon extended and elaborated by the studies of Ballard and Bamford (20). They showed that some of the complex features of the kinetics of sarcosine NCA polymerization, which were reported by the former workers, arose from the catalytic action of carbon dioxide. As the reaction progressed, the pressure of carbon dioxide increased in Waley and Watson's reactor, and hence its catalytic contribution became time-dependent. To avoid the problem of variable carbon dioxide pressure, Ballard and Bamford developed a technique in which the pressure of carbon dioxide was kept constant and the rate of polymerization was then determined by measuring the increase of its volume.

Later, alternative techniques for studies of the kinetics of NCA polymerization were developed. Patchornik and Shalitin (20a) described a method whereby the evolved carbon dioxide is continuously titrated. Nitrogen saturated with the solvent vapor is bubbled through the reacting solution and then passed through a carbon dioxide-absorbing train. The carbon dioxide swept by the gas is then titrated as absorbed. This method allows maintenance of an exceedingly low partial pressure of carbon dioxide in the reactor and thus the complicating features of the reaction caused by its presence may be avoided. Another technique giving similar advantages but fully automatic and allowing studies of reaction with half-life time as short as 1 minute was reported by Block (20b).

In their kinetic studies, Ballard and Bamford (20) followed the procedure of Waley and Watson and initiated the polymerization in nitrobenzene or *o*-nitroanisole by the preformed, low molecular weight "living" polypeptides of DP ~ 4 (DP = degree of polymerization). Since moisture had an undesirable effect, a rigorous drying technique for handling the reagents was recommended. In the absence of moisture, reproducible results were obtained with freshly prepared and rigorously purified anhydrides.

The catalytic action of carbon dioxide, which vitiates the kinetics of sarcosine NCA polymerization was not observed (20) in similar polymerizations of D,L-leucine and D,L-phenylalanine NCA's. The kinetics of the latter processes showed simple behavior. The rates were proportional to the monomer concentration over the whole range of conversion, and a set of experiments performed with different initial concentrations of the starting, living oligomer, see Figure 2, proved the polymerization to be a first order reaction with respect to growing ends. The absence of termination was again demonstrated by two observations: (*1*) the concentration of the terminal amine groups

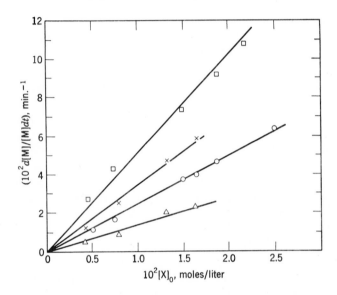

Fig. 2. Polymerization of *N*-carboxy-D,L-leucine anhydride initiated by preformed polymer in different solvents: (□) nitrobenzene, 45°C; (○) nitrobenzene, 25.2°C; (×) *o*-nitroanisole, 45°C; (△) *o*-nitroanisole, 25.0°C.

checked by several analytical techniques, was found to be constant during the whole course of the process, and (*2*) the DP_n of the resulting polymer was given by the equation

$$\overline{DP}_n = p + \text{(total monomer)/(total amount of initiating oligomer)}$$

p denoting, as previously, the degree of polymerization of the initiating oligomer.

The kinetic results obtained by Ballard and Bamford are summarized in Table II. The perfect linear relation of the log M with time observed for the reaction initiated by preformed polymer, provides proof that all the consecutive steps of the propagation proceed with *the same* rate constant, i.e., $k_{p,j}$ is independent of j for $j \geqq 4$. However, the rate constant of the step

$$\text{Base} \cdot \text{M} \cdot + \text{M} \rightarrow \text{Base} \cdot \text{M} \cdot \text{M} \qquad (j = 1),$$

was found to be three or four times greater than the other $k_{p,j}$'s.

TABLE II

Propagation Rate Constants for the Polymerization of NCA's of D,L-Leucine and D,L-Phenylalanine Initiated by Preformed Polymer[a]

Rate $= k$[NCA][growing ends]

	D,L-Leucine $k \times 10^2$, liter/mole sec		D,L-Phenylalanine $k \times 10^2$, liter/mole sec
T,°C	Nitrobenzene	o-Nitroanisole	Nitrobenzene
25	—	2.38	1.56
25.2	4.17	—	—
45	8.72	5.83	3.50
E, kcal/mole	6.9	8.5	7.7
A, liter/mole sec	5.0×10^3	3.8×10^4	6.0×10^3

[a] From D. G. H. Ballard and C. H. Bamford, *Proc. Roy. Soc.* (*London*), *Ser. A*, **223**, 495 (1954).

The first order dependence on monomer and initiator indicates that the bimolecular formation of the adduct is the rate-determining step in the polymerization of D,L-leucine or D,L-phenylalanine NCA's. Apparently, the opening of the ring and decarboxylation follow rapidly thereafter. Inspection of Table II shows that neither the propagation rate constants nor their activation energies are drastically affected by the small change in the structure of the solvent. The very low A factors apparently indicate a greater degree in polarity of the transition state than of the initial one; this observation justifies, therefore, the proposed resonating zwitterion structure for the adduct (see p. 563).

The simplicity of the polymerization kinetics of leucine and phenylalanine NCA's contrasts with the complex behavior of sarcosine NCA. For a constant concentration of growing ends and a constant pressure of carbon dioxide, the propagation was found to be first order with respect to monomer, but the rate increased with the carbon dioxide pressure as shown in Figure 3. Moreover, dependence on the concentration of growing ends was quite complex, and to elucidate this matter it was necessary to investigate the mechanism of the catalytic action of carbon dioxide. This catalysis is clearly demonstrated by Figure 4 which shows that the rate of polymerization which proceed for a while at a high carbon dioxide pressure abruptly decreases upon sudden lowering of its pressure.

Studies by Frankel and Katchalski (31) proved that carbon dioxide reacts reversibly with amino acids and their derivatives giving the respective carbamic acids, viz.,

$$\sim\!\!CO\cdot CH(R)\cdot NH_2 + CO_2 \rightleftarrows \sim\!\!CO\cdot CH(R)\cdot NH\cdot COOH$$

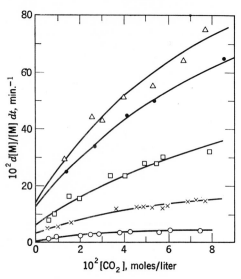

Fig. 3.   Polymerization of N-carboxy-sarcosine anhydride in nitrobenzene at 25°C initiated by preformed polymer. Values of 10^3 $[X]_0$ mole/liter; ($\triangle$) 1.58; ($\bullet$) 1.40; ($\square$) 1.09; ($\times$) 0.748; ($\bigcirc$) 0.900.

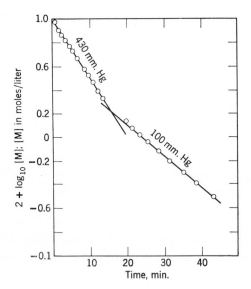

Fig. 4.   Polymerization of N-carboxy-sarcosine anhydride in nitrobenzene at 25°C initiated by preformed polymer: $[M]_0 = 9.5 \times 10^{-2}$ mole/liter; $[X]_0 = 0.748 \times 10^{-2}$ mole/liter. Pressure of carbon dioxide was suddenly reduced from 430 to 100 mm Hg at $t = 15$ min.

and, as suggested by Faurholt (32), who investigated the interaction of carbon dioxide with aliphatic amines, further equilibria may be established in the presence of an excess of amine, i.e.,

$$\sim\!\!\text{CO}\cdot\text{CH(R)}\cdot\text{NH}\cdot\text{COOH} + \text{H}_2\text{N}\cdot\text{CH(R)}\cdot\text{CO}\!\sim\ \rightleftarrows$$

$$\sim\!\!\text{CO}\cdot\text{CH(R)}\cdot\text{NH}\cdot\text{COO}^-,\text{H}_3\text{N}^+\cdot\text{CH(R)}\cdot\text{CO}\!\sim\ \rightleftarrows$$

$$\sim\!\!\text{CO}\cdot\text{CH(R)}\cdot\text{NH}\cdot\text{COO}^- + \sim\!\!\text{CO}\cdot\text{CH(R)}\cdot\text{NH}_3^+.$$

Ballard and Bamford assumed, therefore, that these compounds were formed in the polymerizing mixture in the presence of carbon dioxide. They verified their hypothesis by determining the solubility of carbon dioxide in pure solvent and in the solution of amine-terminated polypeptide. The difference was attributed to the "bonded" carbon dioxide, and the results indicated that in a $0.2M$ solution* of the amines, the bulk of the carbon dioxide is in the form of the ammonium salt (ion-pairs). The equilibrium constant for salt formation in nitrobenzene at 25°C was found to be 22 liter^2/mole^2 for the oligomer of sarcosine and about 60 liter^2/mole^2 for its dimethylamide. The latter seems, therefore, to be more basic than the former.

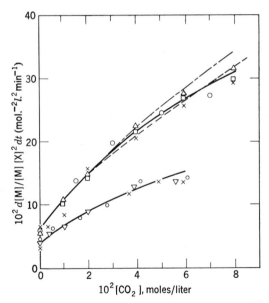

Fig. 5. Polymerization of *N*-carboxy-sarcosine anhydride in nitrobenzene initiated by preformed polymer. Values of 10^2 [X]$_0$: (△) 1.58; (○) 0.400; (□) 1.09; (▽) 1.87; (×) 0.748; Full lines, experimental curves at 25°C and 45°C, respectively; (– — — —) curve calculated from equation 11; (————) curve calculated from equation 12.

* At the concentration of amines corresponding to the conventional composition of the polymerizing mixture the uptake of carbon dioxide is too low to be reliably determined.

Formation of carbamic acids has a dual effect on the polymerization.

1. The acids and their salts do not grow. They behave, therefore, as "dormant" polymers which are in dynamic equilibrium with the living ones. Hence, in deriving the correct order of propagation with respect to growing ends, only the concentration of living species should be used in computations. When this is done, the propagation step of sarcosine NCA polymerization becomes truly *second* order with respect to *living* ends. This may be seen from the data given in Figure 5.

2. The carbamic acids may catalyze polymerization, and indeed, in the polymerization of sarcosine NCA the catalytic effect of weak carboxylic acids, such as α-picolinic acid, was demonstrated (20). The polymerization is inhibited, however, at high acid concentration, or upon addition of a strong acid such as *o*-nitrobenzoic, because the decrease in the concentration of the base, caused by its conversion into salt, outweighs the catalytic effect of the acid. This, indeed, is seen in Figure 6.

Apparently, the rate of polymerization of sarcosine NCA is not entirely determined by the rate of formation of the adduct. The enhancing effect of decreasing temperature on the rate of polymerization, which was reported by

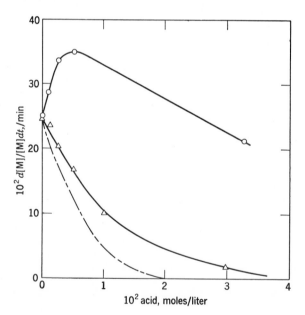

Fig. 6. Polymerization of *N*-carboxysarcosine anhydride in nitrobenzene at 25°C. initiated by preformed polymer: $[CO_2] = 1.05 \times 10^{-2}$ mole/liter; $[X]_0 = 1.60 \times 10^{-2}$ mole/liter. (○) α-picolinic acid added; (△) *o*-nitrobenzoic acid added; (– – – –) calculated curve based on assumption of nearly complete neutralization of base by acid at concentration 4×10^{-2} mole/liter.

Waley and Watson (14) and confirmed by Ballard and Bamford (20), suggests that the addition of NCA to the growing amine is reversible, i.e., k_{-1} (reaction a) is *not* negligible when compared with k_2 (reaction b) (see p. 564). Since adduct formation is exothermic, its stationary concentration *increases* on lowering the reaction temperature and, consequently, the apparent activation energy of the overall process may become negative.* Under these conditions the rate of polymerization is governed by the rate of ring opening (reaction b), and apparently bases and acids catalyze this process. These ideas are rationalized by the following sequence of reactions originally postulated by Ballard and Bamford (20a):

$$\begin{array}{ccc}
R \cdot NMe & & CH_2\!-\!CO \cdot NMeR \\
CH_2\!-\!C\!-\!OH \quad & \xrightarrow{\ k_2'\ } & | \\
\qquad \text{O} + HN(Me)\!\sim & & Me \cdot N\!-\!COO^- + H_2N^+(Me)\!\sim \\
Me \cdot N\!-\!-\!C\!=\!O & &
\end{array} \qquad (b')$$

and

$$\begin{array}{ccc}
R \cdot NMe & & CH_2\!-\!CO\!-\!NMeR \\
CH_2\!-\!C\!-\!OH \quad & \xrightarrow{\ k_2''\ } & | \qquad\qquad H \\
\qquad \text{O} & & \qquad\qquad O \\
Me \cdot N\!-\!-\!C\!=\!O \quad \text{O} & & Me \cdot NH + CO_2 + OC \cdot R \\
\qquad\qquad\quad \| & & \\
\qquad\qquad HO \cdot C \cdot R_1 & &
\end{array} \qquad (b'')$$

The former reaction accounts for the catalytic action of a base and, therefore, also for that of a growing polymer, while the latter explains the catalytic action of an acid. Note that an ammonium ion (or ion-pair) may also act as an acid and catalyze reaction.

Decarboxylation of the zwitterions formed by (b') is very rapid (18), and hence, in the absence of acids, the propagation should be second order with respect to base,† if the spontaneous decarboxylation of the primary adduct is

---

* The negative "activation energy" of polymerization is sometimes observed in other systems. For example, carbonium ion polymerizations often proceed faster at lower temperatures than at higher. This may be accounted for by a decrease in the rate of termination which probably has a higher activation energy than initiation and propagation. However, the polymerization of NCA's proceeds without termination, and such an explanation cannot apply. The acceleration of some ionic polymerizations on lowering the temperature arises from an increased dissociation of ion-pairs into much more reactive free ions (see p. 416). This explanation again cannot be valid, if the addition (step a) is rate-determining.

† Recent studies by Bamford and Block cast doubt on reaction b'. The polymerization appears *not* to be catalyzed by bases; it is nevertheless second order with respect to growing ends. (Private communication from Professor Bamford.)

very slow and k_2' [base] $\ll k_{-1}$. On the other hand, the kinetic relation, $k_{-1} \ll k_2'$ [base], seems to apply to the polymerization of D,L-leucine or D,L-phenylalanine, and indeed their rate of propagation is determined by the rate of formation of the primary adduct. In such a process, no catalysis by acids or bases may be observed, and the overall activation energy of the polymerization must be positive in agreement with experimental findings. We see, therefore, that a change in the sign of the inequality accounts for the entirely different kinetic behavior of sarcosine NCA on the one hand and of D,L-leucine or D,L-phenylalanine NCA's on the other.

The detailed mechanism of sarcosine NCA polymerization may be radically modified by a change of solvent. For example, in dimethylformamide this reaction is first order with respect to growing ends (see, e.g., ref. 44) instead of being second order as in nitrobenzene. The higher acidity of the former solvent reduces the basicity of the dissolved amine and, therefore, destroys its catalytic action. This effect influences also the equilibrium between amine and dissolved carbon dioxide. In nitrobenzene, the carbamic acid produced is associated with the free amine forming the respective ammonium salt (ion-pairs), whereas in the more acidic dimethylformamide it exists as an unionized acid. The uniformity of the mechanism for these apparently diverse types of polymerization is the most appealing feature of the proposed scheme.

The relatively low k_1 values observed in the polymerizations of D,L-leucine and D,L-phenylalanine NCA's are caused by the low basic strength of the terminal amines formed in these processes. The respective k_1 values are probably further decreased by steric hindrance caused by the bulky side chains of leucine and phenylalanine.

The inefficiency of water and of amino acids in initiating the NCA polymerization is now easily understood. The reaction has to be initiated and propagated by a base, whereas the amino acids exist mainly in their zwitterion form. Hence, the initiation must be slow because it involves the unionized amino acid present only at low concentration. As the peptide grows in length, the equilibrium between the zwitterion and its unionized form shifts in favor of the latter, and the rate of growth increases. This phenomenon contributes toward the autocatalytic behavior of that polymerization.

To recapitulate, the mechanism of the "normal" amine-propagated polymerization of NCA's assumes initiation and propagation by primary or secondary amines which act as nucleophiles and add to the 5-CO group of the NCA. The rates of initiation and of propagation increase with the basicity of the respective, initiating or propagating amines, e.g., initiation of D,L-phenylalanine polymerization by the weakly basic p-chloroaniline is 600 times slower than its homopropagation step (30). The presence of bulky substituents in the amine may hamper the rate of its addition and the significance of this phenomenon will be seen later. The reader may appreciate the effect of various structural factors upon the rate of polymerization by inspecting the data collected in Table III.

TABLE III
Rates of Amine-Initiated Solution Polymerizations of NCA's

NCA of	Solvent	Initiator	T,°C	$k \times 10^2$, liter/mole sec[a]	Refs.
D,L-Alananine	Dioxane	n-Hexylamine	35	4.3	(33)
D,L-α-Amino-n-butyric acid	Dioxane	n-Hexylamine	35	1.7	(33)
Glycine	Dioxane	n-Hexylamine	35	34	(33)
Glycine	Dimethylformamide	Diethylamine	25	60	(34)
γ-Benzyl L-glutamate	Dioxane	n-Hexylamine	35	3.9	(33)
γ-Benzyl L-glutamate	Dioxane	n-Hexylamine	34	4.4	—
γ-Benzyl L-glutamate	Dioxane	n-Hexylamine	25	3.2 (0.6)	(17)
γ-Benzyl L-glutamate	Benzene	Diethylamine	25	41.6	(34)
γ-Benzyl L-glutamate	Nitrobenzene	n-Hexylamine	25	11 (2.1)	(17)
γ-Benzyl L-glutamate	Nitrobenzene	Diethylamine	25	7	(34)
γ-Benzyl L-glutamate	Chloroform	n-Hexylamine	25	30 (4.5)	(17)
L-Leucine	Dioxane	Diethylamine	25	1.2	(34)
L-Leucine	Benzene	Diethylamine	25	33	(34)
D,L-Leucine	Dioxane	n-Hexylamine	35	0.6	(33)
D,L-Leucine	Nitrobenzene	Preformed polymer	25	4.2	(20)
D,L-Leucine	o-Nitroanisole	Preformed polymer	25	2.4	(20)
D,L-Phenyl-alanine	Benzene	Diethylamine	25	16.6	(34)
D,L-Phenyl-alanine	Dioxane	n-Hexylamine	35	1.2	(33)
D,L-Phenyl-alanine	Nitrobenzene	Preformed polymer	25	1.6	(20)
D,L-Phenyl-alanine	Dimethylformamide	Diethylamine	25	5.8	(34)
L-Proline	Dimethylformamide	Diethylamine	25	75	(34)
Sarcosine	Nitrobenzene	Preformed polymer	25	25	(14)
Sarcosine	Acetophenone	Preformed polymer	25	26	(14)
D,L-Valine	Dioxane	n-Hexylamine	35	0.2	(33)

k is calculated from the equation: Rate = k [initiator] [M]. E. Katchalski and M. Sela, *Advan. Protein Chem.*, **13**, 249 (1958).

[a] Some of these polymerizations are heterogeneous, viz., the polymer precipitates out as it is formed. Therefore, of the quoted rate constants some may be erroneous.

In principle, the addition is reversible and, if the rupture of the C(5)—O(1) bond of the NCA adduct is sufficiently slow, an equilibrium may be established between adduct, monomer, and initiating or propagating base. The polymerization of sarcosine NCA exemplifies such a case.

If the fission of the C(5)—O(1) bond is much faster than the dissociation of the adduct into its original components, bimolecular formation of the adduct between monomer and growing amine governs then the propagation step. This is observed in the amine-initiated polymerization of D,L-leucine or D,L-phenylalanine NCA in nitrobenzene, or in the polymerization of γ-benzyl L-glutamate NCA in dimethylformamide. Catalysis of the fission reaction cannot affect the rate of those polymerizations for which adduct formation is the rate-determining step. However, if equilibrium is established between the NCA and the amine on the one hand and the adduct on the other, catalysis of the fission step enhances the polymerization. In such systems, weak acids and bases may act catalytically. The self-catalysis of the growing amine, if it occurs, leads to second order dependence of the propagation step on the concentration of growing ends.

Termination is negligible, although not entirely eliminated. A "wrong" monomer addition gives, e.g., urea derivatives which cease to grow. Lack of termination leads to the relation: DP_n (polymer) = monomer/initiator; and for a not too slow initiation the resulting product has a Poisson molecular weight distribution.

The proposed mechanism applies to both N-substituted as well as non-N-substituted NCA's, and in view of the nucleophilic character of the growing ends, the reactions may be classified as anionic polymerizations.

5. DEVIATIONS FROM THE SCHEME OF "NORMAL" POLYMERIZATION

Further studies of NCA polymerization initiated by primary or secondary amines, as well as by other initiators, showed that the mechanism summarized above needed some modification. Blout and Asadourian (35) reported that the polymerization of γ-benzyl L-glutamate NCA initiated by n-hexylamine in dioxane yields two types of polypeptides, recognized today as the helical and the random coil forms. The amount of the latter type decreased as the monomer-to-initiator ratio (M/I) was increased. It appears that the molecular weight of the polymer determines the relative proportions of these two forms, with the high molecular weight material composed of the helical form only.

In the same year, some unusual kinetic features of this polymerization were described (36). As shown in Figure 7, taken from the work of Lundberg and Doty (17), the initial, slow first order reaction seems to be replaced in the later stages of polymerization by a fast and also first order reaction. The rather abrupt change in rate occurs when the average DP of the product exceeds the value of 8–10. In fact, only the fast reaction is observed when the polymerization is initiated by an amine-terminated polypeptide of DP > 10. The molecular weight distribution of the resulting polymer was found (26) to be very broad, a fact established for polymers having molecular weights of about 20,000 or less by applying the Archibald approach-to-equilibrium

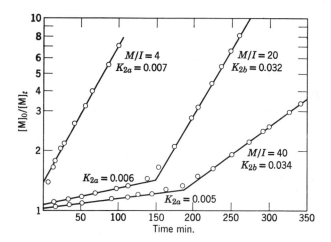

Fig. 7. The polymerization of γ-benzyl L-glutamate NCA in dioxane at 25°C initiated by n-hexylamine. Values of [NCA] initially and at time t are denoted by $[M]_0$ and $[M]_t$, respectively. I is the initial concentration of n-hexylamine; k_{2a} and k_{2b} are the apparent propagation coefficients in liters/mole sec.

ultracentrifugation technique. Had the reaction involved a termination, the anticipated narrow, Poisson-type molecular weight distribution would be broadened; however, as shown by the calculations of Katchalski et al. (37), termination could *not* account for the substantial broadening observed by Doty (26). Moreover, later studies of Sober (93) revealed a bimodal molecular weight distribution of the product.

In searching for a solution to these puzzling features of the polymerization, it was suggested (36) that the polymer starts its growth slowly as a random coil but eventually continues to grow rapidly, after acquiring a helical structure which becomes stable for DP greater than 8–10. The merits of this most interesting suggestion are discussed in Chapter XI, but at this juncture it should be emphasized that the reported molecular weight distributions, as well as the kinetics of these polymerizations, do not conform to the simple mechanism outlined in the preceding section and demand, therefore, new approaches to the problem.

The problem of a two-stage propagation led to an enlightening argument. In a brief communication, Ballard and Bamford (38) disputed Doty's findings. They had noticed that γ-benzyl L-glutamate NCA cannot be rigorously purified by sublimation because of its low volatility. Acceleration of polymerization, they claimed, is observed whenever the prepared anhydride required several crystallizations for its purification; while a compound synthesized by an improved procedure, requiring only one rapid crystallization, polymerized over the whole conversion range in a perfectly normal way—the rate being described by a single first order reaction. Moreover, contrary to Doty's observations (17), Ballard and Bamford (38) insisted that the rate of polymerization of "pure" NCA is the same whether the reaction had been initiated by n-hexylamine or by a performed polymer with a degree of polymerization up to 15.

In their reply, Doty and Lundberg (39) questioned the purity of Bamford's samples which, they pointed out, were not recrystallized from methylene chloride at $-30°C$, a procedure Blout considered as essential to assure the purity of NCA. They correctly stressed that the "simple" kinetic scheme cannot account for the broad molecular weight distribution produced by a two-stage polymerization. This result seems to be indisputable, e.g., a sample produced by such a polymerization had $DP_n = 20$ and $DP_w = 170$. In fact, recent fractionation of the polymer on a Williams-Baker column (66) confirmed this result, the molecular weight distribution of the investigated polypeptide showed two distinct maxima, contrary to the earlier evidence reported by the same workers (67).

Bamford's group admitted in later publications (40) that acceleration *is* observed in the above polymerizations. Samples of NCA's were exchanged between Blout's and Bamford's groups, and the purity of both types were found to be satisfactory. Polymerization of both samples led to acceleration, although the effect observed by Bamford's group was much less dramatic than that reported by Doty (increase in the rate by a factor of 2 rather than by 5). However, a new problem was raised because Bamford's group questioned the homogeneity of the solution and attributed the acceleration to gel formation.

The problem of reaction heterogeneity will be considered in a later section (see p. 614). The writer wishes to stress, however, that this controversy has been reviewed in so much detail in order to impress upon the reader the difficulties and pitfalls of research when the results are so very susceptible to traces of impurities and to slight changes in experimental conditions.

Another anomaly which called for revision of the "simple" mechanism of amine-initiated polymerization of NCA was first reported by Blout and Karlson (49) and stressed recently by Bamford and Block (18). As seen from

Figure 8, taken from their paper, the polymerization of γ-ethyl L-glutamate NCA initiated by diisopropylamine is *faster* and yields a higher polymer than the reaction performed under identical conditions but initiated by *n*-hexylamine. "Normal" initiation by both amines is not *slower* than the propagation step and, therefore, the rate constant for the polymerization yielding peptides of DP = 10 or higher should be *independent* of the initiator's nature if the "simple" mechanism is valid (see, e.g., the data listed in Table III).

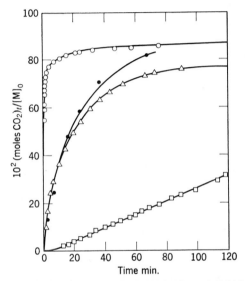

Fig. 8.   Conversion-time curves for the polymerization of NCA's (0.224 mole/liter) with amines (0.915 mole/liter) at 25°C in N,N-dimethylformamide: (○) γ-ethyl L-glutamate NCA (diisopropylamine); (△) γ-ethyl L-glutamate NCA (*n*-hexylamine); (●) sarcosine NCA (*n*-hexylamine); (□) sarcosine NCA (diisopropylamine).

Evidence provided by Blout and Karlson (49) and by Bamford and Block (18) implies, therefore, that there are at least two modes of *propagation* even in a polymerization initiated by primary or secondary amines. Evidence for other modes of propagation is apparent from studies of polymerizations initiated by tertiary amines and other aprotic initiators. We shall, therefore, review these reactions first and then continue the discussion of possible mechanisms for such intriguing polymerizations.

6. POLYMERIZATION OF NCA's INITIATED BY TERTIARY AMINES AND OTHER APROTIC BASES

The initiating capacity of tertiary amines in NCA polymerization was recognized by the earlier investigators of this process. Thus, Wessely (11) reported in 1925 the polymerizations of glycine and phenylalanine NCA's which proceeded spontaneously in pyridine at ambient temperatures, apparently being initiated by this tertiary base. In the following paper (12) a similar polymerization of sarcosine NCA was described. It was believed that this initiator apparently formed *cyclic* polypeptides because no terminal end groups could be detected (41). It is significant that appreciable quantities (a few per cent) of 3-hydantoin-acetic acid derivatives were found in polymers obtained from glycine and phenylalanine NCA's, but *none* were detected in the polymerized sarcosine NCA (12). This evidence suggests that the mechanisms of polymerization initiated by aprotic bases may be different for the non-*N*-substituted and the *N*-substituted anhydrides.

At the beginning of these studies, some controversy arose as to whether traces of water were necessary for initiation induced by tertiary amines. Initiation by primary and secondary amines has been explained by postulating transfer of a labile proton to the monomer. Such protons are not available in a tertiary base, and therefore Coleman (42) suggested cocatalysis by traces of water which would supply the protons needed. However, elaborate and painstaking studies by Wessely's group (41) demonstrated that water is not necessary for pyridine initiation, and more recent studies by Bamford's group (43,44) have confirmed this fact.

Another controversy had been created by the problem of whether tertiary amines really initiate the polymerization of the *N*-substituted NCA's. Wessely (11,12) found pyridine unable to initiate the polymerization of *N*-phenylglycine NCA, but reported the polymerization of sarcosine, also an *N*-substituted NCA, in rigorously dried pyridine (41). This observation was questioned by Bamford's group (43,44) which claims explicitly that carefully purified sarcosine NCA is *not* polymerized by tertiary amines. It is even harder to decide whether L-proline NCA, another *N*-substituted anhydride, or any of its derivatives may be polymerized by aprotic bases. The extremely high reactivity of these monomers makes the experimental results disputable. Pyridine-initiated polymerizations of these NCA's were observed by Katchalski's group (45), whereas Bamford's team (18,46) reported only an extremely slow polymerization of L-proline NCA in the presence of a tertiary base, although very rapid polymerization occurred upon addition of 3-methyl-hydantoin to a slowly polymerizing mixture. It is difficult to visualize an adventitious inhibitor which could slow down the polymerization studied by Bamford, while it is plausible to assume the presence of some adventitious impurities which enhanced the reaction studied by other workers. Therefore, the results of Bamford's group appear to be genuine, and the *N*-substituted NCA's might not be readily polymerized by tertiary amines, if all proton-donating impurities are rigorously excluded. However, Bamford's conclusion should not be generalized as it might not apply to aprotic bases stronger than tertiary amines (see, e.g., p. 612).

The NCA's polymerizations initiated by tertiary amines and by other aprotic bases such as sodium hydroxide, sodium methoxide, triphenylmethyl-sodium, etc., are of the greatest significance because this method has led to truly high molecular weight synthetic polypeptides. Such materials are most valuable as models for extensive studies of the physical and chemical properties of proteins. Their synthesis was reported in 1954 by Blout, Karlson, Doty, and Hargitay (47) who polymerized γ-benzyl L-glutamate NCA in dry dioxane, or in its mixture with tetrahydrofuran, using a methanol solution of sodium hydroxide as the initiator. The polymerization was carried out at 25°C and the reaction was completed in about 4 hr. The molecular weight of the product was found to exceed 100,000 and, in fact, samples having molecular weights of 350,000 were reported (48). Molecular weights were determined by light scattering employing Zimm plots over the angular range of 30–135°. The absence of polymer association was definitely established because the results obtained in dichloroacetic acid agreed with those found in a chloroform-formamide solution. Moreover, the reduced viscosity varied linearly with concentration down to the lowest measurable dilutions.

Rigorous purification of NCA and meticulous exclusion of any traces of moisture, hydrochloric acid, amino acids, etc., appear to be imperative for a reproducible preparation of really high molecular weight materials (47). Further improvements (49) eventually led to polypeptides having molecular weights over 1,000,000.

7. CONDITIONS LEADING TO HIGH MOLECULAR WEIGHT SYNTHETIC POLYPEPTIDES

The discovery of high molecular weight synthetic polypeptides came as a result of systematic studies by Blout and his co-workers of the conditions which limit the degree of polymerization of products formed by polymerization of NCA. Contrary to previous beliefs, it was found that the polymerization initiated by *n*-hexylamine in dry dioxane yields polypeptides of higher molecular weights than expected from the monomer-to-initiator ratio, provided M/I < 100. It should be stressed again that this observation demonstrates that some additional modes of propagation contribute to the reaction, i.e., the "simple" polyaddition mechanism cannot account *in toto* even for the polymerization initiated by primary amines in dioxane.

The increase in the M/I ratio above 50 leads only to a small increase in the molecular weight of the polymer, and its value remains virtually constant when M/I exceeds 400 or 500. Hence, some inherent termination steps limit further growth of the polymeric molecule and, therefore, the *n*-hexylamine-initiated polymerization cannot give products of MW (molecular weight) greater than 100,000, or of DP higher than 450. Searching for a method which would give still higher molecular weight polypeptides, Blout explored the action of other initiators and a methanol solution of sodium hydroxide was the first which produced the desirable result. As shown in Figure 9, taken from the paper by Blout and Karlson (49), at an M/I ratio of about 50, the molecular weight of the product was twice as high for the polymer initiated

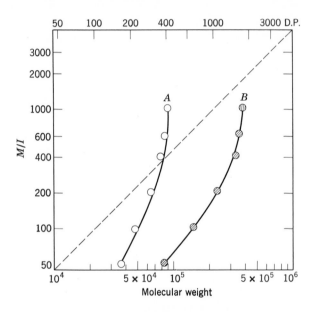

Fig. 9. Poly-γ-benzyl L-glutamate formed in polymerizations in dioxane solutions; (*A*) initiator, *n*-hexylamine; (*B*) initiator, sodium hydroxide. Degree of polymerization (DP) as a function of the anhydride to initiator ratio, M/I. Molecular weights were obtained from reduced specific viscosity at concentrations of 0.2% in dichloroacetic acid.

by sodium hydroxide than for that formed by *n*-hexylamine initiation. A limiting value of about 500,000 was reached when [M]/[NaOH] exceeded 500.

These results stimulated investigation of other alkali initiators. For the ratio M/I ≈ 50, sodium methoxide yielded a polymer having MW in excess of 100,000, and products having MW over 600,000 were obtained when M/I was greater than 1000. Similar results were reported for sodium borohydride initiation. Most unexpectedly, polymerization initiated by diethylamine in a dioxane solution produced a polypeptide having a constant molecular weight of about 200,000 and independent of the value of the M/I ratio which was varied from 50 up to 1000. Initiation by triethylamine produced an even higher degree of polymerization approaching 4500 (MW ≈ 1,000,000). Since tertiary amine initiation was known for a long time, it is most surprising that the synthesis of very high molecular weight polypeptides was not reported earlier. Correlation of the nature of the initiator with the molecular weight of the polypeptide is summarized in Table IV.

TABLE IV

Effect of Initiator on Molecular Weight of Polypeptide[a]
(Monomer: γ-benzyl L-glutamate NCA; M/I = 5. Solvent: dioxane. Temperature: 25°C)

Initiator	Molecular weight × 10^{-3}, from [η] in dichloracetic acid
n-Hexylamine	15
Diethylamine	83
Triethylamine	280
Sodium hydroxide	31
Sodium methoxide	29.5
Sodium borohydride	32.5

[a] From E. R. Blout and R. H. Karlson, *J. Am. Chem. Soc.*, **78**, 941 (1956).

All the polymerizations discussed above were found to proceed much faster (47,49) than those previously observed. The molecular weights of the prepared polypeptides appear to be constant for monomer concentrations ranging from 0.5 to 5%; however, they decrease upon further increases in monomer concentration. Raising the temperature of polymerization above 30°C also had an unfavorable effect on the molecular weight of the polymer, and this might explain the strange effect due to changes of monomer concentration. It may be that in a more concentrated solution, the local temperature increases in the vicinity of the growing polymers because the reaction becomes fast and heat transfer in a viscous medium is very slow. The same explanation might account for the observed decrease in molecular weights caused by the precipitation of the polymer because it is probable that in the precipitated phase, swollen by the monomer, its concentration increases in the vicinity of growing ends (for further discussion of heterogeneous polymerization see p. 614).

The effect of solvent upon the molecular weight of the product is summarized in Table V. High molecular weight polypeptides were obtained even in benzene solution, indicating that the polar nature of the solvent is not imperative for these processes. This is puzzling because there are good reasons to believe that some ionic species participate in these polymerizations and obviously a hydrocarbon solvent does not favor their formation (see, however p. 610). On the other hand, it is possible that the growing end of a high molecular weight polymer is surrounded by its polar chain. Thus, its immediate environment may be polar even when the bulk of the solvent is nonpolar.

TABLE V

Effect of Solvent on Molecular Weight of Polypeptide[a]
(Monomer: γ-benzyl L-glutamate NCA; M/I = 200. Initiator: sodium methoxide. Temperature: 25°C)

Solvent	[M], g/100 cc	Molecular weight $\times 10^{-3}$, from $[\eta]$ in dichloroacetic acid
Dioxane	2	365
Anisole	2.5	345
Benzene	1	340
Chloroform	2.5	290
Chlorobenzene	2.5	230
Dioxane + 0.1% hexaldehyde	5	170
		135
Ethyl acetate	2.5	134
Nitrobenzene	2.5	124
Acetonitrile	2.5	115
Dimethylformamide	5	83
Nitromethane	2.5	59
Methanol	2.5	12
Dioxane + 1% hexaldehyde	5	None

[a] From E. R. Blout and R. H. Karlson, *J. Am. Chem. Soc.*, **78**, 941 (1956).

Data quoted in Table V illustrate also the effect of some impurities on the molecular weight of the product. The presence of 0.1% of hexaldehyde in the dioxane solution was sufficient to reduce the molecular weight of the product by a factor of 2, and only a low molecular weight oligomer, not precipitated by ethanol, was produced when the amount of aldehyde was increased to 1%. The presence of acetone had no influence on the product. It is possible that impurities present in the reacting mixture, or formed by some side reactions, impose an upper limit on the maximum molecular weight of the polypeptide formed in the system.

Blout and Karlson (49) tried to explain the high DP values of polymers produced at relatively low ratios of M/I by assuming that the interaction of the initiator with the monomer involved two or more simultaneous reactions, of which only one gave a product leading to high polymers, while the products produced by the other reactions did not grow, i.e.,

Product I → Polymer

Initiator + Monomer

Product II → no Polymer.

Such a scheme might account for the inequality DP $\gg$ M/I. However, it requires also, contrary to observations, a proportionality between DP and M/I.

TABLE VI

Post-Polymerization of γ-Benzyl L-Glutamate NCA[a]

(Initiator: sodium methoxide. Momomer: constant concentration (4%).
Solvent: dioxane)

Experiment	Total M/I	Added M/I	DP_w
1	20	—	650
	40	20	750
2	20	—	550
	40	20	650
	60	20	650
	80	20	650
	100	20	700
3	100	—	1100
	200	100	1100
4	400	—	2150
	800	400	2250
5	1000	—	3350
	2000	1000	3500

[a] From M. Idelson and E. R. Blout, *J. Am. Chem. Soc.*, **80**, 2387 (1958).

Further peculiarities of the strong-base initiated polymerization were revealed by the work of Idelson and Blout (50). Polymerization occurred whenever the monomer, γ-benzyl L-glutamate NCA, was added to a solution containing products resulting from a previously completed polymerization initiated by sodium methoxide in dioxane. Each post-polymerization resulted in quantitative conversion of the added monomer to polymer. At each stage of this repetitive process, the product had a constant molecular weight for a constant amount of added monomer. Typical experimental results are shown in Table VI and it is obvious that the observed phenomenon cannot be explained by the growth of living polymers.

Idelson and Blout also demonstrated that the molecular weight of the product formed in a single-stage polymerization did *not* increase, or increased only slightly as the conversion of monomer rose from 25 to 100%. This is shown in Table VII taken from their paper. Since actually all of the monomer was converted to polymer, we see again that the concept of living polymers cannot account for these observations.

TABLE VII

Degree of Polymerization of Polybenzyl L-Glutamate at Various Conversions[a]

M/I	NCA consumed, %	DP_w
100	25	1100
	50	1250
	75	1250
	100	1250
200	31	800
	40	1000
	53	1050
	62	1050
	89	1100
	100	1200
400	40	1650
	51	1750
	73	2100
	100	2400
1000	21	2550
	51	3050
	76	3450
	100	3350

[a] From N. Idelson and E. R. Blout, *J. Am. Chem. Soc.*, **80**, 2387 (1958).

This result, and the occurrence of post-polymerization, are highly significant. They indicate:

1. A continuous creation and destruction of growing chains probably leading to their stationary concentration in the course of polymerization.

2. A similar kinetic order with respect to monomer for propagation and termination, becuase the molecular weight of the polymer formed remains nearly constant notwithstanding varying concentrations of monomer. Termination arises perhaps from a chain transfer to monomer or a wrong monomer addition which prevents further propagation (23).

3. The initiator, or any product formed from it which in turn initiates the reaction, is not actually consumed during polymerization because on each addition of fresh monomer the process can be repeated over and over again.

The following alternative explanations may be considered: (*a*) the initiating species are regenerated during polymerization; (*b*) the initiator's role is to activate the monomer rather than to initiate a growing polymeric chain. Upon its incorporation into the chain, the activated monomer regenerates the "initiator" or activates another molecule of monomer. This fruitful idea has been advocated and substantiated by Bamford's school (18,51,52).

On first inspection, the effect of the M/I ratio on the $\overline{\text{DP}}$ of the polymer produced appears to be puzzling. For example, data shown in Table VI indicate a higher $\overline{\text{DP}}$ if the product results from the addition of *all* the monomer at the onset of the reaction, while its stepwise addition in small portions, albeit in the same total quantity, gives a slightly lower DP. However, as shown in Table VI, the *initial* concentration of monomer was kept constant (4%) in Blout's experiments (50), and the increase in the M/I ratio was achieved by decreasing the concentration of the initiator. Hence, the results given in Table VI indicate a slight decrease of $\overline{\text{DP}}$ upon increasing the initiator concentration, implying, e.g., that the termination, or chain transfer, involves some reactions which are of second (or still higher) order in initiator. Be that as it may, one should stress that these effects are small and in some systems, e.g., in a polymerization initiated by diethylamine or triethylamine (49), the $\overline{\text{DP}}$ of the polymer is *independent* of the initiator concentration.

8. KINETICS OF NCA POLYMERIZATION INITIATED BY APROTIC BASES

Most kinetic studies of NCA polymerizations used the evolution of carbon dioxide as a measure of the degree of conversion. An alternative method, the determination of the amount of reacted NCA, was developed by Idelson and Blout (53). The infrared absorption spectra of NCA show two characteristic bands at 1860 and 1790 cm^{-1} which are attributed to the C=O stretching vibrations. On conversion of monomer into polymer they disappear and, therefore, the progress of the reaction may be followed by a decrease in optical density at these wavelengths. An additional band at 1735 cm^{-1}, present in the infrared spectrum of γ-benzyl L-glutamate NCA, arises from the stretching vibration of the carbonyl group of the benzyl ester moiety, and because its intensity is not affected by polymerization, it serves conveniently as an "internal standard." The infrared spectrophotometric technique, developed along these lines, was used by Idelson and Blout in their kinetic studies of γ-benzyl L-glutamate NCA polymerization. Their results were checked by the ordinary gasometric technique and the agreement was most satisfactory. Alternatively, the nonpolymerized monomer may be titrated with sodium methoxide (92).

The basic kinetic features of the "simple" polymerization of NCA were described and discussed in Sections 2 and 3. It has been shown that the "simple" or "normal" polymerization of the non-N-substituted as well as of the N-substituted NCA's proceeds by the same amine-propagated mechanism. The propagation initiated by aprotic bases and similar agents reveals different kinetic features, and the two types of NCA monomers, the N-substituted and the non-N-substituted, do not behave similarly in such polymerizations.

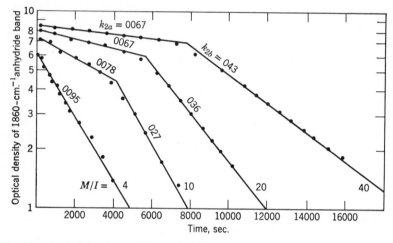

Fig. 10.   Optical density at 1860 cm^{-1} of NCA's C=O band measured as a function of time during the n-hexylamine-initiated polymerization of N-carboxy-γ-benzyl L-glutamate anhydride. M/I values 4, 10, 20, and 40. Initial rate constants k_{2a} are shown below the symbol; final rates k_{2b} are shown in a horizontal line with the symbol.

Studies by Idelson and Blout revealed an important kinetic feature of the polymerization initiated by strong bases, namely that an initial accelerating period is followed by a pseudo-first order reaction (see Figure 10). It seems that this behavior accounts for the previously mentioned findings of Lundberg and Doty (17) who reported two first order reactions, a slow one observed initially and a fast one at later stages. The accelerated portion of the conversion curve could easily be mistaken for a slow first order reaction. The initial acceleration, confirmed by other workers (e.g., see refs. 59 and 60), results most probably from a relatively slow approach of the system to its stationary state of the growing species. Such a situation is characteristic of those polymerization processes in which the rates of initiation and termination are slow compared with the rate of the overall growth.

The high rate of NCA polymerization initiated by aprotic bases is its most remarkable kinetic feature. This is due to the high rate of propagation because the concentration of growing species appears to be low as indicated by the relatively high degree of polymerization of the product. It is apparent, therefore, that the *propagation* mechanism of this reaction differs from that postulated for the primary amine-initiated polymerization.

Detailed kinetic studies of some aprotic systems were reported by Bamford's group. The kinetics of the polymerization initiated by the sodium salt of dihydrocinnamic acid in *N*-methylformamide exhibits a few interesting features. Typical conversion curves for D,L-phenylalanine and γ-benzyl

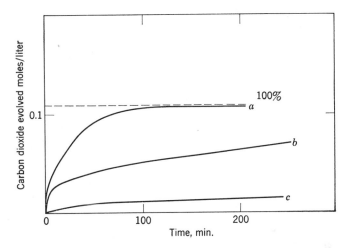

Fig. 11. Conversion-time curves for the polymerization of NCA's in 80% nitrobenzene–20% *N*-methylformamide mixtures at 25°C. $(NCA)_0 = 0.109$ mole/liter in all cases. (*a*) γ-benzyl L-glutamate NCA (sodium dihydrocinnamate) $= 4.6 \times 10^{-3}$ mole/liter; (*b*) D,L-phenylalanine NCA (sodium dihydrocinnamate) $= 4.6 \times 10^{-3}$ mole/liter; (*c*) D,L-phenylalanine NCA (sodium 3-hydantoin acetate) $= 9.2 \times 10^{-3}$ mole/liter.

L-glutamate NCA's are shown in Figure 11 as taken from the paper by Ballard and Bamford (54). For a low salt concentration, the reaction is self-inhibited, i.e., a relatively rapid polymerization ceases after a while (curve *C*) although a substantial fraction of the still unreacted NCA remains in solution. With a larger salt concentration, complete conversion is observed (curve *A*). It is apparent that an inhibitor is formed in the process, and Ballard and Bamford (54) proved that the formation of 3-hydantoin-acetic acid is responsible for this behavior. The yield of this acid is larger in the polymerization of D,L-phenylalanine than in the reaction of γ-benzyl L-glutamate.

The initial rate of polymerization appears to be first order with respect to monomer for a low salt concentration. The order of the reaction increases, however, for higher concentrations of the initiating salt. Sarcosine NCA, which is generally *more* reactive than the other two NCA's, reacts very slowly, and probably even this slow reaction is induced by some impurity present in the solvent. This N-substituted NCA was also found to be inert with respect to polymerization initiated by lithium chloride in dimethylformamide (44), although this sytem initiates rapid polymerization of non-N-substituted NCA's.

The kinetics of the relatively fast polymerization of the non-N-substituted NCA's initiated by lithium chloride in dimethylformamide was thoroughly studied by Ballard, Bamford, and Weymouth (44). Addition of this salt speeded enormously, by a factor of at least 1000, the extremely slow, spontaneous polymerization of glycine NCA. (In the absence of salt, no reaction was detectable even after a few hours.) Other NCA's behaved similarly, although in the absence of the salt their spontaneous polymerizations were slightly faster than that of glycine NCA. The initial rate of reaction was found to be proportional to the square of the monomer concentration, indicating the combined effect of the monomer on the rates of initiation and propagation.

A trace of water had no effect upon the reaction rate. This was demonstrated by changing the method of drying the lithium chloride and by deliberate addition of small amounts of water to the system containing dry lithium chloride. In this respect, a very different behavior is shown by the N-substituted NCA's, e.g., sarcosine anhydride. Although its solution in dimethylformamide does not react in the presence of dry lithium chloride, a fast reaction results upon addition of a trace of water. This again is strong evidence for different mechanisms of initiation and, probably also, of propagation for the non-N-substituted and the N-substituted NCA's.

Amongst major products of the reaction of non-N-substituted NCA's are 3-hydantoin-acetic acid, which is responsible for the previously mentioned autoinhibition of polymerization and for the formation of a cyclic low molecular weight polypeptide. The latter was isolated from the polymerization products of glycine NCA and identified as a cyclic hexamer. Apparently, the high proportion of salt speeds up intramolecular termination. On the other hand, polymerization of sarcosine NCA initiated by lithium chloride *in the presence* of 3-methylhydantoin produced linear polypeptides possessing a terminal amine group at one end of the chain and a hydantoin group probably attached to the other end. It should be noted that this system requires a cocatalyst if any trace of water is absent. The significance of this fact will be apparent when we discuss the mechanisms of these polymerizations (see the following section).

Ballard and Bamford (43) investigated also the kinetics of the polymerization of D,L-phenylalanine NCA initiated by tributylamine in nitrobenzene.

This reaction does not occur* in pure tertiary amine (11), but it proceeds rapidly in polar solvents. (For a tentative explanation of this observation, see p. 611.) However, polarity does not appear to be a necessary condition for rapid reaction because a fast polymerization initiated by triethylamine in benzene was observed by Blout (49). It seems, therefore, that the role of the solvent in these reactions needs further and extensive examination (see also p. 610). In the tributylamine-initiated polymerization, sarcosine NCA again proved to be inert (44) under conditions which led to rapid polymerization of non-N-substituted NCA's.

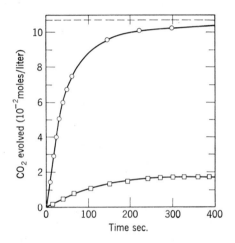

Fig. 12. Conversion-time curves for reactions of N-carboxy-D,L-phenylalanine anhydride with tri-n-butylamine in nitrobenzene at 25°C. $[M]_0 = 0.107$ mole/liter. (○) $[T]_0$ (initial total base concentration) = 9.103 moles/liter; (□) $[T]_0 = 1.6 \times 10^{-3}$ moles/liter. (------) corresponds to evolution of 1 mole of CO_2 per 1 mole of NCA.

The time-conversion curves for the polymerization initiated by tributylamine are shown in Figure 12. They resemble the curves of Figure 11, indicating the basic similarity in the processes initiated by tertiary amines, by salts of carboxylic acids, and by lithium chloride in dimethylformamide. Auto-inhibition was observed at a low base concentration and its cause was traced again to the formation of substituted 3-hydantoin-acetic acids. The order of the initial rates with respect to base is slightly lower than one, and slightly higher than one with respect to monomer. At constant monomer concentration (about 1 mole/liter), the intrinsic viscosity of the final polypeptide increases with increasing M/I ratio, attains a maximum at M/I ≈ 50 and then decreases. These changes are shown in Figure 13.

* It might be desirable to reinvestigate this claim.

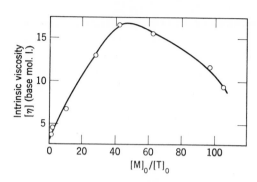

Fig. 13. Intrinsic viscosities of polymers prepared in nitrobenzene at 25°C as a function of $[M]_0/[T]_0$; $[M]_0 = 1.04$ mole/liter in all cases.

9. THE MECHANISM OF POLYMERIZATION INITIATED BY APROTIC BASES

Any proposed mechanism of polymerizations initiated by aprotic bases must account for the following observations:

1. Initiation proceeds without proton transfer to monomer and the initiator is not consumed in the reaction.

2. Polymerization initiated by aprotic bases proceeds *faster* and gives products of *higher* molecular weight than the reaction initiated by primary amines. Hence, the propagation mechanism of the former must differ from that of the latter, its rate constant being higher than that of the "simple" amine-propagated reaction.

3. Increase in initiator concentration enhances the polymerization, although the concentration of growing species does not necessarily increase.

4. The degree of polymerization of the resulting polypeptide may be substantially higher than the M/I ratio. Moreover, for a constant M/I ratio the $\overline{\mathrm{DP}}$ depends on the nature of the initiator.

5. The molecular weight distribution of the polymer may be very broad,* e.g., the ratio $\overline{\mathrm{DP}}_w/\overline{\mathrm{DP}}_n$ could be as large as 10.

6. No long-lived polymers are formed, although polymerization is resumed each time upon addition of fresh monomer to the polymerized mixture.

7. Cyclic polypeptides, or polypeptides devoid of terminal amine groups are among the products of, at least, some polymerizations initiated by aprotic bases.

* However, the high-molecular weight polypeptides resulting from initiation by sodium methoxide did not show unusually broad distribution (unpublished results of Blout and Doty).

8. 3-Hydantoin-acetic acid or its derivatives are frequent byproducts of these reactions. They were not found, however, among products of reactions initiated by sodium methoxide (Dr. Blout, private communication).

9. It seems that the polymerization of N-substituted NCA's initiated by aprotic bases takes a different course than the reaction involving non-N-substituted NCA's.

The various mechanisms which were proposed to account for these processes will now be reviewed. An amine is isoelectronic with a carbanion,

$$\diagdown N \diagdown \qquad \diagdown \overset{\ominus}{C} \diagdown ,$$

hence, its attack on a C=O double bond may resemble a typical propagation step in anionic polymerization:

$$\diagdown C \diagdown + C=C \rightarrow \diagdown C-C-C \ominus$$

$$\diagdown N \diagdown + C=O \rightarrow \diagdown \overset{+}{N}-C-O \ominus.$$

It was suggested, therefore (50), that this mechanism operates in the initiation and propagation steps of NCA polymerization induced by tertiary amines and other aprotic bases, viz.,

$$
\underset{\substack{\text{R·CH—CO}\\ \,\,\,|\qquad\quad \diagdown O\\ \text{NH—CO}\diagup}}{}
\quad + \text{NR}_3 \quad
\underset{\substack{\overset{+}{N}R_3\\ \text{R·CH—C—O}^-\\ \,\,\,|\qquad\quad \diagdown O\\ \text{NH—CO}\diagup}}{}
\quad \rightarrow \quad
\underset{\substack{\text{R·CH·CO·}\overset{+}{N}R_3\\ \,\,\,|\\ \text{NH·COO}^-}}{}
\quad \rightarrow
$$

or

$$
\underset{\substack{\text{R·CH—CO}\\ \,\,\,|\qquad\quad \diagdown O + CH_3O^-, Na^+\\ \text{NH—CO}\diagup}}{}
\quad \rightarrow \quad
\underset{\substack{\text{R·CH—CO·OCH}_3\\ \,\,\,|\\ \text{NH·COO}^-Na^+}}{}
$$

The propagation step had to be visualized as some kind of decarboxylation, which would lead to a terminal —COCH(R)NH⁻ group, followed by attack of an *amide* ion on another NCA molecule. However, it is known that the salts of carbamates, e.g., —CO·CH(R)·NH·COO⁻,Na⁺, do not decarboxylate. Hence, it would be necessary to postulate mediation by some proton-donating substance, e.g., a non-N-substituted NCA which converts carbamate to carbamic acid and then regains the proton from the terminal amine (produced by decarboxylation) by converting it into an amide ion (or ion-pair). This is, however, improbable because the terminal amine is a weaker acid than, e.g., non-N-substituted NCA and it is difficult to visualize a rapid reaction which would tend to shift the equilibrium in the required direction.

Alternatively, one may consider a direct reaction between carbamate and NCA. Walker (58b) suggested that the carboxylate ion activates the adjacent NH group and thus the reaction

$$—CO \cdot CH(R) \cdot NH \cdot COO^- + NCA \rightarrow$$
$$—CO \cdot CH(R) \cdot NH \cdot CO \cdot CH(R) \cdot NH \cdot COO^- + CO_2$$

perpetuates the propagation step. Idelson and Blout (53) slightly modified this proposal by postulating the formation of an intermediate linear anhydride, viz.,

$$—CO \cdot CH(R) \cdot NH \cdot COO^- + NCA \rightarrow$$
$$—CO \cdot CH(R) \cdot NH \cdot CO \cdot O \cdot CO \cdot CH(R) \cdot NH \cdot COO^-.$$

Linear anhydrides are known to undergo rapid decarboxylation with the formation of a peptide bond (7).

The main objection to mechanisms which assume attack by an aprotic base, e.g., CH_3O^- or tertiary amine, on NCA leading eventually to the formation of a carbamic acid is provided by the recent work of Goodman and Arnon (55). Such an initiation should retain the base, e.g., CH_3O^-, in the polymer; however, Goodman and Arnon demonstrated the absence of initiator fragments in the resulting polymer. In their studies, polymerization of γ-benzyl L-glutamate was initiated in dioxane by 9-fluorenylpotassium or by radioactive sodium methoxide. The polymer produced by the former initiator was precipitated and examined spectrophotometrically. For a M/I ratio of 60, the degree of polymerization was found from the intrinsic viscosity of the product to be about 180, and the solution of the purified polymer did not absorb at $\lambda = 300$ mμ where a strong absorption band ($\epsilon = 10^4$) of the fluorenyl moiety should appear. On the other hand, all the fluorenyl residues were found in the supernatant solution remaining after precipitation of the polymer.

Similar results were obtained with radioactive sodium methoxide.* The activity of the precipitated polymer was only 0.7% of that expected had each polymer molecule contained one methoxy group. Again, actually all the activity was found in the residual solution free of polymer.

Subsequent studies confirmed these conclusions. Goodman and Hutchison (68) repeated the work of Goodman and Arnon and found no $C^{14}H_3O$ moiety in the polymer initiated by radioactive sodium methoxide or radioactive triethylamine. On the other hand, activity was found in the polymer when polymerization was initiated by radioactive benzylamine or hexylamine. The same observations were made independently by Peggion, Terbojevich, Cosani, and Colombini (69). No activity was detected in a polymer initiated by C^{14} methyl diisopropylamine, while all the initial activity was found when C^{14} isopropylamine was used as the initiator (see also ref. 94).

---

* In their original work the radioactive sodium methoxide was diluted by non-active methanol. This mistake was rectified in later work (68).

Absence of initiator fragments in the polypeptide implies regeneration of the initiator during polymerization or an initiation process arising from a transfer of some monomer fragment, e.g., a proton, to the initiating base.

Two of the published mechanisms visualize regeneration of the initiator. Wieland (56,57) suggested that tertiary amines convert NCA molecules into zwitterions, i.e.

$$
\begin{array}{ccc}
\mathrm{R \cdot CH{-}CO} & & \mathrm{R \cdot CH{-}CO{-}\overset{+}{N}R_3'} \\
| \quad\quad\;\; \diagdown & & | \\
| \quad\quad\;\;\; \mathrm{O + NR_3' \;\rightleftarrows} & & | \\
\mathrm{NH{-}CO} & & \mathrm{NH{-}COO^-.}
\end{array}
$$

Their interaction then leads to a linear anhydride which on decarboxylation gives a peptide bond, viz.,

$$
\begin{array}{c}
\mathrm{R \cdot CH \cdot CO \cdot {}^+NR_3' \quad {}^-O \cdot CO \cdot NH} \\
| \qquad\qquad + \qquad\qquad | \qquad\qquad\; \rightarrow \\
\mathrm{{}^-O \cdot CO \cdot NH \qquad\qquad R \cdot CH \cdot CO \cdot {}^+NR_3'}
\end{array}
$$

$$
\begin{array}{c}
\mathrm{R \cdot CH \cdot CO \cdot O \cdot CO \cdot NH + NR_3'} \\
| \qquad\qquad\qquad\qquad | \qquad\qquad \rightarrow \\
\mathrm{{}^-O \cdot CO \cdot NH \qquad\qquad R \cdot CH \cdot CO \cdot {}^+NR_3'}
\end{array}
$$

$$
\begin{array}{c}
\mathrm{R \cdot CH \cdot CO \cdot NH} \\
| \qquad\qquad\qquad | \qquad\qquad + \; CO_2 + NR_3' \\
\mathrm{{}^-O \cdot CO \cdot NH \quad R \cdot CH \cdot CO \cdot {}^+NR_3}
\end{array}
$$

and repetition of these steps should yield a polypeptide.

Wieland's mechanism accounts for the formation of diketopiperazines, and this is an interesting aspect of his hypothesis. The above product results, according to his scheme, from an interaction of two "activated" monomers involving both of their ends, namely,

$$
\begin{array}{c}
\mathrm{{}^+NR_3'} \\
| \\
\mathrm{R \cdot CH \cdot CO \qquad\quad {}^-O \cdot CO \cdot NH \qquad\quad R \cdot CH \cdot CO \cdot O \cdot CO \cdot NH} \\
| \qquad\qquad + \qquad\; | \qquad\qquad \rightarrow \quad | \qquad\qquad\qquad\quad | \qquad\qquad + \; 2NR_3' \rightarrow \\
\mathrm{NH \cdot COO^- \qquad\quad CO{-}CH \cdot R \qquad\quad NH \cdot CO \cdot O \cdot CO \cdot CH \cdot R} \\
\qquad\qquad\qquad\qquad | \\
\qquad\qquad\qquad\;\; \mathrm{{}^+NR_3'}
\end{array}
$$

$$
\begin{array}{c}
\mathrm{R \cdot CH \cdot CO \cdot NH} \\
| \qquad\qquad\qquad | \qquad\qquad + \; 2CO_2 + 2NR_3'. \\
\mathrm{NH \cdot CO \cdot CH \cdot R}
\end{array}
$$

For the sake of brevity, this reaction was written in one step, while it probably proceeds through several stages, first forming a linear anhydride which then decarboxylates, and finally cyclizes. It seems significant that diketopiperazine is formed in the reaction of tertiary amines with sarcosine NCA* and its yield may be as high as 35% [p. 318 of (3)]. It should be stressed, however, that this process involves an *N*-substituted NCA, whereas in the polymerization of the non-*N*-substituted NCA's no substantial amounts of diketo-

---

* It is suggested that this product is formed *only* in the presence of water. No reaction was observed in anhydrous systems. (Private communication from Professor Bamford.)

piperazines, if any, were observed. This again might indicate different mechanisms of polymerization for these two types of NCA's.

The separation of terminal charges required by Wieland's mechanism is its most unsatisfactory feature, i.e., the form involving an open chain

$$R_3'N^+ \cdot CO \cdot CH(R) \cdot NH \cdot \{CO \cdot CH(R) \cdot NH\}_n \cdot CO \cdot CH(R) \cdot NH \cdot COO^-$$

is most unlikely to exist in solvents of low dielectric constant. Contact between terminal charges should lead to cyclization, as exemplified by the formation of diketopiperazines, and it might be possible that cyclic polypeptides are partially formed by this route. It is also improbable to expect interaction between two "activated" monomers in the initiation step. The interaction of a non-activated molecule with an "activated" one would be much more plausible, and such a step was indeed proposed by Ballard and Bamford (43).

In their original suggestion, which was modified later, they visualized the formation of a "complex," or of an "activated" monomer between NCA and the aprotic base, or its positive ion if the initiator is an ion-pair. Such a complex resembles conceptually the one proposed by Wieland. For example, initiation by lithium chloride in dimethylformamide solution was represented by the equation (44):

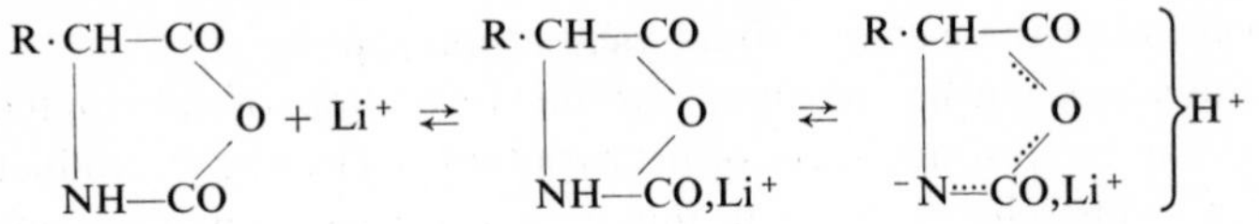

and a similar species was proposed for initiation by alkali salts of weak carboxylic acids (54). The activated NCA then reacts with a non-activated monomer, regenerates the initiator, and gives a dimeric carbamic acid, viz.,

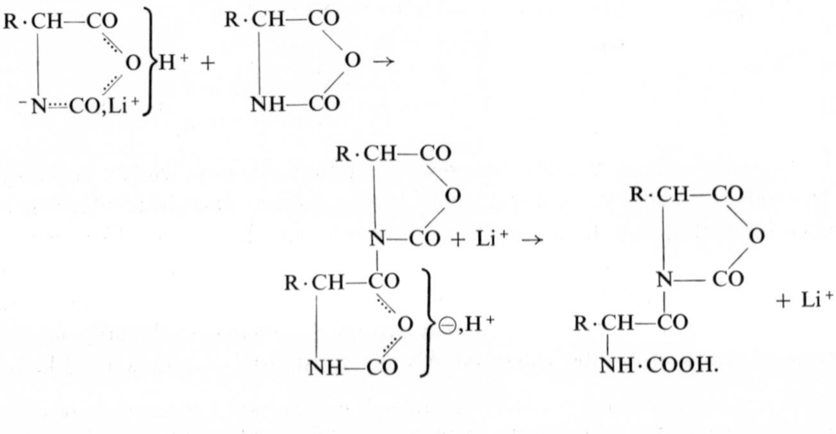

Finally, the latter decarboxylates into a bifunctional, polymerization-propagating species

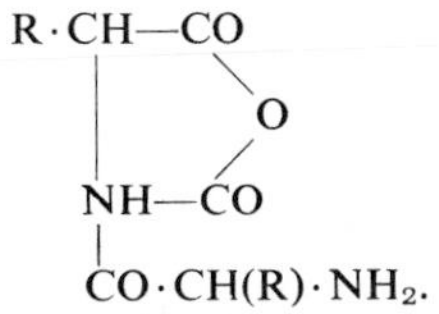

$$
\begin{array}{c}
\text{R·CH—CO} \\
\quad\big| \qquad \diagdown \\
\qquad\qquad \text{O} \\
\quad\big| \qquad \diagup \\
\text{NH—CO} \\
\quad\big| \\
\text{CO·CH(R)·NH}_2.
\end{array}
$$

It is assumed that the labile proton (denoted symbolically by H^+) is attached to a base associated with the primary positive ion, e.g., to Cl^- if lithium chloride is the initiator or to a carboxylate ion if initiation results from the action of the sodium salt of a weak carboxylic acid (54).

Initiation by tertiary amines is proposed to proceed similarly and to give (43)

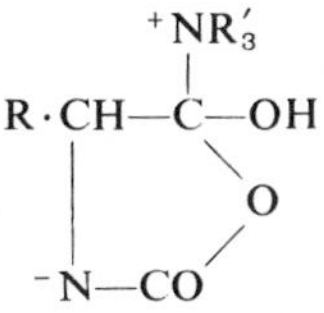

$$
\begin{array}{c}
{}^+\text{NR}'_3 \\
\big| \\
\text{R·CH—C—OH} \\
\quad\big| \qquad \diagdown \\
\qquad\qquad \text{O} \\
\quad\big| \qquad \diagup \\
{}^-\text{N—CO}
\end{array}
$$

i.e., a tautomeric form of Wieland's zwitterion. However, the labile proton is assumed in this case to be attached to the "activated" monomer rather than to another molecule of NR'_3 and, as previously proposed, the interaction with a non-activated NCA regenerates the initiating base and gives the dimer which decarboxylates into the difunctional amine, i.e.,

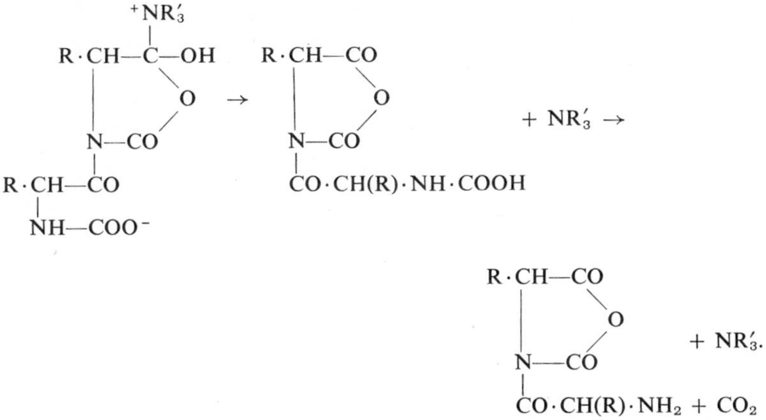

$$
\begin{array}{ccc}
{}^+\text{NR}'_3 & & \\
\big| & \text{R·CH—CO} & \\
\text{R·CH—C—OH} & \quad\big| \quad \diagdown & \\
\quad\big| \quad \diagdown & \qquad \text{O} & \\
\qquad \text{O} & \quad\big| \quad \diagup & + \text{NR}'_3 \rightarrow \\
\quad\big| \quad \diagup & \text{N—CO} & \\
\text{N—CO} & \quad\big| & \\
\quad\big| & \text{CO·CH(R)·NH·COOH} & \\
\text{R·CH—CO} & & \\
\quad\big| & & \\
\text{NH—COO}^- & &
\end{array}
$$

$$
\begin{array}{c}
\text{R·CH—CO} \\
\quad\big| \qquad \diagdown \\
\qquad\qquad \text{O} \qquad\qquad + \text{NR}'_3. \\
\quad\big| \qquad \diagup \\
\text{N—CO} \\
\quad\big| \\
\text{CO·CH(R)·NH}_2 + \text{CO}_2
\end{array}
$$

Perhaps it was unfortunate that these basic ideas evolved from Bamford's studies of NCA polymerization initiated by lithium chloride in dimethylformamide—certainly not the most typical polymerization system initiated by an aprotic base. The presence of lithium chloride in the solution raised doubts in the minds of other investigators as to whether some specific effects, caused by the relatively high concentration of electrolyte, are not imparting a peculiar character to this polymerization. Moreover, the lithium chloride system drew Bamford's attention to the positive ion, rather than to the base, and led him to the above formalism. The essential features of his mechanism may be preserved, and presented in an even more convincing manner, if one postulates initiation involving a transfer of proton from monomer to base instead of addition of the base to the monomer. This, indeed, was proposed by Bamford and Block (51) in their paper published in 1961. The initiation was represented by the equation

$$
\begin{array}{ccc}
\text{R·CH—CO} & & \text{R·CH—CO} \\
| \quad\quad\quad \diagdown & & | \quad\quad\quad \diagdown \\
| \quad\quad\quad\quad \text{O + base} \rightleftarrows & & | \quad\quad\quad\quad \text{O + base·H}^+ \\
| \quad\quad\quad \diagup & & | \quad\quad\quad \diagup \\
\text{NH—CO} & & {}^-\text{N}\cdots\text{CO}
\end{array}
$$

followed by the conventional steps of Bamford's mechanism, leading to the dimeric carbamates

$$
\begin{array}{c}
\text{R·CH—CO} \\
| \quad\quad\quad \diagdown \\
| \quad\quad\quad\quad \text{O} \\
| \quad\quad\quad \diagup \\
\text{N——CO} \\
| \\
\text{CO·CH(R)·NH·COO}^-.
\end{array}
$$

The latter is neutralized by the protonated base and thus the initiating proton-acceptor is regenerated. Finally, decarboxylation of carbamic acid yields the dimeric bifunctional species possessing an amine at one of its ends and the active oxazolidone ring at the other.

In fact, this interpretation became imperative when it was found that lithium perchlorate in dimethylformamide does *not* initiate the polymerization of systems for which lithium chloride is an efficient initiator. This proves that the reaction involves the negative ion, i.e., Cl^- or ClO_4^-, and *not* the positive Li^+ ion, and while Cl^- in dimethylformamide is a sufficiently strong base capable of accepting a proton and initiating the process, the ClO_4^- apparently is not. Actually, one may question the extent to which these salts are dissociated in dimethylformamide. It is possible, therefore, that the reaction involves ion-pairs rather than free ions, and the Li^+, Cl^- ion-pair may be a more powerful proton acceptor than Li^+,ClO_4^-.

To prove their hypothesis, Bamford and Block (51) applied the diagnostic test previously devised by Gold and Jefferson (58) in their studies of the hydrolysis of carboxylic anhydrides catalyzed by tertiary bases. The technique employed involved the use of a series of tertiary bases having different relative abilities to associate with Lewis acids and to act as Brönsted bases. Pyridine, α-picoline, and 2,6-lutidine form such a series. The *o*-methyl group in α-picoline, or two such groups in the lutidine, enhance the proton acceptability of the ring *N*-atom, making 2,6-lutidine the strongest Brönsted base and pyridine the weakest. On the other hand, the shielding of the *N*-atom by the adjacent methyl groups inhibits the addition of 2,6-lutidine to the carbonyl group and, therefore, for such an addition pyridine is the most reactive while 2,6-lutidine

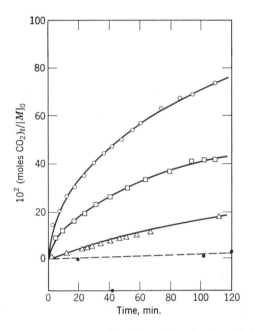

Fig. 14. Conversion-time curves for the polymerization of *N*-carboxy-γ-ethyl L-glutamate anhydride (0.2 mole/liter) catalyzed by pyridine and its homologs (2.0 mole/liter) in *N*,*N*-dimethylformamide at 25°C: (△) pyridine; (□) α-picoline; (○) 2,6-lutidine. (------) shows a comparative experiment with *N*-carboxy-sarcosine anhydride (0.2 mole/liter) and 2,6-lutidine (2.0 moles/liter).

is the least. In the studies of Gold and Jefferson (58), pyridine was the most powerful catalyst to enhance the hydrolysis of anhydrides. Hence, in that reaction the catalytic action was caused by the addition of tertiary base to the carbonyl group of the anhydride. The reverse order of activities was observed by Bamford and Block (51) (see Figure 14) and this result proves conclusively that the initiation of polymerization by aprotic bases arises from proton abstraction and not from the addition of base to an NCA molecule.

The concept of proton-transfer to the base leads to an important generalization (18,52), namely, this type of polymerization may be initiated not only by aprotic bases but also by primary or secondary amines. The latter may then act in a dual way—as a direct proton-acceptor, or as a Lewis base and, indirectly, as a proton-donor. Hence, initiation by primary or secondary base may lead to two simultaneous reactions competing for the base—one which irreversibly removes the base from the system

$$
\begin{array}{ll}
\mathrm{R \cdot CH-CO} & \mathrm{R \cdot CH \cdot CO \cdot NR_1R_2} \\
\quad | \qquad \diagdown & \quad | \\
\quad | \qquad \quad O + HNR_1R_2 \rightleftarrows & \quad | \qquad\qquad\qquad \rightarrow \text{polymer} \\
\quad | \qquad \diagup & \quad | \\
\mathrm{NH-CO} & \mathrm{NH \cdot COOH}
\end{array}
$$

and the other which is only catalyzed by the base but does not affect its ultimate concentration, viz.,

$$
\begin{array}{ll}
\mathrm{R \cdot CH-CO} & \mathrm{R \cdot CH-CO} \\
\quad | \qquad \diagdown & \quad | \qquad \diagdown \\
\quad | \qquad \quad O + HNR_1R_2 \rightleftarrows & \quad | \qquad \quad O + H_2NR_1R_2{}^+. \\
\quad | \qquad \diagup & \quad | \qquad \diagup \\
\mathrm{NH-CO} & {}^-\mathrm{N}\text{.......}\mathrm{CO}
\end{array}
$$

Therefore, two types of polymerization arise and this explains, e.g., the results of Blout and Karlson (49) who found, for the reaction initiated by *n*-hexylamine in dioxane, a degree of polymerization *higher* than expected on the basis of the "simple" amine-propagated mechanism and the bimodal distribution (93).

The different behavior of, e.g., diisopropylamine and *n*-hexylamine becomes now intelligible. The sterically unhindered *n*-hexylamine functions mainly as a Lewis base which adds to the Lewis acid (the NCA) and initiates the "simple" amine-propagated polymerization. On the other hand, the sterically hindered diisopropylamine is inefficient as a Lewis base but, being more powerful Brönsted base than *n*-hexylamine, it initiates a rapid polymerization resulting from proton abstraction and "activation" of the monomer. This is seen in Figure 8.

The bifunctional dimer, postulated in Bamford's initiation, is assumed to grow by two distinct mechanisms. Its amine-terminated end may propagate in the "normal" fashion, but since this mode of growth is relatively slow, its contribution to the overall process is expected to be of *minor* significance. According to Bamford and Block (18,51) the bulk of NCA polymerizes by the mechanism involving an "activated" monomer and it is the latter mode which is assumed to be fast. The "monomer-activation" mechanism is analogous to the formation of the initiating dimer, viz.,

$$
\begin{array}{llll}
\mathrm{R \cdot CH-CO} & \mathrm{\bar{N}\text{----}CH(R)} & \mathrm{R \cdot CH \cdot CO-N\text{----}CO} \\
\quad | \qquad \diagdown & \quad | \qquad | & \quad | \qquad\qquad | \qquad \diagdown \\
\quad | \qquad \quad O + CO \quad CO & \rightarrow & \quad | \qquad\qquad | \qquad\quad O \\
\quad | \qquad \diagup & \quad \diagdown \quad \diagup & \quad | \qquad\qquad | \qquad \diagup \\
\text{\textasciitilde CO \cdot N\text{----}CO} & O & \text{\textasciitilde CO \cdot N \cdot COO}^- \ \mathrm{CH(R)CO}
\end{array}
$$

$$\sim\text{CO}\cdot\text{N}\!-\!\text{CH(R)}\!-\!\text{CO}\!-\!\text{N}\!-\!\text{CO} \quad\quad \text{R}\cdot\text{CH}\!-\!\text{CO}$$

with pendant COO^- and $\text{R}\cdot\text{CH}\!-\!\text{CO}$ groups forming $\text{O}+$; $\text{HN}\!-\!\text{CO}$ forming $\text{O}\rightarrow$ (non-activated)

$$\sim\text{CO}\cdot\text{N}\!-\!\text{CH(R)}\!-\!\text{CO}\!-\!\text{N}\!-\!\text{CO} \quad\quad \text{R}\cdot\text{CH}\!-\!\text{CO}$$

with pendant COOH and $\text{R}\cdot\text{CH}\!-\!\text{CO}$ groups forming $\text{O}+$; $^-\text{N}\!-\!\text{CO}$ forming O (activated)

The newly formed activated monomer then continues the chain process and decarboxylation of carbamic acid produces the previous type of polypeptide but one unit longer.

Increasing the concentration of initiator has a dual effect on the rate of such a polymerization. It increases the stationary concentration of growing species —a trivial effect expected in nearly all polymerization processes—but in addition it greatly enhances the rate of *propagation* of each growing polymer by increasing the concentration of activated monomers. This situation is not encountered in conventional polymerization processes where the rate of growth of each chain is independent of the initiator concentration.

Bamford's mechanism offers an interesting possibility of coupling polymeric molecules, i.e., the conventional reaction of the terminal amine of an *n*-mer with the cyclic group of an *m*-mer should produce an (*n* + *m*)-mer. This process, if it occurs, gives some polycondensation character to this essentially polyaddition type of polymerization. However, although such a condensation may rapidly increase the molecular weight of the polypeptide produced, it contributes relatively little to carbon dioxide evolution or to the rate of monomer consumption. It should be interesting, therefore, to investigate the change in molecular weight of polypeptide in the last stages of the reaction. Idelson and Blout reported (50) that the molecular weight of the polypeptide remained constant as the conversion, viz., carbon dioxide evolution or monomer consumption, increased from 25 to 100% (see Table VII). This indicates that the coupling is slow and does not occur to any significant extent during polymerization. On the other hand, it has been reported recently (70,71) that precipitation of the polymerization product, followed by its dissolution, leads to a substantial increase in its molecular weight. This observation appears to be the first indirect evidence for the bifunctional nature of the polypeptide formed by the action of an aprotic base. The "normal" addition of an amine end of one chain to the oxazolidone end group of another takes place only in a concentrated medium, and such conditions prevail in the swollen precipitate.

The coupling process cannot account for the high rate of reaction because the number of growing chains is exceedingly small even at the onset of the reaction. Therefore, the statement by Ballard and Bamford [see p. 387 of (43)] attributing the rapid rate of reaction to the coupling process, is probably erroneous.

The intramolecular coupling of the terminal amine with the cyclic end of the same polypeptide represents an interesting termination process (44,54). Such a reaction results in the formation of cyclic polypeptides, and indeed, a hexameric cyclic polypeptide was isolated (44,54) from the products of polymerization initiated by some aprotic bases. The rate of such a termination should depend on the molecular weight of the terminated polymer because the probability of ring closure is the greatest for a relatively low degree of polymerization, i.e., 4–8, and thereafter it decreases with increasing chain length (72,73). Hence, a substantial fraction of growing polymers should be terminated by cyclization when their degree of polymerization is low, whereas those which survived this critical stage should attain considerable size before being terminated, probably by some other processes. This feature of the termination step may account for the large $\overline{M}_w/\overline{M}_n$ ratio of polypeptides produced by polymerization initiated by aprotic bases* and for the constant DP of the polymer produced in consecutive "after" polymerizations (see p. 586).

The most probable mode of termination of long-chain polymers may result from a wrong addition of an activated monomer, e.g.,

$$
\begin{array}{c}
\text{R} \cdot \text{CH}{-}\text{CO} \qquad\quad \text{O} \qquad\qquad\qquad \text{R} \cdot \text{CH}{-}\text{CO} \cdot \text{O} \cdot \text{CO} \\[2pt]
\big| \qquad\qquad \text{O} + \text{CO} \quad\ \text{CO} \;\rightarrow\; \quad \big| \qquad\qquad \big| \qquad\qquad \rightarrow \\[2pt]
{\sim}\text{CO} \cdot \text{N}{-}\text{CO} \quad {}^{-}\text{N}{-}\text{CHR} \qquad {\sim}\text{CO} \cdot \text{N} - \text{CO} {-} \text{N} \cdot \text{CH(R)COO}^{-}
\end{array}
$$

$$
\begin{array}{c}
\text{R} \cdot \text{CH}{-}\text{CO} \\[2pt]
\big| \qquad\qquad \text{N} \cdot \text{CH(R)COO}^{-} + \text{CO}_2 \\[2pt]
{\sim}\text{CO} \cdot \text{N}{-}\text{CO}
\end{array}
$$

and not from a spontaneous termination such as cyclization. The former type of termination, being first order with respect to monomer, leaves the $\overline{DP}_w$ of the product independent of initiator concentration, whereas a spontaneous termination demands a decrease in $\overline{DP}_w$ upon decreasing the initiator concentration at a constant monomer concentration (i.e., on decreasing the concentration of activated monomer).

It is also possible that the termination of the long chain polymers may be

---

* Intermolecular coupling does not lead to a high $\overline{M}_w/\overline{M}_n$ ratio. The high $\overline{M}_w/\overline{M}_n$ ratio arises from the presence of a substantial mole fraction of very low molecular weight product and not from a process forming of a small mole fraction of high molecular weight polymer.

caused by some impurities present in the system because it has been frequently observed (private communication from Dr. A. Berger) that the polymerization induced by tertiary bases becomes faster when monomer and solvent are more rigorously purified.

The intramolecular coupling of the terminal amine with the oxazolidine ring attached at the other end of the polymeric chain involves the C-5 carbonyl group because the C-2 CO group appears to be much less reactive (see, e.g., p. 561). However, for steric reasons the reaction with the C-5, carbonyl group is prevented in the initiating dimer. Hence, the amine of the dimer may react only with the 2-carbonyl group, viz.,

$$
\begin{array}{ccc}
\text{R·CH—CO (5)} & & \text{CH(R)·COOH (or COO}^-\text{)} \\
\end{array}
$$

R·CH—CO (5)   CH(R)·COOH (or COO⁻)
 | \ |
 | O |
 | / |
N— CO (2)   →   N— CO
 | NH₂ (or NH⁻) | \NH
 | / | /
CO—CH(R)   CO—CH(R)

giving a derivative of 3-hydantoin-acetic acid. This isomerization competes for the dimer with the propagation step of polymerization and this occurrence reduces, therefore, the efficiency of the initiating process. Such description is preferred to that used by Bamford's group who classified the isomerization of the dimer to 3-hydantoin-acetic acid as a termination process. Dilution of the polymerizing solution increases the yield of the 3-hydantoin-acetic acid (43), and this result, predicted by Bamford's mechanism, provides additional evidence in its favor.

Two additional propagation mechanisms should be considered. In the presence of strong bases such as sodium methoxide or triphenylmethyl-sodium the terminal amine may be converted, at least partially, into the respective sodium amide. Hence, the following propagation step may arise under these conditions,

$$\sim\!\!\text{NH}^-,\text{Na}^+ + \text{NCA} \rightarrow \sim\!\!\text{NH·CO·CH(R)·NH·COO}^-,\text{Na}^+.$$

Carbamate salts are stable and do not decarboxylate. Therefore, to account for polymerization one must assume either a proton transfer to carbamate, e.g., by reversing the process creating the amide or by providing protons from another source. Alternatively, one may postulate (58b) activation of the $>$NH by the adjacent carboxylate ion and represent the propagation by the equation

$$\sim\!\!\text{NH·COO}^- + \text{NCA} \rightarrow
\begin{array}{l}
\sim\!\!\text{NH} + \text{CO}_2 \\
| \\
\text{CO·CHR·NH·COO}^-.
\end{array}$$

This is in essence the mode of propagation postulated by Idelson and Blout

(53) who assumed the formation of a linear anhydride as an intermediate and then its eventual decarboxylation, viz.,

$$\sim\!NH\cdot COO^- + O\!\!\begin{array}{c} CO\!-\!CH(R) \\ | \\ CO\!-\!NH \end{array} \rightarrow \sim\!NH\cdot CO\cdot O\cdot CO\cdot CH(R) \\ \qquad\qquad\qquad\qquad\qquad\quad | \\ \qquad\qquad\qquad\qquad\qquad\quad NH\cdot COO^- \rightarrow$$

$$\sim\!NH\cdot CO\cdot CH(R)NH\cdot COO^- + CO_2.$$

At least two approaches should allow us to decide which is the correct propagation step in the NCA polymerization initiated by an aprotic base. The following line of argument was chosen by Bamford's group.

Propagation involving the activated monomer-anion is possible only for the non-N-substituted NCA's; the N-substituted anhydrides cannot form such ions. Hence, if the polymerization initiated by aprotic bases proceeds entirely by this route, no N-substituted NCA may be polymerized by these initiators (see, however, refs. 74–76 where such polymerizations are claimed to take place). Therefore, Bamford's group spared no effort to prove the validity of this statement and their elegant experiments deserve detailed discussion.

As has been seen in Figure 8, sarcosine NCA, the N-substituted anhydride, polymerizes faster than γ-ethyl L-glutamate NCA if the polymerization is initiated by n-hexylamine in dimethylformamide (18,52). However, sarcosine NCA is found to be extremely unreactive when diisopropylamine is used to initiate the reaction, while the polymerization of γ-ethyl L-glutamate becomes even faster under these conditions (18,52) (see again Figure 8). Steric hindrance inhibits the "simple" amine-addition initiation of diisopropyl amine, and hence the slowness of sarcosine NCA polymerization is not surprising. However, the high basicity of this initiator may result in activation of the monomer by proton-transfer and this fast mode of polymerization is operative for γ-ethyl L-glutamate NCA but not for sarcosine NCA which lacks the necessary proton. Thus, the results shown in Figure 8 are fully rationalized.

An even more convincing argument is provided by further studies (46) results of which are shown in Figure 15. Tri-n-butylamine in dimethylformamide induced only a slow polymerization of L-proline NCA whereas the polymerization of γ-ethyl L-glutamate NCA was very fast under these conditions, even if the concentration of the base was reduced by a factor of 10 (18,46). This is a striking observation because proline NCA is an extremely reactive monomer and polymerizes very fast on addition of a primary amine. However, addition of 3-methylhydantoin

$$\begin{array}{c} CH_2\!-\!CO \\ | \qquad\quad \diagdown \\ \quad\qquad\qquad N\cdot CH_3 \\ | \qquad\quad \diagup \\ NH\!-\!CO \end{array}$$

to the investigated mixture enhanced substantially the polymerization of

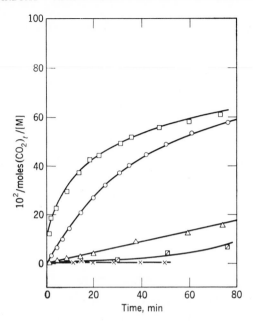

Fig. 15. Conversion-time curves for reactions of NCA's (M) with tri-n-butylamine (B) in N,N-dimethylformamide at 25°C: (□) γ-ethyl L-glutamate NCA, $[M]_0 = 0.1$ moles/liter, $[B] = 0.0126$ moles/liter; (○) sarcosine NCA, $[M]_0 = 0.332$ moles/liter, $[B] = 0.042$ mole/liter + 3-methylhydantoin, 0.178 mole/liter; (△) L-proline NCA $[M]_0 = 0.291$ mole/liter, $[B] = 0.129$ mole/liter; (▨) sarcosine NCA $[M]_0 = 0.289$ moles/liter, $[B] = 0.262$ mole/liter; (×) sarcosine NCA $[M]_0 = 0.323$ moles/liter, $[B] = 0$ mole/liter + 3-methylhydantoin, 0.175 mole/liter.

L-proline NCA as may be seen in Figure 15. Bamford therefore drew the obvious conclusion, namely, that 3-methylhydantoin, which is expected to behave similarly to the non-N-substituted NCA's, reacts with the base and produces the anion

$$CH_2\!-\!CO$$
$$\diagdown$$
$$N\cdot CH_3.$$
$$\diagup$$
$$^-N\!-\!\!-\!CO$$

This, in turn, interacts with the N-substituted NCA and gives the respective dimeric carbamic anion

$$CH_2\!-\!CO$$
$$\diagdown$$
$$N\cdot CH_3$$
$$\diagup$$
$$N\!-\!\!-\!CO$$
$$|$$
$$CO\cdot CH\!-\!N\cdot COO^-$$
$$|\qquad\quad|$$
$$\!-\!(CH_2)_3\!-\!$$

which is neutralized by the protonated base, viz., 3-methylhydantoin, as shown by the equation:

$$Dimer \cdot COO^- + Base^-, H^+ \rightarrow Dimer \cdot COOH + Base^-.$$

Decarboxylation of the acid then produces the amine group which continues the "normal" propagation. Of course, the hydantoin ring, being stable, cannot contribute to the growth, and furthermore, no "activated" monomer may be formed from proline-NCA. The argument was substantiated (46) by isolating and identifying the postulated intermediate, which was formed in the reaction of N-phenylglycine NCA with a tertiary base and 3-methylhydantoin. The substituted aniline is too weak a base to sustain polymerization and, therefore, the reaction produced only a decarboxylated dimer

$$
\begin{array}{l}
CH_2\!-\!CO \\
\quad\qquad\searrow \\
\qquad\qquad N\cdot CH_3 \\
\quad\qquad\nearrow \\
N\!-\!-\!CO \\
\mid \\
CO\!-\!CH_2\!-\!NH(Ph).
\end{array}
$$

The exceptionally high capacity of 3-methylhydantoin to initiate such a polymerization (see ref. 46) calls for some further explanations* and weakens the argument presented by Bamford. Moreover, because rapid polymerization of N-substituted NCA's initiated in an anhydrous medium by aprotic bases has been observed by many investigators, Bamford's evidence might not be sufficiently convincing to prove the validity of the "activated monomer" mechanism. Therefore, an alternative proof will be outlined but, before this is done, we may consider in greater detail the significance of the facts presented up to now.

The results discussed above raise the following questions:

1. Is initiation of N-substituted NCA's polymerization by aprotic bases prevented by the lack of intermediates having the structure of the NCA anion, viz.,

$$
\begin{array}{l}
R\cdot CH\!-\!CO \\
\mid \qquad\qquad\searrow \\
\qquad\qquad\quad O \\
\mid \qquad\qquad\nearrow \\
{}^-N\!-\!CO
\end{array}
$$

or

$$
\begin{array}{l}
R\cdot CH\!-\!CO \\
\mid \qquad\qquad\searrow \\
\qquad\qquad\quad N\cdot CH_3 \\
\mid \qquad\qquad\nearrow \\
{}^-N\!-\!CO
\end{array}
$$

---

* According to Professor Bamford the 3-methylhydantoin anion is indeed exceptional in its ability to initiate the polymerization of N-substituted NCA's in the presence of tertiary amines.

or is decarboxylation of the resulting carbamic acid hindered by the lack of a suitable proton donor.

2. Is the fast polymerization of the non-*N*-substituted NCA's, which leads to high molecular weight polymer, the result of propagation involving the "activated" monomer attacking the terminal ring, or is it caused by a mechanism involving the terminal $\sim$NH$^-$ or $\sim$NH·COO$^-$ group (Blout's or Walker's mechanism).

There is no obvious reason why the $\bar{\text{N}}$ ion (or the corresponding ion-pair) derived from NCA (or hydantoin) should be an exceptional base, more efficient in attacking an NCA molecule and opening its ring than any other sufficiently basic ion (or its ion-pair). Hence, reactions such as

$$\text{NCA} + {}^-\text{OCH}_3 \rightarrow \text{CH}_3\text{O}\cdot\text{CO}\cdot\text{CH(R)}\cdot\text{NH}\cdot\text{COO}^-$$

or

$$\text{NCA} + {}^-\text{C(Ph)}_3 \rightarrow \text{C(Ph)}_3\cdot\text{CO}\cdot\text{CH(R)}\cdot\text{NH}\cdot\text{COO}^-$$

could also be efficient in initiating polymerization, if the carbamate can be protonated and then decarboxylated, or if the growth results from Blout's type of propagation. Such a propagation may be faster than that resulting from the "ordinary" amine-propagated mechanism. However, under normal conditions leading to high-molecular weight polymer, the concentration of NCA is much greater than that of any base used in initiation. Because of this, and because proton-transfer is faster than base addition, the anions derived from NCA are expected to be essentially the *only* bases capable of initiating polymerization, the originally added base having been converted to its conjugated acid. Therefore, the evidence presented by Goodman and Arnon (55), who did not find fragments of initiator in the polymer, is to some extent misleading. This observation does not disprove the intrinsic possibility of initiation by the addition of CH_3O^- ions. It merely shows that these ions are essentially absent under the conditions of their experiments, as well as under conditions prevailing in other conventional polymerization systems. In fact, Zilkha (77) has shown recently that polypeptide chains may be grafted on cellulose containing —ONa groups and this result implies that an anionic residue may be incorporated into a polypeptide chain.

There is no contradiction between the findings of Goodman and those of Zilkha. By the choice of his study, Zilkha observed only those polypeptide chains which were grafted onto cellulose; all the remaining ones, which formed probably the bulk of the product, were removed in the isolation procedure which separates the homopolymer from the graft polymer. On the other hand, Goodman's results show that the bulk of methoxy ions, i.e., 98 or 99%, do not appear in the polypeptide formed in the process, and the accuracy of his technique cannot permit us to know whether 1 or 2%

are genuinely incorporated or whether the sample is contaminated with radio-activity.

A basic problem is raised by the second question, namely, which propagation mechanism operates in the polymerization giving the *high molecular weight* polypeptide. The writer proposed, therefore, the following experiment to resolve the controversy (78). A radioactive salt of carbamic acid is prepared, e.g., $CH_3O \cdot CO \cdot C^{14}H(R) \cdot NH \cdot COO^-, Na^+$, and to its solution is added the investigated NCA. If Blout's type of propagation operates, a fast reaction should occur immediately and yield a high molecular weight polymer containing one radioactive group per chain. On the other hand, if Bamford's scheme operates, one may anticipate polymerization but no radioactivity in the *high* molecular weight product. A polymerization which proceeds slowly and yields radioactive, *low* molecular weight polypeptide might also occur, the process being initiated by the sequence

1. $CH_3 \cdot O \cdot CO \cdot C^{14}H(R) \cdot NH \cdot COO^- + NCA \rightleftarrows$
$\qquad CH_3O \cdot CO \cdot C^{14}H(R)NH \cdot COOH +$ activated NCA^- ion.

2. The carbamic acid decarboxylates displacing the above given equilibrium to the right.

3. The activated monomer then initiates the "aprotic base" type polymerization leading to the high molecular weight polymer containing *no* radioactivity.

4. The radioactive ester of the amino acid may initiate an "amine"-propagated polymerization, but this reaction should be of little importance, producing only a *low molecular* weight polymer endowed with radioactivity, if the conversion is limited to ca. 20%.

Hence, the presence of activity in the *high* molecular weight polymer should prove Blout's type of propagation, while its absence supports Bamford's type of propagation involving attack by an activated monomer on a polymer possessing a diketo-oxazolidone ring on its end.

Goodman and Arnon (74) investigated recently the ability of carbamates to initiate NCA polymerization and reported a successful initiation of γ-benzyl L-glutamate NCA polymerization by sodium *N*-benzyl carbamate. This experiment, however, is still inconclusive because no evidence was provided as to whether the carbamate residue is incorporated in the high molecular weight product.

Goodman and Arnon's goal was to disprove the thesis that polymerization initiated by a strong base is propagated by the free amine (the "simple" amine-propagated mechanism) or by the amide ion, viz.,

$\sim CO \cdot CH(R) \cdot NH^- + NCA \rightarrow \sim CO \cdot CH(R) \cdot NH \cdot CO \cdot CH(R) \cdot NH \cdot COO^-$
$\qquad\qquad\qquad \rightarrow \sim CO \cdot CH(R) \cdot NH \cdot CO \cdot CH(R) \cdot NH^- + CO_2$, etc.

This, indeed, they demonstrated. The addition of methyl iodide to the polymerizing mixture had *no* effect upon the process. Since the iodide methylates the free amine, a polymerization propagated by such a group should be inhibited. Indeed, if methyl iodide is added to a mixture being polymerized by the "simple" amine-propagated mechanism [experiments conducted by J. Hutchison and quoted in ref. 74] inhibition is observed but no effect whatsoever is found in a polymerization initiated by a strong base.

Again, addition of methyl methacrylate had no effect upon the rate of polymerization or upon the structure of the resulting polypeptide. Since amide groups initiate the polymerization of methyl methacrylate, the absence of such a reaction proves the absence of the amide groups.

The crucial experiment was eventually performed by Goodman and Hutchinson (68,79) who initiated the polymerization of γ-benzyl L-glutamate NCA by sodium-*N*-benzyl carbamate possessing a C^{14} label in its benzyl group. Previous results, which showed the kinetics of this polymerization to be similar to that initiated by sodium methoxide, were confirmed and in addition the complete absence of radioactivity in the polymer was established. Traces of activity, amounting to 2.5–3%, only, may be easily accounted for by the "ordinary" slow polymerization initiated by the radioactive amine produced in the decarboxylation of carbamic acid, viz.,

$$Ph \cdot CH_2 \cdot NH \cdot COO^- + NCA \rightarrow Ph \cdot CH_2 \cdot NH \cdot COOH + NCA^-$$

$$Ph \cdot CH_2 \cdot NH \cdot COOH \rightarrow Ph \cdot CH_2 \cdot NH_2 + CO_2.$$

Although, the equilibrium of the above proton-transfer process lies far to the left, decarboxylation of the acid shifts it slowly to the right. This characteristic of the reaction was emphasized earlier; it may lead to an induction period predicted by this writer (78) and observed by Goodman and Hutchison. Moreover, the degree of polymerization was significantly higher than the M/I ratio. These findings disprove the anhydride mechanism.*

In NCA polymerization initiated by a radioactive, secondary amine, the specific activity of the resulting polymer should depend in a complex manner on the concentration of the initiator. Its increase accelerates the rate of propagation (per growing chain) of those molecules which grow by an "activated" monomer mechanism, without affecting the propagation rate of those growing by the "simple" amine propagation. Both types of polymers are formed simultaneously in a constant proportion, but only the latter contain the radioactive label. Hence, the specific activity of the polymer formed is expected to *increase* as the concentration of the radioactive initiator decreases.

* Determination of the direction of the chain growth may distinguish between the two mechanisms. Growth takes place from C to N in the anhydride mechanism, while the direction is reversed in the "activated" monomer mechanism. (Suggested by Dr. Y. Shalitin.)

The autocatalytic character of the strong base initiated polymerization (17,53) which has been confirmed by some recent observations (59,60), calls for further comments. It seems, that slow attack by an activated monomer, NCA^-, on the nonactivated anhydride

$$
\begin{array}{c}
\text{R·CH—CO} \\
\Big| \qquad \diagdown \\
\qquad\qquad O \\
\Big| \qquad \diagup \\
\text{NH—CO}
\end{array}
+
\begin{array}{c}
\bar{\text{N}}\text{—CO} \\
\Big| \qquad \diagdown \\
\qquad\qquad O \\
\Big| \qquad \diagup \\
\text{R·CH—CO}
\end{array}
\xrightarrow{\text{slow}}
\begin{array}{c}
\text{R·CH — CO}\text{———}\text{N— CO} \\
\Big| \qquad\qquad\qquad \Big| \quad \diagdown \\
\qquad\qquad\qquad\qquad\qquad O \\
\Big| \qquad\qquad\qquad \Big| \quad \diagup \\
\text{NH·COO}^- \quad \text{R·CH—CO}
\end{array}
$$

is followed by a rapid attack of NCA^- on the ring of the growing polymer, i.e.,

$$
\begin{array}{c}
\text{R·CH—CO} \\
\Big| \qquad \diagdown \\
\qquad\qquad O \\
\Big| \qquad \diagup \\
\text{⁓CO·N— CO}
\end{array}
+
\begin{array}{c}
^-\text{N— CO} \\
\Big| \qquad \diagdown \\
\qquad\qquad O \\
\Big| \qquad \diagup \\
\text{R·CH—CO}
\end{array}
\rightarrow
\begin{array}{c}
\text{⁓CO·N—(R)CH—CO—N— CO} \\
\Big| \qquad\qquad\qquad\qquad\quad \Big| \quad \diagdown \\
\qquad\qquad\qquad\qquad\qquad\qquad O \\
\Big| \qquad\qquad\qquad\qquad\quad \Big| \quad \diagup \\
\text{COO}^- \qquad\qquad \text{R·CH—CO}
\end{array}
$$

The former reaction is the initiation, while the latter represents the propagation step of the polymerization. The large difference in their rates suggests that the additional carbonyl group linked to the nitrogen in the growing polymer activates the ring, making the

$$
\begin{array}{c}
\text{R·CH—CO} \\
\Big| \qquad \diagdown \\
\qquad\qquad O \\
\Big| \qquad \diagup \\
\text{⁓CO—N— CO}
\end{array}
$$

ring, much more reactive than that of the monomer, viz.,

$$
\begin{array}{c}
\text{R·CH—CO} \\
\Big| \qquad \diagdown \\
\qquad\qquad O. \\
\Big| \qquad \diagup \\
\text{NH—CO}
\end{array}
$$

This phenomenon appears, even in a more striking form, in other similar polymerization processes, e.g., polymerization of pyrolidone (95).

It is not surprising that the rate of aprotic NCA polymerization and the molecular weight of the resulting product are extremely susceptible to the nature of the base, counterion, added electrolyte, solvent, etc. (see, e.g., refs. 68 and 71). These factors affect the equilibrium

$$
\text{NCA} + \text{base} \rightleftarrows \text{activated NCA}^- + \text{base·H}^+
$$

and hence, the rate of polymerization if it indeed proceeds by the activated NCA mechanism. Moreover, if the termination involves a "wrong" addition of the activated NCA^- then the nature of the solvent and of the counterion (hence, also of $Base·H^+$, if it is an ion) should affect the ratio k_p/k_t. This in turn affects both the molecular weight of the product and the rate of polymerization (by affecting the stationary concentration of growing ends). It is interesting to recall at this stage the peculiar fact, namely, that the polymerization of some NCA's does not take place in tri-*n*-butylamine as a solvent, but it proceeds in dimethylformamide in the presence of a small amount of the amine. One may wonder whether the equilibrium

$$\text{NCA} + \text{Bu}_3\text{N} \rightleftarrows \text{activated NCA}^- + \text{Bu}_3\text{NH}^+$$

is not displaced too much to the right when Bu_3N is the solvent, leaving too low a concentration of free NCA to allow an efficient initiation.

Finally, we should consider some of the objections recorded in the literature and raised against the activated monomer hypothesis. There is some confusion about the significance of the bifunctional dimer proposed by Bamford. For example, Blout (61) remarked that the fast polymerization cannot be ascribed to the free amine end. This is correct, but the activated monomer mechanism operates on the ring-terminated end and this is the supposedly fast reaction which yields a high molecular weight product. Blout emphasized also that in the polymerization initiated, e.g., by triphenylmethylsodium, the free amine end cannot exist as such but should be converted into the respective sodium amide. Since the contribution of free amine propagation is insignificant when compared with that of activated monomer propagation, its conversion to another species is of little importance. Moreover, under conditions which lead to a high molecular weight polymer, the free amine is most probably unaffected by the base. Since the concentration of NCA exceeds that of the free amine, and the former is a stronger acid than the latter, the equilibrium in the reaction

$$\sim\!\!\text{NH}^-,\text{Na}^+ + \text{NCA} \rightleftarrows \sim\!\!\text{NH}_2 + \text{activated NCA}^-,\text{Na}^+$$

is displaced far to the right.

Another misconception arises from the statement that a strong base, e.g., $\text{C(Ph)}_3{}^-,\text{Na}^+$ will be converted into an extremely weak acid, i.e., Ph_3CH, and will not provide a proton for the decarboxylation of carbamate ions. This again is true, but the proton need not be provided by the conjugated acid, in fact it is given by the non-activated NCA which is converted in this process into an activated NCA. This reaction cannot take place with the *N*-substituted NCA's and therefore the activated monomer mechanism is not operative for these monomers.

The most serious objection is the lack of experimental evidence for the existence of the ring compound at the end of the polymer (62). Apart from the general difficulty of detecting and identifying end groups of high molecular weight polymers, an additional problem arises in this system. The reaction does not involve a termination step which could destroy the terminal ring group responsible for growth. There is, however, indirect evidence for the existence of an activated monomer. Goodman (90) prepared NCA of the benzyl ester of α-amino-adipic acid. Its polymerization in dioxane initiated by sodium methoxide produced a by-product

$$
\begin{array}{c}
\text{CH}_2 \\
\diagdown \\
\text{CH}_2 \quad \text{CH—CO} \\
\diagdown \\
\quad\quad\quad\quad O \\
\text{CH}_2 \quad \text{N——CO} \\
\diagdown \diagup \\
\text{C} \\
\| \\
\text{O}
\end{array}
$$

which obviously arose from the reaction

$$
\begin{array}{c}
\text{CH}_2 \\
\diagup \diagdown \\
\text{CH}_2 \quad \text{CH—CO} \\
\quad\quad\quad\quad O \rightarrow \\
\text{CH}_2 \quad {}^-\text{N——CO} \\
\diagdown \\
\text{CO(OBz)}
\end{array}
\qquad
\begin{array}{c}
\text{CH}_2 \\
\diagup \diagdown \\
\text{CH}_2 \quad \text{CH—CO} \\
\quad\quad\quad\quad O \\
\text{CH}_2 \quad \text{N——CO} \\
\diagdown \diagup \\
\text{C} \\
\| \\
\text{O} \quad + \text{BzO}^-
\end{array}
$$

In conclusion, the complexity of NCA polymerization arises from the numerous modes of reaction of such monomers. Collating all the available evidence, we find that nucleophiles and bases may interact with NCA in two fashions: through addition whereby the monomer behaves as a Lewis acid or by proton abstraction whereby it acts as a weak proton-donating acid. Growth takes place either by addition of nonactivated monomer to a free amine group, or perhaps to $\sim$NH$\cdot$COO$^-$, or, much more efficiently, by addition of an activated monomer to an oxazolidine ring formed at the end of a growing polymer and activated by the adjacent CO group. The latter mode is fast and gives a high molecular weight polymer. The variety of termination steps add to the complex character of this polymerization.

10. POLYMERIZATION OF N-SUBSTITUTED NCA's

The mechanism of polymerization of N-substituted NCA's is still in doubt. The "simple" amine-propagated polymerization may operate for N-substituted as well as for non-N-substituted NCA's. This was fully discussed in earlier sections of the present Chapter and no further clarification of the subject is needed. The "activated monomer" mechanism, which involves NCA anions possessing the negative charge on the nitrogen atom, is, of course, inoperative for the *N*-substituted NCA's. This is obvious and undisputed, but the real problem lies whether the *N*-substituted NCA's are polymerized by aprotic bases in the *absence* of proton donors? Bamford's school strongly maintains that such a polymerization is intrinsically impossible, while many other investigators claim this polymerization to be feasible.

Let us recall, first of all, that even in an "aprotic" medium containing an *N*-substituted NCA, protons may be available. Goodman and Arnon (55) reported that the addition of a large excess of sodium methoxide to an aprotic solution of L-proline NCA leads to its racemization. They interpreted this observation in terms of the following reactions:

$$CH_3O^- + H(R)C^*\!\!-\!\!CO \qquad\qquad CH_3OH + R\bar{C}\!\!-\!\!CO$$

with the bridging structures (Racemization):

$$R_1\cdot N\!\!-\!\!CO \qquad\qquad R_1\cdot N\!\!-\!\!CO$$

Hence, if protons are needed for the decarboxylation of carbamates, the above process may supply them. However, there are other modes of propagation which may apply to the *N*-substituted NCA's. For example, it has been pointed out that the reaction:

$$CH_3O^- + CO\!\!-\!\!CHR \quad\rightarrow\quad CH_3O\cdot CO\!\!-\!\!CHR$$

$$CO\!\!-\!\!NR_1 \qquad\qquad {}^-O\cdot CO\!\!-\!\!NR_1$$

is feasible and may be observed, although only to a small extent, for the non-*N*-substituted NCA's. In the latter case, proton transfer

$$CH_3O^- + NCA \rightleftarrows CH_3OH + \overline{N}CA$$

is greatly favored and, therefore, direct addition is hardly significant. This is not the case for the *N*-substituted NCA and hence, for these monomers, direct addition may be important.

During polymerization of the non-*N*-substituted NCA, the carbamates become rapidly protonated, and eventually they decarboxylate, because the presence of a large excess of the proton-donating NCA favors such reactions. This again is not so for the *N*-substituted NCA's. The lack of protons makes the carbamates stable and permits them to undergo the Blout-Walker type of reaction

$$\sim\!\!CO\cdot CH(R)\cdot NR_1\cdot COO^- + CO\!\!-\!\!CH\cdot R \quad\rightarrow$$

$$CO\!\!-\!\!NR_1$$

$$\sim\!\!CO\cdot CH(R)\cdot NR_1\cdot CO\cdot O\cdot CO\cdot CH(R)\cdot NR_1\cdot COO^-$$

giving linear anhydrides, and eventually, after their decarboxylation, a new peptide bond may be formed. It is important to note that this reaction was shown (68,79) to be insignificant in the carbamate-initiated polymerization of the *N*-non-substituted NCA's, because another mode of propagation (involving protonation and formation of an "activated" NCA) was so much more efficient. Because this course is impossible for the *N*-substituted NCA alternative modes take over, e.g., formation of linear anhydrides. We may test this idea by initiating, under strictly aprotic conditions, the polymerization of *N*-substituted NCA, for example, sarcosine with a radioactive *N*-benzylamine carbamate. The presence of radioactivity in the polymer, a result contrary to that found for the *N*-non-substituted NCA's, would prove the proposed mechanism.

Finally, it should be noted that the feasibility of polymerization of *N*-substituted NCA initiated by aprotic bases in an aprotic medium does not constitute an argument against the "activated" monomer mechanism. This mechanism is now well proved by the experiments of Goodman and Hutchison, and the negative proof, sought by Bamford's school, is no longer necessary.

11. PHYSICAL FACTORS IN NCA POLYMERIZATION

The complexity of NCA polymerization is often magnified by some phenomena of a physical rather than of a chemical nature. Thus, the state of aggregation of the polymerizing system or the shape of the polymeric molecule may play a profound role in determining the course of these reactions.

It is well known (80,81) that many radical polymerizations are enhanced by an increase in the viscosity of the polymerizing system, and this phenomenon was explained by a decrease in the rate of the diffusion-controlled termination. In fact, the effect of viscosity should be observed at any state of radical polymerization, and this problem has been discussed recently by Benson and North (82,83). Of course, this type of acceleration cannot be observed when growth involves living polymers and, therefore, such an explanation does not apply to the polymerization of NCA, because no termination resulting from a bimolecular interaction involving two active ends takes place in these processes.

In some systems, the polymer may precipitate in the course of a reaction and this again greatly affects the kinetics of polymerization, e.g., in a radical polymerization the precipitation may lead to the formation of "trapped" radicals (84). Moreover, separation into two phases affects the concentration of monomer around the growing centers and this may either speed up the propagation, if the polymer is a better solvent for the monomer than the original medium, whereas a retardation is expected in the reverse case. These phenomena are particularly important in processes involving living polymers or long-lived polymers and, therefore, they should be observed in NCA polymerization. In fact, in several NCA polymerizations a marked acceleration of the rate was noted when the systems became heterogeneous (85). These reactions were investigated in benzene or nitrobenzene, and the precipitated polymer appears to act as a better solvating medium for the monomer than the original solvents. Hence, the local increase in monomer concentration seems to account for the observed phenomenon. An acceleration was also observed in the polymerization of sarcosine NCA.

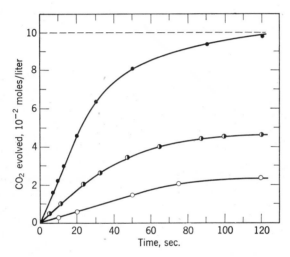

Fig. 16. Polymerization of D,L-phenylalanine NCA in nitrobenzene solution at 25°C with various initiators: (○) sarcosine dimethylamide; (◑) polysarcosine dimethylamide, $n = 5$; (●) polysarcosine dimethylamide, $n = 10$; (------) theoretical yield. $[M]_0 = 0.10$ mole/liter; $[X] = 8.6 \times 10^{-3}$ mole/liter.

A peculiar and intriguing effect of polysarcosine, which was used as an initiator in block polymerization, was observed by Ballard and Bamford (86). Their results are shown in Figure 16 and indicate that the initial rates of D,L-phenylalanine NCA polymerization are affected by the degree of polymerization of the initiator. For a constant monomer concentration and a constant concentration of the initiating amine, the rate increases significantly as the degree of polymerization of polysarcosine rises from 1 to 10.* These findings are summarized in Figure 17 which shows that the initial rate of polymerization approaches a constant value for $n \geq 20$.

* These effects were never observed in the polymerization of γ-benzyl L-glutamate NCA. (Private communication from Professor Blout.)

The following facts should be noted: (*1*) Polymerization produces a block polymer, polysarcosine linked to the newly-formed polypeptide. (*2*) The effect was observed in the polymerization of four different NCA's all initiated by polysarcosine. (*3*) The solvent plays an important role in the process. The

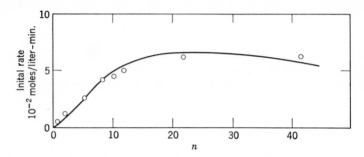

Fig. 17.   Polymerization of D,L-phenylalanine NCA initiation by polysarcosine dimethylamides with different degrees of polymerization in nitrobenzene solution at 15°C. $[M]_0 = 0.100$ mole/liter; $[X] = 5.4 \times 10^{-3}$ mole/liter. The solid line represents theoretical curve, circles represent experimental points.

TABLE VIII

Effect of Polysarcosine and Acetylated Polysarcosine on Polymerizations[a]
(Monomer: $[M]_0 = 0.105$ mole/liter. Solvent: nitrobenzene. Temperature: 15°C)

Initiator	Concentration, 10^{-3} mole/liter	Initial rate, 10^{-3} mole liter^{-1} min^{-1}
Polysarcosine dimethylamide, $n = 20$	5	5.3
N-Acetyl polysarcosine dimethylamide, $n = 20$	5	0
N-Acetyl polysarcosine dimethylamide, $n = 20$ + phenylalanine dimethylamide	Each 5	0.4
Phenylalanine dimethylamide	10	0.13

[a] From D. G. H. Ballard and C. H. Bamford, *Proc. Roy. Soc. (London), Ser. A,* **236,** 384 (1956).

reaction is greatly accelerated in chloroform and nitrobenzene and much less in dimethylformamide. (*4*) As shown in Table VIII, acetylation of polysarcosine destroys the effect and the addition of such a polymer to phenylalanine dimethylamide accelerates the process only slightly. (*5*) No effect is observed when sarcosine NCA is block polymerized with D,L-polyphenlalanine and the active growing end is located on the latter.

Ballard and Bamford proposed an interesting explanation for this phenomenon to which they refer as "chain effect". The molecules of NCA are preferentially assumed to be adsorbed on the polysarcosine chain and hence, they are kept in the vicinity of the growing end. This increase in the "monomer supply" leads to a fast growth rate. Their argument is strengthened by the observation that addition of "dead" (acetylated) polysarcosine inhibits the polymerization. This, they argue, removes the monomer which becomes bonded to a useless polymer.

The "chain effect" deserves further study, and, indeed, some new facts were reported recently by Ballard (87). It is strange that it appears only with polysarcosine, because Bamford's explanation suggests that a similar mechanism may also operate for other polypeptides.

Further investigations (88) verified the "adsorption" through infrared studies. Because this interaction is possible only with non-*N*-substituted NCA's (it is the hydrogen bonding between NH of NCA and CO of polysarcosine that leads to "adsorption") it has been concluded that in a mixture of *N*-substituted and non-*N*-substituted NCA's only the latter would be susceptible to the "chain effect". Therefore, this monomer should polymerize first, if the polymerization is initiated by poly-sarcosine, and the expected product should be a block polymer. On the other hand, if an ordinary initiator is employed a random copolymer should be formed. This idea has been tested (89) and fully verified by experiments performed on mixtures of β-phenylalanine and sarcosine NCA's.

References

(1) H. Leuchs, *Ber.*, **39**, 857 (1906).
(2) (a) H. Leuchs and W. Manasse, *ibid.*, **40**, 3235 (1907). (b) H. Leuchs and W. Geiger, *ibid.*, **41**, 1721 (1908).
(3) E. Katchalski and M. Sela, *Advan. Protein Chem.*, **13**, 243 (1958).
(4) P. D. Bartlett and R. H. Jones, *J. Am. Chem. Soc.*, **79**, 2153 (1957).
(5) E. Miller, I. Fankuchen, and H. Mark, *J. Appl. Phys.*, **20**, 531 (1949).
(6) M. Sela and E. Katchalski, *J. Am. Chem. Soc.*, **76**, 129 (1954).
(7) W. Dieckmann and F. Breest, *Ber.*, **39**, 3052 (1906).
(8) K. D. Kopple, *J. Am. Chem. Soc.*, **79**, 662, 6442 (1957).
(9) K. Heyns and R. Brockmann, *Z. Naturforsch.*, **9b**, 21 (1954).
(10) F. Fuchs, *Ber.*, **55**, 2943 (1922).
(11) F. Wessely, *Z. Physiol. Chem.*, **146**, 72 (1925).
(12) F. Wessely and F. Sigmund, *ibid.*, **157**, 91 (1926).
(13) W. E. Hanby, S. G. Waley, and J. Watson, *J. Chem. Soc.*, **1950**, 3009.
(14) S. G. Waley and J. Watson, *Proc. Roy. Soc. (London)*, Ser. A, **199**, 499 (1949).
(15) J. H. Fessler and A. G. Ogston, *Trans. Faraday Soc.*, **47**, 667 (1951).
(16) M. T. Pope, T. J. Weakley, and R. J. P. Williams, *J. Chem. Soc.*, **1959**, 3442.
(17) R. D. Lundberg and P. Doty, *J. Am. Chem. Soc.*, **79**, 3961 (1957).
(18) C. H. Bamford and H. Block, "Polyamino Acids, Polypeptides and Proteins," Wisconsin University Press, Madison, Wis., 1962, p. 65.

(19) J. L. Bailey, *J. Chem. Soc.*, **1950**, 3461.

(20) D. G. H. Ballard and C. H. Bamford, *Proc. Roy. Soc. (London)*, *Ser. A*, **223**, 495 (1954).

(20) (a) A. Patchornik and Y. Shalitin, *Anal. Chem.*, **33**, 1887 (1961). (b) H. Block, *J. Sci. Instruments*, **41**, 370 (1964).

(21) K. Heyns, H. Schultze, and R. Brockmann, *Ann. Chem.*, **611**, 33 (1958).

(22) K. Heyns and H. Schultze, *ibid.*, p. 40.

(23) M. Sela and A. Berger, *J. Am. Chem. Soc.*, **77**, 1893 (1955).

(24) K. D. Kopple, J. J. Katz, and L. A. Quarterman, *ibid.*, **78**, 6199 (1956); *J. Org. Chem.*, **27**, 1062 (1962).

(25) W. E. Hanby, S. G. Waley, and J. Watson, *J. Chem. Soc.*, **1950**, 3239.

(26) J. C. Mitchell, A. E. Woodward, and P. Doty, *J. Am. Chem. Soc.*, **79**, 3955 (1957).

(27) F. Wessely and M. John, *Z. Physiol. Chem.*, **170**, 38 (1927).

(28) (a) R. B. Woodward and C. H. Schramm, *J. Am. Chem. Soc.*, **69**, 1551 (1947). (b) L. A. Sluyterman and B. Labruyere, *Rec. Trav. Chim.*, **73**, 347 (1954).

(29) D. Coleman and A. C. Farthing, *J. Chem. Soc.*, **1950**, 3218.

(30) J. W. Breitenbach and K. Allinger, *Monatsh. Chem.*, **84**, 1103 (1953).

(31) M. Frankel and E. Katchalski, *J. Am. Chem. Soc.*, **65**, 1670 (1943).

(32) C. Faurholt, *J. Chim. Phys.*, **22**, 1 (1925).

(33) H. Weingarten, *J. Am. Chem. Soc.*, **80**, 352 (1958).

(34) Y. Shalitin, Ph.D. Thesis, Hebrew University, Jerusalem, Israel (1958).

(35) E. R. Blout and A. Asadourian, *J. Am. Chem. Soc.*, **78**, 955 (1956).

(36) P. Doty and R. D. Lundberg, *ibid.*, p. 4810.

(37) E. Katchalski, Y. Shalitin, and M. Gehatia, *ibid.*, **77**, 1925 (1955).

(38) D. G. H. Ballard and C. H. Bamford, *ibid.*, **79**, 2336 (1957).

(39) P. Doty and R. D. Lundberg, *ibid.*, p. 2338.

(40) D. G. H. Ballard, C. H. Bamford, and A. Elliott, *Makromol. Chem.*, **35**, 222 (1960).

(41) L. Bilek, J. Derkosch, H. Michl, and F. Wessely, *Monatsh. Chem.*, **84**, 717 (1953).

(42) D. Coleman, *J. Chem. Soc. (London)*, **1950**, 3222.

(43) D. G. H. Ballard and C. H. Bamford, *ibid.*, **1956**, 381.

(44) D. G. H. Ballard, C. H. Bamford, and F. J. Weymouth, *Proc. Roy. Soc. (London)*, *Ser. A*, **227**, 155 (1955).

(45) J. Kurtz, G. D. Fasman, A. Berger, and E. Katchalski, *J. Am. Chem. Soc.*, **80**, 393 (1958).

(46) C. H. Bamford, H. Block, and A. C. P. Pugh, *J. Chem. Soc.*, **1961**, 2057.

(47) E. R. Blout, R. H. Karlson, P. Doty, and B. Hargitay, *J. Am. Chem. Soc.*, **76**, 4492 (1954).

(48) P. Doty, A. M. Holtzer, J. H. Bradbury, and E. R. Blout, *ibid.*, p. 4493.

(49) E. R. Blout and R. H. Karlson, *ibid.*, **78**, 941 (1956).

(50) M. Idelson and E. R. Blout, *ibid.*, **80**, 2387 (1958).

(51) C. H. Bamford and H. Block, *J. Chem. Soc.*, **1961**, 4989.

(52) C. H. Bamford and H. Block, *ibid.*, p. 4992.

(53) M. Idelson and E. R. Blout, *J. Am. Chem. Soc.*, **79**, 3948 (1957).

(54) D. G. H. Ballard and C. H. Bamford, *Chem. Soc. (London)*, *Spec. Publ.*, **2**, 25 (1955).

(55) M. Goodman and U. Arnon, *Biopolymers*, **1**, 500 (1963).

(56) T. Wieland, *Angew. Chem.*, **63**, 7 (1951).

(57) T. Wieland, *ibid.*, **66**, 507 (1954).

(58) (a) V. Gold and E. G. Jefferson, *J. Chem. Soc.*, **1953**, 1409, 1416. (b) E. E. Walker, *Proc. Intern. Colloq. Macromol. Amsterdam*, **1949**, 381.

(59) E. Peggion, A. Cosani, A. M. Mattucci, and E. Scoffone, *Biopolymers*, **2**, 69 (1964).

(60) R. E. Nyland and W. G. Miller, *Biopolymers*, **2**, 131 (1964).
(61) E. R. Blout and M. Idelson, "Polyamino Acids, Polypeptides and Proteins," Wisconsin University Press, Madison, Wis., 1962, p. 79.
(62) E. Katchalski and M. Sela, reference 3, p. 362.
(63) S. Okamura, K. Hayashi, and T. Natori, *Ann. Rep. Japan. Assoc. Radiation Res. Polymers*, **3**, 161 (1961).
(64) G. Kovacs, E. Kovacs, and H. Morawetz, *J. Polymer Sci. A*, **4**, 1553 (1966).
(65) T. Yoshida, S. Sakurai, T. Okuda, and Y. Takagi, *J. Am. Chem. Soc.*, **84**, 3590 (1962).
(66) A. Cosani, E. Peggion, E. Scoffone, and A. S. Verdini, *J. Polymer Sci. B*, **4**, 55 (1966).
(67) E. Scoffone, E. Peggion, A. Cosani, and M. Terbojevich, *Biopolymers*, **3**, 535 (1965).
(68) M. Goodman and J. Hutchison, *J. Am. Chem. Soc.*, **88**, 3627 (1966).
(69) E. Peggion, M. Terbojevich, A. Cosani, and C. Colombini, *ibid.*, 3630.
(70) E. Peggion, E. Scoffone, A. Cosani, and A. Portolan, *Biopolymers*, **4**, 695 (1966).
(71) A. Cosani, G. D'Este, E. Peggion, and E. Scoffone, *ibid.*, p. 595.
(72) R. N. Haward, *Trans. Faraday Soc.*, **46**, 204 (1950).
(73) B. H. Zimm and J. K. Bragg, *J. Polymer Sci.*, **9**, 476 (1952).
(74) M. Goodman and U. Arnon, *J. Am. Chem. Soc.*, **86**, 3384 (1964).
(75) G. D. Fasman and E. R. Blout, *Biopolymers*, **1**, 3 (1963).
(76) G. D. Fasman and E. R. Blout, *ibid.*, p. 99.
(77) Y. Avny and A. Zilkha, *Israel J. Chem.*, **3**, 207 (1966); see also *Europ. Polymer J.* **2**, 367 (1966), and Y. Avny, S. Migdal and A. Zilkha; *ibid*, 355.
(78) M. Szwarc, *Advan. Polymer Sci.*, **4**, 1, 44, 45 (1965).
(79) M. Goodman and J. Hutchison, *J. Am. Chem. Soc.*, **87**, 3524 (1965).
(80) R. G. W. Norrish and R. R. Smith, *Nature*, **150**, 336 (1942).
(81) E. Trommsdorff, H. Köhle, and P. Lagally, *Makromol. Chem.*, **1**, 169 (1947).
(82) S. W. Benson and A. M. North, *J. Am. Chem. Soc.*, **81**, 1339 (1959).
(83) A. M. North and G. A. Reed, *Trans. Faraday Soc.*, **57**, 859 (1961).
(84) C. H. Bamford and A. D. Jenkins, *Proc. Roy. Soc. (London), Ser. A*, **216**, 515 (1953).
(85) D. G. H. Ballard and C. H. Bamford, *J. Chem. Soc.*, **1959**, 1039.
(86) D. G. H. Ballard and C. H. Bamford, *Proc. Roy. Soc. (London), Ser. A*, **236**, 384 (1956).
(87) D. G. H. Ballard, *Biopolymers*, **2**, 463 (1964).
(88) C. H. Bamford and R. C. Price, *Trans. Faraday Soc.*, **61**, 2208 (1965).
(89) C. H. Bamford, H. Block and Y. Imanishi, *Biopolymers*, **4**, 1049 (1966).
(90) M. Goodman, unpublished results.
(91) V. V. Korshak, S. V. Rogozhin, V. A. Dazankov, Yu. A. Davidovich, and T. A. Makarova, *Russ. Chem. Rev.*, **34**, 329 (1965).
(92) A. Berger, M. Sela, and E. Katchalski, *Anal. Chem.*, **25**, 1554 (1953).
(93) H. A. Sober; in "Polyamino Acids, Polypeptides and Proteins," Univ. of Wisconsin Press, 1962.
(94) M. Terbojevich, G. Pizziolo, E. Peggion, A. Cosani, and E. Scoffone, *J. Am. Chem. Soc.* **89**, 2733 (1967).
(95) See e.g., pp. 226 and 227 of Review by M. Szwarc and J. Smid; *Progr. Reaction Kinetics*, **2** (1964).

Chapter XI

Helical Polymerization

1. HELICAL CONFORMATION OF POLYMERIC MOLECULES IN THE SOLID STATE

The first convincing evidence of a polymeric molecule having a helical conformation in the solid state was provided by the classic work of Pauling and Corey (1). Their x-ray studies of keratin proved that the molecules of this fibrous protein form an array of 13-membered rings linked together through NH$\cdots$OC hydrogen bonds into spirals with a pitch of 5.4 Å. Each turn of the spiral, known now as an α-helix, contains 3.6 amino acid residues, the consecutive units being axially transposed by 1.50 Å. Such a structure seems to produce the most stable helix; it leaves no empty space in the middle of the spiral, does not impose any strain on the polymeric chain, and satisfies the linearity of the hydrogen bonds. The α-helix hypothesis rapidly assumed a dominant role in subsequent studies when it was demonstrated that such a shape is acquired by the molecules of many other proteins and polyamino acids. Nevertheless, the possibility of other helical structures, anticipated by Pauling, must not be renounced. For example, Luzzati and his colleagues (2) reported transitions of α-helices into more elongated 3_{10} helices in which the amino acid residues are displaced by 1.95 Å.*

Prior to the work of Pauling, many attempts were made to describe the structure of proteins in terms of helices stabilized by intramolecular hydrogen bonds. The pioneering work of Astbury and his school (3) indicated that molecules of proteins, especially of wool, may exist in an extended or folded configuration, the latter being considered as a helix. These ideas were applied to polyamino acids by Elliott, Bamford, and their associates (4–6) who were concerned with the different structures of films and fibers. The folded form of the protein molecules, reported by Astbury, was identified with the α-helix, the extended one was recognized as an intermolecularly hydrogen bonded array of chains, designated in earlier literature as the β-form. In films made from synthetic polyamino acids the α- and β-forms could be interconverted. The folded structure is formed when the films are cast from m-cresol solutions, whereas in films cast from formic acid solutions the polymer

*Their conclusions are, however, disputed.

takes the extended β-form. These observations show that the solvent may affect the shape of polymeric molecules.

The helical structure of stereospecific vinyl polymers was demonstrated by the extensive work of Natta and his associates (7). Polypropylene prepared in tetralin with a $TiCl_4 : Al(Et)_3$ catalyst system contained an insoluble fraction which was identified as isotactic polypropylene. X-ray studies of this crystalline polymer (8) showed that its molecules fit into a monoclinic cell ($a = 6.65$ Å; $b = 20.96$ Å; $c = 6.50$ Å, and $\beta = 99° 20'$) containing 12 monomeric units. The chain is folded into a helix having 3 monomeric units per turn of the spiral. Subsequent studies showed that helical forms are adopted by many stereospecific poly-α-olefins, the type of helix depending on the structure of the investigated polymer and the conditions encountered during preparation of the sample. Continuation of such investigations proved that helical polymers may be formed from other classes of vinyl and vinylidene monomers, as well as from dienes.

A most interesting study of helix stability has been completed recently by Liquori (25). The total repulsive energy of a helix arising from interactions between nonbonded atoms can be calculated, by using a computer program technique, as a function of two or three angles which define the chain configuration. For example, in studies of polyamino acids, the inclination angles of the CH(R)—NH—CO and NH—CO—CH(R) planes with respect to the helix axis may be chosen for this purpose. For R = H or CH_3, these two angles are sufficient to describe fully the conformation of the helix but, if R contains 2 or more carbon atoms, an additional angle has to be introduced to define its spatial position. Conventional bond lengths and bond angles, as well as planarity of the —NH—CO group are assumed in the calculation. The results show that the energy, represented as a function of the variable angles, has exceptionally deep and sharp minima. The respective angles, which define the pitch of the helix, may therefore be accurately determined. This treatment shows unambiguously whether the right-handed or left-handed helix is preferred, and agreement with observations is, indeed, remarkable.

Similar calculations were previously performed for helical, isotactic polymeric hydrocarbons possessing the —CH_2—CR_1R_2— recurring units (26). In this case, the chosen variable angles give the inclinations of the —CH_2—$C(R_1R_2)$—CH_2 and $C(R_1R_2)$—CH_2—$C(R_1R_2)$ planes relative to the helix axis. The minima are not as sharp as those calculated for polyamino acids and the symmetry of these chains makes, of course, the right-handed and left-handed helix equally probable.

2. HELICAL CONFORMATION OF POLYMERS IN SOLUTION

Doty and his colleagues (9) were first to recognize that the helical structure of polyamino acids may be retained in solution. This fact, which had far-reaching consequences on further developments, was established on the basis of striking differences in the hydrodynamic behavior in various solvents of synthetic, high molecular weight poly-γ-benzyl L-glutamate. Studies of viscosity, light scattering, infrared spectra, etc., revealed that these polymer molecules behave in dioxane or chloroform as rigid, rod-like particles, whereas in dichloroacetic acid they acquire the characteristic random coil conformation of ordinary polymer chains in solution. The rod-like molecules were recognized as α-helices. In dichloroacetic acid, the intramolecular hydrogen bonds which stabilize the helix are replaced by intermolecular bonds between the amino acid residues and the powerfully solvating $CHCl_2 \cdot COOH$ molecules. Therefore, in this medium the chain acquires the random coil shape.

Further work led to the realization that, even in a non-destructive solvent, a perfect helical conformation cannot be maintained along the entire length of a high molecular weight polymer chain. Helical units of variable length have to be linked by disoriented, randomly coiled chains. The degree of disorientation being determined by temperature, the nature of the solvent and of the polymer, and perhaps also its molecular weight (10). The coexistence of oriented and disoriented portions of the chain poses the interesting problem of helix–random coil transition (11). While the atoms in the coiled portion of the polymeric molecule perform an endless Brownian dance, their positions become fixed in the helical stretches which are one-dimensional crystals. The spontaneous "crystallization" of the disoriented chains presents a fascinating aspect of the physical chemistry of polyamino acids, and this phenomenon has been observed only in one other system, namely, in polynucleic acids and polynucleotides. The question arises, therefore, as to what is the smallest number of residues which permits spontaneous formation of a helix.

To answer this question, Goodman and his team (12) synthesized by classical techniques a series of polypeptides of γ-methyl L-glutamate ranging from 3 to 11 units, all composed of the same enanthiomorphic form of the amino acid residue. The specific optical rotation of a polymeric molecule possessing asymmetric centers is enhanced when it acquires a helical structure (13) and therefore the shape of an oligomer may be recognized from the degree of optical rotation of its solution. In view of this, Goodman's group examined the optical rotation of their oligomers in a variety of solvents (14), the results being shown in Figure 1. It may be seen that the specific optical rotation in dichloroacetic acid continuously *decreases* with increasing degree of polymerization, n (see the full circles, Figure 1). In dioxane, rotation was observed to decrease with increasing n only for the first 3 oligomers ($n = 2$, 3, and 4), whereas the optical rotation of the pentamer solution was considerably higher, and even higher values were found for the solutions of hexa-, hepta-, and nonamers.

An increase in the specific rotation of low molecular weight polypeptides may also be caused by their intermolecular association (9). Such an effect should be concentration dependent, and indeed in dioxane the specific rotations of the penta- and hexamer were found to increase with concentration. Moreover, ultracentrifugation studies (15) proved that these oligomers are associated. However, the specific optical rotations of the dimer,

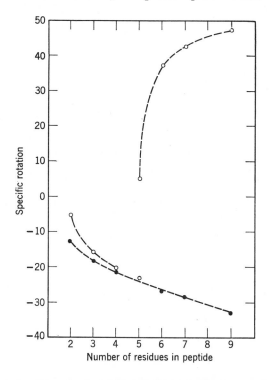

Fig. 1. Optical activity of polypeptide as a function of solvent and number of amino acid residues. Optical rotations were measured in dioxane (○) and dichloroacetic acid (●) at 2% concentration except those for the hepta- and nonapeptides. These rotations were measured in dioxane solution at concentrations of 1.43 and 0.22%, respectively.

trimer, and tetramer, as well as of the hepta- and nonamer were concentration independent, and hence the abnormally high specific rotations of the hepta- and nonamers confirm their helical form.

To avoid any complications caused by intermolecular association, Goodman (16) reinvestigated the optical rotation of the peptides in dimethylformamide, since in this medium the specific rotation is independent of concentration. From the latter study it was concluded that at 25°C the

spontaneous helix formation of poly-γ-methyl L-glutamate in dimethylformamide takes place at the critical range of 7–9 units. Extension of these
studies (12b,17) led to a better understanding of temperature and solvent
effects upon the helix–coil transition of oligomeric polypeptides.

Alternatively, the problems of helix–coil transition may be investigated by
determining the ultraviolet absorption of a polypeptide in the 190 mμ region,
because the extinction coefficient of this band is markedly increased by the
formation of a helix (18). This criterion was used in the studies of Goodman
and Listowsky (19).

Recently, studies of the conformation of oligomers were extended to peptides derived from β-methyl L-aspartates. Their synthesis ($n = 2$ up to 14) was
described by Goodman and Boardman (20), and subsequently the specific
rotations of their solutions in dimethylformamide, dichloroacetic acid, and in
chloroform were reported (21). In the first two solvents these oligomers
exist in a random coil form, but the helical form becomes stable in chloroform for $n = 11$ and 14. These peptides are unusual in one respect, because
their L-amino acid residues produce a left-hand helix (22,23), whereas most
of the investigated polyamino acids "crystallize" in right-hand spirals (24).

Stability of the helical structure of polyamino acids is, to a great extent,
due to the strength of the intramolecular hydrogen bonds which are formed in
those polymeric molecules. Polynucleic acids (DNA and RNA) provide
another striking example of systems in which the shape of the polymer
is profoundly determined by hydrogen bonds. However, other factors contribute also to the preferential conformation of polymeric molecules, e.g., in
aqueous solutions hydrophobic bonds (27) may play an important role,
shaping the configuration of polyamino acid chains. This type of interaction
should be significant in other solvents. Its importance is magnified as the
solubility of the polymeric molecule in the investigated medium decreases.

The helix forming tendency is greatly repressed in solutions of polymers
deprived of groups which may form intramolecular hydrogen bonds. Nevertheless, it is not repressed entirely. For example, extensive studies by Pino
and his colleagues (see for a review ref. 28) revealed a preference for the
helical conformation in solutions of tactic poly-α-olefins.

Evidence for coiling of polymeric hydrocarbons in solution was obtained
from studies of the optical rotation of optically active polymers. Such polymers have been recently synthesized in a high degree of optical purity, mainly
through the combined efforts of Natta, Pino, and their colleagues (28), and
these investigations have been eventually extended by other workers to nearly
all classes of synthetic polymers (29,30).

The chains of optically active, head-to-tail linked vinyl polymers possess
one type of asymmetric carbon atom in their backbone and another in the
side groups. The former, which may be classified as pseudo-asymmetric

carbon atoms, do not contribute much to the rotary power, and therefore, the optical activity of the polymer is determined mainly by the atoms in the lateral side chains.

The preferential conformational equilibria of high molecular weight polymers in solution or in the melt are revealed by differences in the sign and value of the optical rotation found for optically active polymers and their low molecular weight model compounds. A remarkable difference was noted in these polymers in which the asymmetric carbon atoms of the lateral side chains are in a position α or β with respect to the principal chain. This is shown by a few examples collected in Table I. The observations listed in Table I have

TABLE I

Comparison of Optical Activity of Some Aliphatic Addition
Polymers with the Respective Low Molecular Weight Models

n	Model, optical rotation	Polymer, optical rotation	References
	$\begin{array}{cc} CH_3 & C_2H_5 \\ \mid & \mid^* \\ CH-(CH_2)_n-C-H \\ \mid & \mid \\ CH_3 & CH_3 \end{array}$	$\begin{array}{cc} CH_2 & C_2H_5 \\ \mid & \mid^* \\ CH-(CH_2)_n-CH \\ \mid & \mid \\ CH_2 & CH_3 \end{array}$	
0 (α)	-11.4	$+161$	(a), (b)
1 (β)	$+21.3$	$+288$	(c), (b)
2 (γ)	$+11.7$	$+68.1$	(d), (b)
3 (δ)	$+14.4$	$+20.4$	(e), (b)
	$\begin{array}{cc} & C_2H_5 \\ & \mid^* \\ C_2H_5O(CH_2)_n-C-H \\ & \mid \\ & CH_3 \end{array}$	$\begin{array}{cc} CH_2 & C_2H_5 \\ \mid & \mid^* \\ CHO(CH_2)_n-C-H \\ \mid & \mid \\ CH_2 & CH_3 \end{array}$	
0 (β)	$+34.5$	$+312$	(f), (g)
1 (γ)	$+1.1$	$+5.6$	(h)
2 (δ)	$+11.0$	—	(i)

(a) K. Freudenberg and W. Lwowski, *Ann. Chem.*, **587**, 224 (1954).
(b) P. Pino et al., *Makromol. Chem.*, **61**, 207 (1963).
(c) P. A. Levene and R. E. Marker, *J. Biol. Chem.*, **92**, 456 (1931).
(d) P. A. Levene and R. E. Marker, *ibid.*, **95**, 1 (1932).
(e) F. E. Ciardelli and P. Pino, *Ric. Sci.*, **34**, 694 (1964).
(f) P. Pino et al., unpublished results.
(g) P. Pino et al., *Ric. Sci.*, **34**, 193 (1964).
(h) P. Pino and G. P. Lorenzi, *Makromol. Chem.*, **47**, 242 (1961).
(i) P. Salvadori and L. Lardicci, *Gazz. Chim. Ital.*, **94**, 1205 (1964).

been supported by semi-empirical calculations of the rotary power of the polymers, using a method developed by Brewster (31) to account for the effect of conformation on optical activity. Thus, it could be concluded that even non-hydrogen bonded polymers have some residual organization in solution, or in melt. Small sections of their chains tend to turn in a unique way in an attempt to form a helix (32).

3. HELICAL POLYMERIZATION OF NCA*

Conversion of a randomly coiled growing polymer into a helix affects the surroundings of its active end, and, therefore, such a transformation is expected to modify the rate constant of propagation. This most interesting situation was visualized by Doty and Lundberg (34), and a model, taken from a paper by Weingarten (33) and shown in Figure 2, illustrates the possible

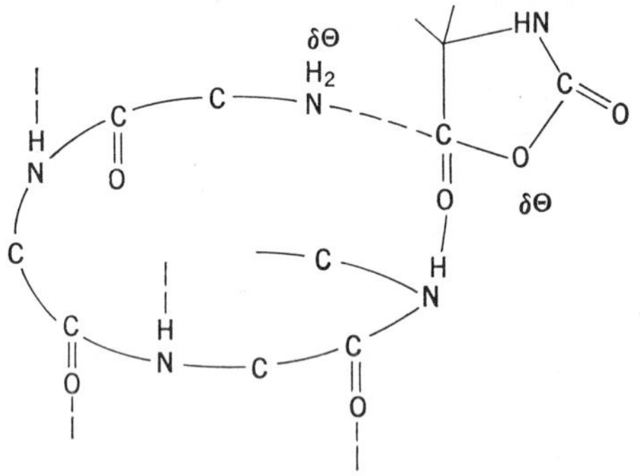

Fig. 2.   Hypothetical effect of the structure of the growing helix of polyamino acid on the NCA polymerization.

influence of helical structure on monomer addition. A molecule of NCA is assumed to be hydrogen bonded to the terminal active amine of a growing helix and linked simultaneously to the NH group of a residue separated by three amino acid units from the end of the polymer. The resulting orientation of the NCA is assumed to be beneficial for its addition. Hence, the destruction of the helix, or even replacement of one of the two H-bonding amino acid residues by its enanthiomorph, destroys the favorable configuration and, according to this hypothesis, slows down polymerization. The model of Weingarten may be also adapted to account for the fast growth of a helix terminated by an oxazolidine ring. For example, it could be argued that a hydrogen bond between the terminal ring and a suitable unit of the helix activates the oxazolidine residue and thus enhances propagation.

* This is a speculative subject. Its discussion is justified because it clarifies some concepts important for polymer chemistry.

Although the details of this mechanism still need verification, other available evidence seems to support the hypothesis of helical growth. The first indication of such a phenomenon came from the detailed kinetic study (35) of γ-benzyl-L-glutamate polymerization initiated by n-hexylamine in dioxane. The initiation was found to be very rapid ($k_{in} \approx 0.2$ liter/mole sec), but the propagation involved two steps described in terms of two second order rate constants: k_{2a}—referring to the first slow stage of the process, and k_{2b}—characterizing the second fast period. The rate constants were calculated from the slopes of the respective straight lines assuming that the concentration of growing ends is given by the initial concentration of n-hexylamine. This procedure was apparently justified, because a 10-fold change in the monomer-to-initiator ratio (from 4 to 40), left k_{2a} and k_{2b} reasonably constant. Addition of monomer to a polymerized solution led to further polymerization, the rate constant being identical with that of the preceding fast reaction. For example, when the polymerization, initiated at $M/I = 20$, was followed to 98% conversion, k_{2b} was determined as 0.035 liter/mole sec. Subsequent addition of monomer and solvent, to give an M/I ratio of 12, reinitiated the reaction which proceeded in a perfectly pseudo-unimolecular fashion with a second-order rate constant k_2 of 0.038 liter/mole sec. It seems, therefore, that the concentration of growing polymers, and the rate constant of their propagation, were unchanged after the polymer had reached a critical size of 8–11 units. On the basis of this evidence, Lundberg and Doty concluded that the slow reaction results from the growth of a randomly coiled, low molecular weight peptide, while the fast polymerization is caused by the propagation of a helix. In agreement with this hypothesis, the polymerization initiated at an M/I ratio of 4 showed only the first slow stage, $k_{2a} = 0.007$ liter/mole sec. Obviously, no helices can be formed at such a low DP, and therefore, the fast reaction could not possibly ensue.

Two additional facts apparently support the hypothesis of helical growth: (1) In dimethylformamide, only the first slow stage of the reaction was observed. Helices are less stable in this solvent and therefore their spontaneous formation could be prevented or inhibited. (2) The infrared studies of Idelson and Blout (36) demonstrated that the initially formed polymer possesses a random coiled structure in dioxane (described by the author as the β-form). As the reaction proceeds a new material appears, the latter being identified through its infrared spectrum as the α-helix. This stage of the process coincided with the onset of the fast reaction. By using a deuterated n-hexylamine as initiator, a labeled β-peptide was prepared and used to initiate further polymerization which yielded an α-peptide. The isolation of the latter and its analysis proved that the α-peptide contained the expected percentage of deuterium, demonstrating that the β-polymer was the precursor of the α-peptide.

Although the phenomenon of "two-stage polymerization" is well established and had been observed by other investigators (see Chapter X), it seems that this process does not involve really two pseudo-first order reactions, as suggested by Lundberg and Doty. Apparently, its kinetics arise from a relatively slow approach of the system to the stationary state, imposing an autocatalytic character on the overall process. Moreover, it is likely that other factors, cooperating with the helical growth, contribute to its acceleration. The following observations lead to this conclusion:

1. The growth of random coils is uniform and the onset of helix stability should occur rather abruptly at some critical size of the polymer (10). In view of the absence of termination this should lead to narrow molecular weight distribution, whereas the experimental results indicate $\overline{M}_w/\overline{M}_n$ ratio ranging from 3 to 8 (37). This implies that only a small fraction of the initially-formed polymers yield high molecular weight material and, hence, k_{2b} was incorrectly calculated. Its constancy appears, therefore, to be fortuitous.*

Lundberg and Doty tried to explain the high $\overline{M}_w/\overline{M}_n$ ratio by assuming that a small fraction of polymers, namely, those which reached the critical size, consumed all the available monomer in their rapid growth. This depleted the monomer supply preventing the remaining chains from attaining the critical length. Such an idea was formalized mathematically by Coombes and Katchalski (38), who derived the appropriate expression for the molecular weight distribution of the resulting polymer. Unfortunately, their calculations do not agree with the experimental data.

2. The kinetics of γ-benzyl D,L-glutamate NCA polymerization (35) resembles that of the L-isomer. Again, two stages are discerned in the process, although the onset of fast growth appears later than in the polymerization of the L-isomer. Lundberg and Doty suggested that in order to form a stable helix, the D,L-polymer has to be longer than the all L-polymer. However, it was shown by Huggins (39) that the presence of a wrong enantiomorph in sequence of four amino acid residues makes the helical structure improbable and the calculations of Liquori (40) fully confirm this prediction. Hence, the self-acceleration of racemic polymerization should be attributed to factors other than helical growth.

3. The self-acceleration of γ-benzyl L-glutamate NCA polymerization initiated by sodium methoxide was reported by Idelson and Blout (41). This reaction produces a high molecular weight polymer and acceleration, although less pronounced, is still observed when fresh monomer is added to a polymerized solution (see Figure 3). It is significant that the straight portions

---

* Unpublished results of Dr. Blout indicate that the high molecular weight polypeptide formed by sodium methoxide initiation does not show abnormally broad molecular weight distribution.

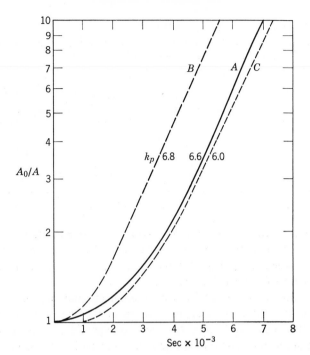

Fig. 3. Curve A (———): plot of anhydride consumption (A_0/A) measured by the CO_2 evolution method for $A/I = 400$. Sodium methoxide initiated polymerization of γ-benzyl L-glutamate-N-carboxyanhydride in dioxane solution (4 g/100 ml). Curve B (- - -): the rate curve obtained upon the addition of an equal quantity of L-NCA to a completely polymerized L-monomer. Curve C (- - - -): the rate curve obtained upon the addition of an equal quantity of D-NCA to a completely polymerized L-monomer. The k_p values are shown in a horizontal line with the appropriate symbol.

of curves A and B, shown in Figure 3, are parallel. Apparently, in both reactions the same stationary state is eventually attained.

The above arguments do not disprove the reality of helical polymerization: they show only that the kinetic evidence may not be sound. A stronger proof for such a growth is provided by the study of the effects exerted by the presence of one enantiomorph on the propagation of the other. This approach was also explored by Lundberg and Doty, and their results are summarized in Table II. They show that the racemic monomer, as well as the mixture of both isomers, polymerizes more slowly than either of the pure enantiomorphs. The identical rates found for DL- and D- + L-mixtures prove that the observed effect is not caused by impurities, because different procedures were followed in the preparation of the respective samples. It is important to note that the reduction in rate is small for $M/I = 4$, as well as for the slow stage of the process when $M/I = 20$. In the fast stage of the

TABLE II

Propagation Constants of Isomeric and Racemic γ-Benzyl Glutamate NCA.
Initiated by n-Hexylamine in Dioxane at 25°C[a]

Monomer	$M/I = 4$, $k \times 10^3$ liter/mole sec	$M/I = 20$ $k_{2a} \times 10^3$ liter/mole sec	$k_{2b} \times 10^3$
L-	7.4	5.6	32
D-	7.1	7.4	34
D,L-	5.1	4.5	13
D- + L-	5.5	4.2	15–17

[a] From Lundberg and Doty, *J. Am. Chem. Soc.*, **79**, 3961 (1957).

reaction, however, the D- – L-mixture polymerized twice as slow as the pure enantiomorph. Apparently, the growing end shows a preference for its own enantiomorph when the polypeptide is long, and this suggests that helical growth is responsible for such a behavior.

The effect caused by the addition of a small amount of D-isomer to an L-monomer is even more striking. In sodium methoxide initiated polymerization, which yields a high molecular weight polypeptide, the addition of the wrong enantiomorph results in a pronounced retardation of polymerization (41) as shown in Figure 4. Even 5% of D-isomer added to L-monomer reduced the polymerization rate by a factor of 3, and the 50:50 mixture polymerized 17 times more slowly than either isomer. The cooperative effect of this phenomenon is therefore indisputable.

The NCA polymerization is not expected to be 100% stereoselective, and therefore there is a chance that a wrong enantiomorph may be incorporated in an otherwise optically pure chain. The rate of the fast stereospecific polymerization is determined not only by the last unit of the chain, but also by some other segments, say, the fourth one from the end (see Figure 2). To allow for a fast reaction, both units must possess the same configuration and form a regular helix. Therefore, incorporation of a D-unit into an all L-polymer inhibits the polymerization of *both* isomers. The fact that the effect was greatly magnified when high molecular weight polymer was formed is another strong indication of helical growth. Further proof of the effect of a remote unit on chain propagation was furnished by another elegant experiment of Lundberg and Doty (35). Polymerization of each, the L-isomer, L,D-racemic mixture, and D-isomer, was initiated by preformed, monodispersed L-polymer of about $DP = 20$. The results are shown in Figure 5 and demonstrate a "normal" fast growth of the L-isomer. The growth in the racemic mixture started slowly and then decelerated, indicating the delaying effect of a wrong monomer on the growth involving either NCA. In a solution of the D-isomer,

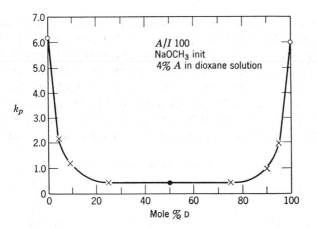

Fig. 4.   The polymerization constants, k_p, for sodium methoxide initiated polymerizations of γ-benzyl-L-glutamate-N-carboxyanhydride in dioxane solution (4 g/100 ml) as a function of mole per cent of added D-anhydride isomer: (×) polymers made with DL-NCA and optically active NCA; (○), polymers made with mixtures of pure D- and pure L-NCA.

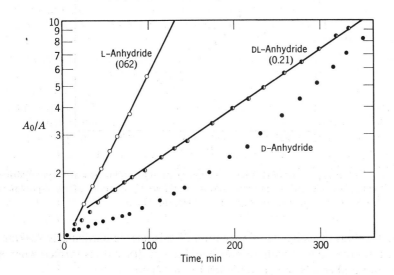

Fig. 5.   Initiation by aliquot of preformed polymer ($DP = 20$) of polybenzylglutamate on L-(○), DL-(◐), and D-(●) carboxyanhydrides of γ-benzyl-glutamate in dimethylformamide; $[A]_0 = 0.058$; $[I]_0 = 0.048$.

the initial growth was very slow but polymerization accelerated as D-segments were accumulated on the L-helix. It appears that on addition of a sufficient number of D-units to an L-polymer a new helical spiral is formed. Thus, the end of the resulting block copolymer (D- preceding L-) behaves in the same way with respect to D-monomer as the previous all L-polymer behaved with respect to an L-monomer. This conclusion is verified by studies of changes in the optical rotation of the polymerizing solutions. These are shown in Figure 6. The increase and then decrease in optical rotation when the poly-

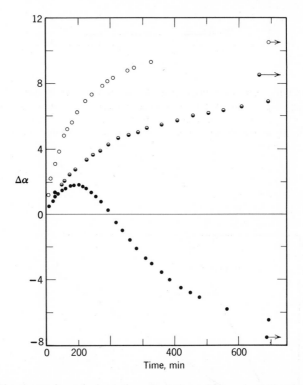

Fig. 6.   The change in optical rotation of L-(○), DL-(◗), and D-(●)carboxyanhydride of γ-benzyl-glutamate during polymerization, initiated by aliquots of an L-polymer; $[A]_0 = 0.076$; $[I]_0 = 0.0034$; polarimeter tube length, 20 cm.

merization was investigated in a solution of D-isomer is particularly striking. It indicates that a helix with an opposite sense of direction is formed when a sufficient number of D-units are added to an L-helix.

In conclusion, the helical growth in NCA polymerization seems to be probable, although not proved finally. It is interesting to know whether this phenomenon may be significant in other polymerizing systems.

4. POSSIBLE HELICAL POLYMERIZATION IN OTHER ANIONIC SYSTEMS

The concept of helical polymerization has been extended by Szwarc (42) to the anionic polymerization of vinyl monomers. It was reported by Williams et al. (43) that triphenylmethylpotassium polymerizes styrene to an amorphous polymer, if the reaction proceeds in benzene, whereas a crystallizable product is formed in hexane. The crystallinity of a vinyl polymer is usually associated with its stereoregularity and hence, Williams et al. assumed that isotactic polystyrene was formed in the latter solvent.

There was a popular belief that only heterogeneous catalysts are capable of producing stereoregular polymers. Therefore, the above workers rationalized their findings by arguing that triphenylmethylpotassium, which is soluble in benzene, represents in this medium a homogeneous initiator and yields then an amorphous polymer. The organometallic compound is, however, insoluble in hexane, and therefore in the latter medium the system is heterogeneous and consequently forms a stereoregular polymer.

This writer believes that heterogeneity of the initiator is not essential for its stereoregulating power, albeit most stereospecific catalysts happen to be heterogeneous. The stereospecificity is determined by the conditions which prevail within a distance of probably not more than 20 Å from the reaction center and, therefore, it is not imperative to involve a macroscopic surface to account for the stereo-regulating phenomena. In fact, a similar opinion was expressed by Patat and Sinn (44).

The correctness of this view has been demonstrated by the studies of R. J. Kern (45) and by the independent investigations of Braun, Betz, and W. Kern (46). These workers showed that homogeneous solutions of butyllithium initiate an isotactic polymerization of styrene, if the reaction proceeds in hydrocarbons at a sufficiently low temperature ($-30°C$ or lower). It should be stressed that solutions of butyllithium may often contain a trace of insoluble lithium chloride and this heterogeneous impurity could be considered the cause of the stereospecificity. However, the lithium chloride-free butyllithium was found to be even a *more* stereospecific catalyst than the compound contaminated by lithium chloride (46) and actually, the stereospecificity was destroyed when larger quantities of lithium chloride were added (46). Hence, the homogeneity of this system seems to be indisputable.

In view of the above discussion, one must seek an alternative explanation for Williams' observations. In some solutions, stereospecific polymers may acquire a helical structure, particularly when the solvent is poor. If Doty's idea applies to such systems, one may expect the helically organized polymers to grow faster than the randomly coiled ones, and moreover their growth may then be stereospecific. This approach was advocated by Szwarc (42) who pointed out that polystyrene is insoluble in hexane but soluble in benzene, and, therefore, a helical structure may be stable in the former solvent, but not in the latter. If by chance, 4 or 5 monomeric molecules become linked together in an isotactic fashion, a helix may be formed in hexane but not in benzene, and the latter may then grow relatively fast, retaining its tacticity. This could account for the formation of an isotactic polystyrene in hexane.

The results of Kern (45) and of Braun, Betz, and Kern (46) may be interpreted in the same way. The stability of a helix increases with decreasing temperature, thus butyllithium yields an isotactic polystyrene at low, but not at high, temperatures. Segment–segment interaction is responsible for helix stability, and therefore it is not surprising that lower temperatures are required for the isotactic polymerization of styrene in toluene than in hexane (46).

The idea of helical growth transcends the domain of ionic polymerization. Various strange phenomena observed in the radical polymerization of acrylonitrile and vinyl chloride have recently been explained by Ham (47) in terms of a stereoregular helical propagation. This hypothesis accounts for the initial induction period followed by a rapid autocatalytic reaction which takes place at the onset of precipitation. The helical growth may, perhaps, provide a more satisfactory explanation for all these observations than the alternative hypothesis invoking "buried" radicals.

5. HELICAL POLYMERIZATION OF METHYL METHACRYLATE

The best documented study of the helical polymerization of vinyl monomers is that reported by Glusker and his associates (48) who investigated the kinetics of the anionic polymerization of methylmethacrylate in a toluene–ether mixture (90:10) at $-60°C$. The reaction, initiated by fluorenyllithium, produces living polymers. The absence of termination was demonstrated by a post-polymerization that ensued upon the addition of fresh monomer to the polymerized material (49). [See also in this connection Wenger (50).] An additional, although only qualitative proof, for the presence of living polymers was provided by experiments in which tritiated acetic acid, CH_3COOT, was used as a killing agent. It was shown that C—T bonds were formed, demonstrating the presence of carbanions in the polymer. It was arbitrarily assumed that the isotope effect is the same for the reaction of fluorenyllithium as for the lithiated polymer, and on this basis the concentration of living ends was calculated. Unfortunately, the observed isotope effect was influenced by moisture making the above described analysis rather unreliable.

The initiation step was found to be very fast, i.e., more than 98% of fluorenyllithium was consumed within the first 5 sec of the polymerization (51). The concentration of fluorenyl end groups was established by a spectrophotometric technique, and for the same samples the number average molecular weight, $\overline{M}_n$, was determined by osmometry. The combined results showed that each polymeric molecule possessed one fluorenyl end group, i.e., $\overline{M}_n$ (spec)/$\overline{M}_n$ (osm) = 0.86–1.19 for $\overline{M}_n$ varying from 4000 to 500,000. Hence, no chain transfer occurs in this reaction. Fractionation of the product showed a surprisingly broad molecular weight distribution; e.g., 90% of the initiator fragments were found in the polymer having a molecular weight less than 2000, while the remaining 10% was found in a material of an average molecular weight of about 27,000. The experiments with tritiated acetic acid demonstrate that at completion of the polymerization the low molecular weight polymer is still living, and hence the presence of the low molecular weight fraction in the final product cannot be explained by its preferential termination.

The polymerization was shown to be a pseudo-first order reaction with respect to monomer. The conventional linear plot extrapolated, however, to a finite conversion at zero time, indicated an extremely rapid formation of a trimer. The same conclusion was drawn from the analysis of the curve representing the time-dependence of the $\overline{DP}$ of the produced polymer. This curve extrapolated, also, to a value of 3 for zero time. Both of these results were found to be independent of the initial ratio of monomer to initiator.*

On the basis of this information, the following mechanism was postulated. The initiation was represented by the reaction

$$Fl^-,Li^+ + CH_2\!:\!C(CH_3)\cdot COOCH_3 \longrightarrow Fl\!-\!CH_2\cdot \overset{\overset{\textstyle Li^+ \cdots\cdots O}{\textstyle \vdots \qquad \|}}{C^-(CH_3)\cdot CO\cdot CH_3}$$

where Fl denotes the fluorenyl residue. The activated monomer is assumed to add very rapidly two monomeric molecules forming a stable, and presumably cyclic, trimer denoted by Tr,

Apparently, this trimer polymerizes very slowly, retaining its cyclic structure in the growth process. Its existence is suggested by the results of Goode et al. (52), who polymerized various methacrylates at 0°C by Grignard compounds. It was shown that each polymer formed by this process possesses one terminal phenyl group derived from the initiator, PhMgBr (tested by using a tritiated Ph), and the molecular weight distribution of the product is very

* Recent studies of Bywater indicate that the initially formed oligomer is rather a penta- or hexamer (private communication).

broad. Hence, Goode's polymerization resembles that studied by Glusker, although in the experiments of the former a termination reaction was observed. The termination produced a 6-membered ring compound identified as a cyclic ketone,

which presumably is formed from the analog of the trimer, Tr, by the elimination of $BrMgOCH_3$. It seems that the elimination reaction producing a cyclic ketone proceeds relatively fast at $0°$ but is extremely slow at $-60°C$. Moreover, this reaction might be faster for the Mg counterion than for Li.

Examination of models demonstrates that ring closure is hindered when three consecutive monomeric units are linked through isotactic placements, because appreciable steric hindrance arises from the axial conformation of the bulky groups (52). Hence, whenever such an event takes place an open chain may result. The latter acquires probably a helical structure which apparently favors its rapid isotactic growth. The open chains are assumed to produce high molecular weight polymer. However, since the probability of a syndiotactic placement is not negligible, a mistake may occur and a new unit could be added in a syndiotactic way to an isotactic chain. Such a chain then acquires the inert cyclic structure and reverts to its extremely slow growth. This process therefore represents a pseudo-termination and explains the first order character of the polymerization resulting from a stationary concentration of the fast growing chains.

It follows from the proposed mechanism that two simultaneous polymerizations proceed in this system—a slow, syndiotactic growth which involves many centers present in the form of oligomeric rings, and a fast, isotactic growth involving a few open chain structures. The resulting polymer should have a binodal molecular weight distribution which indeed was observed. Moreover, it was proved by x-ray studies (53) as well as by NMR investigations (54) that the high molecular weight polymer is isotactic. All these facts fit well into the model proposed by Glusker and demonstrate the ability of an isotactic chain to retain its tacticity during its polymerization, presumably due to its helical structure.

In spite of the agreements this theory is still questionable. Much more work is needed to unravel the mechanism of this polymerization. However, the proposed ideas are interesting and deserve discussion.

References

(1) L. Pauling, R. B. Corey, and H. R. Branson, *Proc. Natl. Acad. Sci. U.S.*, **37**, 205, 235 (1951).

(2) V. Luzzati, M. Cesari, G. Spach, F. Masson, and J. M. Vincenti, "Polyamino Acids," Wisconsin University Press, 1962, p. 121.

(3) (a) W. T. Astbury and A. Idveet, *Phil. Trans. Roy. Soc. (London)*, Ser. *A*, **230**, 75 (1931). (b) W. T. Astbury and H. J. Woods, *ibid.*, Ser. *A*, **232**, 333 (1933).

(4) E. J. Ambrose and A. Elliott, *Proc. Roy. Soc. (London)*, Ser. *A*, **205**, 47 (1951).

(5) C. H. Bamford, W. E. Hanby, and F. Happey, *ibid.*, Ser. *A*, **206**, 407 (1951).

(6) A. Elliott, *ibid.*, A **221**, 104 (1954).

(7) (a) G. Natta et al., *Atti Accad. Nazl. Lincei, Mem. Classe Sci. Sez. II^a*, **4**, 61 (1955). (b) G. Natta et al., *J. Am. Chem. Soc.*, **77**, 1708 (1955); *J. Polymer Sci.*, **16**, 143 (1955).

(8) G. Natta, P. Corradini, and M. Cesari, *Atti Accad. Nazl. Lincei, Rend. Classe Sci.*, **21**, 365 (1956); **22**, 1 (1957).

(9) P. Doty, A. M. Holtzer, J. H. Bradbury, and E. R. Blout, *J. Am. Chem. Soc.*, **76**, 4493 (1954).

(10) C. G. Schellman and J. A. Schellman, *Compt. Rend. Trav. Lab. Carlsberg, Sér. Chim.*, **30**, 465 (1958); *J. Polymer Sci.*, **49**, 129 (1961).

(11) B. H. Zimm and J. K. Bragg, *J. Chem. Phys.*, **28**, 1246 (1958); **31**, 526 (1959).

(12) (a) M. Goodman and K. C. Stueben, *J. Am. Chem. Soc.*, **81**, 3980 (1959). (b) M. Goodman, E. E. Schmitt, and D. A. Yphantis, *ibid.*, **84**, 1283 (1962).

(13) (a) P. Doty and J. T. Yang, *ibid.*, **78**, 498 (1956); **79**, 761 (1957). (b) W. Moffitt and J. T. Yang, *Proc. Natl. Acad. Sci. U.S.*, **42**, 596 (1956).

(14) M. Goodman and E. E. Schmitt, *J. Am. Chem. Soc.*, **81**, 5507 (1959).

(15) D. A. Yphantis, *Ann. N.Y. Acad. Sci.*, **88**, 586 (1960).

(16) M. Goodman, E. E. Schmitt, and D. A. Yphantis, *J. Am. Chem. Soc.*, **82**, 3483 (1960).

(17) M. Goodman, E. E. Schmitt, and D. A. Yphantis, *ibid.*, **84**, 1288 (1962).

(18) K. Rosenheck and P. Doty, *Proc. Natl. Acad. Sci. U.S.*, **47**, 1775 (1961).

(19) M. Goodman and I. Listowsky, *J. Am. Chem. Soc.*, **84**, 3770 (1962).

(20) M. Goodman and F. Boardman, *ibid.*, **85**, 2483 (1963).

(21) M. Goodman, F. Boardman, and I. Listowsky, *ibid.*, **85**, 2491 (1963).

(22) R. H. Karlson, K. S. Norland, G. D. Fasman, and E. R. Blout, *ibid.*, **82**, 2268 (1960).

(23) A. Elliott, E. M. Bradbury, A. R. Downie, and W. E. Hanby, *Proc. Roy. Soc. (London)*, Ser. *A*, **259**, 110 (1960).

(24) See, e.g., E. R. Blout, in "Optical Rotatory Dispersion," McGraw-Hill, New York, 1961, chap. 17.

(25) A. M. Liquori, et al., in the course of publication (1966).

(26) P. De Santis, E. Giglio, A. M. Liquori, and A. Ripamonti, *J. Polymer Sci. A*, **1**, 1383 (1963).

(27) H. A. Scheraga, S. J. Leach, and R. A. Scott, *Discussions Faraday Soc.*, **40**, 268 (1965).

(28) P. Pino, *Fortschr. Hochpolymer.-Forsch.*, **4**, 443 (1966).

(29) C. Schuerch, *Ann. Rev. Phys. Chem.*, **13**, 195 (1962).

(30) R. C. Schulz and E. Kaiser, *Advan. Polymer Sci. (Fortschr. Hochpolymer-Forsch.)*, **4**, 236 (1965).

(31) J. H. Brewster, *J. Am. Chem. Soc.*, **81**, 5475 (1959).

(32) P. Pino, F. Ciardelli, G. P. Lorenzi, and G. Montagnoli, *Makromol. Chem.*, **61**, 207 (1963).

(33) H. Weingarten, *J. Am. Chem. Soc.*, **80**, 352 (1958).

(34) P. Doty and R. D. Lundberg, *ibid.*, **78**, 4810 (1956).

(35) R. D. Lundberg and P. Doty, *ibid.*, **79**, 3961 (1957).

(36) M. Idelson and E. R. Blout, *ibid.*, **79**, 3948 (1957).

(37) J. C. Mitchell, A. E. Woodward, and P. Doty, *ibid.*, **79**, 3955 (1957).

(38) J. D. Coombes and E. Katchalski, *ibid.*, **82**, 5280 (1960).

(39) M. L. Huggins, *ibid.*, **74**, 3963 (1952).

(40) A. M. Liquori, unpublished data.

(41) M. Idelson and E. R. Blout, *J. Am. Chem. Soc.*, **80**, 2387 (1958).

(42) M. Szwarc, *Chem. Ind.* (*London*), **1958**, 1589

(43) J. L. R. Williams, T. M. Laakso, and W. J. Dulmage, *J. Org. Chem.*, **23**, 638 (1958).

(44) F. Patat and H. Sinn, *Angew. Chem.*, **70**, 496 (1958).

(45) R. J. Kern, *Nature*, **187**, 410 (1960).

(46) D. Braun, W. Betz, and W. Kern, *Makromol. Chem.*, **42**, 89 (1960).

(47) G. E. Ham, *J. Polymer Sci.*, **40**, 569 (1959).

(48) D. L. Glusker, I. Lysloff, and E. Stiles, *ibid.*, **49**, 315 (1961).

(49) R. K. Graham, D. L. Dunkelberger, and E. S. Cohn, *ibid.*, **42**, 501 (1960).

(50) F. Wenger, *Chem. Ind.* (*London*), **1959**, 1094.

(51) D. L. Glusker, E. Stiles, and B. Yoncoskie, *J. Polymer Sci.*, **49**, 297 (1961).

(52) W. E. Goode, F. H. Owens, and W. L. Myers, *ibid.*, **47**, 75 (1960).

(53) J. D. Stroupe and R. E. Hughes, *J. Am. Chem. Soc.*, **80**, 2341 (1958).

(54) F. A. Bovey and G. V. D. Tiers, *J. Polymer Sci.*, **44**, 173 (1960).

Chapter XII

Termination of Polymerization

1. TERMINATION PROCESSES OF ADDITION POLYMERIZATION

Propagation of addition polymerization continues until the activities of the growing ends of polymeric molecules are lost or the supply of monomer exhausted. The reactions which deprive the growing polymeric molecules of their ability to add further monomer are known as *termination processes*. They are classified as *proper terminations* if they take place irreversibly and destroy the growth capacity of individual polymeric molecules without creating centers from which new polymeric molecules evolve. *Reversible termination* processes form dormant polymers; and those which produce some active centers, one for each deactivated end group of a polymer, are known as *chain-transfer reactions*.

The modes of termination occurring in anionic polymerization are the subject of this chapter. However, it is advisable to review the termination steps of other addition polymerizations in order to see in a proper perspective the problems pertaining to anionic polymerization.

2. BIMOLECULAR TERMINATION

Bimolecular collision of the radical end groups is the main contributor to the termination of radical-polymerization. Radicals readily recombine or disproportionate and these reactions annihilate the growing ends of polymeric molecules. The bimolecular terminations cannot be prevented as long as the growing chains remain mobile, although it is possible to reduce their rates. The combination of radicals is intrinsically fast and in the liquid phase its rate is diffusion-controlled. Hence, an increase in the viscosity of the medium slows down the termination, and indeed large effects are observed in polymerizations proceeding in undiluted monomers after a high degree of conversion has been attained (1,2). Detailed studies of these systems demonstrated that the increased rate observed at high conversion is accompanied by an increase in the molecular weight of the product. The absolute rate constant of propagation remains unchanged (3) but the absolute rate constant of termination is drastically reduced. The effect of diffusion upon the rate of termination of polymerization is more complex than may appear from a superficial inspection of the problem. Its dependence on molecular weight of the product is most interesting; a thorough study of this subject was reported by North and his associates (4).

An interesting situation is encountered in a polymerization in which a gaseous monomer is directly converted into crystallinic polymer. In such a system the polymeric chains are virtually immobile; they grow because the monomer is continually supplied from the gas phase, but the bimolecular termination is essentially eliminated. Polymerization of p-xylylene exemplifies such a case (5) and a somewhat similar situation exists in polymerizations in which the polymer precipitates as it is formed. These are the conditions under which trapped (or buried) radicals are formed (6).

In ionic polymerization, bimolecular collisions between the growing end groups usually do not lead to termination. An exceptional case was described by Chien (7). The polymerization of ethylene was initiated by a soluble biscyclopentadienyltitanium dichloride–dimethylaluminium chloride complex. The termination was shown to be second order in growing ends and the kinetic observations were accounted for by postulating the following reaction,

$$2\text{\raise.4ex\hbox{$\sim$}}CH_2 \cdot CH_2^-, Ti^{4+}\diagdown^{\diagup} \rightarrow \text{\raise.4ex\hbox{$\sim$}}CH{=}CH_2 + \text{\raise.4ex\hbox{$\sim$}}CH_2CH_3 + 2Ti^{3+}\diagdown^{\diagup},$$

i.e., the disproportionation of the two alkyl end groups occurs simultaneously with the reduction of the two Ti^{4+} complexes into Ti^{3+}. The rate constant of this reaction is exceedingly low, although the high concentration of the growing species makes the termination relatively fast.

3. TERMINATION INVOLVING COUNTERIONS

In a radical-polymerization the active end group of a polymeric molecule, viz. a free radical, is a single entity. In contradistinction, two species compose the growing end group of a polymer when propagation involves an ionic mechanism, i.e., a charged terminal atom or a group located at the end of a polymeric chain, and an oppositely charged counterion which is more or less strongly associated with the active end group.

The charged end group of a polymer and its counterion may combine and form a stable covalent bond. Such a process transforms the reactive end group into a stable and unreactive one, terminating, therefore, the growth of the relevant polymeric molecule. This type of termination is frequently observed in carbonium ion polymerization. For example, consider a hypothetical polymerization of a vinyl monomer initiated by hydrochloric acid. The resulting carbonium ion is associated with Cl^- counterions and on their combination a stable C—Cl covalent bond is formed. This destroys the activity of the growing end group. In fact, the combination of carbonium ion with chloride anion is so rapid that it follows immediately the proton addition. The reaction yields, therefore, an HCl adduct to a C=C double bond instead of a polymer or an oligomer. Nevertheless, an oligomeric product may be obtained if the combination is slowed down by powerful solvation. For example, a solution of styrene in nitromethane, but not in a hydrocarbon, can be polymerized by hydrochloric acid (8); in nitromethane the combination of the solvated ions is relatively slow and therefore the monomer addition may compete with termination. Thus, a low molecular weight polymer is produced.

Polymerization of styrene initiated by trifluoroacetic acid (9) provides another example of the same phenomenon. Addition of styrene to trifluoroacetic acid yields polystyrene of molecular weight 20,000 to 30,000 while no polymer is formed on the addition of trifluoroacetic acid to styrene! The first procedure yields trifluoroacetate counterions stabilized by the solvent (trifluoroacetic acid) through powerful hydrogen bonds; hence their capacity for combining with the growing carbonium ions is greatly reduced. The alternative order of mixing the reagents forms the counterions in a poorly solvating hydrocarbon milieu (styrene). Their combination with carbonium ions remains, therefore, unhindered and, consequently, the formation of a polymer is prevented.

Termination arising from the combination of the oppositely charged ions may be prevented if the counterion becomes coordinated with a powerful binding agent. For example, $BF_3 \cdot H_2O$ complex initiates cationic polymerization of isobutene, and then $(BF_3OH)^-$ ion becomes the companion of the growing carbonium ion. The complex anion may be treated as a hydroxyl ion coordinated with BF_3. One may visualize, therefore, a termination caused by the combination of OH^- with the growing polyisobutene, resulting in the formation of a stable C—OH bond. This is not the case because the binding power of BF_3 is too great (10). The structure of the complex $(BF_3OH)^-$ ion is such that its direct combination with the growing polymer is also precluded. Hence, termination by ion recombination is impossible in this system.

The solvation affects termination as well as propagation. For example, it was shown in Chapter VII that anionic propagation involving alkali cations is greatly enhanced when the cation becomes coordinated with solvent molecules or with some other suitable agent (see pp. 429 and 443). Consequently, specially designed experiments are needed to elucidate the effect of counterion solvation on the rate of termination. It may be misleading therefore to draw any conclusions if they are exclusively based on the overall behavior of polymerization. An interesting system which permits a direct investigation of termination was described by Pepper (11). Polymerization of styrene was initiated by sulfuric or perchloric acid. The initiation was shown to be virtually instantaneous and quantitative,* all the growing chains were present at the onset of the reaction and their initial concentration was therefore known. The course of polymerization is determined then by propagation and termination only; any further initiation is eliminated. Such experiments permit one to determine the absolute rate constants of termination and propagation and, therefore, this technique may be profitably used to investigate the influence of coordinating agents on the termination. It is hoped that future studies will shed more light on this problem.

Initiation of ionic polymerization by reactions such as

$$XY + monomer \rightarrow Y\text{-}(monomer)^+,X^-$$
$$or\ Y\text{-}(monomer)^-,X^+,$$

usually leads to ion-pairs. Termination arising from the collapse of a pair is a first-order reaction. Formation of the free ions is feasible only if their combination into inactive product is not taking place at every collision; otherwise the two oppositely charged species have hardly a chance to separate after being formed. The termination in systems involving free ions is, of course, a second-order reaction. A recent example of such a process was provided by studies of Penczek et al. (12). This group investigated the polymerization of oxetane initiated in nitrobenzene by nitroform. The data shown in Figure 1

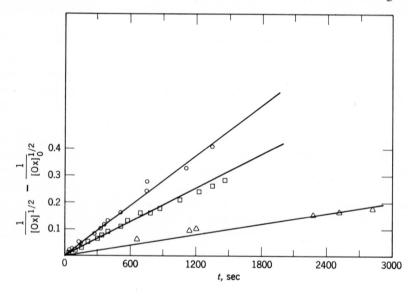

Fig. 1. Bimolecular termination of oxethane polymerization initiated by nitroform.

*Recent, although still unpublished, studies of Pepper raise some doubts about this point.

confirm the second-order kinetics of the termination; most probably the reaction results from the combination of the oppositely charged ions:

$$\sim CH_2 \overset{+}{O} \begin{matrix} CH_2 \\ \diagup \quad \diagdown \\ \quad \quad \quad CH_2 \\ \diagdown \quad \diagup \\ CH_2 \end{matrix} + C(NO_2)_3^- \rightarrow \sim CH_2O \cdot (CH_2)_2 \cdot CH_2 \cdot C(NO_2)_3.$$

The solvation of the $C(NO_2)_3^-$ anion and the necessity of ring opening account for the slowness of this process.

Even more interesting is a chain-transfer process terminating propagation of the cationic polymerization of vinylcarbazole also initiated by nitroform or tetranitromethane in nitrobenzene (12). Detailed kinetic argument shows that the transfer does not produce an intermediate nitroform molecule, i.e., the reaction

$$\sim CH_2 \cdot \underset{\underset{/\,N\,\diagdown}{|}}{CH} \cdot CH_2 \cdot \underset{\underset{/\,N\,\diagdown}{|}}{CH}_\oplus + C(NO_2)_3^- \rightarrow \sim CH_2 \cdot \underset{\underset{/\,N\,\diagdown}{|}}{CH} \cdot CH : \underset{\underset{/\,N\,\diagdown}{|}}{CH} + CH(NO_2)_3$$

does not take place. Apparently the transfer proceeds through the *intramolecular* hydrogen shift,

$$\begin{matrix} C(NO_2)_3^- \\ \sim H_2C \overset{\frown}{} CH_2 \\ \diagdown \quad \quad \diagdown \\ \underset{\underset{/\,N\,\diagdown}{|}}{CH} \quad \underset{\underset{/\,N\,\diagdown}{|}}{CH}_\oplus \end{matrix} \longrightarrow \begin{matrix} C(NO_2)_3^- \\ \sim HC \quad + CH_3 \cdot CH_\oplus \\ \diagdown \quad \quad \quad | \\ \underset{\underset{/\,N\,\diagdown}{|}}{CH} \quad \quad \underset{/\,N\,\diagdown}{} \end{matrix}$$

which is catalyzed by the approach of $C(NO_2)^-$ anion. The presence of this negative center facilitates dissociation of the proton and its subsequent transfer. Indeed, this termination was found to be a second-order reaction and not a unimolecular decomposition.

It is well known that high-energy radiation, such as γ-rays, ionizes the irradiated molecules, producing positive radical-ions and free electrons. Theoretical arguments (38) had indicated that the electron cannot, on the average, escape the parent ion—even in a solvent of such high dielectric constant as water. However, studies of pulse radiolysis (see p. 347) and of the conductivity of organic liquids induced by radiation (39) led to the conclusion that the Compton recoil electrons are capable of being separated from their parent ions by rather large distances, even in media of low dielectric constant. Hence, free ions may be formed under conditions of irradiation and therefore they may induce ionic polymerization propagated by *free* ions if suitable monomers are available. In those reactions the counterion is not judiciously chosen as in a conventional ionic polymerization; it is formed directly, or indirectly, by the irradiation and its collision with the growing chain is probably always effective and terminates its growth (13). Hence, the rate of such ionic polymerization is expected to be 1/2 order in respect to the intensity of radiation.

Irradiation of conventionally purified monomers, under conditions which prevent radical propagation, led to ionic polymerization (40) which was found to be *first* order in respect to radiation intensity, as shown in Figure 2 (14).

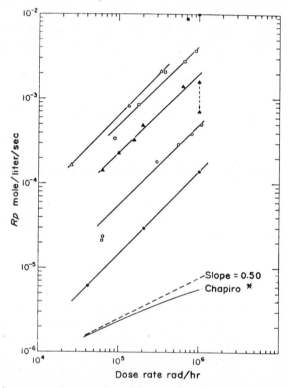

Fig. 2. Rate of polymerization of styrene at 0°C vs. dose rate (14b): (○) S-5; (▲) S-6; (□) S-7; (■) S-8; (△) S-10 (at 40°C).

Order Dependence of Rate of
Polymerization on Dose Rate $(R_p I^n)$

Sample	Order dependence (n)
Ampoules	0.98
S-5	0.94
S-6	0.92
S-7	0.93
S-10	0.97

See reference 40e.

In most of the studied systems the level of impurities, especially of moisture, was sufficiently high to make the impurity-terminated process, which exhibits a pseudo-first order kinetics, much faster than the bimolecular combination of ions. However, meticulous and painstaking purification of monomers and purging of the reaction vessel led to a substantial *increase* in the rate of polymerization, and under these conditions the expected 1/2 order relation was observed (15). Because impurities were eliminated in the purging process their reaction did not obscure the bimolecular termination arising from the combination of the oppositely charged ions. Detailed analysis of the results indicates that the propagation of styrene and α-methylstyrene involves free carbonium ions. Contribution of carbanions, if any, is negligible.

The cationic character of the polymerization explains why a minute amount of moisture efficiently quenches the reaction by a pseudo-first order process. Apparently water acts as a base, e.g.,

$$\sim\!CH_2\cdot CH(Ph)^+ + H_2O \rightarrow \sim\!CH\!:\!CH(Ph) + H_3O^+.$$

The positive H_3O^+ ion subsequently reacts with the solvated electron, or with another negative ion formed through its reaction, and produces a H atom or a product of protonation. Thus, H_2O is regenerated and its concentration remains constant.

An example of anionic polymerization induced by γ-radiation was provided by the studies of Dainton and his associates (41). Polymerization of acrylamide was initiated by the radiation in liquid ammonia at $-65°C$. The NH_3^+ ions are converted in about 10^{-14} sec into NH_4^+ by the reaction

$$NH_3^+ + NH_3 \rightarrow NH_4^+ + NH_2^{\cdot}.$$

The electrons are trapped by the monomer and subsequently anionic polymerization ensues. The latter is terminated by the bimolecular reaction

$$\sim\!M^- + NH_4^+ \rightarrow \sim\!MH + NH_3$$

or

$$\sim\!M^- + NH_2^{\cdot} \rightarrow \sim\!M\cdot + NH_2^-.$$

Thus, the rate of polymerization is first order in monomer and the dose-rate exponent is close to one-half.

A rather unconventional case of termination caused by the combination of the oppositely charged ions was described by Berger, Levy, and Vofsi (16). They reacted living polystyrene prepared by anionic initiation with living cationically grown polytetrahydrofuran. The carbanions combined with the cyclic oxonium ions yielding a block polymer $(S)_n(THF)_m$. The same procedure was employed in the preparation of a block polymer composed of anionically grown polyisoprene and cationically produced polytetrahydrofuran. The identity of the resulting block polymers was established by suitable tests.

The above type of termination differs from the others described in this section in that the termination does not compete with propagation. The two processes take place at different times, the propagation being completed before the termination could occur. Therefore, such a reaction is classified as a "killing" process, a type of termination to be discussed in later sections of this chapter.

4. TERMINATION INVOLVING PROTON OR HYDRIDE ION TRANSFER

Proton transfer is a most common mode of termination in carbonium ion polymerization. For example, cationic polymerization of isobutene initiated by $BF_3 \cdot H_2O$ complex is propagated by $\sim\sim CH_2 \cdot C(CH_3)_2^+,(BF_3OH)^-$ ion-pair. The termination of this polymerization involves a proton transfer from a C—H bond adjacent to the carbonium ion to the negative counterion. The transfer is coupled with the formation of a C=C double bond, i.e.,

$$\sim\sim CH_2C(CH_3)_2^+,(BF_3OH)^- \rightarrow \begin{array}{c} \sim\sim CH_2C(CH_3)=CH_2 \\ \text{or} \\ \sim\sim CH=C(CH_3)_2 \end{array} + BF_3 \cdot H_2O,$$

and regenerates the protonic acid ($BF_3 \cdot H_2O$) which may initiate further polymerization. Therefore, such a termination is classified as chain-transfer to the initiator and a growing carbonium-ion acts as a Brönsted acid in transfer reactions but as a Lewis acid in propagation.

The rate of proton transfer might be reduced by associating the counterion with a suitable complexing agent. In such a case the termination would be hindered and the molecular weight of the resulting polymer should increase. Some unsuccessful attempts to find such an agent were made in the writer's laboratory; compounds such as tetracyanoethylene, tetraphenylethylene, and ninhydrin were tested in polymerizations initiated by trifluoroacetic acid. In spite of the failure of these experiments it is believed that this approach deserves further exploration.

The transfer of hydride ion from a suitable hydride donor terminates the cationic polymerization if the resulting new positive ion is inert and incapable of initiating further polymerization. The solvent in which the polymerization takes place, or some impurities present in the system, often act as the donors. In some industrial processes judiciously chosen donors, known as *moderators*, are added to the polymerized solution to control the molecular weight of the product. Of course, the use of moderators is not restricted to cationic polymerization. Mercaptans are the most common moderators employed in many radical polymerizations—e.g., in the production of synthetic

styrene–butadiene rubber, hydrogen gas acts as a moderator in the polymerization of ethylene initiated by heterogeneous Ziegler-type catalysts,
etc.

The loss of hydride ion from a C—H bond adjacent to carbanion is the
counterpart of a proton loss in the cationic systems. Such a reaction which
leads to the formation of metal hydrides is common in the heterogeneous
polymerization of α-olefins involving transition metals. Its rate is not much
slower than the rate of propagation and therefore this reaction competes with
the growth of the polymeric molecule and influences the kinetics of polymerization.

Hydrides may also be formed in anionic polymerization involving alkali
counterions, but this process is too slow to compete significantly with the
propagation and, therefore, this mode of termination does not affect the
kinetics of polymerization. However, such a termination may be observed
when solutions of living polymers are stored. Under these conditions the
hydride formation leads to spontaneous but slow destruction of growing
ends.

A thorough study of this process was reported by Spach et al. (17). A change
in the absorption spectra of living polystyrene has been often noted when
their solutions were left at room temperature for a few days. The 340 mμ
absorption peak characteristic of the $\sim\sim$CH(Ph)$^-$,Na$^+$ end group gradually
disappears and a new absorption band appears at about 535 mμ. These
changes are illustrated by Figure 3. The isosbestic point, clearly revealed in
the figure, proves that the reaction has a constant stoichiometry and that the

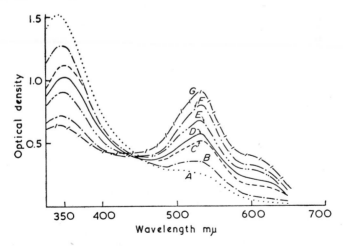

Fig. 3. Aging of "living" polystyrene of low molecular weight (17). Time (hr):
(A) 0, (B) 4.5, (C) 29, (D) 49, (E) 97, (F) 222, (G) 318. Note the isosbestic point indicating
a fixed stoichiometry of the reaction.

colored species produced in the process is stable and does not react further. Although the results shown in Figure 3 were derived from experiments performed in tetrahydrofuran, similar spectral changes were observed in other solvents, e.g., dioxane, dimethoxyethane, or benzene.

Sometimes the spectral changes occur extremely rapidly. For example, it was observed in this writer's laboratory that some batches of commercial tetrahydrofuran induced a color change within minutes after preparation of the living polymer. Various modes of purification of such a solvent did not improve its quality; the cherry-red solution still turned purple within less than an hour. Apparently some unknown impurity, which is not removed in the course of solvent purification, catalyzes the reaction. A similar behavior was often noted with dioxane. It should be stressed, therefore, that in any quantitative study involving living polymers it is desirable to check their absorption spectrum whenever the solution had been stored. The presence of new species not only reduces the concentration of the original growing ends but it may affect, in various ways, the reactivity of those genuine living polymers which still remained in the solution.

The stability of living polymers depends on the nature of the counterions. The spectral changes discussed above, as well as "fading" of the color of living polymers solution, are relatively fast for the sodium salts but slow for the cesium. The cesium-containing polymers appear to be more stable if they possess two growing ends per chain than do the polymers endowed with one growing end only. Perhaps the formation of triple ions is responsible for this behavior (see p. 440).

Sodium hydride is formed in aged solutions of living polystyrene or living poly-α-methylstyrene, but only the former polymer yields the species absorbing at 535 mμ. This indicates that the reaction involves the α-hydrogen of polystyrene. The formation of hydride is evident from the fact that hydrogen gas is evolved on addition of water to "aged" (converted) solutions but not when water is added to fresh (unconverted) aliquots. Moreover, addition of D_2O yields HD and not D_2, excluding colloidal sodium as a possible reagent yielding the hydrogen. Formation of sodium hydride is coupled with the appearance of a terminal $\sim\!\!CH\!\!=\!\!CH(Ph)$ double bond in the product; its presence was established through spectrophotometric analysis. Hence, the reaction seems to follow the equation

$$\sim\!\!CH_2\!\!-\!\!CH(Ph)^-,Na^+ \rightarrow \ \sim\!\!CH\!\!=\!\!CH(Ph) + NaH.$$

The formation of the terminal C=C double bond makes the penultimate —CH(Ph)— group of polystyrene extremely acidic because the resonance stabilization of the

$$\sim\!\!\overset{-}{C}\!\!-\!\!CH\!\!=\!\!CH$$
$$\quad | \qquad\qquad |$$
$$\quad Ph \qquad\quad Ph$$

allylic group reduces the strength of the relevant C—H bond. Hence, the reaction

$$\text{—}\overset{-}{\text{C}}\text{H(Ph)} + \text{—}\overset{\overset{\displaystyle H}{|}}{\underset{\underset{\displaystyle Ph}{|}}{\text{C}}}\text{—CH=CH} \rightarrow \text{—CH}_2\text{(Ph)} + \text{—}\overset{\ominus}{\text{C}}\text{---}\underset{\underset{\displaystyle Ph}{|}}{\overset{\overset{\displaystyle}{|}}{\text{C}}}\text{H---CH}$$

rapidly converts the benzyl type of anion into the 1,3-diphenylallyl anion, and it is the latter that absorbs at 535 mμ. Of course, such a reaction is impossible in the poly-α-methylstyrene system.

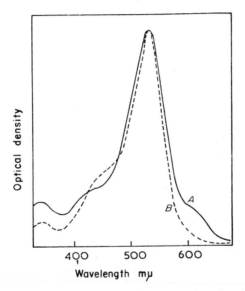

Fig. 4.   Spectra of (A) aged polystyrene and (B) 1,3-diphenylbut-1-ene plus "living" polystyrene (counterion K⁺) (17). The optical densities refer to arbitrary concentrations and were chosen so that the two peaks coincide.

Further evidence for the formation of the 1,3-diphenylallyl anion was provided by the following experiment. A dimer CH$_3$—CH(Ph)—CH=CH(Ph) was prepared and its solution was added to living polystyrene. The absorption spectrum changed nearly instantaneously, the new absorption band being shown in Figure 4. For the sake of comparison, the spectrum of the aged polystyrene is shown in the same figure. Both spectra are virtually identical.

It was noted that dilution of the polymer solution increases the rate of conversion. Since the reaction requires an approach of Na⁺ ion to the β-hydrogen,

$$\text{—}\overset{\overset{\displaystyle H}{|}}{\text{C}}\text{H—CHPh}^-\text{,Na}^+ \rightarrow \overset{\overset{\displaystyle \text{Na}^+}{\displaystyle \text{H}^-}}{} \text{—CH=CHPh,}$$

an increase in the degree of dissociation of the ion-pair may possibly affect the rate of the process. The necessity of partial or complete dissociation of the ion-pair in the transition state of hydride formation not only could account for the effect of dilution but it may also explain why the process proceeds faster in ethers than in benzene and it occurs nearly instantly in hexamethyl phosphor amide.

Quantitative observations show that the formation of sodium hydride is accelerated when the solution is in contact with alkali metal. It was suggested (17) that the following surface reaction may catalyze the reaction:

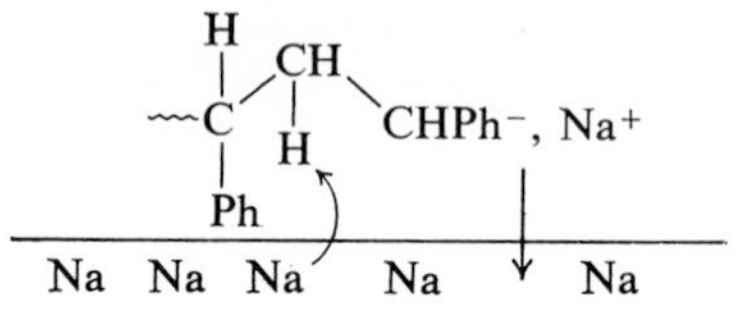

The generation of 1,3-diphenylallyl ions provides an example of a reaction converting an active living polymer into inactive ions. Other types of reactions, of unknown nature, were also observed. For example, Asami et al. (18) described spectral changes which take place on storage of benzylsodium but which are not observed in aged cumylpotassium. The new species, absorbing at 485 mμ, is inactive as an initiator of styrene polymerization. Obviously, this reaction could not produce the 1,3-diphenylallyl anion.

The transfer of proton to carbanions is the counterpart of the termination of cationic polymerization by hydride ion transfer. A classic example of a proton transfer which competes with propagation was described by Sanderson and Hauser (19), who reported polymerization of styrene in liquid ammonia initiated by sodium amide. The molecular weight of the isolated polymer was found to be constant, about 3000, independent of the concentration of sodium amide. The termination was described by the equation

$$\sim\overset{-}{C}H(Ph) + NH_3 \rightarrow \sim CH_2(Ph) + NH_2^-$$

and such a mechanism leads to the relation

$$\overline{DP}_n = k_p[M]/k_t,$$

in which the symbols have their conventional meaning.

Subsequent very thorough investigation of this polymerization by Higginson and Wooding (20) confirmed these conclusions. Potassium amide was used as an initiator. The molecular weight of the polymer again was found to be independent of the amide concentration, but it increased with increasing concentration of styrene. The relation $\overline{DP}_n \sim [M]$ was established quantitatively.

A similar situation was encountered in the polymerization of styrene initiated in dimethyl sulfoxide by the $CH_3SO \cdot CH_2^-, Na^+$ anion (21). A rapid chain-transfer terminates the propagation, viz.

$$\sim\sim CH(Ph)^-, Na^+ + CH_3SO \cdot CH_3 \rightarrow \sim\sim CH_2(Ph) + Na^+, {}^-CH_2 \cdot SO \cdot CH_3.$$

Consequently, only low molecular weight products were formed under those conditions.

A comparison of polymerization in liquid ammonia with the Birch reduction (22) may be of some interest. Some olefins, such as styrene, are rapidly reduced by solutions of alkali metals in liquid ammonia, e.g., styrene yields ethylbenzene. Alkali-metal solutions contain solvated electrons, $e^-(NH_3)_n$, hence one may expect the reduction to be induced by an electron transfer process,

$$e^-(NH_3)_n + CH_2:CH(Ph) \rightarrow nNH_3 + (CH_2:CHPh)^{\overline{\cdot}},$$

which produces the respective radical-anion. The latter could then react with ammonia

$$(CH_2:CHPh)^{\overline{\cdot}} + NH_3 \rightarrow CH_3 \cdot CH(Ph) \cdot + NH_2^-,$$

or it could dimerize and eventually polymerize styrene. Proton transfer to living polystyrene is not too rapid; propagation adds a score of styrene molecules before the termination takes place. The $(CH_2:CHPh)^{\overline{\cdot}}$ radical-ions are expected to be less basic than the $\sim\sim CH_2CH(Ph)^-$ carbanions and, therefore, their dimerization, and the eventual initiation of styrene polymerization, appear to be likely events which could compete with the protonation. Surprisingly, this is not the case. Ethylbenzene is produced in a high yield and only a small amount of polymer is found in the product.

These findings could be rationalized by assuming a direct reaction:

$$e^-(NH_3) + CH_2:CHPh \rightarrow NH_2^- + CH_3 \cdot CH(Ph) \cdot.$$

Alternatively, one may argue that the radical-anion is not paired with its counterion, whereas the carbanion in liquid ammonia forms ion-pairs, viz. $\sim\sim CH_2 \cdot CH(Ph)^-, Na^+$. The above findings may be then accounted for if free ions are protonated faster than ion-pairs. It is significant that under the conditions which lead to protonation of styrene some other olefins that form radicals of higher electron affinity than $CH_3 \cdot CH(Ph) \cdot$ yield carbanions. For example, 1,1-diphenylethylene yields $CH_3 \cdot C(Ph)_2^-$ carbanions as well as the dimeric dianions ${}^-C(Ph)_2 \cdot CH_2 \cdot CH_2 \cdot C(Ph)_2^-$. This may indicate that the course of the reactions

$$e^-(NH_3) + olefin \rightarrow radical + NH_2^-$$

and

$$e^-(NH_3) + R \cdot \rightarrow RH + NH_2^-$$

depend on the relative electron affinities of $NH_2 \cdot$ and of the formed radical R.

The work of Wooding and Higginson (20) provides further evidence for the fast hydrogenation of styrene by a solution of potassium in liquid ammonia; the protonation proceeds more readily than the electron transfer process. These workers found that the anionic polymerization of styrene initiated by a solution of potassium in liquid ammonia proceeds like the polymerization initiated by an equivalent amount of potassium amide. They concluded, therefore, that the hydrogenation of styrene to ethylbenzene rapidly converts solvated electrons into NH_2^- and the latter then initiate the polymerization. Indeed, the blue color of the potassium solution disappears almost instantly, while the polymerization proceeds rather slowly (half-lifetime about 15–30 minutes).

The rate of proton transfer depends on the basicity of the anion. For example, polymerization of styrene in liquid ammonia yields a polymer having a molecular weight of about 3000. Under comparable conditions polymerization of methacrylonitrile results in a polymer of molecular weight of 100,000 (22). Although the increase in molecular weight may be caused by faster propagation, it is highly probable that the lower rate of protonation is, at least partially, responsible for this result.

In the anionic polymerization of styrene proceeding in liquid ammonia the addition of α-methylstyrene lowers the molecular weight of the product (19). Apparently the more basic carbanion resulting from the addition of α-methylstyrene enhances the termination. Since the addition of butadiene to styrene leads to a similar effect (19), one might be inclined to believe that the carbanion formed from butadiene is more basic than the polystyryl$^-$ ion.

5. CHAIN-TRANSFER AND TERMINATION BY MONOMER

The termination caused by the monomer is well known in a radical polymerization. It usually results from a hydrogen transfer from the growing polymer to the monomer, or vice versa. If a reactive radical is formed in the process, one refers to this phenomenon as a chain-transfer to the monomer, but whenever an unreactive radical is produced one deals with self-inhibition of polymerization. Polymerizations of some allyl derivatives (23,24) provide examples of such phenomena.

Chain-transfer to monomer is very plausible in some anionic polymerizations. It usually involves a proton transfer from the monomer to the growing anion. It is suggested, for example, that such a reaction might contribute to termination of anionic polymerization of isoprene. In a system producing "living" polymers the chain-transfer to monomer changes the molecular weight distribution of the product (25); the resulting distribution was described by Litt and Szwarc (25a).

A self-retardation (or inhibition) was observed in the anionic polymerization of vinylmesitylene (26). This subject was discussed in Chapter VII, Section 13. The reaction produces a more stable primary carbanion from the reactive growing secondary carbanion, viz.

$$\sim CH_2-CH^- \quad + \quad CH=CH_2 \quad \longrightarrow \quad CH_2-CH_2 \quad + \quad CH=CH_2$$
$$CH_2^-$$

6. INTRAMOLECULAR PROTON (OR HYDRIDE) TRANSFER

In some systems an intramolecular proton (or hydride ion) transfer may lead to isomerization, which competes with propagation. A spectacular example of such a reaction was reported by Kennedy and associates (27), who investigated the cationic polymerization of 3-methyl-1-butene. At low temperatures, the intramolecular hydride transfer viz.,

$$\sim CH_2-\overset{+}{CH}-\overset{\overset{\displaystyle CH_3}{|}}{CH}-CH_3 \longrightarrow \sim CH_2-CH_2-\overset{\overset{\displaystyle CH_3}{|}\,+}{\underset{\underset{\displaystyle CH_3}{|}}{C}}$$

is faster than propagation, and therefore the polymerization yields

$$poly(CH_2-CH_2-\overset{\overset{\displaystyle CH_3}{|}}{\underset{\underset{\displaystyle CH_3}{|}}{C}}-)$$

instead of

$$poly(CH_2-\overset{|}{\underset{\underset{CH_3 \quad CH_3}{C}}{CH}}-).$$

An extensive review of similar isomerization processes was published by Kennedy (45).

A somewhat similar phenomenon was reported by Breslow, Hulse, and Matlack (28), who studied the anionic polymerization of acrylamide. The propagation step

$$\sim CH_2-\underset{\underset{CO-NH_2}{|}}{CH^-} + CH_2=\underset{\underset{CO-NH_2}{|}}{CH} \rightarrow \sim CH_2-\underset{\underset{CO-NH_2}{|}}{CH}-CH_2-\underset{\underset{CO-NH_2}{|}}{CH^-}$$

competes with intramolecular proton transfer

$$\sim CH_2 \cdot \overset{\displaystyle \cdot}{CH} \longrightarrow \sim CH_2 \cdot CH_2 \cdot CO \cdot NH^-$$
$$\qquad | \quad \backslash$$
$$\qquad CO \cdot NH_2$$

followed by propagation

$$\sim CH_2 \cdot CH_2 \cdot CO \cdot NH^- + CH_2 : CH \rightarrow \sim CH_2 \cdot CH_2 \cdot CO \cdot NH \cdot CH_2 \cdot CH^-$$
$$\qquad\qquad\qquad\qquad\qquad | \qquad\qquad\qquad\qquad\qquad\qquad\qquad\qquad |$$
$$\qquad\qquad\qquad\qquad\qquad CO \cdot NH_2 \qquad\qquad\qquad\qquad\qquad\qquad\qquad CO \cdot NH_2$$
$$\rightarrow \sim CH_2 \cdot CH_2 \cdot CO \cdot NH \cdot CH_2 \cdot CH_2 \cdot CO \cdot NH^-.$$

Thus, the 3-nylon is formed instead of polyacrylamide. Unfortunately, the reaction is not following this course uniquely and consequently both segments,

$$-CH_2 \cdot CH_2(CONH_2)- \quad \text{and} \quad -CH_2 \cdot CH_2 \cdot CO \cdot NH-,$$

appear randomly distributed in the polymer. In addition, some intermolecular transfer of proton from the dangling $-CO \cdot NH_2$ groups leads to an appreciable degree of branching (46).

The conventional vinyl type of propagation may be prevented by introducing bulky substituents on the α-carbon. Thus, for a sufficiently bulky substituent X, the vinyl polymerization of $CH_2 : CX \cdot CO \cdot NH_2$ may be slowed down to such an extent that only proton transfer may be observed. The polymerization should yield, therefore, a pure polyamide:

$$-(CH_2 \cdot CH_2 \cdot CXH \cdot CO \cdot NH)_n-.$$

In anionic polymerization of isoprene or vinylmesitylene (26) the intramolecular proton transfer yields a primary carbanion from a secondary one. Thus

Such reactions are too slow to affect significantly the kinetics of polymerization. They reveal themselves by some slight spectral shifts observed in aged solutions of the relevant polymers (see also pp. 647–649).

7. TERMINATION INVOLVING ISOMERIZATION

The high reactivity of carbonium ions and carbanions is responsible for their numerous isomerization reactions. Those resulting from proton (or hydride) transfer were discussed in the preceding section, but many other reactions are possible which convert the reactive growing ion into an unreactive one. Numerous examples of such a termination of carbonium ion polymerization are known, and a most thorough study of this subject was undertaken by Overberger and his students (29). Isomerization of carbanions are not uncommon, e.g., the rearrangement takes place in the Michael condensation of α,β-unsaturated ketones with malonic ester derivatives due to migration of COOEt groups (30). This is illustrated by the equations:

$$Ph \cdot CH : CH \cdot CO \cdot Ph \; + \; CH_3 \cdot CH(COOEt)_2 \xrightarrow{\text{1. eq. NaOH}}$$

$$Ph \cdot CH \cdot CH^- \cdot CO \cdot Ph \xrightarrow{\text{isomerization}}$$
$$\overset{|}{CH_3 \cdot C(COOEt)_2}$$

$$Ph \cdot CH \cdot CH(COOEt) \cdot CO \cdot Ph \xrightarrow{H^+}$$
$$\overset{|}{CH_3 C \cdot COOEt}$$

$$Ph \cdot CH \cdot CH(COOEt) \cdot CO \cdot Ph$$
$$\overset{|}{CH_3 CH \cdot COOEt}$$

Investigations of termination processes caused by isomerization of carbanions are rather scanty. It was suggested, e.g., that the anionic polymerization of methyl methacrylate may be terminated by isomerization of a carbanion into a carboxylate ion through migration of a CH_3^+ ion (31):

$$\begin{array}{ccc} CH_3 & & CH_3 \\ | & & | \\ \text{\textasciitilde\textasciitilde\textasciitilde}-C^- & \rightarrow \text{\textasciitilde\textasciitilde\textasciitilde}-C-CH_3 \\ | & & | \\ COO-CH_3 & & COO^- \end{array}$$

However, subsequent studies seem to indicate that this rearrangement process requires a high activation energy and, therefore (32), this reaction does not take place, at least at low temperatures. Nevertheless, a slow termination process is possible in such a system. Schreiber (33) draws attention to the reaction of phenyllithium with esters, e.g.,

$$CH_2 : C(CH_3) \cdot CO \cdot OCH_3 \; + \; PhLi \rightarrow CH_2 : C(CH_3) \cdot CO \cdot Ph \; + \; LiOCH_3,$$

and suggests that similar processes between a "living" polymethyl methacrylate and the ester group of a monomer or polymer may lead to termination. Consequently, three termination reactions may be visualized—termination by monomer, grafting on the "dead" polymer, or the formation of a loop on the growing polymer.

Anionic polymerization of rigorously purified acrylonitrile involves a termination if the reaction proceeds at higher temperatures (50–70°C). This is manifested by low conversions found at higher temperatures as compared with quantitative conversion achieved at lower temperatures. It has been proposed that the termination is due to an isomerization that resembles the Thorpe reaction:

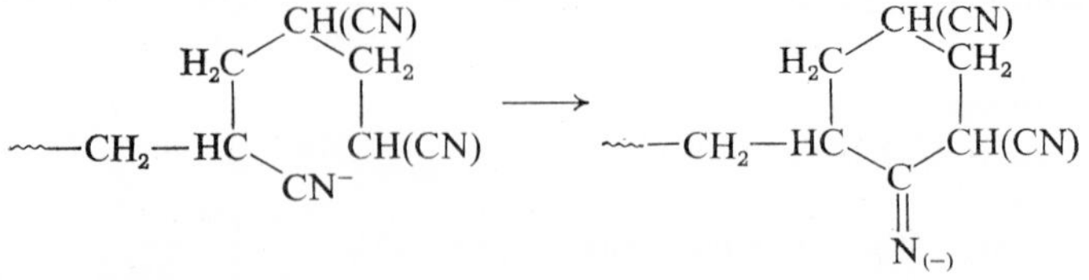

The latter ring-ion may exist either in the above written form or as its tautomer. The ring-ion apparently is not sufficiently reactive to initiate further polymerization.

Solution properties of polyacrylonitriles produced by anionic and by radical-polymerization are different. It seems that the former is a branched polymer while the latter has a linear structure. The amount of branching apparently increases with rising temperature of polymerization, suggesting a termination due to a proton transfer from the polymer to the growing anion.

Another example of termination resulting from the formation of cyclic trimers is reported by Shashoua et al. (34), who investigated anionic polymerization of monoisonitriles to 1-nylons. The trimerization is favored by a higher temperature of polymerization.

8. TERMINATION CAUSED BY WRONG MONOMER ADDITION

Propagation involves an addition of monomer to the growing end of a polymeric molecule in the course of which another, identical end group is formed. There are monomers which may be added in two, or more, different ways to the growing polymer. One mode of addition re-forms the growing end group and therefore represents the propagation, while the other mode, or modes, form new terminal groups of low reactivity. Since those cannot propagate such an addition terminates the polymerization.

A few examples of this termination will be discussed. Sela and Berger (35) pointed out that a wrong addition of Leuchs' anhydride (NCA) may lead to termination (see also p. 564). The propagation follows the course

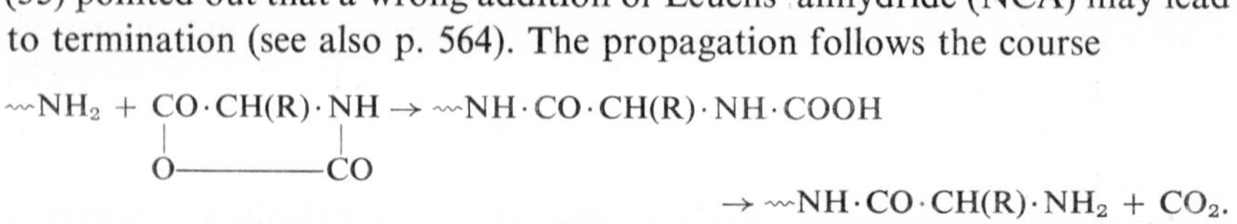

$$\sim NH_2 + CO \cdot CH(R) \cdot NH \rightarrow \sim NH \cdot CO \cdot CH(R) \cdot NH \cdot COOH$$
$$O\text{————}CO$$
$$\rightarrow \sim NH \cdot CO \cdot CH(R) \cdot NH_2 + CO_2.$$

However, the addition to the other keto group results in the formation of a stable end group that cannot propagate the reaction, viz.,

$$\text{---NH}_2 + \underset{\underset{\text{CO}\text{----}\text{O}}{\big\lfloor}}{\text{NH}\cdot\text{CH(R)}\cdot\text{CO}} \longrightarrow \text{---NH}\cdot\text{CO}\cdot\text{NH}\cdot\text{CH(R)}\cdot\text{COOH}.$$

The second mode of addition is less favorable and requires higher activation energy. Consequently, it becomes more significant at higher temperatures of polymerization.

Cationic polymerization of vinylcarbazole may be also terminated by wrong monomer addition (36). The normal reaction is given by

$$\underset{\underset{N}{|}}{\text{---CH}_2\text{---CH}^+} + \underset{\underset{N}{|}}{\text{CH}_2\text{=}\text{CH}} \to \underset{\underset{N}{|}}{\text{---CH}_2\text{---CH}}\text{---}\underset{\underset{N}{|}}{\text{CH}_2\text{---CH}^+},$$

while the terminating addition probably involves an attack on the N-center,

$$\underset{\underset{N}{|}}{\text{---CH}_2\text{---CH}^+} \quad \underset{N}{\text{N}}\text{---CH=CH}_2 \to \underset{\underset{N}{|}}{\text{---CH}_2\text{---CH}}\text{---}\underset{N}{\text{N}}^+\text{---CH=CH}_2,$$

and produces an unreactive quarternary ammonium ion.

It is possible that the anionic polymerization of 9-vinylanthracene is terminated by a wrong monomer addition, viz., the reaction

produces an anion more stable than the "normal" anion resulting from vinyl addition. However, another resonance form of that anion, i.e., may be capable of continuing the propagation, giving rise to a 1,6- instead of 1,2-linkage.

There are, of course, many additions which introduce irregularities in the chain without causing termination. For example, in many radical polymerizations a head-to-head addition introduces a new type of end groups (37), but these are usually *more* reactive than the "normal" end groups. Consequently, a disordered chain is formed but the propagation is not interrupted.

9. OTHER MODES OF TERMINATION OF ANIONIC POLYMERIZATION

Many anionic polymerizations were studied in ethereal solvents, e.g., in tetrahydrofuran. The growing carbanions may react with the ethers, e.g.,

$$\sim CH(Ph)^- + \underset{\underset{H_2C \longrightarrow CH_2}{|}}{\overset{\overset{O}{\overset{H_2C}{\diagup}\diagdown CH_2}}{|}} \longrightarrow \sim CH(Ph)\cdot CH_2\cdot CH_2\cdot CH_2\cdot CH_2O^-,$$

and such a reaction destroys their activity. On the whole, these terminations are slow and therefore are classified as "killing" processes. Quantitative studies, carried out in this writer's laboratory, demonstrated that the free carbanions are responsible for the process; the half-lifetime of the active ends, a majority of which form ion-pairs, increases with their concentration.

Conversion of carbanions into alcoholate led to some confusing observations. It was reported (42) that lithium polystyryl in tetrahydrofuran solution slowly becomes associated at room temperature. Over a period of a day the viscosity of the solution greatly increases, although the intrinsic viscosity of the dead polymer is unaffected by such a change. The aliquots, even after being kept for a long time at $-78°C$, retain their original viscosity when the solutions are warmed up to the ambient temperature; however, when the sample is left for a day at this temperature its viscosity increases again. It was erroneously concluded that the lithium-containing ion-pairs become slowly associated, the association requiring activation energy. The correct explanation was given by Fetters (43), who pointed out that the alcoholates, but not the carbanions, are associated in tetrahydrofuran (44). The observed changes in viscosity are slow because the conversion of $\sim CH(Ph)^-,Li^+$ into $\sim CH_2O^-,Li^+$ proceeds slowly; this process is arrested by low temperature.

Finally, in some polymer systems, dormant polymers may be formed through reversible cyclization or formation of macro rings. These phenomena were discussed previously (see pp. 23 and 636). Other examples of dormant polymers were reviewed in Chapter VII, pp. 440, 446, and 449 and in Chapter VIII.

References

(1) E. Trommsdorff, H. Köhle, and P. Lagally, *Makromol. Chem.*, 1, 169 (1947).

(2) R. G. W. Norrish and R. R. Smith, *Nature*, 150, 336 (1942).

(3) M. S. Matheson, E. B. Bevilacqua, E. E. Auer, and E. J. Hart, *J. Am. Chem. Soc.*, 71, 497 (1949).

(4) (a) A. M. North and G. A. Reed, *Trans. Faraday Soc.*, 57, 859 (1961). (b) S. W. Benson and A. M. North, *J. Am. Chem. Soc.*, 81, 1339 (1959).

(5) L. A. Errede and M. Szwarc, *Quart. Rev. (London)*, 12, 301 (1958).

(6) G. J. Minkoff, *Frozen Free Radicals*, Interscience, New York, 1960, pp. 130–134.

(7) J. C. W. Chien, *J. Am. Chem. Soc.*, 81, 86 (1959).

(8) D. C. Pepper, *Quart. Rev. (London)*, 8, 88 (1954).

(9) J. J. Throssell, S. P. Sood, M. Szwarc, and V. Stannett, *J. Am. Chem. Soc.*, 78, 1122 (1956).

(10) F. S. Dainton and G. B. B. M. Sutherland, *J. Polymer Sci.*, 4, 37 (1949).

(11) M. J. Hayes and D. C. Pepper, *Proc. Chem. Soc.*, 1958, 228; *Proc. Roy. Soc. London A*, 263, 58, 63 (1961).

(12) S. Penczek, R. Gumbs, J. Jagur-Grodzinski, and M. Szwarc, to be published.

(13) M. Szwarc, *Makromol. Chem.*, 35A, 123 (1960).

(14) (a) K. Ueno, K. Hayashi, and S. Okamura, *J. Polymer Sci. B*, 3, 363 (1965). (b) R. C. Potter, C. L. Johnson, D. J. Metz, and R. H. Bretton, *J. Polymer Sci. A-1*, 4, 419 (1966). (c) R. C. Potter, R. H. Bretton, and D. J. Metz, *ibid.*, A-1, 4, 2295 (1966).

(15) (a) K. Ueno, F. Williams, K. Hayashi, and S. Okamura, *Trans. Faraday Soc.*, 63, 1478 (1967). (b) F. Williams, K. Hayashi, K. Ueno, K. Hayashi, and S. Okamura, *ibid.*, 63, 1501 (1967). (c) D. J. Metz, R. C. Potter, and R. A. Bretton, *J. Polymer Sci.*, in press.

(16) G. Berger, M. Levy, and D. Vofsi, *J. Polymer Sci., B*, 4, 183 (1966).

(17) G. Spach, M. Levy, and M. Szwarc, *J. Chem. Soc. (London)*, 1962, 355.

(18) R. Asami, M. Levy, and M. Szwarc, *ibid.*, 1962, 361.

(19) J. J. Sanderson and C. R. Hauser, *J. Am. Chem. Soc.*, 71, 1595 (1949).

(20) W. C. E. Higginson and N. S. Wooding, *J. Chem. Soc.* 1952, 760.

(21) G. E. Molan and J. E. Mason, *J. Polymer Sci. A-1*, 4, 2336 (1966); see also C. E. H. Bawn, A. Ledwith, and N. R. McFarlane, *Polymer*, 8, 484 (1967).

(22) C. B. Wooster and J. F. Ryan, *J. Am. Chem. Soc.*, 56, 1133 (1934).

(23) P. D. Bartlett and R. Artschul, *ibid.*, 67, 812, 816 (1945).

(24) E. D. Morrison, E. H. Gleason, and V. Stannett, *J. Polymer Sci.*, 36, 267 (1959).

(25) (a) M. Litt and M. Szwarc, *J. Polymer Sci.*, 42, 159 (1960). (b) W. T. Kyner, J. R. M. Radok, and M. Wales, *J. Chem. Phys.*, 30, 363 (1959).

(26) D. N. Bhattacharyya, J. Smid, and M. Szwarc, *J. Polymer Sci. A*, 3, 3099 (1965).

(27) J. P. Kennedy and A. V. Langer, *Adv. Polymer Sci.*, 3, 508 (1964).

(28) D. S. Breslow, G. E. Hulse, and A. S. Matlack, *J. Am. Chem. Soc.*, 79, 3760 (1957).

(29) C. G. Overberger and G. F. Endres, *J. Polymer Sci.*, 16, 283 (1955).

(30) C. K. Ingold, *Organic Chemistry*, Cornell Press, 1953, pp. 690–699.

(31) A. Rembaum and M. Szwarc, *J. Polymer Sci.*, 22, 189 (1956).

(32) (a) D. L. Glusker, E. Stiles, and B. Yoncoskie, *ibid.*, 49, 297 (1961). (b) F. Wenger, *Chem. Ind. (London)*, 1959, 1094. (c) T. J. R. Weakley, R. J. P. Williams, and J. D. Wilson, *J. Chem. Soc.*, 1960, 3963. (d) D. M. Wiles and S. Bywater, *Polymer (London)*, 3, 175 (1962).

(33) H. Schreiber, *Makromol. Chem.*, 36, 86 (1959).

(34) V. E. Shashoua, W. Sweeny, and R. F. Tietz, *J. Am. Chem. Soc.*, **82**, 866 (1960).

(35) M. Sela and A. Berger, *J. Am. Chem. Soc.*, **77**, 1893 (1955).

(36) J. Pac and P. H. Plesch, *Polymer*, **8**, 237 (1967).

(37) P. J. Flory and F. S. Leutner, *J. Polymer Sci.*, **3**, 880 (1948); **5**, 267 (1950).

(38) A. H. Samuel and J. L. Magee, *J. Chem. Phys.*, **21**, 1080 (1953).

(39) A Hummel and A. O. Allen, unpublished data.

(40) (a) W. H. T. Davison, S. H. Pinner, and R. Worrall, *Chem. Ind. (London)*, **1957**, 1274. (b) R. Worrall and S. H. Pinner, *J. Polymer Sci.*, **34**, 229 (1959). (c) W. H. T. Davison, S. H. Pinner, and R. Worrall, *Proc. Roy. Soc. (London) A*, **252**, 187 (1959). (d) E. Collinson, F. S. Dainton, and H. A. Gillis, *J. Phys. Chem.*, **63**, 909 (1959); *J. Polymer Sci.*, **34**, 241 (1959). (e) A. Chapiro and V. Stannett, *J. Chim. Phys.*, **56**, 830 (1959).

(41) F. S. Dainton, D. Smithies, T. Skwarski, and E. Wezranowski, *Trans. Faraday Soc.*, **60**, 1068 (1964).

(42) K. S. Das, M. Feld, and M. Szwarc, *J. Am. Chem. Soc.*, **82**, 1506 (1960).

(43) L. J. Fetters, *J. Polymer Sci.*, B, **2**, 425 (1964).

(44) H. Brody, D. H. Richards, and M. Szwarc, *Chem. Ind. (London)*, **1958**, 1473.

(45) J. P. Kennedy in *Encyclopedia of Polymer Science and Technology*, Vol. 7, Mark and Gaylord, Eds., Interscience, New York, 1967, p. 754.

(46) L. W. Busch and D. S. Breslow, *Macromolecules*, **1**, 189 (1968).

Author Index

Numbers in parentheses are reference numbers and show that an author's work is referred to although his name is not mentioned in the text. Numbers in *italics* indicate the pages on which the full references appear.

Subject Index

A

Acrylamide, 645, 653, 654
  radical anion of, 175
Acrylic acid, radical anion of, 175
Acrylonitrile, 655
  anion copolymerization of, 532
  radical anion of, 175
"Activated" monomer, 587, 595–597, 600–614, 635
Addition polymerization, 1–4, 8, 9
  thermodynamics of, 104–109
D,L-Alanine NCA 559, 566, 575
Alkali metals, 153–156
  equilibrium with aromatic hydrocarbons, 327–334
  initiation by, 9, 10, 80, 513–516, 521, 527
  in hexamethylphosphoramide, 352, 353
  in liquid ammonia, 348–352, 521–523, 651, 652
Allyllithium, 477, 488
Amylose, monodisperse, 69
Anionic copolymerization, absolute rate constants of, 532–544
  classic studies of, 526–532
  kinetics of, 520–556
  reversible, 552–555
Anionic polymerization, early studies of, 9–16
  experimental techniques, 151–209
  initiation of, 487–508
  propagation of, 405–472
  with lithium counterion, 476–516
Anthracene, complex with living polystyrene, 23, 446–450, 551
  conductance of, 282, 317
  dianion of, 175, 354, 424
  equlibrium with radical anion, 320, 384–386, 389, 390
  half-wave potential of, 322
  radical anion of, 153, 315, 316, 338, 339, 347
  radical cation of, 344
  spectrum of, 174, 339, 340
Arrhenius parameters, of anionic propagation, 14, 416–419

by free ions, 419–424
by ion pairs, 424–434
"negative" value of, 417, 426–430

B

Batch polymerization, 37, 540
  in kinetic studies, 181–188
  molecular weight distribution in, 37–44
Beer's law, in living polymer systems, validity of, 170–180
Benzanthracene, radical anion of, 336, 338
Benzene, radical anion of, 174, 310
  radical cation of, 343
Benzophenone, dianion of, 354
  equilibrium with radical anion, 382, 383
  radical anion of, 153–155, 335, 354
  dimerization of, 179, 180
  disproportionation of, 366, 367
  removal from diphenylethylene, 154
Benzoquinone, radical anion of, 384
3,4–Benzpyrene, radical cation of, 344
Benzylamine, 594
γ-Benzy-D-glutamate NCA, 559, 560, 629–632
γ-Benzyl-D,L-glutamate NCA, 559, 560, 628–632
γ-Benzyl-L-glutamate NCA, 184, 559–565, 575–589, 594, 608, 609, 615, 622, 627–632
Benzyllithium, 156, 157, 389, 477, 488
  complexes with lithium salts, 443
  exchange with lithium salts, 483
Benzylsodium, 156, 157, 650
Biphenyl complexes of, 268, 269
  dissociation of, 315, 316
  equilibrium with lithium, 329–333
  equilibrium with sodium, 268, 269, 329–333
  isomerization of, 333, 334
  polarography of, 322
  proton transfer to, 347
  radical anion of, 311–313, 335–338, 460
  radical cation of, 344
  removal of from diphenylethylene, 154
  spectrum of, 155, 174, 263